DICTIONARY OF SHIPPING INTERNATIONAL BUSINESS TERMS AND ABBREVIATIONS

Fifth Edition

By

Professor Alan E Branch FCIT FIEX

International Business Consultant – University Lecturer – Chief Examiner in Shipping, Export Practice, International Marketing and Buying.

And

David Branch BSc Econ (LSE) MBA (IMD Lausanne)

International Business Consultant

Books by the same Author

Elements of Shipping, 8th Edition. Routledge London (2006)
Export Practice and Management, 5th Edition. Thomson Learning, London (2005)
International Purchasing and Management. 1st Edition, Thomson Learning, London (2001)
Shipping and Air Freight Documentation for Importers and Exporters and Associated terms., 2nd Edition. Witherby (2000)
Maritime Economics, Management and Marketing, 3rd Edition. Routledge London (1998)
Elements of Export Marketing and Management, 2nd Edition. Chapman & Hall (1990)
Dictionary of Multi-lingual Commercial/International Trade/Shipping Terms in English-French-German-Spanish (over 10,000 entries). Witherby, London (1989)
Dictionary of English-Arabic Shipping/International Trade/Commercial Terms and Abbreviations (4,000 entries). 1st Edition. Witherby, London (1988)
Elements of Port Operation and Management. 1st Edition, Chapman & Hall, London (1986)
Dictionary of Commercial Terms and Abbreviations (6,000 entries). 1st Edition, Witherby, London (1984)

First Published in 1976
Second Edition 1982
Third Edition 1986
Fourth Edition 1995
Fifth Edition 2005
Reprinted 2008
© Witherby Seamanship International Ltd.

ISBN 1 85609 236 4

British Library Cataloguing in Publication data
A catalogue record for this book is available from the British Library.

Terms of Use

All rights reserved. No part of this publication may be reproduced, stored in a retrieval system, or transmitted in any form or by any means, electronic, mechanical, photocopying, recording or otherwise, without the prior permission of the publisher and copyright owner.

While the principles discussed and the details given in this book are the product of careful study, the authors and publisher cannot in any way guarantee the suitability of recommendations made in this book for individual problems, and they shall not be under any legal liability of any kind in respect or arising out of the form or contents of this book or any error therein, or the reliance of any person thereon.

Printed & bound in Great Britain by Bell & Bain Ltd. Glasgow

Published in 2008 by
Witherby Seamanship International
4 Dunlop Square
Deans Estate
Livingston EH54 8SB
United Kingdom
Tel No: +44(0)1506 463 227
Fax No: +44(0)1506 468 999
Email: info@emailws.com
www.witherbyseamanship.com

[5665]

Cover illustration supplied by permission of Allan & Bertram (Corporate Calendars) 01727 876677 and John Lines RSMA (artist) 01788 578522

Preface to the Fifth Edition

Nearly 30 years have elapsed since this established internationally sold dictionary, was published. Today, it is a standard work and world market leader in its field. It is found on every bookshelf across the world which focuses on International Trade/Shipping/Seaports/Finance/Insurance ranging from the student to the International Executive and the Multi National Enterprise.

The fifth edition contains 18,000 entries – compared with 14,000 in the previous edition – and has six appendices. It is a reference work to many in the industry of international trade and distribution/shipping/air/seaport and related integral businesses/activities. This new edition reflects the enormous changes which have emerged in recent years. Overall, it includes all the ingredients for conducting International Business which has become more complex and diverse. One must remember that the international contract embraces four elements finance, carriage, insurance and export sales contract.

The dictionary had been completely up-dated and covers the wide spectrum of International Business, Distribution -Shipping/Air and Trade embracing terms found in Advertising, Airports, Air Transport, Buying goods overseas (International Purchasing), Cargo handling, Computerisation, Containerisation, Customs, Documentation, E-mail, Export Practice, Export Marketing and Selling, Import Practice, International Banking, International Business, International Marketing, Logistics, International Road Haulage, Marine Engineering, Marine Insurance, Multi-modalism, Seaports, Shipping, Tourism and World Currencies featuring separately Euro currency countries. Such enrichment to the fifth edition will further its popularity in international business in the industrial and service sectors globally. Equally, it will remain increasingly popular amongst students, undergraduates, libraries involving numerous institutes, colleges and universities throughout the world. The book will be especially useful for students taking courses sponsored by the Institute of Chartered Shipbrokers, British International Freight Association, Chartered Institute of Marketing, Chartered Institute of Bankers, Chartered Institute of Purchasing and Supply, Chartered Institute of Logistics and Transport, Institute of Marine Engineers, Institute of Commercial Management and Institute of Export. It will also prove a popular dictionary for Chambers of Commerce, Trade Associations and Training Agencies conducting short courses and seminars on International Trade and Shipping. Degree level undergraduates studying International Business, Finance, Management, Marketing, Logistics, Maritime Transport, Seaports, International Physical Distribution/Tourism will find the dictionary an essential reference work. Such university courses are situated not only in the United Kingdom but globally especially throughout Europe, United States, Singapore, Hong Kong, Australasia, South Africa, sub-continent, Nigeria, Cyprus, Malta and the Middle East.

The fifth edition contains a much enlarged input from organisations and practices around the world. This not only enriches the dictionary but also enables the reader to have better understanding of world culture and related terms and abbreviations used in the conduct of International Business. The extent of such input is exemplified in the many organisations which have helped us so enthusiastically with this work as recorded in the acknowledgements. In particular we would mention WTO, IATA, BIMCO, INTERTANKO, numerous banks, IPE, International Underwriting Association of London, Lloyd's Register of Shipping, HM Customs and Excise, UNCTAD, and IMO. We acknowledge with gratitude their help and likewise those we featured in the acknowledgements.

Finally, we would like to acknowledge with grateful thanks the generous secretarial help from Mr & Mrs Splarn, Mrs Jane Salter, and as always my dear wife Kathleen Branch. Such a quartet have provided us with encouragement, forbearance and above all complete professionalism in enabling this much enlarged and enriched fifth edition to be possible, for which we are greatly indebted.

Professor Alan Branch
David Branch
19 The Ridings
Emmer Green
Reading
Berkshire RG4 8XL
England
International Tel: +44 (0) 118 947 6291
International Fax: +44 (0) 118 947 6291
E-mail kathleenbranch@hotmail.com

March 2005

Acknowledgements

American Bureau of Shipping
Associated British Ports

Baltic and International Maritime Council
Baltic International Freight Futures Exchange Ltd.
Baltic Mercantile & Shipping Exchange Ltd.
Barclays Bank PLC
BIFFEX
British Airways PLC
British International Freight Association
British Maritime Technology
British Shipbuilders Ltd.
British Waterways Board
Bureau Veritas

Chamber of Shipping
Chartered Institute of Purchasing and Supply
Containerisation International
Corporation of Lloyd's

Data Ship (UK) Ltd.

Euler Hermes
Export Credits Guarantee Department
Exportmaster Ltd.

Freight Transport Association

GE SeaCo Services Ltd.
Glaxo Export Ltd.

HM Customs & Excise
HSBC Bank plc

International Air Transport Association
International Association of Classification Societies Ltd.
International Cargo Handling Co-ordination Association
International Chamber of Commerce

International Maritime Organisation
International Petroleum Exchange of London
International Underwriting Association of London
Institute of Bankers
Institute of Chartered Shipbrokers
Institute of Export
Institute of Marine Engineers
Institute of Packaging

Kay O'Neill-logistics professionals
LEP Transport
LIFFE
Lloyd's of London
Lloyd's Register of Shipping
London Chamber of Commerce & Industry

Overseas Containers Ltd.

Marine Safety Agency
MAT International Group Ltd.

National Waterways Transport Association
National Westminster Bank
Nautical Institute

P & O Cruises (UK) Ltd.
Port of Singapore Authority

Road Haulage Association

Simpler Trade Procedures Board
Suez Canal Authority

Thailand Board of Investment
TNT

UNCTAD

World Trade Organisation

Contents

Preface to the fifth edition — iii

Acknowledgements — v

Dictionary of Shipping/International Trade Terms and Abbreviations — 1

Appendix A	Progressing Export Consignment – Flow Charts, Diagrams A and B	373
	Export Process Improvements Cycle Diagram C	375
	Export Master Sales Process and Shipping Process Flow Chart Diagram D	376
Appendix B	Major Ports of the World and Their Location	377
Appendix C	(i) World Currencies	387
	(ii) Euro Based Currency Countries	393
Appendix D	Conversion of Weights and Measures	395
Appendix E	IMO Dangerous Goods Labels	397
Appendix F	Further Recommended Reading	399

This book is dedicated to
Anna and Richard

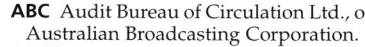

A1 Highest Hull classification for American Bureau of Shipping.
A3 Aegean Airlines S.A.-Aviation Limited Company.
A4 Southern Winds (Southern Winds S.A.) – IATA associate member airline.
A Aft; or automation.
a Alloy containers; or aft.
aa Always afloat – indicates that the vessel will either load and/or discharge as specified in the charter party.
AA Average Adjuster; or Automobile Association.
AAA Association of Average Adjusters.
AAAA Always afloat always access-ible.
AAC Aeronautical Administrative Communications – IATA term.
AACC Airport Associations Co-ordinating Council – IATA term.
AACI Airport Associations Council International – IATA term.
AACO Arab Air Carriers Organisation.
AAD Administrative Accompanying Document.
AAE Asia Australia Express. An Associated company P & O containers trading between Far East and Australia.
AAF Asian Automotive Federation
AAG AIS Automation Group (of EANPG) – IATA term.
AAL Asia America Line. A Shipping company.
AAPA American Association of Port Authorities.
AAR Against all risks.
AARTC Anchorage Air Route Traffic Control Centre – IATA term.
AAS Annual automated controls survey (of a ship).
AASA Airline Association of South Asia – IATA term.
AASMM Associated African States, Madagascar and Mauritius.
AASO Association of American Shipowners – Based in New York.
AAT Airports Authority of Thailand
AATF Airworthiness Assurance Task Force – IATA term.
AATG Airline Airport Technical Group (ATA) – IATA term.
AAWG Airmiss Analysis Working Group (IATA) – IATA term.
AB Air Berlin (Air Berlin GmbH & Co Luftverkehr KG) – IATA airline member; able bodied seaman; or Above Bridges.
AB (or ABS) American Bureau of Shipping – American Ship Classification Society.
Abacus Asian/Pacific computerised reservations system owned by Cathay Pacific, China Airlines, Malaysia Airline, Philippine Airlines, Royal Brunei, Singapore Airlines and Worldspan.
Abaft Towards the stern of a vessel.
Abandon To allow an option to expire worthless – an International Petroleum Exchange term.
Abandoned The option is not exercised as the price is against the buyer (or token) in a declaration day – a BIFFEX term.
Abandonment Leaving a ship as unseaworthy; or in marine insurance terms the process of giving up the proprietary rights in insured property to the underwriter in exchange for payment of a constructive total loss. It is the right a marine assured has to abandon property in order to establish a constructive total loss. An underwriter is not obliged to accept abandonment, but if he does he accepts responsibility for the property and for any legal liabilities attaching thereto, in addition to being liable for the full sum assured.
ABC Audit Bureau of Circulation Ltd., or Australian Broadcasting Corporation.
ABCC Association of British Chambers of Commerce; or Arab-British Chamber of Commerce.
ABCU Automatic bridge control system for unattended engine room.
Abeam In a line at right angles to the vessel's length – opposite the centre of vessel's side.
ABECOR An association of European banks seeking to improve the services that each individual member is able to offer to its customers by developing banking and financial objectives on the basis of co-operation.
ABH Vessel classed by American Bureau of Shipping (Hellas).
ABI Association of British Insurers.
Abinitio (Latin) From the beginning.
Able seaman Holder of the UK Department for Transport, local Government and the Regions certificate of competency; a fourth year apprentice. It is now termed Seaman Grade I; or RN rating.
ABMEC Association of British Mining Equipment Companies.
Abnormal invisible load Any load which cannot without undue expense or risk of damage, be divided into two or more loads for the purpose of carriage on any transport mode and which owing to its weight and dimensions cannot be carried on the transport unit which complies with the relevant regulation i.e. Construction and Use regulations for the road vehicle.
Abnormal load A consignment necessitating special arrangements. Such loads may be heavy lifts, out of gauge, or indivisible consignments.
ABOI Association of British Oceanic Industries.
About When referring to the amount or the quantity of items or the unit price of the goods, "about" means plus or minus 10%. See Article 39 of UPC – ICC publication.
Above par At a premium.
Above the line advertising Advertising for which a commission or fee is payable to an Agency and usually arise with press, TV, radio, cinema and posters.
Above the line promotion See above the line advertising entry.
Above the norm Above the average.
ABP Associated British Ports.
Abrasion Resistance The ability of a fibre or rope to withstand surface wear and rubbing due to motion against other fibres of rope components (internal abrasion) or a contact surface such as a fairlead (external abrasion).
ABS American Bureau of Shipping – the USA ship classification society.
Absolute advantage An individual has an absolute advantage in the production of two goods if, by using the same quantities of inputs, that person can produce more of both goods than another individual. A country capable of producing specific goods or services using fewer resources (per unit of output and times) than other countries. An economic term
Absolute net claims Claims paid less premium returns less reinsurance recoveries.
Absolute net premium Premiums plus additional premiums less return premiums less reinsurance's, brokerage and commission.

Absolute pressure The absolute pressure is the total of the gauge pressure plus the pressure of the surrounding atmosphere.

Absolute temperature The fundamental temperature scale with its zero at absolute zero and expressed in degrees Kelvin. One degree Kelvin is equal to one degree Celsius or one degree Centigrade. For the purpose of practical calculations in order to convert Celsius to Kelvin add 273. It is normal for the degree Kelvin to be abbreviated in mathematical formulae to 'K' with the degree symbol being omitted.

Absolute zero The temperature at which the volume of a gas theoretically becomes zero and all thermal motion ceases. It is generally accepted as being – 273.16°C.

Absorption Acceptance by the air freight carrier of a portion of a joint rate; charge which is less in amount than that which it would receive for the service in the absence of such joint rate or charge; or the process by which incident radiation is taken into a body and retained without reflection or transmission, thereby increasing the internal or kinetic energy of the molecules or atoms composing the absorbing medium.

Absorption pricing A pricing policy based on assigning all cost – fixed and variable – to the company's products/services to determine the selling price.

ABS SAFENET Software package for ship management industry and trademark of ABS.

Abt About.

ABTC Australian Competition Trade Association.

AC Account current; alternating current; account; Assistant Controller; authorisation under consideration; air conditioning; Air Canada. The Canadian National airline and a member of IATA; Assistant Collector – HM Customs post; or Arbitration Clause.

Ac Airfreight containers.

ACA Accession Compensatory Amount. A charge or refund which compensates for differing price levels between new member states and the rest of the European Community; or Agricultural Credit Administration (Philippines).

ACAC Arab Civil Aviation Council – IATA term.

ACAP Advanced Composite Airframe Programme.

ACARS ARINC Communications Addressing & Reporting System – IATA term.

ACAS Airborne Collision Avoidance System (connected with SSR Mode S) – IATA term.

Acc Acceptance; or accepted.

ACC American Chamber of Commerce; Automatic control certified; Average claims cost; Acceptable container condition; Airport Consultative Committee (IATA) – IATA term; or Area Control Centre – IATA term.

ACCA Air Charter Carriers Association (European Charter Carriers) – IATA term.

ACCC Australian Competition and Consumer Commission.

ACE Group A Group of Shipping Companies including Consortium of Franco Belgian Service; K Line, Korea Shipping Corporation, Neptune Orient Line and Orient Overseas Container Line.

Acceptance The signing (usually across the face) by the drawee of a tenor (usance) bill of exchange which then binds him to make payment of the bill at maturity. A banking term; or notice given by the insurers of their willingness to give cover – an insurance definition. This embraces notification by a party to a proposed contract that he accepts the offer of the other party of the terms proposed, as where an underwriter initials a slip. On receiving an offer of cover from assurer's, signifies his acceptance by paying the premium; or an IPE definition with the agreement of intention to carry out a transaction. A similar criteria obtains for a client accepting any legal constituted offer such as agreed sale price of a product. See Acceptance of Goods entry.

Acceptance bills A form of export finance. A bank establishes an acceptance credit in the exporter's favour allowing bills of exchange to be drawn by the exporter. They usually run in parallel with bills drawn on the overseas buyer by the exporter and lodged with the bank for collection.

Acceptance certificate Document issued by the forwarding agent on request to certify that the goods to be despatched in accordance with the contract of carriage have been actually handed over for transport.

Acceptance credit The process of specific Banks arranging acceptance credit involving the acceptance of a Bill of Exchange drawn on any of its members. The 'Bank Bill' as it is called permits funds to be drawn for a set period and amount, and is particularly useful to importers of raw materials or components who have to meet their costs before the finished goods are sold. Usually a 'minimum amount' is imposed when financing the scheme. It operates on the basis the importing company draws a bill of exchange up to an agreed value for payment at a specified date in the future, and the bank arranges for the bill of exchange to be 'discounted' in the money market to one of the discount houses that specialise in this business. The sale proceeds are credited to the importer, making funds immediately available to him. When the bill of exchange matures, the bank pays it and debits the importing company's accounts for the amount.

Acceptance of the goods Under the terms of the Sale of Goods Act 1979, acceptance of the goods occurs where the buyer (i) intimates to the seller he has accepted the goods; (ii) does any act inconsistent with the seller's ownership of the goods, or (iii) keeps the goods for more than a reasonable time, without informing the seller he has rejected them, or in transport terms when the goods/cargo has been accepted by the carrier from the consignor and a cargo receipt has been issued acknowledging the carriers liability at that point thereon.

Accepted Draft A draft accepted by the drawee by putting his signature on its face. The draft thus is subject to the strict provisions of the law regarding bills of exchange if the drawee is publicly registered as a firm. A banking term.

Accepting Bank/Paying Bank A bank that accepts a Usance Bill of Exchange payable upon a stated or determinable future date. The bank is nominated in the documentary letter of credit to accept or pay drawings under that credit. It can be either the issuing or the advising bank.

Accepting House Financial house – often a Merchant Banker specialising in financing foreign trade.

Accepting party The party accepting the goods to be handed over to the carrier. A shipping term.

Access code A combination of characters that enables a computer user to gain entry to a field of information.

Accessories For the purpose of applying to specific air freight commodity rate descriptions, such as additional objects which are not essential to the normal use of the main article or are not an integral component thereof but which are intended for use with the main article.

Accident year The calendar year in which the accident or loss occurred.

Accompanied load The process of the road trailer being accompanied by the driver throughout the international transit. This is very common in the European market with the vehicle operating between Spain and Germany or Italy and UK.

Accord international sur le Transport des denrées Périssables Agreement on the International carriage of perishable foodstuffs. See ATP entry.

Account A record of transactions during a specified period, or the periods into which Stock Exchange dealings are divided – a London Stock Exchange definition.

Accountancy culture An attitude/focus adopted by Government/Company which focus primarily on the financial evaluation in the cost/benefit analysis.

Account day The day for settlement of bargains done during the account – a London Stock Exchange definition.

Account Executive An executive/manager in an entity responsible for all aspects of a sales account client business. It may be in a manufacturing business, service industry such as Airlines, Liner Cargo Companies or an Advertising Agency.

Account sales A term used when goods imported on consignment are valued for customs purposes by reference to the price they achieve when sold in the Community. Account sales commonly apply to importations of perishable goods. An EU customs definition.

Acct Account

ACCU Automatic control certified for unattended engine room.

ACCU – OS Automatic control certified for unattended engine room – open seas.

Acc Yr Accident Year – insurance abbreviation.

ACEP Approved Continuous Examination Programme.

ACETF Airport Capacity Enhancement Task Force – IATA term.

ACF Annual Cash Flow.

Achievers A social/economic category term which identifies successful, work oriented people who derive complete satisfaction from their family and work/job environment.

Acidic In general, an acidic substance is one which contains hydrogen and which dissolves in water to produce one or more hydrogen ions. Such solutions turn litmus dye red and cause other indicator dyes to change to characteristic colours. They also react with certain metals and bases/alkalis to form salts. Acidity is commonly measured using the pH scale; on this scale water has a 'neutral' pH, i.e. neither acidic nor basic, of 7 and acids have a pH lower than 7. Some examples of acidic substances are hydrochloric acid, sulphuric acid, hydrogen sulphide (inorganic acids) and acetic acid, e.g. vinegar and citric acid (organic acids). A dangerous cargo term.

ACIS Advance Cargo Information System.

ACK Acknowledgement.

Acknowledgement of order Document acknowledging undertaking to fulfil the order and confirming conditions or acceptance of conditions.

ACMI contract Aircraft, crew, maintenance, insurance.

ACMEL Associated Continental Middle East Line.

ACO Aerosat Co-ordinating Office (located in The Netherlands) – IATA

ACOP Approved code of practice.

ACORN A classification of residential neighbourhoods. A marketing term used to classify residential areas in socio economic terms.

ACOS Automated Container Operations System – a PSA facility.

ACP African, Caribbean and Pacific countries. Group of 71 countries with preferential trading relation with the EU under the former Lomé Treaty now called the Contonou Agreement.

ACPC Average Cost per Claim – insurance abbreviation

ACPs African, Caribbean and Pacific States (parties to the Lome Convention).

Acquisition The process of one company buying another for a variety of reasons. It maybe to increase market share; raise more capital for investment; develop technology; obtain more research and development as in the pharmaceutical industry; or extend the product range.

ACS Australia Container Service – Shipping Company or Air Cavity System.

ACSC Australian Coastal Surveillance Centre.

ACT Associated Container Transportation Ltd; a consortia of container ship operators.

Actio personalis moritur cum persona (Latin) A personal action dies with the person.

Actio 'in personam' (Latin) An action against a person.

Actio 'in rem' (Latin) An action for the recovery of a particular thing e.g. a seaman's action against the ship's owner for wages or an action for the specific recovery of land.

Actio 'in rem' A legal term involving a court seeking judgement against property.

Action plan A plan drawn up with a view to identifying various areas for action/attention and in so doing ensure the plan objective is realised. For example to improve a container service, the operator may list twenty-five points which need attention such as improved ship punctuality, introduce new container control techniques, develop new distribution arrangements in port area and so on. Such an action plan would have a time scale and be delegated to various parts of the organisation to ensure its overall execution/completion.

Activated Alumina A desiccant (or drying) medium which operates by absorption of water molecules.

Active drive The disk drive on which the computer looks for files. Sometimes called the current drive or default drive

Active exporter An exporter who adopts a proactive as compared to a reactive strategy. In so doing he/she initiates an overseas promotion of the product/service. See passive exporter entry.

Active fleet The world fleet minus surplus/laid up tonnage.

Active importer An importer who adopts a proactive as compared to a reactive strategy. In so doing he/she researches the overseas market place for a particular product/service. See passive importer entry.

Active inventory The provision of raw material, finished products, which will be sold or used within a specified period without extra cost or loss.

Active underwriter In relation to a Lloyd's syndicate, the person at, or deemed by the Committee to be at, the underwriting box with the principal authority to accept risks on behalf of the members of the syndicate.

Activity (Radioactive Material Only) is a measure of the quantity of radioactivity emitted by a radioisotope and is used to determine the amount of radioactive material which may be transported in various types of packagings. A dangerous cargo definition.

ACTS Alfred C Toepfer Schiffahrt, MB Hamburg – a shipowner.

Act of God Any fortuitous act which could not have been prevented by any amount of human care and forethought such as storms, floods, lightning, earthquakes, etc.

Actual flight number See flight number.
Actual Gross Weight The weight of product plus packing (and/or container, trailer or wagon).
Actual pay load (net weight) The difference between the actual gross weight and the gross tare weight of a container, trailer or wagon; or the weight of goods unpacked.
Actuals The physical commodity – an International Petroleum Exchange Term.
Actual total loss A situation when one of the following circumstances exist relative to a marine insurance cover: (i) the assured is irretrievably deprived of the insured property; (ii) the insured property is completely destroyed; (iii) the vessel is posted at Lloyd's 'as missing' in which situation both the ship and her cargo are regarded to be an actual total loss; (iv) the cargo changes in character so that it is no longer the thing that was insured i.e. cement becomes solidified.
Actual voyage number See voyage number.
Actuator pole Long pole of hollow construction used for locking twistlocks in inaccessible positions.
Actus reus (Latin) A prescribed act.
ACU Administration of the customs union – a part of the EEC commission with responsibility for administration of the customs union.
ACV An air cushion vehicle – a hovercraft.
AD Above deck (of a ship); or Art Director a Marketing term and post found in Advertising Agency.
a/d After date.
ADAG Air Traffic Demand and Analysis Group (of EANPG) – IATA term.
ADAPT ATS Data Acquisition, Processing and Transfer (Panel – ICAO) – IATA term.
ADB Asian Development Bank.
ADD Address or In PAM, identifies an application and places its name on the PAM Main Menu so that it can be run from PAM. A computer definition.
Added value The process of providing an input into product/service thereby increasing its realisable value. This may be processing, assembly, packaging, etc. An example is the importation of the oriental cane furniture and the importer providing cushions to enhance its buyer's appeal and value in accord with the market need. See also value added benefit.
Additional Extra premium due from the insured either because the insurance is more hazardous than normal or arising from an endorsement because additional benefits are added to the policy. An insurance definition.
Add on A term equivalent to proportional rate or arbitrary involving an additional charge on the basic rate to cover a particular circumstance e.g. currency surcharge.
Add-ons Additional arrangements that can be purchased in conjunction with a tour.
ADDR Address (commission).
Address Commission A percentage of commission sometimes specified in a charter party due to charterers based on the amount of freight.
ADEAN Common Market An economic trading bloc comprising of Bolivia, Colombia, Ecuador, Peru and Venezuela. It was introduced in 1967.
ADELT Automatically Deployed Emergency Locator Transmitter.
ADG Adaptive Digital Gyropilot.
Ad hoc (Latin) For that purpose.
Adiabatic Describes an ideal process undergone by a gas in which no gain or loss of heat occurs.
Ad idem (Latin) In perfect agreement.
Ad infinitum (Latin) Without end.

Adjoining rooms Two or more hotel rooms side by side but not necessarily with connecting doors.
Adjustable policy An insurance where a provision premium is payable at inception and adjusted after expiry.
Adjustable shore ramp A roadway providing an intermediate connection between the shore and the ship, and on which the shore end of a ship ramp can rest. The roadway is suspended vertically and is adjustable and is usually hinged at the inshore end and supported near the outer end independently of the ship
Adjuster An independent professional who negotiates loss settlements on behalf of the insurer.
Adjusting loading wall Sheet lining which allows dunnage bars to be fitted into the ISO container at any point.
Adjustment Settlement of a loss incurred by insured.
Adk Awning deck.
Ad lib. (ad libitum) (Latin) According to pleasure: Freely.
ADM Asian Dollar Market
Administrative Accompanying Document A customs term involving the documentation required within the European Community embracing Registered Excise dealers and shippers. It operates outside the bonded warehouse scheme and the administrative accompanying document describes the document which enables Customs controls at both ends of the movement. It maybe an official form or a commercial document providing all the data requirements. It involves goods subject to excise duty. A customs document which must accompany all goods subject to excise duty involving an intra community movement which requires Customs approval for such a transit.
Administrative driven A company/entity/department which is motivated and managed by administrative strategies and tactics/techniques.
ADN European provisions concerning the International Carriage of Dangerous goods by Inland Waterways.
Ad nauseam (Latin) Until disgusted, to a sickening degree.
ADNR Accord European Relatif au Transport International des Merchandises. Dangereux par Voie de Navigation Interieure Rhin.
ADP Automatic data processing.
ADR Accord European Nation du transport international des merchandise Dangereux par Route. The European agreement concerning the international carriage of dangerous goods by road. ADR concerns the dangerous substances which are listed in the convention and specifies the nature of the tanks in which these substance may be carried. These tanks are subject to inspection and must carry an ADR certificate issued by the Department of Transport.
ADRIATIC Adriatic Container Service – a shipowner.
Adrift Floating at random.
ADRs American Depository Receipts.
ADS Automatic Dependent Surveillance. An air navigation system.
ADTF Airline Designator Task Force – IATA term.
ADV Advice enclosed.
(Adv) Advance of Special Survey (of a ship).
Ad val Ad val – in proportion to the value.
Ad Valorem 'According to Value'. An Ad Valorem freight rate is one where freight is based on the value of goods.
Ad valorem Bill of Lading An ad valorem Bill of Lading is one where the value of the goods is shown on the face of the document, which value becomes the carrier's limit of liability, in return for the payment of a freight surcharge.

Ad valorem duty Duty evaluated on percentage of cargo value.

Ad valorem freight Freight rate based on percentage value of goods shipped.

Advance arrangement The shippers' obligation to contact the carrier prior to tendering the consignment with a view to initiating the despatch arrangements.

Advanced charge/disbursements A charge paid by a carrier to an agent or to another carrier, which the delivering carrier then collects from the consignee. Such charges are usually for agents forwarding fees and incidental expenses paid out of pocket for account of the shipment by an Agent or other carrier.

Advanced countries Advanced countries are those with high living standards and their economies fully developed yielding high incomes. Examples include Japan, USA, Germany, Switzerland, UK, France and Canada. Literacy and education is to a high standard and may be regarded as the leading nations in economic, trade, political, technical and education terms/influence.

Advance deposit Partial payment in advance for goods or services.

Advanced interline An interline carrier which picks up cargo from the shipper and delivers it to another carrier for shipment to the consignee.

Advanced payment letter of credit An advanced payment arrangement featured in the letter of credit terms. See advanced payment.

Advance freight Freight payable (by Charterer/Shipper etc.) at time goods accepted for despatch.

Advance note A draft on a shipowner for wages, given to a seaman on signing Articles of Agreement and redeemable after the ship has sailed with the seaman on board.

Advance notice A chartering term whereby the shipowners or the master shall give notice of expected readiness to load. See notice of readiness entry.

Advance of freight Payment 'on account' for disbursements.

Advance payment Partial or complete payment prior to completion of contract. (E.g. cash with order).

Advance payment guarantee A guarantee to the import (buyer) that if the exporter (seller) fails to complete the contract any advance payments can be reclaimed. Basically as an exporter advance payment from the buyer of the full contract price is greatly to be preferred. However, cash with order, as this is often known, is uncommon since the buyer is effectively extending credit to the exporter and bears the risk that the goods will not be despatched in accordance with the contract terms. Nevertheless, provision for partial advance payments in the form of deposits (normally between 10 per cent and 20 per cent of the contract price), or progress payments at various stages of manufacture (particularly for capital goods) is often included in the contract terms.

Advance purchase excursion fare (APEX) Special round-trip air or rail fare calling for restrictions and advance payment and subject to penalties for alterations.

Advances against shipping documents The process whereby the bank may be prepared to make advances against shipping documents with or without bills of exchange in course of collection or payable.

Advertising The process of taking paid space/time in a media outlet which maybe a publication-newspaper/journal/magazine; cinema, television or radio. In so doing the advertisement is essentially a form of communication between the seller and buyer containing a specific message and usually persuasive in terms.

Advertising Agency An Agent specialising in Marketing and Advertising techniques and who undertake such work on behalf of a principal.

Advertising Budget The process of formulating forecast and objectives relative to advertising expenditure during a specified period for a particular area of the business activity of the company. It is usually aligned to the Marketing and Sales plan.

Advertising Budget review The process of reviewing critically the advertising budget especially in the light of changed circumstances compared with when the budget was compiled, and the situation obtaining at the time of the budget review and likely events/needs in the remaining period of the budget.

Advertising code The code of ethics/standards/criteria on which advert is conducted in all its areas including press, TV, commercial radio, brochures, posters and so on. The prime considerations involved are authenticity of presentation whereby the contents must be factually correct, and not to mislead the client through misrepresentation of the facts; avoidance of false/unconfirmed statements; not to be abusive or offensive; and so on.

Advertising conversion rate The number of coupons completed by respondents in an advertisement feature requesting for product/service details related to the number of actual orders received. For example, 1,000 coupons are received requesting product details which are duly despatched, and ultimately 300 orders are received thereby having a 30% conversion rate.

Advertising cost per product sale The cost incurred in advertising to secure a product sale. For example, during a year a manufacturer/retailer sold 100,000 washing machines and spent $1.0 million on advertising embracing TV, commercial radio, sponsorships, brochures, newspaper/journal advertisements etc., incurring an advertising cost per washing machine sale of $10.

Advertising criteria The basis on which an advertising product is devised/formulated. This may be a brochure, TV advertisement, press advertising, poster and so on. Individual Advertising Agencies and their clients will have a varying basis on which their advertising criteria is formulated bearing in mind the product, advertising budget, degree of competition and other factors.

Advertising Expenditure Expenditure incurred to promote a product(s) or service(s) by an entity/company. This includes, TV, commercial radio, press advertising, brochures, poster sites, sponsorship, sales conferences, general promotions and so on.

Advertising Manager The executive responsible for developing/planning an entity advertising strategy/execution in accordance with Company strategy.

Advertising message Prime/salient information found in a company advertising material such as brochures, press adverts, TV adverts and so on promoting a product(s) or services(s). It may be features/advantage of the improved product design; strength of after sales service; price competitiveness and so on. See also advertising slogan entry.

Advertising Policy The advertising policy of a particular company or in a particular set of circumstances. It maybe for example on the advertising budget which will be limited to five per cent of the total retail value of the product promoted thereby permitting only €12 to be spent on every €240 budgeted income.

Advertising programme objectives The aims/objectives of a particular advertising programme for a specific period involving a particular product(s) or service(s). This may include the improving of the market share,

Advertising research

successfully launching the new product, the developing of new markets overseas, improving product image and so on.

Advertising research The process of evaluating the effectiveness of the advertising campaign and the actual and potential buyers' readership trends. The latter will specify what newspaper/journal/magazine the buyer reads.

Advertising Revenue Income from advertising. It maybe from a journal, poster site, newspaper, television, commercial radio and so on.

Advertising revenue by media An analysis of advertising income by source such as TV, newspaper, magazine, commercial radio, posters, cinema and so on.

Advertising schedule The advertising timetable plan for a particular product(s)/service(s) of a specified company/entity. The schedule is likely to be between one and twelve months' duration and feature on specific dates TV, journal, press and advertising. This will include cost and the name of the TV network/journal/newspaper, etc.

Advertising slogan A phrase devised to facilitate selling a particular Company product(s) or service(s) and in so doing will feature in publicity material. It maybe 'never knowingly undersold' indicating the retail store will match any competitors' price on documentary evidence for any item they retail.

Advertising strategy A course of action adopted by a company/entity/advertising agency to realise an objective. It maybe market penetration, increase market share, competitive pricing, etc.

Advice note A document describing the condition of the cargo/goods.

Advice of arrival (goods) Notification from the carrier to the consignee in writing, by telephone or by any other means (express letter, message, e-mail etc) informing him that a consignment addressed to him is being or will shortly be held at his disposal at a specified point in the place of destination.

Advising Bank An international trade term. It is the bank through which the issuing bank advises its credit. It may be a bank in the beneficiary's country or a bank in another country. Also see transit credit entry.

AEA Agricultural Engineers Association (UK); or Association of European Airlines – IATA term.

AECMA Association Europeenne des Constructeurs de Material Aerospatial – IATA term.

AECS Australia Europe Container Service.

AELE Association Europeenne de libre Exchange (French). Based in Geneva European Free Trade Association.

AEMERO African, Eastern Mediterranean and European – Regional Office of UNICEF.

AER Accord European Relatif au Travail des Equipages des Vehicles effectuant des Transports Internationaux par Route; or Annual Equivalent Rate: an interest rate defined in the Code of Conduct for the Advertising of Interest Bearing Accounts published by the BBA and the Building Societies Association in the UK.

AERA Automated Enroute Air Traffic Control (US) – IATA term.

Aerating Aerating means the introduction of fresh air into a tank with the object of removing toxic, flammable and inert gases and increasing the oxygen content to 21 per cent by volume.

Aer Lingus The National Airline of the Republic of Ireland.

Aero Lineas Argentinas The Argentine National Airline and a member of IATA.

Aeromexico The Mexican National Airline and a member of IATA.

Aeronautical Radio, Inc. (ARINC) A company owned primarily by the airlines that provides internal domestic US computer communication services among them and others.

AEROSAT Aeronautical Satellite – IATA term.

Aerosol Any non-refillable receptacle made of metal, glass or plastic and containing a gas compressed, liquefied or dissolved under pressure, with or without a liquid, paste or powder, and fitted with a self-closing release device allowing the contents to be ejected as solid or liquefied particles in suspension in a gas, as a foam, paste or powder, or in a liquid or gaseous state. A definition associated with dangerous cargo shipments.

Aerosols Particles, other than water or ice, suspended in the atmosphere. They range in radius from one hundredth to one ten-millionth of a centimetre – or 10″ to 10Ø≥micrometres(μm). Aerosols are important as nuclei for the condensation of water droplets and ice crystals and as participants in various atmospheric chemical reactions.

AESC Australia to European Shipping Conference.

AEW Airborne Early Warning.

AF Advance freight – payable at time goods are accepted for despatch; or Air France – National Airline of France and member of IATA.

AFAICT As far as I can tell – cyberspace term.

AFC Airflex clutch; or ARINC frequency Committee – IATA term.

AFCAC African Civil Aviation Commission – IATA term.

AFCS Automatic Flight Control System – IATA term.

AFDB African Development Bank.

AFEA AFEA Line – a shipowner.

AFFF Aqueous Film Forming Foam.

Affinity group An organisation formed for any purpose and subsequently sponsoring group travel arrangements; clubs, schools and trade associations are examples.

Affreightment A contract to carry goods by ship. Charter parties and bills of lading are contracts of affreightment

Affidavit A written declaration on oath.

AFI Africa(n) – IATA term.

AFIC Asian Finance and Investment Corporation

AFIM PSG AFI ILS/MLS Transition Planning Sub-Group – IATA term.

AFL – CIO American Federation of Labour – Congress of Industrial organisation based in Washington DC.

AFNOR The French Standards Institute.

A Fortiori (Latin) With stronger reason; all the more.

AFPIG Automated Fraud Prevention Interest Group – IATA term.

AFPTC Agricultural Farm Produce Trade Corporation (Burma).

AFRA Average Freight Rate Assessment Panel representing London freight brokers. See world scale entry.

AFRAA Association of African Airlines – IATA term.

Aframax A vessel which is capable of using the African ports having regard to its specification embracing draught, beam, shipboard handling equipment and cargo transhipment facilities. It is a crude oil tanker of 80,000-120,000, and identified with the average freight rate assessments. See AFRA.

Aframax tankers The tankers are between 80,000 and 120,000 dwt.

AFRASEC Afro-Asian Organisation for Economic Co-operation.

Africa Travel Association Group of governments and travel suppliers promoting travel to and within the Africa continent.

AFS Aeronautical Fixed Service – IATA term.
Aft The rear portion/area of an aircraft or ship.
AFTA Asian Free Trade Area established in 1992 in order to encourage progress towards freer intra-ASEAN trade.
Aftermarket The period immediately following a new issue – a London Stock Exchange definition.
After peak Compartment at stern, abaft, aftermost water tight bulkhead.
After Peak Tank Situated below Steering Gear and Deck Store and used either for sea water ballast or filled with fresh water for the daily use of the officers and crew.
After Sales Manager A managerial position responsible for looking after customer interest following purchase of the company's product(s) or service(s). Sometimes it is called Customer Relations Manager.
After Sales Service A department within a Company/ Entity whose prime task is to look after Customer needs following purchase of the Company's product(s) or service(s). In an Engineering Company, for example, this may involve provision of technical servicing manuals, spares replacements, equipment servicing, disputed invoices and so on.
AFTN Aeronautical Fixed Telecommunication Network – IATA term.
AG Hunting Cargo Airlines; or Arabian Gulf.
AGA Aerodromes. Air Routes and Ground Aids (ICAO) – IATA term.
Age multiplier The process of applying the age of a product to a tariff. For example, for vessels registered in the Republic of Cyprus the fees are in various bands according to the age of the ship: up to 10 years, 11-20 years and over 20 years.
Agency administrator IATA conference official responsible for overseeing travel agency relations.
Agency Agreement A contract between the Company and a Broker, which lists all the conditions under which the Company grants an agency upon appointment, i.e. commission terms. Types of Business, etc. – insurance definition. Also relevant to Exports, Imports, Travel, etc.
Agency by ratification A situation/circumstance whereby an Agent commits an act for which he has no authority whatsoever but acquaints his Principal after the occurrence with this act.
Agency Commission Clause A clause in the standard hull clauses (1983) whereby no claim will be admitted under the policy to remunerate the assured for expense incurred, or time expended, in obtaining information or providing documents to support his claim.
Agency Compliance Board A Board responsible for investigating and controlling pretended or actual breaches of IATA Agents.
Agency fees Charges payable by the shipowner to the Agents for services at loading or discharging port. In the UK these fees are set out in a scale issued by the Institute of Chartered Shipbrokers.
Agency Law The legislation dealing with Agents who are people appointed by another to make contracts for him.
Agency manager The person who manages a retail travel agency; in an airline-appointed agency in the US, must have at least three years of experience selling travel to the public.
Agency management system Computerised system that performs various agency functions such as accounting, ticketing, itinerary preparation, report preparation and mailing list control.
Agency rep Sales representative of a hotel, airline or other travel industry segment who calls on travel agents.
Agency tour See familiarisation trip.

Ageing society A populace where the average age is increasing annually.
Agent One who represents a principal or buys or sells for another. For example, in insurance terms, a person who provides advice on the appropriate policy but does not generally provide the same service as a broker or a representative covering the territory to which you are exporting, who may have exclusive rights to your product. An Agent may be employed by an insurer, exporter or a shipper. Types of Agents include a Travel Agent, Forwarding Agent, Receiver's Agent, Clearance Agent, Ships' Agent and so on. (See separate entries).
Agent, IATA Cargo Sales An Agent recognised and approved by IATA appointed by a carrier and authorised by the respective carrier to receive and prepare shipments and to collect charges.
Agent – Lloyd's Underwriting A company, firm or person approved by Lloyd's authorised to manage the affairs of syndicates or Names at Lloyd's; (a) Managing Agent – The agent who manages a syndicate, appointing the underwriter and other staff, investing syndicate funds and preparing accounts. A managing agent frequently also acts as a members' agent; (b) Members – Agent who looks after the interests of the Name to ensure he derives maximum benefit from his participation in syndicates; and (c) Co-ordinating – The agent appointed by a Name to co-ordinate his affairs at Lloyd's where more than one agent is involved. See separate entries.
Aggregate excess of Loss Reinsurance A form of reinsurance providing excess of loss cover for losses arising from any one event (or vessel) in excess of the reinsured's retention up to an agreed limit, but only when the aggregate of claims otherwise recoverable under the excess of loss treaty exceeds a stated amount.
Aggregate inventory The inventory of any group of items or products involving more than one stock keeping unit.
Aggregate measure of support (AMS) Method of converting all forms of support which are trade distortive into a total monetary value in order to have a clearer idea of the level of support for each commodity. An AMS can be used to determine targets for support reductions or as a monitoring tool. A WTO term.
Aggregate Stop Loss Reinsurance This is a wider application of the ordinary Stop Loss treaty in that it applies to the entire portfolio of one branch of the reinsured's activities. See Stop Loss Reinsurance entry.
Aggressive advertising An attitude adopted by a company/entity/advertising agency to put more persuasion/pressure on the client to make a purchase through advertising. See aggressive selling.
Aggressive Marketing The process of having a very active and strong marketing strategy directed towards the client with a view to generating a sale and in so doing placing the customer under pressure to make a purchase. Also in so doing through advertising expose the weaknesses of the competitor's product(s)/services(s) and the strengths of the particular company product(s)/service(s).
Aggressive selling The process of having a very active and strong selling strategy thereby putting a lot of pressure on the client to make a purchase.
Agio The difference in value between one kind of currency and another. This can be a significant balance sheet item when, for example, borrowings are in a different currency to that in which the accounts are prepared, often giving rise to substantial unrealised profits or losses on outstanding debt balances.
AGM Annual General Meeting.

Agora The currency of Israel. See Appendix (C).
Agorot The currency of Israel. See Appendix (C).
Agrarian base Agricultural based (economy).
Agribulks Bulk shipments of agricultural products such as grain and animal feed.
Agricultural Community The work force and their families engaged in agriculture activities in a particular area/region. It would also extend to the supporting industries and the general social fabric of the area/region.
Agri port A seaport handling agricultural products/commodities.
Aground Vessel laying (touching) sea bed.
AGS Annual general survey (of a ship).
AGV Automated Guided vehicles – a PSA facility.
AGVS Automated guided vehicle system – shore based container equipment.
AGW All going well.
Ah After hatch.
AH Range of ports between and including Antwerp/Hamburg; or Air Algerie – An IATA airline.
AHASC Airport Handling Agreements Sub-Committee – IATA term.
AHC Airport Handling Committee – IATA term.
AHESC Airport Handling Equipment Sub-Committee – IATA term.
AHIP Army Helicopter Improvement Programme.
AHM (IATA) Airport Handling Manual – IATA term.
AHPSC Airport Handling Procedures Sub-Committee – IATA term.
AHR Antwerp-Hamburg Range (of ports).
AHS Annual hull survey; or Assistant Head of Section.
AHST Anchor Handling Salvage Tug.
AHT Anchor Handling Tug.
AHTS Anchor Handling Tug Supply (Vessel).
AI Air India – National Airline of India; or artificial intelligence.
AIA Aerospace Industries Association (of America) – IATA term; American Insurance Association or Automotive Industry Association (in Thailand).
AIAA American Institute of Aeronautics and Astronautics – IATA term.
AICES Association of International Courier and Express Services – a trade organisation representing Express parcels operators.
AID Agency for International Development – a US Governmental Organisation.
AIDA International Association for the distribution of Food Products; or Attention, Interest, Desire, Action – a marketing term.
AIDAB Australian International Development Assistance Bureau.
AIDC Automotive Industry Development Committee.
Air donor An organisation/entity/person providing foreign aid to a country.
Aid programme The provision of foreign aid to a country such as the USA despatching in US vessels shipments of grain to India.
AIDS Aircraft Integrated Data Systems – IATA term.
Aid scheme A scheme which provides aid to a specified country/region. It may be financial, human resources, services such as transport and so on.
AIFA American International Freight Association.
AIFSG AFI IFR Flight Study Group – IATA term.
AIFTA Associate Institute Freight Trades Association; or Anglo-Irish Free Trade Area Agreement.
AIG Accident Investigation and Prevention – ICAO term.
AIGSS Annual inert gas system survey.
AIJV ASEAN Industrial Joint Venture

AIL Abnormal indivisible load.
AIM Automatic Identification Manufacturers.
AIMS American Institute of Merchant Shipping – based in Washington DC.
AIO Activities, interest and opinions – a marketing term.
AIR Airworthiness – ICAO term.
AIRAC Regulation and Control of Air Information – ICAO term.
Air Algerie The Algerian National Airline and a member of IATA.
Air Botswana The Botswana National Airline and a member of IATA.
Air Bridge Part of an overall transit in which an air transit features and integrated with at least one other transport mode such as maritime transport thereby forming an intermodal transit. A example is the sea/air bridge with the container/goods being conveyed by the sea transport and the next leg by air transport thereby providing a fully dedicated integrated and co-ordinated schedule.
Air Bus A passenger airline service which tends to operate on short duration flights but having the salient feature of no pre booking requirement. This permits the passenger to simply 'turn up' at the airport terminal for a prescribed flight and purchase the ticket at the time of flight embarkation. It is renowned as a low cost airline service offering frequent services between major industrial cities of relative short flight transit time on an internal or international basis. All the accommodation is one class and no 'in flight' facilities are usually provided such as video entertainment.
Air Canada The Canadian National Airline and a member of IATA.
Air charter The process of a merchant/agent chartering an aircraft for a particular flight or period of time usually the former to convey an abnormally large or awkward consignment. See also Split Charter.
Airline codes The systems of abbreviations for airlines, cities and fares used by the travel industry throughout the world
Air Commuter A person who makes regular and frequent journeys between one's place of residence and work location travelling by air.
Air Consignment Note An Air Waybill.
Air consolidators An Agent usually who offers regular service on scheduled flights and in so doing despatches as one overall consignment under one document the Master Air Waybill a number of individual compatible consignments from various consignees to various consignors. The consolidator tariff is lower than the airline rate for the individual consignments. The air consolidator retails the service to the market, formulates their own rates and usually undertakes the physical consolidation in airline containers.
Air container Any unit load device primarily intended for transport by air having an internal volume of one cubic metre or more incorporating restraint provisions compatible with an aircraft restraint system and an entirely flush base bottom to allow handling on rollerbed cargo handling systems.
Aircraft Kilometres The sum of the products is obtained by multiplying the number of flights performed by the stage distance in kilometres.
Aircraft Manifests A list of individual consignments conveyed on a particular flight.
Aircraft operator A person/Company who has the management and overall control of an aircraft.
Aircraft turn round time The duration of a particular aircraft stay at an airport terminal based on the time the aircraft is received at the terminal to commence

passenger disembarkation and/or cargo discharge, until the aircraft duly loaded leaves the terminal to proceed on her flight or other specified assignment. The aircraft may arrive in a loaded or empty condition and likewise depart on a similar basis.

Aircrew Personnel who man an aircraft such as a pilot, navigator, flight engineer, air hostess and so on.

Air France The French National Airline and member of IATA.

Air Freight Cargo conveyed by airline services.

Air Freight consolidation The process of despatching as one overall consignment under an Agent/Freight Forwarders sponsorship by Air involving a number of individual compatible consignments from various consignees to various consignors.

Air Freighter Aircraft carrying exclusively air freight consignments.

Air Freight insurance The process of insuring air freight consignments which is usually cheaper than by surface transport.

Air Guinee A National Airline and member of IATA.

Air hostess Part of the aircrew team and primarily responsible for 'in flight' passenger welfare.

AIRIMP ATC/IATA Reservations Interline Message Procedures Manual – an IATA term.

Air India The Indian National Airline and a member of IATA

Airline Includes the air carrier issuing the air waybill and all air carriers that carry or undertake to carry the cargo under the air waybill or to perform any other services related to such air carriage. IATA definition.

Airline Delivering The carrier who delivers a consignment to the consignee or his agent.

Airline, First The participating airline over whose air routes the first section of carriage under the air waybill is undertaken or performed.

Airline, issuing The airline who issues the air waybill – IATA definition.

Airline, Last The participating airline over whose air routes the last section of carriage under the air waybill is undertaken or performed; or for the purposes of determining the responsibility for collecting charges and disbursement amounts, the airline which delivers the consignment to the consignee whether or not the airline has participated in the carriage. IATA definition.

Airline Stewards and Stewardesses Association The trade union of US airline flight attendants.

Airline, Participating An airline over whose air routes one or more sections of carriage under the air waybill is undertaken or performed.

Airline Pilots' Association (ALPA) Collective bargaining unit for airline Pilots, co-pilots and engineers in the US.

Airline, Receiving A participating airline that receives the consignment from a transferring airline at a transfer point.

Airlines Reporting Corp. (ARC) Non-profit airline-owned corporation that accredits US travel agencies and processes their air sales remittances; also sets and enforces standards for agency bonding and handling and storage of tickets.

Airline, transferring A participating airline that transfers the consignment to a receiving airline at a transfer.

Airlock A separation area used to maintain adjacent areas at a pressure differential. For example, the airlock to an electric motor room on a gas carrier is used to maintain pressure segregation between a gas-dangerous zone on the open deck and the gas-safe motor room which is pressurised.

Air Malawi The National Airline of Malawi and member of IATA.

Air Malta Airways The Maltese National Airline and member of IATA.

Air mass A widespread body of air that is approximately homogeneous in its horizontal extent, particularly with reference to temperature and moisture distribution; in addition, the vertical temperature and moisture distributions are approximately the same over its horizontal extent.

Air Mauritius The National Airline of Mauritius and member of IATA.

AIRMIC The Association of Insurance and Risk Managers in Industry and Commerce.

Air mode container An ISO series container suitable for international exchange and for conveyance by road, rail and sea as well as air including interchange between these forms of transport.

Air Navigation (Dangerous Goods) Regulations 1986 See Air Navigation Order 1985 entry.

Air Navigation Order 1985 UK legislation in conjunction with the Air Navigation (Dangerous Goods) Regulations 1986 which require compliance by Air freight operators/shippers. The legislation specifies the Technical Instructions for the Safe Transport of Dangerous Goods by air as published by the International Civil Aviation Organisation.

Air New Zealand The National Airline of New Zealand.

Air Pacific A national airline and member of IATA

Airport A terminal and an area within which aircraft are loaded with and/or discharge of cargo and likewise passengers embark or disembark. A modern airport as found in Singapore or Paris is fully integrated with the transport network and has an extensive range of facilities for both passengers and freight including fuelling and servicing/victualling of aircraft.

Airport and Obstacle Data Base Information on obstacles near airports necessary for aircraft performance analysis – an IATA resource.

Airport Association Council International/North American North American branch of world organisation of governmental operators of airports, formerly Airport Operators Council International.

Airport infrastructure All the facilities provided at an airport embracing air traffic control, passenger lounge, shops, rail/road networks, aircraft maintenance/servicing resources, customs, immigration and so on.

Airport lounge Accommodation provided in a passenger terminal at an airport whereby, passengers are accommodated/seated/assembled following their final 'check in' prior to proceeding to board the aircraft.

Airport tax Airport Authority tax on passengers passing through it. See head tax entry.

Air Rights A term which relates to the use of 'free space' within a building which can be used for conveying, storage or lifting purpose.

Air/sea Cruise or travel arrangement in which air and sea transportation are combined.

Air taxi Small operator of non-scheduled or on-demand air transportation for short distances.

Air time The period during which an advertisement is transmitted on television or radio.

Air Traffic Controller A Managerial position usually located at an Airport responsible for all aircraft movements both take off and landing at the Airport. This involves very close liaison with air crew at all times during such periods which is greatly facilitated by modern telecommunication systems and computerisation. See also Traffic Control entry.

Air Transport Association (ATA) The trade association of US scheduled airlines, based in Washington.

Air Transport Association of Canada (ATAC) A trade association of Canadian airlines.

Air Transport Document An authenticated receipt, indicating that goods have been accepted for carriage, naming the airport of departure and the airport of destination as stipulated in the Documentary Credit.

Air Travel Card A credit card owned and administered by the airlines; formerly called the Universal Air Travel Plan.

Air Waybill An air freight consignment note made out by or on behalf of the shipper which evidences the contract between the shipper and carrier(s) for carriage of goods over routes of the carrier(s).

Air Waybill, neutral A standard air waybill without classification of issuing carrier in any form. An IATA term.

Air Waybill, substitute A temporary air waybill which contains only limited information because of the absence of the original air waybill, and is the document issued to cover the forwarding of cargo in the absence of the original air waybill. An IATA definition.

Airworthiness certification Equipment/unit has been certified by a governmental airworthiness authority that it meets the safety requirements for the aircraft in that country.

Air Zaire The National Airline of Zaire and member of IATA.

Air Zimbabwe The National Airline of Zimbabwe and member of IATA.

AIS Aeronautical Information Service – ICAO term; Automatic Identification Systems – navigation radar aide; or Advance Info Service Co., Ltd. (in Thailand).

AISG AIS Group – see EANPG entry.

Aisle space The distance between rows of parked or stacked containers which allows access for lifting or transporting equipment.

AIT Alliance Internationale de Tourisme.

AITAL International Association of Latin American Air Transport.

AITS Automated Information Transfer system – a USA resource providing an information service available to exporters.

AJCL Australia/Japan/East Asia container service.

AL Aluminium cover (hatch covers on a ship); or Albania – customs abbreviation.

ALADI Latin American Integration Association.

ALALC Asociacion Latin Americana de Armadores (Latin American Association of Shipowners).

ALAMAR Latin American Association of Shipowners.

Albedo The proportion of the radiation falling upon a non-luminous body which it diffuses.

ALEX The computerised reservation and national distribution system of Iceland; partner of Amadeus.

Alias (Latin) Otherwise named – a fictitious name.

Alibi (Latin) Elsewhere.

Alien insurer or re-insurer An insurer or re-insurer organised under the laws of a non-United States jurisdiction.

Aligned Export Documentation system Method whereby as much information as possible is entered on a 'master' document so that all or part of this information can be reproduced electronically onto individual forms of a similar design.

Aligned forms A series of forms so designed that items of information common to all forms appear in the same relative positions on each form.

Aligned series original document One of a number of documents reproduced from a master in an Aligned System, designated by means of contrasting into stamping etc. to be valid as the original, the other being similarly designated as copies.

Aligned systems Method whereby as much information as possible is entered on a 'master' document or EDP record so that all or part of this information can be reproduced electronically on individual forms of similar design.

Alitalia The Italian National Airline and a member of IATA.

Aliter (Latin) Otherwise; on the contrary.

ALL Admiralty List of Lights.

All cargo carrier An airline conveying only air cargo.

All expense tour A fixed-price tour including transportation, meals, lodging, porterage and sightseeing; be aware that the terms 'all-expense' and 'all-inclusive' are often misused.

Alliance of Canadian Travel Associations (ACTA) Federation of travel industry associations in the provinces.

All-in All-inclusive.

All-inclusive tour The time saved to the ship from the completion of loading/discharging to the expiry of the laytime excluding any notice time and periods excepted from the laytime. A charter party term and also referred to as 'all working time saved'.

Allocated material Stock which has been allocated, earmarked, or set on one side for a particular purpose.

Allocatee A third party to whom, in certain defined circumstances, either a part or all of the proceeds obtained under a Documentary Credit may be paid. An international trade financial definition.

Allocation The process of allocating costs, cargo, cargo space, passenger seats, facilities, personnel and so on to a specified activity. For example, an assignment of rooms to a specific organisation for them to sell exclusively.

Allotment A method by which a seaman may allot part of his wages regularly to a near relative or savings bank; or the issue of shares to persons who have subscribed to an offering or placement. The subscriber may receive his allotment in full, or, if the issue has been oversubscribed, he may be scaled down.

Allowance free baggage The baggage which may be carried without payment of a charge in addition to the fare.

All Risks Insurance cover providing for an extensive cover; special risks, however, being usually not covered by it (e.g. risks of war and strikes or perishing of goods). It includes all fortuitous causes of loss. It does not embrace inevitable loss, such as wear and tear. In practice, such a policy always specifies certain risks that are excluded from cover.

All Risks Clause Protection against all losses and damage caused by external and fortuitous events.

All time saved The time saved to the ship from the completion of loading/discharging to the expiry of the laytime including periods excepted from the laytime. A charter party term.

All told Charter party term indicating that deadweight capacity of vessel embraces bunkers, water dunnage, stores and spare parts.

All working time saved The time saved to the ship from the completion of loading/discharging to the expiry of the laytime excluding any notice time and periods excepted from the laytime. A charter party term and also referred to as 'all laytime saved'.

Aloft Above or overhead.

Alongside The process of accommodating a vessel berthed at a quay/wharf usually for the purpose of discharging and/or loading cargo or passenger disembarkation/embarkation.

ALPA Air Line Pilots' Association (US).

ALPAC Alpac Line – a shipowner.

Alphanumeric Keys The keys on the keyboard that produce either letters or numbers when pressed.

Alpha stocks Part of a former, but popular, classification of share marketability, alpha stocks were shares of large companies most actively traded on the Stock Exchange. There were also beta, gamma and delta stocks in diminishing order of tradeability.
ALRS Admiralty List of Radio Signals.
Alt Altered.
Alu Aluminium (hatch cover material).
Aluminium powder The unpolished powder may evolve hydrogen in contact with water and finely divided dust may be ignited by naked lights or sparks. Polished aluminium powders which have been treated with oils or wax for printing or paint purposes are not generally dangerous.
Aluminium processing by-products These are the materials, consisting of skimming of virgin aluminium, rising to the surface of impure molten aluminium metal.
A l'usine (à la mine, ex magasin en magasin etc.) French Cargo delivery term – Ex-works (ex factory, ex-mill, ex-plantation, ex-warehouse etc.).
Always accessible or reachable on arrival In such a situation the charterer undertakes that when the ship arrives at the port there will be a loading/discharging berth for her to which she can proceed without delay.
AM Aerovias de Mexico S.A. de C.V. (AEROMEXICO). The Mexican National Airline and a member of IATA; assistant Manager; or America – customs abbreviation.
AMA American Marketing Association; Area Minimum Altitude; or Air Marketing Associates.
Amadeus The European computerised reservations system whose founding owners are Air France, Lufthansa and Iberia; it is based in Madrid.
Ambient temperature The temperature of a substance surrounding a body. Thus the ambient temperature of a container holding refrigerated cargo would be the temperature of the air to which it is exposed outside.
AMCM Airborne Mine Counter Measures.
AMCP Aeronautical Mobile Communications Panel – ICAO term.
AMD Advanced Multi-Hull Design.
Amendment An international trade term found in banking. It is an alteration to the credit. Advice of any alteration comes from the issuing bank only and must be advised to the beneficiary through the advising bank if there is one. Irrevocable credits cannot be altered without the consent of all parties to the credit.
American Automobile Association (AAA) A not-for-profit federation of motor clubs in the US and Canada that provides members with travel information, highway services, insurance and other auto-related services. It also operates travel agencies throughout its network.
American Bureau of Shipping The USA Ship Classification and Survey Society.
American Bus Association (ABA) Trade group of intercity and charter bus operators that has strong relations with other segments of the travel industry.
American Chamber of Commerce The organisation based in London representing the USA, whose basic role is to develop, promote and facilitate international trade between the UK and USA and vice versa.
American Deposit Receipts Certificates issued by American Agents for shares of UK Companies dealt in USA.
American Hotel & Motel Association (AH & MA) A federation of state and regional lodging industry associations.
American Petroleum Institute This is a major national trade association representing the entire oil and natural gas industry in the USA. With its headquarters in Washington DC and councils in 33 States, it is a forum for all parts of the oil and natural gas industry to pursue priority public policy objectives and advance the interest of the industry.
American plan (AP) A hotel rate that includes a room and three meals a day; see full pension.
American selling price A system of US Customs evaluation whereby duty is levied not on the price of imported goods but on the price at which the comparable domestic article is freely offered for sale in the USA.
American short ton A total weight of 2000lbs.
American Society of Travel Agents (ASTA) Principal US travel industry trade association representing agents and tour operators with airlines, hotels and other industry segments as allied members.
American Style An option which can be exercised by the buyer (holder) at anytime during its life. An IPE definition.
AMG Airspace Management Group – IATA term.
Amidships Centre of the ship – mid point area between the forward and aft portions.
AMI Ex Associate Member of the Institute of Export.
AMIME Associate Member of the Institute of Marine Engineers.
AMM Asia Merchant Marine (Hyundai Group) – a ship-owner.
AMMI American Merchant Marine Institute – organisation of shipowners in the United States.
AMM SYS Ammonia system.
Ammunition Generic term related mainly to articles of military application consisting of all kinds of bombs, grenades, rockets, mines, projectiles and other similar devices or contrivances.
Ammunition, Incendiary Ammunition containing incendiary substance which may be a solid, liquid or gel including white phosphorus. Except when the composition is an explosive per se, it also contains one or more of the following: a propelling charge with primer and igniter charge; a fuse with burster or expelling charge. The term includes: ammunition, incendiary, liquid or gel, with burster, expelling charge or propelling charge; ammunition, incendiary, with or without burster, expelling charge or propelling charge; and ammunition, incendiary, white phosphorous, with burster, expelling charge or propelling charge.
Ammunition, illuminating With or without burster, expelling charge or propelling charge. Ammunition designed to produce a single source of intense light for lighting up an area. The term includes: illuminating cartridges; grenades; projectiles and illuminating and target identification bombs. The term excludes Cartridges, signal; signal devices, hand; Signals, distress; Flares, aerial and Flares, surface.
Ammunition, Practice Ammunition without a main bursting charge, containing a burster or expelling charge. Normally it also contains a fuse and a propelling charge.
Ammunition, Proof Ammunition containing pyrotechnic substance(s) used to test the performance or strength of new ammunition or weapon components or assemblies.
Amortisation The redemption of bonds or loans by annual payment from a sinking fund. In the case of ship mortgage loans, the lender is looking to the contractual or anticipated operating cash flow of the mortgaged vessel to amortise the loan.
Ampoule A small receptacle sealed after filling by fusing the glass neck.
AMPS Advanced Mobile Phone System.
AMS Annual machinery survey (of a ship); American Bureau of Shipping-highest classification of machinery;

Automated Manifest system or Aeronautical Mobile Service.
AMSA Australian Maritime Safety Authority.
Amt. Amount.
AMT Air Mail Transfer – a remittance purchased by the debtor from his banker in international trade.
Amtrak The name under which the National Railroad Passenger Corp. operates US intercity passenger trains.
AMVER Automated Mutual-Assistance Vessel Rescue (System).
Amwelsh 93 Code name of BIMCO approved Americanised Welsh coal voyage charter party.
AN Ansett Airlines of Australia – an IATA airline.
ANA All Nippon Airways; Airline Network Architecture Development and Specification Ad Hoc Group (ATA/IATA); or Article Number Association.
Analogue The representation of the performance of a system by continuously variable physical entities, for example currents, voltages. Contrasted with Digital – a computer term.
Analysis The process of evaluating/analysing critically a particular situation/circumstance of all its component/constituents parts with a view to forming a conclusion/recommendation/decision. In the business sector, economics and statistical techniques can play a significant role in the facilitation of the analytical process.
ANC Absolute net claims – an insurance abbreviation; Air Navigation Commission or Air Navigation Conference – ICAO term.
ANCAT Abatement of Nuisance Caused by Noise – see ECAC entry.
Anchor Implement by which a ship is rendered stationary.
Anchorage dues Dues raised on a vessel by the port authority involving ships anchored in the port environs.
Anchor bracket Corner casting device used in some railcars.
Anchor handling tugs A vessel specially designed to handle offshore oil and gas platforms and drilling rigs involving maritime towage and anchor handling situations.
Anchor position A navigational term indicating the anchored location of a specified vessel.
ANCOM Andean Common Market including Bolivia, Colombia, Ecuador, Peru and Venezuela.
And arrival Used in relation to return premiums allowed on a hull policy. No such return premium is payable by the Underwriters until expiry of the policy and subject to the condition the vessel has arrived i.e. she is safe in port.
ANDEAN A trading bloc embracing Bolivia, Colombia, Ecuador, Peru and Venezuela.
Andean Community Bolivia, Colombia, Ecuador, Peru and Venezuela.
ANEAC Aircraft Noise and Emissions Advisory Committee – IATA term.
ANETAF Aircraft Noise and Emissions Task Force – IATA term.
ANF Arrival notification form.
ANFOOD Animal Food (type of Cargo)
Angled stern ramp A ramp that extends from the stern of a Ro-Ro vessel at an angle to the centre line of the vessel. Fixed stern ramps are customarily located on the starboard.
Angle of Repose The angle between the horizontal and maximum slope that loose material assumes when it is stockpiled.
Anglo French Trade Trade between France and Great Britain and vice versa.

Anglo German Trade Trade between Germany and Great Britain and vice versa.
Anglo Japanese Trade Trade between Japan and Great Britain and vice versa.
Angola Airlines Luanda Angola National Airline and member of IATA.
ANHSING Anhsing Overseas Line – a shipowner.
ANL The Australian National Line – a shipowner.
ANM Air Traffic Management Notification Message – IATA term.
Annecy round The second round of GATT multi trade negotiations which were held in Annecy in 1949.
Annual Sales Conference An annual conference usually sponsored by a Company/entity which presents its latest products and related developments during the preceding twelve months to the market place. The invited audience would be the Company's major clients (existing-potential). Such a Conference is an important event in the Company Sales calendar as it maintains close contact with the market and thereby facilitates development of Company sales.
Annual Sales Review The process of reviewing/analysing the sales results of a particular Company, Product etc. It will cover a particular period and could be compared with budget and previous years results. Remedial measures to improve sales is likely to be discussed.
Annul To cancel or render void.
ANP Absolute net premiums – an insurance abbreviation; or Air Navigation Plan – IATA term.
ANRO The consortium of Australian National Lines including Australia Straits Container Line, Nedlloyd Lines and Neptune Orient Lines.
ANSI American National Standards Institute (formerly ASA) – IATA term.
Ante (Latin) Before.
Anticipation equivalence An extension to equivalence, the relief earning exportation taking place before the relieved importation. A customs term.
Anticipatory breach An anticipatory breach arises where one party announces his/her intention of not performing a contract which is due to be performed in the future. A legal term.
Anti-dumping See anti-dumping tariff duty entry.
Anti-dumping duties Article VI of the GATT 1994 permits the imposition of anti-dumping duties against dumped goods, equal to the difference between their export price and their normal value, if dumping causes injury to producers of competing products in the importing country. WTO term.
Anti-fouling system A technique whereby to counter marine growth on the hull of a vessel special measures are taken such as the application of a special type of hull coating or self-polishing co-polymer paint.
Anti-jack-knife device A device provided for road haulage units for preventing by one of two methods: (a) the wheels from locking so as to avoid any skidding action, or (b) by locking the fifth wheel under braking to keep the vehicle in a rigid state.
Anti-nosedive leg Support provided at the front end of a container chassis used to support that end during loading operations.
Anti-rack device Hardware that can be attached to container doors to provide additional strength and stiffness to the door and frame assembly, enabling ISO containers to withstand greater racking forces.
Anti-roll suspension A suspension system which incorporates a torsion bar between the axle and the vehicle chassis to resist the rolling action which occurs when a vehicle negotiates a corner.

Anti-theft device System for preventing vehicle theft by immobilising the vehicle or causing warning buzzers to sound if an attempt is made to enter the vehicle or drive it away. Immobilisers include steering column locks, fuel cut out devices and electrical cut outs as well as warning devices.

ANTS Automatic Navigation and Track keeping System. The name adopted by the Furuno Electric Company to describe its navigation system for use with integrated bridge design.

AN (US) Arrival Notice: US equivalent of an ANF Advice to Consignee of goods coming forward.

Any one bottom (or any one policy) A term applied by a reinsurer to a limit in marine reinsurance treaty to avoid a potential accumulation of liability (bottom hull).

ANZECS Australia, New Zealand, Europe, Container Service.

ANZ/GUL Australia, New Zealand/Gulf container service.

ANZMSA Australian & New Zealand Merchants' & Shippers' Association.

ANZUS Mutual Aid Pact between Australia, New Zealand and the USA – a Pacific Pact.

AO Aviación y Comercio SA (AVLACO) – an IATA member Airline.

A/o Account of.

AOB Any one bottom. A marine insurance term.

AOC Aeronautical Operational Control – IATA term.

AOCI Airport Operators Council International (US).

AOCTs Associated overseas countries and territories.

AODB Airport and Obstacle Data Base – IATA term.

AODC Association of Offshore Diving Contractors.

AOLOC Any one location.

AOP Aerodrome Operations – IATA term.

AOPA Aircraft Owners and Pilots Association (US).

AOPG Aerodrome Operations Group – see EANPG.

AOPTF Aerodrome Operations Task Force.

AOS Any one steamer.

AOV Any one vessel.

AP Angkutan Pertambangan – a shipowner; American Plan; or Air One SpA – IATA associate Airline member.

AP or A/P Additional premium or aft perpendicular (of a ship).

APAA Asean Port Authorities Association.

APANPIRG Asia/Pacific Air Navigation Planning & Implementation Regional Group – IATA term.

APATSI Airport/Air Traffic System Interface – see ECAC entry.

APBH After peak bulkhead.

APC Administrative Policy Committee; or Aeronautical Passenger Communications – IATA term.

APEC Asian Pacific Economic Co-operation forum is made up of 15 Countries including Australian, Brunei, Canada, China, Chinese Tapei, Hong Kong, Indonesia, Japan, Malaysia, New Zealand, the Philippines, Singapore, South Korea, Thailand and the United States. A WTO term.

APEX Advance purchase excursion fare.

APG Automation Policy Group – IATA term.

API American Petroleum Institute.

APIRG AFI Planning and Implementation Regional Group – ICAO term.

Apollo Marketing name in the US, Mexico and Japan for the Galileo International computerised reservations system.

APP Application.

Apparel The vessels outfit embracing rigging, life boats, anchor and so on.

Apparent good condition The statement relative to the condition of the goods which have been shipped and are so described on the Bill of Lading. The shipowner having no right of inspection of the contents, receives goods and acknowledges receipt only according to their external condition. See clean bill of lading.

Apparent good order of the goods An indication the merchandise/consignment is in a satisfactory condition.

Appellate body An independent seven-person body that, upon request by one or more parties to the dispute, reviews findings in panel reports. A WTO term.

Appellations of origin Indications of where goods originate with characteristic qualities which are due exclusively or essentially to the geographical environment (for example, 'Bordeaux' or 'Roquefort'). A WTO term.

Appellate review Proposal for an independent appeals body to reappraise the findings in panel reports when requested by one or more of the parties. A WTO term.

Applicant An international trade term found in banking. It is the person, usually the buyer, on whose behalf the credit is issued. Also sometimes known as the 'opener'.

Application A computer program that performs a specific task or set of tasks for the user. Sometimes called an application program. Applications can be 'added' to PAM so that they can be started (or executed) from the PAM Main Menu. A computer term.

Application label The label on the PAM Main Menu with the name of an application. You can select an application to run by moving the pointer to the label so that it becomes highlighted. A computer term.

Applications software Software used for a specific purpose, e.g. textual document processing (word processing), text and graphic printed presentation (desktop publishing), numerical analysis processing (spreadsheets), record and transaction processing (databases), computer-aided design and graphic drawing (vector graphics), graphics/artwork processing (bitmap graphics), slide/picture presentation (bitmap graphics), accounts processing. Some software integrates several of the above uses in one package. A computer term.

Applications software routines Facilities providing user automation facilities such as macros, styles and templates. A computer term.

Appointment Procedure by which travel agencies obtain rights to sell products on behalf of travel industry suppliers. Travel agency term.

Appreciating currency A floating currency which is tending to appreciate in value in terms of its exchange rate in comparison with major convertible currencies.

Appro Approval.

Approval Marks Official markings on certain motor vehicles or motor vehicle parts which show that they conform with a type approved by an international agreement to which the UK is a party. See CE entry.

Approved continuous examination programme An agreement between the equipment owners and classification society to permit continuous examination of the equipment such as a container.

Approved Equipment Equipment of a design that has been type-tested and approved by an appropriate authority such as a governmental agency or classification society. Such an authority will have certified the particular equipment as safe for use in a specified hazardous atmosphere.

Approved or held covered A vessel which is approved or held covered is one which the Underwriters consider adequate to carry the insured cargo at the agreed premium rate. In circumstances when the vessel is not approved the risk is still covered but subject to a reasonable additional premium.

Approved premises Premises approved by the Commissioners of Customs & Excise for purpose specified.

APR Annual Percentage Rate: The interest rate which has to be quoted in advertisements and quotations for credit regulated by the Consumer Credit Act 1974 in the UK.

A priori (Latin) From the cause to the effect.

Apron That portion of a wharf or pier lying between the waterfront edge and the (transit) shed, or that portion of the wharf carried on piles beyond the solid fill; or a defined area of land at a aerodrome intended to accommodate aircraft for the purpose of loading or unloading passengers or cargo, refuelling, parking or maintenance.

Apron Wharf That portion of a wharf or pier lying between the waterfront edge and the (transit) shed, or that portion of the wharf carried on piles beyond the solid fill.

APS Arrival Pilot station.

APSA Australian Port Shippers Association.

APSC Aircraft Performance Sub-Committee – IATA term.

APT Aft peak tank; Automated Pit Trading or actual pumping time – a chartering term.

AQ Aloha Airlines, Inc. An IATA associate member Airline.

A quai (dédouané ... port convenu) (French) Cargo delivery term. Ex quay.

A/R All risks (insurance term).

AR All Risks – insurance abbreviation; Aerolineas Argentinas. The Argentine National Airline and a member of IATA; average revenue or Arrival Report.

ARA Antwerp-Rotterdam-Amsterdam ports. These include ports of Terneuzen, Flushing, Scheveningen, Moerdijk and Dordrecht.

Arab-British Chamber of Commerce The organisation based in London representing the Arab States whose basic role is to develop, promote and facilitate international trade between UK and the Arab states and vice versa.

Arab Common Market This agreement came into force on 1st January 1965 to eliminate tariffs and provide for eventual (a) freedom of movement of persons and Arab capital; (b) exchange of local and foreign goods and products; (c) freedom of residence, work, employment and any economic activity; (d) freedom of transport, transit, and the use of means of transport, civil harbours and airports within the territory of all members of the Arab League.

Arabian Light Crude Oil The world's most prolific crude oil frequently used as a 'market crude' by which the prices of other crude oils are fixed by comparing their quality with Arabian light.

Arab Maghreb Union An economic group formed in 2003 embracing Libya, Morocco and Tunisia to exploit/develop the opportunities for inter-regional trade.

Arab Organisation for Standardisation and Metrology Founded in 1965 to serve as a specialised technical body for the League of Arab States in the fields of standardisation, metrology and quality control.

ARAG Antwerp, Rotterdam, Amsterdam, Ghent (seaports).

Aramid A manufactured fibre consisting of very long molecular chains formed by rearranging the structure of aromatic polyamides.

Arbiter See travel agent arbiter.

Arbitrage The switching of funds from one financial market to another to take advantage of higher yield or capital gain opportunities as a result of interest or exchange rate differentials prevailing between two or more centres. This involves the buying of securities in one country and selling them in another with the object of making a profit and thereby taking advantage of price variations in parallel markets e.g. differences in prices between LIFFE and Eurex (the German/Swiss futures exchange), or price differences between any two OTC products or their constituent parts.

Arbitrageurs A person who takes advantage of pricing anomalies to make risk free profits in the futures market.

Arbitration Method of settling disputes which is usually binding on the parties concerned – clause usually found in charter parties.

Arbitration clause A clause found in a contract, such as a charter party, which binds both parties to settlement of any dispute arising from the contract, to an arbitration procedure; or in insurance terms a clause providing a means of resolving differences between the reinsurer and the reinsured without litigation. Usually, each party appoints an arbitrator. The two thus appointed selected a third arbitrator, or umpire, and a majority decision of the three becomes binding on the parties to the arbitration proceedings.

ARC Applied Research Corporation (Singapore); Airlines Reporting Corp. or Committee for the Review of the Application of Satellites and Other Techniques to Civil Aviation.

ARCADIA Arcadia Shipping Lines Pte. Ltd. – a shipowner.

Arc elasticity of demand The value of elasticity of demand between two points calculated on the basis of the average price method. An economic term.

ARCH Decks supported by web frames or diagonal bracing with pillars.

Archie An internet resource for finding files online – a cyberspace term.

ARCP Aerodrome Reference Code Panel – ICAO term.

Arcquip system A range of equipment in a container embracing, for example, the attachment rails on the inside walls to provide lashing and separation of the cargo.

ARCS Admiralty Raster Chart Service.

Arctic Pollution Prevention certificate A certificate required by the Arctic Waters Shipping Pollution Prevention Regulations for shipowners making Arctic voyages north of latitude 60° N in a shipping safety zone. It is issued by a Canadian inspector or ship classification society.

Area off hire lease Geographical area where a leased container becomes off hire.

Area off hire sublease Geographical area where a subleased container becomes off hire.

Area settlement plan The mechanism through which US travel agents report and remit ticket sales to the airlines and Amtrak, operated by ARC.

ARFA Allied Radio Frequency Agency – NATO term.

Argentine Peso The currency used in Argentina. See Appendix 'C'.

Arian Afghan Airline A national airline and member of IATA.

ARINC Aeronautical Radio Inc.

Arm chair research An alternative description of desk research. See desk research entry.

Armitage Report An inquiry into lorries, people and environment published in December, 1980. Its terms of reference were 'to consider the causes and consequences of the growth in the movement of freight by road and, in particular of the impact of the lorry on people and their environment; and to report on how best to ensure that future development serves the public interest'. Subsequently legislation was introduced in 1983 permitting commercial vehicles to have a gross laden weight of 38 tonnes spread over a five axle vehicle unit with two axles on the tractor unit. Additionally the maximum permitted length of articulated vehicles was increased by half a metre to a total of 15.5 metres and the trailer unit from 12 metres to 12.20 metres in the UK. In 2004 the gross laden weight permitted was 44 tonnes.

ARN ATS Routes and Associated Navigation Means – IATA term.

ARO Average rate option.

ARPA Automatic Radar Plotting Air; or Agent Reporting Plan (wholly owned Subsidiary of IATA).

Arranged total loss A compromised total loss.

Arranged total loss (comprised total loss) A compromise settlement where a total loss has not in fact occurred, but is deemed to have done.

Arrangement fee A fee usually charged by a bank over and above the interest payable on any loan advance.

Arrangement with creditors or members In the event of a company experiencing serious financial difficulties, the directors may propose to both shareholders and creditors a composition of its debts or a scheme of arrangement of its affairs. Such arrangements, involving capital reconstructions and rescheduling and waivers of debt, have not been uncommon in the shipping industry since the mid-1970s.

Arrest Detention of a vessel with a view to her ultimate release when the purpose of her arrest has been fulfilled. An arrest may take place when a ship has contravened some port or national regulation or when the ship is held during the process of exercise of a maritime lien. The arrest of vessels in port has been a measure to which creditors have had frequent recourse. In the event of subsequent non-payment of liabilities, the consequence is generally the auction of the vessel in compliance with the legal procedures of the country in which the vessel was seized. Where this is a peril in the policy but is part of a judicial process it is not covered by the policy. See detention.

Arrest of vessel for time charterer debt The process of arresting a vessel for debts incurred by time charterers.

Arrival date The date on which the consignment arrived/was received by the specified consignee.

Arrival notice See arrival notification form entry.

Arrived notification form A document which advises a consignee or a container operator that goods or containers have arrived at the port of discharge.

Arrived ship Vessel arrived at berth or port or off the port – according to charter party terms.

ARS Annual refrigerated machinery survey (of a vessel).

ARSA Airport Radar Service Area (US).

ARSO The African Regional Organisation for Standardisation. Founded in January 1977 under the auspices of the UN Economic Commission for Africa (ECA). Membership open to the national standards bodies of African countries who are members of ECA and Organisation of African Unity.

Art Director A senior position found in an Advertising Agency responsible for all the artwork in the preparation of advertisement and promotional merchandise.

ARTI Advanced Rotorcraft Technology Integration.

Article Number Association Organisation responsible for establishing the bar-code seen on consumer products.

Articles of Agreement Contract of employment between shipowner and crew.

Articulated vehicle A combination of tractive unit and semi trailer in which the trailer is resting on and attached to the tractive unit, and 20 per cent or more of the weight of the trailer when it is uniformly loaded is superimposed on the tractive unit. The maximum permitted overall length of an articulated vehicle is 15.50 metres with a gross laden weight of 44 tonnes spread over five axles.

Artificial exchange rate A currency exchange rate which does not have any necessary alignment with the current market exchange rate.

Artificial trade barriers This represents a barrier to the conduct of free trade between countries and features non tariff regulations, industrial standards and favouring local industry rather than foreign companies in awarding contracts. See trade barrier entry.

Artificial tweendeck An artificial tweendeck is a container without endwalls, sidewalls or roof, also known as a platform container. These units are used for oversize and overweight cargo which cannot otherwise be containerised. By combining several artificial tweendecks on board a vessel it is possible to obtain a very high payload.

ARYA Arya National Shipping Lines – a shipowner.

AS Alaska Airlines Inc. An IATA member Airline.

A/S After sight; alongside; or account sales.

ASA Aircraft Separation Assurance – IATA term; or Australia Shipowner's Association.

ASAP As soon as possible.

ASBA Association of Shipbrokers and Agents (USA) Inc.

ASC Austrian Shippers' Council; or Autoship systems corporation.

ASCARS ASEAN Sub-Committee on Aviation and Related Subjects: see ASEAN entry.

ASCII American Standard Code Interchange – the standard format for computer coding.

ASCL Australia Straits Container Line. P & O containers associated company trading between Australia and the Malacca Straits area.

ASCs Automatic Stacking Cranes – shore based container lifting equipment.

ASD Accounting Services Division.

ASDIC Anti-Submarine Detection Investigation Committee.

ASE Athens Stock Exchange.

ASECNA Agence pour la Securite de la Navigation Aerienne en Afrique.

ASEAN Association of Southeast Asian Nations. The seven ASEAN members of the WTO – Brunei, Indonesia, Malaysia, Myanmar, the Philippines, Singapore and Thailand – often speak in the WTO as one group on general issues. The other ASEAN members are Laos and Vietnam. Its aim is to accelerate economic growth, social progress and cultural development of the region.

ASEAN-CCI The ASEAN Chambers of Commerce and Industry.

ASF Asian Shipowners Forum.

As fast as the vessel can Receive/Deliver The laytime is a period of time to be calculated by reference to the maximum rate at which the ship in full working order is capable of loading/discharging the cargo. A Charter Party term.

Asia Australia Express Overseas Containers Ltd., associate company trading between Far East and Australia.

Asian Shipowner Forum Formed in 1992 and has 85 representatives of shipowners' associations from seven Asian regions. It features shipowners' associations of ASEAN, plus China, Japan, Australia, Hong Kong and Korea. ASF Associations control or manage about 40 per cent of the world's ocean-going cargo carrying fleet.

Asia Pacific-Economic Co-operation Group Formed in 1989 by the Australian initiative, it comprises Australia, New Zealand, Canada, United States, Japan, South Korea, China, Taiwan, Hong Kong, Thailand, Malaysia, Singapore, Indonesia, the Philippines and Brunei. Its objective is to provide a loose economic federation in the

Asia-Pacific Region, designed to prevent the creation of trade barriers in the region and deter any that might be established in Europe or elsewhere.

ASIAS Airline Schedules and Interline Availability Study – IATA term.

A similibus ad similia (Latin) From like to like.

ASK An indication of willingness, to sell a specified amount of a commodity at a specific price (also known as offer). An IPE definition.

Ask price The price at which a seller is prepared to trade – an International Petroleum Exchange Term.

ASM After Sales Manager, or Area Sales Manager.

ASMO Arab Organisation for Standardisation and Metrology. Founded in 1965 to serve as a specialised technical body for the League of Arab States in the fields of standardisation, metrology and quality control.

ASOA Australasian Steamship Owners Association.

ASP Accelerated surface post; or American selling price. A system of US Customs evaluation whereby duty if levied not on the price of imported goods but on the price at which the comparable domestic article is freely offered for sale in USA; Aeronautical Fixed Service Systems Planning for Data Interchange Panel (ICAO); or active server pages – a computer term.

ASPA Association of South Pacific Airlines – IATA trerm.

ASPAC Asian and Pacific Council. Organisation of several Far East states for the discussion of economic, technical, cultural, social and political questions.

ASRRI Indonesian Association of Industries.

Assay The independent evaluation of the physio-chemical composition of a metal to determine its degree of purity. LME-registered brands must not exceed stated levels of impurity and are assayed to establish this; OR ACTIVE SERV

Assembler A readable form of low-level program language which converts an English-like code program into a machine code program which can be executed. A computer term.

Assembly cargo The separate reception of parcels or packages and the holding of them for later dispatch as one consignment.

Assembly of National Tourist Office Representatives (ANTOR) Professional association of executives of foreign tourist offices based in New York City.

Assessor Person who estimates the value of goods for the purpose of apportioning the sum payable by the underwriters to settle claims.

Asset backing overall A company's land, building, plant, cash, etc. minus its debts. It is usually calculated on an assets per share basis and described sometimes as asset value or net asset value.

Asset based The infrastructure of a country or company which contributes to the generation of wealth. In a country it is transport, commercial services, computer technology and so on.

Asset fund An investment fund committed to the purchase of assets which are expected to increase substantially in value. During the mid-1980's, when ship values were exceptionally depressed, a number of funds of this type were set up to take advantage of the recovery potential in ship prices. The term has also been loosely and somewhat misleadingly applied to investments funds which concentrate more on the cash flow generating potential of the ships acquired rather than their scope for price appreciation.

Asset play A term which has become highly popular in Norway to describe highly-geared investments in second-hand bulk vessels with the intention of reselling such vessels at higher prices within a relatively short period, e.g. 6-18 months. The usual vehicle for these investments has been the K/S partnership (q.v. Kommandittselskap) which has offered equity investors a combination of high internal rates of return and tax deferral benefits. The asset play concept, however, is dependent for success on a period of consistently rising second-hand vessel prices. Investors run the risk of substantial losses in stagnant or falling markets. Banks look to the resale value of ships as one protection in the event of a borrower defaulting on a ship mortgage loan and will attempt to limit the value of their loan to a percentage of the estimated asset value

Asset productivity The productive use of an asset such as a ship, port equipment, etc.

Asset protection clause A banking term associated with ship investment. It is designed to protect the bank in the event that the value of the asset – the ship – becomes too low in relation to the outstanding loan. The clause stipulates that should the second-hand value of the ship fall below a given amount or a given per cent of the outstanding principal, the bank would have the right to demand additional security either in the form of extra mortgages on the owners' fleet or increased corporate or personal guarantees. Also termed security maintenance clause.

Asset Swap A transaction that attaches a swap contract to an asset in a way that changes one or more attributes of the asset's cash flow.

Asset value The realisable value of the assets upon sale

Assignee One who receives rights from an assignor.

Assignment This is the passing of beneficial rights from one party to another. A policy or certificate of insurance cannot be assigned after interest has passed, unless an agreement to assign was made, or implied, prior to the passing of interest. An assignee acquires no greater rights than were held by the assignor, and a breach of good faith by the assignor is deemed to be a breach on the part of the assignee. The Institute Time Clauses (Hulls) incorporate a clause which terminates the policy automatically if, amongst other things, the ship changes ownership or management unless the underwriters agree in writing to continue the insurance. This effectively restricts free assignment of a hull time policy. If underwriters do agree to the assignment an endorsement signed by the assignor, showing the date of assignment must be attached to the policy. The assignment is still subject to the restrictions above.

Assignment of requisition compensation This security consists of the borrower assigning to the bank, any compensatory payments in the event of the vessel requisitioning, either in terms of absolute title or for hire purposes during times of war, hostilities or nationalisations.

Assignor One who assigns his rights to another.

Association Français de Normalisation The French Standards Institute.

Association of British Travel Agents (ABTA) The principal trade association of travel agents and tour operators in the UK.

Association of Conference Executives (ACE) UK trade association of executives who arrange conferences.

Association of Corporate Travel Executives (ACTE) A US group of travel management and corporate executives dealing in business travel.

Association of Group Travel Executives (AGTE) A US trade association of executives responsible for the promotion, sale, operation or purchase of group travel programs.

Association of Southeast Asian Nations See ASEAN entry.

Association of Retail Travel Agents (ARTA) A trade association of travel retailers in the US.
Association of Retail Travel Agents Consortia (ARTAC) UK trade association of travel agents' associations.
Association of Special Fare Agents London-based organisation of international air consolidators from around the world.
ASPPR Arctic Shipping Pollution Prevention Regulations.
ASST Anti-ship, Surveillance and Targeting.
Assured Party indemnified against loss by means of insurance.
Astern The process of a vessel travelling in the reverse direction or the area behind a vessel/aircraft moving in a forward direction.
Astern power Power available for driving a ship astern.
ASTM American Society for Testing and Materials (US).
ASV Anti-surface Vessel.
ASW Anti-submarine Warfare.
AT Austria – customs abbreviation; Air Transport (Committee of ICAO); or Royal Air Maroc – IATA member airline.
A/T American terms (grain trade).
ATA Actual time of arrival; Air Transport Association of America; or Admission Temporaire – Temporary Admission.
ATAC Air Transport Association of Canada; or Airport Terminals Advisory Sub-Committee – IATA term.
ATA (CARNET) Admission Temporaire/Temporary Admission. A simplified import and export documentary procedure, backed by an international guarantee chain, for temporary importations. It is administered by the ICC.
ATAG Air Transport Action Group – IATA term.
ATARS Automatic Traffic Advisory and Resolution Service – IATA term.
ATB Automated Ticket and Boarding Pass.
ATC The WTO Agreement on Textiles and Clothing which integrates trade in this sector back to GATT rules within a ten year period; Air Traffic Conference of America; Automatic Train Control; or Air Traffic Control.
ATCO Air Traffic Control Officer – IATA term.
ATD Actual time of departure.
ATE Automatic Test Equipment – IATA term.
ATFM Air Traffic Flow Management – IATA term.
Athens Convention International convention governing carriers' liability for passengers and their baggage by sea.
Athwart Across.
Athwartships Across the vessel/ship: port to starboard or vice versa. A term used in cargo stowage. Conversely the cargo could be stowed lengthwise such as bow to stern.
ATI The Association of Thai Industries.
ATIS Air Traffic Information Service -ICAO term.
Atlantic Container Line Ltd A major liner operator.
ATLAS Atlas Lines, Chittagong – a shipowner. – IATA term.
ATM Automated Teller Machines; or Air Traffic Management – IATA Term.
ATMG Air Traffic Management Group (EANPG) – IATA term.
ATN Aeronautical Telecommunications Network – IATA term.
ATO Automatic Train Operation.
ATOL Air Travel Organisers Licence.
ATOM Automated Trade Order Matching.
ATOMOS Advanced Technology to Optimise Manpower Onboard Ships – A Lloyd's Register of Shipping service.
ATP Aid and Trade Provision; convention covering the international transport of perishable goods by rail and road vehicles and containers on international journeys, and the Special Equipment to be used for such carriage. It requires that the equipment to be used must be subjected to certain tests and issued with a certificate; Accord Relatif aux Transports Internatinaux de Denrees Perissables: or Automatic train protection.
ATPA Andean Trade Preference Act.
At risk A situation where the particular circumstances are at risk or subject to change. This may be due to a variety of situations including technical, political, commercial, industrial relations and so on. For example, an increase in oil fuel tax may cause many businesses to examine the economics of a gas fired central heating system in preference to their oil fired system thereby putting at risk the latter facility continuing on a permanent basis.
Atrum deck A deck situated amidships or thereabouts on a vessel.
ATRM Airport Terminals Reference Manual (IATA) – IATA term.
ATS All time saved – a chartering term; Assistant Traffic Supervisor; Arrival time of ship; Automatic Train Supervision; or Air Traffic Service.
Ats The currency of the Peoples Democratic Republic of Laos. See Appendix 'C'.
ATS-DS ATS – Direct Speech – IATA term.
At sight drafts Drafts payable on demand and take no days of grace.
ATS-RTG AFI Air Traffic Services Route Task Force – IATA term.
ATT Admiralty Tide Tables.
Attachment Date The date entered on the broker's slip by the leading Underwriter to provide a point from which the period allowed in the terms of credit scheme will operate.
Attestation Legal act of witnessing a deed by affixing one's signature thereto.
Attestation clause That part of the policy in which the underwriter is bound to the policy conditions.
At-the-money A call/put option where the exercise price is approximately the same as the current market price of the underlying security.
Attitude research The process of identifying through research buyer's behaviour and especially attitude towards a particular product/service or response to a promotional campaign.
Attitude survey A market research technique whereby buyer behaviour/attitude is determined /researched relative to a particular circumstance such as a change in colour, design of a product; a price variation; new product launch; and so on.
Attribute In relation to database entities, an attribute is a single data item representing an individual property of the object (entity). A computer definition.
AU Austral Líneas Aéreas SA – an IATA member Airline.
Audi alterem partem (Latin) Let the other party be heard.
Audience coverage The extent to which a media plan or other marketing initiative reaches the potential customers.
Audience Research The process of evaluating the effectiveness of the media plan or other marketing initiative from the potential or existing customers. It could take the form of face to face interviews, or postal questionnaires. A marketing term.
Audience response The level of response to which a media plan or other marketing initiative gains reaction from the potential or existing customers.

Audio visual

Audio visual A piece of salesman's equipment enabling him to present to his client(s) a short film or slides with commentary featuring the product he has to sell.

Audit reserves Reserves created according to rules laid down by Lloyd's and approved by the DTI. They are made on open years of account to ensure that Names have sufficient funds to wind up their accounts after 36 months have elapsed. This solvency must be on an individual Name basis (i.e. aggregating the Name's participation in all relevant syndicates) and any deficiency must be covered by the lodgement of additional assets. Insurance definition.

Au Fait (French) Well acquainted with.

Aurar The Currency of Iceland. See Appendix 'C'.

AUS The Australia dollar – a major currency.

AusREP Australian Ship Reporting system.

AUSTRA Australian Committee on Trade procedures and Facilitation based in London.

Austral Code name of Chamber of Shipping approved charter party for Australian Grain trade – agreed pre-war with Australian Grain Shipping Association

Australian Dollar The currency of Australia and Tuvalu together with the Republic of Naura. See Appendix 'C'.

Australia Japan Container Line Overseas Containers Ltd. Associate company trading between Australia and Japan.

Australia New Zealand Europe Container service The Consortium in which Overseas Containers Ltd. trades in the Australia/New Zealand-Europe service.

Australia Straits Container Line An Overseas Container Ltd., associated company trading between Australia and the Malacca Straits area.

Austrian Airlines The National Austrian airline and member of IATA

Austrian Shippers Council The organisation in Austria representing the interest of Austrian shippers in the development of international trade.

Austwheat 1990 Code name of BIMCO approved Australian Wheat voyage charter 1990 and amended in 1991.

Austwheatbill Code name of BIMCO approved Bill of Lading used with shipments under Austwheat charter party.

Authentication The process of legalising a document through the means of a signature and sometimes an official seal/stamp. An example is the certificate of origin approved by an accredited Chamber of Commerce.

Authorisation The notification by Customs and Excise to an exporter allowing him to import goods under inward processing relief arrangements for processing and subsequent export from the Community (EU) in the form of compensating products or the appointment of an Agent/person/Company to act on behalf of his/her principal. It may involve issue of a house airwaybill or house Bill of Lading, collect cargo, deliver cargo and so on.

Authorised Capital The maximum volume of shares at par value which a particular company is allowed to issue.

Authorities to purchase The process adopted chiefly for shipments to the Far East including Malaysia and Hong Kong together with Mauritius whereby the authority to purchase the bill are drawn on the buyer of the goods and payable at sight, or at some later date as specified in the Authority. It is an alternative to Documentary Credits.

Auto container A container designed/equipped to convey vehicles.

Auto drop PNR A passenger name record that automatically appears in the appropriate queue when it needs to be attended to; for example, when space becomes available for a wait-listed client.

Autoexec.BAT File A MS-DOS Batch file created by the user that is automatically executed when the operating system is first started, or restarted. This file contains MS-DOS commands that the user wants executed every time the system is started. A computer term.

Auto ID Automatic identification technology. A computer facility which provides automatic identification of bar codes, optical character recognition and radio frequency. It speeds up the handling of air cargo consignments.

Auto-ignition Temperature The lowest temperature to which a liquid or gas requires to be raised to cause self-sustained spontaneous combustion without ignition by a spark or flame.

Automated guided vehicle system Vehicles equipped with automatic guidance equipment which follow a prescribed path, stopping at each necessary station for automatic or manual loading or discharging. Such vehicles will be found in an airport, seaport, container yard, warehouse and so on.

Automated Pit Trading A fully operational computer based trading system for futures. It enables LIFFE members to execute trades in designated contracts at specified times from workstations located in members offices and at the exchange.

Automated Teller Machines A cash point facility operated by the holder of an accredited card and using the specified allocated PIN number.

Automated ticket/boarding pass Ticket stock that can include a flight coupon and boarding pass on one document.

Automatic coupling Coupling system occasionally used on articulated road vehicles in which the tractive unit is equipped with guide ramps to lift the front off the trailer from the ground when the unit is reversed under the trailer. The trailer landing support wheels are released and fold away, and the rollers, mounted on the trailer, move up the ramps until the trailer turntable locks into place on the tractive unit. Automatic coupling also exists for railway rolling stock both for passenger and freight wagons.

Automatic Exercise Where the Clearing House automatically exercises all in-the-money options. A financial definition.

Automatic fare collection A system so devised which obviates the need to engage staff to examine/collect tickets on a passenger transport system. This primarily involves for example on a railway system the provision of coin operated ticket issuing machines, and automatic ticket barrier control whereby passengers may only gain access to the platform/train by processing their ticket through the automatic barrier equipment. It is used primarily on railway systems especially in conurbation areas such as the Paris metro, London Underground and City bus services with one man operation.

Automaticity The 'automatic' chronological progression for settling trade disputes in regard to panel establishment, terms of reference, composition and adoption procedures. A WTO term.

Automatic Radar Plotting Aids A method of obtaining and displaying target data onto the radar screen. The advantage of ARPA is that multiple target information can automatically be acquired so relieving the observer of lengthy manual plotting techniques.

Automatic watch keeper Equipment embracing monitoring and alarm surveillance on a vessel.

Automation The repetitive production, assembly and/or inspection of manufactured products by the use of a programmed control mechanism.

Auto pilot A navigational device found in sea and air transport.

Aux Auxiliary (sailing ship with auxiliary machinery).

Aux B Boiler(s) supplying steam to auxiliary essential services of a ship.

AV Aerovias Nacionales de Colombia S.A. (AVIANCA). An IATA member Airline.

Availability Usually refers to seat or bed capacity or its availability for booking. A tourist term.,

Available seat miles One aircraft seat flown one mile whether occupied or not.

Aval An unconditional guarantee for each Bill of Exchange or promissory note from an internationally recognised major bank. See also forfeiting entry.

Avalisation The guaranteeing of payment by a party, usually a bank in the buyer's country, which has no other involvement in the transaction.

Avalised bills Avalised bills is the specific endorsement on a Bill of Exchange by a bank which guarantees payment should the drawee (the importer) default on payment of the bill at maturity.

Avalising The process of dealing in Bills of Exchange guaranteed by the buyer's bank.

Average Loss – usually termed general or particular average. In marine insurance a partial loss (a) General average is a loss which arises in consequence of an extraordinary sacrifice made, or expenses incurred for the preservation of the ship and cargo. All underwriters involved in the ship and cargo will participate to indemnify the insured against the actual loss incurred, and (b) Particular average is partial loss of the subject matter insured, caused by a peril insured against, and which is not a general average loss.

Average Adjuster Independent expert who assesses the liabilities of the various parties to a common maritime adventure -when a claim arises – and to marine insurance contracts, and to classify the various items of expenditure between general and particular average, viz. ship, freight and cargo. The adjuster may also adjust claims on hull insurance policies for submission to underwriters.

Average Bond An agreement signed by all interested parties acknowledging their liability to pay a share of the loss under general average. A 'general average guarantee' is sometimes referred to (particularly in the USA) as an 'average bond'.

Average Clause A clause in a policy, whereby partial losses are subject to special conditions (e.g. a franchise or deductible is to be applied to claims).

Average cost per-claim reserve Total claims outstanding divided into the total reserves (less payments) equal an average claim reserve. Reserves in this context means all amounts provided for all known claims less any deductions for subrogation – An insurance term.

Average cost pricing The pricing policy where an average price is established over a product range based on average cost.

Average deposit Cash security deposited by consignee pending assessment of general average contribution.

Average Disbursements Expenditure incurred by the shipowner in connection with a general average act or an act of salvage. Such expenditure, when properly incurred, is recoverable from the GA or salvage fund created by the adjuster. Hull underwriters are not liable directly for GA expenditure. The assured must recover his expenditure from the GA fund. Underwriters' liability for GA contribution, if any, will incorporate their proportion of the GA expenditure that is included in the contribution paid by the insured.

Average (General) Partial loss of the whole adventure deliberately made to prevent loss of the adventure. It may be sacrifice of property or expenditure incurred to save the adventure. Parties who benefit from a general average loss are required to make good that loss by contributing in the proportion that the saved value of the party's property bears to the saved value of all interests involved in the adventure. General average is a rule of the sea and is implied in all contracts of carriage by sea. Parties to the contract are liable to contribute in GA whether or not their property is insured.

Average laytime This is determined by the number of days saved/lost at the load port(s) added to/subtracted from those allowed at the discharge port(s), but only after a separate discharge port(s) calculation had been made. For example, the results were two days demurrage accruing at load port and two days despatch at discharge, the results would cancel out each other, leaving no demurrage and no despatch on the basis 'two days saved minus two days lost equals nothing'.

Average life The total of the amounts outstanding at the end of each year of the loan for its entire life, divided by the total principal sum borrowed to give the average life of the loan in year. An International Banking term.

Average (Particular) A fortuitous partial loss of insured property proximately caused by an insured peril, but which is not a general average loss.

Average rate The average rate which is produced from conveying a given volume of traffic during a specific period or calculating the range of rates offered by a transport company. For example, during the month of April 10,000 tons of merchandise was conveyed yielding $100,000 or $10 per ton. Alternatively an analysis of all types of parcel rates available to the market by a particular Company indicated the average rate per kilogramme is $2. It is used extensively in budgetary control systems.

Average rate option Where an exporter or importer has a series of regular payments or receipts, and has budgeted for a particular exchange rate over the period, the average rate option will provide the same protection as a currency option but because of the averaging process is achieved with lower premium cost. The Company selects the rate it wishes to protect (the strike price) and agrees a mechanism by which the bank can calculate the 'average' rate. See currency option entry.

Average statement Average adjusters statement.

Average total cost Total cost divided by the quantity of a good or service produced in a given time period.

Averaging The practice of buying more of the same shares on a fall or selling on a rise, in the hope of gaining financial gain advantage by the fluctuations.

AVIA Aviation.

Avianca An International National Airline and a member of IATA.

Aviation Consumer Action Project (ACAP) Organisation that promotes consumer interests before governmental bodies dealing with air travel.

Aviation hull insurance In aviation insurance, hull insurance (i.e. the aircraft's structure) is distinguished from cargo, passenger and liability insurance.

Aviation spirit A specially blended light hydro carbon intended for use in aviation piston-engine power units.

Aviation turbine fuel A specially refined kerosene intended for use in aviation gas turbine power units.

Avogadro's Law Avogadro's Hypothesis states that equal volumes of all gases contain equal numbers of molecules under the same conditions of temperature and pressure.

Avoidance The right of an underwriter in the event of a breach of good faith or delay in commencement of an insured voyage to treat the insurance contract as null and void and thereby cancelling it. This can occur in the event of a breach of good faith by the assured or by his broker or, in the case of a voyage policy, where the voyage does not commence within a reasonable time after acceptance of the risk by the underwriter.

Avos The currency of Macao. See Appendix 'C'.

AVR Agent's Vehicle Record.

Award The decision in arbitration. A decision given by a court of law or by an arbitrator to conclude a dispute. The term can be used also to define the amount of damages allowed, if any, in the award. A judge in court will often state the reasoning for the decision, whereas this is not the case, usually, with an arbitrator's award.

Awareness The process of informing the market place (consumers/buyers) the existence of a product or service and its perceived benefits it offers to the potential buyer/end user. In marketing terms the process of making aware the target audience of a product/service. Alternatively, the product/service which may be long established and has a brand image is likely to have a high level of awareness in the market place through the continuous interaction of seller/buyer.

AWB Air Waybill – air freight consignment note.

AWD Awning deck.

AWES Association of Western European Shipbuilders.

AWFC America-West Africa Freight Conference.

Awkward cargo Merchandise which is difficult to handle and usually requires special handling and stowage arrangements. Also see out of gauge and heavy lift entries.

AWOP All Weather Operations Panel – ICAO term.

AWOS Automated Weather Observing Stations – IATA term.

AWPG ARN Workshop Preparation Group (ICAO) – IATA term.

AWRA Asia Westbound Rate Agreement – Liner Conference facility relative to the Japan/Europe Freight Conference.

AXB Auxiliary boiler (of a ship)./

AXBS Auxiliary boiler survey (of a ship).

AXFTB Auxiliary fire tube boiler (of a ship).

AXFTBS Auxiliary fire tube boiler survey (of a ship).

Axial Compression Fatigue The tendency of a fibre to fail when it is subjected to cyclic loading, which exerts compression along its axis.

Axial-flow turbine A turbine in which the direction of steam flow is parallel to the rotor shaft, with the blades fitted perpendicularly to the shaft.

Axle loading The total downward pressure exerted by a vehicle through any given axle. This may be transmitted through two or four wheels.

Axle weight The sum of the weights transmitted to the road surface by all the wheels of the axle. A road transport term.

AXWHB Auxiliary waste heat boiler (of a ship).

AXWHBS Auxiliary waste heater boiler survey (of a ship).

AXWTB Auxiliary water tube boiler (of a ship).

AXWTBS Auxiliary waste tube boiler survey (of a ship).

AZ Alitalia – Linee Aeree della Sardegna. An IATA airline.

Azcon Code name of grain charter party.

AY Finnair Oy. The Finnish National Airline and a member of IATA.

AZ Alitalia – Linee Aeree Italiane S.p.A. The Italian National Airline and a member of IATA; or Azerbaidzhan – Customs abbreviation.

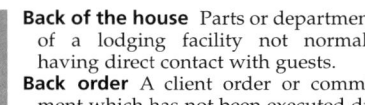

B2 Belavia -Belarusian Airline – IATA member Airline.
B3 Bellview Airlines (limited) – IATA member Airline.
B4 Before.
B6 Britannia Airways AB – an IATA member airline.
B26 A bulk carrier of 26,000 dwts capacity with seven holds and of single deck construction ideal for world wide tramp operation especially bulk grain shipments with a speed of 15 knots.
B30 A bulk carrier of 39,000 dwts capacity with five holds and single deck construction ideal for world wide tramp cargo operation with a speed of 15 knots.
B Bale capacity in cubic feet or metres; Box vans; Bridge (of a ship); or Beam (of a ship).
b Bin containers. See Bin type container entry.
B & D Bad and doubtful debt (a debt that is unlikely to be repaid).
BA Buenos Aires; or British Airways plc. – UK National Airline and member of IATA.
BAA British Airports' Authority.
BAC Berlin Agreement Carriers – IATA term; Burma Airways Corporation and member of IATA; or Bunker adjustment charge.
BACAT Barge aboard catamaran.
BACC British-American Chamber of Commerce.
Back Freight Freight (additional) incurred through cargo being returned from destination port, usually because its acceptance was refused or beyond the contract port owing to circumstance beyond the control of the shipowner.
Background music Music which is pre-recorded and played almost continuously in a public place to provide entertainment.
Back haul See Back load entry.
Back haul voyage The return load in a ship thereby giving a loaded voyage in both directions.
Back letter An addition to a contract which is specific in its rights and obligations between two contracting parties. Also termed letter of indemnity.
Back load The return load in a transport unit, such as a road vehicle, thereby giving a loaded journey in both directions.
Backloading Subsidy reduction scaling for smaller subsidy cuts in the beginning of an agreed period and steeper cuts at the end of a specified time. Frontloading calls for the opposite process. A WTO term.
Back log A build up/arrears of work. It may be in orders received but not despatched, or documents received but not processed.
Back pressure Resistance created in oil flow through cargo pipes on VLCC.
Back-pressure turbine A steam turbine after which the exhaust steam is not immediately condensed but led away and used for heating or in some industrial processes. The exhaust pressure of this type of turbine is higher than that of the condensing turbine (q.c.), and the name derives from the fact that it is often a little above atmospheric pressure.
Back office The area dealing with all the back up resources of a company such as research, planning, product development and so on.
Back-office automation Systems that computerise agency functions other than reservations; most often used to refer to automated accounting.

Back of the house Parts or departments of a lodging facility not normally having direct contact with guests.
Back order A client order or commitment which has not been executed due to insufficient stock or inadequate resources.
Backpricing A consumer may establish a contract with a producer which commits him to receive a certain specific tonnage, usually for the next year, in periodic shipments. The price used by producers currently the LME settlement price (qv). Each day's settlement price is valid until noon the following day, and the consumer has the option of 'fixing' a proportion of his quota on that price up to that time i.e. some 23 hours later. This price-fixing operation is referred to as backpricing or 'pricing on the known'.
Back to Back The process of buying and selling on basically the same terms; or a program of multiple air charters between two or more points with arrivals and departures co-ordinated to eliminate deadheading – sending an empty aircraft – and waiting.
Back to back credits An international banking term whereby two credits are established completely independently of each other. The buyer establishes its credit in the exporter's favour. The buyer arranges for a second credit (usually based on the terms of the original credit) to be established in favour of the ultimate supplier of the goods or the supplier of raw materials.
Back to back loan Companies with surplus liquidity in one currency may wish to obtain funds in another, for investment or expansion, by employing their own surplus without conversion or incurring exchange exposure, or without incurring increased interest costs by borrowing unmatched funds; this may be arranged by means of a parallel, or back to back, loan. In such a situation a form of financing whereby money borrowed in one country or currency is covered by the lending of an equivalent amount in another.
Back up service The provision of ancillary services. A good example is the 'after sales service' which looks after customer needs following completion of the sale.
Backwardation A circumstance/situation whereby nearly 'future prices' are at a premium to 'forward prices'. Overall, it is a payment per share or unit stock made by a (seller) to a bull (buyer) for the loan of securities for which the bear wishes to defer delivery. A BIFFEX term. A futures market where the priced of near delivery months trades at a premium to more distant months – an International Petroleum Exchange term. The situation when the cash or spot price of a metal is greater than its forward price. A backwardation occurs when a tight nearby situation exists in a metal. The size of the backwardation is determined by differences between supply/demand factors on the nearby positions compared with the same factors on the forward position. There is no official limit to the backwardation. The backwardation is also referred to as the 'back'.
Baco Liner An ocean going barge carrier with a capacity of 21,000 dwt's and may carry 500 TEU's plus twelve barges below deck. The barges are transhipped from the mother vessel on arrival at the port area to be towed along the various inland waterways thereby providing a door to door service.
BACS Ben Asia Container Services – a shipowner; or Bankers Automated Clearing Services.

Bad and doubtful debt A bad debt is regarded as not recoverable and therefore written off. A doubtful debt is one concerning the full or partial repayment of which there is uncertainty. Banks make provision for these by setting sums aside out of profits.

Bad stowage The inadequate stowage of cargo on a vessel having regard to the ship specification, the prescribed voyage, the nature of the cargo and any particular relevant circumstances including nature of packaging (if any) and maritime legislative obligations. Overall, it will focus ultimately on ship safety at risk.

BAF Bunker adjustment factor – a bunkering surcharge.

Bag A package made of paper, plastic, film or any woven material to be closed by such means as stitching, gluing or heat sealing or by a valve.

Baggage Articles, effects, and other personal property of a passenger as are necessary or appropriate for wear, use, comfort or convenience in connection with his/her trip/journey. Unless otherwise specified it includes both checked and unchecked baggage. Also termed luggage. An IATA definition.

Baggage allowance Weight or pieces of baggage that may be carried by a passenger with no extra charge.

Baggage check Those portions of the ticket which provide for the carriage of passenger's checked baggage and which are issued by the carrier as a receipt for passenger's checked baggage. An IATA definition.

Baggage checked Basically 'registered baggage' which is baggage of which the carrier takes sole custody and for which the carrier has issued a baggage check. An IATA definition.

Baggage excess That part of baggage which is in excess of the baggage which may be carried free of charge. An IATA definition.

Baggage handling The process of handling passenger baggage at an airport or seaport.

Baggage reclaim area An area at an airport/seaport where passenger baggage is placed following discharge/transhipment from a ship/aircraft. In so doing the passenger collects his/her baggage and accompanies it when presenting it to customs for baggage examination and customs clearance.

Baggage tag A document issued by a carrier solely for identification of checked baggage. The baggage (strap) tag portion is attached by the carrier to a particular article of checked baggage and the baggage (identification) tag portion is given to the passenger. An IATA definition.

Baggage trolley A trolley accommodating passenger baggage.

Baggage vehicle A vehicle accommodating/conveying accompanied or unaccompanied passenger baggage.

Bagging of cargo The process of placing into bags specific merchandise. This may arise in the environs of a seaport where the merchandise arrives in a bulk carrier and on discharge the commodity is bagged for distribution purposes.

Bags Method of packing cargo consisting of paper, plastic, film, cotton, textiles, woven material, or jute bags.

Bahamas Maritime Authority A ship registration authority based in Nassau in the Bahamas.

Bahamian Dollar The currency used in the Bahamas. See Appendix 'C'.

BAHARI Perusahaan Pelayaran Nusantara 'Bahari' – a shipowner.

Bahrain Dinar The currency used in Bahrain – See Appendix 'C'.

Baht The currency of Thailand. See Appendix 'C'.

Bailee A person or entity/company responsible for goods whilst in his care.

Bailment A contract by which one person (bailer) entrusts personal property to another (bailee) for some specific purpose. All contracts for the carriage of goods are bailments.

Baiza The currency of Oman. See Appendix 'C'.

Bal Balance.

Balance of advantages The principle whereby advantages derived from the trading of concessions in trade negotiations should be broadly balanced among the participants.

Balance of payments Financial statement of balance of a country's visible and invisible trade exports and imports.

Balance of trade Financial statement of balance of a country's visible trade exports and imports.

Balance of trade deficit See trade deficit entry.

Balance of trade surplus See trade surplus entry.

Balboa The currency of Panama. See Appendix 'C'.

Bale A bundle or package of merchandise.

Bale clamp Specialised attachment which enables a fork-lift to pick up wool or other bundled commodities.

Bale space Total amount of under deck cargo hold capacity expressed in cubic feet or metres available for shipment in bags, bales, or boxed cargo, and excludes space between frames and beams and odd corners which would be inaccessible to baled cargo,.

Baling Method of packing cargo for shipment.

Ballast Liquid or other material loaded or pumped into ship's hold's tanks to vary the draught, regulate stability and improve the trim.

Ballast distance The total distance the vessel steams in ballast.

Ballast Line Used to carry sea-water for ballast completely separate from cargo lines.

Ballast sailing A vessel sailing empty other than ship's stores, bunkers and water ballast.

Ballast space Portions of vessel (which may be loaded or empty) available to be filled with water, or heavy material to improve ship's stability.

Ballast tank A double bottom peak, or deep tank for either fuel or water ballast to alter the trim of a vessel thereby improving the stability.

Ballast trip A vessel sailing empty other than ship's stores, bunkers and water ballast.

Balloon Credit terms under which the final repayment is larger than previous remittances. An international trade banking term. Overall, it involves a contractual payment of specific amount, intended primarily to reduce the amount of periodic instalments of principal on a loan. The balloon is added to the final instalment, at which time it is not infrequently refinanced. Balloons are widely used in ship mortgage loans because (a) shipping cash flows may be neither large nor reliable enough to amortise the full value of the loan in the ordinary way, and (b) the scrap value of a ship provides ample security for such a deferred payment.

Balloon repayment When interest payments are 'rolled up' and not paid until the end of the loan period, at which time the loan plus interest is repaid in one lump sum. An international banking term.

Ballot A small bale of about 70 to 120 lbs (35 to 60 kgs).

Baltcon Code name of charter party for coal trade issued by the Baltic and International Maritime Council, Bagsvaerd, Denmark adopted by the Chamber of Shipping of the UK.

BALTIC Baltic Enterprises – a shipowner.

Baltic and International Maritime Council An International organisation of shipowners whose object is to unite the industry worldwide and develop it to the long-term benefit of the members. It is based in Bagsvaerd, Denmark. It is the world's largest private association of shipping companies, with nearly 2,600 members in 122 countries. The owner-members of BIMCO control a fleet of 510 million DWT (dead-weight tonnes) thereby representing 65 per cent of the world's merchant fleet. Among its many activities, the organisation provides strictly professional, non-political information and counselling services for the maritime community including owners, brokers, agents and club members. BIMCO is an official observer at the International Maritime Organisation.

Baltic Capesize Index An index devised for the Capesize tonnage chartering market. See Baltic Index entry.

Baltic Conference Ice Premium clause 1947 A BIMCO approved charter party clause which states 'any premium for insurance of the vessel against ice risks shall be borne by the charterers'.

Baltic Dry Cargo Index See Baltic Freight Index entry.

Baltic Exchange The Baltic Mercantile & Shipping Exchange Ltd. – The Baltic – place in London where chartering transactions are predominantly undertaken.

Baltic Freight index An index compiled by taking a specified number of dry cargo voyages; each voyage is weighted according to its importance in the market and historical data. Each day a panel of Baltic Exchange Members submits the spot rate which they consider applicable to each voyage. The secrecy of this information allows known fixture rates to be included and as the Panel Members are unaware of each others contribution a random sort computer program ensures total impartiality in assessing the information provided. Thus the daily index accurately and rapidly reflects each day's 'spot market' freight rate movements.

Baltic International Freight Futures Exchange Situated on the Baltic Exchange its objective is to provide a means by which many elements of the international freight and shipping industry can protect themselves against adverse price movements. The futures market is one on which one can buy or sell on a series of conditions, standardised, except for price and time of delivery. Hence, it provides the trade with an effective form of price protection by reducing the financial risk of buying or selling large quantities of forward freight.

Baltic Panamax Index An index devised for the Panamax tonnage chartering market. See Baltic Freight Index entry.

Baltic Pool Liner conference between North Atlantic ports of the United States to Baltic Ports.

Baltime 1939 Code name of time charter party approved by BIMCO. Various editions exist including those in French, Italian, Spanish and English.

Baltime Form C Code name for charter party for grain trade from USA/Canadian ports.

Baltpulp Code name of charter party for pulp and paper from Finland issued by BIMCO.

Baltwar The code name of a BIMCO approved charter party clause relative to 'war' and entitled 'Baltic Conference War Risks clause for voyage charters 1938'.

BAM Bulk air mail.

Bani The currency of Romania. See Appendix 'C'.

BANK The Bank Line Ltd. – a shipowner.

Bank bill A bill of exchange which permits funds to be drawn for a set period and amount. It is particularly useful to importers/buyers of raw materials or components who have to meet their costs before the finished goods are sold. Se also acceptance credit entry.

Bank Commission A commission charge is levied by banks for all remittances. An overseas bank may also make a charge for payment of mail or telegraphic transfers (bank drafts are usually paid by overseas banks free of charge). It is desirable agreement be reached between the importer (buyer) and the exporter (seller) as to who is to pay any charges.

Bankers' guarantee This is a written instrument usually issued on behalf of a customer in favour of a third party. In international trade, the customer is normally the exporter, and the third party the foreign importer/buyer. Can also be called a bond. This instrument; indemnifies the third party for a stated amount against the failure or the exporter or contractor to fulfil obligations under the contract, and may relate to tender, performance, retention monies, advance payments etc. Bankers' guarantee can also be provided as security for many types of transaction/debt, which need not necessarily relate to international trade.

Bank for International Settlements, Basle Established in 1930, it now acts as agent for the European Monetary Agreement, but more important are its periodic meetings of European and US central bankers. The Bank's present stockholders are the central banks of 25 European countries and US commercial banks which are represented by the Federal Reserve Bank of New York.

Banking channels The international banking system/matrix through which international trade payments are effected.

Banking operations The transfer of funds by bankers for investment or deposit in foreign centres.

Banking system The practice of always keeping more than one pallet on the wharf or in the ship so that the hook does not have to wait for a load. A cargo handling term.

Bank rate The minimum rate at which the Bank of England or a 'national' central bank will discount first class Bills of Exchange. It is usually called Minimum Lending Rate.

Banker's draft A draft drawn by one bank on another bank in favour of a third party, or draft drawn by a branch of a bank on its head office (or vice versa) or upon another branch of the same bank, in favour of a third party. The draft should comply with the specifications laid down for cheques in the country in which it is to be payable.

Bank selling rate The rate at which a bank agrees to sell to its customer a specified amount of one currency in exchange for another currency. With a few exceptions a bank will always sell currency at a rate lower than that at which it will buy.

Bank settlement plan Airline-operated system for payment of travel agents accounts.

BAOVIET Vietnam Insurance Company.

BAP Bankers Association of the Philippines.

BAPPENAS Indonesian Development Planning Board.

Banqueting rooms Rooms set aside for functions.

Bar Barrel.

BAR Board of Airline Representatives (at different airports) -IATA term.

Barbados Dollar The currency used in Barbados. See Appendix 'C'.

Barbados Ships Registry A ship registration authority.

Bar Channel A channel which has been dredged to maintain a minimum depth to enable vessels to journey through it to gain access to and from a particular area such as a port/berth/estuary.

Bar coding A method of encoding data for fast and accurate electronic readability. It is used extensively in cargo sorting particularly by major airlines and shipowners handling consolidated consignments (LCL) and warehouse distribution.

Bar draught The draught to which a vessel is restricted by the presence of a bar at a port entrance.

Bare Boat Charter Charterer hires a vessel for a long period (time charter) appoints the master and crew and pays all running expenses.

Barecon Code name of a Bareboat Charter Party issued by BIMCO. There are 'A' and 'B' versions.

Barge A cargo carrying inland waterway vessel. There are three categories of barges: self-propelled, dumb and push tow. See separate entries.

Barge aboard catamaran Specially constructed semi-catamaran barge carrier which can be used as a feeder vessel.

Barge carrier A vessel capable of conveying barges such as the LASH, BACO liner.

Barge-Carrying Vessel Specialised sea-going ships, which can carry barges. This permits the separation of the ship, as a motive unit, from the barge, as a cargo-carrying unit, thus allowing the ship to proceed on its way while the actual cargo is handled. The barges are loaded/unloaded to/from the barge-carrying vessel mother ship by means of elevators, gantry cranes, or floatation, depending upon the types of barge-carrying vessel involved. The barges carried on barge-carrying vessels are known as Shipborne Barges and are dumb push-tow craft. Four standard types of shipborne barges have been adopted by the International Standards Organisation, as below:-

	Length	Beam	Max. Draught
LASH	18.7m	9.5m	2.7m
SEABEE	29.7m	10.7m	3.2m
DANUBE-SEA	38.2m	11.0m	3.3m
HALF-LASH	18.7m	4.7m	2.7m

The most popular type of barge-carrying vessel has so far proved to be the LASH concept. This type of vessel is in decline.

Barge forwarding The offloading of containers from vessel to barge for forwarding by river or canal.

Barratry All wilful acts or misdemeanours committed fraudulently and with criminal intent by the Master and/or Crew against the vessel or cargo without the knowledge or consent of the owners to the detriment of the shipowner or charterer. The Institute hull clauses cover loss of or damage to the ship when it is proximately caused by barratry; except when it is excluded by the paramount exclusions in the clauses. For example, the latter would exclude barratry involving the use of an explosive or weapon of war; also barratry carried out by a seaman on strike.

Barrel A wooden receptacle made of natural wood of round cross section with bulging walls, constructed with staves, heads and hoops; or 42 US gallons (34.7 Imperial gallons).

Barrel handler Fork lift truck. Fork lift truck equipped to handle barrels.

Barrel hooks A piece of cargo handling equipment used by dockers to hoist barrels or drums between the ship's hold and the quay.

Barriers to trade in shipping A situation in the shipping industry which acts as a barrier(s) to a Particular National Maritime fleet trading freely for traffic in the open market. It maybe reservation of cargoes; flag discrimination in its various forms; undue burdens on a particular national fleet operating in a particular trade; or directly prohibiting a particular national fleet operating in a particular trade and so on.

Bars Special devices mounted on container doors to provide a watertight locking.

Barter A counter trade technique which involves a direct exchange of goods. A single contract covers both barter flows and in the simplest cases no cash is involved; or the exchange by an airline or other supplier of its product for goods and services provided by another entity.

BAS Base.

Base The home depot of a container, road haulage fleet, equipment and so on.

Base business The core of a business. See core business entry.

Base date The day, month, year or period on which the starting date of the index is based usually at 100. Ideally the base date should be representative or normal in the sense that the date chosen is not subject at that time to any irregular or abnormal influences.

BASEEFA British Approvals Service for Electrical Equipment in Flammable Atmospheres (HSE).

Basel Convention An MEA dealing with hazardous waste.

Base load What traffic can be potentially generated as, for example, a seaport, by taking into account the hinterland and competing terminals.

Base-load power station Power demand varies during the day and during the year. In most countries the daily demand is least at night and has two peaks during the day. Base-load power plant runs continuously to supply the 'base' of the load demand, i.e., that which is present all or most of the time.

Base Metal Any metal other than gold, silver, or a member of the platinum group. All the LME metals except silver are base.

Base port A seaport providing regular services and featuring in liner companies sailing schedules all the year round.

Base port for acceptance Name of port, accepted by freight conference or carrier which for tariff purposes is used as a base port for delivery from port of loading.

Base port for delivery Name of port, accepted by freight conference or carrier which for tariff purposes is used as a base port for delivery from port of discharge.

Base product The product with which a particular Company/Entity is best known/renowned and on which the marketing strategy is primarily based.

BASIC Beginner's all purpose Symbolic Instruction Code (Computer Language).

Basic (Alkali) In general, a basic substance dissolves in water to produce one or more hydroxyl ions. Such substances have the ability to turn litmus blue and to cause other indicators to take on characteristic colours. They also react with (neutralise) acids to form salts. Basicity is commonly measured using the pH scale; on this scale water has a 'neutral' pH (neither acidic nor basic) of 7 and bases have pH higher than 7. Some examples of basic substances are sodium hydroxide, (e.g. caustic soda or lye; calcium hydroxide (e.g. lime); potassium hydroxide and ammonium hydroxide.

Basic currency The currency in which rates of an IATA Area are specified by IATA.

Basic service port A port to which cellular container vessels and/or vehicle deck container vessels of multi-modal transport operators will call for transportation of cargo in containers.

Basic stock Items of an inventory intended for issue against demand during the re-supply lead time.

Basic rate The Air Freight rate based on either a specified or a constructed rate expressed in the basic currency as defined under Basic currency – an IATA definition; or the rate of interest on which bank lending rates are founded and determined by the central bank.

Basing point A master point to and from which fares are established.

Basing point pricing system A pricing system which ensures that final selling prices in an industry are identical irrespective of the retail outlet location and freight/distribution charges involved.

Basis The difference between the physical market and futures market within the Freight market – a BIFFEX term, or the difference between the futures price for a given commodity and the comparable cash or spot price for a commodity. Normally quoted as cash price minus futures price i.e. a positive number indicates a futures discount, a negative number indicates a futures premium – an International Petroleum Exchange term.

Basket hitch A method of slinging in which the chain forms a loop secured at its extremities to the upper terminal fitting, the load being cradled in the loop of the chain.

Basket of rates A collection of rates involving for example various commodities or the same product from a number of different companies. This enables one to evaluate the similarity of the rates and the overall range. It can also be used as a basis for subsequent comparison at a later date.

Basle framework A set of rules for minimum capital adequacy standards for international banks, so-called because they were drawn up by the Basle committee on banking supervision under the auspices of the Bank for international Settlements. These rules have been adopted by the G10 countries and the European Economic Community and will effectively apply to all banks engaged in international business as from the end of 1992. The central feature of the rules is that banks will be obliged to hold capital equivalent to at least 8% of their risk-weighted assets, with half of that capital being in the form of core capital. Given the losses which many banks have been sustaining on their past lending and the excessive pace of credit expansion in the 1980s, the task of meeting the Basle rules is presenting problems for numerous institutions, notably in the United States, Japan and Norway. A number of the banks facing the greatest squeeze on their new lending capacity are institutions which have been active lenders to the shipping industry. This fact has contributed to a marked tightening of credit availability to the shipping industry since the latter part of 1990.

BAT Bureau of Air Transport (Philippines).

BATAN Indonesian Atomic Energy Agency.

Batch A collection of products or data which is treated as one entity with respect to certain operations e.g. processing and production

Batch file An operating system program file which is created to automate a processing activity.

Batch lot A definite quantity of some product manufactured or produced under conditions which are presumed equivalent and for production control purposes passing as a unit through the same series of operations.

Batch processing This type of processing involves collecting jobs or material to be processed over a period of time, and creating a schedule followed by one complete processing session. The user has very little interaction with the process.

Battens Strips of wood added to a box, crate, barrel, etc., to strengthen it. Also members protruding from the inside walls of a vessel's hold or a (thermal) container to keep away the cargo from the walls to provide an air passage.

Batteries, containing sodium Articles consisting of a series of Cells, containing sodium that are secured within, and fully enclosed by a metal casing so constructed and closed as to prevent the release of dangerous goods under normal conditions of transport. Although designed and intended to provide a source of electrical energy, these batteries are electrically inert at any temperature at which the sodium contained in the battery is in a solid state.

Batteries, dry These are sealed, non-vented batteries of the type used in flashlights/torches or for the operation of small apparatus. They contain zinc salts and other solids, or may be of the nickel cadmium type or other combinations of metals.

Batteries, dry, containing solid potassium hydroxide Storage batteries filled with potassium hydroxide, solid which are shipped from the factory in their original dry state and filled with the dry alkali. Water would be added to the battery before first being used.

Batteries, wet, electric storage These consist of series of metal plates immersed in an electrolyte. The electrolyte is a dilute sulphuric acid, but for a certain type of battery the electrolyte is a solution of potassium hydroxide. Both of these electrolytes are corrosive liquids. The casing for the acid containing batteries is commonly plastic. Storage batteries of either of these types, when contain-ing electrolyte, are classed as corrosive liquids. Storage batteries in transit may cause damage by leakage of the electrolyte or may produce fire by accidental short circuiting of the terminals. Non-spillable batteries are designed and constructed so as to positively prevent leakage of the electrolyte, irrespective of the position of the battery. This is achieved by the use of jelly type electrolyte or porous absorbent separators or by specially designed filling and venting devices.

Batteries wet, without electrolyte These are usually wet type batteries which have been shipped from the factory in their original dry state with the intention that electrolyte would be added just before placing the batteries in service. They may also be wet batteries from which the electrolyte has been removed. In this latter instance the cells should be thoroughly flushed with water and allowed to drain before shipping.

Bay A vertical division of a vessel from stem to stern used as a part of the indication of a stowage place for containers. The numbers run from stem to stern, odd numbers indicate a 20 foot position, even numbers indicate a 40 foot position. A Nedlloyd term; or an area in a transit shed or warehouse between posts or columns or the area between lateral ceiling beams or trusses projected downward to wharf or warehouse floor.

Bay plan A plan which shows the locations of all the containers on the vessel.

B2B Business to Business.

B & B Bed and breakfast.

BB Bulbous bow; or Below Bridges – vessel will load or discharge below bridges i.e. London – below London Bridge.

BBA British Business Association (Singapore).

BBB Before breaking bulk; or Bankers Blanket Bonds.

BBC Brand-to-Brand Complementation.

BB Certificate Certificate for clearance inwards of a vessel retained by the Master for inward clearance by the Customs Authorities.

BBD Break bulk delivery.

Bbl Barrel.

BBP Built up non ferrous propeller (of a ship).

BBS Barber Blue Sea – a shipowner; or Bulletin board system; a computer based information source operated for the general public by the United States Coast Guard Navigation Information Service.

BC Budgetary Control; British Council; Bristol Channel; bale cubic metres; or bulk carrier.
B2C Business to consumer.
BCAS Beacon-derived Collision Avoidance System – IATA term.
BCB British Consultants Bureau.
BCC Brazilian Chamber of Commerce; or British Chamber of Commerce.
BCCB British Chamber of Commerce Bangkok.
BC Code Safe working practice code for solid bulk cargo.
BCG Boston consulting Group – marketing term.
B/Ch Bristol Channel.
BCIC Birmingham Chamber of Industry & Commerce.
BCP Built up cast iron propeller (of a ship).
BCS British Calibration Service.
BCSP Built up cast steel propeller (of a ship).
BCV Barge carrying vessel.
BD British Midland Airways Ltd. – a UK airline; below deck; Billing day – the despatch of an account to a customer; Bar draft; below deck (of a ship); or Bonded Distributor – a person who has given bond and has been individually approved for the purpose of supplying oil in the course of trade for eligible use.
B/Dft Bank draft.
bdi Both dates inclusive.
Bdl Bundle.
Bdls Bundles.
BDT Bureau of Domestic Trade (Philippines).
BE Belgium – customs abbreviation; Bill of Exchange; Bill of entry; British Embassy (HMG); both ends; British European (Jersey European Airways (UK) Limited) – an IATA member airline.
BEA Break even analysis – accountancy term – see separate entry.
Beaching Voluntary stranding of a vessel.
Bead A circumferential convex (outward) rib mechanically expanded or pressed out.
Beam The breadth of a vessel at its widest part.
BEAMA British Electrical and Allied Manufacturer's Association Ltd.
Beams Athwartship steel rolled sections supporting the deck on a ship.
Bear An investor or speculator who expects prices to fall in the market. A pessimist: one who has sold a security in the hope of buying it back at a lower price, often in the same Account.
Bear covering The process of closing shot positions. A BIFFEX term.
Bearer bond The security of which ownership by the holder is presumed.
Bearer stocks/shares Securities for which no register of ownership is kept by the company. Ownership of bearer securities passes by delivery, so they are very attractive to thieves. Dividends are not received automatically from the company but must be claimed by removing and returning 'coupons' attached to the certificates.
Bearing surface (ramp) The underneath part of the ship ramp and which touches the quay or shore ramp, and on which the load is taken.
Bearish The belief that the market price will fall.
Bearish market A competitive market place whereby interest is displayed in the product/service on offer thereby generating a situation whereby speculators sell their products in the expectation of a fall in prices. Usually associated with financial markets. A market in which prices are declining.
Bear market A market in which bears would prosper – that is, a falling market.
Bear Position One where the trader had sold metal for forward delivery with the view to buying back at a lower price; his profit being the difference between the sale and repurchase prices.
Bear straddle The process of selling nearby months and buying distant months – a BIFFEX term.
Beaufort scale Wind force expressed numerically on a scale generally from 0 to 12. Overall, a classification system determining sea states and wind speeds.
Beavertail A low loading platform found on a road vehicle which has a sloping tail beyond the rearmost axle to facilitate loading.
Becquerel The Becquerel is the standard unit of measure for the specific activity of a radionuclide used in these Regulations; it is represented by the symbol 'B'. Because the Becquerel is a very small unit, larger multiples are used. The Becquerel replaces the older unit for specific activity, the 'Curie'(ci). One Ci is equal to 37 GBq.
BED Bureau of Energy Development (Philippines).
Bed-and-breakfast An arrangement for breakfast to be included in guest accommodations; usually a private house or a small hostelry that services breakfast as the only meal, including the price in the room rate.
Bedplate Structure forming base of a machine.
Bedienung German term meaning the tip is included on the bill.
Bed night One person spending one night in a hostelry; 10 persons staying 10 nights would be 100 bed nights; a standard measure of overnight tourist traffic.
Bee Business enabled electronically.
BEEA British Educational Equipment Association.
Behavioural variables A marketing term whereby consumers/buyers/passengers etc. behavioural attitudes are measured/recorded/evaluated. This may for example record the buying pattern of customers using a supermarket or clients using a retail shoe outlet.
Belgian Shippers Council The organisation in Belgium representing the interest of Belgian shippers in the development of International Trade.
Belgo – Luxembourg Chamber of Commerce The organisation based in London representing Belgium and Luxembourg whose basic role is to develop, promote and facilitate international trade between UK and Belgium/Luxembourg and vice versa.
Belly Lower portion of an aircraft.
Belly container Unit load device often with contoured profile, suitable for stowage in the belly holds of aircraft.
Belly hold A confined space below the main deck of an aircraft used for carrying baggage, mail or cargo.
Below the line The process of advertising/promoting product/service which does not attract any commission or fee payable to the Advertising Agency. See below the line advertising entry.
Below the line advertising The process of advertising a product/service which does not attract any commission or fee payable to the Advertising Agency. Examples include exhibitions, sponsorship, direct mail coupon promotions, sales demonstrations and point of sale material. – IATA term.
Below the line budget See below the line entry.
Below the line promotion See below the line advertising entry.
Belship An ocean going heavy lift vessel capable of conveying heavy or bulky cargoes such as locomotives, transformer etc.
Benacon Code name for Chamber of Shipping approved charter party for softwood from Eastern Canada; adopted by BIMCO.
BENBULK Ben Bulk Timber Service – a shipowner.

Benchmark The predetermined level/standard. This may embrace price, quality, design, efficiency and cost effectiveness. It is against the benchmark that all comparisons are made. The base line of a product/service cost.

Beneficial owner The ultimate owner of a security, regardless of the name in which it is legally registered.

Beneficiary The party in whose favour the credit is to be issued and the party who must comply with the terms and conditions of the credit in order to be entitled to receive its proceeds, normally the seller or exporter. An international banking term; or the natural or legal person established in the member state of departure who carries out the procedure – customs definition.

Benefit of Insurance Clause A clause in a contract of carriage by which the bailee of goods claims the benefit of any insurance policy effected by the cargo owner on the goods in care of the bailee. Such a clause in a contract of carriage, issued in accordance with the Carriage of Goods by Sea Act, is void at law.

BENELUX A Customs Union established in 1948 in Brussels – an important fore-runner of the Common Market. Its members included Belgium, Luxembourg and Holland, which was then superseded by the larger Customs Union of the EEC.

BENOCEAN Ben Ocean -a shipowner.

BEP Booked Earned Premium – an insurance abbreviation.

Berdiscon The code name of a BIMCO approved charter party clause relative to a 'ready berth' and entitled 'Baltic Conference Waiting for Berth clause (discharging) 1964'

Bergy bit A large piece of floating glacier ice, generally showing less than 5m above sea level, but more than 1m and normally about 100-300m" in area.

Bergy water An area of freely navigable water in which ice of land origin is present in concentration less than 1/10. There may be sea ice present, although the total concentration of all ice shall not exceed 1/10

BERI Index Business environment risk index. First published in 1972 assesses 48 countries on 15 economic, political, and financial factors and allocates them a 'score' of 0-4, with 0 indicating unacceptable conditions and 4 equalling optimum conditions.

Bermuda Dollar The currency used in Bermuda. See Appendix 'C'.

Bermuda plan (BP) Hotel accommodations with full American breakfast included in the rate.

Berne Convention Treaty, administered by WIPO, for the protection of the rights of authors in their literary and artistic works. These rights relate to the expression of the ideas of the author when this expression constitutes an original creation. These rights are referred to as 'copyright' in the Anglo-American legal system and as 'author's rights' in other legal systems. This is more than just a matter of terminology. Copyright protects authors against imitations of their creations by others without their authorisation. Author's rights, in addition, can prevent any distorted reproduction of those creations. A WTO term.

Berne Union The International Union of credit and investment insurers founded in 1934.

Berth The specific place within a port where the vessel is to load or discharge. If the word 'BERTH' is not used, but the specific place is (or is to be) identified by its name, this definition shall still apply. A BIMCO term; a location/place with a seaboard where vessels are moored, loaded or discharged; or sleeping accommodation on a ship or train.

Berth Accommodation Sleeping accommodation on a ship or train; or area allocated at a quayside to a particular Shipping Service, Company, or other specified purpose for berthing; or a point to which a chartered ship is ordered to load or discharge, or operate a service.

Berth charter A voyage charter party which specifies the vessel to discharge her cargo at a specified berth and in so doing the vessel will not have officially arrived until she actually reaches the named berth where cargo operations are to be performed.

Berth hire Charges raised by a port authority to use/hire a particular berth to enable a vessel to dock and tranship cargo. It is usually based on the ships GRT and time in port. Some ports have an integrated berth hire charge such as Bombay and Mormugao (India) which includes crane charges; night allowance to labour; overtime charge for cargo work during second and third shift and Sundays and holidays; and gate and shed opening charges during the second and third shift and Sundays and holidays.

Berth management strategy The process of a Port Authority or Port Terminal operator deciding on the most acceptable/optimum allocation of berths in the light of shipping demands and the resources needed at each berth to handle the traffic passing through it.

Berthing of ships The process of a vessel entering a port/harbour and subsequently being moored/berthed/tied up alongside a particular quay/berth.

Berth No Berth If the location named for loading/discharging is a berth and if the berth is not immediately accessible to the ship, a notice of readiness can be given when the ship has arrived at the port in which the berth is situated. Alternatively it may be termed 'whether in berth or not'. A charter party term.

Berth strategy See Berth management strategy.

Berth terms A chartering term whereby the shipowner may agree to his vessel's loading or discharging operation being subject to the custom of the port where the cargo handling is taking place, or he may be agreeing that the vessel will load or discharge as fast as can or under customary dispatch or any or all of this type of term. See 'custom of the port' and 'as fast as can entries'.

Berth user The degree of utilisation/occupation of a particular berth at a port, or a berth in a cabin on a ship/train.

BERI Business Environment Risk Index – a marketing term.

Bespoke charters British usage for special group charters, as for sports teams and musical organisation.

Bespoke service The process of the manufacturer or service industry entity providing a product/service with a differing specification specially formulated to meet the needs of each buyer on an individual basis thereby generating the strategy of empathy.

Best available A pledge by a travel supplier to furnish the top accommodation possible to a client.

Best orders Orders to buy or sell which the LME Ring – dealing broker executes on the market at what the dealer judges to be the best available price. It is also termed buying (or selling) 'at best'.

Best practice The classification description of an acceptable code of practice adopted to execute/undertake/perform a specific activity. It will have regard to prescribed quality standard, environment of activity, resources available, competition, quality control/assessment techniques, actual work place location; and any financial/environmental disciplines. A useful measure would be BS 5750, ISO 9001.

BETA Business Equipment Trade Association (UK).

Beta Factor A factor used to calculate the degree of correlation between the (BIFFEX) index and a particular route. A term used by BIFFEX.

BEUC Bureau Europeen des Unions Consommateurs (French) (European bureau of consumer unions based in Brussels); or Bureau of Energy Utilisation (Philippines).
BExH British Exporters Association.
BF Bridge/Forecastle.
Bf Brought forward; or bring forward.
BFC Baltimore Berth Grain Charter Party 1913.
BFCD Bureau of Flood Control and Drainage (Philippines).
BFD Food and Drugs Bureau (Philippines).
BFEC British Food Export Council.
BFI Baltic Freight Index.
BFO Bunker fuel oil.
BFT Bureau of Foreign Trade (Philippines).
Bg Brig.
BG Bulgaria – customs abbreviation; or Biman (Biman Bangladesh Airlines) – member IATA airline.
Bgd Bagged.
Biman Bangladesh Airlines An IATA member airline.
BH Bill of Health, or Bulkhead(s).
BHC British High Commission.
Bhp Brake horsepower (oil engines).
BH Range Range of ports between and including Bordeaux and Hamburg.
BI Royal Brunei Airline. The National Airline of Brunei and member of IATA.
BIA British Insurance Association.
Bias Preference given to flights of one carrier over those of others in the displays of choices on computerised reservations systems, a practice forbidden in the US, Canada and most European countries if the listed carriers pay to appear in a system.
Bias Belted A pneumatic tyre, the structure of which is such that the ply cords extend to the bead so as to be laid at alternate angles of substantially less than 90 degrees to the peripheral line of the tread, and are constrained by a circumferential belt comprising two or more layers of substantially inextensible cord material laid at alternate angles smaller than those often ply cord structure.
BIBA British Insurance Brokers Association
BIBO Bulk in – bags out.
BIC British Importers' Confederation; or Bureau International des Conteneurs.
Bid The price the buyer is prepared to pay. Hence, an offer to purchase at a specified price – a BIFFEX term; or an extensive crescent-shaped indentation in the ice edge, formed by either wind or current; or a proposal at a given time on the trading floor to buy a specific amount of a commodity at a specified price – an IPE definition; or Bureau of Industrial Development (Philippines).
Bid bond A bond provided to support the exporter's offer to supply the goods or services. It is an indication to the buyer of the serious intent of the exporter in bidding for the contract. The bond is usually required for amounts of two to five per cent of the tender. Also termed tender bond. An International banking term.
Bidding theory Pricing based on quantifying of factors influencing purchasing and the probability of their acceptance at different price levels.
Bid price The price at which a buyer is prepared to trade.
Bid rate The lower side of interest rate quotations; it is the rate of interest a bank is prepared to pay for deposits or to acquire securities
BIE Bureau International des Expositions (French) (Organisation for International Exhibitions) – based in Paris.
BIEC British Insurers European Committee.
BIFA British International Freight Association.

BIFA NET British International Freight Association software package involving Electronic Data Interchange designed for the freight forwarding industry.
BIFFEX The Baltic International Freight Futures Exchange.
Big bang The change in the rules of the London Stock Exchange which occurred on 27th October, 1986. It resulted in the abolition of fixed commission charges which caused a complete alteration in the structure of the market.
Big figure Foreign exchange dealers' term for the major digits of an exchange rate e.g. $1.90 to £1.00. Quotes between dealers assume the 'big figure' i.e. $1.90 and are only negotiated on decimal points smaller than that e.g. '40/50' meaning $1.9040 to $1.9050 to the £1.00.
BIGHT A loop formed by doubling back a rope upon itself.
Big ticket lease The leasing of capital equipment for sums above a certain amount, e.g. in the United Kingdom Pounds 1 million. Effectively all ship leases are big ticket leases, which may involve a consortium of lessors. The term 'big ticket' may also be used of large medium-and long-term loans.
Bilateral Agreement between two countries respecting the designation of air carriers to fly routes between them and the limitations, if any, of schedules and fares on those routes.
Bilateral aid Agreement between two countries to provide some form of aid to an overseas territory. It may be food, financial-soft loans, line of credit, technology, medical, etc. See donor government.
Bilateral donor Financial aid or other provision donated to an overseas territory by two countries operating together. See donor aid entry.
Bilateralism The conduct of trade negotiations between two countries leading to material privileges not extended to others.
Bilateral procedure A procedure whereby the central bank deals directly with one or only a few counterparties, without making use of tender procedures. Bilateral procedures include operations executed through stock exchanges or market agents.
Bilateral road agreements A situation/circumstance whereby two countries formulate an agreement for the movement of merchandise by goods vehicles. In the main the agreement permits the conveyance of goods to, from and in transit through the country concerned. It embraces the acceptance of a return load and applies to hauliers involved in the 'hire and reward' operations.
Bilateral Trade Agreement An agreement between two trading countries. It may be over a specified period involving particular goods and services up to a specific overall value such as £900 million over five years, or simply in general terms to facilitate trade between the two countries.
Bilateral Trade Treaty Trade treaty between two countries.
Bilateral Transport Agreement An agreement between two nations concerning their transport relations. See also Bilateral road agreements entry.
Bilge The curved portion often circular between the bottom and side shell plating of a ship; or the largest circumference of a barrel.
Bill The shortened term for the Bill of Exchange.
Billing of customers The process of despatching accounts to clients for payment.
Billing participant A party who is neither a CASS airline nor a port participant and who submits in an electronically readable form, to the Settlement Office Air Waybill data of transactions made on its behalf by Agents (aircargo).

Bill of Exchange An unconditional order in writing, addressed by one person to another signed by the person giving it, requiring the person to whom it is addressed to pay on demand or at a fixed or determinable future time, a sum of money to, or to the order of, a specified person, or the bearer.

Bill of Health Document certifying health of every passenger/crew member is free of any contagious disease.

Bill of Lading A receipt of goods shipped on board a ship signed by the person (or his agent) who contracts to carry them, and stating the terms on which the goods are carried. It is a document of title and as such is required by the importer to clear the goods at the port of destination. The documentary letter of credit will specify what type of bill of lading is required. The bill of lading information includes the name of the shipping company; the name of the shipper (usually the exporter); the name and address of the importer (consignee) or order; the name and address of the notify party (the person to be notified on arrival of the shipment usually the importer); the name of the carrying vessel; the names of the ports of shipment and discharge; the marks and numbers identifying the goods; a brief description of the goods (possibly including weights and dimensions); the number of packages; whether freight is payable or has been paid; the number of originals in the set; the signature of the ship's master or his agent; the date on which the goods were received for shipment and/or loaded on the vessel (this must not be later than the shipment date indicated in the credit); the signature of the exporter (or his agent) and his designation if applicable.

Bills of lading are usually made out in signed sets of two or three original copies known as negotiable copies, any one of which can give title to the goods. The number of copies in a set is shown on each copy. There may also be non-negotiable (unsigned) copies which are not documents of title and are normally used for record purposes. The credit will indicate how the various copies of the bill of lading are to be distributed.

The reverse of the bill of lading bears the terms and conditions of the contract of carriage. The clauses on most bills of lading will be similar in effect if not in wording. A bill of lading should be 'clean,' i.e. contain no superimposed clause recording a defective condition of the goods or their packing.

The goods can be consigned 'to order' which means the importer can authorise someone to collect the goods on his behalf. In this case the bill of lading will be endorsed, normally on the reverse side, by the exporter. If the importer (consignee) is named, the goods will only be released to him, unless he transfers his right by endorsement. The bill of lading must, however, provide for this.

Bill of Lading frauds The forgery of a Bill of Lading for non-existent goods, or by the shipowner against the charterer. It may arise through fraud by the seller, or fraud by the buyer. It can also arise through a general average fraud.

Bill of Lading (house) A document of carriage issued by a forwarding agent to his principal or a domestic document between two ocean carriers when they are carrying each other's cargo.

Bill of Lading issued to a named party (and not to order) A bill of lading issued to the name of a certain party and which cannot be transferred by endorsement.

Bill of Lading ton The greater of weight or measurement of goods where one tonne is either 1,000 Kilos or one cubic metre.

Bill of Materials A list showing all raw materials or components required to make a final product.

Bill of Sale A registered transfer of goods to a person for some consideration empowering him to dispose of them upon non-fulfilment of certain conditions.

Bill of Sight A custom import form used when the importer is unable to make a complete customs entry owing to insufficient information from the shipper. A Customs Officer opens and 'sights' the goods and the information thus provided enables a normal entry to be made.

Bill payable A bill of exchange or promissory note payable, prepared by a drawer to a specified payee, and often subject to acceptance by a third party as a guarantee or as a discounter.

Bill receivable A bill of exchange or promissory note receivable.

Bimchemtime Code name of BIMCO approved time charter party for the bulk shipment of chemicals in tankers. It was introduced in 1984.

Bimchemvoybill Code name of BIMCO approved Bill of Lading for shipments on the Bimchemvoy Charter Party.

Bimchemvoy Code name of approved BIMCO voyage charter party for the shipment of chemicals in tank vessels.

Bimcosale Code name of BIMCO approved standard bill of sale relative to a vessel.

BIMCO The Baltic & International Maritime Council with headquarters in Bagsvaerd, Denmark. An International Association of Shipowners. See separate entry.

Bimetallism Adoption of two metals (silver and gold for example) as legal tender.

Bimodal trailer A trailer which is able to carry different types of standardised unit loads e.g. a chassis which is appropriate for the carriage of one FEU or two TEUs.

Binding Tariffs fixed in schedules to the WTO which can only be raised if compensation is negotiated. Also known as Bound tariffs. A WTO term.

BINTANG Bintang Lines – a shipowner.

Bin type container An ISO container designed for the shipment of bulk grain/coal/fertiliser/powder type cargoes.

Biological Products These are either finished biological products for human or veterinary use manufactured in accordance with the requirements of national public health authorities and moving under special approval or licence from such authorities; or finished biological products shipped prior to licensing for development or investigational purposes for use in humans or animals; or products for experimental treatment of animals that are manufactured in compliance with the requirements of national public health authorities. They also cover unfinished biological products prepared in accordance with procedures of specialised governmental agencies. Live animal and human vaccines are considered biological products and not infectious substances. Importation of human and animal vaccines may be subject to authorisation by the country of destination.

BIP Baggage improvement programme – IATA term.

BIRD Banque Internationale pour la Reconstruction et le Development (French). International Bank for Reconstruction and Development.

BIRPI United International Bureau for the Protection of Intellectual Property based in Geneva.

Birr The currency of Ethiopia. See Appendix 'C'.

BIRU Perkapalan Lautan Biru – a shipowner.

BIS Bank for International Settlements, Basle. Established in 1930, it now acts as agent for the European Monetary

Agreement, but more important are its periodic meetings of European and US central bankers. The Bank's present stockholders are the central banks of 25 European countries and US commercial banks which are represented by the Federal Reserve Bank of New York; or Built for in-water surveys.

Biscoilvoy 86 Code name of BIMCO standard voyage charter party for vegetable/animal oils and fats.

Biscoilvoy bill Code name of BIMCO approved Bill of Lading for shipments on the Biscoilvoy charter party.

BISRA British Iron & Steel Research Association.

BIT Bureau International du Travailler (French). (International Labour Organisation).

Bitmap graphics A graphic image or text formed by a pattern of dots or pixels. Examples include scanned documents and printed newspaper pictures. An electronic graphic file where each minute item (dot) in the graphic picture is represented by a single (or several for colour) bit of information in the file. Thus a picture with 8,000 bits of information would produce a 1,000 byte file. (1 byte =8 bits). These files cannot be scaled in the way that vector-based images can.

BITR Baltic International Tanker Routes.

BK Bar Keel.

Bk Bulk containers. See bulk container entry.

BKK Bangkok IATA definition.

BKKA Foreign Contractors Co-ordinating Body in Pertamina (Indonesia)

BKPM Indonesian National Co-ordinating Investment Board.

B/L Bill of Lading.

BL Bale

Black body radiation The radiation that is emitted by a surface which absorbs all incident radiation at all wavelengths. The wavelength-dependence of this radiation is defined by the temperature of the surface.

Blacked out A time when special lower fares and other prices do not apply.

Black lists A very high risk. See grey and white list entries.

Black oils Fuel oils. See also white products.

Black plate A term covering many types of uncoated mild steel sheet.

Black powder (gun powder) Substance consisting of an intimate mixture of charcoal or other carbon and either potassium nitrate or sodium nitrate, with or without sulphur. It may be meal, granular, compressed or palletised. Black powder may be readily ignited by a spark. The finer the grains the easier the powder may be ignited. It is the peculiar susceptibility to sparks which renders the transportation of black powder hazardous. A term associated with dangerous cargo transport movement.

Blackseawood Code name of charter party to cover timber shipments from USSR, Rumanian Black Sea ports and Danube ports.

Blackseawoodbill Code name of BIMCO approved Bill of Lading for shipments on the Blackseawood charter party.

Blades (Chisel forks) Extremely thin wide forks on a fork lift truck used for sliding beneath loads which are not on pallets – often termed chisel forks.

Blair House A preliminary agreement between the US and the EC, reached on 20th November 1992 at Blair House in Washington DC, regarding bilateral differences on agricultural issues including domestic support, export subsidies and market access. The agreement also contained a resolution for the longstanding oilseeds dispute between the two parties. A WTO term.

Blank bill of lading A bill of lading where the name of the receiver or consignee is not inserted and is replaced by the word bearer.

BLC Ben Line Containers Ltd; – a shipowner.

BLCC Belgo-Luxembourg Chamber of Commerce.

Bleeding The act of cutting full bags of bagged cargo.

Blending The process of mixing differing aggregate powders or ores according to a pre-determined formula or ratio.

BLEU Belgian and Luxembourg Economic Union.

BLEVE Boiling Liquid Expanding Vapour Explosion – it is associated with the rupture, under fire conditions, of a pressure vessel containing liquefied gas.

Blind Sectors Areas not readily observable from Navigating Bridge for lookout purposes.

BLK Bulk.

BLK CAR Bulk carrier.

Block booking The process of an entity having a contractual arrangement with a retailer for the former to provide a specified capacity on specified dates in return usually for a form of guarantee. This facility may be on transport services such as a ship, airline, train; a hotel for a particular season; a theatre and so on. Such an arrangement is usually undertaken by an Agent.

Block Coefficient (C) Printer: where space after C inferior small cap B please. A measure of the fullness of a vessel: volume of displacement/length x breadth x draught.

Blocked space Reservation, often subject to forfeiture of deposit, made with suppliers by travel agents or wholesalers in anticipation of resale.

Blocking A phenomenon, most often associated with stationary high pressure systems in the mid-latitudes of the northern hemisphere, which produces periods of abnormal weather.

Blocking off Making the cargo tight when the holds are not full to prevent cargo shifting during the voyage.

Blocks, chains and shackles Ships equipment.

Block train load A complete train of wagons which operate between specified terminals with no intermediate marshalling of wagons en route. It may be a train load of containers or car transporters operating between Birmingham and Milan.

Bloodstock Horses – usually with the conveyance of live horses by air or sea transport.

Bloodstock rate The rate applicable to the movement/conveyance of live horses by air or sea transport.

BLR(S) Boilers.

BLT Bureau of Land Transport (Philippines); or built.

B/L Ton Bill of Lading ton. The greater weight or measurement of goods when one ton is either 1,000 kilos or one cubic metre. Also called Freight ton.

Blue Asbestos See Asbestos.

Blue book A booklet issued by HM Government Department of Transport and indicates requirements for Dangerous Goods aboard British ships or any ship in British Ports. It largely refers to the International Maritime Dangerous Goods Code for the carriage of dangerous goods in ships. The Blue Book deals with recommendations for dangerous cargo packing/shipment.

Blue Certificate A document issued by the International Transport Workers Federation which identifies vessel complying with the seafarers' agreement. It confirms the vessel has an ITF standard collective agreements which reflects seafarers' negotiable employment conditions.

Blue chip A term for the most highly-regarded industrial shares. Originally, this was an American term derived

from the game of poker, where blue is the colour of the highest value chip.
Blue chip company A well established large company which is prosperous and attractive to investors.
Blue chips The shares of companies, usually large, well-established and prosperous concerns which have a high status in investments.
BLUEFUNNEL Blue Funnel (SEA) Centaur – a shipowner.
Blue Peter A rectangular flag, blue with a white square in the centre, which may be displayed to indicate that a ship is ready to proceed.
BLUESTAR Blue Star Line – a shipowner.
Bm Board measure (timber).
BM Boom.
BMA Bangkok Metropolitan Administration.
BMEC British Marine Equipment Council.
BMEG Building Materials Export Group (UK).
Bmep Brake mean effective pressure.
BMIF British Marine Industries Federation.
BMITA British Malaysian Industry & Trade Association.
BMLA British Maritime Law Association.
BMR Bangkok Metropolitan Region.
BMSOA British Motor Ship Owners Association.
BMTA The Bangkok Mass Transit Authority.
BN Brussels Nomenclature – now known as Customs Co-operation Council Nomenclature; or Forward Air (Forward Air International Airlines, Inc.) – IATA associate member Airline.
B/N Booking note.
BNA British North Atlantic – defines North Atlantic Institute Warranty Limits.
BNCEL National Bank for Foreign Trade (Laos).
BNC/ICC British National Committee of the International Chamber of Commerce.
BNFC Brazil-Nigeria Freight Conference.
bo Brokers order; buyer's option; or branch office.
BO Boiler manufacturer.
Boarding gate The departure gate at an Airport Passenger terminal through which passengers assemble and ultimately are processed prior to joining/boarding their specific flight/aircraft.
Boarding pass A form of ticket enabling/authorising the holder to board a specific flight or sailing. In the case of an airline it is usual to specify the seat number, the flight number and name of the passenger. With regard to the shipowner, usually no details are given of the sailing departure time. The boarding pass is obtained when the passenger 'checks in' for a specified flight or sailing. It is a form of passenger control.
Boarding priority The standardised order in which airlines board passengers holding different types of tickets.
Boatage Taking of mooring lines ashore.
Boatman A shore based employee at a seaport responsible for the mooring and unmooring of vessels.
Boatswain Able seaman 20 years of age or over with at least four years service on the deck.
BOC Bulk oil carrier.
Body flange The outward flare of the body of a metal drum in preparation for a double seaming.
Body language Broadly a non-verbal form of communication and expressed in attitudes, dress and gestures. This is particularly relevant to different cultures and very evident in sales negotiation.
BOEA British Offshore Equipment Association.
BOGIE A set of wheels specially constructed for use under a container, railway wagon or chassis.
Bogie lift Device which enables one axle of a bogie to be raised from the ground usually by a pneumatic system on a road vehicle.

BOI Board of Investment (Philippines); or Office of the Board of Investment.
Boiler survey Examination of ship's boiler by surveyor.
Boiling Point The temperature at which the vapour pressure of a liquid is equal to the pressure on its surface (the boiling point varies with pressure).
Boil-off Boil-off is the vapour produced above the surface of a boiling cargo due to evaporation. It is caused by heat ingress or a drop in pressure.
Bolero International Ltd An electronic trade joint venture between the transportation insurance mutual Through Transport Club (TTC) and the international settlement system (SWIFT) launched in 2000. It is a worldwide system that provides secure transmission of electronic global trade documents. This paperless trade network also fully supports the EAN.UCC toolkit for electronic trade and document exchange between importers, exporters, shipping companies, banks and customs.
Bolivar The currency of Venezuela. See also Appendix 'C'.
Bolivian Peso The currency of Bolivia. See also Appendix 'C'.
Bollard Cast-steel posts, with a swelling head, and secured to a wharf or pier and used for mooring vessels. A vertical post to which the eye of a mooring line can be attached.
Bolster Component of a skeletal trailer that supports twistlocks.
Bombs Explosive articles which are dropped from aircraft. They may contain a flammable liquid with bursting charge, a photo-flash composition or a bursting charge. The term excludes Torpedoes (aerial) and includes bombs, photo-flash; bombs with bursting charge and bombs with flammable liquid, with bursting charge. A dangerous cargo term definition.
Bona fide (Latin) In good faith, without fraud or deceit.
Bona vacantia (Latin) Ownerless or heirless property.
Bond Guarantee to Customs of specified amount of duty to be paid; or security in registered or bearer form usually with coupons attached bearing interest at a fixed or variable rate or an interest bearing certificate of debt (traditionally long-term) that is offered to the public for subscription during a specified period. Overall, a security issued by a borrower as receipt for a loan longer than twelve months, indicating a rate of interest and date of repayment.
Bonded Goods stored under Customs seal pending payment of outstanding import duty. See Bonded cargo and Bonded warehouse entries.
Bonded area An area or accommodations which Customs Authorities have authorised for the storage of uncleared consignments.
Bonded cargo Cargo which is in transit passing through a country and usually not subject to any customs examinations at customs Frontier points, or cargo imported into a country and placed in a bonded warehouse pending customs clearance. These goods must be held in a bonded area until cleared by Customs or moved in bond to interior points or re-exported. See Bonded warehouse entry.
Bonded goods Dutiable goods upon which duty has not been paid i.e. goods in transit or warehoused pending customs clearance.
Bonded warehouse Accommodation under Customs surveillance housing highly dutiable cargo i.e. tobacco, spirits which may be stored on importation and withdrawn at importers' convenience on payment of relevant duty. Depending on local customs regulations

Bonding

the goods may be unpacked, repacked, relabelled, cleaned, sorted and tested. In some situations certain modes of assembly are permitted.

Bonding Purchase of a guarantee of protection for a supplier or a customer; certain bonding programs being mandatory in the travel industry.

Bonding Company An organisation that is prepared to undertake an agreement to make good a financial guarantee on behalf of another responsible for such guarantee. Owners of 'arrested' vessels may obtain such a bond to satisfy a court and to obtain release of the vessel.

Bond Note Written authority to remove goods in bond from a bonded store either for export or for transference to another store – a Customs term.

Bonds with warrants A long or medium term certificate of debt to which is attached a document/warrant entitling the holder to purchase equity securities shares or participation certificates within a predetermined period – exercise period at a predetermined price-exercise price.

Bond washing The process of reducing income tax liabilities by buying bonds shortly after payment of a dividend and selling shortly before the next payment.

Bonus See override.

BONVOYAGE Bon Voyage Shipping Line – a shipowner

Booked traffic Traffic which has reserved accommodation earmarked for it on a particular transport mode service. It may be on a passenger train, passenger flight, cruise, container ship and so on.

Booking cargo space The process of booking cargo space on a transport unit which may be a ship, aircraft, lighterage and so on.

Booking office A point of sale dispensing travel and/or theatre etc. tickets.

Booking reference number The number allocated by the carrier or his agent relative to cargo space booked on a particular transport mode – ship – aircraft – wagon – truck. See Booking cargo space entry.

Booking System The process of the client reserving accommodation for a particular facility, service, or product. It may be booking for an appointment to see a doctor or, hairdresser, or reserving a seat on train, theatre or ships cabin.

Book value The value of a company's assets as stated in the balance sheet. Because of the marked fluctuations which may occur in the second-hand prices of ships, estimates of shipping companies' net worth usually involve the adjustment of book value to take account of the market value of their ships.

Boom lift type Forklift truck A fork-lift truck equipped to handle pipes, carpets etc.

Booster Articles containing a detonating explosive and used to increase the initiating power of detonators or detonating cord. A term associated with dangerous classified cargo movement.

Booster Pump A pump used to increase the discharge pressure from another pump (such as a cargo pump).

Boot Topping A protective composition painted round the hull of a ship to prevent corrosion between the load and light waterlines.

BOP Bureau of Post (Philippines); or Balance of Payments.

BORD Bordereau.

Border adjustment measures Exports may be relieved of indirect taxes and imports taxed on an amount equivalent to indirect taxes on similar domestic goods.

Border country A country situated contiguous/adjoining to a particular country. For example, Switzerland has border countries with Italy, France and Germany.

Bordereau A document used in international road transport listing the cargo carried on a road vehicle and often referring to appended copies of the road consignment note.

Border post Customs post/examination point situated at an international border point and manned by customs personnel.

Border protection Any measure which acts to restrain imports at point of entry. WTO term.

Bore A tidal flood which rushes up the mouth of a river and meeting the ebb tide or opposing current, forms a wall of water which increases in height as the river narrows.

BORG Basic Operational Requirements and Planning Criteria – IATA term.

BORO Bulk/oil and ro/ro vessel.

Boston Consulting Group A company portfolio planning technique developed by the Boston Consulting Group to determine the best way an entity can use its strengths and develop its advantages within its business environment.

Bosun Senior deck rating with European crews.

BOSVA British Offshore Support Vessels Association.

BOT Board of Trade of Thailand.

BOTT Board of Trade of Thailand.

Bottle A small narrow necked receptacle usually of glass, stoneware or plastic.

Bottleneck A point of congestion whereby the resources/capacity available is inadequate to handle the input. It may be a warehouse, export documentation office, customs clearance unit and so on.

Bottlescrew A tensioner that is tightened by turning a threaded screw with left and right hand thread. Also termed a turn buckle or screw.

Bottom fittings The special conical shaped devices inserted between a container and the permanent floor on the deck of a vessel in order to avoid shifting of the container during the voyage.

Bottom lift The handling of containers with equipment attached to the four bottom corner fittings (castings).

Bottom line The ultimate final financial result of an enterprise which may be a profit or loss situation.

Bottom out A term associated with a freight rate, currency exchange rate and so on, indicating the rate has reached its lowest level having regard to market conditions.

Bottomry Loan taken out on the ship.

Bottomry Bill (or Bond) The pledge of a ship, or of its cargo, as security for repayment of money advanced to the master in an emergency, and of no avail if the ship be lost.

Bottomry Bond Process of Master pledging his ship or part of it to secure money to enable voyage to continue.

Bottoms-up An approach to national services schedules in which national commitments are only those positively identified. These would be added to over time. A WTO term.

Bottom Treatment Clause A clause in the Institute hull clauses, whereby underwriters specify the extent to which they shall be liable for surface preparation and primer painting of plates in repairing the bottom of a ship damaged by an insured peril.

Bound tariffs Tariffs fixed in schedules to the WTO which can only be raised if compensation is negotiated. Also known as Binding. A WTO term.

Bourse A stock exchange.

Bow door A watertight barrier which seals an opening in the forward end of a roll-on roll-off ship through which rolling cargo is wheeled or driven along a ramp into or out of the ship. The door is often made up of the ramp itself.

Bows Extreme forward end of vessel.

Bow loading An oil tanker fitted with a facility to load cargo over the bow usually at an offshore mooring buoy. Likewise a Ro/Ro vessel equipped with a bow visor allows bow loading with vehicles passing over the shore based ramp.

Bow ramp An inclined plane which connects the forward end of a roll-on roll-off ship to the shore or quay on which rolling cargo is wheeled or driven into or out of the ship. The ramp is designed to make a watertight door to cover the opening in the ship.

Bow Stopper A device for securing the mooring between offtake vessel and installation. It is fitted near the bow of an offtake vessel and is used to secure the chafe chain, found at the end of the mooring hawser.

Bow thruster Reversible propeller to give movement athwartships in either direction thereby aiding ship manoeuvrability especially in confined waters. Hence, a propeller used to provide a transverse thrust to the bow of the ship and to assist movement in confined spaces.

Bowtime Code name of BIMCO approved uniform time charter party for container vessels.

Bow visor Upward hanging portion of the bow to allow vehicle access to and from the vessel via the shore based ramp.

Box A complete package with rectangular sides, of wood, plywood, reconstituted wood, fibreboard, metal, plastics, or other suitable material and without orifice; or a container; or a portion of a strip or a portion of a column contained within imaginary or printed lines found on export or import documentation; or category of domestic support – Green box: supports considered not to distort trade and therefore permitted with no limits – Blue box: permitted supports linked to production, but subject to production limits and therefore minimally trade-distorting – Amber box: supports considered to distort trade and therefore subject to reduction commitments. WTO term.

Boxer class A container type vessel.

Box K Box keel (on a ship).

BOXNET The code name of an interactive link fully computerised between PSA and port users relative the handling of containers at the container terminals.

Box one piece A fibreboard box with one manufacturing join, constructed as a complete 'sleeve', the top and bottom each being formed by four flaps.

Box pallet A pallet with side (or sides and top) mostly of wire mesh or grilles. Standard box pallets on 800mm or 1,200mm bases are exchangeable pallet pool equipment in parts of Europe.

Box pull through A fibreboard box consisting of one or two creased sheets placed within an outer sleeve to which it is (they are) stitched or taped.

Box rate A container rate.

Box ships Container vessels conveying ISO containers.

Box terminal A terminal handling containers.

Box three piece A fibreboard box consisting of a body sleeve or envelope sometimes having flanges at the openings, a lid and a bottom which are usually identical components in the shape of trays.

Boxtime The standard BIMCO time charter for container ships. The latest edition is dated 2001.

Box trailer A trailer whose body consists of a metal box similar to a container which can be sealed and which permits goods to be shipped with maximum security.

Box wood frame end A fibreboard box in which the body section consists of a single sheet forming the sides, bottom and lid, with an overlap at the joint: the two ends consists of a fibreboard panel fastened to the wooden frame of the ends.

bp Basis point – economic/financial term.

B/P Bills payable.

BP Bollard pull in long tons; Air Botswana Corporation – International Airline of Botswana and a member of IATA; or Bermuda plan.

BPA British Ports Association.

BPF British Plastics Federation.

BPHR Bureau of Ports, Harbours and Reclamation (Philippines).

BPI Burma Pharmaceutical Industry.

Bpm Best Practical Means.

BPOP Business Process Outsourcing Programmes – a computer logistics term.

BPSS Barge-mounted production and storage system.

BP Worldwide Marine Distance Table The internationally recognised nautical mileage distance between seaports.

BQS Bunker Quality Survey.

BR Bulgarian Register of Shipping – the Bulgarian Ship Classification Society; boiler room; vessels classed by the Brazilian Register – Ship Classification Society; or EVA Air (EVA Airways Corporation) – an IATA member Airline.

B/R Bills receivable or builders risk.

Braided Rope A rope produced by intertwining a number of strands.

Brain storming The process of a group of people pooling ideas on a common theme and in so doing generating a thrustful discussion to evolve fresh strategies/proposals. It may be a new advertising slogan, improved product/service specification or new products/services. Freedom of expression and commitment to the brain storming session are essential to stimulate thinking and ideas.

Brake-horsepower The amount of horsepower produced by an engine on a bench test. Gross brake horsepower is the total power produced, whilst net brake horsepower is the total power, less losses caused by driving auxiliaries (water pump, engine fan, alternator) and other auxiliaries such as air cleaners, and silencers, all of which reduce the usable horsepower.

Brand The process of an entity developing a name for some or all of its products/services and thereby develop a brand image such as 'M & S' involving Marks and Spencer. A product with a set of characteristics which differentiate it from other products.

Brand awareness The extent to which the Brand is known in the market place and its perception, including value added benefit, as viewed by existing and potential clients.

Brand building The development of a brand by adding value to it such as extending the warranty period or making it multi-purpose.

Brand development The process of developing and making more competitive the Brand products/service in the market place and the perceived benefits in the eyes of the buyer.

Brand exposure The process of exposing the brand product/service in the market place thereby generating more market awareness of the commodity to potential clients.

Brand image The perceived image created by the brand in the competitive market place as perceived by competitors and potential clients/buyers and particularly the value added benefit the brand offers.

Brand label The label/logo on a product/service identifying the company brand.

Brand leader A product which holds the greatest share of the market.

Brand loyalty The process of the buyer remaining loyal to the seller's brand through continuous/repeat buying. The extent to which users of a brand re-purchase it. The buying cycle is likely to be shorter for the consumer product and longer for the industrial product. Accordingly, different strategies are required to retain the buyer's patronage.

Brand Management The process of the brand management effectively managing, developing and sustaining its brand in the market place. This is primarily achieved through advertising and the buyer evaluation of the product/service. It is essential that strong empathy be developed with the buyer.

Brand Marketing The process of marketing the brand image of a product/service and in so doing extolling the perceived benefits of the brand to the user/customer.

Brand positioning The position of a branded product or service in the market place. It may be a market leader or a laggard. See also positioning in the market place entry.

Brand share The market share a particular brand product/service commands/holds in the market place.

BRAS Building Research Advisory Service (BRE).

Brash ice Accumulations of floating ice made up of fragments not more than 2m across, the wreckage of other forms of ice.

Brazilian Chamber of Commerce The organisation based in London representing Brazil whose basic role is to develop, promote and facilitate international trade between UK and Brazil and vice versa.

Brazilian Register Ship Classification Society of Brazil.

BRE Building Research Establishment (DOI).

Breach of Warranty The Marine Insurance Act (1906) demands that the assured comply literally with any warranty in the policy, whether it be an express warranty or an implied warranty. Non-compliance, where not excused, is termed 'breach of warranty' and discharges the underwriter from all liability under the policy, whether or not it relates to the breach, as from the date of the breach; but without prejudice to any valid claims arising from accidents occurring prior to the date of the breach. Breach of warranty is excused where circumstances have changed so that the warranty has become unnecessary; or where it would be illegal to comply with the warranty; or where the policy conditions waive breach of warranty.

Breach of Warranty Clause A clause in a policy where underwriters waive breach of certain specified warranties. In hull policies, the clause relates only to warranties as to cargo, trade, locality, towage, salvage services or date of sailing, and requires prompt notice, and amended conditions and an additional premium of underwriters so desire. There is no similar clause in cargo policies as a standard, but cargo underwriters usually waive breach of the implied warranties of sea worthiness and cargo worthiness of the overseas carrying vessel, provided the assured are not privy thereto.

Break A sharp often unexpected movement in price in either direction. A BIFFEX term.

Breakage Money earned by a travel supplier when client does not use all the services paid for as part of a package.

Break-away Coupling A system fitted in the cargo transfer hose, designed as a weak-link, which can part under over-tension but which seals each open end to minimise spillage to the sea.

Break bulk General cargo conventionally stowed as opposed to bulk, unitised, or containerised cargo; goods that have been stripped from containers (or other form of bulk carriage) for forwarding to final destination; or the process of commencing discharge.

Break bulk cargo Cargo packed in separate packages (lots or consignments) or individual pieces of cargo, loaded, stowed and unloaded individually.

Break even analysis A profit/volume planning resource, whereby the volume and related sales volume is correlated to fixed and variable cost on a graphical presentation to determine at what point cost equates to revenue. See Break even point entry.

Break even load factor The load factor on a transport mode which may be a train, road vehicle, aircraft, ship at which the total income from the journey/voyage/flight equals the total expenditure for that particular transit and thereby yield no loss/profit but merely equates income with cost to produce a breakeven situation. Overall, the load factor necessary for scheduled traffic revenue to cover operating costs. See also load factor.

Break even point The situation at which total revenue covers the total cost.

Break even pricing Pricing policy based on an analysis of break even point.

Break even weight The weight at which it is cheaper to charge the lower Air Freight rate for the next higher weight break multiplied by the minimum weight indicated, than to charge the higher rate for the actual weight of the shipment.

Breaking bulk To open hatches and commence discharge.

Breaking load Minimum load applied at which fracture occurs.

Breaking strength For cordage, the nominal force (or load) that would be expected to break or rupture a single specimen in a tensile test conducted under a specified procedure. On a group of like specimens it may be expressed as an average or as a minimum based on statistical analysis.

Break of journey A condition of carriage term permitting passengers to break their journey en route on a full fare paying ticket provided all travel is completed within the validity of the appropriate portion of the ticket.

Break point/weight The chargeable weight of an air freight consignment beyond which it is cheaper to charge for a higher weight which has a cheaper rate per kilo.

Break-up value The value of a company's assets on the basis that it is no longer a going concern. This involves estimating the realisable value of the assets, less payment of taxes which would no longer be deferrable.

Break-up Voyage A voyage where the ship's final destination is a break-up yard, for the purpose of breaking up the ship. The 1983 ITC provides that where the insured ship undertakes such a voyage during the currency of the policy, any claim occurring subsequent to sailing on the break up voyage shall be limited to the scrap value of the ship. This does not apply if prior arrangements have been made with the underwriters regarding the break up voyage. The break-up voyage clause does not affect claims recoverable under the three fourths collision clause, nor under the GA and salvage charges clause.

Breakwater An above deck structure designed to protect containers from heavy seas.

Breasthook A place bracket joining the starboard and port stringers at the bow of a ship.

Breast lines Mooring lines leading ashore as nearly perpendicular as possible to the ship's fore and aft line.

Breech base A low profile dock socket working on the same principle as the breech of a cannon.
BREEMA British Radio and Electronic Equipment Manufacturers Association.
Bretton Woods In 1944 an International Monetary and Financial Conference of the United and Associated nations was held in Bretton Woods, New Hampshire, USA, to discuss alternative proposals relating to post-war international payment problems put forward by the USA, Canadian and UK Governments. It resulted in the establishment of the International Monetary Fund in 1944 and later the International Bank for Reconstruction and Development.
BRG Bridge.
Bridge Navigating centre of a ship.
Bridge Box Sandwiches and tea/coffee drinks container placed by stewards in Wheelhouse.
Bridge felting Felting that may be tensioned for rigidly connecting the corner castings of adjacent containers or tiers.
Bridging loan A short-term advance pending the borrower's receipt of funds from another source.
Briefing The process of appraising/informing a person(s) of a particular situation. It may be prior to a meeting, a sales conference to outline selling techniques and so on.
Brings an action against him The process of the plaintiff suing the plaintiff – a legal term.
Britcont Code name of Chamber of Shipping approved charter party for the home trade 1928; or Code name of BIMCO approved Bill of Lading for shipments on the Chamber of Shipping General Home Trade Charter, 1928.
British Airways PLC A UK International Airline and a member of IATA.
British and Continental Shipping Lines Joint Container Service The consortium in which P & O Containers trades in the Europe-East Africa service.
British American Chamber of Commerce The organisation based in London representing USA whose basic role is to develop, promote and facilitate international trade between UK and USA.
British dependent territories A term used in shipping to identify tonnage which has been registered in countries regarded as the British dependent territories. These include Montserrat, St. Helena, Anguilla, Gibraltar, Turks and Caicos Islands, Bermuda, Caymon Islands, Falkland Islands and British Virgin Islands.
British East Africa Containers The consortium in which the Overseas Container Ltd trades in the Europe – East Africa service.
British Importers Confederation An organisation in the UK who aims to look after the interest of parties concerned with importing goods into Britain. Membership is comprised of individual businesses and import trade associations.
British International Freight Association The organisation which represents the freight forwarding industry and incorporates the Institute of Freight Forwarders Ltd.
British Midland A United Kingdom airline.
British Shippers Council The organisation in the UK representing the interest of UK shippers at National and International level in the development of international trade. It now forms part of the Freight Transport Association.
British Soviet Chamber of Commerce The organisation based in London representing USSR whose basic role is to develop, promote and facilitate international trade between UK and USSR and vice versa.

British Standards Institution A recognised body in the UK for the preparation and promulgation of national standards. It represents the UK in the International organisation for Standardisation (ISO) in the International Electrotechnical Commission and in other Western European organisations concerned with harmonisation of standards. Its work governs the preparation of glossaries of terms, definitions and symbols, methods of test, specifications for quality, safety, performance or dimensions, preferred sizes, and codes of practice.
British Waterways Board The nationalised authority in Great Britain responsible for the maintenance of and navigation on much of the British inland waterway network. British Waterways Board also operate their own freight handling facilities, including docks, inland depots and barge fleets.
BRKRS Brokers.
BROA British Rig Owners Association.
Brochure A printed folder describing a tour, package, attraction, or designation and specifying the arrangements.
Broken stowage Space wasted in a ship's hold, container, trailer, railway wagon, by stowage of uneven cargo i.e. irregular shaped consignments.
Broker One who puts buyer and seller in touch with one another for a fee/commission. In a broader sense the term is used for an organisation which is a member of an exchange and which acts as the means through which a trader has access to the market. In these circumstances the broker acts as principal, but does not share in the risk as each deal is offset, with other brokers, in the market. In maritime terms a person who transacts the business of negotiating between merchants and shipowners respecting cargoes and clearances; one who effects insurances with the underwriters.
Brokerage Percentage of freight payable to broker on completion of charter party negotiations of sale or purchase of vessel – the time of payment being specified in the contract; or the commission charged by a broker to his clients – a BIFFEX term.
Browser The software you use to surf the World Wide Web – a cyberspace term.
Brunei Dollar The currency in Brunei. See also Appendix 'C'.
Brussels Nomenclature International agreed tariff classification system containing 21 sections, subdivided into 99 chapters with 1095 headings for all goods in international commerce. It is now known as the Customs Co-operation Council Nomenclature.
B/S Bales or barrels.
BS Bill of Sale; hull and equipment or iron and steel ships classed according to former British Corporation Rules; boiler survey; balance sheet; bill of store; bow and stern thruster; Bunker surcharge; British Standard; or Bahamas.
BS 5750 The national standard for quality management systems. The standard lays down procedures which help a company create a cost effective quality conscious working environment including customer service. Overall, it is a set of guide lines for the professional management of any commercial venture including transport and distribution. Companies complying with this quality management system are awarded BS 5750 accreditation which is subject to routine annual surveillance. See also EN 29000 and ISO 9000.
BSC British Shippers Council; Belgian Shippers Council; or Bangladesh Shipping Corporation – a shipowner.
BSCC British-Soviet Chamber of Commerce; or Brunei State Chamber of Commerce.

BSCE Bird Strike Committee Europe – IATA term.
BS (Comp) Composite ships classed according to former British Corporation Rules.
BSE Bovine spongiform encephalopathy, or 'mad cow disease'.
BSEA Black Sea.
BSI British Standards Institution – authorised body for the preparation of National Standards in the United Kingdom.
BSJC British Seafarers' Joint Council.
BSMI Bureau of Small and Medium Industries (Philippines).
BSP Brunei Shell Petroleum Co Ltd.
BSRA British Ship Research Association.
BSS British Standards Specification.
BST British Summer Time.
BT Bow thrusters (ship propulsion unit); British Telecom; Wagen Behalter Tragwagen; or Air Baltic (Air Baltic Corp. (SIA)) – An IATA member Airline.
Bt Berth terms; or billion tons.
BTA Border tax adjustment.
BTG British Technology Group.
BTI British Trade International.
btm Billion ton miles.
Btm Bottom (of a ship).
BTMA British Textile Machinery.
BTMS Bottoms (parcels).
BTO Build-Transfer-Operate.
BU Bulgarian Register of Shipping – vessel classed by Bulgarian Ship Classification Society; Braathens S.A.F.E. – an IATA member airline; or Bonded User – a person who has given bond and has been individually approved for the purpose of using oil for eligible use.
Bubble car See observation car.
Bucket elevator Cargo handling equipment used for grain.
Bucket shop Travel agency that sells to the public discounted air tickets obtained for consolidators.
Buck the market To adopt a policy/strategy which goes directly against the market trend. For example, a government may decide to avoid devaluation of its currency and buys substantial quantities of its currency in the hope the exchange rate will improve, but in reality the market forces are too strong and the government intervention only delays slightly the currency devaluation eventuality.
Budget The process of formulating forecast and objectives of either expenditure and/or revenue including traffic volume during a specified future period for a particular service/trade or Company.
Budget analysis The process of analysing a budget compared with for example actual results during a specific period and area of the business.
Budgetary control A management aid whereby a particular Company, or entity will prepare through its most senior Management a budget reflecting Company aims/policy for the forthcoming year, or in some cases a three or five year period. The one year budget would be the more detailed and all sponsoring Executives involved in its compilation would be committed to its realisation. It would embrace income (sales), expenditure, investment and cash flow. Budget statements compared with actual results would be issued each month, quarterly or other convenient period, and significant variations discussed by Management and any remedial measures instituted. It is a modern Management aid of financial discipline on an individual participating sponsoring Executive basis.

Budgeted cost Those costs as predicted in the budget.
Budgeted income The income or revenue as predicted in the budget.
Budgeted profit The profit as predicted in the budget.
Budget Officer The post in a particular organisation responsible for devising the specified budget and instituting the appropriate measures and systems to ensure Management are able to compare the actual results of the Company with the Budget. In so doing Management would need to have explanations of any significant budget variations and details of any remedial measures in activities which need such treatment. The budget may be on income (sales), expenditure, investment, cash flow and so on. A monthly or quarterly statement may be prepared to record actual results against the budget. Each Sponsoring Officer would submit to the Budget Officer a commentary on the basis of the Budget submitted. Likewise, the Budget Officer would circulate to all Senior Management requisite guidelines on how the budget is to be compiled and other requisite data such as economic forecasts, exchange rates, etc.
Budget review The process of reviewing the constituents of a budget by the sponsoring officer(s) within an entity. This may be due to a large order being obtained. In such a situation production will need to be increased in the factory and a fresh look taken at the unit production cost in the light of the desire to optimise economic resources within the production plant.
Buckling To crumple under longitudinal pressure.
Buffer A cushioning device to reduce shunting shocks on rail cars.
Buffer stocks The process of storing a specific commodity/merchandise to meet an unpredictable market need and thereby enabling such a product to be distributed to the consumer/retailer as soon as the market demand dictates. In such a situation the buffer stocks acts as a distribution centre for the product either worldwide, nationally, or on a regional/area basis. Overall, a quantity of goods/articles kept in store to safeguard against unforeseen shortages or demands which acts as a buffer between adjacent operations in the cycle of movement and use of materials allowing each operation to carry on at its own speed.
Buffer storage capacity To provide above the norm storage capacity to meet an unforeseen market need/demand. See buffer stocks entry.
Building destinations The process of a company, usually a global one, with an international brand image developing a site in a remote part and thereby identify-ing it on the global commercial map. It maybe an oil jetty/terminal, industrial plant or leisure centre/park.
Building subsidies A Government subsidy granted to shipowners to encourage them to have built new vessels and so displace usually less modern tonnage.
Built in obsolescence See obsolescence entry.
Built up vehicles Vehicles (cars/coaches/lorries etc) in finished/complete condition. A term used in the conveyance of such vehicles by sea transport.
Bulgarian Register of Shipping The Bulgarian Ship Classification Society.
Bulbous bow Protruding bow below the waterline designed to reduce the vessel's resistance to motion under certain circumstances, such as to aid more economical operation through less pitching of the vessel
Bulk buying The process of buying in large quantities. See discounted bulk buying entry.
Bulk cargo Cargo usually a homogeneous raw material shipped in bulk i.e. complete shipments such as grain,

timber, chemicals and oil. It is not shipped in drums, containers or packages.

Bulk cargo fleet See dry bulk carrier fleet entry.

Bulk cargo vessel Single deck ship designed to carry homogeneous unpacked dry cargoes such as grain, iron ore and coal. Such ships have large hatchways to facilitate cargo handling.

Bulk carrier Purpose built vessel specialising in the shipment of bulk cargoes i.e. iron ore, grain.

Bulk container Containers of ISO specification designed for the carriage of any powders and granular substances in bulk. To facilitate top loading, three circular hatches are fitted in the roof and to discharge the container a hatch is fitted in the right hand door of the container. Overall, a container designed for the carriage of free-flowing dry cargoes which are loaded through hatchways in the roof and discharged through hatchways at one end of the container.

Bulk container ship Vessel that carries either bulk cargoes or containers simultaneously or consecutively.

Bulk dry and wet cargo berth A berth situated at a seaport which accommodates/handles bulk dry and wet cargo shipments such as cement, fertilisers, grain, rice, wine and so on.

Bulk dry cargo berth A purpose built/designed berth situated at a seaport which accommodates/handles bulk dry cargo shipments such as cement, fertiliser, phosphate and so on. A modern bulk dry cargo berth would be fully automated.

Bulk fare Fare available only to tour organisers or operators who buy a block of seats from a carrier.

Bulk freight container Any container, which by its own peculiar design i.e. roof loading hatches and door front wall discharge hatch will allow bulk handling of commodities.

Bulkhead Vertical partition which subdivides interior of ship or aircraft into compartments; or sometimes used to denote the front wall, i.e. opposite the doors of a container.

Bulkhead seat Usually roomier seat on an aircraft right behind the bulkhead.

Bulk licence Licence issued to manufacturers and exporters to cover their requirements for a certain bulk quantity or period.

Bulk liquid berth A purpose built/designed berth situated at a seaport which accommodates/handles bulk liquid cargo shipments such as oil, liquid propane gas and so on. A modern bulk liquid berth would be fully automated.

Bulk liquid bag A form of collapsible container made of rubber and ideal for bulk liquid shipments.

Bulk liquid container An ISO container designed for the shipment of bulk wine/spirits/oil/chemicals type cargoes.

Bulk purchase See bulk buying entry.

Bulk shipping Maritime bulk cargo fleets such as oil tankers, chemical carrier, LNGs, ore carriers and so on.

Bulk transhipment centre A terminal linking two modes of transport where bulk cargo is loaded and/or discharged by mechanical means. It is usually situated on a berth at a seaport and cargo handled includes coal, ore, cement, phosphate, gravel and so on. A modern terminal would be computer operated.

Bulk unitisation Using a container, road trailer or railway wagon to group a quantity of small consignments together. Often called consolidation and sponsored by a Freight Forwarder.

Bulk unitisation charge A charge which applies to consignments carried from the airport of departure to the airport of arrival entirely in Unit Load Devices (air cargo). See Unit Load Device entry.

Bull A speculator who buys for future delivery hoping to sell at a profit before he has to take delivery. One who believes prices will rise.

Bulldog bond A bond denominated in sterling but issued by a non UK resident.

Bullet A loan or placing for which no 'managed' sinking fund is established and repayment of the loan is made at maturity in a lump sum. An international Banking term.

Bullet loan A loan, the principal of which is repayable in its entirety upon maturity, with interest only being paid in the interim. It is analogous to a corporate debt issue without a sinking fund.

Bullet points Major/salient points in a document/discussion/case study/debate, etc.

Bullet train High-speed Japanese train.

Bullion market A market where dealers both buy and sell gold and silver on behalf of their clients.

Bullish The belief that a market price is going to rise.

Bullish market A competitive market place whereby interest is displayed in the product/service on offer thereby generating a strong aggressive market demand in the expectation of rising prices. Usually associated with financial markets. A market where speculators/buyers anticipate a rise in price and thereby make a purchase with a view to a sale later at a profit.

Bull markets A market in which prices are rising.

Bull position One where the trader has bought metal with the view to selling it later after prices have risen to higher levels, his profit being the difference between the purchase and sale prices.

Bullrings Rings for lashing the cargo in containers.

Bull straddle The process of buying nearby months and selling distant months – a BIFFEX term.

Bulwark The raised woodwork running alongside the vessel above the level of the deck or a breakwater.

Bum-boat A boat which supplies provisions to a ship in a harbour or in a roadstead.

Bump To displace a passenger or guest in favour of a reservation or sale to a person with higher priority or supposed importance.

BUNDES Austrian Committee/Organisation on Trade Procedures and Facilitation based in Vienna.

Bundling yard A place where timber is bundled. It is usually situated in the environs of a port.

Bung A plug used to close a barrel bung hole.

Bunker Accommodation on board a vessel to store fuel and usually in the form of tanks.

Bunker Adjustment Factor A surcharge expressed either as a percentage of the freight or a price per cargo tonne W/M, added when fuel prices increase above the level allowed for by air or shipping lines in their freight rates. Overall, bunkering surcharge on the freight rate.

Bunker broker A broker which deals with the bunker market. See broker and bunker entries.

Bunker clause Charter party clause obliging the owners and the charterers to pay for the vessel's remaining bunkers at the port of delivery (by the charterers) and re-delivery (by the owners).

Bunker contract See Fuelcon entry.

Bunker Fuel Any fuel which is loaded in the bunker tanks for use on powering a vessel. Often referred to as 'bunkers'.

Bunker fuel supply ship services The provision of supply vessels containing bunker fuel available at a seaport to bunker ships at the seaport. Such supply vessels embraced bunker barges and standard tankers.

Bunkering The process of refuelling a vessel.

Bunkering plan Schedule of ports at which to bunker a vessel on a specific voyage.

Bunkering barge A barge operative usually in the environs of a port which provides a bunkering service to those vessels at anchor.

Bunkering surcharge A technique whereby a surcharge is raised by the shipowner on the basic freight rate to reflect increased fuel cost incurred due to the upward price of fuel oil which may be attributable to currency variations or simply an increase in the basic bunker rate. The bunkering surcharge may be later consolidated with the basic rate etc. It is sometimes called a fuel surcharge.

Bunker port A port which is available for bunkering ships.

Bunker Quality Survey The process of conducting a survey/audit of fuel oil delivered in any given bunker stem. A service provided by Ship Classification Societies.

Bunkers The quantity of fuel on board a vessel.

Bunker safety surplus The additional bunkers conveyed on board a vessel. This may be some 15 per cent in addition to the actual bunkers required for normal consumption for that leg of the proposed voyage.

BUP Bulk Unit Programme.

Burden of Proof Responsibility for proving some fact.

Bureau of Consular Affairs An agency within the US State Department charged with administration of US passports.

Bureau Veritas French Ship Classification and Survey Society.

Burning cost A loss ration determined from the statistics of a number of preceding years in order to assess the premium to be charged to the reinsured in connection with excess loss reinsurance.

Burning oil (Kerosine) A refined petroleum distillate intermediate in volatility between motor spirit and gas oil, used for lighting and heating.

Burster Cases charged with detonating or deflagrating explosive used to burst projectiles or bombs in order to project or disperse their contents. The burster is no larger than is required to burst the case and disseminate the contents. A term associated with dangerous classified cargo transport movement.

Burundi Franc The currency of Burundi. See also Appendix 'c'.

Bushing A circular Hush or low profile deck socket (e.g. post socket, or flush ISO hole deck socket).

Business acumen The ability of a person to identify, create and cultivate maximum advantage of business opportunities which maybe self created or simply arise through a particular circumstance. In so doing the person will develop through his own enterprise the business on a profitable long term basis improving the overall market share.

Business card A card depicting the name and position held in the company together with full address details. For overseas business many are double sided bi-lingual with the seller's language on one side and the buyer's language on the other. For example, English and German.

Business centre A place where Business facilities are provided. It includes, conference facilities, secretarial resources, dictating and reprographic services, reference library, telecommunications including telephone and facsimile, restaurant, computerised facilities including computerised press facilities, conference hall and meeting rooms, language service and so on.

Business class A grade of airline seat and service usually between first class and economy/tourist/coach, with better seating, food, service and check-in facilities; called Club class by some carriers.

Business Community The entrepreneurs in a community/competitive market place.

Business culture The business protocol, and culture operative in a particular country in which business is conducted. It will be different in India and Japan for example. It could also vary by Company and business activity.

Business cycle The process of the business environment going through various phases of its development. This involves the peak, decline, trough and growth.

Business day Any day except Saturday, Sunday or public/bank holiday in England or a day designated not to be by the management.

Business Development Manager A key managerial post responsible for the development/growth, strategies/ initiatives in a Company which may be a manufacturing or service sector. The post calls for (a) an understanding of strategic business embracing evaluation of all factors, consideration of all perspectives, conceptional approach, and take a broad long term view; (b) develop creativity and flexibility involving innovation, anticipation and initiation of change, developing new ideas and multi-disciplinary interest; (c) analysis and decision making embracing identification of key issues, clarification of objective, evaluate options, consider implications, initiate action and take decisions; (d) numeracy and financial understanding of financial accounts, profit motivation, costing and quantitative data; (e) results orientation, featuring organisation and planning, proactive strategies, exploitation available resources, overcome objections, deliver promises and achieve results; (f) management of relationships embracing listening to others, teamwork, highly persuasive, direct and open, confident in views and promotion of effective working relationships, and (g) computer literate, embracing applying new technology, performance analysis and data assessment.

Business ethics The ethics adopted by a company/ entity/government when undertaking business. This will vary by country and company. It will reflect culture and focus on morals. See culture entry.

Business Houses A retail outlet specialising in the Business Executive's needs in the travel sector embracing all its infrastructure such as concluding the arrangement of an overseas trip involving hotel, travel, entertainment and so on.

Business interruption An insurance covering the loss of earnings of a business following fire or other specified damage. It is also called consequential loss.

Business logistics Logistics within a business system, or the co-ordinating function of material management and physical distribution which executes the integral control of the goods flow. See also logistics entry.

Business luncheon A lunch attended by two or more people which has been arranged for the purpose of discussing/negotiating business matters.

Business Management The technique(s) of managing a business.

Business organisation A company organisation structure usually based on the business markets which the specific company serves.

Business Partner An entrepreneur in a business partnership relationship situation. See Business relationship entry.

Business plan A plan or programme covering a period usually between three to five years featuring the Company intentions/objectives/policy and identifying annually by service/product/activity etc., the predicted level of income, expenditure and capital investment.

Business Process Re-engineering The process of re-engineering the processes in order to simplify and streamline low key objectives such as the supply of products and services are delivered. An example is concurrent engineering.

Business Re-engineering A company/entity organisation which is structured around the concept of continuous business processes being designed to get the company's products to the customer. This concept abandons the old practice of companies dividing themselves into bureaucratic empires focused on discrete task such as sales, marketing and accounts. Overall, the concept bypasses middle managers and devolves responsibility to multi-disciplined teams of shop floor workers.

Business relationship A relationship between two parties strictly on a business plateau with each party retaining/maintaining the relationship on the value added benefit it yields.

Business to Business Process of conducting business between two companies direct without any intermediary such as an agent or broker. It includes inter-organisational information system transactions and electronic market transactions between organisations.

Business to consumer See consumer to business entry.

Business travel A passenger whose journey purpose is to fulfil a business engagement such as to attend a meeting, meet a client, attend a conference and so on. This forms a large part of the passenger market in many sectors and special arrangements are usually provided to meet the business executive's needs travelling on an expense account.

Business travel agent See commercial travel agent.

Business travel market That segment of the passenger market whose purpose of journey is to fulfil a business engagement.

Bustout Purchase of an agency to steal and/or fraudulently use its ticket stock and perhaps other assets before abruptly closing down.

Butane Hydrocarbon containing four carbon atoms, gaseous at normal temperature, but generally stored and transported under pressure as a liquid. Used for industrial purposes, some domestic heating and cooking and as a constituent of motor spirit to improve volatility and as a chemical feedstock.

BUTEL Bureau of Telecommunications (Philippines).

Butterworth store Situated on Main-deck housing tank cleaning equipment and deck stores.

Buttress A rigid above deck container securing system where containers are supported on frames which in turn are supported by buttresses at the end of the hatch covers.

Buttweld The buttweld is used to join two plates in the same plane and is considered to be the strongest welded joint produced in plates of between 6mm-18mm thick. It is often necessary to 'vee' the plate edges.

Bututs The currency of Gambia. See also Appendix 'C'.

Buy back A counterstrade technique being in the form of a barter in which suppliers of capital, plant, or equipment agree to payment in the future from output of the investment concerned. For example, exporters of equipment for a chemical works may be paid with part of the resulting output from that chemical plant.

Buyer The person/company/entity which purchases a product(s)/service(s).

Buyer-centric portal A market place set up by several sellers with the objective of offering the potential buyer an integrated product line. An E commerce term.

Buyer credit The process of providing finance direct to the buyer (or to a borrower e.g. a bank in his country) in the form of buyer credit. In so doing a loan agreement is negotiated between for example an acceptable bank in the United Kingdom and the overseas borrower under which finance is made available in connection with the supply contract between the UK supplier and the overseas buyer. Payment is normally made direct to the UK supplier by the lending bank against agreed documentation. The responsibility for the payment of interest and the repayment of principal is that of the borrower and the UK supplier is largely in the position of having a cash contract.

Buyer's behaviour The reaction of the buyer in the market place to a change in circumstances in which the buying decision takes place. It maybe economic, social, legal, technical or political. Usually, buyer behaviour is measured in terms of a change in price and/or product specification. A marketing term.

Buyer on behalf of foreign firms A buying office in the UK used by overseas importing firms to find UK sources of supply.

Buyer's market A situation in the competitive market place where an over supply of a commodity places buyers in a very favourable negotiating situation as a result of an imbalance of supply and demand coupled with buying power and generally lower prices. This often involves discount pricing to stimulate volume sales and a very active selling effort to secure any business.

Buying agents Also called indent agent. Acts for an overseas' buyer to place orders or indents with manufacturers.

Buying cycle The sequence of events/the process of the client buying a product/service. The consumer buying cycle is shorter than the industrial buying cycle in terms of the decision making process and frequency of purchase.

Buying House An indirect trading technique whereby numerous large retail organisations in any country establish offices in other countries in order collectively to purchase from manufacturers in those territories. The extension of the system relates to the establishment of independent houses who buy on behalf of well known foreign department stores. An ideal outlet for manufacturers of consumer products and often referred to as Indent House.

Buying overseas The process of purchasing goods/services which cross international boundaries.

Buying pattern The consumer habits in buying terms over a period of time relative to the purchase of a particular product or service. It may be consumers make their purchases when volume discounts are available and hold back when prices are at their peak. See also buying cycle entry.

Buying power The strength/ability of the consumer/industrialist to fund the goods/service sought. It is primarily determined on a national basis on the Gross National Product and Gross Domestic Product. This is high in developed markets like USA, Japan, Germany, but low in underdeveloped countries like Gambia, Tanzania and Sudan. Developed countries have greater buying power than less developed countries.

Buying rate of exchange The price at which a Bank, or other specific organisation/company is prepared to buy any particular currency.

Buying syndicate A number of persons who form themselves into an integrated group to form a negotiating and buying group.

Buy-out The purchase of a company by its former management team, who have usually obtained substantial outside capital backing for this purpose. There have been a few examples of this type of transaction in shipping, with the ratio of asset values to cash flow frequently posing serious obstacles.

BV Bureau Veritas – French Ship Classification Society; or Blue Panorama (Blue Panorama Airlines S.p.A.) – an IATA member Airline.

BW BWIA International Airways (BWIA West Indies Airways Ltd.) – an IATA member Airline.

BWB British Waterways Board.

BWS Bureau of Water Supply (Philippines).

bx Box Containers.

BY Belarus – customs abbreviation.

Bypass Practice by some suppliers of selling travel products directly to the public rather than through travel retailers.

12c's environmental analysis model
A marketing information framework system model designed to produce reliable data on which international marketing/business strategic decisions can be made. It embraces analysis of country, choices, concentration, culture/consumer behaviour, consumption, capacity to pay, currency, channels, commitment, communication, contractural obligations and caveats.

c Centre; cage containers; or Cede – see separate entry.

C Compound expansion (marine) engines.

C/- Case packaging.

C88 The single administrative document form – see separate entry.

CA Controlled atmosphere; Air China International Corp. – an IATA member.

CAA Civil Aviation Authority; current account; or central accounting – as used by the main bureau, Lloyd's ILU and PSAC.

CAAC Civil Aviation Administration of China.

CAACE Comite des Associations d'Armateurs des Communautes Européenes – the Committee of Shipowners Associations of the European Community.

CAAG Civil Aviation Assistance Group (US) – IATA term.

CAB Civil Aeronautics Board of the USA; or Civil Aeronautics Board (Philippines).

CABAF Currency and bunker adjustment factor.

Cabana A room or hut in a beach or pool area, usually separate from a main building and maybe having beds.

Cabin Passenger compartment of an aircraft; standard bedroom (berth) on a ship; a small, isolated building for rent to travellers.

Cabin Allocation The process of allocating cabins on a ship e.g. thirty may be allocated to agents for their clients and the residue to direct bookings to the shipowner by the client/passenger.

Cabin Charges The tariff applicable to cabins/berths on a particular vessel. It may be a one, two, three, or four berth cabin. The tariff may be based on the cabin complete or, individual berths within the cabin. Hence, a two berth cabin tariff may be €30, or €19 per berth in a two berth cabin on an overnight voyage between two ports. The passenger fare would be extra. On a cruise however, the cabin tariff and passenger fare are combined, and would also include meals to produce an overall cruise tariff of for example €2050, for a seven day cruise in a two berth cabin. Cabin accommodation is also provided on an aircraft in seated conditions.

Cabin service The steward service offered to passengers during the course of a flight or on a voyage.

Cabin staff Airline crew employees responsible for 'in flight' passenger welfare – this applies specially to the executive/first class passenger accommodation, or shipboard stewards responsible for passenger/crew cabins.

Cable A measurement equal to about 100 fathoms or the tenth part of a nautical mile.

Cable rate The cable rate arises when the 'forward rate' is lower than the 'Spot rate'. The term emerges from cable transfer in dollar/sterling deals. An international trade banking term.

Cabotage Air Freight traffic originating at a point in one country and destined to another point within the territory of the same country; or the reservation by law – in some countries – of the national coastal trade to national flag ships. This results in foreign operators being excluded from the coastal trade; or the process of leasing of containers to other operators to reposition them; or the transport of cargo in a country other than the country where the vehicle is registered (road cargo) or the carriage of a container from a surplus area to an area specified by the owner of that container in exchange of which and during which the operator can use this container.

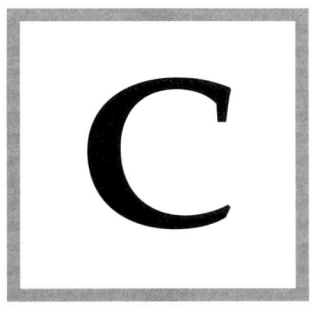

CAC Customs Additional Code; Central Arbitration Committee; or Currency adjustment charge.

Cache The storage of some often used parts of a disc file in an area of memory to provide rapid access (disc cache). The storage of some portions of RAM or ROM in very fast RAM devices to increase speed of access (memory cache).

CACIV Convention additionelle a la CIV.

CACM Central American Common Market.

CAD Cash against documents; Civil Aviation Department (Brunei); or Computer aided design.

CADD Changi Airport Devt. Division.

CADSB Cargo Automation Data Standards Board – IATA term.

C & D Collection and delivery.

C & D Canal Chesapeake & Delaware Canal.

CADDIA Co-operation in automation in data in import and export documents.

CADV Cash advance.

CAEE Committee on Aircraft Engine Emissions – ICAO term.

CAeM Commission for Aeronautical Meteorology – WMO term.

CAEP Committed on Aviation Environmental Protection – ACAP term.

C & E Customs & Excise.

CAF Currency Adjustment Factor, a surcharge usually expressed as a percentage added or subtracted on freight.

CAFAL Cargo Facilitation – IATA term.

CAF (French-co`t, assurance, Fret) Cargo delivery term: cost, insurance and freight.

C & F Cost and Freight (named Port of Destination) Under his cargo delivery term the seller (exporter) pays the cost and freight necessary to convey the goods to the specified port of destination. The buyer (importer) risk commences when the goods pass the ship's rail at the port of shipment and bears all the risks from this point onwards. The buyer (importer) would bear all the cost and risk of insurance, unloading of cargo at destination port, customs import charges, import licence and so on. The seller (exporter) would bear the cost of packing; pay loading cost; obtain any export licence, pay Customs/Export dues; provide commercial invoice, Bill of Lading, Certificate of Origin, Consular Invoice and so on. The seller (exporter) may agree to bear all the unloading cost by adding the word landed – 'C & F Landed'. The buyer's (importer's) risk commences on acceptance of the 'clean on board Bill of Lading'. See CFR entry.

CAFFA-ICC Commission on Asian and Far Eastern Affairs of the International Chamber of Commerce.

Cage-tainer Container with sides and roof, or sides, roof and ends, consisting of grids, meshing and lattice.

C & I Cost and insurance. See Incoterms 2000 entry.

CAI A major container leasing company.

Cairns Group Group of agricultural exporting nations lobbying for agricultural trade liberalisation. It was formed in 1986 in Cairns, Australia just before the beginning of the Uruguay Round. Current membership: Argentina, Australia, Bolivia, Brazil, Canada, Chile, Colombia, Costa Rica, Guatemala, Indonesia, Malaysia, New Zealand, Paraguay, Philippines, South Africa, Thailand and Uruguay. WTO term.

Calculable laytime A chartering term which indicates that the laytime requires to be calculated in accordance with the specific factors/data as found in the charter party clauses.

Calcutta conference The first liner conference formed in 1875.

Calcutta – Pacific Conference Liner conference from Calcutta to the United States through transhipment in Hong Kong.

Call The amount due to be paid to a company by the purchaser of nil-paid, or partly paid shares; the demand for payment if an instalment is due; the trading period in which a price for each Baltic Freight futures month is established in sequence – a BIFFEX term; or the visit of a vessel to a port.

Call account A deposit account, usually interest bearing, from which funds may be withdrawn 'at call' (on notice being given).

Call analysis An analysis of a sales persons visits to customers embracing person seen, object of the visit, outcome of the visit, future calls, cost effectiveness of the call and so on.

Call cycle Organisation of a sales person calling/visiting routine on customers/clients.

Calling forward notice Instructions for release or delivery of goods.

Call Option An option which gives the holder the right to buy metal from or sell it to the seller of the option at the basis or strike price. A call option is bought in the expectation that prices will rise, and limits the buyer's risk to the premium paid should his view of the market prove to be wrong.

Call report A sales report on a customer which will form the client portfolio. A significant aspect of the report will be the object of the visit, business secured/lost, action plan, problem/risk areas, future developments and so on.

Calls The amount still payable after allotment in order to make part paid securities fully paid. Failure to meet calls may mean forfeiture of the security.

Call sign Assigned by the National Authority with which the ship is registered e.g. Lloyd's Register of Shipping.

CALM Catenary Anchor Leg Mooring; or Computer Aided Level Measurement system used in tanker cargo control.

Calor Gas It is a liquefied flammable hydrocarbon gas or a mixture of any of the liquefiable petroleum gases. A term associated with the movement of dangerous classified cargo.

CALS Computer Aided Acquisition and Logistics Support. An EDI research organisation designed to speed up and lower the costs of the supply chain; or Continuous Acquisition and life cycle support.

CALSEL Selective Calling (air-ground) – IATA term.

CAM Computer aided manufacturing.

Cam and lever steering Simple design of steering system in which a cam on the steering shaft moves against a peg or lever on the steering.

Camber The rise or crown of a deck above a horizontal line connecting the ends of a beam.

Cameron Airlines The Camerons international airline and member of IATA

Campaign plan The plan devised by the seller to realise a particular marketing objective within the competitive market place. It will be closely controlled and monitored.

Campaign planning The process of formulating a plan for a particular product(s)/services(s) in a specified area/region/country relative to its promotion in a competitive environment. It will have regard to existing/future competition; methods of promotion in the market place; pricing strategy; product specification and so on. Campaign planning tends to aim for increased product sales. A marketing term.

Camshaft A shaft on a vessel carrying the cam(s) which operate valves.

CAN The Canadian dollar – a major currency; or Committee on Aircraft Noise – ICAO term.

Canada-United Kingdom Chamber of Commerce The organisation based in London representing Canada whose basic role is to develop, promote and facilitate international trade between UK and Canada and vice versa.

Canadian Dollar The currency of Canada. See also Appendix 'C'.

Canadian dollar invoice An invoice payable in Canadian dollars relative to an export sales contract.

Canal dues Dues raised by the Canal Authority on vessels passing through the canal. It is usually based on the tonnage of the vessel.

Canal Passage Process of voyaging through a canal.

Cancelcon The code name of a BIMCO approved charter party clause relative to the cancelling clause entitled 'Baltic Conference cancelling clause 1970.'

Cancellation insurance Insurance policy for travellers guarding against loss or funds in case traveller cannot take trip or the arrangements are cancelled by the supplier.

Cancellation policy Travel supplier policy cancelled bookings and amount of notification necessary for granting of refunds.

Cancelling date Charter party term giving charterer option to cancel the contract in event of vessel not being ready to load by the specified date.

Cancelling return A return of premium paid because a hull policy has been cancelled before the specified expiry date.

Cancl Cancelling.

'C and U' regulations Construction and Use regulations relative to road transport vehicles – their construction and use.

Can hooks Cargo handling equipment suitable for barrel transhipments.

CANISA Compania Argentia de Navegacion Intercontinental SACIF (French) – a shipowner.

Canister Filter Respirator A respirator consisting of mask and replaceable canister filter through which air mixed with toxic vapour is drawn by the breathing of the wearer and in which the toxic elements are absorbed by activated charcoal or other material. A filter dedicated to the specific toxic gas must be used. Sometimes this equipment may be referred to as cartridge respirator. It should be noted that a canister filter respirator is not suitable for use in an oxygen deficient atmosphere.

CANR Chamber of Agriculture and Natural Resources (Philippines).

Cantlings Upright ribs at shipside to strengthen plates.

Canvas sling Cargo equipment suitable for handling bagged cargo.

CAP Common Agricultural Policy; The EUs comprehensive system of production targets and marketing mechanisms designed to manage agricultural trade within the EU and with the rest of the world. WTO term; The Continuing Airworthiness Panel – ICAO term; Civil Aviation Purchasing Scheme – ICAO term; or conditional assessment programmes.

Cap A light gauge metal or plastic cover used to seal or protect the neck of opening of a receptacle.

Capability Plots A series of paper charts showing the performance of an offtake vessel's dynamic positioning system in varying weather condition.

Capacity The ability of an insurance market to absorb risks.

Capacity planning The process of correlating evaluating available capacity to market demand and in so doing optimise capacity utilisation and its infrastructure. An example arises within transport whereby individual services are allocated particular streams of traffic such as twenty customers each having an allocation of five per cent on one particular service. Additionally, the transport operator can vary the service frequency/capacity availability to market conditions. This can be realised by evaluating market forecasts and translating them into service provision. It also arises in a manufacturing plant and similar situations.

CAP charges Those charges arising from the Common Agricultural Policy levies, variable charges, countervailing duties and monetary compensatory amounts.

Capesize Vessels operating in the Cape market and for a dry bulk carrier the capacity is over 80,000 dwt. Such vessels cannot negotiate the Panama Canal as they are too wide.

Capesize bulk carrier A bulk carrier of over 80,000 dwts permitted to ship a wide range of bulk cargoes embracing grain, phosphate rock, iron ore, coal and toxic chemicals.

Cape Verde Escudo The currency of Cape Verde Islands,. Se also Appendix 'C'.

Cape Verde Escudo invoice An invoice payable in Cape Verde escudo currency relative to an export sales contract.

CAP goods Goods liable to agricultural levies and other import and export charges arising in a third country trade from Common Agricultural Policy.

Cap issue The process whereby money from company's reserves is converted into issued capital which is then distributed to shareholders as new shares in proportion to their original holdings in a capitalisation issue which is often abbreviated to 'cap issue'. The correct term is 'capitalisation issue' but is sometimes called 'Free', 'Scrip', or 'Bonus' issues because no extra money is payable by the shareholder at the time.

Capital The equity or shares (authorised and/or issued in a company/entity); equity plus reserve plus profit retained plus loan and debenture stock in a company/entity; or a factor of production, being a stock of resources which have been produced in the past, or involve a sacrifice of present consumption, to enhance productivity and in so doing increase utility. The word in economics means 'real capital', physical goods, ships, bridges, ports, roads, etc.

Capital Allowances The United Kingdom term for deduction allowed against tax for amounts expended on capital equipment. The acquisition of ships typically creates large capital allowances, enabling the shipping company to defer significant amounts of taxation. Similar tax depreciation rates exist in all countries where corporate income tax is payable, and they have sometimes had an undue influence on the capital expenditure policies of shipping companies.

Capital Budget The process of formulating forecast and objectives relative to capital investment projects during a specified period for a particular area of the business activity or overall company results.

Capital employed The aggregation of capital employed in a particular situation/business usually related to the income and level of profitability.

Capital expenditure budget A plan for capital expenditure in monetary terms.

Capital funding planning The process of selecting suitable funds to finance long term assets and working capital.

Capital International World Index An index which is a rough measurement of the performance of the world's major stock markets in aggregate. It contains some 1,100 stocks listed – each of which are weighted according to their market value on stock exchanges in US, Europe, Canada, Australia and the Far East.

Capital investment appraisal The process of evaluating proposed investment in specific field assets and the benefits to be obtained from their acquisition. The techniques used in the evaluation can be summarised as non-discounting methods (i.e. simple payback), return on capital employed, and discounted cash flow methods (i.e. yield, net present value, and discounted pay back).

Capitalisation issue The process whereby money from a company's reserves is converted into issued capital which is then distributed to shareholders as new shares, in proportion to their original holdings. The correct term is capitalisation issue often abbreviated to 'Cap issue' but sometimes is called 'Free', 'Scrip' or 'Bonus' issues because no extra money is payable by the shareholder at the time.

Capitalisation of Reserves The process whereby money from a company's reserves is converted into issued capital, which is then distributed to shareholders as new shares, in proportion to their original holdings in a capitalisation issue. Also known as Bonus, Free or Scrip issue.

Capital lease A major container leasing company.

Capital markets The market on which financial instruments, or established financial contracts (securities and derivatives as opposed to commodities) are bought and sold.

Capital projects Investment projects within a Company/Entity which feature on the investment budget.

Capital risk In the foreign exchange market the danger of not receiving payment from a counterpart on a bargain is known as the capital risk.

Capital surplus The surplus distributed amongst shareholders in accordance with their rights under the articles of association after discharge of all outstanding costs and liabilities following their liquidation of a company.

Capital turnover Turnover of the year. Average capital employed in a year. It measures the number of times the capital is turned over in the year or alternatively the turnover generated by each Euro of capital employed.

CAP levy Common Agricultural Policy levy.

Caps, toy (amorces) Articles consisting of a small quantity of an explosive substance between two strips or discs of paper or contained in a plastic cup or covered by varnishing or other means. These caps should not contain more than 16 milligrams (0.25 grain) of the explosive mixture each. Term associated with dangerous cargo shipments.

Capstan Revolving device on a vertical axis on a vessel used for the heaving in of mooring lines or anchor cable.

Captain Ship's officer.

Captain of Industry A very outstanding person in industry involving a good track record of industrial success as an 'entrepreneur'.

Captain protest See ship protest.

Caption Term used to signify 'heading' on letter/document etc.

Captive audience An audience which for example is almost compelled to take notice of a particular audio visual presentation. See captive market.

Captive market A market in which virtually no competition exists relative to the product offered such as in most countries where the Railway is State owned. In such a situation no competition exists on alternative railway services, but only on substitute transport modes such as road and air.

Captive price A price which is valid and not subject to change over a period of time. It is usually found in a contract as a time charter fixture rate which is locked into the charter party and guarantees the shipowner the fixture rate in the period specified. Also found in the IPE dealings. See International Petroleum exchange entry.

CAR Computer aided routing; or Caribbean – IATA term.

CARAT Cargo Agents Reservation Air Waybill Issuance and Tracking system – British Airways computerised air freight reservation system.

Caravan The provision of sleeping/eating accommodation totally self-contained and reliant on a vehicle/car to tow it.

Caravanette The provision of a motorised caravan. This involves the combined facility of sleeping/eating accommodation forming an integral part of the driving/motorised unit with integrated cab accommodation for the driver and passengers.

Carbamates A white powdery substance produced by the reaction of ammonia with carbon dioxide.

Carbon Dioxide In approved cylinders, not exceeding 230g (8oz) net weight, and worn by passengers for the operation of artificial limbs or orthopaedic appliances, is not considered to be a dangerous article under these Regulations, providing the cylinders are approved for the purpose and pressures do not exceed the maximum working and service pressures for which they are designed. Spare cylinders of the same size and make may be carried by the passengers if required to insure an adequate supply for the duration of the journey. Term associated with dangerous cargo shipments.

Carbon dioxide, solid (dry ice) Carbon dioxide, solid (dry ice) is produced by expanding liquid carbon dioxide to vapour and 'snow' in presses that compact the product into blocks. It is used primarily for cooling and due to its very low temperature (about -79° C) can cause severe burns to skin upon direct contact. When Carbon dioxide, solid (dry ice) converts (sublimates) directly to gaseous carbon dioxide it takes in heat from its surroundings. The resulting gas is heavier than air and can cause suffocation in confined areas as it displaces air. Venting of packages containing Carbon dioxide, solid (dry ice) is required to avoid pressure build-up. Term associated with dangerous cargo shipments.

Carboy A large narrow necked receptacle with a capacity usually of 10 to 60 litres made of glass, earthenware, porcelain, plastic or metal other than steel, may be with or without protection.

Car carrier Vessel built to convey cars or other vehicles.

Carcinogen A substance capable of causing cancer.

Car deck Deck of a ship on which cars are carried. See car carrier.

CAR DI SYS Carbon Dioxide system (ship machinery).

Career development The process of developing Management resources within a Company. This usually involves staff appraisals and other techniques to facilitate a person progressing through a series of job appointments – may be at two yearly intervals in a Company. It is very much dependant on his/her job performance relative to the next promotional appointment.

Career vitae A history/inventory of a person's career to date. It will include age, address, sex, education, and the salient details of career progression on a date order basis, but giving details of work experience. It is usual to mention qualifications and any particular attainments in work performance such as a research chemist who would mention details of new products developed for which he/she has been responsible or associated with. The career vitae is used primarily in job application and obviously the content will vary by job specification.

Car ferry A ship designed to carry automobiles, coaches and passengers.

Cargo Merchandise conveyed on a ship, aircraft or hovercraft, other than mail or other property, carried under the terms of an international postal convention, or baggage. The IATA definition is that 'cargo' is equivalent to the term 'goods' means any property carried or to be carried in an aircraft, other than other property carried under the terms of an international postal convention, baggage or property of the carrier provided that baggage moving under an air waybill is cargo.

Cargo accounting device A document prepared and forwarded by the accounting department of an airline carrier to the accounting department of another carrier notifying of additional charges incurred en route for collection from the shipper or consignee or that charges were collected by the last air line carrier. An IATA definition.

Cargo Accounts Settlement Systems An IATA facility which exists in 19 countries. Its role is to settle monies due on air cargo consignments between carriers and their intermediaries. It covers both exported and imported consignments and developed jointly by airlines and representatives of IATA Cargo Agents. It is fully computerised.

Cargo agent A person or organisation authorised by an airline to receive shipments, execute Air Waybills and collect charges. An IATA cargo agent is one that is recognised by IATA as having met its requirements for an IATA registered cargo agent.

Cargo aircraft Any aircraft, other than a passenger aircraft, which is carrying goods or property.

Cargo area That part of the ship which contains the cargo containment system, cargo pumps and compressor rooms, and includes the deck area above the cargo containment system. Where fitted, cofferdams, ballast tanks and void spaces at the after end of the aftermost hold space or the forward end of the forwardmost hold space are excluded from the cargo area.

Cargo assembly An area earmarked for the reception of parcels or packages and holding them for later dispatch as one consignment.

Cargo Booking Service The process of the Shipper/Agent using the cargo reservation service thereby ensuring space/capacity is allocated to a consignment on a particular flight/sailing/road haulage truck schedule. In the case of a container vessel it would involve booking the container and allocating space on the specified sailing. It is fully computerised.

Cargo bridge An airbridge, sea airbridge or land bridge. See separate entries.

Cargo Charges Correction Advice (CCCA) The document used for the notification of changes to transportation charges or other charges.

Cargo Claim A demand by the purchaser of goods for monetary compensation because the goods delivered are damaged, short delivered or not of the specified quality.

Cargo claims prevention The process of devising a system to reduce or eliminate the level of claims emerging from the carriage of goods on an international transit.

Cargo claims procedure The procedure adopted on the progressing of a cargo claim. This briefly involves the receiver notifying the agent of the damage/loss; submitting claim details to the agent/shipowner/P & I club; evaluation of the claim by the surveyor; progressing surveyor's report with interested parties with a view to a claims settlement.

Cargo clearance The process of clearing cargo through customs which may be at a seaport, airport, lic/lec, or inland clearance depot.

Cargo configuration The composition of the total cargo capacity on a particular transport unit. For example a vessel may have a cargo configuration of 2050 TEU's involving a cargo mix of 20 feet (6.10m) and 40 feet (12.20m) containers, break bulk and unitised cargo, cars, lorries, buses, earth moving and construction vehicles; and heavy machinery of all descriptions.

Cargo Containment Systems The arrangement for containment of cargo including, where fitted, primary and secondary barriers, associated insulations, interbarrier spaces and the structure required for the support of these elements.

Cargo declaration Document established in the convention on 'Facilitation of International Maritime Transport' to be the basic document on arrival and departure providing information required by public authorities relating to the cargo.

Cargo delivery term The precise terms under which the cargo is delivered to the buyer (importer) under an export sales contract and usually expressed in one of the thirteen Incoterms 2000 terms devised by the International Chamber of Commerce such as, EXW, CIF, CPT, CIP and so on.

Cargo derrick Boom on board ship used for loading and discharging cargo.

Cargo disassembly The separation of one or more of the component parts of a consignment (from other parts of such consignment) for any purpose other than that of presenting such part or parts to customs authorities at the specific request of such authorities.

Cargo documents Documents appertaining to the shipment of a consignment including the Bill of Lading, Air Waybill, Cargo Manifest, Commercial/Export Invoice, CMR/CIM Consignment Note, Dangerous Goods Note and so on.

Cargo dues Charges raised by port authority on cargo passing over the quay usually based on tonnage, or a specified unit method i.e. per vehicle.

Cargo handling All the elements/procedures involved in the handling of cargo.

Cargo handling charges The charges raised on handling cargo which may be at a port, warehouse, airport, depot etc. and based on tonnage, commodity type, and method of distribution such as palletised, unitised, or containerised merchandise.

Cargo handling equipment Equipment/facilities provided at a port or freight depot to handle cargo (such as cranes, pallet trucks, pallets, fork lift trucks, straddle carriers and so on).

Cargo – IMP The IATA official source for message specification covering space allocation, air waybill, information flight manifest, accounting, status, discrepancy, embargo and proposed airline customs systems. It also includes an encoding and decoding of all approved codes and abbreviations. An IATA computer term.

Cargo Imp Code A standard system of coding for cargo message elements. Imp codes are used by airlines in data exchange in order to minimise transmission time.

Cargo in isolation The process of isolating cargo away from other merchandise/traffic for a variety of reasons but usually attributable to its being classified as dangerous. Cargo isolation is especially common at airports and seaports where a dangerous cargo compound is provided under strict surveillance.

Cargo insurance The insurance of cargo in transit. It maybe on a vessel, aircraft, or combined transport operation.

Cargo insurance rating Factors influencing the cargo insurance premium rates such as nature of cargo; type of packing; type of cover required; volume of cargo; previous claims, experience and nature of transit.

Cargo Liners Vessels engaged on regular services usually under liner conference conditions.

Cargo Manifest Inventory of cargo shipped.

Cargo marking symbols A label affixed to a consignment signifying it is dangerous classified cargo, or 'use no hooks', 'this way up', etc. Such symbols are devised by IATA and IMO.

Cargo mix The constituents of the cargo commodities conveyed in a particular transport unit. For example, it may be a ship conveying 20 TEUs, 40 cars, 200 tonnes break bulk and unitised cargo, and 15 earth moving and construction vehicles. See also commodity mix.

Cargo net Cargo handling equipment suitable for mail bags, cartons, bagged cargo etc. transhipment, or alternative equipment to make the merchandise more secure on the vehicle, pallet etc.

Cargo (or mail) tonne kilometres used A metric tonne of cargo carried one kilometre. Cargo tonne kilometres used, equal the sum of products obtained by multiplying the number of tonnes of cargo on each stage of the flight, or other transport mode, by the stage distance.

Cargo plan Plan depicting space in a ship occupied by cargo.

Cargo preference A form of flag discrimination whereby under a bi-lateral agreement which insists the bulk of the cargo will be carried on the trading nations' fleets and imposes a penalty of up to 50 per cent of the total freight for violation of this stipulation.

Cargo receipt A receipt for a cargo acceptance. The form of receipt may take several forms. See Bill of Lading, Certificate of Shipment, mates receipt and CMR consignment note.

Cargo reservation The process of reserving cargo space on a vessel or aircraft.

Career route The progression of an employee's career through various jobs/posts which may be plotted by an employer or initiated by the individuals.

Cargo restriction code A code indicating that the use of a certain container is restricted to a particular cargo.

Cargo sharing The Liner code devised by UNCTAD whereby all shipping traffic between two foreign countries is to be regulated as far as the quantities of shipments are concerned on the following percentages: (a) owners of the country of origin 40%; (b) owners of the country of destination 40% and (c) owners of the country which is neither the origin nor the destination 20%.

Cargo shed Accommodation handling imported/exported cargo.

Cargo Ship Safety Construction Certificate A document/certificate issued by a ship classification society usually on behalf of a Maritime Government/National Administration. It is required by any ship engaged on international voyages except for passenger ships, warships and troopships, cargo ships of less than 500 tons gross tonnage, ships not propelled by mechanical means, wooden ships of primitive build, pleasure yachts not engaged in trade and fishing vessels. The classification survey ensures the SOLAS 1974 convention for hull, machinery and relevant equipment are complied with.

Cargo Ship Safety Equipment Certificate A document/certificate issued by a ship classification society usually on behalf of a Maritime Government/National Administration. It is required by any ship engaged on international voyages except for passenger ships, warships and troopships, cargo ships of less than 500 tons gross tonnage, ships not propelled by mechanical means, wooden ships of primitive build, pleasure yachts not engaged in trade and fishing vessels. The classification survey ensures the ship complies with Chapters II – 1, II – 2, III and IV of SOLAS, 1994 along with any other relevant requirements.

Cargo Ship Safety Radio Certificate A document/certificate issued by a ship classification society on behalf of a Maritime Government/International Administration. It is required for all cargo ships of 300 gross tonnage and upwards confirming the vessel is fitted with a radio station. For keels laid before February 1995 the radio station should be either a radio telephone station (only applicable for Ships of 300 to 1599 gross tonnage), a radio telegraph station, or a Global Maritimes Distress and Safety System for operation in specific sea areas. For keels laid after 31 January 1995 a GMDSS installation must be fitted.

Cargo 'shut out' Cargo not carried on transport unit – which may be a ship, air freighter and so on due to the non availability of space/capacity. See also 'shut out' entry.

Cargo stowage The technique of stowing cargo in a ship's hold, ISO container, TIR vehicle, International Road Haulage vehicle, aircraft hold etc.

Cargo Superintendent A senior managerial position at a port, responsible to ensure all booked or manifested cargo – including break bulk, containers, bulk cargo etc – is loaded or discharged in an efficient manner; to ascertain the condition of the cargo on receipt or outturn; to comply with the requirements of the Master in regard to ship stowage; and finally to ensure that the stevedores responsibilities are satisfactory discharged. The Cargo Superintendent may be employed by a shipping company, stevedores, or port authority.

Cargo surveyor A person who examines/evaluates the condition of cargo primarily for insurance purposes – usually following an insurance cargo claim.

Cargo sweat Condensation which occurs when a ship sails from a cool to a relatively warm climate.

Cargo tag A form of labelling, often seen on constructional steel, and applied or affixed to a package or packages at the point of manufacture of origin.

Cargo tank Ship's tank used for the carriage of cargo and not used as a ballast tank.

Cargo terminal A terminal receiving/handling freight traffic which may be road/rail/maritime/air served.

Cargo tonnage It may be expressed in weight – usually based on the metric ton of 1000kg – or measurement – based on the cubic metre being equivalent to one (metric) tonne. See also weight/measurement ships option entry; or vessels designed/built to convey cargo.

Cargo tracer A document sent by the Agent to all relevant parties stating that certain cargo is either missing or overloaded.

Cargo trucking The process of monitoring the cargo throughout its transit to identify its whereabouts at any one time as required. It is particularly evident in logistic driven supply chain transits and also much in evidence in the international container business together with air freight.

Cargo transfer Cargo arriving at a point by one service such as rail, air, road, ship, and continuing therefrom by another service of the same or different transport mode.

Cargo transit Cargo equipment suitable for handling cartons, bags, small wooden cases.

Cargo tray The movement of a consignment by a specified transport service.

Cargo unit A vehicle, container, pallet, flat, portable tank, or other entity or any part thereof which belongs to the ship, aircraft, railway wagon or barge, but is not permanently attached to that ship, aircraft, railway wagon or barge.

Cargo valuable Any consignment which contains one or more of the following articles: any article having a declared value for carriage of USD1000 (or equivalent) or more, per gross kilogram; gold bullion (including refined and unrefined gold in ingot form), dore bullion, gold specie and gold only in the form of grain, sheet, foil, powder, sponge, wire, rods tube, circles, mouldings and casting, platinum, platinum metals (palladium, indium, ruthenium, osmium and rhodium) and platinum alloys in the form of rain, sponge, bar, ingot, sheet, rod, wire, gauze, tube and strip (but excluding those radioactive isotopes of the above metals and alloys which are subject to Dangerous Goods labelling requirements); legal banknotes, traveller's cheques, securities, shares, share coupons and stamps (excluding mint); diamonds (including diamonds for industrial use), rubies, emeralds, sapphire, opals and real pearls (including cultured pearls); jewellery consisting of diamonds, rubies, emeralds, sapphires, opals and real pearls (including cultured pearls); Jewellery and watches made of silver and/or gold and/or platinum, articles made of gold and/or platinum, other than gold and/or platinum plated.

Cargo Whip A rope or chain used in connection with a derrick for handling cargo. One end is provided with a heavy hook and the other end is rope through the derrick block and taken to the winch.

Cargo Width Maximum width of loadable cargo in metres on a Ro/Ro vessel.

Cargo worthiness The vessel is sufficiently strong and equipped to carry the particular kind of cargo which she has contracted to carry and the cargo must be so loaded that it is safe for the ship to proceed on her voyage.

Car hire The process of hiring a car for a specified period. It may be for a day, one week, one month, one year and so on.

Caribbean Common Market An economic trading bloc consisting of Antigua, Barbados, Dominica, Grenada, Guyana, Jamaica, Montserrat, St. Lucia, St. Vincent, Trinidad and Tobago.

Caribbean Development Bank It was formed in 1970 and located in Barbados. The DCB can make loans to the private sector without government guarantee and may finance project loans and technical assistance as well as invest in equity. Membership comprises of Anguilla, Antigua and Barbuda, the Bahamas, Barbados, Belize, the British Virgin Islands, Canada, the Cayman Islands,

Colombia, Dominica, France, Grenada, Guyana, Jamaica, Mexico, Montserrat, St Kitts and Nevis, St Lucia, St Vincent and the Grenadines, Trinidad and Tobago, the Turks and Caicos Islands, the United Kingdom and Venezuela.

Caribbean Hotel Association (CHA) Group of individual hotels and hotel groups formed for promotion of travel to the area.

Caribbean Tourism Organisation (CTO) Association created for promotion of travel to and around the area.

CARICOM Caribbean Community and Common Market. A regional community whose principal objective is economic integration. It introduced a single line tariff on 1 January, 1976, and at present has 15 members. WTO term.

Car insurance The provision of insurance coverage for a car. It may be comprehensive, covering the motorist and vehicle, or merely limited to third party coverage.

Carlisle leasing A major container leasing company.

Carnet A customs document usually obtained for commercial samples, exhibition material, and similar goods with which temporary importation is allowed without payment of duty. See ATA Carnet entry.

Carnet de passage en douane A document required by certain countries to accompany the vehicle to cover its temporary importation into the country.

Carousel A baggage reclaim facility usually found in the passenger terminal at an airport. Overall, it is a moving belt into which passenger accompanied baggage is placed by baggage handlers and claimed by the passenger in the baggage reclaim area.

Car pallet Flat tray with corner post for transporting cars on container ships

CARPE DIEM (Latin) Live for the day. Literally 'Pluck the day'.

Car park Accommodation provided for the parking of cars. It may be a one level (ground level) facility or a tiered basis providing various levels of car parking space. Usually, a scale of charges exists relative to users, and based on the period the vehicle is parked in the accommodation.

Carr fwd. Carriage forward.

Carriage The transportation of passengers, cargo, freight or post office mail and the accompanying charges.

Carriage and insurance paid to (named point of destination) – CIP This term is identical to carriage paid to (named point of destination) except that the seller also funds the cargo insurance. The seller is required to clear the goods for export. Again it is ideal for the combined transport operation and the seller's liability ceases when the cargo has been accepted by the first carrier at the named place and any requested Bill of Lading, Waybill, or carrier receipt as been handed over. An Incoterm 2000. See Carriage paid to (named place of destination) entry.

Carriage, Interline The carriage over the routes of two or more air carriers.

Carriage International The conveyance of traffic in a circumstance/situation whereby the placed of departure and the place of loading/arrival/destination are situated in more than one State/Country and thereby crossing international boundaries. It may be passengers/cargo/parcels, etc.

Carriage of Goods by Road Act, 1965 The UK statutory provisions containing the International Convention concerning the Carriage of Goods by Road (CMR) which was introduced in October, 1967 and permits the carriage of goods by road under one consignment note under a common code of conditions applicable to 26 countries primarily in Europe.

Carriage of Goods by Sea Act, 1924 UK national legislation implementing Hague Rules setting out minimum conditions under which cargo shipped under a Bill of Lading. It is now succeeded by Carriage of Goods by Sea Act, 1971.

Carriage of Goods by Sea Act, 1971 UK national legislation which contains the Hague Rules as amended by the Brussels protocol 1968 embracing the Hague-Visby Rules, and outlines minimum conditions under which cargo shipped under outward Bill of Lading or similar document of title from a UK port.

Carriage paid to (named place of destination) – (CPT) Under such a term the seller pays the freight for the carriage of the goods to the named destination. The seller is required to clear the goods for export. The buyer's risk commences when the goods have been delivered into the custody of the first carrier. Moreover, at this point the buyer accepts full liability for an additional cost incurred in the conveyance of the goods. On request the seller may have to provide a Bill of Lading, Waybill or carriers receipt to the buyer or other person prescribed at which stage the seller's obligations are fulfilled. In common with the FCA term, it is ideal for the multimodal transport operation which includes roll on/roll off and container movements. An Incoterm 2000. See CIP entry.

Carrier The operator company/entity who contracts to provide the transport service which may be by rail, road, ship, air, hovercraft or canal. For example, it includes the air carrier issuing the Air Waybill and all air carriers that carry or undertake to carry the cargo; thereunder or to perform any other services related to such air carriage.

Carrier International This involves (except for the purpose of the Warsaw Convention) carriage in which, according to the contract of carriage, the place of departure and any place of landing are situated in more than one State. As used in this definition, the term 'State', includes all territory subject to the sovereignty, suzerainty, mandate, authority of trusteeship thereof.

Carrier last The participating carrier over whose air routes the last section of the carriage under the Air Waybill or ticket is undertaken/performed. An IATA definition.

Carrier point Carrier point is an online transportation market that allows shippers and carriers to offer shipments, communicate prices and information. Its founders aim at permitting carriers and shippers a true market environment free of middlemen or hidden charges.

Carriers' Bill of Lading Ports. Terminal, pre-terminal port, or post terminal port as per tariff indicated on the Bill of Lading and which is not the port physically called at by carriers ocean vessels.

Carriers' haulage A pre-shipment or post shipment charge in the ocean carriers' tariff for collection or delivery haulage of LCL or FCL traffic. Usually termed zone charge.

Carrier's lien The right of the carrier to retain the cargo as a security for the collection of the freight and other charges

Carr Pd. Carriage paid.

Carry The process of making a purchase of a commodity for delivery on a given date combined with the sale of an identical amount for delivery at a later date. A BIFFEX term.

Carry forward When an exporting country uses part of following year's MFA quota during the current year. A WTO term.

Carrying Carrying is the general term used for both borrowing (qv) and lending (qv) on the LME. It is also used to describe a borrowing operation which results in physical metal being held in an LME warehouse by the borrower.

Carrying charge The cost, including interest, insurance and storage charges of holding a commodity for a given period of time. A BIFFEX term.

Carrying temperature The required cargo temperature during transport and storage.

Carry over When an exporting country utilises the previous year's unutilised MFA quota. A WTO term.

Cars knocked down Car kits partly assembled and usually crated.

Cartage The collection or delivery of goods by road from a warehouse, goods depot, airport, seaport, etc.

Cartel A market situation in which a number of companies producing/supplying a product or service agree on price and control supply. A group of firms that act as if they were a 'single seller', by entering into a collusive agreement to control or restrict output so as to increase profit, or price.

Cartons A popular form of packaging made of cardboard, strawboard, or fibreboard which is relatively inexpensive and ideal for a wide range of consumer goods shipped in containers or as palletised cargo.

Cartouche A non refillable receptacle, packed under pressure, having to be punctured for removal of its contents.

Car Transporter A road haulage vehicle or railway wagon capable of conveying cars.

Cartridges Generic term applied to any explosive article designed to deliver combustion gases, under pressure, with a view to performing a given mechanical action, for example, to propel a projectile. In particular, it applies to assembled ammunition consisting of a case fitted with a primer, filled with propellant powder with or without projectile. The term cartridge is used to indicate a unit charge of blasting explosive, wrapped in a thin paper, plastic or other envelope, the shape of which is ordinarily a cylinder. However, cartridged blasting explosives are not considered as articles but as substances. A term associated with the movement of dangerous classified cargo. See separate entries.

Cartridges, actuating for fire extinguisher or apparatus valve, explosive Contrivances containing a small explosive charge with a primer, the functioning of which ruptures a metal piece (for example, a bursting disc) and thereby actuates a fire extinguisher or either opens or loses a valve. See Cartridges, power device. Term associated with dangerous cargo shipments.

Cartridges, flash Articles consisting of a casing, a primer and flash powder, all assembled in one piece ready for firing. Term associated with dangerous cargo shipments.

Cartridges for weapons Fixed (assembled) or semi-fixed partially-assembled ammunition designed to be fired from weapons. Each cartridge includes all the components necessary to activate the weapon once. The name and description should be used for small arms cartridges that cannot be described as 'Cartridges, small arms'. Separate loading ammunition is included under this name and description when the propelling charge and projectile are packed together; or incendiary smoke, toxic and tear-producing cartridges are described as Ammunition incendiary, etc. Term associated with dangerous cargo shipments.

Cartridges for weapons, blank Articles which consist of a cartridge case with a centre or rim fire primer and a confined charge of smokeless or black powder but no projectile. Used for training, saluting or in starter pistols, etc – Term associated with dangerous cargo shipments.

Cartridges for weapons, inert projectile Ammunition consisting of a projectile without bursting charge but with a propelling charge. The presence of a tracer can be disregarded for classification purposes provided that the predominant hazard is that of the propelling charge – Term associated with dangerous cargo shipments.

Cartridges, oil well Articles consisting of a casing of thin fibre, metal or other material containing only propellant which projects a hardened projectile. The term excludes charges, shaped, commercial which are listed separately – Term associated with dangerous cargo shipments.

Cartridges, power device Articles designed to accomplish mechanical actions. They consist of a casing with a charge of deflagrating explosive and a means of ignition. The gaseous products of the deflagration produce inflation, linear or rotary motion or activate diaphragms, valves or switches or project fastening devices or extinguishing agents – Term associated with dangerous cargo shipments.

Cartridges, signal Articles designed to fire coloured flares or other signals from signal pistols, etc – Term associated with dangerous cargo shipments.

Carving note Statement confirming that ships name, port of registry, official number, registered tonnage and draft measurement is inscribed on the ship in the appropriate places.

CAS Continuous survey cycle – automated controls; currency adjustment; currency adjustment surcharge or collision avoidance system. – IATA term.

CASA Casablanca

CASAS Computer Aided Slot Allocation System – IATA term.

Cascade Reliquefaction cycle A process in which vapour boil-off from cargo tanks is condensed in a cargo condenser in which the coolant is a refrigerant gas such as R22 or equivalent. The refrigerant gas is then compressed and passed through a conventional sea water-cooled condenser.

Case history A record of a particular experience or specific item usually over a period of time or throughout the life of the item. It may be a case history of a piece of equipment spanning seven years or a prototype product. The case testing would reveal the general performance of the item and its adequacy/reliability/maintenance cost and so on; or a detailed record of a person's career progression including promotion, salary, discipline, courses attended, sickness details, within a particular company. It may be recorded on a card or computer file. Alternatively, it may be a medical record of a patient at a doctor's surgery or a record of an item of machinery/equipment detailing periods of maintenance and by whom; any modification, dates of any certificate of inspection was undertaken and so on. Overall, a portfolio devised on a particular product/service/circumstance with a view to it being analysed and used as 'in house' market research data.

Case oil Rectangular metal cans holding five US gallons usually packed in pairs in wooden cases, which superseded barrels.

Cases, cartridge, empty, with primer Articles consisting of a cartridge case made from metal, plastics or other non-flammable material, in which the only explosive component is the primer.

Cases, combustible, empty, without primer Articles consisting of cartridge cases made partly or entirely from nitrocellulose.

Case study The study/evaluation of a particular set of circumstances relative to a particular discipline. It may be a case study of a particular patient medical-social history and associated treatment; an investment programme for a fleet of vehicles; export contract award and the overall environment which led to the contract award embracing competition, product design, price, delivery terms, market research, etc.

Cash against documents Usually relates to a sight collection, whereby documents – which may include title documents to goods – are only released to customer/importer against payment.

Cash and carry An arbitrage transaction involving the simultaneous purchase of a cash commodity with borrowed money and the sale of the appropriate futures contract – an International Petroleum Exchange term.

Cash and new (contango) A price paid by a seller for being allowed to postpone delivery of stocks or shares.

Cash collateral security Cash securities usually comprise a special cash collateral account, at a bank which may be established on an initial lump sum or monthly payment basis. Cash collateral accounts are normally interest bearing, subject to negotiation.

Cash covered credit Cash cover is normally involved as a form of security for a transaction. Essentially, it is a cash deposit, usually for the full amount of the transaction, which may be taken/charged as security, or held informally. It may be used when one party is otherwise unwilling to accept the risk without the provision of cash 'up front' from which funds may be disbursed. It can also be used as a means of reducing charges. Cash cover may be provided from one bank to another or from an importer to issuing bank. Cash cover is not exclusive to letters of credit.

Cash cow A marketing term indicating a well established product/service in the competitive market place is generating income well above associated cost thereby maintaining a high cash flow, sustainable market share and developing a favourable portfolio of the company. A low growth product.

Cash discount A situation in a competitive market place whereby the seller offers discount on the price within a certain criteria such as a cash purchase and volume sale. It provides the entity with improved cash flow, lower working capital, develops market share and increases the throughput of stock.

Cash flow The total amount of monies received by a Company during a specified period. The pattern of the flow of cash within a business over a period of time. It is usually established by adjusting the net profit for all non-cash items (principally depreciation) and profits/losses on capital transactions (e.g. ship sales), thereby arriving at operating cash flow, and then adding and deducting items relating to the inflow of capital, the purchase of assets for cash and the repayment of debt. Cash flow is the banker's chief concern in assessing the probability of a loan being repaid on schedule. The ratios of net cash flow to debt service, interest charges and total capital are critically important measures of the creditworthiness of a business.

Cash Loss A payment made outside the normal accounting procedure for treaty reinsurance. An insurance definition.

Cash Management The process of a Company or Group of Companies managing its cash inflow and/or cash outflow.

Cash market The market is the instrument which underlies a futures or option contract – an International Petroleum Exchange term.

Cash on delivery An arrangement between the shipper and carrier whereby the latter upon delivery of the consignment is to collect from the consignee the amount indicated on the Air Waybill as payable to the shipper. An IATA definition.

Cash rich A company/entity/nation/person etc with an excessive liquidity.

Cash settlement The settlement of futures and options contract for money (as in BIFFEX) or when the physical delivery of futures or options is cumbersome then contracts are settled by attaching a monetary value – an IPE definition.

Cask A barrel.

CASO Council of American Flag Ship owners.

CASPPR Canadian Arctic Shipping Pollution Prevention Regulations.

CASS Cargo Accounts Settlement System (IATA).

Casting off The process of a vessel departing from either her berth or mooring place and in so doing releasing the mooring ropes.

Castor Beans or Castor Meal or Castor Pomace or Castor Flake The residue from extraction of oil from the castor seed.

Casualty insurance US term for non-life insurance other than fire, marine or surety business.

Casualty report A report by Lloyd's or Institute of London Underwriters Agents of accidents to vessels occurring within the agents' area: when possible the report will give a detailed specification of the damage together with the costs involved. Edited versions of such reports are published daily in Lloyd's List.

CAT The Communications Authority of Thailand.

Catastrophe cover A form of reinsurance that indemnifies the ceding company for the accumulation of losses in excess of a stipulated sum arising from a catastrophic event such as a conflagration, earthquake or windstorm. Catastrophe loss generally refers to the total loss of an insurance company arising out of a single catastrophic event.

Catastrophe risk A risk in which the potential loss is of exceptional magnitude e.g. nuclear power station.

Catchment area The area which is receptive/responsive to the promotion of a particular product. For example the provision of a new newspaper launch which may be an evening, weekend or daily paper for a particular town, in which case the catchment area would be primarily that particular town.

Catering Officer A ship's officer in charge of catering on a passenger vessel and usually responsible to the Purser.

Cathodic protection An electrolytic system for the protection of a metal from corrosion by providing an electric current to neutralize the current naturally produced in the corrosion process; use of cathodes/anodes protecting hull/tanks against corrosion.

Cathode A flat rectangular piece of metal which has been refined by electrolysis. Copper and nickel are commonly traded and delivered in this form.

Cathode ray tube (CRT) Screen of a computer or computer system that allows the user to see what he or she is entering into the system and what is being sent to it from outside; cathode ray refers to the way light is projected on the screen; more general terms for the TV-like screen used with computers are video display unit and monitor.

CATSE Task Force on Capacity of the Air Transport System. – IATA term.

Cattle container Partly open container equipped with rails, boxes, and cribs for livestock transport.

Cattle float A purpose built road vehicle designed for the conveyance of cattle.
Catwalk Safety passage ways from Accommodation to Foc'sle Head.
Cauris The currency of Republic of Guinea. See Appendix 'C'.
Cauris invoice An invoice payable in Guinean Cauris relative to an export sales contract.
Causa Causans The cause of a cause of loss.
Causa Proxima non remota spectatur (Latin) The proximate and not remote Cause (determines whether loss covered by marine insurance policy).
Causa sine qua non (Latin) An indispensable cause or condition.
Caveat emptor Latin and legal meaning 'let the buyer beware'.
Cavitation A process occurring with the impeller of a centrifugal pump when pressure at the inlet to the impeller falls below that of the vapour pressure of the liquid being pumped. The bubbles of vapour which are formed collapse with impulsive force in the higher pressure regions of the impeller. This effect can cause significant damage to the impeller surfaces and, furthermore, pumps may lose suction.
CAVN-FMG Campania Anonima Venezolana de Navegacion Flota Mercante Graneolombinana SA – a shipowner.
CAXB Composite auxiliary boiler (of a ship).
CAXBS Composite auxiliary boiler survey (of a ship).
Cayman Dollar The currency of Cayman Islands. See Appendix 'C'.
Cayman dollar invoice An invoiced payable in Cayman dollars relative to an export sales contract.
C2B Consumer to Business.
CB Container base; or Central Bank (Philippines).
cb Collective bargaining.
CBD Convention on Biological Diversity – a WTO term.
cbd Cash before delivery.
CBI Confederation of British Industry.
CB & H Continent between Bordeaux and Hamburg.
cbm Cubic metre.
CBM Conventional Buoy Mooring; condition based maintenance; or cubic bulk metres.
CBOT Chicago Board of Trade.
CBR Common Base Radio.
CBS Cyprus Bureau of Shipping.
CBSWG Joint Cross Border Selling W/G (ATC/IATA).
CBT Clean Ballast tanks (on a ship); or computer based training.
CBTG China Britain Trade Group.
CBU Completely built up.
CBW Crude oil washing (in tankers).
C2C Consumer to consumer.
CC Civil commotions; canvas covers; continuation clause; Coal Corporation (PNOC) (Philippines); cargo capacity (in cubic metres); container control; or Cement Carrier.
cc Current cost; charges collect; cubic capacity; closed containers; carbon copy; catastrophe cover – insurance abbreviation; Ceding Company; Contingent Covers – insurance abbreviation; or Cost Centre.
(cc) A Lloyd's Register approved system of corrosion control installed in the tanks indicated.
CCB Commercial Crime Bureau – part of the ICC.
ccc Customs clearance certificate; Goods declaration (outwards) – document used for the Customs declaration of goods at exportation.
CCC Customs Co-operation Council. The international organisation responsible for the simplification and standardisation of Customs formalities; Communication Consultative Committee (India and other countries) – IATA term.
CCDB Corporate Claims Data Base – insurance abbreviation.
CCECA Consultation Committee on Electronics for Civil Aviation (UK).
CCF Central Control Function – IATA term.
CCFF Compensatory and Contingency Financing Facility – IMF facility.
CCFS Container control and follow up system – a computer term.
CCCN Customs Co-operation Council nomenclature – the standard classification of goods for customs tariff purposes, now used by most countries.
CCI Chambre de Commerce Internationale (International Chamber of Commerce).
CCIR International Radio Consultative Committee – ITU term.
CCIT International Telegraph Consultation Committee – ITU term.
CCITT International Telegraph & Telephone Consultative Committee – ITU term.
CCJ The European Committee of Legal Co-operation. Consultative organ for legal co-operation under the Council of Europe, established in 1963. CCJ co-ordinates and watches the development of the legal programme of the Council of Europe.
CCITT International Telegraph & Telephone Consultative Committee.
CCL Customs clearance.
CCLINE CC Line – a shipowner.
CCLN Consignment Note Control Label Number.
CCMS Container control and management system – a computer term.
CC/O Certificate of consignment/origin.
CCP Cargo clearance permit.
CCR Commodity Classification Rates; Customs Clearance Request or Commission Centrale pour la navigation du Rhin.
CCS China Classification Society – the Chinese Ship Classification Society, Consolidate-cargo (container) service; Customs Clearance Status – a code associated with an entry at CHIEF; Cargo Community Systems – Computer Term (IATA); Combined container service, or it denotes that the ship can be operated with the machinery spaces unattended and that the control engineering equipment has been arranged, installed and tested in accordance with the Lloyd's Register of Shipping rules.
CCSC Cargo Community Systems Council (IATA term).
CCT Common Customs Tariff. The Tariff of the European Community.
CCTO Central Community Transit Office. A central office in the UK to which Community transit documents are returned by Customs Authorities in other member states and EFTA countries. The CCTO also despatches Community transit documents to other member states and EFTA countries on behalf of UK Community transit offices.
CCTV Close Circuit Television.
c/d carried down.
CD Clutch drive (on a ship); commercial dock; consular declaration; customs declaration; creative Director; Canal dues; or Candela.
C & D Collected and delivered. Carriage from/to Customer's premises to/from CFS.
CDB Central Data Bank or Caribbean Development Bank.
CDC Commonwealth Development Corporation.
CDCP Construction and Development Corporation of the Philippines.

CDIC Container damage inspection criteria.
CDIS Curriculum Devt. Institute of Singapore.
CDK Carrying capacity in numbers and length(s) of laden containers on weather deck.
CD Certificates of deposit – a marketable bank deposit.
CDP Container data processing (tracking).
CDR Conflict Detection & Resolution – IATA term.
CDS Continuous Discharge Certificate. A seaman's document.
CDSB Cargo data standards board – IATA term.
CDV Current domestic value. Information often required for customs/insurance purposes for a particular product.
CDW Collision damage waiver.
CE The council of Europe, domiciled in Strasbourg, is a parliamentary forum for discussion by all West European nations. It was founded in 1949, and its present members are Austria, Belgium, Cyprus, Denmark, France, Germany, Greece, Iceland, Ireland, Italy, Luxembourg, the Netherlands, Norway, Sweden, Turkey and the United Kingdom. The council works for economical, social, cultural, legal, scientific, legislative and administrative co-operation of the member countries; Chief Executive – a senior Management position; Customs & Excise – HM Government department; Claim expense (loss expense or adjustment expense) – insurance term; or marking for EU products – see separate entry.
CEA Confederation Europeene de l'Agriculture. It is based in Switzerland and is an organisation for European agricultural co-operation.
CEAC European civil aviation conference; or Committee for European Airspace Co-ordination – NATO term.
CEBIS European Community Edifact Board Information System.
CEC Clothing Export Council (UK).
CECA Communaulte Europeenne du Charbon et de l'Acier (French) (European Coal and Steel Community).
CECC CENELEC Electronic Components Committee. European system to facilitate international trade by harmonisation of specification and quality assessment procedures for electronic components, and by the granting of an internationally recognised mark and/or certificate of conformity. Founded 1970. Comprises national electrotechnical committees of 11 West European countries. About 20 active working groups.
CECLA Special Committee on Latin American Co-ordination.
CECOP Council of Engineering Consultants of the Philippines.
CECs Certificate of Equivalent Competency.
CED Customs & Excise Department.
Cedant One who cedes a reinsured risk.
Cede To advise a reinsurer of risk written which is attached to the reinsurance contract. When a company reinsures its liability with another.
CEDEL A clearing system for Eurobonds based in Luxembourg, where Eurobonds are physically exchanged and stored.
CEDEX ISO container equipment data exchange code. It is used for container inspection, repair and repair authorisation communications developed by ISO inputs from the shipping and leasing industry for incorporation into the UN/EDIFACT system. The standard is ISO 9897.
Cedi The currency of Republic of Ghana. See Appendix 'C'.
CEDIC Container EDI Council.
Ceding Company The original or primary insurer, the insurance company which purchases reinsurance.

Cedi invoice An invoice payable in the Ghanaian currency of cedis relative to an export sales contract.
CEE International Commission for Conformity Certification of Electrical Equipment. Founded in 1946. Comprises national electrotechnical committee of 23 European Countries, with Australia, Canada, Hong Kong, Iceland, India, Japan, South Africa and the USA as observer members; 'CBN' certification board for domestic electrical equipment, operating 'E' mark; Communaute Econmique Europeenne (European Common Market) – see EEC; or Central Eastern Europe.
CEEA Communaute Europeenne de l'Energie Atomique (French) (European Atomic Energy Community).
CEEC Central and East European Countries.
Ceefax Trademark of the teletext service of the British Broadcasting Corporation (BBC).
CEFIC European Chemical Industry Council.
CEFTA A trading bloc embracing Bulgaria, Czech Republic, Hungary, Poland, Romania, Slovenia and Slovak Republic.
CEICAP Centre of Export Inspection and Certification for Agricultural Products.
CEIF Council of European Industrial Federations, Paris. Organ for co-operation of the European industrial federations.
Ceiling A mechanism for limiting the amount of goods imported into the Community under preference arrangements.
Ceiling air duct A passage in a thermal container located in proximity to the ceiling to direct air flow.
Ceiling price The maximum price of a particular product/service the market will support. For example it may be a container rate; fixture rate; or passenger fare.
Celarix A web based transport exchange that combines the ability to buy and sell transportation services and also to manage and monitor the entire logistic process. Through the exchange, information is provided from sellers, carriers, customs brokers, freight forwarders and consolidators (NVCCs).
Cell Location on board of a container vessel where one container can be stowed.
Cell guides The guidance system enabling containers to be carried in a vertical pile in the ship, each container supporting the one above it, the guides providing horizontal restraint.
Cell Ho Cellular Hold(s) – of a container ship fitted with fixed cell guides for the carriage of laden containers.
Cell position A code indicating the position, by means of successively the bay, the row, and the tier of a container on a vessel.
Cells The guidance system enabling containers to be carried in a vertical pile in the ship; each container supporting the one above it; the guides providing horizontal restraint.
Cells, containing sodium Articles consisting of hermetically sealed, metal casings which fully enclose the dangerous goods and which are so constructed and closed as to prevent the release of the dangerous goods under normal conditions of transport. In addition to sodium, cells covered by this entry may also contain sulphur, but no other dangerous goods. Although designed and intended to provide a source of electrical energy, these cells are electrically inert at any temperature at which the sodium contained in the cell is in a solid state. Term associated with dangerous cargo shipments.
Cellular Term used to describe hold configuration of purpose built container ships equipped with cell guides into which containers fit.

Cellular container ship A container vessel so designed to have each hood fitted with a series of vertical angles guides adequately crossbraced to accept the container. This system obviates any lashing of the container.

Cellular service A fully containerised service.

Cellular vessel See cellular container ship entry.

CELTICBULK Celtic Bulk Carriers – a shipowner.

CE Marking The CE marking is the legal requirement for trade/manufacturers selling their products within the European Union. The CE marking must feature on EU products and/or packaging and accompanying literature.

Cemenco Code name of Chamber of Shipping approved charter party for cement trade.

Cement The fine grey powder composed of lime, alumina and silica which sets to a hard product when water is added. Also known as hydraulic cement or Portland cement. It is used to make concrete. This product is not restricted for transport by air.

Cement Boxes Temporary reinforcement to minor plating cracks. Also sealing anchor spurling pipe in small ships.

Cement, flammable This product, properly called an adhesive, usually contains rubber or a rubber-like substance and a solvent. It is used to band other substances together such as paper or leather. The solvent may be flammable. Term associated with dangerous cargo shipments.

Cementvoy Code name of BIMCO approved voyage charter party for the bulk cement and cement clinker trade.

Cementvoybill Code name of BIMCO approved bill of lading used with shipments under cementvoy charter party.

Cement Wagon A railway wagon designed to convey cement in bulk.

CEMT Conference Europeenne des Minstres des Transports. (French) It is based in Paris and also deals with transport matters.

CEN European Committee for Standardisation. Founded in 1991. Comprises national standards bodies of 16 EEC and EFTA countries, plus Spain. Prepares European Standards that are published without variation of text as national standards in the countries approving them. Each member has one vote and there are about 55 active technical committees.

CENE Commission on Energy and the Environment.

CENELEC European Committee for Electrotechnical Standardisation. Electrotechnical counterpart of CEN. Founded in 1973 from union of CENEL and CENELCOM. Comprises national electrotechnical committees of EEC and EFTA countries. Prepares European standards for identical publication nationally and also Harmonisation Documents as the basis of technically equivalent standards. Weighted voting system. About 35 technical committees.

CENFRE Sri Lanka Committee/Organisation on Trade Procedures and Facilitation based in Colombo.

CENSA It was closed down in 2001 and the shipping policy functions taken over by the International Chamber of Shipping.

Centavoc A currency used by a number of countries including Argentina, Bolivia, Brazil, Columbia, Cuba, Nicaragua and so on. See Appendix 'C'.

Centesimos The currency of Panama. See Appendix 'C'.

Centimos The currency of Costa Rica. See Appendix 'C'.

CENTO Central Treaty Organisation.

Central American Common Market Set up in 1960, its members are Costa Rica, El Salvador, Guatemala and Nicaragua; Honduras has temporarily withdrawn and Panama is not a member although represented on some of the subsidiary bodies of the organisation. The aims of CACM are to broaden the members' internal markets, supplement their economies, co-ordinate their development plans and make changes in their productive structure until a complete economic union can be achieved which would eventually lead to a political union of the countries in a federation of Central American states.

Central Bank A Bank usually under Government control/influence controlling/determining/influencing a country's banking activities especially in the role of lending, deposits, interest rates, borrowing limitations etc.

Central counterparty Sometimes referred to as 'CCP': is legally interposed between buyer and seller in member-executed transactions. Although it is more common for CCPs to accept transactions that take place on exchanges, e.g. derivatives and equities, they are increasingly willing to accept OTC products, including interbank interest-rate swaps, repos, cash bonds and commodity derivative. An ECB term.

Central Freight Booking Office A situation/circumstance which exists in some countries whereby National Government legislation compels exporters to process their shipments/consignments through one 'central freight booking office' for sea transport merchandise. This eliminates the competitive nature of the Freight Forwarding business as usually all the tariffs are government controlled. Also called Central Freight Bureaux.

Central Freight Bureaux Offices situated worldwide who are involved in the process of allocating cargoes to vessels on the basis of individual instructions, or by following general guidelines. Such offices tend to reserve cargo to specific trades and national flag tonnage. It also controls the level of sea freight rates.

Centralised organisation A company organisation which centralises all its decision making processes and permits very limited delegated powers/decision making to subordinate areas of the company organisation structure such as local/regional/district management teams.

Centrally planned economies There are some countries in Asia and Latin America that have economies whose foreign trade is centrally managed according to annual and medium term plans. The State has a monopoly of foreign trade in these particular countries and trade must conform to the plans. Imports, exports and the use of foreign exchange are regulated accordingly. Firms wishing to export to these countries will have to deal through foreign trade organisations. Trade can sometimes involve compensation agreements which link the goods a country wishes to import to the types of goods it wishes to export.

Central processing unit (CPU) The single electronic silicon chip which contains the arithmetic logic unit (ALU), the memory registers and the control unit (CU). Some computers (parallel processors) have more than one CPU.

Central reservation office An office whereby all the reservations for a particular Company's activity are dealt with in one office. This may be for hotel bookings, train seat reservations, flight bookings and so on. Computers feature prominently in the modern central reservation system with a data base and online booking facilities.

Central Reservation System A reservation system such as hotel booking, flight, train, where all the bookings are undertaken in one central point. See central reservation office.

Central securities depository (CSD) An entity which holds securities and which enables securities transactions to be processed by book entry. Physical securities may be immobilised by the depository or securities may be dematerialised (i.e. so that they exist only as electronic records). In addition to safekeeping, a central securities depository may incorporate comparison, clearing and settlement functions. An ECB term.

Centre girder Continuous girder in a ship's double bottom that runs fore an aft on centre line.

Centre of gravity The point at which a load will balance or is in equilibrium, through which it can be said the whole of the weights acts vertically downwards.

Centrocon Code name of charter party for River Plate grain trade.

Cents A currency used by a number of countries including Australia, Hong Kong, Singapore, South Africa and so on. See Appendix 'C'.

CEO Chief Executive Officer.

CEOC Colloque Europeen des Organisations de Controle. Comprises leading independent technical inspection organisations in most European countries. Exchanges information and aims to harmonise practical standards for inspection of plant and machinery.

CEP Central East Pacific.

CEPAC Confederation Europeenne de l'industrie des pates, papiers at cartons (European confederation of pulp, paper and board industries), 14 rue de Crayer, 1050 Brussels.

CEPT Conference of European Postal and Telecommunication Administrations. Organ for co-ordination of and co-operation between the European post and telegraph administrations.

CERN Conseil Europeen pour la Recherche Nucleaire (European Council for Nuclear Research), Geneva.

Cert Certificate.

Certificate A document by which a fact is formally or officially attested and in which special requirements and conditions can be stated.

Certificate of analysis A certificate issued by an independent surveyor of international repute which specifies that the samples used to determine quality of, or to analyse the cargo were taken from the cargo being loaded on board the vessel (stating the name of the vessel). Several such samples should be obtained during loading. After the analysis the certificate should be sent directly to the paying bank by the surveyor and not given to the shipper. These points should be made one of the conditions of the documentary credit. Also termed certificate of quality.

Certificate of Classification A certificate issued by the classification society under which supervision (new Building) has been carried out and stating the class under which a vessel is registered.

Certificate of Class (of a ship) The Certificate of Class assigned to a ship classification society for a particular vessel.

Certificate of competency (Deck Officer) A certificate issued to deck officers who satisfy the requisite Standards of Competency as prescribed by the Department of Transport.

Certificate of competency (Marine Engineer Officer) A certificate issued to Marine Officers who satisfy the requisite Standards of Competency as prescribed by the Department of Transport.

Certificate of compliance The certificate of compliance is issued in accordance with the ISM code on a five year basis and confirms that the Shipping Company complies with the ISM Code for the safe operation of ships and for the pollution prevention (ISM Code) adopted by resolution A741 (18) as amended for the types of ships listed.

Certificate of compliance – Crew Management Document issued for example by the ISMA relative accreditation of quality crew management system. See certificate of compliance entry.

Certificate of compliance – Ship Management Document issued for example by the ISMA relative accreditation of quality ship management system. See certificate of compliance entry.

Certificate of Conditioning A certificate issued by a competent officer in which, on the basis of the ascertained humidity factor, the dry weight of wool or silk is reckoned and certified.

Certificate of delivery A certificate, indicating the condition of a vessel upon delivery for a charter including ballast available and fresh water; or a document confirming delivery of cargo by a carrier.

Certificate of Deposit An interest bearing negotiable bearer certificate which evidences a time deposit with the bank. These are issued by banks and building societies for large deposits that cannot be withdrawn on demand. CDs can be traded in the discount market until the final buyers receive full payment plus bank interest at the end of the period.

Certificate of first entry of classification Confirmation the specified vessel had been accepted by the Classification Society e.g. Lloyd's Register of Shipping.

Certificate of Fitness A document /certificate issued by a Classification Society following survey of specified ship types. See International Society following survey of specified ship types. See International Liquefied Gas Carrier Code and International Bulk Chemical Code entries.

Certificate of Free Sale A certificate required by some countries as evidence that the goods are normally sold on the open market and approved by the regulatory authorities in the country of origin.

Certificate of Health A certificate of health is normally required when agricultural/animal products are imported. The certificate is issued and signed by the health authority in the sellers' supplier's country. It states that the health requirements were satisfied at the time of shipment.

Certificate of Inspection A certificate issued by an overseas Chamber of Commerce or independent inspector stating that the goods were inspected/ examined prior to despatch and were in accordance with the specifications laid down in the sales contract. It includes details of the identifying marks, weight and number of packages, the result of the qualitative and analytical examination of a specified sample of the goods; the shipping details and the signature of the inspector and seals of the inspecting organisation together with the date. See PSI entry.

Certificate of Insurance Document generally issued where goods which are the subject of a c.i.f. sale are insured by a floating policy covering other goods as well as the particular goods in question.

Certificate of Manufacture A certificate of a producer that the goods have actually been produced and are available.

Certificate of mortgage on sale of ship A document issued by the registrar of the port at which a ship is registered to a registered owner who wishes to mortgage or sell the ship to a foreign mortgagee or buyer.

Certificate of Origin Specific form identifying the goods, in which the authority empowered to issue it, certifies

Certificate of posting

expressly that the goods to which the certificate relates originates in a specific country. It thereby usually enables the importer to obtain concessionary import duties. The certificate may also include a declaration by the manufacturer, producer, supplier, exporter or other competent person. A certificate of origin will normally give the following details: the name and address of the exporter; the name and address of the importer; a description of the goods; the country of origin of the goods; the signature and seal or stamp of the certifying body.

Certificate of posting A document issued by the Post Office confirming the packages specified thereon have been posted on the date and the place indicated.

Certificate of pratique Document verifying that every passenger/crew member on board a vessel on port arrival is free of any contagious disease.

Certificate of quality A certificate issued by an independent surveyor of international repute which specifies that the samples used to determine quality of, or to analyse the cargo were taken from the cargo being loaded on board the vessel (stating the name of the vessel). Several such samples should be obtained during loading. After the analysis the certificate should be sent directly to the paying bank by the surveyor and not given to the shipper. These points should be made one of the conditions of the documentary credit. See PSI and Certificate of analysis entries.

Certificate of redelivery A certificate indicating the condition of a vessel upon redelivery from a charter including ballast available bunkers and fresh water.

Certificate of registry Document containing full details of ship including port of registry, name of vessel, particulars of ownership and registration number.

Certificate of shipment A certificate issued by a shipping agent to confirm consignment of goods has been shipped on a date recorded on a specified vessel/sailing from a particular berth/dock/port.

Certificate of survey Document detailing the GRT & NRT of the ship as assessed by the surveyor when vessel surveyed.

Certificate of weight Document stating the weight of the goods mentioned thereon.

Certified cheque A cheque the payment of which is guaranteed by the bank in which it is drawn.

Certified declaration or origin Declaration of origin certified by an authority or body empowered to do so.

Certified gas free A tank or compartment is certified to be gas-free when its atmosphere has been tested with an approved instrument and found in a suitable condition by an independent chemist. This means it is not deficient in oxygen and sufficiently free of toxic or flammable gas for a specified purpose.

Certified invoice A commercial invoice bearing a detailed statement of the value and origin of the merchandise described thereon and signed by the exporter. Some overseas countries require commercial invoices to be legalised by the buying nation's consulate in the exporter's/seller's country before the goods can enter the importer's/buyer's country.

Cert.Invs Certified invoices.

Cert Origin Certificate of origin.

CES Committee of European Shipowners; or Coast Earth Station – term used with GMDSS communications.

CESI Centre for Economic and Social Information (UN).

Ceskoslovenske Aero Line The National Airline of Czechoslovakia and member of IATA.

Cesser clause Charter party clause granting shipowner lien which he would not otherwise possess on the cargo.

Cession A ceded line is known as a cession – insurance term.

Cestui que trust (plural: Cestuis que trustent) (Latin) The person for whom the trust funds are held i.e. the beneficiary.

CET Common External tariff; or Central European Time.

Ceteris paribus (Latin) Other things being equal. The assumption of nothing else changing, all other relevant things remaining the same.

Ceu Car equivalent units.

CEU Central Executive Unit – IATA term.

CEYLON Ceylon Shipping Corporation – a shipowner.

cf Cubic feet; carried forward; or corrugated furnaces.

c/f Carried forward.

c & f Cost and freight.

C & F (Named port of destination) A cargo delivery term under which the seller (exporter) pays the cost and freight necessary to convey the goods from the port of shipment to the port of destination. The seller is required to clear the goods for export. The buyer (importer) will be responsible for any damage or loss to the goods when the goods pass the ship's rail at the port of shipment. The seller bears the cost of freight from his factory to the port of shipment; sea freight from the port of shipment to the port of destination including cost of packing, and loading in port of departure. The buyer (importer) bears the cost of preparing the certificate of origin, consulate documents and any other documents that may be needed to clear his goods in the port of arrival and unloading cost. See CFR entry.

CFB Central Freight Bureaux.

CFBR Customs Freight Business Review.

CFC Chlorofluorocarbon usually abbreviated to CFC; or Currency fluctuation clause.

CFEL Continental Far East Lines – a shipowner.

CFIT Controlled Flight Into Terrain – IATA term.

CFM Container Flow Management

CFMU Central Flow Management Unit – IATA term.

CFO Container flow operation.

cfo Channel for orders; or coast for orders.

CFR Cost and freight – cargo delivery term. Incoterm 2000. See separate entry.

CFSP Customs Freight Simplified Procedures. A paperless simplified clearance procedure. See separate entry.

CFT Controlled flight into terrain.

cft Cubic feet.

CFTB Cylindrical Fire Tube Boiler (of a ship).

CFTBS Cylindrical fire tube boiler survey (of a ship).

CFS Container (ISO) Freight Station.

CG Clearance Group. A group of AVRs and associated manifest items (usually only one); or coast guard.

CGA Cargoes proportion of general average.

CGDI Conditions generales pour les demenagements Internationaux.

CGFPI Consultative Group on Food Production and Investment in developing Countries (UN/IBRD).

CGLAR Consultative Group on International Agricultural Research.

CGO Cargo.

CGPM General Conference of Weights and measures. Membership drawn from those 41 nations who are signatories to the Metre Convention. The Conference meets at approximately four-year intervals responsible for implementing decisions of the CGPM and preparing for each Conference is the International Committee of Weights and Measure (CIPM). The International Bureau of Weights and Measures (BIPM), a metrological laboratory under the responsibility of the CIPM, can arrange for the measurement standards of any country to be compared with internationally agreed standards.

cgrt Compensated gross registered tonnage.
cgt Compensated gross tonnage.
C-GULF Central Gulf Line – a shipowner.
Ch Channel.
CH Carriers haulage; or Switzerland – customs abbreviation.
CH & H Continent between Havre and Hamburg.
Chafing Longitudinal stress point where rough seas create movement between cargo pipe and cushioning.
Chain conveyor A conveyor consisting of two or more strands of chain running in parallel tracks with the loads carried directly on the chains.
Chain Lashing Lashing constructed out of chain.
Chain Locker Housing for ship's anchor cable leading from windlass to spurling pipe.
Chain Sling Cargo handling equipment used for heavy lifts particularly.
Chain stopper A fitting for securing a chain, consisting of two parallel vertical plates mounted on a base, with a pivoting bar or pawl which drops down to bear on a chain link.
Chain tensioner A type of tensioner designed for use with no chain.
Chairman Person in overall charge of a Company. The precise responsibilities will vary by Company size and nature of the business, but usually responsible for overall policy in consultation with a Board of Directors of which he is the Chairman. In the USA, the position is called President. It usually involves the overall policy of the Company.
Chamber of Commerce An organisation representing industry/businesses/commerce in the area served whose two main functions are (i) to protect the domestic trade interests of commerce and industry in the area and (ii) to develop the international trade of all British Companies. A large Chamber of Commerce as found in most UK major industrial cities should have a number of departments within its organisation including International, Export Services, Home Affairs, Economic, Press Administration, Membership, Finance and Trade Sections. Its basic theme is to develop wealth within the area served through its members and represent them at all levels.
Chamber of Shipping British shipowners trade association formerly General Council of British Shipping.
Chandlers A port victualling retail outlet.
Change of class The process whereby a shipowner decides to change the Registration of vessel from one classification society to another such as changing from The Korean Register of Shipping to Lloyd's Register of Shipping. Also termed transfer of class.
Change of voyage This applies only to 'voyage' policies (hull or cargo). A change of voyage occurs when the destination of the ship is voluntarily changed after commencement of the voyage. Unless arrangements have been made with the underwriters to continue cover, prior to the change, underwriters are discharged from all liability as from the time the decision to change the voyage is manifested; but without prejudice to any valid loss occurring prior to such time. Underwriters hold covered change of voyage; subject to prompt notice to underwriters, payment of an additional premium and change of conditions, if required.
Changerep Change report.
Channel Passage of water leading to the port that is normally dredged and policed by the port authority.
Channel money Partial payment of seaman's outstanding wages in ship 'pay off' arrival at UK port.

Channel Navigation and Information Service A monitoring facility operative from the radar stations at Dover and Griz which serves the Dover Strait area and monitors the daily flow of over 400 vessels which transit the area.
Channel of communication The system adopted/selected to establish and maintain effective communication between two parties. It may be ship to shore, air-traffic control, or exporter and importer or their intermediaries. High tech systems are now common methods of communication especially electronic data interchange. See EDI.
Channel selection The criteria and selection process of the channel of distribution adopted by the seller or an intermediary in the product supply chain. See channel of distribution.
Channels of distribution The route followed to enable the goods to reach the buyer/end user from the point of origin in the supply chain. This may involve the Malaysian produced car Proton being bought through a distributor in Birmingham. Hence, the stages and organisations through which a product must pass between its point of production and consumption.
Channel Structures The supply chain structure in the market place to enable the goods to reach the buyer/end user from the point of origin/supply. See Channels of Distribution entry.
Channel Tunnel A 31 mile route subterranean rail tunnel between Folkestone (UK) and Frethun near Calais (France). It conveys international through passenger and freight trains, and between the two portals coaches, motorist and road haulage vehicles which are conveyed on purpose built railway wagons. See Euro Tunnel.
Channel Tunnel Intermodal Services Rail services using the Channel Tunnel and integrated with other transport modes such as the deep sea Container Services which are rail borne to and from the seaport through the Channel Tunnel to the Inland Distribution Centre/Inland Clearance Depot.
Chapter eleven (xi) A situation relevant/operative in the USA which represents the equivalent of calling in the Administrator. It also allows the Company to continue operations but provides legal protection from creditors.
CHAR Chartered to third party.
Charcoal A product made by the destructive distillation of wood in iron retorts or ovens. The valuable volatile products, principally wood alcohol, and acetic acid, are also recovered in this process. This charcoal, unless properly cooled, aired and packed, is liable to spontaneous ignition in transit. A term associated with the movement of dangerous classified goods
Charge An amount to be paid for carriage of goods or excess baggage based on the applicable rate for such carriage, or an amount to be paid for special or incidental service in connection with the carriage of a passenger, baggage, or goods. An IATA definition.
Charge card A card bearing the holder's name and registration number permitting accounts to be settled at specified retail outlets on presentation of the card. This includes hotels, travel tickets, car hire, restaurants etc. The aggregated accounts each month must be settled monthly and no credit facilities are permitted. It is used primarily by business personnel. Examples of a charge card include Diners Club International and American Express. See credit card entry.
Charge excess baggage A charge for the carriage of excess baggage.
Charge local A charge which applies for carriage over the lines of a single carrier.

Charge Minimum The lowest amount which will be charged for the transportation of a consignment between two points irrespective of weight or volume.

Chargeable Kilogramme A freight rate term whereby each kilogramme or part of a kilogramme is charged. Hence, a package of 7.5 kilogramme would be charged as 8.0 kilogrammes. Also rate for goods where volume exceeds six cubic metres to the tonne.

Chargeable tonne A freight rate term whereby each tonne or part of a tonne is charged. Hence, merchandise of 10.5 tonnes would be charged at 11 tonnes.

Charge published A charge the amount of which is specifically set forth in the carrier's fares or rates tariff.

Charges collect Charges which are entered for example on the air waybill for collection from the consignee. This is equivalent to the term 'freight collect' or 'charges forward'.

Charges, Combination of An amount which is obtained by combining two or more charges.

Charges, demolition Articles containing a charge of a detonating explosive in a casing of fibreboard, plastics, metal or other material. The term excludes bombs, mines, etc. which are listed separately. Term associated with dangerous cargo shipments.

Charges, depth Articles consisting of a charge of detonating explosive contained in a drum or projectile. They are designed to detonate under water. Term associated with dangerous cargo shipments.

Charges, explosive, commercial, without detonator Articles consisting of a charge of detonating explosive without means of initiation, used for explosive welding, jointing, forming and other metallurgical processes. Term associated with dangerous cargo shipments.

Charges Forward Goods/merchandise despatched on the basis that the charges/freight/cartage/export fees are payable at the destination by the consignee.

Charges, Forwarding Charges paid or to be paid for preliminary surface of air transportation for example the airport of departure by a surfaced or air transportation agency, not a carrier under the Air Waybill.

Charges Prepaid Under such consignment forwarding terms, the Shipper is responsible for payment of the charges specified prior to or at the time of shipment by air, sea or international combined transport. Hence, for example, the charges entered on the Air Waybill for payment by the shipper.

Charges, propelling Articles consisting of a propellant charge in any physical form, with or without a casing, for use as a component of rocket motors or for reducing the drag of projectiles. Term associated with dangerous cargo shipments.

Charges, propelling for cannon Articles consisting of a propellant charge in any physical form, with or without a casing, for use in a cannon. Term associated with dangerous cargo shipments.

Charge, Volume A charge for carriage of goods based on the volume of such goods.

Charge, Weight The charge for carriage of goods based on the weight of such goods.

Charta partita (Latin) Document in duplicate.

Charter Originally meant a flight where a shipper contracted hire of an aircraft from an airline. It has usually come to mean any non-scheduled commercial service.

Charter airline A carrier with an operating certificate or permit that allows charter service only; formerly called supplemental airline.

Charter back To charter the vessel back usually to the original owner.

Charter Broker A shipbroker who fixes the whole ship or a whole series of ships for one shipowner with one charterer for the carriage of a specific quantity in full ship loads.

Charter coach Motorcoach or bus to be used by a specific group for a specific trip.

Charter contract An agreement/contract for a period of time or voyage/flight whereby the shipowner or airline places the entire vessel or aircraft capacity at the disposal of the merchant or other person (known as the charterer) for the conveyance of goods between specified seaports or airports for an agreed negotiable rate. It can also apply to a road haulage international transit.

Charterer Person who hires a vessel/aircraft either on voyage, flight, or time basis.

Charterer's account A chartering term which indicates that all charges/overtime to be described/identified in the charter party, are to be paid by the charterer.

Charterer's voyage orders Orders initiated by the charterer regarding the execution of the voyage.

Chartering Process of hiring vessel/aircraft under charter party terms for a voyage/flight or period of time.

Chartering Agent Person who acts for the merchant seeking a vessel to carry his goods.

Charter Party A contract whereby a shipowner agrees to place his ship or part of it at the disposal of a merchant or other person (known as a charterer) for the carriage of goods from one port to another on being paid freight or to let his ship for a specified period, his remuneration being known as hire money. Also applies to aircraft.

Charter Party frauds The process of frauds arising from a charter party which can be minimised by checking on the credit worthiness and reputation of the charterer.

Charter party freight Negotiated prior to shipment at an agreed rate (e.g. per tonne carried).

Charter rate Payment by charterer (such as cargo owner) to shipowner for the charter of the vessel. It is determined by market conditions and terms of charter.

Charter revenue Income from the chartering of ships or aircraft.

Charter Sharing Process of allocating between two parties a chartered vessel in terms of sailings, cost, fixture rate, time and so on.

CHC Cargo handling charges.

CHD Container height division; or Container Height dimensions.

Checker A member usually of a stevedoring gang at a port involved in the loading/discharging of cargo to/from the ship and quayside/berth/lighterage. The checkers job is to record details of the cargo including marks and numbers of each package to ensure the data agrees with the cargo information on the Ship Manifest/Bill of Lading/consignment note etc. Additionally, the checker must record and draw attention to any merchandise the condition of which is unusual such as damaged, inadequately packed, packages missing, stained packages and so on. In a modern port with modern cargo technology the checker would operate under a Dock foreman/supervisor. A checker performs a similar role at an airport.

Check-in time The hour at which a room is ready for occupancy by a guest; also, the time for pre-flight ticket checking and baggage handling.

Check list A list of points etc. to be observed when processing a particular situation. It is usually devised on a cost efficiency basis with a view to completing the task in the most proficient/professional way practicable.

Check-out time The hour by which a guest must vacate his accommodations, although the management may permit continued use of facilities.

Check point In transport terms, a place where a vehicle is 'checked in' relative to details of the vehicle, its contents and so on. It may be at the factory gates, or a car ferry terminal involving examination of the tickets, reservation details and so on.

CHEI Commission for Heavy Engineering Industries.

Chemical carrier Vessel specially designed to carry chemicals; many of such products are highly corrosive, poisonous and volatile.

Chemical tanker A purpose built transport unit capable of conveying chemicals in bulk. It may be a ship, road vehicle, railway wagon, or container.

Chemtankvoy Code name of BIMCO approved standard voyage charter party for the transportation of chemicals in tank vessels.

Chemtankvoybill Code name of BIMCO approved bill of lading used with shipments under the chemtankvoy charter party.

Chemtankwaybill 85 Code name of BIMCO approved non-negotiable chemical tank waybill.

CHENGSOON Cheng Soon Co – a shipowner.

Cheque A bill of exchange drawn on a banker payable on demand. (Bill of Exchange Act 1882).

Cheque Book journalism A form of journalism whereby 'informants' are persuaded to give information on payment of a significant sum of money relative to a particular cogent press story/information.

Cherry picking The process of selection with particular emphasis on identifying elements/areas which favour strongly the selector.

CHESS Chess Line Pte. Ltd – a shipowner.

Ch Fwd Charges forward.

ChIA Chemical Industries Association.

CHIEF Customs Handling Import and Export Freight. See entry processing units entry.

Chief Engineer Officer in charge of ship's machinery and related equipment.

Chief Executive A most senior Managerial position within a Company. It is usually the Management level below the Directorate and tends to be involved in the 'Day to Day' running of the business. A post usually found in the larger organisation both in the Public and Private sectors.

Chief Executive Officer The most senior executive officer in a Company and a member of the Directorate. Usually the position is immediately below the Chairman in seniority terms.

Chief Mate Officer in charge of the ship's deck department.

Chief Petty Officer (Deck) A post in the Deck department on a ship.

Chief Petty Officer (Motorman) A post in the Marine Engineer's department on a ship.

Chief Steward Person responsible to the Catering Officer in a ship for catering.

Chief Traffic Controller A person responsible for the Traffic Control operations of a particular transport mode. See Traffic Control entry.

Chiksans Cargo loading/discharging cranes situated on oil jetty.

Chile Airline The Chilean National Airline and member of IATA.

Chime The protecting edge rim or brim at the ends of a drum or of a barrel; or the depth of the end stamping of a drum.

China Classification society The Chinese Ship Classification Society based in Beijing, China.

China Corporation Register (Taiwan) Chinese Ship Classification Society.

China Ocean Shipping (Group) Company (COSCO) A major liner shipowner and container operator and registered in China.

Chinese walls Barriers – physical or otherwise – between different departments of a financial institution or bank to prevent the exchange of price-sensitive information that could lead to a conflict of interest.

Chipboard Low grade fibreboard made mainly from waste paper, or reconstituted wood.

Chisel Forks Extremely thin wide forks on a fork lift truck used for sliding beneath loads which are not on pallets – often termed blades.

CHo Carrying capacity in number and length(s) in holds (of a ship).

Chock A guide for a mooring line which enables the line to be passed through a ship's bulwark or other barrier (See also FAIRLEAD).

Choke hitch A method of slinging in which the chain forms a sliding noose around the load.

Choke point An area of intense congestion such as a narrow channel of a busy shipping lane resulting in excessive delay to proceed through it.

Chon The currency of Democratic People's Republic of Korea. See Appendix 'C'.

CHOPT Charterers option.

Chose in action A thing capable of recovery only in legal proceedings.

CHOYANG Cho Yang Line – a shipowner.

CHRTS Charterers.

CHS Continuous hull survey cycle (of a ship).

CHUNGLIEN Chung Lien Navigation Co. Ltd. – a shipowner.

CI Consular Invoice; Chief Inspector; or Compulsory Insurance.

c/i Certificate of insurance.

c & i Cost and insurance – Cargo delivery term whereby the seller (exporter) pays the cost and insurance whilst the importer (buyer) funds the freight necessary to bring the goods to the named destination. See Incoterms 2000 entry.

cia Cash in advance.

CIA Cement Industry Authority (Philippines); Chemical Industries Association; or Credit Insurance Association.

CIAS Changi International Airport Services Pte Ltd.

CIAT International Centre of Tropical Agriculture (IARI)).

CIB Conseil International de Ble (French) – (International Wheat Council based in London); or Counterfeiting Intelligence Bureau – part of the ICC.

CIC Committee for Industrial Co-operation (EEC/ACP); Capital Issues Committee (Malaysia); Ceramic Industries Corporation (Burma) or Container inspection criteria.

CICA Convention on International Civil Aviation.

CICOS Computer Integrated Conventional Operations System – a PSA facility.

CICR Comite International de la Croix – Rouge (French) (International Red Cross Committee based in Geneva).

CICT Commission on International Commodity Trade, United Nations, New York. A commission established by ECOSOC with the object of investigating international raw material problems.

CID Centre for Industrial Development (EEC/ACP).

CIDA Canadian International Development Agency – IATA term.

CIDIN Common ICAO Data Interchange network – IATA term.

CIE Customs Input of Entry; or Committee on Invisible Exports.

CIEM Conseil International pour l'Exploration de la Mer. An International Council for the exploration of the sea, based in Denmark.

CIF Cost, insurance and freight – cargo delivery term – Incoterms 2000.

CIF & C Cost Insurance, freight, commission – cargo delivery term.

CIF & I Cost, insurance, freight and interest – cargo delivery term.

CIFC & E Cost, insurance, freight, commission and exchange – cargo delivery term.

CIFC & I Cost, insurance, freight, commission and interest – cargo delivery term.

CIF Contract Contract for the sale of goods for overseas shipment under CIF terms – cargo delivery term.

CIFI & E Cost, insurance, freight interest and exchange – cargo delivery term.

CIF Landed Cost, insurance, freight – cargo delivery term identical to CIF except the cost of unloading including lighterage and wharfage are borne by the seller at the destination port and not by the importer (buyer).

CIFLT Cost, insurance and freight London terms – cargo delivery term.

CIFOS Computer Integrated Office system – a computer facility available in the Port of Singapore.

CIFW Cost, insurance and freight/war – cargo delivery term.

CIHBB Cast iron hub non-ferrous blades propeller (of a ship).

CIHOS Computer Integrated Hydrographic Operations System – a PSA facility.

CIHSB Cast iron hub steel blades propeller (of a ship).

CII Chartered Insurance Institute.

CILA Chartered Institute of Loss Adjusters.

CILC Confirmed irrevocable letter of credit.

CIM Convention Internationale Concernant le Transport des Marchandises par Chemin de Fer. The international convention for conveyance of goods by rail operative in Europe signed in Berne 1970 covering the conditions and performance of the contract of carriage (the CIM consignment note) the carriers' liability, carriage of dangerous goods and so on. It's valid through Euro tunnel and operative in 17 countries most of which are in Europe. Chartered Institute of Marketing or Computer Integrated Manufacturing.

CIM/COTIF Consignment Note The international railway consignment note relative to the conveyance of goods. It is not a negotiable or transferable document or a document of title. It is a cargo receipt and contract of carriage under the international convention. See CIM and COTIF entries.

CIMOS Computer Integrated Marine Operations System – a computer facility available in the Port of Singapore.

CIO Chief Immigration Officer; or Chief Information Officer.

CIP Carriage or Freight and Insurance paid to (named point) – cargo delivery term. This is the container equivalent of cost insurance and freight, see separate entry; Cargo Investigation Panel (IATA); or EATCHIP Convergence & Implementation Programme – IATA term; see EATCHIP entry.

CIR Cage Inventory Record. The Dover computer record of a consignment of goods off-loaded to a security cage.

Circa (Latin) About (usually with dates).

Circle fare A special fare lower than the sum of the point-to-point fares for a circular trip.

Circle trip A journey with stopovers that returns to the point of departure without retracing its route.

CIRM International Radio-Maritime Committee.

Circuit working A transport term involving the exclusive disciplined operation of specified transport units between two points. This may be for example some fifty railway wagons carrying iron ore operating only between port A and steel plant B.

Circulation The total number of papers/journals/magazines distributed. This usually only includes subscribed copies and excludes free copies. See also readership entry.

Circumvention Measures taken by exporters to evade anti-dumping or countervailing duties. An example might be shifting assembly operation to a third country or to the importing country itself. Also avoiding quotas and other restrictions by altering the country of origin of a product.

CIRR Commercial Interest Reference Rate – ship building term.

CIS Commonwealth of Independent States.

CISG UN Convention on Contracts for the International Sale of Goods – Vienna convention.

CITES Convention of International Trade in Endangered Species of wild fauna and flora. An MEA WTO term.

CITOS Computer Integrated Terminal Operations System – a computer facility available in the Port of Singapore.

City pair The origin and destination points of an air trip.

City terminal An airline ticket office, not an airport, where a passenger can check in, check baggage, receive a seat assignment and get transportation to the airport.

City ticket office (CTO) Airline ticket office or counter not located at an airport.

City tour Sightseeing tour through a city to view its main attractions, usually by bus and usually lasting a half day.

Civil Aviation authority Government body that regulates civil aviation in the UK.

Civil law Legislation involving claims or damages by one individual against another for compensation. Also referred to as Private Law.

CK Cask.

CKD Completely knocked down – consignments which are assembled at destination as distinct from being transported as complete unit; or cars knocked down.

CL Corporation of Lloyd's; continuous liner; or Lufthansa City Line GmbH – an IATA associate airline member.

c/l Craft loss.

Cladding Coating of one metal with another.

Claim The process of lodging/presenting a claim with a view to obtaining some measure of financial compensation; or a request by the policy holder or his representative for payment under the policy. An insurance definition.

Claim Expense (loss expense or adjust expense) Expenses of adjusting claims, such as court costs, fees and expenses of independent adjuster, lawyers, witnesses, and other expenses which can be charged to specific claims.

Claim made basis An insurer only meets claims if they occur in the insurance year which has been agreed upon i.e. from 1st January – 31st, December, 1989. If a claim is made on 1st, January, 1990 it is not covered under this basis.

Claim occurring basis Claims are net if they occurred (happened) during the insurance year even though a claim might not arise for a number of years e.g. an insurance company which covered an asbestos risk 5 years ago which now results in a claim. The insurance company is forced to pay out under a claims occurring basis.

Claim run off An underwriting year of account all of whose benefits and liabilities have been transferred to a succeeding year or another Lloyd's syndicate by means of a reinsurance to close. This is normally after 36 months (occasionally 48 months with special permission from Lloyd's). The ascertained profits or losses of a closed year are available for distribution to or are collected from the Names.

Claims co-operation Where the reassured agrees to consult and co-operate with reinsurer's on claims handling and settlement.

Claims department A department dealing with clients' claims. If may be an insurance company honouring claims submitted within the terms of the policy; a transport operator such as an airline, shipowner, railway company processing claims payment on goods damaged, missing in transit and so on.

Claims equalisation Amount set aside for the purposes of preventing abnormal fluctuations in claims due to the occurrence or exceptional events.

Claims expenses The expenses incurred in the investigation, adjustment and settlement of claims. There are two types of claim expenses: allocated and unallocated. Allocated claims expenses are generally considered to cover expenses incurred in settling claims that can be attributed to specific cases including such items as actual court costs, lawyer's fees, medical examinations, independent adjuster's fees, etc.

Claims frequency The number of claims divided by the number of units of exposure, expressed as a percentage.

Claims incurred The claims for which liability exists and which have been advised to the Company.

Claims Inspector A post which is responsible for the examination/inspection of merchandise etc. on which a claim has been made. The claims inspector would submit a report on the result of his evaluation which would be progressed by the Insurance Company/Underwriter who were progressing the claim.

Claims Manager A managerial position in a Company/entity such as a Shipowner, Freight Forwarder, Receiver/Agent responsible for settling claims submitted and contributing to the overall policy on claims matters in consultation with the Directorate. This includes claims prevention; may be premium levels; liaison with Insurance Companies/Underwriters and so on.

Claims notified/reported Claims resulting from accidents or events which have taken place and on which the insurer has received notices or reports of claim.

Claims outstanding Amount set aside for claims that have been devised but have not been settled.

Claims Prevention The process of devising a policy/circumstances/code of practice which will lessen the incidence of claims. This involves conducting transit test to determine damage etc. and institute remedial measures; improve documentation to specify circumstances giving rise to claims; quality control of product; analysis of packaging; improved security to counter pilferage; develop staff training; code of reporting procedure for claims and so on.

Claims procedure See cargo claims procedure entry.

Claims ratios Expression in terms of ratio of the relationship of claims to premiums. Two ratios in common usage are: (a) paid claims ratio – paid claims divided by written or earned premiums; (b) incurred claims ratio – incurred claims divided by earned premiums.

Claims reserve The provisions made by an insurer for outstanding claims and IBM. Under a reinsurance contract an amount is retained by the reassured against payment of claims which will be recoverable under the contract in due course. This is called a 'Claims Reserved' (see also Outstanding Claims Advance, Outstanding Claims Reserved and Premium Reserve). The difference between subsequent claims experience versus reserves established as of a given date.

Clamp A fork lift truck attachment associated with the handling of bales, cartons and drums.

Clamping force The horizontal compressive component of the forces produced when lifting by means of a multi-leg sling. Also the force resulting from the use of fork lift clamps and similar equipment.

Class A category of insurance business defined by the nature of the cover provided or by the nature of the object at risk – insurance definition. See tourist class entry; all options of the same type pertaining to the same underlying commodity. Calls and puts form two separate classes. An IPE definition; or a category passenger transport travel. See classification entry.

Classification A listing of commodities/articles into classes for freight rate purposes.

Classification certificate A document issued by a Ship Classification and Survey Society indicating when the vessel was last surveyed and the class attributed to a ship.

Classification Clause ICC Insurance clause listing criteria for vessels on which automatic cargo insurance cover applies at standard rates.

Classification Clause – Cargo An Institute clause used in cargo open cover contracts to indicate the age and class of overseas carrying vessels acceptable at the premium rates specified in the contract. Basically, the ship must be iron or steel and mechanically self-propelled. It must be fully classed with any of the listed classification societies. It must be no more than 15 years of age. The age limit is raised to 25 years for liners. Goods carried by ships not attaining the required standard are held covered subject to payment of an additional premium.

Classification Clauses – Hulls
A set of London market clauses available for attachment to hull and machinery policies. They incorporate an 'existing class maintained' warranty and various restrictive conditions relating to the ship's classification.

Classification Rates An IATA freight rate applicable to highly specialised commodities such as gold, bullion, live animals, newspapers and human remains.

Classification society An organisation which surveys and classifies ships such as Lloyd's Register of Shipping and Bureau Veritas. Also publication issued by a classification society which lists all the ships classed by that society. Against each ship is recorded such information as place and date of build, the tonnages and capacities, dimensions, number of decks, holds and hatches and details of engines and boilers.

Classification survey Survey carried out by a surveyor of a classification society, either a periodical survey, or one required, for example, after a collision to ensure that the ship meets the minimum standards for continued trading set by the classification society.

Classification surveyor Classification surveyors are those appointed to supervise and inspect the building and repairing of vessels classed by a register of shipping and to examine them periodically or when damaged. See marine surveyor.

Classified question A question found in a market research questionnaire which is so worded to classify the respondent by age, group, sex, social/economic group, nature of job, marital status, income and so on.

Class rate A rate applicable to a specifically designated class of goods. It may be air freight, intermodal market, consolidated shipments in trucks or containers and so on.

Claused Bill of Lading Bill of Lading which has been endorsed by the shipowner as the goods described thereon do not conform to what is offered for shipment e.g. package missing, damaged, stained, inadequately packed.

Clause paramount A clause in a bill of lading or charter party which stipulates that the contract of carriage is governed by the Hague Rules or Hague-Visby Rules or the enactment of these rules of the country having jurisdiction over the contract.

Claw A hook especially designed to hold a chain, often used in cargo securing i.e. on chain tensioners.

Clawback A charge payable to IB by exporters from Great Britain of sheep and sheep meat and goats and goat meat. The amount payable is equivalent to the variable slaughter premium payable in the week of export.

CLC Civil Liability Convention (For oil pollution damage).

CLCLD Cycloidal propeller.

Cld Cleared.

CLD Container length dimension; or Container length division.

Clean bill of health All ships leaving port or harbour and entitled to be furnished with a clean bill of health. This is an official certificate to the effect that the place or port of departure enjoys good public health and no infectious diseases are present.

Clean Bill of Lading A Bill of Lading which has no superimposed clause(s) expressly declaring a defective condition of the packaging or goods. Hence, one in which there is nothing to qualify the admission that the goods are shipped in good order and condition.

Clean B/L Clean Bill of lading.

Clean collection A Bill of Exchange which has already been accepted by the importer and is being presented for payment on its maturity.

Clean floating The process of currency value being determined by the market forces found in international trade. See also entries of floating currency and dirty floating.

Clean payments The process of effecting money transfers internationally in a specified currency for payment of the goods and in so doing unconditionally to the payee (except possibly against a simple receipt). The importer must, therefore, rely on the supplier to ship the goods and despatch the relative documents to him.

Clean period charter A product/chemical trades involving refined products time charter.

Clean petroleum products Refined products such as aviation spirit, motor spirit and kerosene. See dirty petroleum products.

Clean receipt A receipt for cargo delivered and in an acceptable condition and overall as prescribed in quantity/specification on the receipt.

Clean report of findings The document issued by the Inspection Agency under the Pre-shipment Inspection arrangement. It will be issued following final inspection of the goods and specifies the goods are in order and accord with the commercial export invoice. This document will be sent to the Bank to arrange payment of the goods to the exporter.

Clean single charter A product/chemical trades involving refined products voyage charter from Port A to Port B.

Clearance Process of clearing cargo/ship/aircraft/vehicle/container through Custom House Procedures.

Clearance Agent One who represents a principal and in so doing undertakes all the requisite documentation work and associated items to process a consignment through Customs thereby enabling the goods to be released and continue on their transit. The Clearance Agents may be situated at a seaport, ICD, or airport. Most clearance agencies specialise in imported cargoes, but some do both export and import goods.

Clearance depot See ICD.

Clearance label Denotes that a vessel has complied with all the regulations for clearance outward. It is attached to the Victualling Bill by the Customs Officer who clears the vessel, and is then known as the Outward Clearance.

Clearance of cargo goods The process of clearing cargo through customs which may be at a seaport, airport, or inland clearance depot.

Clearance of ships The process of a ship being 'cleared inwards' and in so doing enable the vessel to accept cargo for the outward sailing.

Clearance outward of a vessel The process of obtaining customs clearance of a vessel and her cargo/passengers prior to departure from the port on an international voyage.

Clearance terminal A terminal where customs facilities exist for the clearance of goods.

Clear Days Consecutive days commencing at 0000 hours on the day following that on which a notice is given and ending at 2400 hours on the last of the number of days stipulated. A BIMCO term.

Cleared without examination A customs term indicating the goods have been cleared by Customs without examination/inspection.

Clear height Maximum deck head clearance between adjacent fixed or moveable decks in metres on a Ro/Ro vessel.

Clearing The process of trade match, registration, settlement and provision of a guarantee of exchange traded contracts – an International Petroleum Exchange term.

Clearing Agent An Agent who represents a 'principal' and in so doing undertakes all the requisite documentation work and associated times to process a consignment through customs. See Clearance Agent.

Clearing fee A sum charged by the clearing house for clearing trades through its facilities. An IPE definition.

Clearing House An accounting settlement centre where all participants submit their debits or credits for settlement as prescribed within the group of Companies within; or the Agency responsible for contract fulfilment and financial settlement between the Members relative to BIFFEX.

Clearing member
A member of a clearing house. All trades executed on the Floor must be registered and held with a Clearing Member – An IPE definition.

CLECAT The European Communities Freight Forwarders Association.

Client account Any individual or entity being serviced by an agent (broker) for a commission. Servicing generally includes advice, accounting, and order execution. A customer's business must be distinguished from the broker's principal in-house business and may be in a physically segregated account at the clients option. An IPE definition.

Client base A list of an entrepreneur's clients. See Mailing list.

Client led marketing policy See consumer led marketing policy entry.

Client/server computing The process of one element of a computer network (the client) sending a request to another element of the network (the server).

Climate The long-term statistical average of weather conditions. Global climate represents the long-term behaviour of such parameters as temperature, air pressure, precipitation, soil moisture, runoff, cloudiness, storm activity, winds, and ocean currents, integrated over the full surface of the globe. Regional climate, analogously, are the long-term averages for geographically limited domains on the Earth's surface.

Climatic packaging The process of packaging the goods to counter effectively the various climatic conditions to be experienced during the transit such as humidity, etc.

Clip-on units Portable refrigeration units designed to clip onto insulated containers which normally rely on a central refrigeration system for their cold air supply.

CLO Classification Liaison Officer. An officer appointed in a Collection or EPU to be the focus for tariff classification expertise.

CLORA/CLLORA Consequential loss list of rating adjustments.

Close The final published official price of the day. A BIFFEX term.

Closed berth A restrictive user berth. See open berth entry.

Closed box container A container that can only be packed through one or more doors in the end of side walls.

Closed conference Liner conference in which member lines vote on the admission of a new line. The purpose of this is to restrict the number of ships in a particular trade.

Closed container A container which totally encloses the contents by permanent structures.

Closed height The height of the mast of a fork lift truck when the forks are completely lowered.

Closed out A futures purchase within the Freight market that has been matched by a corresponding sale or vice versa – a BIFFEX term.

Closed position This arises where the dealer's assets and liabilities in any currency do not balance, thereby resulting in an open position. If the dealer then makes a deal to cover that balance, other than by buying or selling, the position is closed.

Closed shop A situation in a particular Company/Entity whereby all employees must be members of a Trade Union which may or may not be specified by the employer.

Closed type transport units Transport units in which dangerous goods are totally enclosed by sufficiently strong boundaries such as a closed container or vehicle, a demountable or portable tent. Fabric sides or top of a unit are not considered suitable to bring an otherwise open top within this definition relative to maritime dangerous cargo conveyance.

Closed vehicle deck A deck or part of a deck without the free access of air so that additional ventilation is required.

Closed ventilated container A container of a closed type, similar to a general purpose container, but specially designed for carriage of cargo where ventilation, either natural or mechanical (forced) is necessary.

Close ice Floating ice in which the concentration is 7/10 to 8/10, composed of floes mostly in contact.

Close out A finalising transaction by which an equal and opposite trade closes an open position. An IPE definition.

Closing date Latest date cargo accepted for shipment by (liner) shipowner for specified sailing.

Closing prices Final prices at the close of an exchange. A London Stock Exchange term.

Closing purchase A transaction in which a writer purchases an equity or index option having the same terms as an option which he has previously sold, thus terminating the liability as a writer. A financial definition.

Closing range A range of closely related price cuts transactions which take place at the close of a market – an International Petroleum Exchange term.

Closing reinsurance Reinsurance of an entire portfolio of outstanding liabilities and outstanding claims to close a syndicate's underwriting account by transfer of the outstanding liability to the next open underwriting account, normally by the end of year 3.

Closing sale A transaction in which the holder of an equity or index option sells in the market an option identical to one which he holds thus extinguishing his rights as a holder; or the process of closing/concluding a sale.

Closing time Latest date and time cargo accepted for shipment by (liner) shipowner for specified sailing. Usually termed closing date.

CLP Container load plan.

Cloverleaf deck fitting A type of deck fitting capable of accepting a lash from any direction.

Cloverleaf 'D' ring A 'D' ring designed to accept two lashings.

CLU Central Logistics Unit.

Club In marine insurance terms the Protection & Indemnity Association or P & I Clubs. In Export terms the Export Club whose membership is drawn from all those involved in International Trade viz, Freight Forwarders, Exporters, Banks, Insurance Companies, etc.

Club class See business class.

Cluster sampling The process of organising sampling based on local groups such as using two streets in a town rather than cover every road in the town.

Clutch hook A hook especially designed to hold a chain, often used in cargo securing i.e. on chain tensioners.

cm centimetre(s).

CM Condition Monitoring – ship maintenance technique; or Compania Panamena de AviaciÛn SA (COPA) – an IATA airline member.

CMA Cash Memorandum Account: maintained by CREST in the name of a member. The member makes cash posting to the account, which records the cumulative (net) balance, at any time in the course of the settlement day, of payments made and received in the relevant designated currency by that member.

CMA-CMG Group A major 'liner' shipowner and container operator and registered in France.

CMB Claims made basis.

CMCR Cie Maritime Des Chargeurs Reum's – a shipowner.

CMG Course made good.

CMI Comite Maritime International. An International committee of maritime lawyers based in Antwerp.

CMI Rules for Electronic Bills of Lading The rules applicable when the parties agree to conduct trade in this manner relative Electronic transmission of Bills of Lading as laid down by the CMI.

CMN Convention on the contract for the Carriage of Goods by Inland Waterway.

CMP Chamber of Mines of the Philippines; or core messaging platform – computer term – see separate entry.

CMPL Completed

CMR Convention relative au contrat de transport international des Marchandises par route. Convention agreed in Geneva in 1956 by the Economic Commission for Europe (Transport Division) in the contract for the International Carriage of Goods by Road.

CMR Consignment Note The CMR consignment note document is not a negotiable or transferable document or a document of title. It is a cargo receipt and contract of carriage under the international convention concerning the carriage of goods by road. See CMR sentry.
CMR Form Lloyd's goods in transit (CMR) policy.
CMS Continuous machinery survey (of a ship).
C/N Consignment note; cover note; or credit note.
CN International Convention on the Carriage of Passengers and Luggage by railway; or Combined Nomenclature: the combined CCT and statistical nomenclature which form the post 1.1.88 Tariff of the European Community.
CNAN CNAN (Algerian National Line) – a shipowner.
CNC Compagneurs Nationales des Containers. An affiliate of the French National Railways for container traffic.
CNCO China Navigation Co – a shipowner.
cnee Consignee.
CNIS Channel navigation information service.
Cnmt/consgt Consignment.
CNOO China National Offshore Oil.
CNRS Container Number Recognition System – a PSA facility.
CNS Communications Navigation & Surveillance – IATA term; or Cargo Network services.
CNSC Cargo Network Services Corporation (wholly owned Subsidiary of IATA).
CNT Containers.
CNUCED Conference des Nations Unies sur le Commerce et le Development (French). The United Nations Conference on Trade and Development based in Geneva.
Co Course.
Co Case oil; certificate of origin; cast order; cargo oil; or cash order.
c/o Certificate of Origin.
CO Continental Airlines Inc – an IATA member Airline.
CO₂ Carbon Dioxide.
COA Contract of affreightment.
Coal berth A purpose built/designed berth situated at a seaport which accommodates/handles bulk shipments of coal. A modern coal berth is fully mechanised/automated. It is usually loaded by a conveyor system and discharged by a grab up to a lifting capacity of 85 tons.
Coal carrier Vessel built to convey coal.
Coal gas This is the gas obtained by the destructive distillation of bituminous coal. It is shipped in steel cylinders and classed as a poisonous flammable compressed gas.
Coalice '1954' The code name of a BIMCO approved charter party clause relative to 'ice' and entitled 'Baltic Conference Ice Clause for Coal and Coke cargoes 1954'.
Coal-Orevoy The standard dual commodity iron ore and coking coal voyage charter party devised by BIMCO and introduced in 2002. It displaced the Orevoy charter party.
Coal propulsion Vessel built with the propulsion unit coal fired.
Coaming The supporting structure of a hatch cover.
Coaming bridge fitting A bridge fitting designed for the connection of containers to the coaming.
Coaming chock A bridge fitting designed for the connection of containers to the coaming.
Coastal state A country which has a seaboard and coast line.
COASTCON Code name of Chamber of Shipping approved charter party for coal trade.
Coastduty The code name of a BIMCO approved charter party clause relative to customs duty in the coastwise trade and entitled 'Baltic Conference clause in respect of Customs duty in Coastwise trade 1962'.
Coaster All purpose coastal cargo carrier.
Coasthire Code name of Chamber of Shipping approved time charter party in the coasting and short sea trades.
Coasting broker A shipbroker concerned with chartering vessels found in coastal voyages usually primarily conveying bulk cargoes.
Coating Solutions Materials such as automobile undercoating, drums or barrel lining materials, etc. which cannot properly be described as cements, but present similar hazards in transportation. They usually contain flammable solvents. A term associated with the movement of dangerous classified cargo.
COB Clean on board or claims occurring basis.
COBOL Commercial and Business Orientated Language (computer language).
COBRA Continent, Britain, Asia Container Service. The consortium in which P & O Containers trade in the Europe/India/Pakistan/Bangladesh Service.
CoC Certificate of compliance.
C/OC Certificate of origin and consignment.
COCOON A raised hatch cover on which containers are stacked.
COD Cash on delivery, an arrangement between the shipper and the carrier whereby the latter upon delivery of the consignment is to collect from the consignee the amount indicated on the Air Waybill as payable to the shipper; or country of destination.
Code Box An area within a data field designated for a coded data entry and found in export/import documentation processing and chartering.
Coded data entry Data entry expressed in code.
Code of conduct The trading procedures and etiquette of a market. An IPE definition.
Code of conduct on ships A code of conduct for observance by seamen on vessels, breach of which would involve disciplinary action through an agreed procedure as found in the Merchant Shipping Act, 1979.
Code of practice A framework devised within which to operate such as the procedure to follow in the event of a claim being received by the Shipowner for a passenger injury, or technique of handling at a port a particular cargo which is classified as dangerous. Alternatively, it may be termed as code of procedure.
Code of practice on ship recycling This was introduced in 2001 by seven major shipping associations including IMO, INTERTANKO, Inter Cargo. It is a voluntary code to create the 'best practice' framework so that responsible owners can prepare their ships for recycling and disposal in an environmentally friendly way.
Code of procedure See code of practice entry.
Codes Additional instruments to the General Agreement during the Tokyo Round and open for signature by all Contracting Parties which wish to adhere. They cover non-tariff measures (e.g. subsidies and anti-dumping) as well as dairy products, bovine meat and civil aviation. A WTO definition; or standard abbreviations and designations used by the travel industry to indicate elements such as airports, fares, classes of service and cities.
Code-sharing An arrangement between two airlines that allows one carrier to list its flights under the two-letter designation of another, usually to obtain the benefits of on-line status for connections displayed in reservations systems; often part of a broader marketing pact such as feeder agreements between regional and major carriers, or between carriers that co-operate on international routings.

Code – UN A code devised by the United Nations identifying commodities with a number which is inserted in the consignment document.

Codex Alimentarius FAO/WHO commission that deals with international standards on food safety. WTO term.

Coefficient of Friction A factor for calculation that relates to the reluctance to slip or slide. Generally, the higher the value, the lower the tendency to slide over an adjacent surface.

C of C Chamber of Commerce; Container on flat car.

Cofferdam Transverse double bulkheads are at least three feet apart, extending from the keel to the upperdeck, to separate one part of the ship from another. Cofferdams are located either side of the engine-room space and are used as part of the collision bulkhead. Overall, empty spaces separating cargo tanks from engine-room aft and fore-part of tanker.

COFR Certificate of Financial Responsibility required under US oil Pollution Act of 1990.

COG Course over Ground. A term generally employed, but used more so now with the advent of electronic chart systems.

COGSA Carriage of Goods by Sea Act. In the UK the 1924 version (Hague Rules) now superseded by the 1971 version (Hague Visby Rules).

Coherence Reference to the third item in the FOGS negotiating mandate to increase the contribution of the WTO to achieving greater coherence in global economic policy-making through strengthening its relationship with other international organisations responsible for monetary and financial matters. A WTO term.

COI Central Office of Information.

COIE Committee on Invisible Exports (UK).

COINS Co-Insurance.

Co-Insurance Where a number of insurers each cover part of a risk, or where a policy requires the insured to bear a part of each loss, there is said to be co-insurance except in certain circumstances. A co-insurer is not obliged to follow the decision of another co-insurer. Each co-insurance is a separate contract with the insured. In the USA, the term is used to describe the application of an average condition (see Average) whereby if the sum insured is inadequate the insured has to bear a proportionate part of a claim. Reinsurance on the same terms as the original insurance (USA). Overall, where two or more parties share the same insured risk. A re-insurer is not co-insurer with the original insurer.

COL Column.

Col Collision.

Cold call analysis The process of analysing all the cold calls undertaken during a particular period and identifying/reviewing their results, success rate and cost.

Cold calling The process of a sales person making a telephone call or visit without any invitation with a view to obtaining an interview hoping to gain a sale.

Cold front The boundary line between advancing cold air and a mass of warm air under which the cold air pushes like a wedge.

Cold mailing The process of despatching 'mail shots' to potential clients with a view to ultimately securing a sale. Many companies use a rented mail list as a basis for a mail shot.

Cold storage The storage of merchandise under refrigerated conditions.

Cold store Shore based accommodation usually found at a port accommodating refrigerated cargo. It is usually computer controlled and houses perishable foodstuffs including banana, meat, vegetables, citrus fruit and so on.

Cold store terminal Purpose built/designed accommodation at a terminal such as a seaport, road depot etc. which handles/stores foodstuffs under a controlled temperature situation. A modern one would be computer controlled under refrigerated conditions.

Coll Collision.

Collapsible container A container with the main parts hinged or removable so that its effective volume may be reduced for transportation in an empty condition.

Collapsing stresses The primary static force on a ship floating in still water is due to water pressure on the hull. It is resolved in two lines of pressure. The water presses inwards on every submerged part of the ship in a direction perpendicular to the skin surface i.e. either vertical or horizontal. The force of the collapsing stress increases uniformly at a rate of 100kg/sq cm per metre of immersion.

Collateral In United Kingdom parlance, additional security distinct from the primary security. In United States parlance, a synonym for security. In ECB terms assets pledged as a guarantee for the repayment of the short-term liquidity loans which credit institutions receive from the central banks, as well as assets received by central banks from credit institutions as part of repurchase operations. Overall, additional security distinct from the primary security. In many cases though it is a synonym for security.

Collateralised Financing One of the four main business areas of Barclays Capital, Collateralised Financing is dependent on both the credit and risk side of their business, and characterised by the steady integration of the world's bond and loan business. It is comprised of five main product offerings: Futures, Equities, Lending, Prime Brokerage and Repos.

Collection Obtaining the proceeds of a foreign currency item for a customer, prior to giving credit. An International Banking term.

Collection and Delivery The process of conveying the merchandise from/to the customer's premises and to/from the Container Freight Station or other specified distribution point.

Collection charges Those charges raised by the carrier to collect the goods/merchandise/consignment from the consignee.

Collection on wheels The collection of goods by road from the consignor's premises involving consignment(s) up to 3,000 kilogrammes in weight or up to nine cubic metres in volumetric measurement. Such a consignment is usually despatched under LCL/consolidation/groupage arrangement.

Collective bargaining An industrial relations term whereby both unions, representing the work force, and management, negotiate on issue of pay without any constraints being imposed on either party especially by Government.

Collector H.M. Customs Officer.

Collectors office Customs accommodation where declaration(s) (entries) are scrutinised and amounts payable collected.

Collier Coastal coal carrier.

Colliery guarantee terms A chartering term indicating the ship is to be loaded in accordance with the terms of the colliery guarantee. This is a contract between the colliery and the charterer – occasionally between the colliery and the ship owner – describing applicable laytime, excepted periods, holidays and demurrage.

Colliery Working Days A laytime term describing days on which a colliery normally works.

Collision In maritime terms a situation of two ships coming into contact so that damage is suffered to a greater or lesser extent by both of them. It also covers contact between a ship and many and varied objects such as wrecks, piers, dock walls, breakwaters, etc., with anchors of other vessels, or with fishing nets attached to other vessels; or in marine insurance terms the physical impact between two or more ships or vessels used for navigation. The term does not include contact of the insured vessel with anything other than a ship or vessel.

Collision bulkhead Foremost transverse watertight bulkhead which extends to the freeboard deck in a vessel.

Collision Clause A clause in a hull policy, extending cover to embrace collision liability incurred by the assured (See also 'Running down Clause').

Collision damage waiver (CDW) Protection offered, for an extra fee, by a car rental company against liability for damage to a rental car.

Coll Rdr Collision Radar.

Co-loading The loading, on the way, of cargo from another shipper, having the same final destination as the cargo loaded earlier.

Colombo Plan for Co-operative Economic and Social Development in Asia and the Pacific An economic and social plan embracing 28 countries.

Colon The currency of El Salvador. See Appendix 'C'.

Colon invoice An invoiced payable in the currency of El Salvador colons relative to an export sales contract.

ColRegs International regulations for the Prevention of Collision at sea.

Coltainer Collapsible container.

Column inches The amount of space taken up by an article in a particular newspaper, journal and so on. For example, an article on a new shipping service may attract six column inches.

COM Communications; or Commission.

COMAC Communications Advisory Sub-Committee – IATA term.

Co-makership The long-term relationship between e.g. a supplier or a carrier and a customer, on the basis of mutual confidence.

Co-manager The manager of an issue may invite other financial institutions to join the management team as co-managers. An International Banking Term.

COMB Combination, combinable.

Comb Combined (holds).

Combi An aircraft carrying passengers and cargo upperdeck (air cargo). See also Combi Carrier and Combination Carrier entries.

Combi Carrier An 'all purpose' general cargo vessel engaged in deep sea liner trades capable of conveying containers, loose cargo, indivisible loads, vehicular traffic, and equipped with vehicular decks, container accommodation, derricks of 36 tonnes and 126 tonnes lifting capacity, and ramps for vehicular traffic transhipment both at the stern and amidship. Jumbo capacity lifts are provided in the vessel to move cargo between decks. Thus, the vessel is equipped with roll on/roll off, lift on/lift off and side cargo transhipment facilities. Sometimes it is called an Omni carrier.

Combiconbill Code name of approved BIMCO Bill of Lading used for combined transport.

Combidoc Code name of BIMCO approved combined transport document.

Combi King 45 A flexible modern single deck container/bulk carrier ship with an overall length of 194.30m, breadth of 32.20m, draught of 12.20m, deadweight tonnage of 45,500 tonnes and speed of 14.0 knots. The vessel has a cargo hold grain capacity of 58,700m$\geq$ or container capacity of 2127 TEUs. Cargo handling equipment embraces electro-hydraulic deck cranes of one single 25 tonnes, and two twin 25 tonnes – all equipped with grabs. The vessel is suitable for bulk shipments of grain, coal, ore, bauxite and phosphates. She can also convey containers, packaged timber, and standard pipe lengths. The vessel is built by British Shipbuilders.

Combination The establishment of a rate or charge by addition of sectional rates or charges (air cargo).

Combination carrier The provision of a transport service forming part of an overall through transport service involving a combination of transport modes. Under such circumstances the goods are conveyed in a transport unit throughout and thereby usually provide a door to door service. Examples include the road haulage unit involving a road/sea/road transit; a container being conveyed on rail/sea/road service, or palletised cargo being conveyed on road/rail/road carriers.

Combination charge An amount which is obtained by combining two or more charges relative to a consignment such as the normal tariff and currency/fuel surcharge.

Combination joint rate A joint rate which is obtained by combining two or more published rates (air cargo).

Combination of charges The amount which is obtained by combining two or more charges relative to a consignment/goods/merchandise involved in a particular transit. It may be the tariff charge of carriers 'A' and 'B' plus handling charges for a consignment of machinery.

Combination rate A rate which is obtained by combining two or more rates or charges and which is not published as a separate rate.

Combined cycle power station (Gasteam) A combined cycle power station consists of interconnected gas and steam turbines. The hot exhaust gases from the gas turbine(s) raise steam in an exhaust gas boiler which is used to drive the steam turbine(s).

Combined economies of scale in product development. A situation whereby two organisations/companies/entities unify their skills/focus to develop a product and in so doing realise economies of scale to realise this objective.

Combined Nomenclature The Combined Tariff and Statistical nomenclatures of the Community, represented by the first 8 digits of the Commodity Code. A Customs term.

Combined Package Port Tariff A port tariff which embraces and consolidates a collection of port charges such as stevedoring, cranage, cargo dues and so on.

Combined Ratio The combination of an insurer's (or reinsurer's) Loss Ratio and Expense Ratio. Another name for Operating Ratio or Trade Ratio.

Combined Tickets The provision of a ticket which involves more than one transport mode or related facility. Examples include the rail and bus/coach ticket which may be for a rail/coach combined tour; the package holiday involving flight and hotel accommodation tickets; or the road/ship ticket embracing the rail and ship journeys. Each element would be separately ticketed, usually so that the rail, coach, ship or hotel has a ticket to confirm the passenger travelled etc. Sometimes parts of the combined tickets are called coupons.

Combined Transport The carriage by more than one mode of transport against one contract of carriage.

Combined Transport Bill of Lading A document issued by a Combined Transport Operator (CTO) for the carriage of goods by at least two modes of transport service such as rail/sea road.

Combined Transport Document The Combined Transport Document's Bill of Lading.

Combined transport document (ICC) Negotiable or non-negotiable document evidencing a contract for the performance and/or procurement of performance of combined transport of goods and bearing on its face either the heading 'Negotiable combined transport document' issued subject to Uniform rules for a Combined Transport Document (ICC Brochure No 298) or the heading 'Non-negotiable Combined Transport Document' issued subject to Uniform Rules for a Combined Transport Document (ICC Brochure No 298).

Combined transport Operation A multi-modal transport service involving the use of more than one mode of transport on a particular journey offered by a single operator acting as principal for the entire journey. It is thus a through transport service from door to door or warehouse to warehouse; it is orientated to the needs of the user in that it caters for the entire consignment journey required, in contrast to a traditional single-mode/transport service covering only the leg of the journey performed by one mode.

Combined Transport Operator A through transport operator involving a multi-modal transport operation such as rail/sea/road.

Combi ship A vessel designed to carry both containers and conventional cargoes.

COMBITERMS Delivery terms for international groupage traffic (among forwarders).

COMBO Como Line – a shipowner.

COMESA Common Market for Eastern and Southern Africa.

Comex The New York Commodity Exchange, on which copper, aluminium, gold and silver are traded. Comex is often referred to as meaning the prices of these metals in New York e.g. 'Comex is 66 cents', meaning the Comex copper price is 66 cents a lb.

Comfort letters These are in the form of intermediate, less strict guarantees usually containing assurances and intent. Although in cases legally binding, such letters are generally less likely to satisfy the stricter legal requirement of a guarantee.

COMG EANPG Communications Group – see EANPG entry.

Com/I Commercial Invoice.

Comite Maritime International An international committee of maritime lawyers based in Antwerp.

COMITEXTIL Comite de co-ordination des industries textiles de la CEE (French) (Co-ordination committee for the textile industries in the EEC).

COMLOSA Committee of Liner Operators to South America.

Command A request, usually typed at the keyboard, to have a task performed. A computer term.

Command advertising The process of adopting a policy to advertise in the immediate future and do not delay it.

COMMAND.COM The MS-DOS file that contains the MS-DOS internal commands and that processes both internal and external commands. Also, called the Command Processor. A computer term.

Command economy An economy centrally controlled by a government. This involves extensive regulations in all sectors of the business and was found in former Communist bloc countries. It is in complete contrast to a market economy. See market economy entry.

Commercial acumen The ability of a person to identify, create and cultivate maximum advantage of commercial opportunities in the market place which may be self-created or simply arise through a particular circumstance. In so doing the person will develop through his own enterprise the commercial situation or business on a profitable long-term basis improving the overall market share.

Commercial advantage The commercial advantage gained in the market place by a producer over its competitor. This may be a price advantage; larger network of sales outlets; more favourable after sales service and so on.

Commercial Agent An intermediary who brings the exporter into direct contractual relationship with buyers in his or her territory. Agents are normally paid by flat rate commissions. Advantages of using the Agent from the Exporters standpoint include the exporter has many clients which means a spread of small credit risks, and he has effective control over his sales policy. Guide lines regarding the appointment of Agents are found in the Commercial Agency – publication No 410 of the International Chamber of Commerce.

Commercial analysis The process of analysing a particular investment or general business situation in a strictly commercial environment with a view to determining the commercial element involved and the implications which flow from them. This may be price, product specification, competition, market profile, distribution and so on.

Commercial and Development Manager A managerial post responsible for devising and implementing policies to aid the development of the entity/company business on a profitable basis.

Commercial bank A bank involved in international trade/corporate banking.

Commercial cargo Cargo conveyed on a particular transport operators schedule such as container shipment, air freight, international road transport and so on, for which the shipper remunerates the carrier for the conveyance of the goods, and the cargo does not constitute a consignment under a foreign aide/gift programme for which no freight is raised or a substantial discounted tariff is applicable. See also non commercial cargo entry.

Commercial considerations The commercial aspects involved in a situation which may involve the decision making process.

Commercial development The process of developing a product(s)/services(s) for a market. It may be the expansion of a transport service; the provision of a new range of products; or commercial site of offices and flats.

Commercial Director The person at Director level within a Company responsible for all the Commercial policy of the business. He is responsible to the Managing Director and his duties include charging policy, marketing and selling, development of the business and so on. An important post within a Company requiring a person of high business acumen calibre.

Commercial documents The range of commercial documents associated with international trade such as the commercial invoice.

Commercial insurance Insurance against losses arising from international commercial risk including insolvency of the purchasee, default on payment by a private purchasers and non acceptance of goods delivered to the buyer.

Commercial Invoice It is an accounting document prepared by the exporter (seller) in the name of the importer (buyer) or Agent. It evidences the actual dispatch of the goods specified or provision of services, and indicates their quantitative and qualitative characteristics, price, value and other information necessary for the preparation of accounts.

Commercially active population A population of entrepreneurial flair and responsive to marketing initiatives by Companies/Government/individuals.

Commercial Manager The person responsible for the commercial aspects of a business, which in an airline or shipping company can embrace rates, fares, claims, marketing, commercial development etc.

Commercial market A segment of the market embracing the commercial sector. This tends to involve all areas of commerce/business and its infrastructure.

Commercial Market Research The process of conducting market research in the commercial sector. See market research and commercial market entries.

Commercial paper Short-term unsecured promissory notes issued by major companies. Only large shipping companies with the first-class credit ratings have access to the commercial paper market. Overall, a short-term security issued to raise money, typically under one year.

Commercial Policy The commercial policy of a particular company or in a particular set of circumstances. It may be for example on pricing whereby the company will raise prices by five per cent on goods up to €1,000 and four and a half per cent on goods over €1,000.

Commercial presence A foreign company setting up subsidiaries or branches to provide services in another country such as foreign banks setting up operations in a country officially. WTO term

Commercial rate A discount rate offered to a company or to a special customer.

Commercial research The process of researching into the commercial field such as evaluating the merits of an existing rates structure amongst shippers.

Commercial risk A risk undertaken translated into financial terms by an entrepreneur in the process of executing a commercial policy or concluding a contract or other circumstance. It may be a property developer in the development of a site for residential development, or a transport operator starting a rates war by reducing rates to increase market share and in so doing hopes the volume will increase – albeit with a lower average unit yield – but with much improved revenue production.

Commercial Risk analysis The process of analysing a particular investment or general business situation in a strictly commercial environment with a view to determining the commercial risk areas and how they can be best minimised. The risk maybe increased competition, general market vulnerability to decline and control by government of prices and so on.

Commercial Storage The storage of commercial merchandise as distinct from household effects in a warehouse.

Commercial Strategy The commercial strategy adopted by any Company/entity relative to all the commercial aspects of the business. This includes both short-term and long-term policies. It embraces marketing, charges, profit levels, market share, market participation i.e. home and overseas including countries attitude towards competition, and so on.

Commercial Tariff Concessions System A concessionary import tariff facility available in Australia.

Commercial travel agent One whose clients are companies rather than members of the public; also a corporate agent.

Commingling The process of mixing/blending together oils. A clause found in oil tanker charter party restricting such a practice to be only undertaken with the agreement of all shippers concerned.

Commissary goods Goods of little or no duty significance such as catering supplies and equipment.

Commission Remuneration paid to an intermediary for the introduction or servicing of business. For example, an amount, usually based on a percentage, paid to a travel agent for the sale of a supplier's product or service; or the basic fee charged to a customer by a broker for a purchase or sale transaction. Commissions are negotiable. An IPE definition.

Commission Agent An agent who earns his/her living from the commission he/she obtains from selling the product(s)/service(s) of his/her Principal. Hence, a product sold for €100 on a ten per cent commission would earn the Agent €10.

Commission House An essentially American term, referring to a company introducing client business to a ring dealing broker. Many companies introducing business in London as commission houses are themselves members of exchanges in the USA and vice-versa. Individuals acting in this capacity are often referred to as half-commission agents because they are rewarded for introducing the business and maintaining local contact with the client by a half share of the Ring dealer's commission.

Commission on International Commodity Trade: United Nations, New York A commission to investigate international raw material problems.

Commixture A mixture of two or more cargoes which cannot be separated into the relevant consignments.

Commodities Agricultural, mineral or other primary products and raw materials contracts.

Commodity A type of goods.

Commodity analysis The process of analysing a range/selection of commodities. This may arise in transport whereby an analysis of commodities conveyed on a particular route is undertaken during a specified period and the revenue each commodity generates.

Commodity Box rate A rate classified by commodity and quoted per container.

Commodity Classification Rates An Air Freight rate based on commodity classification type. Also applies to sea transport.

Commodity code The term used to denote the code (either at 8 digit or 10 digit level) which identifies a line of tariff nomenclature. The term replaces TTCN (Tariff/Trade Code Number).

Commodity Exchanges A trading centre for the commodity markets. Main commodity markets are Comex in New York, Nymex, the New York mineral exchange, the Chicago Board of Trade, the Baltic Exchange-London, the International Petroleum Exchange-London and the London Metal Exchange.

Commodity item number Specific description number required in air transport to indicate that a specific rate applies. See Commodity Rate entry.

Commodity market economy A commodity market driven economy focusing on commodities. See commodity markets entry.

Commodity markets These embrace both industrial material and foodstuffs. They include oil, aluminium, zinc, copper, vegetable oilseeds, rubber, cotton, lead, wool, cocoa, soya beans, coffee, sugar, maize and wheat. Overall, they represent a substantial volume of international trade.

Commodity mix Details of the commodities conveyed on a consignment which usually is a groupage consignment embracing a number of commodities. For example, it could be a consolidated/groupage consignment conveyed in an ISO container.

Commodity rate A rate devised for particular commodities which maybe subject to minimum weight restrictions, availability on certain services and operative between specified points. It is particularly established in IATA Air Freight services.

Commodity trades Any primary product or raw material marketed internationally either in its original state, for example, mineral ores, corn, cotton, etc., or after the initial state for example mineral ores, corn, cotton, etc.; or after the initial process which makes it acceptable as an industrial raw material, for example metal ingots.

Common account When an original insurer who already has reinsurance effects further reinsurance with another reinsurer it may be either wholly for his own protection or also for the protection of the first reinsurer. In the latter case it is said to be reinsurance for the common account.

Common Agricultural Policy (CAP) Comprehensive system of production targets and marketing mechanisms designed to manage agricultural trade within the EC and with the rest of the world. The system aims at guaranteeing farmers' incomes by bridging the gap between world market prices for major commodities and the normally higher prices set by the EC, through a complex system of price support mechanisms, export restitutions, social and other measures.

Common Agricultural Policy levy A charge on imported foodstuffs into UK from countries other than EEC.

Common Carrier One who carries any type of goods, other than a carrier of special goods. The shipowner is liable for all events except Act of God, war, kings/queens enemies, general average, inherent vice, etc. The common carrier is therefore liable for other causes on the grounds that he has taken on the goods for reward; responsible for the safety and security of the cargo having been paid to look after the cargo; there are no witnesses other than servants/crew of the shipowner – there is need to avoid fraud and falsehood e.g. that the goods were stolen; liable even without fault or with fault e.g. cases of negligence; and he is like an insurer of the goods.

Common code of practice A uniform code of practice/procedure which apples to all concerned within a particular situation/category. For example, the type of ticket found on all IATA Airlines is of similar layout and a common code of practice is adopted regarding the inter-line accountancy settlement arrangements through a clearing house.

Common external tariff A common custom duty imposed on non members of a customs union. See customs union entry.

Common law lien A claim based on common law.

Common Market Countries which are members of the European Economic Community permitting free trading amongst themselves. See EEC entry.

Common rated Two or more relatively adjacent destinations to which the fare from the point of origin is identical.

Common-rating system A European practice enabling long-haul travellers from outside the continent to move freely on a single ticket at no extra cost.

Common Short Form Bill of Lading An abridged form of Bill of Lading which covers shipper/forwarder from port to port and through transport consignments including containerised shipments but excluding combined transport Bills of Lading. It is fully negotiable and has the normal Bill of Lading lodgement and presentation procedures.

Common stocks A United States and Canadian term for equity or ordinary shares. They are often of 'no par value'.

Common Transit A procedure, similar to Community Transit for moving goods without payment of duty and tax between the Community and the EFTA and Visegrad countries.

Common user facility A facility available to all participants within a scheme/group of Companies. An example is found in the facilities available at an airport for passengers of all Airlines using it. Such common user facilities include restaurant, lounge, luggage trolleys, etc.

Commonwealth A group of 82 countries.

Commonwealth of Independent States Formed in 1991 consequent on the breaking up of the Soviet Union involving Glasnost.

Commonwealth Preference System in UK whereby goods from Commonwealth countries imported at concessionary tariffs.

Communal economy An economy/State/Country which is controlled entirely by the State/Government as obtains in former Communist countries and today in China.

Communication mix It embraces advertising, sales promotion, personal selling and public relations. See promotion mix entry.

Communications plan A marketing plan embracing the communication mix. See communications mix entry.

Communications software Software is required for all types of computer-based communication. The main types of software are: system, user interface and communication, e.g. Terminal in Windows.

Community Any or all of the European Communities, the European Economic Community; the European Economic Community; the European Coal and Steel Community and Euratom.

Community goods Goods which wholly originate in the Community (EC) and other such goods which are in free circulation in the EC.

Community Member States In 2004 there were 25 member states comprising European Community embracing Austria, Belgium, Cyprus, Czech Republic, Denmark, Estonia, Finland, France, Germany, Greece, Hungary, Italy, Latvia, Lithuania, Luxembourg, Malta, Netherlands, Poland, Portugal, Republic of Ireland, Slovakia, Slovenia, Spain, Sweden and United Kingdom. Total population 500 million.

Community System Providers Appointed by the major ports to operate systems for inputting, storing and processing import and export freight data and then sharing this information with the CHIEF computer which is the repository for all declarations.

Community Transit A procedure for the movements within the Community of goods, without payment of duty and tax. It has been displaced by the new export system – see New Export System Entry.

Community treatment The admission of goods as Community goods into a member state of the EU.

Commutation The finalisation of an outstanding loss by payment of an agreed figure in settlement.

Commutation clause A clause in a reinsurance agreement that provides for estimation, payment, and complete discharge of all obligations including future obligations between the parties for reinsurance losses incurred.

Commuter A person who makes regular daily work journeys to and from work usually to a conurbation area by rail. It can also apply to a much lesser degree to regular work journeys undertaken by coach, sea (ferry), private car or air.

Commuter airline Scheduled carrier operating regional US service, usually with aircraft carrying fewer than 60 passengers; also called regional airline.

Comp Combined (holds); compound; or complement.

Compacted ice edge Close, clear-cut ice edge compacted by wind or current, usually on the windward side of an area of drift ice.

Compagneurs Nationales des Containers See CNC entry.

Compania Sud-American de Vapores A major liner operator.

Companionway A set of steps, leading from one deck to another on a vessel; or ladder used for disembarkation or embarkation of passengers/crew.

Company A legal entity, the life of which is dependent on that of its members. There are three ways of forming a company or corporation namely by charter, by Act of Parliament, or by registration under the Companies Act.

Company audit The process of formulating a company portfolio embracing all its financial aspects; position in the market place; company structure; product/service portfolio; any legal disputes pending; annual report analysis, short and long-term prospects and its overall degree of competitiveness/technology. A marketing term.

Company Business Plan See Business Plan entry.

Company Liquidity The liquid assets of a company such as cash and bank deposits. The UK Department of Trade conducts a quarterly survey of the net financial position of a range of mainly large companies, looking at their liquid assets (cash and bank deposits) and their liabilities (mainly bank borrowings).

Company owned sales force A sales force which is directly employed by the company and thereby is under the direct control of the Sales Director/Manager.

Company portfolio The profile of a company. See company audit.

Company resources This embraces within a company human resources, capital, land, cash flow, technology, fixed assets, production and company infrastructure.

Company Secretary A post responsible for looking after all the Companies affairs within the legal framework of a public limited company.

Company Service contract Contract of employment between seafarer and shipowner.

Company's own field organisation A method of direct export trading whereby the exporter employs agents or distributors to which the Company's technical export salesmen are attached. The agents or distributors undertake all the importing and processing of shipments. The technical export salesmen provide technical assistance to clients and obtain orders.

Company status The general financial status/audit/evaluation of a particular company and usually conducted to ensure the company is viable and has a favourable credit rating when conducting business with such an entity. A task often undertaken by an exporter's bank as a screening exercise.

Company structure The organisational structure of a particular company.

Company technical specialist export salesmen/saleswomen A method of direct trading overseas. This concerns a Company employing its own salesmen/saleswomen to sell its products/services in overseas markets and in so doing visit the countries concerned and become involved in face to face selling between seller (exporter) and buyer (importer).

Comparative Advantage A country has a comparative advantage relative to a second country in the production of the commodity in which it has a lower opportunity cost than the other country. An economic international trade definition. A similar criteria can be applied to other industrial/commercial situations.

COMPAT Computer aided trade.

Compensated gross tonnage A measurement developed by shipbuilders to indicate the approximate work content of a vessel. It is calculated by weighting the gross tonnage by a 'coefficient' which varies from 0.3 for a large tanker to 3.0 for a small service vessel.

Compensation agreement An international trade term which links the goods a country wishes to import to the types of goods it wishes to export thereby endeavouring to strike a balance between the level of imports and exports. It obtains primarily in centrally planned economies and thereby lessens the risk of a trade imbalance. See centrally planned economy entry.

Compensating products Goods imported under inward processing relief arrangements.

Compensatory and Contingency Financing Facility An IMF facility which provides financial assistance to members experiencing temporary export shortfall, and compensatory financing for excesses in cereal import cost, as well as for external contingencies in fund arrangements. See reserve tranche and credit tranche entries.

Compensator levy A payment made on exportation, whereby compensating products resulting from a manufacturing process may be exported to Turkey without losing entitlement to the reduced rates of duty, even though goods from which these products are made have not borne the Community customs charges applicable to them.

Competition The circumstance whereby a product(s) or service(s) is competing with another product(s) or service(s) in the market place. It may be direct competition such as two taxi companies competing in the same market in a particular area, or indirect competition whereby the railway company competes with the coach operator.

Competitive advantage The value added benefit derived from the entrepreneur product/service in the eyes of the buyer when related to competitive products/services in the competitive market place. It may be price, design, warranty, after sales, packaging, quality and so on.

Competitive broker A shipbroker who operates in a competitive environment and endeavours to find the best deal or fixture for their principal.

Competitive market A market in which no buyer or seller has market power. They are all 'Price Takers'.

Competitive position The overall competitive position of a company or product in the market place. It may be a market leader or amongst the top ten.

Competitive pricing The process of the entrepreneur basing the selling price in the market place on those of his/her competitors having regard to their product/service specification and the perceived benefits in the eyes of the buyer of the entrepreneur commodity.

Competitive strategies The strategies adopted/pursued by a company/entity to out manoeuvre competitors in the market place.

Competitive strategy The strategy adopted by the entrepreneur to retain/develop the competitiveness of the product/service in the competitive market place. This may involve quality, price, after sales, warranty, packaging, technical manuals and training.

Competitive strengths The features of a Company's product/service/circumstance/environment which are perceived by the buyer/market place in the most favourable light and constitutes factors influencing the rationale of the buyer's favourable choice.

Computer Processing

Competitive weaknesses The features of a Company's product/service/circumstance/environment which are perceived by the buyer/market place in the most unfavourable light and constitutes factors influencing the rationale of the buyer not to view the situation favourably.

Competitor audit The process of conducting an evaluation of a competitor embracing, position in the market, product/service portfolio, client base, level of profitability, financial resources, research and development programme, market share and so on.

Competitor orientated pricing The process of adopting a strategy in a Company/entity which focuses on the competitors' pricing structure with a view to gaining price advantage in the market place and thereby gain market share/increased sales.

Compiler A program which translates a complete high-level language program in to machine code to create a fully executable version. A Computer term.

Completely knocked down Consignments which are assembled at destination as distinct from being transported as complete unit.

Completion of end use The putting of goods to the use prescribed in the Customs tariff as the condition of their admission to an end use heading or sub-heading, suspension, or quota. A customs term.

Composite company A company transacting most general lines and long-term business.

Composite packaging A package consisting of a plastic receptacle and an outer protection in sheet metal, fibreboard, plywood etc., so constructed that the receptacle and outer protection form an integral packaging for transport purposes. Once assembled, it remains thereafter an integrated single unit.

Composite ship An iron framed wooden hull ship.

Composite trailer A semi trailer supported on a converter dolly.

Compound deferment An imports accounting system relative to VAT operation in some countries.

Compradore A local adviser or agent employed by a foreign party or company who acts as an intermediary in transactions with local inhabitants.

Comprehensive Bank Guarantee A guarantee given to a bank as security for finance made available to an exporter for goods covered by a comprehensive insurance. It may be a guarantee up to 100 per cent of the invoice value and this is the amount the bank will lend. The guarantee assures the bank of unconditional reimbursement of any payment not received within three months of due date, plus interest. There are two types of Comprehensive Bank Guarantee depending on the method of trading adopted. (i) Comprehensive Bank Guarantee: Bills or Notes: This is available when the payment is obtained by means of bills of exchange or promissory notes, and the credit period ranges from sight to up to two years. To obtain the finance, the exporter presents the bills or promissory notes to the bank, together with evidence of shipment and a warranty that the terms of insurance cover have been complied with. (ii) Comprehensive Bank Guarantee: Open Account: This is available when the exporter receives payment direct from the buyer under open account and the credit period does not exceed six months. To obtain the finance the exporter presents a schedule of debts, evidence of export and a warranty that the terms of insurance have been complied with. The exporter issues a promissory note to the bank to cover the repayment.

Compression The process of a package bulging, such as a carton with side bulges caused by over-tightening of straps is bulged by compression.

Compression ratio The ratio of the absolute pressure at the discharge from a compressor divided by the absolute pressure at the suction.

Compressive Stresses Forces of sea on ships sides and bottom leading to hull distortion.

Compromised Total Loss This term is used where hull underwriters agree to a compromised settlement for total loss of ship, in circumstances where neither an actual loss, nor a constructive total loss may be claimed, but the value of the ship when repaired does not justify the cost of repairs. This type of settlement is not subject to any basic rules, but usually applies to policies where the insured value of the ship is higher than her market value.

COMPROS The forum in which the European Communities Trade Facilitation organisations meet. It deals with simplification of documentation and trade procedures, electric data processing and so on.

Compulsory licensing For patents: when the authorities licence companies or individuals other than the patent owner to use the rights of the patent – to make, use, sell or import a product under patent (i.e. a patented product or a product made by a patented process) – without the permission of the patent owner. Allowed under the TRIPS Agreement provided certain procedures and conditions are fulfilled. See also government use. A WTO term.

Computer A data processor that can perform substantial computation, including numerous arithmetic operations or logic operations, without intervention by human operator during a run.

Computer-aided design/computer-aided manufacturing The application of computer aids to design, drafting, planning, estimating, and manufacture primarily in the mechanical and electrical engineering sectors of industry. The system speeds up the design process and allows the design to be fed directly, in digital form, into a numerically controlled machine tool for automatic manufacture of the product.

Computer driven company A company which may be a service or manufacturing based entity which is managed and operated extensively by computer technology.

Computerised Accounting System A computerised accounting system embracing all the salient accounting processes including income, expenditure, customer billing, account payments, management information and so on.

Computerised reservation system (CRS) Computer system through which many travel products, including airlines and major hotel chains, are booked; several systems, all started and controlled by airlines, compete for travel agency business.

Computer literate The ability of an individual to operate efficiently a computer.

Compute literate market A market which is able to operate and understand fully the benefits of computerisation.

Computer output to microfilm A technique where a mass of text is filmed and stored in a small space.

Computer passenger codes The codes used to identify particular passengers for the purpose of compiling an invoice for billing to a customer.

Computer print out The provision of data in tabulated form from a computer.

Computer Processing The mechanism of processing international trade documents through a computer resource/network.

Computer rate codes The codes used to identify particular rates for the purpose of compiling an invoice for billing to a customer.

Computer Reservation System A computerised reservation system which exists for airline seats/flights, theatre seats, train seats, allocation of cargo space on flights/sailings and so on. In many situations the Freight Forwarder/Travel Agent/Exporter has direct online computer access to the computer reservation system thereby enabling the client to have immediate confirmation of the booking.

Computer software The infrastructure of a computer featuring all the resources and needs of the user embracing computer programs, floppy disk.

Computer system (hardware) Includes: main processor unit, input devices, storage devices, output devices, communication facilities and devices.

Computer virus A program that can infect other programs by modifying them to include a possibly evolved copy of itself.

COMS Commissions.

Conair Container A thermal container served by an external cooling system (e.g. a vessel's cooling system or a Clip on Unit) which controls the temperature of the cargo. Conair is a brand name.

Conbill Code name of approved BIMCO uniform Bill of Lading used when no charter party is signed.

Conbulker crane A device which enables ship to shore container gantry cranes to load bulk carriers; a rotator attachment fitted to the spreader rotates open-top containers around the axis to tip out the dry bulk cargo.

Con-bulkers Container/bulk carriers.

Concentration The ratio expressed in tenths describing the amount of the sea surface covered by ice as a fraction of the whole area being considered. Total concentration includes all stages of development that are present, whereas partial concentration may refer to the amount of a particular stage or a particular form of ice and represents only a part of the total.

Concierge A person or desk in a hotel in charge of porters and bellboys; handles personal services for a guest, including ground arrangements.

CONCORP Construction Corporation (Burma).

Concurrent Engineering The process whereby individual components of a particular product can be manufactured simultaneously in different locations and even countries as part of a single process.

Condensate Reliquefied gases which collect in the condenser and which are then returned to the cargo tanks.

Condensing turbine After passing through the turbine and expanding to a low pressure, the exhaust steam is discharged to a condenser where it is cooled and condensed into water before being pumped back to the boiler. (c.f. Back-pressure turbine).

Conditional bond A bond which specifies documentary evidence or one which contains no requirement for documentary evidence termed a 'Default Bond'. Bonds which do not require documentary evidence are not therefore fully conditional and provide little or no protection to the exporter (seller). An international banking term.

Condition monitoring Ship maintenance technique to determine when it would be carried out instead of undertaking the task on a regular basis without regard to its condition, for example the ship's machinery.

Condition precedent A condition imposing liability only on an event first happening.

Conditions The contractual stipulation/obligations agreed/accepted by two or more persons/companies to a contract and printed/recorded on the contractual document; or in travel/tourism terms the language of a travel contract which specifies what the customer is or is not being offered and which may specify the circumstances under which the contract may be invalidated: also called terms and conditions.

Conditions of carriage The terms and conditions by which the carrier contracts to convey the cargo and referred to on the consignment note or charter party.

Conditions of contract The terms and conditions by which the carrier contracts to convey the cargo and referred to on the consignment note, air waybill, bill of lading, CMR note or charter party. The terms and conditions found in a contract.

Cone Top part of stacking fitting that fits into bottom hole of ISO corner casting.

Cones Devices for facilitating the positioning and lashing of containers. The cones insert into the bottom castings of the containers.

Confederation of Latin American Tourist Organisation (COTAL) Association of official tourism bodies promoting travel to Central and South America.

Conference Organisation whereby number of shipowners often of different nationality offer their services on given sea route on conditions agreed by members; or whereby a company/entity calls a sales conference to promote its product/service. See Sales Conference entry; or an association of carriers formed to establish standards and rules governing others, such as travel agents, who do business with members of the association.

Conference Lines Deep sea Shipping Companies who are members of a Liner Conference.

Conference operator A shipowner who is trading under liner conference conditions. See liner conference system entry.

Conference Ship A ship operated by a signatory to a shipping conference. See conference entry.

Conference system A conference system is an association of liner shipowners operating in the same trade who agree to abide by a set of regulations, and to quote the same rates of freight, to the mutual benefit of shippers and themselves.

Conference terms Qualification to a freight which signifies that it is subject to the standard terms and conditions of the particular liner conference. These are normally set out in the conference tariff.

Conferencing Video conferencing and tele-conferencing enable two or more people to have a conference. Such facilities are available through several network service providers.

Confidential tariff A schedule of wholesale rates, to be marked up to include commission, distributed in confidence to travel agents and wholesalers.

Configuration The interior arrangement of a vehicle, such as an aircraft used for transporting passengers in various classes of service.

Configuration (software) In software, this means setting variables to enable software to run efficiently and effectively for the user. In hardware, it means setting up and connecting hardware devices to ensure that they operate efficiently and effectively for the user. A computer term.

Confirmation The written acknowledgement of a firm deal involving a placing or deposit or sale or purchase of funds or securities; for reasons of bank security no dealer should be permitted access to confirmations; or the process immediately following a transaction whereby the traders confirm the details of the trade. An IPE definition.

Consignment instructions

Confirmation of renewal Acknowledgement sent to the insured after payment of the renewal premium confirming the continuation of the policy for a future term.

Confirmed Letter of Credit/Unconfirmed Letter of Credit An Unconfirmed Credit is where the advising bank has not been asked by the issuing bank to add any financial commitment by way of confirmation.

Notwithstanding that the beneficiary may have presented documents in full compliance of the credit, the advising bank has no obligation to pay the beneficiary until reimbursement has been obtained from the issuing bank.

When trading with a country experiencing political or economic difficulties a beneficiary may desire the additional security of the legal and binding undertaking of the advising bank in his own country.

If this additional undertaking is given by the advising bank, the Documentary Credit becomes a Confirmed Documentary Letter of Credit, and the advising bank does have a financial interest.

Under this arrangement, the confirming bank undertakes that payment will be made irrespective of what may happen to the applicant or issuing bank regardless of any changed political or economic situation in the country of the applicant, providing that the documents submitted by the beneficiary comply fully with the terms of the credit. Confirmations involve payment of an additional fee, and the advising bank may not always be willing to enter into such arrangement.

Confirmed reservation Oral or written statement by a travel industry supplier acknowledging receipt of a reservation and promising to honour it within specified limitations.

Confirming bank A bank, other than the 'issuing bank', which undertakes on its own responsibility to pay, accept, or negotiate under the credit. The bank (usually the Advising Bank) which, at the request of or with the permission of the Issuing Bank, adds its own irrevocable undertaking to pay/accept or negotiate without recourse, documents presented in compliance with the Credit terms either to itself or (if permitted within the terms of the Credit) to another bank. Such an undertaking is in addition to, and not instead of, that already given by the Issuing Bank to the Beneficiary of the Credit. In simple terms, the Confirming Bank is conditionally guaranteeing payment to the Seller, even if the Issuing/Buyer's Bank or the Buyer is unable or unwilling to pay, providing the Seller has satisfied all the terms of the Credit and the documents submitted are exactly as specified by the Credit.

Confirming House An indirect export trading technique whereby confirming house acts as the overseas buyers' agent and will either find sources of supply or negotiate with suppliers nominated by the buyer. The exporter deals directly with the confirming house who will normally pay the exporter for the goods. They will usually be responsible for financing the contract on the buyers' behalf, arranging the necessary documentation and shipping the goods to their final destination. The Confirming House enters into two legal relationships; one with the overseas customer who asks to produce certain goods for him, and one with the seller in the home market with whom it places the order. The relationship with the overseas customer is normally that between principal and agent, whereas the relationship with the seller in the home market depends upon the nature of the contract which the Confirming House concludes with him. However, whilst selling through independent organisations is relatively easy, the exporter has no control over marketing in the country of consumption and the success or failure of the exporter's product(s) depends entirely on someone else.

Confirmed irrevocable credit A confirmed irrevocable credit is an irrevocable credit to which the 'advising bank' (at the request of the 'issuing bank') has added its confirmation that payment will be made. The 'advising bank' thus confirms that it will honour drawings which conform to the terms of the letter of credit. For example, through the documentary letter of credit system in a situation providing credit confirmation in London, and subject to the exporter complying with all the terms of the credit, the actual payment is guaranteed: there is no risk of non payment, either in the event of the insolvency of the buyer or because of exchange control regulations in the country of the importer.

Confiscation and Expropriation risk Legal seizure of a vessel by the authorities.

Congenbill Code name of approved BIMCO Bill of Lading used with Charter Parties 1994 Edition.

Congestion surcharge A surcharge raised by the shipowner payable by the shipper on cargo in circumstances where excessive congestion/delay is experienced in the port. The surcharge is to recoup the additional cost incurred by the shipowner consequent on the delay experienced.

Conglomerate The collection/association/formulation of a number of companies to form one overall entity.

Conline bill Code name of BIMCO approved Liner Bill of Lading. Various editions exist including English, Spanish, German and French.

Conline booking Code name of BIMCO approved liner booking note to be used with the Conlinebill Bill of Lading.

Connecting carrier A carrier with services meeting those of another carrier so that traffic can be interchanged between the carriers at the connecting point.

Consecutive days Day which follows one immediately after the other relative to a charter party terms. Alternatively, it may be termed running days.

Consecutive voyages A chartering term whereby a vessel may be chartered for a whole series of voyages.

Consensus ad idem (Latin) Mutuality of consent and understanding of all the material elements.

Consequential loss Insurance covering the loss of earnings of a business following fire or other specified damage. It is also termed Business interruption.

Conservancy dues Charges raised by a port authority on vessels using a port and based on the ship NRT. In Nigeria it applies to both inward and outward ship movements and includes seven days free berthage.

Consign To forward goods from one place to another.

Consignee Name of agent, company, person receiving consignment. For example, the person whose name appears on the Air Waybill as the party to whom the goods are to be delivered by the carrier.

Consignment One or more pieces of goods accepted by the carrier from one shipper at one time and at one address, receipted for in one lot and moving on one Air Waybill to one consignee at one destination address. A similar criterion obtains for other transport modes such as rail, sea, and combined transport.

Consignment instructions Instructions from either the seller/consignor or the buyer/consignee to a freight forwarder, carrier or his agent, or other provider of a service, enabling the movement of goods and associated activities. It may involve movement and handling of goods (shipping, forwarding and stowage), customs formalities, distribution of documents, allocation of documents (freight and charges for the connected operations) and special instructions (insurance, dangerous goods, goods release, additional documents required).

Consignment note A document which acknowledges the terms under which the goods are accepted and conveyed and usually accompanies the cargo throughout the transit.

Consignor Name of agent, company, person sending consignment – the shipper. For example, the person whose name appears on the Air Waybill as the party contracting with the carrier(s) for carriage of the goods.

Consolidate The process of grouping/stuffing several shipments together of compatible cargo in one container, swap body, truck, wagon and pallet. See also groupage entry.

Consolidated accounts The accounts of a company together with all its subsidiary companies, presented in one set of figures.

Consolidated container Container stuffed with several shipments (consignments) from different shippers for delivery to one or more consignees.

Consolidated data plate Plate affixed to a container giving details of gross and tare weights, external dimensions, owner, serial number, etc.

Consolidated ice Floating ice in which the concentration is 10/10 and the floes are frozen together.

Consolidated rate A rate devised by an Agent/Freight Forwarder for consignments despatched under consolidation arrangements involving an air or surface international transit. It usually includes both the collection and delivery charge, customs clearance, plus of course, the air or surface rate element.

Consolidation
The process of despatching as one overall consignment under an Agent/Freight Forwarders sponsorship by Air, Rail, Container or Road haulage unit, a number of individual compatible consignments from various consignees to various consignors; or selling the same tour with identical departure dates through a number of wholesalers, co-operatives or other outlets so as to increase sales and reduce the possibility of tour cancellations; cancellation by a charter tour operator of flights associated with a specific departure or departure period with transfer of passengers to another flight or flights to depart on or near the same day.

Consolidation depot A depot where parcels of cargo are grouped and packed into containers.

Consolidation point See Consolidation depot entry.

Consolidator A person or Company (Freight Forwarder) undertaking consolidation of consignments, or wholesaler selling discounted scheduled air tickets to travel agents or to the public.

Consortia A company formed from a group of companies to manage a business such as found in some major deep sea container services serving a group of ports situated in various countries, and emerging from individual National Shipping Lines formerly engaged in break bulk/tween deck tonnage operations.

Consortium A group of combined operators who agree to rationalise sailings in a trade and in so doing carry each others containers; or entity formed by individuals or companies to acquire travel products they could not acquire individually or to negotiate better prices or commissions; in the UK, a chain of generally retail travel outlets.

Constr Construction.

Constructed rate A rate other than a specified rate (air cargo).

Construction aid This is financial assistance from the government to its national fleet. It provides funds for shipowners to build new vessels, either as a percentage of the total building cost or as a fixed sum. A form of flag discrimination.

Construction and use The title of legal regulations which govern the construction and use of vehicles in United Kingdom.

Construction rate A non saleable 'add on' amount for Tariff publication purposes to be used only in combination with other rates for carriage from, to, or through a specified point.

Construction subsidies Government financial contribution towards building costs of a specified vessel. See 'construction aid' entry.

Constructive total loss This arises when the assured abandons the subject matter insured such as the cargo or vessel, to the underwriter and claims a total loss. He may do so when he is deprived of the insured property and is unlikely to recover it, or in the following situations: (i) – the ship – the cost of recovery and repair would exceed the insured value; and (ii) for cargo – the cost of recovery, reconditioning and forwarding would exceed the value of the goods at destination.

Consul Commercial representative of one country residing officially in another whose duties are to facilitate business and represent the merchants of his nation.

Consulage Duty paid to a consul for protection of goods.

Consular documents The preparation/legislation of the consulate documents relative to the processing of an international consignment. Examples include the consular invoice, certified invoice, accredited chamber of commerce, certified certificate of origin. Documents required by customs at the time of importation depending on the overseas territory.

Consular fees Fees raised by the consulate relative to the importation of a consignment and passenger immigration. This involves, for example, the authentication of the export invoice by the consulate; the accreditation of documents involving an international consignment and issue of the visa for passenger immigration.

Consular invoice A document representing confirmation of the consulate of the respective country in the country of the exporter that the annotations of the exporting company and the signature of the authorised person of the exporter are authentic. It confirms that the respective company really performs the function of a foreign trade enterprise in the country of the exporter, and that the declaration concerning the price, the total amount, the place of despatch as well as the number of packages and the weight of the goods are correct. Such a document is mandatory when despatching goods to certain parts of the world particularly to those countries which enforce ad valorem import duties.

Consumer Association A national association, which may be under the aegis of government or the private sector, and formed to look after and safeguard the needs of the consumer. The association may be for a specific industry or range of industries. Overall, the consumer association represents the consumer.

Consumer behaviour Consumer response or reaction to a change in price, design, new product promotion, etc. for a particular commodity or range of commodities in a particular area, region, county or countries, etc.

Consumer buying power The strength of the consumer buying power in the competitive market place. This will depend on taxation, employment levels, GDP, social-economic classification/distribution, exchange rates and so on.

Consumer commodity A consumer Commodity is defined as a material which is packed and distributed in a form intended or suitable for retail sales for the purposes of personal care or household use.

Consumer demand The market demand for a product or service emerging from the consumer market base who may be the housewife for washing machine, or the 'man in the street' for men's shoes.

Consumer durables Commodities in the market place which form the infrastructure of the average house holder and include furniture, carpets, cars and furnishings. Such products have a mass market and are replaced when they become obsolete, out of fashion, worn out, out of date technology, etc. Hence, the buying cycle is related to the buyer's behaviour and buyer's purchasing power.

Consumer goods A wide variety of commodities used by the consumer in the market place which require replenishment regularly and frequently such as food, clothing, detergents, footwear and confectionery.

Consumer led marketing policy A marketing policy adopted by a Company/entity which reflects/accords with the consumers needs in the market place. Also termed client led marketing policy.

Consumer market That segment of the market involved in the demand for products emerging from the average household. This involves both food and merchandise but can extend to travel, household appliances and so on.

Consumer marketing The process of promoting a consumer product such as soap, perfumes, washing powders in the consumer orientated market place especially towards the household and through packaged advertising/promotions.

Consumer market research See consumer research entry.

Consumer Panel A group of consumers who may be recruited on a random basis and formed with the objective of evaluating consumer product/service acceptability. Overall, the Panel may be recruited on a Company, Market Research Agency or Consumer Association basis.

Consumer products Goods produced for the consumer market such as household products, clothing, cars etc. See consumer goods entry.

Consumer profile The composition or make up of the consumer(s) identified with a particular product or service provision. It may be, for example, men of 17/35 years of age who are attracted to a particular sporty designed motor car.

Consumer protection legislation of sale The aim of the consumer protection legislation in the field of goods is to prevent individuals being taken advantage of by large businesses imposing on them harsh contractual terms, against which they have neither the financial power, nor the freedom of choice to resist. It involves private, business and consumer sales.

Consumer research The process of researching into the consumer profile within the market place of a particular product or range of products. For example, research may be undertaken into owners of a particular brand of washing machine, the reason of choice, general satisfaction, age group and so on.

Consumer Relations Manager A managerial position responsible for maintaining good relations and satisfactory public image between the Company and its clients.

Consumers This features individuals selling in classified advertisements in a newspaper/journal, residential property, cars and so on. Advertising personal services on the internet and selling knowledge and expertise is another example. Additionally, many individuals are using intranets and other organisational internal networks to advertise items for sale or services.

Consumer to Business This category includes individuals who sell products or services to organisations, as well as individuals who seek sellers, interact with them and conclude a transaction.

Consumer to consumer In this category a consumer sells directly to the consumer.

Consumer services Services provided in the market place for the consumer such as restaurants, hotels and entertainment.

Consumption abroad The process of consumers or companies making use of a service in another country as found in tourism. WTO term.

CONT Containership.

Cont Continent (of Europe).

Contact Damage Contact – when the ship strikes something other than another ship, the latter being 'collision'.

Cont. (AH) Continent, Antwerp – Hamburg range.

Container A container is a transportable unit permitting inter-modal unitised merchandise distribution which may be national or international. This may be involved on FCL or LCL type of consignment. Most container types are built to ISO standards. The usual container modular size is 2.45m X 2.45m with a varying length of 3.05m, 6.10m, 9.15m, or 12.20m. A wide variety of containers exist including covered dry, top loader, bulk liquid, bulk powder, bin type, skeleton, refrigerated and so on.

Container Aircraft A certified container that interfaces directly with an aircraft restraint system and meets all restraint requirements without the use of supplementary equipment. As such it can be either a combination of components (e.g. an aircraft pallet plus non-structural igloo), or one complete structural igloo; or one complete structural unit (e.g. a lower hold cargo/baggage container).

Container Base The depot for packing and unpacking 'less than container load' consignments. See Container Freight station entry.

Container berth Berth or quay at which containerised cargo is handled.

Container Bills Usually a 'received for shipment' bill of lading issued by the Container operator to cover multimodal transport from an inland container depot to final destination. The Bill of Lading is endorsed 'Shipped Bill of Lading' at the time the container is stowed on board the vessel at the port of departure.

Container block A number of horizontal stacks inter connected and secured horizontally using bridge stackers.

Container Bolster A container floor without sides or end walls which does not have the ISO corner fittings and is generally used for Ro/Ro operations.

Container chassis A vehicle specially built for the purpose of transporting a container so that when container and chassis are assembled, the produced unit serves as a road trailer.

Container control The process of controlling the ISO container fleet. In so doing it is desirable the optimum use be made of them and adequate policing is undertaken at all times. Usually computers feature prominently in container control.

Container depot The depot for packing and unpacking less than container load consignments. See Container Freight Station entry.

Container dues Charges raised by Port Authority on containers passing over the quay based on the individual container.

Container Flow Management

Container Flow Management Management personnel responsible for intercontinental logistic services container transport over sea and land between different continents with an important role for differentiation and added value. This embraces Container Management thereby ensuring the most ideal container is available to meet the shipper's needs on a global basis. Overall, it involves transport, forwarding, storage and distribution adapted to the specific requirements of the product, the supplier and his customers. Overall, it involves the organisation responsible for the marketing and management of container services and related products to the market.

Container frame The strength members of the container.

Container Freight Station The depot for packing and unpacking 'less than Container load' consignments. Such a depot is called in the UK a container base; a container depot in Australia, and inland clearance depot in UK and the Indian Sub Continent.

Container head Sometimes used to describe the end opposite the doors of the container.

Container import cargo manifest A document detailing/listing the cargo contents of all the containers which will be unloaded/discharged from a particular ship on a specified sailing at a named port and berth.

Container inter-modality The process of a container capable of being conveyed/interchangeable in two or more transport modes such as rail/sea or sea/air. The inter-modality of ISO containers being conveyed by sea/air is on the increase using the Boeing 747F and sea/air-bridge dedicated services.

Containerisation A method of distributing merchandise in a unitised form thereby permitting an inter-modal transport system to be evolved providing a possible combination of rail, road, canal and maritime transport.

Containerised Indication that goods have been stowed in a container.

Container lease The contract by which the owner of containers (lesser) gives the use of containers to a lessee for a specified period of time and for fixed payments.

Container leasing The facility to lease from a Container Company ISO containers. It may be for one, two, three years or other mutually acceptable periods to the two parties. Two types of container leasing exist; a master lease available for one or two trips, or a long-term lease of up to five years involving several thousand containers.

Container load Consignment which, in a container, fully occupies the internal capacity, or conversely reaches maximum payload for that particular unit.

Container logistics The efficient distribution globally of containers by a carrier to accommodate the needs of the shipper on an added value basis. The controlling and positioning of containers and/or other equipments.

Container market The market demand for the container. This embraces shippers wishing to convey their merchandise therein either on FCL, or LCL basis. See container user analysis.

Container moves The number of actions performed by one container crane during a certain period.

Container on flat car Carriage of containers on conventional railway flat cars, as opposed to specially designed rolling stock.

Container operator A company/entity involved in the provision and operation/control of ISO containers. In so doing the container operator would market the container service encouraging shippers to use it as a FCL or LCL (consolidated/groupage consignments), in liaison with shipowner(s) for a trade(s). The container operator may own the container or have them on lease.

Container owner A party who has a container at his/her disposal and who is entitled to lease or sell the container.

Container packing certificate A document involved in the containerised shipment of dangerous cargo by sea transport. It certifies that the container was clean, dry, and fit to receive the cargo; no cargo known to be incompatible has been stowed therein; all packages have been inspected for damage and only dry and sound packages loaded; all packages have been properly stowed and secured and suitable materials used; the container and goods have been properly labelled; and finally a dangerous goods declaration had been received/completed for such dangerous goods consignment packed in the container.

Container Park A place where containers are accommodated. It may be in a port, inland clearance depot, container depot and so on. The containers may be stacked up to four high.

Container part load A consignment which neither occupies the full capacity of a container nor equals the maximum payload and will, therefore, allow the inclusion of another or other part load.

Container platform A container floor without sides or end walls which can be loaded by spreader directly and is generally used for Lo-Lo operations.

Container pool Agreement between various transport carriers and/or leasing companies concerning the exchange of containers.

Container rail cars Dedicated rail wagon involved exclusively in conveyance of ISO containers.

Container, Registered (IATA) A container registered by IATA and published in the IATA Register of Containers and Pallets. Such containers are owned either by an IATA Member Airline or non-Member i.e. shipper, container manufacturer or leasing concern. The IATA registered container, on behalf of a non-Member, had to meet the IATA minimum design criteria and standard dimensions and entitles the shipper or consignee, to the applicable IATA container discount.

Container safety convention International convention governing container safety standards.

Container service charges Charges to be paid by the shipper as per tariff.

Container ship Vessel built to convey containers.

Container size/type Description of the size and type of a freight container or similar unit load device as specified in ISO 6346.

Container slot Allocation of container space on a vessel or container yard.

Container Slot Management Management personnel involved exclusively in the container fleet management and vessel capacity control. This embraces schedules, ship deployment, and allocation of container capacity on specific sailings relative clients/commodities etc. It particularly involves the operation of container vessels or ports of the capacity of third party vessels for the movement of containers.

Container space allocation The process of allocating container space on a particular vessel. For example, it may be undertaken on a port of call basis with specific numbers for loading and unloading based on the TEUs market needs.

Container Stack Two or more containers placed one above the other forming a vertical column.

Container stuffing The process of stowing a container with merchandise. In so doing one must bear in mind types of commodities, including their characteristics weight/dimension/compatibility; and distribution by weight/volume of merchandise throughout the container.

Container sublease The contract by which a carrier gives the use of containers to another carrier for a specified period of time and for fixed payments.

Container Terminal A container terminal is accommodation where (FCL) full and empty containers are received and/or delivered. See Container Yard entry.

Container use The degree of utilisation of a container or fleet of containers throughout a specific period. Usually it is an annual user analysis identifying the period the containers are loaded or empty; transits; unserviceable awaiting repair; or 'laid up' awaiting employment.

Container unstuffing The process of unloading the contents of a container.

Container user analysis The process of analysing over a specified period of maybe one month, three months, etc. the actual use of the container fleet on a particular route/service/trade or throughout the Company/Entity container business. This may involve, average load; average journey length; average revenue per container movement; availability of containers discounting those non operational such as awaiting repair; average demurrage/standage period; average period operational containers available for traffic but not used because no demand; average period containers in operational revenue production use; and finally any financial detail which may be injected into such an analysis as are meaningfully available. It is likely the analysis will segregate container type and any other special feature.

Container utilisation The utilisation of the ISO container especially in regard to the period of its operational availability and the time the container is being used for revenue earning purposes. See also container user analysis entry.

Container Yard A container yard is accommodation where (FCL) and empty containers are received from or delivered to merchants.

Containment system (Radioactive Material Only) The assembly of components of the packaging specified by the designer as intended to retain the radioactive material during transport. A term associated with dangerous cargo shipments.

Contango The situation when the price of a metal for forward or future delivery is greater than the cash or spot price of the metal. Contangos occur when their metal is in plentiful supply. The size of the contango does not normally exceed the cost of financing, insuring and storing the metal over the future delivery period. An International Petroleum Exchange term.

Cont (BH) Continent Bordeaux – Hamburg range.

Contcoal Code name for Charter Party used in coal trade published by BIMCO.

Cont (HH) Continent Havre – Hamburg range.

Continental breakfast Rolls or toast and a beverage

Continental plan (CP) Hotel rate for room and continental breakfast.

Continental trade International trade between Great Britain and European seaboards.

Contingency arrangements Those arrangements available to cope with any disruption to a scheduled service such as diverting the vessel to another port.

Contingency insurance cover This arises when the exporter is seeking special cover rather than rely on the consignee's insurance. Examples arise when a consignee refuses to accept the goods knowing that fire has damaged the carrying vessel. Moreover, consignees have refused acceptance of sight draft when a ship has sunk on the voyage. Contingency cover effected for the benefit of the exporter will protect the exporter in such circumstances.

Contingency planning The process of planning to cope with an unexpected event/situation. It may be industrial disputes, accidents, shortage of raw materials and so on.

Contingent covers Back-up cover which operates if an original policy e.g. public liability policy, fails.

Contingent interest A marine insurance term indicating as the defeasible interest of the seller ceases, the interest of the buyer commences. See Defeasible interest entry.

Continuous billing The process of billing the customer at the time the purchase is made and not aggregating them until the month end or other specified date in each month. Continuous billing aids cash flow but can increase administration costs.

Continuous forms Forms in continuous length ('endless' web) which can be fed automatically through the machine on which they are processed.

Continuous Research An ongoing research which enables the entrepreneur to build up a profile portfolio which can be monitored and thereby develop trends. Continuous research is very essential in a fast moving consumer goods market to determine the buyer's behaviour and to monitor competition.

Continuous survey
Vessel undertaking only portion of her survey annually i.e. 20 per cent or 25 per cent.

Contr. Contractor.

Contra (Latin) Against.

Contraband Import/export of prohibited traffic.

Contract This involves an offer and acceptance. Contracts are normally but not necessarily confirmed in writing. Also, the standardised unit of trading futures and options (also called a 'lot').

Contract for the Sale of Goods A contract for the Sale of Goods is one whereby the seller agrees to transfer the title to goods for a money consideration called 'the price'.

Contract hire The process of hiring a product under contract terms. This may be a road vehicle for six months.

Contracting Party A country which has fulfilled all the necessary requirements to become a member of WTO, and which has therefore accepted to abide by the contractual rights and obligations which WTO embodies. A WTO term.

Contract licence Authority to import/export material and equipment supplied under a large contract.

Contract logistics The contracting out of all the warehousing transport and distribution activities or a part thereof by manufacturing companies.

Contract Maintenance A situation whereby a Company undertakes to do on a contract basis the maintenance of specified equipment, vehicles, etc.

Contract manufacturer An overseas contract with the manufacturer under contractual obligations to manufacture and deliver a product usually in a time scale basis and a precise specification.

Contract of affreightment A contract whereby the shipowner undertakes to carry a specific quantity of cargo on a particular route or routes over a given period employing vessels of his choice with specified restriction.

Contract of suretyship A contract of suretyship establishes an accessory obligation towards a creditor. The accessory is dependent on the existence and substance of the obligation of the principal debtor. If the debt should for any reason be extinguished (e.g. on the basis of a statute of limitations), the contract of suretyship also becomes void. If a claim is made, the surety is obliged to use all such defences against the creditor as are available to the principal debtor. The surety's obligation extends only as far as that of the principal debtor.

Contractor Land operator providing services to tour operators, wholesalers and travel agents; also called ground operator, independent contractor, land operator, receiving agent, reception agency, receptive operator.

Contracts Manager A managerial position within a Company/Entity responsible for negotiation and terms including price of a particular product or service relative to a particular contract.

Contract Officer A position within a Company/Entity responsible for the negotiation and terms including price of a particular product or service relative to a particular contract.

Contract specification The terms of the contract as found, for example, in a charter party. Also the terms of the futures and options contract (e.g. size, trading times, etc.)

Contract system Situation whereby shipper signs a contract to forward all his goods by conference line vessels either in the general course of business or with a specified project over certain period.

Contract term A term in the contract which defines what is included in the price.

Contractual relationship The relationship between two parties as defined within the terms of the contract.

Contractual salvage Agreed payment for efforts made to save properly (whether successful or not).

Contra preferentem (Latin) Against the one who it should favour.

Contribution In non-marine insurance this term refers to a sharing of loss between the assured and the insurer or between insurers. In marine insurance and shipping practice it refers to salvage charges and general average adjustments. The contribution is the amount paid or payable by the insured to a salvage award or a general average fund, or if claimant has more than one policy in respect of a claim, insurer only liable for the proportion of claims to premium paid. See Indemnity entry.

Contribution pricing The process of adopting a pricing strategy whereby the price is set at a level which is above variable cost and makes a contribution to fixed cost. See fixed and variable cost entries.

Contribution pricing analysis The process of examining one or more products retailed by the entrepreneur to determine the performance of each product financially on the basis the product selling price and the yield it produces above variable cost contributes to fixed cost and profit. See contribution pricing.

Contributory value The value on which a contribution to a general average loss or salvage award is assessed

Control devices Devices connected either within the main processor housing (e.g. disc controller, video card, input/output cards, serial and parallel output cards) or connected to the input/output part of the main processing unit via a buffer (e.g. heaters or lights (output), sensors (input)).

Controlled atmosphere Sometimes used in addition to temperature control to prolong the storage life of fruit.

Controlled economy A situation where the complete economy is State run including all the trading activity except for tiny pockets of free enterprise such as the peasant who is permitted to grow a few vegetables and sell them for cash. This sort of economy is generally communist/state controlled and its international trade is carried on by the State Trading enterprises.

Controlled rates Rates of currency exchange which are not allowed to exceed an upper and a lower limit by means of exchange control.

Controlled transit A consignment which is monitored at specified points of its transit.

Controlling hold The hold – usually the largest – which will take the longest time to load or to discharge. This governs the time taken to complete cargo operations and may affect the turn round time of the ship in port.

Control procedure The program created to operate a process control system. The procedure is designed to read input data, process the data and send output signals according to preset rules, e.g. read light level, compare with limit set, adjust output if necessary. A computer term.

Control system A computer system which automatically controls a process or mechanical device by sensing the need to vary the output. Examples of sensors are light, heat, humidity and Ph. A control system is said to have feedback when it is the output of the controlled device which is sensed and fed back to the computer. A computer term.

Control ticket A coupon/ticket issued by or on behalf of a transport operator with the aim to regulate/control/limit the number of passengers per schedule/sailing(s)/flight(s) etc.

Contship Container Ltd A major container operator.

Contwood Code name of charter party used in timber trade published by BIMCO.

Conv. Conveyance or converted (from).

Convection A type of heat transfer which occurs in a fluid by the vertical motion of large volumes of the heated material by differential heating (at the bottom of the atmosphere) thus creating, locally, a less dense, more buoyant fluid.

Convenience foods Commodities which are pre-prepared usually available in the supermarket or hypermarket and capable of being served/consumed with little further preparation and in very limited time other than heating/cooking the product often involving a microwave oven.

Convenient speed Voyage charter party term indicating that vessel must proceed with due diligence in the usual and customary manner.

Conventional cargo Cargo not carried in containers, flats, etc., but stowed in normal packages.

Conventional cargo liners Vessels which convey containers both below and above deck, with self discharging capability using the heavy deck cranes. Additionally, such tonnage ships have the traditional break bulk cargo.

Convention Internationale Concernant le transport de Marchandises par chemin de fer (French) International convention on Carriage of Goods by Rail applied by 19 European railway Administrators setting out the conditions for the international carriage of goods by rail.

Convention on a code of conduct for Liner Conferences See Cargo sharing.

Convention on the contract for the International Carriage of Goods by Road An international convention concerning the carriage of goods by road which came into force in the United Kingdom in 1967 and is embodied in the Carriage of Goods by Road Act 1965. It permits the carriage of goods by road under one consignment note under a common code of conditions applicable to some twenty-six countries primarily in Europe.

Convention on the International Regulations for Preventing Collisions at Sea. (COL REGS 72) This Convention deals with steering and sailing rules, lights, ships, and sound and light signals. Compliance is a condition for the issue of Passenger Ship Safety Certificates and Cargo Ship Safety Equipment Certificates.

Convention relative au contrat de transport internationale des Marchandises par Route (French) International convention on Carriage of Goods by Road.

Convergence criteria Criteria established in Article 109j (l) of the Treaty (and developed further in Protocol No. 6). They relate both to performance with regard to price stability, government financial positions, exchange rates and long-term interest rates, and to the compatibility of national legislation, including the statutes of national central banks, with the Treaty and the Statute of the ESCB. The reports produced under Article 109j (l) by the European Commission and the EMI examined the achievement of a high degree of sustainable convergence by reference to the fulfilment of these criteria by each Member State. An ECB term.

Convergence programmes These are medium-term government plans and assumptions provided by non-participating Member States regarding the development of key economic variables towards the achievement of the medium-term objective of a budgetary position close to balance or in surplus as referred to in the Stability and Growth Pact. Regarding budgetary positions, measures to consolidate fiscal balances as well as underlying economic scenarios are highlighted. Convergence programmes must be updated annually. They are examined by the European Commission and the Economic and Financial Committee. Their reports serve as the basis for an assessment by the ECOFIN Council, focusing, in particular, on whether the medium-term budgetary objective in the programme provides for a sufficient safety margin to ensure the avoidance of an excessive deficit. Countries participating in the euro area must submit annual stability programmes, in accordance with the Stability and Growth Pact. An ECB term.

Converging commonality The customers desire and use of a product/service in the same way/manner worldwide such as Levi jeans and Swatch watches.

Conversion rate The process of measuring the number of enquiries into an actual sales campaign/strategy. Such enquiries may emerge from a particular advertising campaign, cold calls or mail shots.

Converter dolly An auxiliary undercarriage assembly consisting of a chassis, fifth wheel and tow bar used to convert a semi-trailer or a container chassis to a full trailer.

Convertible bonds A long or medium term certificate of debt that can be converted at the holder's option into securities specified in the terms of the bond within a predetermined period (conversion period) at a pre-determined price (conversion priced).

Convertible bond and stocks A bond or security (unsecured loan stock, debenture or preferred share) giving the holder the right to convert into ordinary shares of the issuing company on pre-determined terms within a stipulated period of time. Typically, the conversion option is judged to have a value justifying the payment of a lower rate of interest than would be obtainable on straight fixed interest bonds and securities.

Convertible currency A convertible currency is one that is readily and freely transferable/traded on the international currency exchanges, and can be exchanged for another currency in the present or some future date. Certain currencies are subject to restrictions, e.g. exchange control regulations in the country of the currency, or even preclusion from physically taking the currency notes in or out of the country. Such currencies are not readily convertible currencies. Also, a London market term for any currency other than Sterling, US dollars or Canadian dollars. Within Lloyd's, any currency other than Sterling, US dollars and Canadian dollars.

Convertible notes A long or medium term certificate or debt that can be converted at the holder's option into equities i.e. shares or participation certificates specified in the terms of the notes within a predetermined period (conversion period) at a predetermined price (conversion price).

Conveyance Transport of goods from one place to another.

Conveyor A mechanical device in the form of a continuous belt for transporting cargo.

Conveyor belt Form of automatic unloader for freight.

Convolute (fibre drums) winding A style of straight winding by which a tube is formed by having each ply placed directly over the preceding ply.

Conwartime The code name of a BIMCO approved charter party clause relative to 'war' and entitled 'Baltic Conference War Risks clause for Time Charters 1939'.

Conway The code name of a BIMCO approved charter party clause relative to Canals and entitled 'Baltic Conference Stoppage of Canals and Waterways clause 1968'.

C of O Certificate of Origin.

COO Country of origin.

COOC Change of Ownership clause.

Cooperage The repairing of casks or the services of a cooper in repairing damaged cargo. Also the opening of Cargo for Customs examination and repacking.

Co-operating carrier Scheduled line that pays or helps to pay for brochures or other costs of marketing a tour programme.

Co-operation Council for the Arab States of the Gulf A group of seven Arab States.

Co-operative A trade association formed for the same purposes as a consortium but structured differently, with shares held by agency owners; see consortium.

Co-ordinating Lloyd's Agent See Agent – Lloyd's Underwriting entry.

COP Custom of port.

COPANT Pan American Standards Commission. Founded in 1961. Comprises national standards bodies of USA and 11 Latin American countries. A co-ordinating organisation concerned with the regional implementation of ISO and IEC standards and recommendations.

Co-pilot The second pilot on the flight deck of an aircraft.

Copper chrome arsenate A solution used in immunising timber against wood-boring insects.

COPR Centre for Overseas Pest Research.

COPY The MS-DOS command that creates a copy of an existing file. A computer term.

COPY CON A form of the COPY command that lets you create and save a file by 'copying' it to the 'console' device. You can use this command to create an AUTOEXEC.BAT file, or to create a new CONFIG.SYS file. A computer term.

Coras Iompair Eireann Eire's state owned bus and railway undertaking. The CIE are UIC members.

Cord, Igniter Used to transmit ignition from a device to a charge or primer, this cord of textile yarns is covered with black powder or another fast burning pyrotechnic composition and of a flexible protective covering. It may contain a metal core wire or textile threads to improve the strength. It burns progressively along its length with an external flame. A term associated with the movement of dangerous classified cargo.

Cordoba The currency of Nicaragua. See Appendix 'C'.

Cordoba invoice An invoice payable in Nicaraguan Cordoba relative to an export sales contract.

Core business The central business of a company/entity.

Coreinsurance Retention by a reinsured in his own reinsurance.

Core messaging platform A system through which all messages are digitally signed and sent. The CMP is known as the hub and spoke system since all messaging is routed via it. When using the security encryption system in the CMP, the users employ public and private key cryptography which converts every message transmitted into code. The very use of this system means that the correct code identifier must be transmitted from the user to Bolero's core messaging platform for the platform to acknowledge the sender.

Coreper The comité des Representants Permanents de la CEE (Committee of permanent representatives of the Community) which prepares the work of the Council of Ministers and carries out tasks assigned to by the Council.

Core product This is the main/most important product/service in a business. For example, for an airline it is the passenger and income from duty free sales.

Core Strategy The central strategy adopted by an entrepreneur to develop his/her core business. In marketing terms this may be to develop a brand image which yields premium pricing and thereby sustains an acceptable level of profitability. It is market research driven.

Coriolis force The term used to explain the fact that a moving object detached from the rotating Earth appears to an observer on Earth to be deflected by a force acting at right angles to the direction of motion. Deflection of moving objects in the Northern Hemisphere is to the right of the path of motion. Deflection in the Southern Hemisphere is to the left of the path of motion.

Corner A position where one operator owns all or virtually all the market stocks of a commodity. His object is to control the supply of the metal, and thus cause an exceptional rise in price. The LME has rules to avert the worst effects of a corner. See also Squeeze entry;

Corner casting Hardware located on top and bottom of each container corner post used for handling and securing a container.

Corner fittings Fittings located at the corners of containers providing means of supporting, stacking, handling and securing the container.

Corner guide A fitting that guides the container into a position when being loaded on the vessel or lifting spreader guides.

Corner post A vertical structural member at either side of an end frame of an ISO container joining a top and a bottom corner fitting.

CORPAC Corporacion Peruana de Aeropuertos y Aviacion Comercial.

Corporate advertising The process of building up/formulating a marketing strategy through advertising which will develop a favourable image of the company in the market place and thereby enhances the company's reputation/may be brand image with major competitive advantage in the eyes of the potential buyer/consumer.

Corporate agent See commercial travel agent.

Corporate Bonds A security issued by a corporate borrower as receipt for a loan longer than 12 months, indicating a rate of interest and date of repayment.

Corporate card An individual charge or credit card issued through the traveller's employer and typically carrying the names of both.

Corporate culture The manner in which a company conducts its business.

Corporate Finance The policy adopted on funding the business throughout the Group and in so doing ensure the overall objectives corporately are adopted to produce a coherent financial policy. This would include the day to day financial policy of the corporate entity.

Corporate image The generation and sustainment of a company image in the market place. This is achieved through the quality of service offered to the clientele and the supporting advertising extolling the virtues of the company to the discerning buyer in a competitive environment. See also corporate culture and corporate advertising entries.

Corporate Management The Management structure found within the corporate entity which may tend to be imposed as far as practicable within the overall Group. See corporate policy.

Corporate Marketing The marketing strategy adopted by the company overall. It is particularly relevant to Multi National Industries which outline the broad strategy to be adopted on the markets in which they operate.

Corporate Market Segmentation It embraces six methods/elements: consumer location, consumer size, usage role, industry classification, financial structure and nature of business strategy.

Corporate Mission The mission statement of the Company. See mission statement entry.

Corporate objectives The overall objectives of the Company embracing all the elements of the business. See corporate policy entry.

Corporate planning The policy adopted on planning the business throughout the Group and in so doing ensure the planning objectives corporately are adopted throughout to produce a coherent policy.

Corporate Policy The overall policy of the Company embracing all the elements of the business. For example, the Company – which maybe a holding company – could indicate the credit permitted to all clients must not exceed 21 days from which date interest charges will arise. It also extends to investment, buying, and other corporate policies within a company.

Corporate position In marketing terms the corporate position of the group in the competitive market place.

Corporate rate A reduced price applicable to business travellers, sometimes specially negotiated.

Corporate travel manager An employee who makes travel arrangements for other employees of a company; also called a passenger traffic manager.

Corporate video A video which reflects the corporate matters of the group and not individual companies forming part of the group. See corporate Planning and Management entries.

Corporation of Lloyd's More commonly known as Lloyd's of London; it is an association of insurers based in London specialising in the insurance of marine and similar risks. It is world renowned for its marine insurance business.

Correspondent Bank A bank which acts as the agent of another bank to provide specified services. An account relationship between the two banks involved is often an integral part of the agreement, although not essential.

Correspondent banking An arrangement under which one bank provides payment and other services to another bank. Payments through correspondents are often executed through reciprocal accounts (nostro and loro accounts), to which standing credit lines may be attached. Correspondent banking services are primarily provided across international boundaries but are also known as agency relationships in some domestic contexts. A loro account is the term used by a correspondent to describe an account held on behalf of a foreign bank; the foreign bank would in turn regard this account as its nostro account. An ECB term.

Correspondent central banking mode (CCBM) A model established by the ESCB with the aim of enabling counterparties to use underlying assets in a cross-border context. In the CCBM, national central banks act as custodians for each other. This implies that each national central bank has a securities account in its securities administration for each of the other national central banks (and for the ECB). An ECB term.

Corrosion Deterioration of a metal by natural causes.

Corrugated container A container with corrugated walls and ends which gives added strength.

Cory The currency of Republic of Guinea. See Appendix 'C'.

CoS Chamber of Shipping.

Cos Cosine.

COS Cash on shipment.

COSCO China Ocean Shipping Company.

COSCO-K line Major merchandise maritime trade alliance whose members are COSCO, 'K' Line and Yangming Marine.

COST European Co-operation in Science & Technology.

Cost analysis The process of analysing a particular investment or general business situation on a strictly cost environment basis with a view to determining the cost elements involved and the implications which flow from them. This may be labour, distribution, materials, publicity, design and so on.

Cost and Freight (named port of destination) – CFR Under this term the seller must pay the costs necessary to bring the goods to the named port of destination; the risk of loss or damage to the goods, as well as of any additional expenses, is transferred from the seller to the buyer when the goods pass over the ship's rail. This is identical to CIF except that the buyer is responsible for funding and arranging the cargo insurance.

The seller's obligations in CFR include: to supply the goods in accordance with the contract of sale terms; to arrange and pay for the conveyance of the goods to the specified port of destination by the customary route and fund any unloading charges at the destination port; to provide and pay for any export licence or other governmental authorisation necessary to export the cargo; to arrange and pay for, on specified date or period, the cargo loading at the agreed port (if no such loading date or period is quoted, such a task to be undertaken within a reasonable period), to bear all the cargo risk until such time as it passes over the ship's rail at the port of shipment; to supply promptly and pay for the clean shipped negotiable Bill of Lading for the agreed destination port, together with the commercial invoice of the goods shipped or its equivalent electronic message; to provide and pay for the customary packing of the goods unless it is the custom of the trade to ship the cargo unpacked; to fund any cargo scrutiny prior to loading the cargo; to pay any cost of dues and taxes incurred relative to the process of exportation in respect of the cargo prior to shipment; to provide the buyer on request and at the buyer's expense, a Certificate of Origin and consular invoice; to render the buyer on request and at the buyer's expense and risk every assistance to obtain any documents required in the country of shipment or transit countries necessary for conveyance of the cargo to its destination.

A point for the seller to note especially with the CFR term is the need to supply a full set of clean on board or shipped Bills of Lading. If the Bill of Lading contains a reference to the Charter Party, the seller must also provide a copy of the latter. The buyer must ensure these provisions are complied with by the seller.

The factors relevant to the buyer include: acceptance of the documents as tendered by the seller (subject to their conformity with the terms of the contract) and payment of the goods etc., as specified in the contract of sale; to receive the goods at the port of destination and, with the exception of the sea-freight, all the costs and charges incurred during the voyage(s); to fund all unloading expenses at destination port including lighterage, wharfage, etc. unless such costs have been included in the freight or collected by the shipowner at the time freight was paid; to fund any pre-shipment cargo inspection arrangements; to undertake all the risk when the cargo has passed the ship's rail at the departure port; in the event of the buyer failing to give instructions (by the specified date or within agreed period) relative to destination port, all additional costs and risks will be borne by the buyer subject to the goods being duly appropriated to the contract; to pay all the costs to obtain the Certificate of Origin and consular documents; to meet all charges to provide any other documentation specified relative to processing the consignment in the country of shipment or transit countries; to pay all customs duties and other taxes raised at the time of importation; to obtain and pay for any import licence or related documentation required at the time of importation.

In the CFR contract sale the buyer can arrange the cargo insurance in his own country thereby saving foreign currency. Many importers are tending to favour this arrangement. This term should only be used for sea and inland water transport. When the ship's rail serves no practical purpose as in the case of roll-on/roll-off or container traffic, the CPT term should be used. An Incoterm 2000.

Cost and Insurance A cargo delivery term whereby the seller (exporter) pays the cost and insurance whilst the importer (buyer) funds the freight necessary to bring the goods to the named destination.

Costa Rican Colon The currency of Costa Rica. See also Appendix 'C'.

Costa Rican Colon invoice An invoice payable in Costa Rican Colons currency relative to an export sales contract.

Cost benefit analysis The measurement of resources used in an activity and their comparison with the value of the benefit to be derived from the activity.

Cost centre A location, function or items of equipment in respect of which cost may be ascertained and related to cost units for control purposes.

Cost effective The process of determining the general efficiency of a particular service/facility/process etc. in cost or 'value for money' terms. In so doing one must have regard to competition, system and the options presented to improve the system to enable it to become more efficient. An example may be, a company adopting a logistics strategy would produce cost savings in production/distribution of thirty per cent.

Cost escalation cover Insurance cover provision against escalation in processing cost in contracts of, for example, five million euros or more where the manufacturing operation takes at least two years. A facility available in insurance policies.

Cost Insurance Freight (CIF) – named port of destination

Cost Insurance Freight (CIF) – named port of destination Under this cargo delivery term the exporter (seller) undertakes all the risks and costs to deliver the goods to the named destination port embracing both the freight and insurance. The buyer (importer) accepts the goods at the named destination port when they have passed over the ship's rail and accepts all the costs from this point onwards including unloading expenses, any wharfage, customs, documentation, and onward freight costs. Packing costs are borne by the exporter (seller). Insurance is undertaken by the exporter (seller) who bears all the costs but the insurance is transferred to the importer (buyer) at the time of shipment when Bill of Lading is tendered to the importer (buyer). The importer (buyer) pays unloading costs which are not included in the freight. The importer (buyer) must give notice to the shipowner of any resultant loss and bears all the risks of the goods from the time when they have passed the ship's rail at the port of shipment. Under this term the exporter (seller) becomes involved in the maximum distribution arrangements and income therefrom. Where the exporter (seller) and the importer (buyer) have agreed to communicate electronically, the documents detailed in the following paragraphs maybe replaced by an electronic data interchange message.

The salient features of the contract as far as the seller is concerned include: to supply the goods in accord with contract of the sale terms: to arrange and pay for the carriage of the goods to the specified destination port by the customary route and to fund any unloading charges at the destination port; to provide and pay for any export licence or other governmental authorisation necessary to export the cargo; to arrange and pay for, on specified date or within an agreed period, the cargo loading at the agreed port (if no such loading date or period is quoted such a task to be undertaken within a reasonable period); to inform the buyer promptly when loading is completed; to arrange and pay for insurance of the cargo in a transferable form against the risk of loss or damage during the transit – such cover shall be Institute Cargo Clauses and embrace the CIF price plus 10% and the insurance will be in the currency of the contract; to bear all the risk of goods until they have effectively passed ship's rail at the port of shipment; to supply to the buyer promptly, at the seller's expense, a clean negotiable 'shipped on Board' Bill of Lading for the agreed port of destination and insurance policy or certificate of insurance; to provide and pay for the customary packing of the goods unless it is the custom of the trade to ship the goods unpacked; to pay the cost of any cargo scrutiny prior to loading the cargo; to bear any cost of dues and taxes incurred relative to the process of exportation in respect of the cargo prior to shipment; to provide the buyer on request and at the buyer's expense a Certificate of Origin and consular invoice; to render the buyer on request and at the buyer's expense every assistance to obtain any documents required in the country of shipment or transit countries necessary for the conveyance of the cargo throughout the transit.

The responsibilities of the buyer include: acceptance of the documents as tendered by the seller (subject to their conformity with the terms of the contract) and payment of the goods, as specified in the contract of sale; to fund any pre-shipment cargo inspection arrangements; to receive the goods at the agreed destination port and to bear, with the exception of the freight and marine insurance, all costs and charges incurred during the voyage(s); to fund all unloading expenses at the port of destination, including lighterage and wharfage, unless such costs have been included in the freight or collected by the shipowner at the time freight was paid; to undertake all the risks when the cargo has passed the ship's rail at the departure port; in the event of the buyer failing to give instructions, by the specified date or within an agreed period, relative to the port of destination, all additional costs and risks will be borne by the buyer subject to the goods being duly appropriated to the contract; to pay all the costs in obtaining a Certificate of Origin and consular documents; to meet all charges to provide any other documentation specified relative to processing the consignment in the country of shipment or transit countries; to pay all customs duties and other taxes raised at the time of importation; and obtain at his expense any import licence or related documentation required at time of importation. This term can be used only for sea and inland waterway transport. If the seller/buyer do not intend to deliver the goods across the ship's rail, the CIP term should be used. An Incoterm, 2000.

Cost insurance freight landed A cargo delivery term identical to CIF except the cost of unloading including lighterage and wharfage are borne at the destination port by the seller (exporter) and not the buyer (importer).

Cost of funds The term sometimes used as the basis for a loan pricing, particularly when the source of funding is uncertain or includes reserve assets costs. A precise definition of what is meant by this term should be established if it is to be of any practical value; it should be noted that the normal funding cost of a commercial loan is the offered rate, being the rate which the bank has to pay to another bank in the market for the funds obtained for the purpose.

Cost of sales The cost of sales is determined by adding the beginning inventory to the net cost of goods purchased during the period and subtracting the cost of ending inventory. This cost usually embraces the cost of materials, labour and the indirect costs of the products sold.

Cost orientated pricing The process of adopting a strategy in a Company/entity which focuses on the product cost in the formulation of the pricing structure in the market place.

Cost per thousand The cost of reaching a thousand potential buyers by means of a particular advertising method.

Cost per vehicle mile The total cost of operating a vehicle mile. This will include fuel, crew cost, general maintenance, road tax, depreciation, insurance, administration, advertising and so on. It will include both direct and indirect cost.

COSTPRO Canadian organisation/committee for the simplification of Trade Procedures based on Ottawa.

COTIF Convention Concerning International Carriage by Rail of goods/passengers 1980 (CIM/CIV).

COT Customers own transport. Customer collects from/delivers to CFS/CY.

Cot Cotangent.

COU Clip on unit associated with the portable refrigeration units; or central operating unit – an organisation/body set up to co-ordinate consortium operations in a trade such as a container ship consortia.

Couchette Train passenger accommodation which is so designed to enable passengers to be seated on day time services and converted to sleeping accommodation on overnight services.

Council for Mutual Economic Assistance Founded in 1949 to promote co-ordination between the USSR and

East European National standards bodies and to assist the production of common or harmonised standards. Membership includes Bulgaria, Czechoslovakia, German Democratic Republic, Hungary, Mongolia, Poland, Romania, USSR, Cuba and Vietnam. It was abandoned in 1990.

Council of Arab Economic Unity A group of 20 Arab countries committed to economic unity.

Council of Europe Based in Strasbourg and a parliamentary forum for discussion by all West European Nations.

Council of Lloyd's The ruling body of Lloyd's formed under the Lloyd's Act 1982 and consisting of 16 elected working Names, 8 elected external Names, and 3 nominated members not being Names, annual sub-scribers of associate members of Lloyd's. The council has powers to delegate, to make additional by-laws, and to carry out disciplinary measures over members: Lloyd's brokers, underwriting agents and employees thereof.

Council of Ministers of EEC Twenty five Ministers of member governments, the actual Minister present depending on the subject in question. It decides on final policy on the basis of proposals from the Commission. Presidency held for 6 months in turn by each member state. Assisted by a Committee of Permanent Representatives (COREPER) and a general secretariat of 3,000 staff. The Commission of the European Union comprises 25 Commissioners responsible for initiating and executing community policy adopted by the Council of Ministers. Answerable to the European Parliament. Acts as mediator between member governments and as guardian of the Treaties. 9,000 administrative staff located in Brussels and Luxembourg. The President and Vice-Presidents are appointed from the Commissioners for renewable four-year terms. See European Union entry.

Counter agent Ticket agent for a carrier.

Counter credit This is virtually the same as the Back-to-Back Credit. The only difference is that the merchant's bank is involved with both Credits, i.e. as the Advising/Confirming/Paying Bank in respect of the first Credit and as the Issuing Bank of the second Credit.

Counterfeit Unauthorised representation of a registered trademark carried on goods identical or similar to goods for which the trademark is registered, with a view to deceiving the purchaser into believing that he is buying the original goods. A WTO term.

Counterfeiting The manufacture of imitation products usually sold under a name intended to deceive consumers into believing they are the genuine article.

Counterpart The opposite party in a foreign exchange deal is called counterpart, i.e. a dealer selling three months forward will look to find a counterpart who wishes to buy three months forward.

Counterparty The opposite party in a financial transaction (e.g. any transaction with the central bank). An ECB term.

Counter purchase A counter-trade technique whereby concurrently with and as a condition of, securing a sales order, the exporter undertakes to purchase goods and services from the country concerned. There are two parallel but separate contracts – one for the principal order, which is paid for on normal cash or credit terms and another for the counter purchase. The value of the counter purchase undertakings vary in value between ten per cent and 100 per cent (or even more) of the original export order. This is the most common form of counter-trade.

Counter-trade A form of foreign trade – loosely known as 'Barter' – in which the sale of goods or services to a country is linked contractually to an obligation to buy goods or services from that country It is practised in 100 countries. It is particularly common in third world countries and Eastern European territories. Various types of counter-trade exist including counter purchase, barter, buy-back, offset, switch trading and evidence accounts. See also separate entries.

Countervailing duties Duties imposed on imports to counter the effects of export subsidies used by an exporting country.

Countervailing measures Action taken by the importing country, usually in the form of increased duties to offset subsidies given to producers or exporters in the exporting country. A WTO term.

COUNTRY Country Line – a shipowner.

Country analysis See key financial variables.

Country damage Damage to baled or bagged goods such as cotton caused by excessive moisture from damp ground or exposure to weather, by grit, dust or sand forced into the insured property by wind-storm or inclement weather.

Country of departure The country from which a transport mode is scheduled to depart or has departed.

Country of despatch Country from which goods are shipped.

Country of provenance The country from which goods or cargo are sent to the importing country.

Country report A report produced on a particular country's economy and market prospects and especially useful for potential exporters for business opportunities. Such reports are regularly issued on a per country basis by International Banks and features also any trade risk involved.

Country risk analysis See key financial variables.

Country specific advantage A country which has a specific advantage in terms of its profile which maybe economic, technical, social, etc. in the judgement of the buyer.

Coupon In financial terms on Bearer Securities a detachable part of the certificate exchangeable for dividends; also the part of a bond giving the holder the right to be paid predetermined interest on a bond (sometimes also used to mean the fixed rate on a swap); the coupon may be physically cut off the bond and sent to the stated paying agent when due for payment; also commercially a document/ticket which may have a redeemable value in a special product promotion such as soap, washing powder etc.; or part of combined ticket which embraces for example a ferry journey.

Coupon, passenger Portion of airline ticket retained by passenger; should be one coupon for each air segment of a trip.

Coupon, tour See vouchers.

Courier A person who accompanies a group on a passenger journey or accompanies/carries important letters/small packages on a transit which may be an international journey.

Courier service An air transport service conveying document envelope despatches and non dutiable very light items. The service operates from the airport of departure to the consignees premises. Couriers accompany on board the consignments.

Couris The currency of Republic of Guinea. See Appendix C.

Coût, Assurance, fret (port de destination convenu) French Cargo delivery term. Cost and Freight (named port of destination).

Cover The balancing of an open position by buying (if the original position was short) or selling (if the original position was long) on the market.

Coverage In marketing terms the extent to which an advertisement, press release or other form of media reaches the market. It is usually measured in market share terms.

Covered bear A seller of securities he/she owns who hopes to repurchase later at a profit and who hopes to save repurchase expenses by not having to deliver.

Covered gangway A piece of port equipment linking the ship with the shore (quay) which is completely enclosed thereby enabling passengers/crew personnel to embark or disembark under cover.

Covered dry container A completely enclosed ISO container having end doors for easy access with floors fully stressed for mechanical aids, i.e. palletisation, flush panelling throughout, alloy interior skirting boards, a form of ventilation and flush mounted bars for interior roping.

Cover Note Confirmation of cover issued for a limited period pending the issue of a policy and where applicable a certificate of insurance. An interim document evidencing the grant of insurance cover.

Covia Airline-owned company that owned the Apollo system before it merged with Galileo to form Galileo International.

Cov pak See pal box entry.

COW Crude oil washing; or Collection on wheels.

C/P Charter Party; or custom of the port.

CP Canada International Airline. An International Airline of Canada and member of IATA; continental plan; or Chareon Pokaphan (of Thailand).

CPA Critical Path Analysis; or Closest Point of Approach. The term is used extensively when radar plotting.

C/P blading Charter Party bill of lading.

CPC Customs Procedure Code. A code identifying the customs and/or excise regimes to which goods are being entered and from which they have been removed (if applicable).

cpd Charterers pay dues.

CPI Consumer Price Index.

CPO Crime Prevention Officer.

CPP Controllable Pitch Propeller; or Clean Petroleum Products.

CP Ships holding Inc A major liner and container shipowner and registered in Canada.

CPT Cost paid to; Cost per thousand; an example arises in the supply of industrialised componentised products; or carriage paid to.

CPU Central processing unit.

CQD Customary quick despatch.

(cr) Centre.

CR Carriers risk; current rate; crane corrosion resistant material; cargo ramps; cash reserves; or credit.

Cr Credit; creditor; or crane.

Crack Any fracture which has not parted.

Crack spread The buying of crude oil whilst taking the reverse position in refinery products. Any combination of energy futures can be used provided that the number of crude contracts equals the number of product contracts – An International Petroleum Exchange term.

Crane, Fitting-out A crane located and especially arranged for shipyard use to place equipment in a ship after it is in the water.

Crane, fixed A crane whose principal structure is mounted on permanent or semi-permanent foundations.

Crane, floating A crane mounted on a barge or pontoon which can be towed or self propelled from place to place.

Crane, Gantry A crane or hoisting machine mounted on a frame or structure spanning an intervening space.

Crane jib type Fork Lift truck A fork Lift truck equipped to lift cargoes.

Crane, Luffing A crane in which the load may be moved radically, or to or from the centre of the crane, horizontally.

Crane, portal A type of gantry crane with vertical legs with sufficient height and width to permit vehicles or rail road equipment to pass between the legs.

Crane, semi-portal A type of gantry crane with one support on the pier or wharf, and the other on the shed wall.

Crate A wooden or metal receptacle of open construction (or framework).

CRB Raised quarter deck/bridge (of a ship).

Cream off the traffic The process of a carrier taking/abstracting the traffic which attracts the highest level of rate.

Creating an Agency The process of being invited to start an entity/business as an Agent. Overall, there are three methods of having an Agency created: express agreement; implied agency and agency by ratification,. See separate entries.

Creative Director A position at senior managerial level found in an advertising agency responsible for innovation and creative ideas/material in the formulation of the advertising material and strategy.

Creative marketeer A marketeer with a creative/innovative mind/profile generating ideas and a creative/perceptive business product/service profile.

Credit The business of making and pricing loans, and of assessing the ability of the borrower, whether company or country, to repay them. An international banking definition.

Credit Account Voucher A system whereby an accredited client is afforded credit on specific terms involving the issue of a voucher when making a purchase.

Credit control The process of ensuring all clients pay their accounts within the prescribed terms/period. See credit controller entry.

Credit controller A person responsible to ensure all clients pay their accounts promptly within the terms of credit granted and thereby minimise the risk of bad debts arising.

Credit d'enlevement (French) Simple deferment. See separate entry.

Credit des droits et taxes/obligations Cautionnees (French) Extended deferments. See separate entry.

Credit information Data available on the standing and integrity of the buyer (importer) before entering into a business relationship.

Credit institution Refers to an institution covered by the definition contained in Article 1 of the First Banking Co-ordination Directive (77/780/EEC), i.e. 'an undertaking whose business is to receive deposits or other repayable funds from the public and to grant credit for its own account'. An ECB term.

Credit Insurance See Export Credit Insurance entry.

Credit insured debt An insurance cover providing indemnity on sales on credit.

Creditor A person who lends money and who has not yet been paid.

Credit rating The rating assignment to the company in terms of its general creditworthiness and having regard to bad debts, credit control techniques, trading performance, gearing, cash flow, order book, etc. See Status enquiry and credit rating company.

Credit Rating Company A credit rating company such as Standard and Poors which provides a creditworthiness of an obligor with respect to a particular debt security or other financial obligation in the case of an issue rating. The ratings are default risk indicators with recovery after insolvency as a secondary consideration. The long-term rating scale is AAA to D with pluses and minuses in each grade from AA to CCC. The evaluation in the maritime industry includes scale of company – diversification, fleet profile; contracts with customers; management skills; operation-safety record; financial policy; accounting; earnings protection; cash flow; capitalisation-debt leverage and financial flexibility.

Credit status report The process of a seller taking an audit on a potential client/company to determine the credit worthiness. Also see status enquiry entry.

Credit terms A variety of forms of credit granted to a client. It may be payment with 28 days; a 2½% discount on payment within 14 days; a credit note for purchases above a certain level and so on.

Credit tranche policies An IMF term. Credits under regular facilities are made available to IMF members in tranches (segments) of 25 per cent of quota. For first credit tranche drawings, members are required to demonstrate reasonable efforts to overcome their balance of payments difficulties and no phasing applies. See reserve tranche policies entry.

Credit voucher A document prepared by the seller to enable the purchaser to reduce the account/bill by the amount detailed on the voucher. See Credit Account Voucher entry.

CREST Committee on Scientific & Technical Research.

Crew The Master, officers and ratings on a vessel.

Crew Agency An agent which recruits crew members and may undertake their management in accordance with a shipowners instructions. See Crewman 'A' and 'B' entries.

Crew Agreement An agreement between a (UK) shipowner and seaman outlining the conditions of employment.

Crew insurance Insurances against which include but not be limited to death, sickness, repatriation, injury, shipwreck, unemployment, indemnity and loss of personal effects.

Crew list Details of crew members/personnel on a particular vessel.

Crewman The BIMCO approved standard crew management agreement introduced in 1995.

Crewman A – cost plus fee A BIMCO standard crew management agreement

Crew Manager The Company responsible for the provision of suitably qualified crew for a particular vessel. An example arises in the BIMCO Crewman A standard crew agreement.

Crewman B – lump sum A BIMCO standard crew management agreement.

Crew manifest Inventory of crews' baggage.

Crew Manning The level of manning for a particular transport unit. For example, on an inter city passenger train it may be the driver and guard and on board ship six officers and ten seamen.

Crew member A person assigned by a carrier to serve on a ship, aircraft, barge or truck and listed as such.

Crew repatriation Process Process of crew members returning to their own country of origin.

Crew subsidisation The policy of some Governments to subsidise their ships crew cost.

CRF Clean report of findings. See separate entry.

CRISTAL Contract regarding an interim settlement to tanker liability for oil pollution damage.

Critical path analysis The analysis of a large project to identify tasks which can be undertaken in parallel and which of such parallel tasks are critical to the project time-scale. A method of recording timed events within complex projects which depend on multiple processes. Individual activities and their relationship to other activities are displayed, enabling critical items in the schedule to be identified. Also known as programme evaluation and review technique (PERT).

Critical Path Method A network planning system used for planning and controlling the activities of a project by showing each of these activities and their associated times, the critical path can be determined. The critical path is the series of successive activities which take up most time and is therefore decisive for the total lead time of the project.

Critical Pressure The pressure at which a substance exists in the liquid state at its critical temperature. In other words it is the saturation pressure at the critical temperature.

Critical Temperature The temperature above which a gas cannot be liquefied by pressure alone. Also, the temperature at which the properties of a fibre begin to deteriorate.

CRL Current Rate Load – chartering term. See current rate.

CRM Crew resource management (IATA term).

CRN Customs Registered Number. A number given to a trader operating LCP or SDP.

CR No Central Registration Number.

CRNS Cranes.

CRO Central Research Organisation (Burma); Companies Registration Office (HMG); or Current Rate Discharge – chartering term. See current rate.

Croatian Register of Shipping The Croatian Ship Classification Society.

Cronos Group A major container leasing company.

C Ro-Ro DK Carrying capacity in numbers and length(s) of laden containers on internal decks accessed by doors and ramps on a container vessel.

Cross When a floor member or dealer matches buying and selling orders, and 'Crosses' them on the market floor. A BIFFEX term.

Cross Border cheque The process of a cheque effecting payment across an international border.

Cross Border Euro payments The process of conducting Euro payments across international borders.

Cross border factoring The process of undertaking factoring from one country to another thereby crossing international boundaries.

Cross border finance The process of transferring funds/money from one country to another thereby crossing international boundaries.

Cross border investments The process of transferring capital from one country to another thereby crossing international boundaries.

Cross border leasing Leasing directly to a lessee in a foreign country.

Cross border marketing The process of marketing a product/services which cross international boundaries.

Cross Border Portfolio flows See Cross trading entry.

Cross border shopping The process of travelling from one country to another thereby crossing international boundaries for the purpose of shopping.

Cross border supply The process of services being supplied from one county to another crossing international boundaries. WTO term.

Cross border trading The process of conducting international trade which cross international boundaries.

Cross Border Transaction Raising an invoice or its electronic equivalent covering the trade between VAT Registered traders in different member states.

Cross Country investments The process of transferring capital from one country to another thereby crossing international boundaries and effecting an investment in the destination country. It may be a multi national based in UK deciding to invest in South Korea involving industrial plant.

Cross cultural analysis A critical overview of a range of culture and how each differs one from one.

Cross cultural business Conducting cultural across cultures. See cross border marketing entry.

Cross currency account An account which services more than one currency. See cross currency exposure.

Cross currency exposure A measure of the ability of an international company to service the debts it has incurred in any one currency with the revenue it has received in the same currency.

Cross docking The process of scheduling of road haulage vehicles/trucks into a cross docking facility whereby the cargo is unloaded from several trucks and then immediately reloaded into one container for delivery to a final destination. The cargo is never actually inventorised just transferred using the warehouse's load bays. It never incurs storage fees or other traditional costs associated with holding inventory. See merge in transit entry. A logistics shipping term.

Cross-dock operation Sometimes also referred to as stockless warehousing. In this type of operation, goods are moved to a warehouse where they may be pre-palletised, pre-configured or pre-packed and then assembled, usually across the loading docks into consolidated outbound loads. Here, the warehouse is used as a transhipment facility rather than for storage in the traditional sense.

Cross guarantees Guarantees given by other companies in a group in respect of a loan received by one or more companies within that group. Such cross guarantees may be of particular value to a bank lending to a one-ship company which is part of a large group.

Cross holdings between companies The practice is found especially in Japan when companies tend to hold each others shares.

Cross International frontiers The process of merchandise/consignments/passengers crossing/travelling over the border/boundary between two countries. This involves adherence to immigration and customs regulations.

Cross licensing A technique of transfer of technology whereby both companies/parties exchange technology data.

Cross members Transverse components attached to the bottom side rail and supporting the floor of an ISO container.

Cross promotion The process of conducting promotion of a product/service which crosses international boundaries.

Cross rate The rate of exchange arrived at by expressing the quotations for any two currencies in terms of a third.

Cross/self trade The simultaneous purchase and sale by the same broker in the pit for equal amounts of the same month of a contract at a single price. An IPE term.

Cross subsidisation The process of one profit centre/service/route subsidising another such as profitable route 'A' subsidising loss maker route 'B'.

Cross-the-market coverage A market analysis based on a broad market assessment embracing its complete range.

Cross trade A shipping service operated between two seaboard countries by a Shipping Company whose fleet does not fly either of the National flags and therefore is a foreign operator or third carrier.

Crowley Maritime Corporation A major liner operator.

Crown agents A Government owned organisation acting as a buyer in the UK for overseas governments and large companies.

CRPA Claims Related Premium Allocation (Mutualisation of Captive).

CRS Continuous survey cycle – refrigerated machinery (of a ship); Croatian Register of Shipping – Hiyatski Registar Brodova – The Croatia Ship Classification Society and based in Split Croatia; Computer Reservation system; or Coast Radio Station.

CRS rules or codes of conduct Regulations imposed on operators of agency CRSs in the US, Canada and most European countries, includes prohibition against biased displays.

CRT Cathode ray tube. See separate entry; or Croatia.

CR (Taiwan) China Corporation Register (Taiwan) – Chinese Ship Classification Society.

Crude basis The simple difference between the cash price of a commodity quoted in futures equivalent terms and the prevailing futures market price – an International Petroleum Exchange term.

Crude death rate The number of deaths in a year per 1000 population. Also affected by the population's age structure.

Crude oil carrier An oil tanker designed to convey crude oil.

Crude Oil Washing System using crude oil from cargo for tank cleaning.

Cruise department A department in a passenger cruise line company or tour operators/agency handling all aspects of the cruise business relative to processing passenger enquiries/reservations and related needs.

Cruise Director A senior managerial position usually at Company Board level in a Shipping Company or Tour operator/Travel Agency responsible for strategy and development of the cruise business. It may also be a position on a ship – cruise liner – and as a Ships Officer responsible for the management and operation of all aspects of the cruise especially passenger welfare/entertainment and on board passenger facilities resources.

Cruise fleet A dedicated fleet of vessels entirely engaged in the cruise market.

Cruise Line A shipowner operating in the cruise market.

Cruise Liner A passenger ship purpose built to operate in the cruise market having extensive passengers facilities aboard.

Cruise Lines International Association (CLIA) Trade association with promotional and training programs.

Cruise Market Potential clients interested to take a cruise.

Cruise operator See cruise line entry.

Cruise tonnage Passenger liners engaged in the cruise market operation.

Cruise voy The code name of a BIMCO approved voyage charter party between shipowners and tour operators, travel organisations or for corporate or convention purposes. Owners provide and operate the vessel while charters are responsible for marketing, publicity and passenger arrangements. It was introduced in 1999.

Crusader Swire Container Service An Overseas Container Ltd subsidiary company trading between New Zealand and Japan.

Cruzeiro The currency of Brazil. See also Appendix 'C'.

Cruzeiro invoice An invoice payable in Cruzeiro currency of Brazil relative to an export sales invoice.

Cryogenic A freezing mixture used to administer refrigeration.

Cryogenic Liquids Low temperature liquefied gases, such as air, argon, helium, neon and nitrogen. A term associated with the movement of dangerous classified cargo.

Cryosphere The portion of the climate system consisting of the world's ice masses, sea ice, glaciers, and snow deposits. Snow cover on land is largely seasonal and related to atmospheric circulation. Glaciers and ice sheets are tied to global water cycles and variations of sea level. The ice sheets of Greenland and the Antarctic contain 80 per cent of the existing fresh water on the globe.

Cryptography This involves digital signatures operative on digital networks. Security through cryptography is a key area as authenticity and confidentiality of information becomes a greater problem. Cryptographic techniques such as digital signatures and message encryption solve these problems. They ensure to an even greater degree, than comparable paper techniques, that a message has indeed been authenticated by its signatory, has not been altered since then and is not readable by parties not privy to it.

c/s Cases.

CS Continental Micronesia, Inc. – An IATA member airline; Continuous survey; cotton seed; curtain sided trailer; or Czech and Slovak republics – Custom's abbreviation.

CSA China Shipowners' Association or Canada Shipping Act.

CSAV A major container shipowner and registered in Chile

CSC Container safety convention – of ISO containers; Canadian Shippers Council; or cargo service conference – IATA term.

CSC Line Ltd A shipowner.

csd Closed shelter deck – type of vessel.

CSD Central Supplies Dept. (Singapore).

CSFP Customs Simplified Freight Procedures.

CSFR Czech and Slovak Federal Republics.

CSG Consultative Shipping Group – A Government Committee of Western European countries involving EEC, Scandinavia and Japan to deal with Shipping questions.

CSH Cash; Continuous Survey Hull; or Cargo ship.

CSHBB Cast steel hub non ferrous blades propeller (of a ship).

CSIC China Shipbuilding Industry Corporation.

CSM Container slot Management; or Continuous Survey Machinery.

CSP Commencement of Search Pattern.

CSPC Cargo systems and procedures committee – IATA term.

CSPS Community System Providers. See separate entry.

CSS Co-ordinator Surface search; or Computer Customer Service System.

CSSC China State Shipbuilding Corporation.

CSS (Code) IMO Code of Safe Practice for Cargo Stowage and Securing.

CST Container Service Tariff.

CSTC China shipbuilding Trading Co Ltd.

CS-with date Continuous survey of the (ships) hull quoting date.

CT Cubic tonnage of a ship; cargo tank; combined transport – a transporter system where goods are despatched to destination on a container or trailer which may undergo several modes of carriage whilst in transit i.e. road/sea/rail against one contract; counter-trade; conference terms; container terminal; or chemical tanker.

CTC Corn Trade clauses; cargo tank centre; Customs Transaction code – code letter identifying the type of journey (e.g. European Union); or Certified Travel Counsellor.

CTCS Commercial Tariff Concessions System.

CTD Combined Transport Document; The WTO Committee on Trade and Development.

CTETF Cargo Technical Evaluation Task Force – IATA term.

CTG Council for Trade in Goods – overseas WTO agreements on goods, including the ATC. A WTO term.

CTL Constructive total loss – insured interest must be considered a total loss when the cost of retrieving and/or repair would exceed recovered/repaired value.

CTLO Constructive total loss only.

CTO Combined transport operator (usually involving land and sea movements) A Carrier who contracts as a principal to perform a CT operation; container terminal operator; city ticket office; or Caribbean Tourism Organisation.

C-TPAT Customs – Trade Partnership Against terrorism. See separate entry.

CTPC Cargo Traffic Procedures Committee – IATA term.

Ctr Cutter.

CTUs Cargo Transport Units.

CTVM Centre for Tropical Veterinary Medicine (ODA etc.).

CTW Cargo tank wing (of a ship).

CTX Cargo tank common (of a ship).

CU Empresa Consolidada Cubana de Aviación (CUBANA An International Airline and member of IATA.

Cubana Airways An international Airline and member of IATA.

Cuban Peso The currency of Cuba. See Appendix 'C'.

Cuban Peso invoice An invoice payable in Cuban Peso currency relative to an export sales contract.

Cube cutting The practice whereby a freight forwarder obtains money by deception from the shipper and causes loss to the shipowner by giving to the shipping company reduced measurements of cargoes on which freight is charged by volume. When claiming reimbursement from the shipper, the freight is charged on the correct cubic measurement, thereby the freight forwarder obtains more money from the shipper than he actually paid to the shipping company.

Cube out This occurs when the volumetric capacity of a container has been reached in advanced of the permitted weight limit.

cu ft Cubic foot (feet).

cu in Cubic inch(es).

CCKCC Canada-United Kingdom Chamber of Commerce.

Cultural dynamics The ingredients in a society which are driving culture change/development. See culture briefing and cultural goods.

Cultural goods The social, economic and artistic resources in a country embodied in the institutions and other hybrid factions such as a museum,. Overall, the soft infrastructure embracing the judiciary, government, arts, literature, theatre, films, sport, life style, attitudes and so on. Basically a country's culture is all its national elements.

Cultural Specificity Clause Refers to concerns voiced by some participants to allow special treatment under the GATS for audio-visual services (e.g. films, television programmes) because of their 'unique character'. A WTO term.

Culture Culture is defined as the configuration of learned behaviour and results of behaviour whose component elements are shared and transmitted by members of a particular society. It embraces demographics, religion, language, education, values and attitude, aesthetic values, wealth, etc. Overall, culture features in all areas of international trade embracing both service and industrial sectors.

Culture analysis The process of conducting an analysis in culture terms of a particular country/region/company. See culture briefing entry.

Culture briefing The process of briefing a person/group of people in the culture of a particular country/religion/company. This will include language, protocol, religion, dress, entertainment, travel, currency, gifts, sales negotiation and so on. Culture briefings are essential to sales persons or buyers visiting an overseas territory for the first time.

Culture environment All the ingredients of culture found in a particular country/region/area such as religion, language, protocol, social/economic factors, attitudes, education and so on.

Culture environment of international business The environment in which international business is conducted which embraces language, religion, values and attitudes, aesthetics, education, law and politics, technology and material culture and social organisation.

Cum-dividend This is the usual procedure when stocks and shares are sold. The buyer has the right to the next dividend. 'Cum' is the Latin word for 'with'.

Cum (Latin) With or cumulative.

Cumulative preference share A share whose dividend is fixed and guaranteed; if the company cannot pay in any year, the dividend is rolled over for payment the next year.

Currency A medium of exchange representing a monetary value and associated with a particular country or trading/economic bloc.

Currency accounts If an importer has a continual to and fro flow of international business, it may be preferable to open 'accounts' in the currencies he requires. The various balances can then be used to meet any expenses incurred in those currencies and the balances replenished from receivables in the same currencies.

Currency adjustment factor A technique whereby a surcharge is raised by the shipowner on the basic freight rate to reflect a variation on the currency exchange rate. This helps to minimise the shipowner's losses through variations in the exchange rates. The currency adjustment factor is usually consolidated with the basic freight rate.

Currency and Bunker adjustment factor The combination of the currency adjustment factor and bunker adjustment factor.

Currency appreciation A rise in the value/price of a currency in terms of others in an appreciating currency.

Currency area The area over which a particular currency circulates or within which rates of exchange among different currencies are fixed.

Currency at a discount If the forward rate of a specified currency is higher than the spot rate, the currency is not at a premium but at a discount. An international trade banking term.

Currency at a premium If the forward rate of a specified currency is lower than the spot rate, the currency is at a premium. An international trade banking term.

Currency basket A mix of currencies used as a reference unit of account in order to reduce the risks involved in trading in widely fluctuating single currencies by spreading it over several currencies. Also, sometimes termed currency cocktail.

Currency borrowing The process of an importer raising finance in a foreign currency by taking a loan in the same currency as that to be paid to an overseas supplier. Borrowing in a foreign currency may be cheaper than borrowing in the national currency depending on the relative interest rates prevailing. Foreign currency loans can help the importer expand or develop international business and various types of loans are available including fixed rate and floating rate loans.

Currency cocktail See currency basket entry.

Currency code A 3-aplha code (which also identifies countries) to be included on the customs entry to notify the currency in which the customs value is being declared.

Currency deposits If the importer knows that he will need foreign currency in the future, he could buy the currency at today's spot rate, and place it in a deposit account until the time when he requires it to make payment. By buying the currency now the importer lessens the exchange risk and can earn interest on the foreign currency deposit which could be greater than the interest earned on an equivalent national currency deposit account operation in the importer's country.

Currency depreciation A fall in the value/price of a currency in terms of others is a depreciating currency.

Currency fluctuation clause A clause in a non-proportional treaty, the purpose of which is to stabilise the treaty retention and cover at the rates of exchange prevailing at the start of the underwriting year.

Currency linked to the USD A non convertible currency which is linked/aligned to the US dollar such as the Baht of Thailand.

Currency management The process of managing a prescribed currency usually within a specified strategy.

Currency of a Bill The period which must elapse before it is due to be paid.

Currency options It provides the importer or exporter with the right, but not the obligation, to buy or sell a certain amount of currency at an agreed rate (called the strike price) on or between future dates. The purchaser of the option pays a premium, in advance, for the benefit of being able to take advantage of any favourable movement in the exchange rate.

Currency restrictions Restrictions imposed by a Government on the 'out flow' and or 'in flow' of its national currency.

Currency swap An agreement between companies or governments to exchange currencies for a specific period e.g. a UK company may provide a US company with sterling in exchange for dollars to cover a transaction, with both parties agreeing to reverse simultaneously the exchange at some future date.

Current asset Cash or other asset e.g. stock or short-term investment held for conversion into cash in the normal course of trading.

Current cost The calculated cost of acquiring goods for processing or resale, or of assets, at the time when their value is consumed by the entity, usually by some form of averaging or index.

Current cost accounting The conversion of a company's accounting figures to show what the effects of inflation have been in any one period.

Current Directory The directory in which Vectra looks for information when no other directory is specified. Sometimes referred to as the default directory.

Current rate A term whereby the shipowner contracts to pay the current rate costs of loading and/or discharge.

Current ratio One of the basic credit ratios applied by bankers and investment analysts. It is simply the ratio of current assets to current liabilities, indicating the margin of working capital in the business. A variant of the current ratio is the quick ratio which is the ratio of current assets minus stocks to current liabilities. This is the key indicator of the margin of liquidity in the business. Both are regarded as having particular importance in shipping, where fluctuations of income can quickly absorb working capital and liquid funds.

Current transfers Government transfers to enterprises, households and the rest of the world, net of transfers received from the rest of the world, that are not related to capital expenditure; they comprise, among other operations, production and import subsidies, social benefits and transfers to EC institutions. An ECB term.

Current yield The annual return on an investment at current prices as based on the interest rate or dividend.

Curtain sides Moveable soft side curtains running on tracks fitted to open trailers, containers, or rigid vehicles allowing side access and loading.

CUSCAR Customs Cargo Report message (UN EDIFACT International Standard Message)

This permits the transfer of cargo manifest data from a carrier to a Customs administration (or other official organisation) for the purposes of meeting customs cargo reporting requirements (or other governmental requirements).

CUSDEC Customs Declaration message (to CHIEF).

Cusdel clause A charter party clause designed to compensate shipowners for delays waiting for cargo documentation in connection with customs clearance. In 2003 it was changed by BIMCO to read 'delay for Charterers' purposes clause. The free time has been reduced to two hours.

Cushion tyres Solid rubber tyres made of fairly soft rubber or composition.

CUSRES Customs Response message (from CHIEF).

Customary despatch The charterer must load and/or discharge as fast as is possible in the circumstances prevailing at the time of loading or discharging.

Custom built A product/service provided/built in accordance with the specified needs of the client. It may be a computer, ship, aircraft, airport passenger terminal or house.

Customer Asset Management A logistic term whereby the transport operator focuses on the customer needs. This embraces providing a service and customer retention Also, it includes knowing the specific customer requirements.

Customer awareness The image created by the work force/entity/company in the market place as to how well or otherwise the customer needs are met. This may be to keep passengers closely informed on any service disruption situation, or develop through personal contact a courteous helpful attitude by staff at all levels at all times in any particular situation involving contact/communication with the customer.

Customer audit The process of formulating a portfolio on each customer embracing all its financial aspects, any legal disputes and trading relations with the client.

Customer base A data bank of each customer profile.

Customer Billing The process of despatch accounts to clients for payment.

Customer Card The case history of a particular client or Company.

Customer care The processing of developing the strategy of empathy whereby the entrepreneur reflects the customer's needs and objectives in providing a market research led product/service. Hence, the major consideration in the marketing strategy is to target the buyer/consumer/customer and continuously respond to their defined needs and perceptions derived from the entrepreneur product/service.

Customer Confidence The process of potential/existing clients having confidence in the product/economy in terms of quality/buoyant well managed economy. For example, a transport service which offers a high standard of service, provides reliable service and generally creates an image of customer care, will tend to produce customer confidence in the market place.

Customer credit status report The credit rating of a company. See credit rating and status enquiry entries.

Customer focused To develop a marketing strategy focused on the customer needs and expectations.

Customer liaison The process of liaising with the customer. This may be through an Association, through a Consultative Committee, or through a Shippers Club as adopted by some National Airlines Freight clients.

Customer loyalty The process of the buyer remaining loyal to the product/retail outlet and making repeat purchases. See Brand Loyalty.

Customer needs The needs/provision required for the customer market. This may be for rail passengers on long distance inter city service, the provision of an 'on train' buffet/restaurant, or station catering facilities.

Customer orientated A situation whereby the prime consideration is the needs of the customer/client. It maybe a cruise line whereby special provisions are made to provide passenger care in terms of catering, baggage, excursions ashore, entertainment, or a container service where customer liaison officers are appointed to deal with shippers needs/complaints thereby ensuring the customers' needs are progressed quickly and effectively by the department concerned

Customer orientated marketing sales programme
A sales programme for a particular product/service with particular focus on the customer needs with a view to increasing sales.

Customer pick up Cargo picked up by a customer at a warehouse or other specified point.

Customer profile The portfolio of the entrepreneur customer. This includes culture, buying pattern, socio-economic category, buying power and so on. The areas which feature in the profile will depend on the nature of the research and especially its objectives.

Customer Relations The relationship between the client buying the goods/service and the supplier/manufacturer.

Customer Relations Manager A managerial post responsible for maintaining good relations and a satisfactory public company image between the Company and its clients.

Customer Research The systematic gathering/recording and analysing of data of the behaviour of the buyer and reasons for his/her preference in terms of product benefits and brands. Buyer behaviour is influenced by social, psychological and economic variables.

Customer service The service offered to the client/customer by the retailer/entity/company. It maybe in a supermarket with a wide range of products at competitive prices, or a railway service providing quality service throughout embracing punctual services, quality rolling stock, well appointed stations, public information, caring and courteous staff and so on.

Customers own transport A situation whereby the customer collects the merchandise from or delivers to the Container Freight Station/Container Base/Container depot/Inland Clearance depot/Seaport/Airport or other specified point.

Customer user meeting A meeting where the buyer (importer) meets the seller (exporter).

Customer value driven analysis The process of evaluating the customer value in the market place.

Custom fence The provision of a high fence of about three metres capped by barbed wire enclosing an area under customs surveillance. Such fences are usually found at seaports and airports.

Customised logistics network It is based upon the customer service segmentation needs utilising correct facets of logistics for time sensitive groups and reserve this facet for these groups only. A logistics term.

Custom of the port A term indicating a shipowner's agreement for his ship to load or discharge as per the custom of the port i.e. at the shipowner's risk and expense, laytime being indefinite. Sometimes the term is combined with 'fast and can'.

Customs Account Manager Customs officer responsible for overall control of a participating trader's business with customs.

Customs & Excise department H.M. Government revenue collecting service under Treasury control.

Customs and Excise Management Act UK legislation which contains the authority under which the Commissioners of H.M. Customs and Excise control the movement of goods into, and out of, the UK and collect revenue (duties).

Customs and Excise Warehouse A warehouse approved by H.M. Customs and Excise for the deposit of goods liable to customs duty and goods liable to excise duty or goods liable to both duties.

Customs barriers See tariff and trade barriers.

Customs broker An Agent specialising in 'inbound' customs clearance.

Customs cargo clearance Process of clearing import/ export cargo through customs examination, which may be at a seaport, airport, container base, or inland clearance depot.

Customs clearance Process of clearing import/export cargo through customs examination.

Customs Clearance Agent An Agent or customs broker specialising in inbound customs clearance services at a CB, ICD, airport, seaport, or other specified area approved by the Customs Authorities.

Customs consignee Equivalent to the term Customs Clearance Agent, it involves a Customs Broker or other Agent of the consignee designated to perform customs clearance services for the consignee.

Customs Co-operation Council A multilateral body located in Brussels through which participating countries seek to simplify and rationalise customs procedures.

Customs Co-operation Council nomenclature The standard classification of goods for customs tariff purposes now adopted by most countries. It enables Customs to quickly identify the goods for statistical, duty, or clearance purposes, and is used for all goods subject to customs tariffs in international trade.

Customs debt An obligation on a person to pay the amount of the import duties (customs debt on importation) or export duties (customs debt on exportation).

Customs declaration Presentation/statement to customs of the merchandise description in accordance with the National legislation. See Customs entry.

Customs duties Revenue collected by Customs at time of importation from duties imposed on various classes of goods.

Customs Duty An indirect tax which inter alia, provides protection for Community industry, raised on imported goods under the Combined Nomenclature of the Community. An EU term.

Customs Entries These are the means by which traders can pass to Customs details of all goods imported, or exported – now properly termed 'Export Declarations'.

Customs entry Form on which importer/agent gives necessary details for ultimate customs clearance, usually in a computerised format. An importer of goods into any country is required by law to give a declaration in respect of those goods to Customs. The normal method of making the declaration is by the completion of a special form (Customs Import Entry Form) issued by the Customs authorities. There are various types of Customs import entry forms. Information about the supplier of goods, the importer, the customs tariff, the goods, shipment details, duty etc. are required on the form which must be completed by the importer or his authorised agent. Normally, completed entry forms must be presented at the appropriate Customs office within a prescribed time of the arrival of the importing ship, aircraft or vehicle. Customs entries can usually be presented before the expected arrival of the ship or aircraft although they should not be earlier than four working days. However, certain classes of goods are not accepted for entry prior to importation.

Customs Handling of Import and Export Freight A customs system to process export and import entry introduced in 1994 under the simplified clearance procedure and displaced the DEPs. See also SDP.

Customs House report Report prepared by Master or Agent on ship arrival detailing cargo/passengers on board submitted to Customs House.

Customs Invoice It is a document prepared by the exporter in accordance with the requirements of customs authorities in the country of the importer, serving the customs authorities of the importing country as a basis for establishing the customs value of the goods and for the calculation of the customs duty.

Custom Smelter A smelter which depends for its intake mostly on concentrate purchased from independent mines and on scrap metal, rather then its own captive mine sources. This type of operation has become more important in recent years.

Customs planning This involves a wide variety of customs activity including tariff costing outwards, duty preference and origin, freight procurement, VAT, outward processing relief, standard exchange relief, temporarily imported goods, custom warehousing, free zones, carnets, export licensing, and local clearance procedure. Overall, it focuses attention on the customs, implementation of international distribution and whether any alternatives exist.

Customs port A seaport with customs facilities available.

Customs Post A place where customs examination is undertaken.

Customs Pre-entry Exports All goods subject to Customs controls; goods subject to export licensing requirements and all goods subject to official pre-shipment documentation.

Customs Pre-entry – Imports As a concession in many countries traders are allowed to present import entries to Customs shortly before the goods are due to arrive in order to facilitate rapid clearance of the goods.

Customs procedure The process of presenting/lodging a consignment either for export or import with customs with a view to it being cleared either for shipment (export) or import purposes. It may be at a seaport, airport or inland customs depot. A range of customs procedures exists and these vary by individual countries. These embrace release for free circulation, transit, customs warehousing, inward processing,

processing under customs control, temporary admission and exportation. See separate entries.

Customs procedure code Used on entries on import, export and warehouse removal to identify the type of procedure for which the goods are entered and from which they came. Can be found in Volume 3 of the Tariff. See Customs Tariff entry.

Customs Registered number A unique 5 digit number assigned to an exporter or agent approved to use one or more of the following export procedures; full preshipment declaration; low value preshipment advice; non-statistical preshipment advice; SDP, LCP; all of which maybe done electronically.

Custom's tariff The tariff based on the Harmonised Commodity Description and Coding System (HS) which is used worldwide. It classifies cargo and essential for anyone engaged in the preparation of import or export custom entries. The commodity classification will determine any customs duty raised. The tariff consists of three volumes – Volume I – General information; Volume II – Nomenclature, codes and rates of duty and volume III – freight procedures. The tariff contains 97 chapters, and over 15,000 sub-headings. The actual code numbers provide for eight digit code for inter EC trade and an eleven digit code required for imports from outside the EC. A further four digits may be required to identify MCA, ACA, variable charges, certain anti-dumping duties and wine reference prices.

Customs Territory of the Community The single customs area to which Community legislation, particularly customs legislation, expressly refers. For the UK, includes the Channel Islands and the Isle of Man. An EU definition.

Customs – Trade Partnership Against terrorism An USA sponsored agreement which reflects new security regulations and measures adopted by the USA. Customs service in 2002 to protect shipowners and results in increased expenses and delays to vessel calling at USA ports.

Customs Union A situation where a number of countries join forces to be treated as one customs union such as EC with free circulation of goods and a common external tariff established against non members. Hence, members apply a common external tariff.

Customs union: (Article XXIV) Group of countries which (a) have removed duties and other barriers to all or almost all the trade between them and (b) have adopted a common external customs tariff towards non-members of the group. A WTO term.

Customs value The value of an item or group of items expressed in a monetary value within a consignment declared to customs for duty and statistical reasons. The value of imported goods for customs purposes.

Customs warehouse A place authorised by Customs and Excise for the storage of non Community goods without payment of import charges

Cut off date Specific date when final action must be taken on a reservation or blocked space.

CUV Construction unit value.

C/V Certificate of value.

cv Chief value; or career vitae.

CVC Countervailing charge. This is imposed in addition to any other duties or charges on wines, dried grapes and certain other commodities to prevent unfair competition from cheaper non-EC imports. A Customs term.

CVGK Customs value per gross kilogram.

CVGP Customs value per gross pound.

CVMO Commercial value movement order.

CVO Certificate of value and origin; usually combined with the export invoice.

CVR Convention relative au Contrat Transport International de Voyageurs et de Bagages par route.

CW Air Marshall Island – an IATA member Airline; or Continuous Wave.

CWC Country whence consigned.

CWDE Centre for World Development Education.

CWE Cleared without examination – a customs term or Continental Western Europe.

cwo Cash with order.

CWTB Cylindrical water tube boiler (of a ship).

CWTBS Cylindrical water tube boiler survey (of a ship).

CX Cathay Pacific Airways Ltd. An IATA member Airline.

CXBG Comprehensive Extended Term Banker's Guarantee.

CY Cyprus Airways Ltd. The Cypriot National Airline and member of IATA; container yard; or cylinders.

Cy Currency; or cylinders.

Cybernetics The study of control processes in mechanical, biological, electrical and information systems.

CYCLD Cycloidal propeller (on a ship).

Cycle In this context, usually a thermodynamic process, following the progress of the working medium (e.g. water/steam) from its starting point through the entire process. In a closed cycle, the working medium is returned to the starting point and re-used: in an open cycle, it is discharged; in shipping terms the sequence of four events: a trade boom, a short shipping boom during which there is overbuilding, followed by a prolonged slump and growth. See shipping market entry cycle; in business terms the various stages through which a process follows to realize its objective – see also product life cycle, procurement cycle, payment, economic and family life cycles.

Cycle stock That portion of stock available or planned to be available in a given period for normal demand excluding excess stock and safety stock.

Cycle time The maximum number of days a shipper chooses to hold an order to achieve sufficient volume for consolidation. A shipping term. The shorter the cycle time constraint, the lower the volume and hence the higher the cost (opportunity costs for foregoing savings through full container utilisation). See consolidation entry.

Cyclical peak Part of the trade/shipping cycle when international trade is at its most favourable realising a peak condition. See Trade/shipping cycles entries.

Cyclical trough Part of the trade/shipping cycle when international trade is in a depressed condition. See trade/shipping cycles entries.

Cyclical unemployment The peaks and troughs of unemployment with unemployment reaching a peak in a period of deep recession and attaining a low level in times of excessive economic prosperity.

Cyclone Generally a name given to a region of low pressure. In temperate latitudes cyclones are usually spoken of as depressions and the term cyclone is taken to refer to only a 'tropical cyclone'.

CYL Cylinders

Cymogene A petroleum product, gaseous at ordinary temperatures but liquefiable by cold or pressure. It is classified as a hydrocarbon gas, liquefied, n.o.s. A term associated with movement of dangerous classified cargo.

Cyprus Airways The Cypriot National Airline and member of IATA.

Cyprus pound The currency of Cyprus. See Appendix 'C'.

Cyprus pound invoice An invoice payable in Cypriot pounds relative to an export sales contract.

CZ China Southern Airlines – an IATA member Airline.

Czech Crown (Koruna) The currency of Czechoslovakia. See Appendix 'C'.

Czech Crown (Koruna) invoice An invoice payable in Czech Crown (Koruna) currency relative to an export sales contract.

D6 Inter Air (Inter Aviation Services (Pty) Ltd.) – IATA Airline member.
D7 Dinar Lineas AÈreas S.A. (limited) – IATA Airline member.
D Delivery, delivered, diagonal engines, Director or depth.
d Dry-bulk containers.
D/A Deposit account, discharge afloat; days after acceptance; doubled acting (machinery); or documents against acceptance (insurance).
DA Dan Air – a UK airline; Defence Attache (Embassy)/Defence Advisor (High Commission); or disbursement accounts.
DAC Development Assistance Committee – based in Paris under the aegis of OECD; or Departamento de Aviacao Civil.
DAF Delivered at frontier (named point); cargo delivery term. See separate entry.
DAGAS Dangerous Goods Advisory Service.
DAGMAR A marketing term/abbreviation – defining advertising goals for measured advertising response.
DAIICHI Daiichi Chuo Kisen Kaisha (Japan) – a shipowner.
Daily net freight revenue voyage The level of profit accruing from a particular voyage on a daily basis and calculated as follows:–
 Daily net revenue (voyage) – average daily running cost = daily net revenue (voyage, running costs included). Hence, the daily net freight revenue (voyage) embraces the level of daily freight net revenue accruing from a particular voyage and calculated as follows:–

$$\frac{\text{Net freight revenue}}{\text{Number of days (voyage)}} = \text{Daily net freight revenue (voyage)}.$$

Dalasi The currency of Gambia. See Appendix 'C'.
Dalasi invoice An invoice payable in the Dalasi currency of Gambia relative to an export sales contract.
Dalton's Law of Partial Pressures This states that the pressure exerted by a mixture of gases is equal to the sum of the separate pressures which each gas would exert if it alone occupied the whole volume.
Damaged cargo report Written statement concerning established damages to cargo and/or equipment.
Damage report Surveyor's report on ship and cargo damage and its causes.
Damp proof Material which resists penetration by moist air.
Dammum sine injuria (Latin) Injury without breach of a legal right.
DAN Deferment Approval Number. The number allocated to persons approved by HMC & E to defer payment of duty/VAT . A customs term.
Dangerous cargo Merchandise classified as dangerous for which stringent regulations exist regarding its acceptance procedure, packaging, stowage, documentation, and conveyance for both local and international transits by road, rail, air and sea. There are some nine classifications of dangerous goods for international transits by sea and air. The regulations, documentation, acceptance procedures, packaging, and stowage arrangements are laid down by IMO (sea), ADR (road), RID (rail), and IATA (air) in regard to their respective transport modes.
Dangerous cargo compound An area which accommodates/stores dangerous cargo awaiting shipment or inland distribution at a seaport or airport. It is under strict surveillance.

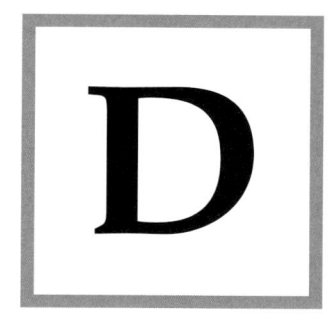

Dangerous Cargo Endorsement Endorsement issued by a flag state administration to a certificate of competency of a ship's officer allowing service on dangerous cargo carriers such as oil tankers, chemical carriers, or gas carriers.
Dangerous Goods Refers to certain commodities such as corrosive substances, explosives, flammable liquid, flammable solids, infectious substances, liquid and compressed gases, magnetised materials, oxidising substances, poisons and radioactive materials.
Dangerous goods authority form A document issued by the shipowner bearing a reference number giving the sailing details including port of departure and destination on which the consignment is authorised; the hazard; UN number labels; key number (for emergency in event of any incident, spillage etc.); and any special instructions (i.e. special Department of Transport approved or restrictions on stowage, i.e. on deck only (passenger ship), freight vessel only etc.). It is often referred to as a 'stowage order'. The Master has the right to refuse to ship dangerous classified cargo if the circumstances warrant such a decision.
Dangerous Goods declaration The documentation involved in the shipper's declaration of the dangerous classified cargo.
Dangerous Goods labels Labels required by IMO in movements of dangerous goods by sea.
Dangerous Goods Note A multi purpose document used for any combination of surface modes of transport valid for the whole journey involving a dangerous classified cargo transit. The document combines the function of application for stowage space on the transport unit(s); the dangerous goods declaration; the special stowage order; the container/vehicle packing certificate; the standard shipping note and finally the 'back up' document for the forwarder/haulier.
Dangerous goods packing certificate A document as part of the dangerous goods declaration in which the responsible party declares that the cargo has been stowed in accordance with the rules in a dry clean container in compliance with the IMDG regulations and properly secured.
Dangerous Goods rate The tariff applicable to the conveyance of dangerous classified cargo which is usually up to fifty per cent above the basic commodity/general cargo rate.
Dangerous Goods regulations Regulations determining the criteria/arrangements under which goods classified as dangerous are conveyed and packaged on a particular transport mode such as by road, rail, air, and sea transport. This involves ADR by road; RID by rail; IATA by air and IMO by sea transport for international transits.
Danish Crown (Krone) The currency of Denmark. See Appendix 'C'.
Danish Crown (Krone) invoice An invoice payable in the Danish Crown (Krone) relative to an export sales contract.
Danish Shippers Council The organisation in Denmark representing the interest of Danish shippers in the development of international trade.
DANPRO Danish Committee on Trade Procedures.
DAR Discharge advice report.
Data Information provided for a particular circumstance.

This term is much used in computerisation and represents the information fed into the computer network. See other entries relative to data carrier, data entry and so on.

Database A file designed to allow the information it holds to be obtained in several ways, for example, information on an employee could be obtained by quoting his/her name, or address, or National Insurance Number, or works number, or age. It is a file not just designed to satisfy a specific, limited application, but forms a pool of cross-referenced information from which the computer can draw on. It is basically an electronic library. Electronic databases usually contain data items (e.g. files) and their relationships (indexes and keys).

Database report The production of output from software such as a database for a specific purpose such as a telephone list, a list of orders, an invoice or a statement of account.

Data carrier Medium designed to carry records of data entries. A means of transferring information from one point to another.

Data dictionary A file containing descriptions of all the data items in a database. The descriptions could include: data type, element size, format of data, validation range of values, entity descriptions, attribute-entity relations, purpose, synonyms. A computer term

Data elements The smallest invisible piece of data e.g. date. An EDI definition.

Data entry Data entered on a data carrier.

Data field (Syn Box) An area designated for a specific data entry.

Data file Group of related data – carriers, clients etc. (Records) etc.

Data flow diagram A diagram which shows the movements of data through a system and of the associated changes taking place in the data which result from processing actions.

Data Freight Receipt A document issued by a shipowner which acts as a receipt for the cargo shipped and evidence of the contract of carriage, but unlike the Bill of Lading it is not a document of title or negotiable instrument. It is often referred to as a Sea Waybill.

Data input The provision (input) of data into a computer system.

Data Interchange in Shipping A system for exchanging data between carriers and merchants electronically.

Data model The purpose of a data model is to identify and present user requirements in a way easily understood by users and computer professionals. It is also important that the model can be easily converted to a technical implementation and provide rules and criteria for implementing the system. A data model may consist of some or all of the following: high-level information flow diagrams, data flow diagrams, data dictionary, entity relationship diagrams.

Data plate Identification plate placed suitably on the container, road trailer, railway wagon to carry detail of cargo and transport unit, e.g. actual gross weight, net tare weight.

Data processing The recording and processing of information for commercial computerised tasks.

Data processor A device capable of performing data processing including a desk calculator, computer or a person.

Data representation When characters are transmitted electronically a code must be used to convert the character into numbers and back again. Most commonly used code is the ASCII (American standard code for information interchange). A computer term.

Data sharing The use by several systems of the same information from the same file. A computer term.

Data storage Ways of storing data and information. Non-electronic ways could be: filing cabinet, account book, card file or cupboard; electronic ways could be data files on tape or disc.

Data type The characteristics of the data used. These may differ for each type of system. Typical data types are: character, number (integer, real), graphic, logic (Boolean), date.

Davits Light cranes on ship's sides for lowering and lifting boats.

"Day" A period of twenty-four consecutive hours running from 0000 hours to 2400 hours. Any part of a day shall be counted pro rata. A BIMCO term.

Days of grace Days allowed by law or custom for payment of Bills of Exchange (except those payable at sight or on demand) after specified day of payment.

Day Order Orders placed during a trading session for execution on the same day – a BIFFEX term.

Day Order (GED) An order placed with a broker which is valid throughout the rings and kerbs during the day on which it is placed. If not executed that day, it is automatically cancelled. If valid for all markets (i.e. pre-market, lunchtime and evenings as well as the rings and kerbs), any instruction should clearly state 'all markets'. The order can be executed at any time of the day.

Day trade A trade opened and subsequently closed on the same day. A BIFFEX term.

DB Double bottom; double-ended (boilers); deals and battens (timber) or Development Bank.

dB Decibel.

dbb Deals, battens, boards (timber).

DBB Deals, battens, boards (timber).

DBC Double bottom centre; or Dunlop Beaufort Canada.

DBE Data Bank (Eurocontrol) – IATA term.

dbe Despatch money payable both ends (loading and discharging ports).

DBEATS Despatch payable both ends (loading and discharging ports) all time saved.

DBELTS Despatch payable both ends (loading and discharging ports) all laytime saved.

Dbk Drawback

DBMS Data Base Management System.

DBP Development Bank of the Philippines.

DBS Distressed British Seaman; Development Bank of Singapore; or Dnieper Bureau of Shipping Ltd – Ukraine.

DC Direct Current or Developed Country.

D/C Deviation clause.

DCA Debt collection agency; Department of Civil Aviation (Burma); Dept. of Civil Aviation (Singapore); or Director of Civil Aviation (various countries) – IATA term.

DCD Data post on demand.

DCF Discounted cash flow.

DCO Debt collection order

DCS Departure control system – IATA term.

DCTPB Dubai Commerce and Tourism Promotion Board.

dd Delivered.

Dd Delivered docks – indicates that the seller of the goods pays all charges to the named docks; or days after date.

DD Delivered at docks; demand draft; dry docking; Duty Deferment; or Conti-Flug – an IATA member airline.

D1/2D Despatch money payable at half demurrage rate.

DDA Duty (Customs) deposit account.

ddc Despatch money payable in respect of time saved at the discharging port only.

DDE Direct data entry

DDO Despatch discharging only – discharge money payable on time saved during discharge.

DDP Delivered duty paid – cargo delivery term applicable to all modes of transport. An Incoterm 2000. See separate entry.

DDR Direct debit.

DDU Delivered duty unpaid – Incoterm 2000. See separate entry.

DE Diesel-electric (machinery); diesel electric engine or deemed/earned (Freight).

Deadfreight Freight to be paid for shipping space by a shipper or charterer on a vessel for which a reservation had been made and which was not used because goods remained unshipped. The shipowner is paid a sum in lieu of the freight earnings lost. Hence, freight rate which is paid on empty space in a vessel when the charterer is responsible for the freight rate of a full cargo. It should be paid before sailing.

Deadhead Aircraft, ship or other transportation vehicle in transit without a payload, also known as an empty leg.

Dead heading The return of empty containers; or a road haulage term representing lost running/light mileage and repositioning of empty units. It can be resolved adopting triback. See triback entry.

Deadload The difference between the actual and calculated ships draft.

'Dead' ship A vessel whose propulsion is immobilised and has to be taken in tow.

Dead time The time period between the exit of one vessel from a lock and entry of another.

Deadweight Vessel carrying capacity viz. fuel, stores, water, crew, based on 1,000 kgs to deadweight tonne.

Deadweight all told Total carrying capacity in weight terms (metric tonnes) of a vessel including cargo, fuel, stores, fresh water and so on. It is expressed in tonnes. See tonnage entry.

Deadweight cargo Cargo which is measured by weight avoirdupois, in contrast to measurement by dimensions/volume. Heavy cargoes such as coal, ore etc. which measure less than 1.133m² per tonne weight (1,000kg) are described as deadweight and freight is paid on the weight in tonnes.

Deadweight tonnage The number of tonnes weight (1,000kg to a tonne) of cargo, stores, bunkers, fresh water, passengers and crew on board a ship permitted to be carried and in so doing bring the vessel down to her load line/marks. In marine insurance practice the term refers to the dimensions of a ship for the purpose of calculating the premium for a hull and machinery policy on full conditions.

Deals Lengths of timber between five feet and three feet in length and between two inches and nine inches thick.

De-ballasting The removal of the ballast from a ship. Water ballast is pumped overboard.

Debenture A formal acknowledgement of a debt, usually incorporating a charge over the unencumbered assets of the company issuing it; the rights of debenture holders rank before those of shareholders and unsecured creditor in the event of the issuer's liquidation. Redeemable debentures give the holder the right to repayment of principal at the stipulated date.

Debit note A document prepared by the purchaser notifying the seller that his (the sellers) account has been reduced by a stated amount because of an allowance, return of goods or cancellation; or a bill issued to advise the insured of the premium due. Sometimes the cover note and the debit note are combined in one document a cover/debit note – an insurance definition.

Debit ratio The process of dividing the total liabilities by the total assets results in the debit ratio. The ratio shows the portion of the total assets financed by debt.

De bonis non administratis (Latin) Property left unadministered.

De-briefing The process of appraising/informing a person(s) the outcome of a particular enquiry meeting such as the Sales Director telling all his sales representatives the outcome of a Board meeting.

Debt – bad A person or entity who owes money, the likelihood of which complete recovery is slight as the person/entity is in the hands of the receiver/liquidator.

Debt Collection Agency An Agency which specialises in the process of collecting debts.

Debt-equity ratio The ratio of a company's ordinary share capital to its fixed interest capital, including debentures, loan stock, and preference shares; calculations are often simply based on the ratio of ordinary shares plus retained reserves to prior charge capital. For Example, a rise from 1.39 per cent to 300 per cent will tarnish badly the credit rating of a company. Usually defined as the ratio of long-term debt, including the current portion, to shareholders' funds. Also known as gearing and, in the United States leverage. This is one of the most important credit ratios, variants of which are the equity ratio (book equity to total capital) and the ratio of debt to total capital. The capital structure of the shipping industry has featured debt/equity ratios higher than the average in most other industries, leading to considerable financial instability during periods of recession.

Debt factoring The sale of debts to a third party (the factor) at a discount, in return for prompt cash.

Debt Finance A legally enforceable loan agreement in which the lender receives interest at predetermined intervals with repayment of principal after a specified period.

Debtor A person or an entity who owes money.

Debt ratio The subject of one of the fiscal convergence criteria laid down in Treaty Article 104c (2). It is defined as 'the ratio of government debt to gross domestic product at current market prices', where government debt is defined in Protocol No. 5 (on the excessive deficit procedure) as 'total gross debt at nominal value outstanding at the end of the year and consolidated between and within the sectors of general government'. General government is as defined in the European System of Integrated Economic Accounts. An ECB term.

Debt service/Exports (%) See international debt entry.

Debt servicing Capital repayments and interest payments due on borrowing.

DECCAN Deccan Shipping Ltd. (India) – a shipowner.

Decentralised organisation An organisation structure of a company whereby control/decision making process is undertaken at the localised managerial level and not centrally controlled.

Decision making unit The person(s) responsible for making the decision in a company/entity. It may be the Board of Directors, Manager, Supervisor or user. Additionally, there exists the specifiers/influencers who have a major influence in the decision making process. Both the decision maker and influencer must be targeted by the Advertising Executive Sales person to secure the business on an ongoing basis. This strategy applies both to the consumer and industrial markets.

Decision Support System An interactive computer based system which generates a number of alternatives to solve an unstructured problem. These alternatives are being interpreted by the manager/decision maker, where after he/she decides which alternative is to be used to solve the problem.

Decision table A table which shows the relationships between variables and the actions which must be taken when certain conditions arise. A computer term.

Decision tree The mechanism/matrix/levels of management involved in the decision making process in a particular company.

Deck Platform on a ship covering the whole or part of the hull area at any level. See also tonnage deck, upper deck, entries; or any extended horizontal structure in a vessel or an aircraft serving as a floor and structural support covering partially or fully a portion of the vessel or aircraft.

Deck boy A boy with less than nine months sea service in ship's deck department.

Deck cargo Cargo shipped on open deck.

Deck cargo certificate Manifest of cargo shipped on open deck.

Deck department Vessel's crew department responsible for navigational aspects of ship operation; or a shore based department responsible for upkeep of vessel and her equipment excluding machinery.

Deck fitting Fitting that connects securing equipment to ship's structure.

Deck line Horizontal line marked amidships on each side of the ship.

Deck load Cargo stowed on open deck.

Deck Log Ship's log recording general details concerning the running of the ship, including accidents concerned with ship or cargo.

Deck log book The deck department log book on a particular ship. See log book entry.

Deck Officer Navigating officer aboard ship with responsibilities also for deck and cargo work.

Deck personnel Vessels crew department responsible for navigational aspects of ship.

Deck plan A diagram or model of a ship's decks; often used by agents to select cabins.

Decl. Declaration.

Declarant A person who signs the valuation declaration or statement. Also a person who completes and signs a SAD. A customs term.

Declaration The details of a consignment of goods, imported or removed from a Customs Warehouse, declared to Customs.

Declaration date The latest date or time by which the buyer of an option must indicate to the seller his intention to exercise or abandon the option. A BIFFEX term.

Declaration of age A ship classification document confirming the age of the vessel.

Declaration of origin Statement as to the origin of the goods made, in connection with their exportation, by the manufacturer, producer, supplier, exporter or other competent person on the commercial invoice or any other document relating to the goods.

Declared value for carriage The value of goods declared to the carrier by the consignor for the purpose of fixing the limit of the carrier's liability for loss, damage, or delay to cargo. It is also the basis for possible applicable valuation charges. It is particularly common to air freight consignments as under the terms of the Warsaw Convention a carriers liability is limited per kilogramme on any air freight consignment. By declaring the higher value, the shipper can increase the carriers liability.

Declared value for Customs The value of the goods for Customs purposes declared by the consignor on the Air Waybill/Bill of Lading/Sea Waybill.

Declassing of a ship The process of removing the ship classification from a Ship Classification Registry.

Declining market A market which is becoming smaller.

Declivity The angle of the launching ways at a shipyard.

Decongestion The process of relieving congestion in a traffic situation.

De-consolidators The process of breaking bulk a consolidated consignment and despatching the individual packages to their prescribed consignees.

Decoupling inventory A stock retained to make the independent control of two successive operations possible.

Decoupling point The point in the supply chain which provides a buffer between differing input and output rates.

De die in diem (Latin) From day to day.

Dedicated fleet A fleet of vehicles or ships purpose built to convey only a specific cargo.

Dedicated rail head A railway terminal purpose built which handles exclusively rail borne traffic and offers scheduled time tabled services. It is usually intermodal feeding into road networks as a distributor.

Dedicated service A transport service which is provided exclusively for a particular shipper or group of shippers offering guaranteed schedules and transit time. Usually the transport units are purpose built and operate on an intermodal system such as the landbridge concept. Alternatively, it may be an ocean going chemical carrier which relies on chemical tank wagons rail services to distribute the merchandise to the chemical farm some 100 miles inland from the chemical jetty at which the chemical carrier discharges.

Dedicated terminals A terminal purpose built for a particular type of traffic and used exclusively for this product such as a container terminal at a seaport operated by a particular shipping company. It would have all the back up resources.

Deductible An amount or percentage, expressed in a policy, that is deducted from partial loss claims. Most commonly seen in policies covering hull and machinery, where the policy deductible is applied to all claims arising out of any accident or series of accidents comprising one occurrence, other than a total loss claim or sue and labour charges relating to a total loss. The cost of sighting the bottom for stranding damage is exempt from the deductible in the Institute hull clauses.

Deep Sea Broker A person who will normally act for either the shipowner or merchant, negotiating with a fellow broker for the other side.

Deep Sea Shipping Lane A charted shipping navigation route specially provided for ocean going vessels. Such vessels would fall within the draught range of 10 metres to 24 metres.

Deep sea trades All trades entailing voyages outside coastal and short sea trading areas.

Deep tank A tank fitted and equipped for the carriage of vegetable oil (e.g. palm oil and coconut oil) and other liquids in bulk on a vessel. By means of oil tight bulkheads and/or decks, it is possible to carry different kinds of liquid in adjacent tanks. Deep tanks may be equipped with heating facilities in order to carry and discharge oil at the required temperature.

Deep water berth A berth at a port accommodating ocean going vessels. It may be an oil tanker, iron ore carrier, passenger liner, container ship and so on.

Deep water harbour A harbour offering deep water facilities including deep water berths thereby being able to accommodate vessels up to 20m draught.

Deep water route A designated shipping lane or route through which vessels of a restricted draft may navigate with safety.

Deepwell Pump A type of centrifugal cargo pump commonly found on gas carriers. The prime mover is usually an electric or hydraulic motor. The motor is usually mounted on top of the cargo tank and drives, via a long transmission shaft, through a double seal arrangement, the pump assembly located in the bottom of the tank. The cargo discharge pipeline surrounds the drive shaft and the shaft bearings are cooled and lubricated by the liquid being pumped.

De facto (Latin) As a matter of fact; really.

Default Failure to repay a loan or an overdraft as promised. A banking definition.

Default bonds A conditional bond which contains no requirement for documentary evidence. An international banking term.

Defaults The settings of 'software configuration' or 'hardware configuration' to a standard set of values for the user. A computer term.

Defeasible interest An interest which may cease for reasons other than the operation of a maritime peril. A marine insurance term.

Defence Mapping Agency (US) An American organisation which is responsible for broadcasting specialised and selective navigation information.

Defendant The person against whom a claim for damages, or other charge is preferred – a legal term.

Deferment account An account underwritten by a bank or insurance company to which import duties due are posted. A customs definition.

Deferred Payment which is deferred such as under a documentary credit.

Deferred Account A system allowing, for example, the shipowner to pay his annual premium by instalments.

Deferred payment credit A documentary credit available by deferred payment nominates the bank which is to effect payment under the credit after presentation to it of the stipulated documents.

Deferred rebate System whereby shippers are granted a rebate (usually up to 10 per cent) on freight for consistent exclusive patronage over a given period (minimum usually six months).

Deficiency payments Paid by governments to producers of certain commodities and based on the difference between a target price and the domestic market price or loan rate, whichever is the less. A WTO term.

Deficit A deficit is brought about by the accumulation of a loss resulting in a debit balance in the 'retain' earning account. This balance is known as a deficit.

Deficit-debt adjustment The difference between the government deficit and the change in government debt. Among other reasons, this may be due to changes in the amount of financial assets held by the government, a change in government debt held by other government sub-sectors or statistical adjustments. An ECB definition.

Deficit ratio The subject of one of the fiscal convergence criteria named in Treaty Article 104c (2). Defined as 'the ratio of the planned or actual government deficit to gross domestic product' at current market prices, where the government deficit is defined in Protocol No. 5 (on the excessive deficit procedure) as 'net borrowing of the general government'. General government is as defined in the European System of Integrated Economic Accounts. A term/definition associated with the Eurozone.

Definite laytime A charter party term which defines precisely the laytime period as a number of days or hours.

Deflation The process of reducing the level of inflation in an economy which may emerge through varying measures such as high interest rates, increased taxation, reduced money supply and so on. The basic theme is to reduce aggregate demand in the market place thereby placing less strain on the economy in terms of utilisation of its resources. This could result in a multiple contraction in the economy.

DEFREP Defect Report.

Degressivity The notion that a measure restricting or distorting trade must be phased out over a finite and stated period of time. Overall, a systematic winding-down of safeguard measures during their duration.

Degroupage The process of splitting up shipments into small consignments.

De-icing fluids Such fluids frequently contain large proportions of alcohol and other inflammable liquids.

De-industrialisation The process of an area/region closing down its industrial base of manufacture/production/assembly.

De jure (Latin) As a matter of law.

Del credere Agent/broker guarantee to principal for solvency of person to whom he sells goods.

Del credere agents Agents which guarantee that on consideration of extra commission will pay bad debts. The difference between them and Confirming Houses is that they do not guarantee the overseas sale itself.

Delegatus non potest delegare (Latin) An Agent cannot delegate his authority.

Delay for Charterers Purposes Clause A charter party clause introduced in 2003. See Cusdel clause entry.

Delay surcharge A tariff/charge raised usually by the shipowner on a shipper consequent on the delay incurred by the shipowner waiting a berth to discharge/load cargo. The delay may extend beyond a week and the charge is on a per container TEU or cargo tonne basis. It arises due to the inability of the seaport to handle the cargo awaiting discharge/shipment.

Delivered at frontier (named point) (DAF) The seller's main obligations include: to supply the goods in accordance with contract of sale terms at the seller's risk and expense; to place the goods at the disposal of the buyer at the specified frontier point within the stipulated period and in so doing provide him with the necessary customary documentation including consignment note, export licence, delivery order, warehouse warrant etc. including the equivalent electronic message – to enable the buyer to take delivery of the cargo; to fund any customs charges and other expenses incurred up to the time when the goods have been placed at the buyer's disposal, and bear all the risk throughout this period, to obtain and pay for all documentation necessary for the exportation of the cargo, including transit countries needs and placing them at the disposal of the buyer as appropriate; to fund all transport cost (including incidental charges raised during the transit) up to the nominated frontier point; if no particular frontier point is quoted, the seller may choose the one which is most convenient to him provided he notified the buyer promptly and it offers adequate customs and other facilities to enable the contract to be executed satisfactorily by both parties; to supply the buyer on request the through consignment note embracing the origin and destination place in the importing country – such a request to be executed by the seller on the condition he incurs the risk, or expense other than is customary to incur; in the event of the goods being unloaded on arrival at the frontier point, such cost to be borne by the seller, including lighterage and handling charges (this would also apply if the seller used his own transport); to notify promptly the buyer that the goods have been despatched; to provide and pay for the customary packing of the goods unless it is the custom of the trade to ship the cargo unpacked; to fund any cargo scrutiny necessary to transport the goods to the specified frontier point; to bear any additional cost incurred to place the goods at the buyer's disposal; and render to the buyer on request and at the buyer's expense all reasonable assistance to obtain documents inherent in the importing process in the destination country.

Under the 'delivered at frontier' term the seller's obligations are concluded when the goods have arrived at the named frontier point place or customs examination border. It is usual, to avoid ambiguity, to quote the two countries separated by the frontier. This term is used primarily for rail or road borne traffic but can be used for other transport modes in varying circumstances. Its use could become more common through the development of the combined transport operation for international consignments. The term requires the seller to clear for export.

The buyer's obligations include: to accept delivery of the goods at the specified frontier point and accept all transportation and handling costs therefrom; to meet the risk and custom duties and other costs incurred from the time the goods have been placed at the buyer's disposal relative to the process of importation of the cargo; to fund incidental expenses incurred to unload the cargo at the specified frontier point; in the event of the buyer not taking delivery of the cargo duly put at his disposal in accord with the sales contract, the buyer will pay all additional costs and bear all the risk resulting therefrom; to obtain at his expense any import licence and other documentation required to process the cargo through importation; to fund any additional expense incurred by the seller to obtain the consignment note for the buyer which could include documentation; and to supply the seller on request with details of the ultimate destination of the goods; to fund any expense incurred by the seller to provide the buyer with any third party certificate of conformity of the goods stipulated in the contract of sale. The cost of pre-shipment cargo inspection is borne by the buyer unless when mandated by the authorities of the country of export. The insurance funding to be mutually decided by the seller/buyer. An Incoterm 2000.

Delivered domicile Merchandise delivered to the specified buyers (importers) premise – cargo delivery term.

Delivered duty paid (named point) – DDP
Under this term the seller is responsible for the conveyance of the goods at his own risk and expense to the named destination located in the buyer's country in the contract of sale. It represents the maximum obligation for the seller. This includes the task of processing the cargo through both exportation and importation including duties and taxes plus customs clearance, loading and unloading, together with the related documentation which the buyer usually obtains as necessary on request but at the seller's expense. The seller may use his own transport throughout the conveyance.

The buyer's role is to accept the goods at the named place of destination and he is responsible for all subsequent movement cost of the goods including handling. Any form of transport can be used. The cost of pre-shipment cargo inspection is borne by the buyer unless when mandated by the authorities of the country of export.

The term should not be used if the seller is unable to obtain directly or indirectly the import licence. If the parties wish that the buyer should clear the goods for import and pay the duty, the term DDU should be used.

Hence, the seller fulfils his obligation when he has delivered the goods, cleared for export, into the charge of the carrier named by the buyer at the named place or point. If no precise point is indicated by the buyer, the seller may choose within the range stipulated where the carrier shall take the goods into his charge. When according to commercial practice the seller's assistance is required in making the contract with the carrier, (such as in rail or air transport with fixed freight rates) the seller will act at the buyer's risk and expense. This term may be used for any mode of transport. An Incoterm 2000.

Delivered duty unpaid (named point) – (DDU) The seller fulfils his obligations when the goods have arrived at the named point or place in the country of importation. The seller has to bear the full cost and risks involved in bringing the goods there to, excluding duties, taxes and other official charges payable upon importation. If the parties wish to include in the seller's obligation some of the cost or official charges payable upon import of the goods (such as value added tax (VAT)) this should be made clear by adding words to this effect (for example, 'delivered duty unpaid, VAT paid'). The term may be used irrespective of transport mode. The buyer must pay all costs and charges incurred to obtain the documents or equivalent electronic messages. An Incoterm 2000.

Delivered Ex Quay (named port of destination) DEQ
Under this term the seller arranges for the goods to be made available to the buyer on the quay or wharf at the destination port detailed in the sales contract. The seller has to bear the full costs and risks involved to bring the goods to the quay whilst the buyer undertakes import clearance and associated costs.

The seller's obligations include: to supply the goods in accordance with the contract of sale terms; to make the goods available to the buyer at the specified quay or wharf within the period given in the sales contract and to bear all the associated risks and costs; to provide and pay for the customary packing of the goods unless it is the custom of the trade to ship the cargo unpacked; to pay the cost of any scrutiny immediately prior to the goods being placed at the disposal of the buyer.

The buyer's task is: to take delivery of the goods as specified in the contract; to clear the goods for import and to pay all formalities, duties and other charges upon import; to bear all the expenses and risks of the goods from the time the cargo had been effectively placed at the disposal of the buyer. The cost of pre-shipment cargo inspection is borne by the buyer unless when mandated by the authorities of the country of export. The buyer must pay all costs and charges to obtain documents or equivalent electronic messages. An Incoterm 2000.

DES Delivered Ex Ship (named port of destination)
This sales contract term is not used extensively. It obliges the seller to make the goods available to the buyer on board the vessel not cleared for import at the destination port as specified in the sales contract. The seller has to bear the full cost and risk involved to bring the goods to the destination port before discharging.

The seller's main obligations include: to supply the goods in accordance with contract of sale terms; to make the goods available to the buyer on board the vessel at the agreed destination port to enable the cargo to be conveniently discharged; to bear all the risk and expense of the cargo conveyance to the destination port until promptly collected by the buyer; to provide and pay for the customary packing of the goods unless it is the custom of the trade to ship the goods unpacked; to pay the cost of any cargo scrutiny prior to collection of the cargo by the buyer; to inform promptly the buyer of the expected date of arrival of the vessel and to provide the buyer with a Bill of Lading and any other documents necessary to enable the buyer to take delivery of the consignment; to provide the buyer on request and at the buyer's expense the Certificate of Origin and consular invoice; to render the buyer on request and at the buyer's expense every assistance to provide the requisite documentation issued in the country of shipment and/or origin required for importation in the destination country or transit countries.

It obliges the buyer to bear all the risks and expenses from the time the cargo had been placed at the disposal of the buyer on board the vessel awaiting discharge at the destination port; to bear all the costs and charges associated with the provision of documentation or equivalent electronic messages obtained by the seller necessary for the importation of the goods both in destination and transit countries; to obtain at the buyer's expense all licences or similar documents necessary for the importing process; to bear all customs charges and other duties and taxes incurred at the time of importation. This term can be used only when the goods are to be delivered by sea or inland waterway or multimodal transport on a vessel in the port of destination. An Incoterm 2000.

Delivered in charge The carriers charge to cover the transfer of cargo (carrier-loaded to container) from container freight station or equivalent to loading port or terminal.

Delivering carrier The carrier who delivers the consignment to the consignee or his agent in accordance with the consignment note terms.

Delivery The settlement of a futures contract by receipt or tender of the physical or cash (cash settlement) – an International Petroleum Exchange term; the tendering of a physical commodity against an open futures contract. A BIFFEX term; or the process of delivering the consignment to the specified consignee or appointed agent.

Delivery charges A charge levied by the delivering party for delivering the goods.

Delivery date The date on which the consignment was delivered/received by the specified consignee.

Delivery instruction The document issued by a buyer giving instructions regarding the details of the delivery of the goods ordered.

Delivery month The month in which a futures contract within the freight market matures and delivery is made – a BIFFEX term.

Delivery note See delivery notice entry.

Delivery notice Notification in writing, sent by the carrier to the sender, to inform him at his request, of the actual date of delivery of the goods.

Delivery on wheels The delivery of goods by road to the consignee premises involving consignment(s) up to 3,000 kilogrammes in weight or up to nine cubic metres in volumetric measurement.

Delivery order A written authority to deliver goods etc. to a named party in exchange for the bill of lading usually at the port of destination. It is issued at the port of destination and is subject to all the terms and conditions of the carrier's bill of lading. It must not contain any reservations or clauses other than those appearing in the bill of lading except where increased obligations or extra cost may be incurred in giving delivery beyond the bill of lading. The document is issued at the port of destination in exchange for an original bill of lading and legally is recognised as a token of an authority to receive possession.

The delivery order should be addressed to the ship's master and its role arises, for example, when the buyer may not wish to know the identity of the supplier abroad for trade reasons. Hence, the delivery order issue may prove a useful document. It is important the document is endorsed by the party to whom it is made out. However, if it is issued in one port for delivery in another, and the freight is payable at destination the order would then be 'consigned' to the carrier's agent to ensure that it would have to be presented and released before collection of the goods is authorised.

Delivery party The party to which the goods are to be delivered.

Delivery receipt A receipt signed by the consignee as proof that the goods have been delivered to him/Company.

Delivery service The carriage of 'in bound' consignments from the airport, seaport, ICD, CB or other specified destination to the address of the consignee or that of the designated or to the custody of the appropriate government agency when required.

Delivery terms of sale The delivery terms under which an export sales contract is concluded. Usually one of the INCOTERMS 2000 as issued by the ICC. See Incoterms 2000 entry.

Delivery versus payment system (or DVP, delivery against payment) A mechanism in a securities settlement system that ensures that the final transfer of one asset occurs if, and only if, the final transfer of (an)other asset(s) occurs. Assets could include securities or other financial instruments.

Delmas A major container operator.

Deluxe Used in travel to suggest highest quality, but much misused and meaningless except as part of a rating system in official use.

De luxe cabin Passengers sleeping accommodation of a superior quality cabin type on a ship and accordingly attracts the highest tariff/fare.

Delta system A ship construction system used for the carriage of bulk oil or similar cargoes. Detachable caissons containing the cargo are locked into bays in the side of the mothership which carries only fuel and ballast. When the ship arrives at her destination port, the ballast is adjusted so that she rises. The caissons on which she actually rides, can be detached and towed into berth for unloading and loading. Subsequently, the caissons are re-positioned in the bays at the side of the ship, which adjusts her ballast to re-engage on the caissons.

Dely Delivery.

Dely & re-dely Delivery and redelivery.

Demand orientated pricing The strategy adopted by a Company/Entity to price a product/service at a level which will generate an increased demand. This will gain market share and increase sales volume.

Demerger The process of dismantling a merger involving two companies or more.

Demijohn Glass bottle packed in protective wicker basket.

Demi-pension Half-pension, a hotel rate, particularly in Europe, including bed, breakfast and either lunch or dinner; also called modified American plan.

Denied-boarding compensation Payment in the form of cash or additional airline ticket made to passenger by carrier when he or she is bumped.

Demise charter A charter party giving the charterer the entire responsibility for operating, crewing, etc. Also known as a bareboat charter. The owner in this case is acting purely as a financier, taking a risk on interest rates and the assumed residual value of the vessel.

Demise charter party Charter party (usually time) under which shipowner provides vessel, and charterer crew and cargo. Hence, the charterer takes over control costs and responsibilities of the vessel for an agreed period. The charterer pays hire for an agreed period.

Demographic data Details of the population and disposition of a particular country/region/company. This includes age distribution/sex/marital status/socio-economic category/residence/literacy. For example, a country may have an ageing population of low literacy. Overall, it represents the social statistics provided in a country/region basis. Basically such data is essential in market research terms to devise an effective marketing strategy.

Demographic data analysis The process of analysing population statistics in respect of births, deaths, marriages, divorces, sex (male/female), age group etc. by specific area/country/region.

Demographic market segmentation The composition of the market place in demographic terms. See demographic data analysis entry.

Demographics The study of human populations in terms of size, density, location, age, sex, race, occupation and other statistics.

Demography The science of population statistics in respect of births, deaths, marriage and divorce in a specific community.

Demolishcon 2001 A BIMCO approved form and described as the standard ship recycling contract. It supersedes the 'salescrap 87' standard contract.

Demolition market The market handling the scrapping and demolition of ships.

Demolition of ships The process of scrapping ships usually for economic reasons.

Demountable system System of interchangeable bodies on rigid road vehicles.

DEMS Demurrage.

'Demurrage' An agreed amount payable to the owner in respect of delay to the vessel beyond the laytime, for which the owner is not responsible. Demurrage shall not be subject to laytime exceptions. A BIMCO term.

Demurrage clause A clause specifying the time and number of days allowed for loading and/or discharging usually found in a charter party.

Denholm Ship Management (Holdings) Ltd, Scotland A ship management company based in Scotland.

De novo (Latin) Starting afresh.

Density Relationship of weight to volume relative to cargo assessment or the mass per unit volume of a substance at specified conditions of temperature and pressure.

Density of commodity See density entry.

DEP Department of Export Promotion (Thailand)

Dep Departure.

Departure tax See head tax, airport tax, and passenger facility charge entries.

De-planning The process of alighting/leaving an aircraft following a flight.

Deposit Part of the purchase price (usually up to 10 per cent) which the purchaser pays on or before the exchange of contracts; or the initial outlay required to open a futures positions – a BIFFEX term.

Deposit facility A standing facility of the Eurosystem which counterparties may use to make overnight deposits at the central bank which are remunerated at a pre-specified interest rate.

Deposit policy Sum required to reserve a service or goods, for example, hotel rooms or tours.

Deposit receipt A receipt given in respect of a general average deposit payment. A marine insurance definition.

Deposit reservation A reservation for which a hotel has received advance payment for at least one night and is usually obliged to hold the room regardless of the guest's arrival time; policies vary and should be checked.

Deposits The current liabilities of a bank in the form of current account funds or monies at call, notice or fixed term, in sterling or foreign currency.

Depot The place designated by the carrier where empty containers are kept in stock and received from or delivered to the container operators or shippers.

Depot charges – Export A charge/tariff raised for receiving cargo and stowing into a groupage unit.

Depot charges – Import A charge/tariff raised for receiving a loaded groupage unit and subsequent delivery to a road vehicle for conveyance to the final destination.

Depreciating currency A floating currency which is tending to depreciate in value in terms of its exchange rate in comparison with major convertible currencies.

Depreciation The measure of the wearing out, consumption or other loss of value of a fixed asset whether arising from use, the passage of time or obsolescence through technology and market changes. Overall, the decrease in the value of capital stock.

Depreciation adjustment The difference between the value of the business of the part of fixed asset consumed during the accounting period and the amount of depreciation charged on an historical cost basis.

Depression A part of the atmosphere where the surface pressure is lower than in surrounding parts – often called a 'low'.

Depth alongside The depth of water alongside a specific berth-quay.

Deputy Chairman Person who deputises for a Chairman of a Company – the latter who has overall control of a Company. The Deputy Chairman position is found more in the larger Companies (PLC) – particularly in the Holding Company – and the duties vary by Company size and nature of the business.

Deputy Managing Director The deputy to the Managing Director. Such a post is usually found in the larger Company and besides being deputy to the Managing Director, the position also carries other specific responsibilities at Director level within the Company.

DEQ (delivered ex quay) Named port of destination. See separate entry.

Der Derricks (Ship cargo lifting equipment).

De-ratting certificate A ship's document issued in accordance with Regulation 19 of the Public Health (Ships) Regulations 1970.

Der/cr Derricks /Cranes.

De-recognition The process of an entity/company/product/person loses its recognition such as opting out of Union recognition or trade association membership.

Deregulated economy An economy where a limited number of government regulations exist relative to the conduct of business and the economy. This embraces labour law, fiscal measures, company law, social benefits and technology. This is found in many African, South American and Far Eastern countries including the sub continent. It particularly applies to less developed countries. It favours strongly inwards investment.

Deregulated market See deregulated economy entry.

Deregulation The process of a entity/company/ country/government contracting out/abandoning a regulatory system/mechanism and offering a free market. See deregulated economy entry.
Derelict A vessel that has been abandoned by the crew but has not sunk.
Derivatives A broad product category which essentially covers instruments and contracts which derive their value from an underlying product – but which aim to reduce its risk profile.
Derived demand Demand for an item not for its own sake but for use in the production of goods and services. Demand which arises from the wish to satisfy the demand for another commodity or service. The demand for a factor of production that depends on the demand for other goods or service. The greater the demand for the final product, the greater the demand for the factor of production. Simply, no one wants capital goods, such as a ship for its own sake, but for the final consumer goods they help to produce: the demand for a tanker is the demand for oil and its products.
Derrick Ships lifting apparatus.
Derv Diesel Engined Road vehicle.
Derv Fuel Gas/diesel oil suitable for use in high speed, compression-ignition engines in road vehicles.
DES (Delivered Ex Ship) (Named port of destination).
Descriptive Data Entry Non variable non changeable data entry expressed in plain language, full or abbreviated.
Designated place A place or area approved by Customs for the receipt and examination of imported goods from third countries. A Customs definition.
Design draft Maximum permitted draft of a vessel as defined in accordance with the ship design.
Design weight Weight for which the manufacturer has designed the vehicle, wagon, container etc; to operate.
Desk planner A monthly calendar with adequate space for each daily entry – accommodated on a desk – which forms a record of forthcoming engagements throughout the month and subsequent ones.
Desk research The process of conducting research using existing available published data such as Government Statistics/reports.
De-slopping barges Barges/lighterage available at a port for the purpose of vessels discharging their oil spillage and refuse therein.
Despatch An agreed amount payable by the owner if the vessel completes loading or discharging before the laytime has expired. Also termed despatch money. A BIMCO term; or the process of sending the goods.
Despatch advice See despatch note entry.
Despatch Bays The point from which containers, vehicles, railway wagons etc., are physically loaded or unloaded.
Despatch department The department of a particular entity/manufacturer/exporter which undertakes all the distribution arrangements of the merchandise.
Despatch half demurrage A chartering term indicating the rate of despatch money will be at half the rate of demurrage.
Despatch Money An agreed amount payable by the owner if the vessel completes loading or discharging before the laytime has expired. A BIMCO term.
Despatch note A document prepared to advise a third party of actual quantity and date of despatch and method of transportation.
Despatch on all time saved (ATS) Despatch money shall be payable for the time from the completion of loading or discharging to the expiry of the laytime including periods excepted from the laytime. A BIMCO term.

Despatch on (all) working time saved (WTS) or 'on (all) laytime saved' Despatch money shall be payable for the time from the completion of loading or discharging to the expiry of the laytime excluding any periods excepted from the laytime. A BIMCO term.
Destination The ultimate named destination place according to the contract of carriage.
Destination airport The specified airport of destination for an aircraft, passenger, or consignment.
Destination port The specified seaport of destination for a ship, passenger, or consignment.
De-stuffing The process of unloading a container. Also termed stripping.
Detachable front end The front end of a step-frame or low-loading road trailer which may be detached to facilitate trailer loading from the front. Both manually and hydraulically operated detachable front ends are available.
Detainment Temporary legal restriction of a vessel.
Detention The holding of a vessel in a port – without resorting to seizure, arrest or capture – by a sovereign power or competent authority or compensation assessed against a shipper for the delayed return of the carriers equipment beyond allowable free time; or detention of aircraft, barge, vehicle, railway wagon or cargo. See arrest entry.
Detention charge Charges raised on detention of goods, equipment or transport units. See detention entry.
Det Norske Veritas The Norwegian Ship Classification Society based in Hovik, Norway.
Detonator Assemblies, non-electric, for blasting Non-electric detonators assembled with and activated by such means as safety fuse, shock tube or detonating cord. This may be of instantaneous design or incorporate delay elements. Detonating relays incorporating detonating cord are included. Other detonating relays are included in 'Detonators, non-electric'.
Detonators Articles consisting of a small metal or plastics tube containing explosives such as lead azide, PETN or combinations of explosives. They may be constructed to detonate instantaneously, or may contain a delay element. The term includes detonators for Ammunition; detonators for blasting, both electric and non-electric and detonating relays without flexible detonating cord.
Detour To follow a route which is not the most direct or customary one.
DETR Department of Environment, Transport and the Regions. UK government department also responsible for shipping.
De-unionisation
The process of a company/entity moving from a work force with union membership to one where union membership has been withdrawn.
DEUPRO Freight Trade Facilitation organisation dealing with the rationalisation and modernisation of inter-national trade procedures in Germany, and development of electronic data processing techniques. It represents Germany at both National and International level.
Deutsches Institut Fu Normung (German) German's Standards Institute.
Deutsche Schiffs – Revision und – Klassifikation (German) German Ship Classification Society of Germany.
Devaluation The process of an exchange rate which is fixed officially by a government and is subject to a change imposed by a government causing the price/value to fall/lower in terms of other currencies. It may be a five per cent devaluation of an exchange rate. Overall, it is the process of reducing the official rate at

which one currency is exchanged for one or a range of other currencies.

Devanning The removal of contents from a container. Also termed stripping, unpacking or discharging.

Developed market A market situation whereby the total market is using/buying the facility (product or service), and any penetration by a competitor may prove difficult other than for strong competitive reasons such as price, design, after sales, specification, etc.

Developed Market Economy A country whose economy is fully developed such as USA, Japan, Germany, France, etc.

Developing countries Developing countries are those states which are experiencing a rapid shift from an agricultural community to an industrial and urbanisation expansion. Overall, literacy is improving but wage scales are low but rising as industrialisation develops and brings more technical skills/training to the work force. The country tends to have rapid economic growth which is export led. Formerly such countries tended to be called less developed countries.

Developing market A market in process of development. See developing countries entry.

Development Assistance Committee Based in Paris – under the aegis of OECD – to work for financial and other assistance to the developing countries.

Development Director A most senior managerial position involved in the Development of product(s) or services(s) within a Company or entity. It usually involves product technical or commercial development and often is linked to the Research and Development of a Company/Entity activities. It can include Market Research. An important position in an era of technological development, keen competition and cost efficiency.

Development Manager The person responsible, usually at senior level, for developing a Company business or a sector of it. This is usually a commercial/technical post and may concern product development, provision of new services, commercial development of existing resources and so on. In some Companies it can embrace market research and/or technical development research.

Development Policy The development policy of a particular company which may be of a commercial or technical nature. It may be for example on product development and confining such development which falls into line with the company overall product range and not outside it. For example, it may be greenhouses and confined to the aluminium specification and not wooden structure.

Deviation A deviation occurs when a ship is on a voyage and departs from the customary or designated route with the intention of returning to that route to complete the voyage. A hull time policy is not affected by deviation, nor is a cargo policy that is subject to the Institute Cargo Clauses. In the case of a hull voyage policy deviation discharges the underwriter from all liability as from the time the vessel leaves the stated or customary course, unless the assured gives the underwriter immediate notice of the deviation upon receipt of advices, agrees to any amended terms of cover required and pays an additional premium, if required.

Deviation frauds The theft of cargo, for example, by the shipowner by deviating to another destination port.

Deviation from a route See deviation entry.

Device drivers The software needed to control the actions of equipment components which are connected to or controlled by the computer. They may need to be set (configured) to operate with the computer by setting certain variables. A computer term.

Devices Equipment components which are used to input or output data or are controlled in some way from a computer port, e.g. a printer. They may need to be set (configured) to operate with the computer by using DIP switches or software settings. A computer term.

De-watered Process of draining water, for example, from a canal lock or enclosed dock usually for maintenance purposes.

Dew point The temperature at which a given sample of moist air will become saturated and deposit dew. Also, the temperature at which condensation will take place within a gas if further cooling occurs.

df Dead freight – payable by shipper for space booked but not occupied.

DF Direction finder – ship navigational aid.

DFC Duty Free Confederation.

DFDR Digital Flight Data Recording – IATA term.

DFIs Development Finance Institutions.

DFT Department of Foreign Trade (Thailand).

Dft Draft.

DGB Dangerous Goods Board – IATA term.

DGCA Director General of Civil Aviation (various countries) – IATA term.

DGN Dangerous Goods Note. A document used in the conveyance internationally of dangerous classified cargo for any combination of surface modes of transport valid for the whole journey.

DGPS Differential Global Positioning System

DGSE Department of Geological Survey and Mineral Exploration (Burma).

DGSM Directorate-Gerneral Scheepvaart en Maritime Zaken (Directorate-General Shipping and Maritime Affairs).

DI Deutsche BA Luftfahrtgesellschaft mbH – an IATA member Airline.

Diagnostic specimens Any human or animal material including, but not limited to, excreta, secreta, blood and its components, tissue and tissue fluids, being shipped for purposes of diagnosis, but excluding live infected animals.

Diagonal ply A pneumatic tyre, the structure of which is such that the ply cords extend to the bead so as to be laid at alternate angles of substantially less then 90 degrees to the peripheral line of the tread, but not being a bias-belted tyre.

DIC Delivered in charge.

Dichotomous questions/questionnaire A questionnaire which features questions requiring only a 'yes' or 'no' answer. It can be extended to include 'don't know' or 'no preference' responses. A marketing term.

Dictum Meum Pactum (Latin) My word is my bond – the Stock Exchange motto.

Die intestate The absence of a will on a person's death.

Diesel/gas oil Petroleum distillate having a distillation range intermediate between kerosene and light lubricating oil.

Differential Design of rear axle which permits one wheel on the axle to turn faster than the opposite wheel. This is necessary because in turning, the outer wheel on the axle needs to turn faster than the inner wheel; or the difference between the physical price and futures price – a BIFFEX term.

Differential Global Positioning System A highly accurate GPS fixing system which employs the known difference (error) between true position and the obtained GPS position. The error difference is then used to calibrate precise position information using the direct GPS signal and the differential data.

Differential lock Device on the differential which enables the two wheels on driving axle to be locked in a solid state to eliminate wheel spin by one wheel when other wheel has grip.

Differential pricing The strategy adopted by a Company/Entity to price a product/service at a level which is different in various Regions/Areas of the market. This strategy will take account of the different market conditions in the sales outlets.

Digital Selective Calling A system which uses digital codes which allows a radio station to communicate with another station or group of stations.

Dillon round The fifth round of multilateral trade negotiations 1961-62 named after Douglas Dillon – under Secretary of State of the United States. See GATT entry.

DIMS Dimensions.

DIN Deutsche International Norm – the German equivalent of the BSI. See BSI entry.

Dinar The currency of Iraq and number of other countries. See Appendix 'C'.

Dinar invoice An invoice payable in dinars relative to an export sales contract for a specified country.

Dine-around plan Provision of meals in a package that allows the client to chose from among a variety of restaurants in the area.

DIP Department of Industrial Works (Thailand).

DIR Direct.

Direct access reservations A CRS feature that offers agencies a direct link via the host's mainframe to a number of airline or other suppliers; simple versions allow agents only to look at but not book in the non-host computer; enhanced ones permit real-time bookings there.

Direct air carrier An airline that actually operates aircraft, as opposed to an indirect air carrier, which charters passenger or cargo space for resale in its own name.

Direct Bill of Lading Acknowledgement in a Bill of Lading that cargo has been shipped between two specified ports of loading and discharge.

Direct booking The process of a client/passenger effecting a booking direct for reserved accommodation on a ship, train, aircraft with the transport operator and not booking through an Agent.

Direct cost of sales The sum of direct materials consumed, direct wages, direct expenses and variable production overheads.

Direct costs Costs which can be directly attributed to a product or service – e.g. raw materials or labour.

Direct cross trade service A shipping service operated between two seaboard countries by a Shipping Company whose fleet does not fly either of the national flags and therefore regarded as a foreign operator or a third carrier. It would provide a service between two seaports each of which would be situated in the seaboard trade country served. Such a service would not involve any intermediate ports of call.

Direct delivery Process of discharging cargo direct from the ship to inland transport system such as a bulk cargo shipload of rice which has been cleared by customs.

Direct expenses Costs other than materials or labour, which can be identified in a specific product or saleable service.

Direct Exporting The process of exporting a product direct to the end user/consumer and not through an intermediary. An example is the overseas Sales Executive or through advertising. Both involve a direct response to the exporter.

Direct exports All transactions involved in exporting are controlled by the supplier.

Direct export trading. (Sales abroad) In international trade export terms, the process of a manufacturer wishing to sell his goods/services abroad and in so doing becomes directly involved in contacting overseas buyers and associated product distribution in overseas markets. Basically, there are six methods including employing Company technical specialist export salesmen; using Agents; engaging Distributors, using the Company's own field organisation; establishing overseas branches to retail/distribute the product and finally develop multinational trading. See also separate entries.

Direct flight Service between two points, with or without intermediate stops.

Direct Interchange Transfer of leased equipment from one leasee to another. Such as a container, swap body, trailer and so on.

Directional tariff A reduced fare, often seasonal and usually round trip, for passengers originating at one end of a route only.

Direction Finder Radio bearing equipment. Included in the statutory navigation requirements for commercially operated vessels.

Direct labour cost The cost of remuneration for employees' efforts and skills applied directly to a product or saleable service and which can be identified separately in product costs.

Directly transported Goods are regarded as directly transported between two member EU States in that they do not pass through the territory of a non Community country, or if they pass through the territory of a non Community country under cover of a single transport document made out in a member state.

Direct mail advertising The process of advertising a product/service by postal mail shot.

Direct marketing The process of marketing through various advertising media that interact directly with consumers/end users. It generally requires the consumer to respond directly to the supplier/marketeer. See also direct mail advertising.

Director Person appointed by shareholders to be responsible for the efficient running of a company.

Director General The most senior full time post in an entity/organisation. Usually such a position is found in an entity involving a representative professional body such as a professional Institute. It is less common in a commercial business. Overall, the post is one of Administration and running the activity in accordance with the defined policy on an optimised resource basis.

Direct or held covered A condition in a marine insurance policy requiring that the insured voyage be direct from one place to another. If the voyage is delayed en route an additional premium is charged.

Direct ports A smaller port served by liner cargo services with the provision of transhipment cargo facilities to other shipping services.

Direct post related job A post at a seaport which is directly related to its activities such as a stevedore, berthing master, pilot, customs officer and so on.

Direct route The shortest operated route between two points.

Direct Sales The process of selling direct to the market and not through an intermediary such as an Agent or Distributor. See also direct marketing entry.

Direct sell British usage for selling tours directly to the public without going through agents.

Direct selling The process of the producer/manufacturer selling direct to the consumer and not relying on a 'middleman'/agent or other third party techniques. An example arises with the transport operator selling direct to the passenger and not relying on the Travel Agent to do so who would earn a commission of may be nine per cent. Overall, the process of selling a product/service between producer/supplier and end user without involving any intermediary such as a retail outlet or wholesaler.

Direct service A service between two terminals without any intermediate stop and usually regarded as the quickest and most direct route between the two terminals and free of any detours.

Direct transhipment The process of loading merchandise from one mode of transport to another without landing the consignment.

Direct yield The discount to yield based on the total life of the asset at simple interest.

Dirham The currency of the United Arab Emirates, including Libya, Qatar, and Morocco. See Appendix 'C'.

Dirham invoice An invoice payable in Dirhams relative to an export sales contract of a specified country.

Dirty Bill of Lading Bill of Lading which has been endorsed by the shipowner as the goods described thereon do not conform to what is offered for shipment i.e. package missing, damaged, soiled, inadequately packed.

Dirty bills A Bill of Lading which has been endorsed by the shipowner as the goods described thereon do not conform to what is offered for shipment i.e. package missing, damaged, soiled, inadequately packed and so on.

Dirty floating The process of a currency value being influenced by its government intervention with a view to levelling out/stabilising short run fluctuations in the market. See also floating currency and clean floating entries.

Dirty money Extra payment made to dockers for handling goods of an objectionable nature.

Dirty period charter A crude oil time charter.

Dirty petroleum products Crude oils such as heavy fuel oils. See clean petroleum products.

Dirty single voyage charter A crude oil voyage charter from port A to port B.

Dis Discount.

Disagio Commission raised for example by an Agent.

Disassembly, cargo The separation of one or other of the component parts of a consignment (from other parts of such consignment) for any purpose other than that of presenting such part or parts to customs authorities from a specific request of such authorities.

Disc Discount; or allowance made for money paid before it is due.

Disbursement Is an amount(s) paid by a carrier to an agent, shipper or another carrier which the delivering carrier then collects from the consignee. Such (an) amount(s) is/are incurred at origin for the provision of services incidental to the air carriage of the consignment and are to be collected on behalf of any person other than an air carrier or an air carrier where such amounts relate to services performed prior to air carriage from the point of departure indicated on the Air Waybill. These services will be limited to prior transportation, handling and documentation.

Disbursement Commission Agents commission earned on payment of disbursements on behalf of shipowners.

Disbursements Payments incurred for a variety of expenses embracing fresh water, bunkers, custom fees, telex, telephone, etc. It may involve a Port, Agent/Transport Operator, etc. Expenses incurred by the shipowner in connection with running a ship.

Disbursements Clause A clause in a policy covering hull and machinery of a ship, which incorporates a warranty prohibiting the assured from effecting additional insurances to include total loss of the ship, other than those listed in the clause. Breach of the warranty is not held covered, and discharges the underwriter from all liability under the hull and machinery policy as from the date of breach of warranty. The clause permits a maximum amount that may be insured in respect of 'disbursements', 'increased value', 'anticipated freight', etc. without breach of warranty taking place.

DISC Domestic International Sales Corporation (USA).

Disc Discount; or allowance made for money paid before it is due, such as five per cent discount on €1,000 bill subject to payment within one calendar month of date of account. In such a situation the client would pay €50.

Disc drive The hardware operates a disc of storage media which enables direct access. A computer term.

Discharge The unloading of a vehicle, vessel, aircraft or barge or landing of cargo.

Discharge of a contract A contract is discharged when it has come to an end. This may arise in four situations: by performance, agreement, breach or frustration. A legal term.

Discharge Port Port at which a vessel is finally unloaded.

Discharging berth Place where cargo is discharged from a ship in a seaport.

Disciplined movement The process of merchandise or passenger(s) being conveyed on a transport mode on a predetermined schedule thereby involving a specific departure and arrival time.

Disclosure It is the duty of the assured and his broker to tell the underwriter every material circumstance before acceptance of the risk. Failure to do so can result in the policy being rendered 'null and void'.

Discount The purchase by a bank or finance house of a bill of exchange at its face value, less interest – an international trade term found in banking; the process of offering a reduction on a published tariff/rate/bill, such as ten per cent discount on €1,000 results in a net payment of €900 ; or the amount by which a future or option is priced below its theoretical or 'fair value' or below the price of the underlying instrument. This term is applied to both fair value and crude basis. An IPE term.

Discount (accounting) The calculation of the present value of a cash flow of future payments.

Discount (bills) The purchase of an accepted term bill of exchange for less than its face value.

Discount (bonds) The under-par value of securities.

Discount coupon A coupon offering the holder a discount on specified purchases.

Discounted bill The process of selling a Bill of Exchange at its face value less a discount charge in the money market usually by a bank.

Discounted bulk buy The process of buying a large quantity of a product and in so doing obtaining a discount on the published public price/tariff.

Discounted cash flow An evaluation of the future net cash flows generated by capital project, by discounting them to their present day value. Two methods most commonly used are: (a) yield method for which the calculation determines the internal rate of return (IRR) in the form of a percentage, and (b) net present value (NPV) in which the discount rate is chosen and the answer is a sum of money.

Discounted rates Any rate below the standard minimum rate.

Discount (forex) The forward margin of a currency that is more expensive than the 'spot rate'.

Discount Houses Specialist financial institutions operating in the Money Market.

Discounting The purchase from the Seller at less than face value, of an Accepted Term Bill of Exchange (Draft) which is retained by the Discounting Bank until the Bill matures when the face value is recovered from the Drawee. An international banking definition.

Discount Market Process of selling and buying Bills of Exchange Treasury Bills, and providing a market for short bonds.

Discount rate A rate which is subject to a discount.

Discount pricing A price which is subject to a discount.

Discount travel Passenger fares marketed below the full fare/tariff. An example arises with the day excursion which permits the passenger to do the outward and return journey within one day. The tariff may be €70 compared with the normal full fare of €100 return thereby offering a thirty per cent discount.

Discriminating customers Clients who are discerning in their choice of product/service.

Discriminating market A market place composed of clients who are discerning in their product/service choice, behaviour, attitudes, taste and so on.

Discriminators A marketing term which identifies factors which influence a client to buy a product/service.

Disembarkation Process of passengers leaving a vessel.

DISH Data Interchange in Shipping. A system for exchanging data between carriers and merchants electronically.

Dishonoured Commercial term for the return unpaid of a cheque/draft.

Dishonoured Bill of Exchange A situation whereby an importer refuses to accept and/or pay a bill of exchange.

Disk/Disc A storage device consisting of a number of flat circular plates each coated on both surfaces with magnetisable material. A computer term.

Disk-head A protective ring used to protect 'D' rings from vehicles.

DISP Displacement.

Dispatch When so agreed in the charter-party, this is paid by the shipowner to the charterer as a result of the vessel completing loading or discharging before the stipulated time.

Displacement light The total weight of a vessel in metric tonnes without her stores, bunker fuel or cargo/passengers.

Displacement loaded
The total weight of a vessel in metric tones embracing the ship and her contents.

Displacement scale Table which shows in columns a set of draughts with a ship's corresponding displacement tonnages when she is lying in salt water and fresh water.

Displacement tonnage This term is chiefly used when referring to warships and is the actual weight of water displaced by the vessel when floating at her loaded draught.

Disponent owner When a charterer takes a vessel on bareboat charter/demise charter from the shipowner, the charterer becomes a disponent owner. Disponent owner is responsible for equipping, maintaining, providing stores, spares, lube oil and victualling for the crew of the vessel. He is also responsible for manning the vessel and arranging for surveys (both statutory and class surveys). Disponent owner takes over the complete operation (commercial as well as technical) of the vessel without involving high capital cost as actual owning the ship. Overall, the Master is responsible for all the financial management of the ship in all the situations presented.

Disposable pallet A pallet intended to be discarded after a single cycle of use.

Disposal of goods The act/process of getting rid of the goods.

Dispositioning All activities relating to the inland movement of empty and/or full containers.

Dispute Settlement Procedure See WTO dispute settlement procedure.

Dispute settlement system System which provides opportunities for individual countries to resolve trade problems through conciliation or with help of a WTO panel of experts which rules on the conformity with WTO of government measures. A panel's report must be adopted by the Council before the panel's rulings have force and effect. A WTO term.

Disseminate To remove shipments from a container, railway wagon or trailer for separate deliveries.

Dist Distance.

Distance freight Cargo discharged beyond original destination port at nearest safe port due to inability of shipowner to discharge it at original port e.g. hostilities, ice formation. Additional freight due termed distance freight.

Distant market A market which is some distance from the suppliers (exporters) location but crosses international boundaries. To the UK exporters distant markets include South America, USA, Africa, Middle East and Far East.

Distant months The far dated delivery months in a futures or option contract. An IPE term.

Distortion When prices and production are higher or lower than levels that would usually exist in a competitive market. A WTO definition.

Distress freight Cargo accepted for shipment in tramp vessel below normal rates in order to fill up vacant space.

Distributed systems A computer system in which the machinery is located at several different places but which operates as one unit. Systems in which processing is spread over several computer systems which are connected and share resources. One computer may have data relating to goods and another may have data relating to customers; as a system they provide facilities for combining this data.

Distribution The process of allocating and transporting goods to various parties. See also distribution centre, distribution analysis, and distribution cost entries.

Distribution analysis The process of conducting an analysis of the distribution network of a product involving cost, reliability, transit time, regularity of service, degree of technology, packaging cost and so on.

Distribution centre A depot which is used as a break bulk point for a region, area, or continent and from which individual shipments will be re-forwarded to final destination.

Distribution Chain The chain/elements/constituents through which the goods pass to reach the end user/buyer from the manufacturer. See logistics entry.

Distribution Channel See Channels of Distribution.

Distribution cost Cost incurred in warehousing saleable products and in delivering products to customers.

Distribution depot A depot which is used as a reception point for example, for general merchandise or other specified product(s), and from which the products are processed and re-forwarded to their final destination.

Distribution hub A seaport/depot which is part of a distribution network where individual consignments are received and re-forwarded to final destination. See hub and spoke entry.

Distribution infrastructure

Distribution infrastructure The transportation infrastructure and its associated resources such as commercial network available to the distributor/shipper.

Distribution Management Distribution management relates to the phase of activity which begins at manufacturing output or final product and ends with a satisfied customer.

Distribution Manager A managerial post responsible for all the distribution arrangements within an entity/company. This post is often called Transport Manager.

Distribution network planning A logistics term. It determines the best facilities, location, proper system configurations and selection of the correct transportation providers.

Distribution of the load The process of distributing the load ideally evenly throughout the length of the transport unit.

Distribution of wealth An economic term indicating the manner in which National or International wealth is distributed. This involves all the ingredients of wealth embracing land, property, income and so on. The degree to which such wealth can be distributed in an industrial country, for example, can be very much determined by the manipulation of taxation involving high taxes on the rich and very low on the poor, which enables a more equitable method of wealth distribution especially as the lower income groups enjoy State funded social benefits obtained through high taxation. This is a subject which occupies the attention of Governments worldwide not only in their internal economies, but also internationally. International conferences are held regularly on the subject and attended by world leaders.

Distribution requirements planning The function of determining the need to replenish stock at branch warehouse.

Distribution research The process of conducting market research into the distribution adequacy for a particular service/commodity/client. It would embrace pricing, frequency of service, reliability, technology, transit time and so on.

Distribution resource planning The set of concepts, procedures and techniques for the effective planning and control of the physical distribution.

Distribution service The separation of one or more of the component parts of a consignment (from other parts of the consignment) for any purpose other than that of presenting such part of parts to customs authorities at the specific request of such authorities.

Distribution system The distribution network.

Distributive ability A transport term. A measure of the scale/area the transport product(s)/service(s) serves. For example, road transport generally has a higher distributive ability than rail simply because the number of route miles of road is much higher than rail in any one country which thereby enables the road operator to have deeper penetration of the market without relying on other transport modes.

Distributor An importer who buys goods direct from a manufacturer and distributes them in the market place to retailers. Hence, the exporter sells direct to the distributor who retails to his customers within his territory. With the distributors, therefore, the exporter has one large credit risk in the area covered instead of a number of small ones. A direct export trading technique; or in transport terms an Entity/Company involved in distribution of merchandise. In such situations the Company may be a wholesaler or Agent who distributes the products to retailers.

Districentre An area dedicated in a seaport environs for the provision of a range of distriparks offering purpose built storage/processing and distribution facilities. It provides value added services.

District Heating Boiler Plant This is a boiler plant used to heat water, which is then pumped out in a network of underground pipes to provide space and process heating for household, commercial and industrial use. After use, the water is recycled for reheating.

District Heating Power Station A power station producing both hot water for district heating and electricity. The energy in the steam or gas is partly converted to electricity by the turbine.

District Manager A post involved in the overall Managerial control of a specified geographical area of a particular Company or service. For example, the District Manager may be responsible for a number of Commercial Bank's retail outlet offices within a specified area.

District sales manager (DSM) Person in charge of sales at a district office, for example, of a travel industry supplier.

Distriparks A dedicated purpose built area in a seaport environs for the storage/processing and distribution of cargo which is usually imported through its port.

Distri port A dedicated purpose built area located in a seaport for the storage/processing and distribution of cargo which is usually imported through its port.

DITADA Directorate of City and Regional Planning (Indonesia).

Diversified economy A very broadly based economy embracing both the service and manufacturing sectors each playing their full part in wealth creation. Also, the economy is likely to have a global focus.

Divest The process of stripping/selling the constituents/assets/intellectual property of a company or group of companies. Also called asset stripping.

Dividend cover The number of times that dividends are covered by available profits.

Divider A vertically mounted partition in a compartment on board an air freighter.

DIY Do it yourself.

DK Denmark – Customs abbreviation.

DK(S) Deck(s).

DKY Donkey boiler.

DL Delta Air Lines Inc. – an IATA member Airline.

D.Lat Difference in Latitude.

DLD Department of Livestock Development. (Thailand).

D-LLOYD Djakarta Lloyd – a shipowner.

dlo Dispatch money payable in respect of time saved at the loading port only.

DLO Dead Letter Office; or dispatch loading only – dispatch money payable on time saved during loading.

D. Long Difference in Longitude.

DM Maersk Air – an IATA member Airline.

DMA Defence Manufacturers Association (UK) or Defence Mapping Agency (US).

Dm/d Depth moulded.

DME Distance Measuring Equipment – IATA term.

DMEC Developed market- economy country.

DMP Difference in Meridional Parts.

DMR Department of Mineral Resources (Thailand).

DMT Dimethyl Terephthalate.

DMU Decision making unit.

DMX Distributed Mobile Exchange.

D/N Debit note.

DNV Det Norske Veritas – the Norwegian Ship Classification Society.

DNV-W1 One Man Operation (DNV requirements).

104

D/O Delivery order.
DO Distribution Office; Duty Officer; District officer; diesel oil; or Dock operations.
DOA Department of Agriculture (Thailand).
Dobra The currency of Sao Tome and Principe. See Appendix 'C'.
Dobra invoice An invoice payable in dobra relative to an export sales invoice for the specified country.
DOC Document of compliance.
Doc Credit Documentary credit.
DOC Credits Documentary credits.
DOCDEX Documentary Instruments Dispute Resolution Expertise.
Dock A purpose built basin – usually of rectangular design – situated in a port, often equipped with cargo handling equipment to permit the loading/discharge of ship's cargo where vessels can lie afloat.
Dockage A port handling tariff. See Dock handling charge entry.
Dockage dues Dues raised on a vessel by the port authority involving ships moored within the dock area.
Dock bumpers The cushioning devices (rubber, plastic, wood etc) mounted at the extreme rear of a chassis or trailer to take the impact when it backs into a loading dock or platform (road cargo).
Dock charges Port/Harbour authority charges raised on cargo/passengers passing over quay.
Dock charter A voyage charter party which specifies the vessel to discharge her cargo at a specified dock and in so doing the vessel will not have officially arrived until she actually reaches the named dock where cargo operations are to be performed.
Dock container parks A specified area in a port/dock area which accommodates containers.
Dock dues Charges made upon shipowners for the use of docks.
Docking survey Vessel placed in dry dock for survey purposes.
Dock levellers Devices to bring the height of the container floor and loading bay to equal proportion during the loading and unloading of lorries and railway wagons.
Dock receipt Document issued by a berth operator (port administrator) acknowledging receipt of goods specified therein on conditions stated or referred to in the document.
Dock socket Fitting that connects securing equipment to ship's structure.
Dock warrant Receipt issued by Dock Authorities for goods warehoused.
Doctrine of strict compliance The legal principle that the bank is entitled to reject documents which do not strictly conform with the terms of the credit.
Doctrine of substantial performance A legal term stating that if the person doing the work has completed it, but done it badly, the person can claim the difference between the price of the finished job and what it cost to put that job right.
Document Anything printed, written, relied upon to record or prove something.
Documentary collection A documentary collection is an operation whereby a bank collects a payment for a customer by delivering documents to the buyer. A payment term which involves the seller submitting documents via a bank, which forwards them to a bank in the buyer's country
Documentary credit The basis of international trade by means of which payment is made against surrender of specified documents. A documentary credit is basically an undertaking given by a bank at the request of a customer to pay a particular amount in an agreed currency to a beneficiary on condition that the beneficiary presents stipulated documents within a prescribed time limit. The basis of international trade by means of which payment is made against surrender of specified documents.
Documentary Credit cycle The various stages/legs through which a documentary credit payment is processed/effected involving seller/buyer advising confirming bank/issuing bank.
Documentary Credits Documentary credits are used as a means of guaranteeing the transaction of both importers and exporters in the conduct of international trade. The seller knows that he/she will be paid as long as he/she hands over predetermined documentation relating to the deal, which the buyer is assured that the bank will not hand over the money unless it receives a Bill of Lading, Air Waybill, Clean Report of Findings, Certificate of Origin, Packing List and so on. The Uniform Customs and Practice for Documentary Credits – UCP 500 – came into effect on 1st January, 1994 and outlines/details the arrangements under which Documentary Credits are handled/processed. See UCP 500 and 600.
Documentary draft Draft drawn by seller on buyer accompanied by shipping documents.
Documentary fraud The process of obtaining money, property in the goods or a pecuniary advantage by the issuance of forged and/or falsified documents used in a transaction. In some cases the documents may be forged or fraudulently altered after their execution, in others they may be genuine documents but with false information.
Documentary Instruments Dispute Resolution Expertise Rules drawn up by the ICC Banking Commission to facilitate the rapid settlement of disputes under the ICC Uniform Customs and Practice for Documentary Credits, the ICC Uniform Rules for Bank to Bank Reimbursements/under Documentary credits, and ICC Uniform Rules for Collections and the ICC Uniform Rules for the Demand Guarantees.
Documentary letter of credit Document whereby at the buyers request the importers bank authorises the exporter to draw a financial amount by a specified date for a particular shipment subject to the detailed documents being forthcoming. It may be revocable or irrevocable. See irrevocable and revocable letter of credit entries.
Documentary remittance Remittance of drafts and/or documents.
Documentation In International trade terms, documents used in the processing of an International consignment embracing Bills of Lading, certificate of origin, consular invoice, Air Waybill, delivery note and so on.
Document Code Document identifier expressed in code Document (Compared with a form) Data Carrier (record) with data entries. A data carrier containing a representation of stored information, such as a form, magnetic tape, i.e. a form is a document, but document need not only be a form. A document is a data medium with data recorded on it that generally has permanence and can be read manually or machine or both.
Documented Management System The documented description of the safety and pollution prevention objectives and the organisation structure, responsibilities, activities and resources laid down by the company to ensure its capability to meet defined objectives under the ISM code.
Document name The title of a document whether printed or typed.

Document of Compliance

Document of Compliance It is issued by an accredited Ship Classification Society e.g. Lloyd's Register of Shipping, London and certifies that the Safety Management System of the specified Shipping Company has been audited by an accredited classification society and that it complies with the requirements of the International Management Code for the Safe Operation of Ships and for the Pollution Prevention (ISM Code) adopted by the organisation by resolution A741 (18) as amended for the types of ships listed. The certified document is valid for five years subject to periodical annual verification as prescribed on page two of the document and in accordance with regulation 6 of Chapter IX of the Convention, the Safety Management System as found to comply with the requirements of the ISM Code.

Documents against Acceptance Used in obtaining payment. The buyers accept and sign the bill of exchange, agreeing to pay on a future date, and are then given the shipping documents.

Documents against Payment Used in obtaining payment. The buyers must pay when they accept the bill of exchange before they are given the documents. Also called cash against documents.

Documents of class Shipboard documents classifying the vessel and usually involving regular inspections of such equipment including boilers, tailshaft, etc.

Documents of flags Shipboard documents issued by the authority of the vessels country of registry.

Documents of movement/receipt Also referred to as Transport Documents, these are documents that provide evidence of the receipt of the goods by the shipping/forwarding party. They include Air Waybills, Bills of Lading, Parcel Post Receipts, Railway Consignment Notes and Forwarding Agents' Receipts.

Documents of Title Documents produced by a consignee as evidence of right to take delivery of goods (e.g. Bill of Lading). A document representing ownership of the goods it covers. A bill of lading may act as a document of title.

DOD Department of Defense (US).

Dog A marketing term which identifies a product/service is retained for non financial reasons as it remains no longer viable. It compares sharply with cash cow. See cash cow entry.

Dog bone bridge The fitting together of two twist locks in adjacent stacks.

Dog hooks Cargo handling equipment suitable for timber or case/barrel transhipment.

Dollar An account which is managed in dollars.

Dollar invoice An invoice payable in Dollars relative to an export sales invoice. Usually it would specify the dollar currency which may be US dollars, Canadian dollars, Australian dollars and so on.

Dollar 'spot' and 'forward' The price of the dollar for immediate delivery and prices on contracts, struck for settlement one month or three months. In the case of the sterling/dollar rate itself the settlement is six and twelve months ahead.

Dollar stocks The American and Canadian stocks.

Dolly Set of wheels set under front of container trailer to provide support when motive unit is disconnected. Or a small platform type vehicle equipped with rollers used to convey or move aircraft unit load devices on the ground – an IATA definition.

Dolphin A mooring buoy; a spar or block of wood with a ring – bolt at each end for vessels to ride by.

Dome car See observation car.

Domestic carriage The conveyance of a commodity which originates and terminates within the same country.

Domestic feeder service A feeder service which may be maritime, surface or air which feeds into a transport hub network and operating within a country. Accordingly, the feeder route does not cross international boundaries/customs posts to service the hub.

Domestic flight A flight undertaken within a country's domain. Hence, they are not subject to customs and immigration needs as the flights are internal and not international.

Domestic issue A loan stock or equity raised in the indigenous capital market and currency of the country of issue: Euros raised in Eurozone, US dollars raised in the USA, etc.

Domestic rate A rate applicable only within a country for a local transit consignment and not usually forming part of an overall through international rate.

Domestic Shipping Shipping services which operate within the territorial waters of a particular country.

Dominican Peso oro The currency of Dominican Republic. See Appendix 'C'.

Dominican Peso oro invoice An invoice payable in Dominican Peso oro relative to an export sales contract.

Dong The currency of the Socialist Republic of Vietnam. See Appendix 'C'.

Dong invoice An invoice payable in Vietnamese dongs relative to an export sales invoice.

DONGSUE Dong Sue Line – a shipowner.

Donkeyman Artisan in the ships engineer's department.

Donor Agency An Agency – usually International such as WHO, IMF, giving some form of aid to a country/region/area, or a country/state which donates financial aid to a country overseas usually a Less Developed Country or one subjected to a severe disaster such as an earthquake.

Donor Agent project A project under a donor agent – sponsorship. See donor agency entry.

Donor Air Financial aid or other provisions/services donated to an overseas territory. It may be financial, food, technical skills, research and so on.

Donor Government A Government giving some form of aid to a country/region/area. It may be some 100,000 tonnes of grain.

Door/Depot The conveyance of a consignment/merchandise from the shippers premises and finishing at the Container Depot or Container Freight station in the importing area.

Door/Door The conveyance of a consignment/merchandise from the shippers premises and finishing at the consignee premises thereby providing a combined transport movement.

Door to door The process of despatching goods from the senders/consignor address direct to the receiver/consignees address without any transhipment en route of the merchandise.

Door to door transport See door to door entry.

DOP Dropping the outbound pilot, or dropping off pilot.

DOT Department of Transport

DOTI Department of Trade and Industry. HM Government department with responsibility for trade matters.

Double Banking Two vessels moored alongside each other at a berth or two vehicles parked alongside each other in a roadway.

Double bottom Series of tanks above the keel of the vessel to afford improved construction rigidity and providing tanks for fuel oil, freshwater or water ballast in a ship.

Double-decked pallet-reversible pallet A flat pallet with similar top and bottom decks, either of which will take the full rated load.

Double decker A two deck or double deck bus having passenger seats on the lower and upper deck. A double decker bus has a seating capacity of about 70 passengers.

Double deck pallet A flat pallet with a top and bottom deck.

Double Hulls A vessel built to a double hull specification which became mandatory in UK from 1993 relative to all oil tankers of 600 ton deadweight and over. It protects cargo against collision, grounding, oil pollution and spillage.

Double insurance Where two or more policies are effected by or on behalf of the assured in the same adventure and interest or any part thereof and the sums insured exceed the indemnity, the assured is said to be over-insured by double indemnity (double insurance).

Double manning The process of two persons manning a piece of equipment or undertaking a particular job assignment where one would be sufficient/adequate.

Double option The buyer (taker) acquires the right either to buy or sell a futures contract at a basic (or strike) price agreed as with a call option usually not more than three months – a BIFFEX term.

Double-pricing System whereby domestic processors get a price advantage in procuring local resource materials. A WTO term.

Double stack container trains The movement by rail of containers stacked two high on purpose built railway wagons and particularly common in North America such as the Canadian Pacific Rail system between Montreal and Toronto. It is usually a block train operating between dedicated terminals. See land bridge entry.

Double stack trailers Cargo handling equipment at a container terminal usually a seaport which is capable of handling four 20 feet or two 40 feet loaded containers. It is a double stack trailer for terminal operations. Such equipment is in operation in the Port of Singapore.

Double stack train See double stack container trains entry.

Double stacker Two stackers rigidly connected to perform the function of a bridge fitting. See also bridge fitting and stackers entries.

Double wall A packaging term relative to the fibreboard carton with a double wall structure comprising of two layers of fibreboard to afford improved protection to the goods.

Double width (passenger) gangway A piece of port equipment linking the ship with the shore (quay) of double passenger width – about two metres – thereby enabling passengers/crew personnel to embark or disembark two abreast.

Dove A person or entity which adopts a conciliatory attitude towards a business situation.

Dove-tail socket A low profile deck socket.

DOW Delivery on wheels.

DOWA Dowa Line Co Ltd (Japan) – a shipowner.

Downgrade To change to a lower class of service or accommodation.

Down market The bottom end of the market embracing primarily 'C' and 'D' social/economic categories per capita. Such personnel are on low incomes.

Down market product/service The bottom end or near bottom end of a product/service range which is regarded as below the norm for quality of service, design, durability, specification and general market appeal.

Downside An unfavourable development. It may be lower profits, increased competition, or more punitive legislation.

Down time The period of time for which a work station is not available for production due to a functional failure.

Down turn The process of a decline relative to a particular situation. This would arise if the volume of business in a company fell resulting in a down turn in sales volume.

dp Direct port.

DP Documents against payment; Dynamic positioning: a position reference system employed to maintain station holding and heading; or direct port.

DP(AM) Dynamic Positioning with centralised remote manual control system.

DP(AAA) Dynamic Positioning with fully redundant automatic control system and emerging automatic control system.

DP (CM) Dynamic Position with centralised remote manual control system.

DP Control System The part of the DP System that calculates position and provides thruster command.

DPO Dynamic Positioning Officer – Watch keeping officer designated as a DP controller; or Documentary proof of origin.

DPP Damage protection plan.

DP System All systems and sub-systems that directly or indirectly effect the dynamic positioning of a vessel, comprising the following sub-systems; power system, thruster system and DP control system.

DPW Department of Public Works.

DR Dead rise; deposit receipt; derrick; or double reduction.

Dr Debit; debtor; drawer; or double reduction gearing.

Draft A written order for the payment of money 'drawn on', or addressed to a person holding money in trust or as an agent or servant of the drawer; or ship's draught depth vessel submerged in the water.

Draft agreement An agreement prepared for consideration by two or more parties. It has yet to be approved by all parties to the agreement and may require amendment in the light of discussion/negotiation before the agreement terms is finally settled.

Draft contract The provisional terms of a contract subject to amendment by either party before final agreement and ultimate signature by both parties.

Drag Chain Used to arrest momentum of ship during launching.

Dragging of an anchor An anchor moving over the sea bottom involuntarily and in so doing no longer prevents movement of the vessel.

Draught The depth from the waterline to the vessel's bottom.

Drawback Repayment of import duty partially or completely usually on re-exportation.

Drawbar combination A rigid vehicle towing a separate load carrying trailer attached by means of a draw bar.

Drawbar trailer Trailer with axles at both front and rear, the swivelling front axle being connected at the rear of the towing vehicle by means of a solid drawbar. No portion of the weight of the trailer is imposed upon the towing vehicle.

Drawdown The taking up of part or all of a term loan or standby credit facility. Until the drawdown is complete, the borrower will usually have to pay the lender a commitment fee on the undrawn balanced.

Drawee The person to whom a bill of exchange is addressed and who is expected to accept and/or pay the bill.

Drawer The party who writes out the bill or draft. The drawer is the creditor of the drawee.

Drawing A presentation of documents under a Documentary Credit requiring payment, acceptance or negotiation. Also sometimes referred to as a drawdown.

Drawn bond A redeemable bearer bond that has been selected for repayment (drawn) and should be returned by the holder to a specified person.

DRC Daily running cost (of a vessel).

Dreading Option general cargo.

Dredger A vessel designed/capable of recovering silt etc. from the bed of a river, port or estuary, etc.

Dredging Process of recovering silt etc., from the bed of a river, port, estuary, etc.

Dredging anchor An anchor moving along the sea bed but under the control of the ship.

Dress code The manner in which a person is dressed. It will reflect the culture and the nature of the occasion leisure/business together with the environment including climatic conditions. See culture entry.

Dressing, leather It may contain liquids or solvents of low flash point and hence be classified as flammable liquids.

DRGR Dredger.

Drift ice/Pack ice Term used in a wide sense to include any area of sea ice other than fast ice no matter what form it takes or how it is dispersed. When concentrations are high, i.e. 7/10 or more, drift ice may be replaced by the term pack ice. Previously the term pack ice was used for all ranges of concentration.

Drilling rig Equipment used in exploration of oil in maritime conditions/environment.

Drill Ship A type of drilling rig used in oil exploration at sea.

Driver The person in sole charge of a transport unit. On an aircraft the position is called a pilot; on a ship a master; on an electric multiple unit train a motorman; on a locomotive a driver and so on.

Drivers cab The accommodation housing the driver in a road vehicle, locomotive, etc. and featuring all the driving controls. It is the place where the driver controls/operates the transport unit.

Drop legs Lifting wires between spreaders and container.

Drop lock A bond with variable interest which switches automatically to a fixed rate if the variable rate falls below a certain point.

Drop off charge The charge made by the container owner and/or terminal operator for delivery of a leased or pool container into depot stock. The drop off charge may be a combination of actual handling and storage charges with surcharges; or fee charged by car rental company when the renter does not return the car to the original rental location.

DRP Distribution Requirements planning or Distribution Resource planning.

'D' ring A fitting connected to the ship's structure to which a tensioner or lash is attached.

(Dr) with date Date when special survey including drilling of shell and deck plates took place on a specified vessel.

Drum A flat ended cylindrical unit made of metal, fibreboard, plastic or plywood or receptacles of shapes when made of metal or plastics, for example round taper neck, barrel shaped or pail shaped receptacles.

Drum handle type fork lift truck Fork lift truck equipped to handle drums.

Dry barrel A barrel constructed to hold powders and commodities other than liquids.

Dry-Break Connection The quick acting disc valves; one in the ship's hose coupler, the other in the hose-end coupler, which close automatically as the hose is released.

Dry bulk cargo Bulk cargo shipped in a dry condition in complete shipments or containers. This includes cement, iron ore, forest products (newsprint, pulp and timber), scrap metal, grain, steel in all its forms, sea-dredged aggregate (sand and gravel) and so on.

Dry bulk cargo freight futures The Baltic Freight index relative to the dry bulk cargo market operative under BIFFEX. See also Baltic Freight index entry.

Dry bulk carrier fleet The bulk carrier fleet carry a wide range of dry bulk cargoes embracing grain, phosphate rock, iron ore, coal and toxic chemicals. Overall, there are four main parts: Capesize (80,000 dwt plus); Panamax (50,000-79,999 dwt); Handymax (35,000-49,999 dwt) and Handy (20,000-34,999 dwt).

Dry bulk container A container designed for the conveyance of dry solids in bulk without packaging. See dry bulk cargo entry.

Dry bulk freight rates See Baltic Dry index entry.

Dry cargo Goods not requiring controlled temperature protection and not of a liquid composition. Merchandise other than liquid carried in bulk.

Dry cargo chartering The chartering of a dry cargo shipment. See dry cargo.

Dry cargo container A container designed for the carriage of goods other than liquids. See dry cargo entry.

Dry cargo freight indices The indices which measure monthly the dry cargo charter rates. See dry cargo time charter index and dry cargo tramp trip charter index entries.

Dry cargo ship A vessel which carries all cargoes in bulk, excluding liquid cargoes.

Dry cargo tramp time charter index A monthly index on the dry cargo tramp time charter rates compiled by the German Ministry of Transport.

Dry cargo tramp trip charter A monthly index on the dry cargo tramp trip charter rates compiled by Lloyd's Ship Manager.

Dry dock Dock in which ships may be repaired or built, the water being pumped out as required.

Dry freight Goods not requiring controlled temperature protection.

Dry freight container A dry freight container is a closed container with one set of doors at the rear end. Dry freight containers are used for general cargo. The container is designed for general cargo which does not require temperature controlled protection.

Dry ice See 'carbon dioxide, solid' entry.

Dry lease Rental of a vehicle, particularly an aircraft, without an operator, crew or service; a pure dry lease does not include fuel, supplies or maintenance.

Dry market cargo cycle The periodic oscillation in the level of business activity that is directly aligned to influence the demand/supply of the dry cargo market especially freight rates and the overall dry cargo market. It has four phases: peak, decline, trough and growth. See shipping cycle entry.

Dry plate clutch (single or double) Clutch which operates in a dry state. A single plate clutch comprises a clutch plate with friction material on both sides sandwiched between the engine flywheel and a spring loaded cover plate which when released compresses the clutch plate against the flywheel, thereby transmitting the drive. A double plate clutch comprises two plates with frictional surfaces, thereby providing extra driving force which has the benefit of longer life.

Dry port A port situated inland serving an industrial/commercial region linked to one or more seaports by rail and/or road offering dedicated services between the dry port and importers destination overseas. Usually the dry port is container and multimodal orientated and has all the logistics and facilities needed by the shipper at the seaport. An example is Ipoh which serves the ports of Klang and Panang in Malaysia.

Dry weight Road vehicle weight exclusive of fuel, oil, water and loose equipment.

DS Days after sight; or depth sounder.

DSB Dispute Settlement Body: when the WTO General Council meets to settle trade disputes.

DSC Danish Shippers Council; Digital selective calling or Dynamically Supported Craft.

DSD Diplomatic Service Department (Brunei) or daily stacking demand.

DSM See district sales manager.

DSR DSR-Lines – a ship owner.

DSRK Deutsche Schiffs – Revision und – Klassifikation. German Ship Classification Society.

DSS Decision support system. A computer term.

DSU The Uruguay Round Understanding on Rules and Procedures Governing the Settlement of Disputes. A WTO term.

DSV Drum safety valve; or Diving Support Vessel.

DT Deep tank; Department of Transport; or TAAGT – Linhas AÈreas de Angola (Angola Airlines) – an IATA member Airline.

DTa Deep tank aft.

DTC Direct Trading Corporation (Thailand).

DTDs Document type definitions.

DTEC The Department of Technical and Economic Co-operation (Thailand).

DTf Deep tank forward.

DTI A trade computer system that supports a network of trade users and provides HCI and EDI access to CHIEF IES. Inventory systems are usually associated with the Direct Trader Input system; or Department of Trade and Industry – UK Government Department.

DTm Midship deep tank.

DTma Midship deep tank aft.

DTmf Midship deep tank forward.

DTO District Training Officer.

DTs Deep tanks.

DTS District Traffic Superintendent. Managerial position in a transport department, or transport company responsible for the day to day commercial and operational aspects of the transport mode for a particular area which may be road or rail.

DU Hemus Air – IATA Airline member.

Dual currency bonds Bonds in which the issue price and interest are payable in one currency and the principal amount is repayable in another currency.

Dual rate Situation whereby the shipper had the option of signing a contract and thus obtain an immediate lower freight rate, or later claim a deferred rebate.

Dual rate contract system An arrangement agreed between members of a liner conference for an immediate rebate to be given to a shipper (by making a deduction from the freight account) if he or she contracts to use exclusively the services of members of the conference for all his or her goods passing on the route it covers.

Dual registry A vessel which is registered in two States. For example: established in Panama by means of law No1 of 1973 3Nct and later altered in law No3 – also in 1973. According to these laws a foreign vessel – bare boat chartered for a period of two years can be registered in Panama for the same period without losing its previous registration and the opposite is also permissible, that is, from Panama to other countries.

Dual Valuation clause Clause in marine insurance hull/machinery policies involving total or constructive total loss.

Dual wheels FLTs Fork lift trucks having two sets of double wheels at the front, thus increasing the load area.

Dubbing The process of superimposing an audio commentary or captions/sub titles on an already completed film. The strategy/technique is used extensively when showing promotional films/sales videos overseas to reflect culture/language needs.

Due date The date on which a usance draft or a deferred payment becomes due for payment. A banking definition.

Due diligence Process of carrying out a task to a satisfactory reasonable standard.

Dumb Barges A dumb barge is a barge without its own mechanical means of propulsion. For movement, one or more such barges are towed by a tug, using a line connection. In the USA the term 'tug-boat' is used to define the towing vessel. The term Lighter is often used as a synonym for a dumb barge. This term originally applied to dumb barges which were used to 'lighten' seagoing ships in harbour areas by means of overside discharge.

Dumb craft A vessel without any motive power/propulsion unit such as a barge.

Dump Lowering the level of the water in a lock.

Dumping A deliberate policy by one nation, which sets out to export goods to another nation, at prices far below their true cost. It is often a strategy designed to earn foreign exchange at any price. Overall, it occurs when goods are exported at a price less than their normal value, generally meaning they are exported for less than they are sold in the domestic market or third country markets, or at less than production cost. A WTO definition.

Dunnage Wood, mats, or other material used in the stowage of cargo for protecting packages from damage, to separate, support or secure cargoes, and to prepare working platform.

Duopoly A market place with only two suppliers of a merchandise or service. It is in contrast with monopoly with only one supplier. See monopoly entry.

Duplex A two-story suite connected by a private stairway.

Duplicate sailing A sailing schedule which operates alongside the main sailing to provide additional capacity. The two sailings are usually referred to as the parent and duplicate sailings.

DUR Duration.

Dutch auction A sale in which the property is initially offered above its estimated value, after which the price is gradually lowered until an offer is accepted.

Dutiable cargo Cargo which attracts some form of duty i.e. Customs, Excise or value added tax.

Dutiable cargo list A list of shipboard cargo which attracts some form of duty i.e. Customs, Excise, Value Added Tax.

Duty Free Zone A designated area where goods or cargo can be stored without paying import customs duties awaiting further transport or manufacturing. A free port or free Trade Zone. See Free Trade Zone entry.

Duty of Assured Clause A clause, in the institute cargo clauses 1982, which specifies the responsibility of the assured to protect the goods from risk of loss or damage during the period of cover (this replaces the sue and labour clause, which appeared in the SG policy).

Duty of Assured Clause (Cargo) This appears in the Institute cargo clauses published for use with the MAR form of policy. It directs the attention of the assured, his agents, etc. to the duty (as required by the MIA, 1906) to take reasonable measures to avert or minimise any loss which is recoverable under the policy; also to ensure that all rights against carriers and others are properly preserved and exercised. Underwriters agree to reimburse the assured for any reasonable expenditure incurred by his compliance with the clause; in practice, these expenses are termed 'sue and labour' charges.

Duty of Assured Clause (Sue and Labour) This appears in the Institute hull clauses published for use with the MAR form of policy. It directs the attention of the assured, etc. to the duty (required by the MIA, 1906) to avert or minimise any loss which is recoverable under the policy and provides that underwriters will pay sue and labour charges incurred in compliance with the clause. The clause incorporates a 'waiver clause' and provisions regarding the application of underinsurance to sue and labour claims.

Duty suspension A suspension of customs duty on imported goods intended for export as compensating products after process; the goods remaining under customs control and out of free circulation.

Duties Taxes levied by Customs on imported or exported goods.

Duty Free Imported cargo not subject to any duty.

DVC Delivery Verification Certificate.

DVLC Driver and Vehicle Licensing Centre.

DVP Delivery-Versus-Payment A mechanism in an exchange-for-value settlement system that ensures that the final transfer of one asset occurs only if the final transfer of another asset occurs simultaneously. Assets may include money, securities or financial instruments.

D/W Dock Warrant.

dw Deadweight.

DWAT Deadweight all told.

dwc Deadweight capacity.

DWCC Deadweight carrying capacity.

Dwell time Period during which an air freight consignment following customs clearance is collected by the importer; or the length of time a container remains in the terminal or cargo in the container freight station.

dwt Deadweight tonnage. Weight in tons of cargo, stores, fuel, passenger and crew etc., carried by the ship when loaded to her maximum summer loadline.

Dy Delivery or dandy.

DY Alyemda Democratic Yemen Airlines – an IATA airline.

'Dye, NOS' or ' Dye intermediate, NOS' These are cyclic or ring compounds, containing an amino, hydroxyl, sulfonic acid, or quinine group or a combination of these groups used in the manufacture of dyes.

Dynamic Stresses Forces superimposed on Static Stresses whilst ship is under way.

E1 Aer Lingus (Aer Lingus Ltd.) – IATA member airline.

E4 Aero Asia (aero Asia International PVT) Ltd.) – IATA member airline.

E5 Samara Airlines (Joint Stock Company Samara Airlines) – IATA member airline.

E8 Alpi Eagles SpA. – IATA member airline.

E The substance has explosive properties – an IATA definition; or East – a point on the compass.

EA European Air Express – member of IATA.

EAB (Since December 1997 the Euro Banking Association, formerly the ECU Banking Association) An interbank organisation which is intended to be the forum for exploring and debating all issues of interest to its members and, in particular, matters pertaining to the use of the euro and the settlement of transactions in euro. Within the EBA, a clearing company (ABE Clearing, SociÈtÈ par Actions SimplifÈe a capital variable) has been established with the purpose of managing the Euro Clearing System as from 1st January 1999. The Euro Clearing System (Euro 1) is the successor system to the ECU Clearing and Settlement System.

EABC European/ASEAN Business Council.

E and OE Errors and omissions excepted.

EAC East African Community or European Agency for Co-operation.

EACON Far East, Colombo, Mauritius, East Africa, Far East Service (Nedlloyd lines).

EAD Entry Acceptance Data.

EAEC European Atomic Energy Community; or European Airlines Electronic Committee (IATA term).

EAEG East Asia Economic Grouping.

EAF/GUL/SAS East Africa/Gulf/South Asia container service (IOS).

EAGGF European Agricultural Guidance and Guarantee Fund. It funds the EECs Common Agricultural Policy.

EANA European Article Numbering Association.

EANPG European Air Navigation Planning Group – ICAO term.

EARC Extraordinary Administrative Radio Conference – ITU TERM; or Elimination of Ambiguity of RT Call Signs Study Group of ICAO.

Early cash recovery pricing The process of a company/entity to price a product/services at a level which will generate adequate income to fund the product investment outlay in a relatively short timescale.

Earned premium Premium in respect of that part of the insurance where the adventure has attached and terminated and during which the insurer was on risk. Where time is of the essence, the premium is 'earned' pro rata to the time on risk. If the policy pays a total loss the whole premium is deemed earned.

Earnings The amount of profits after meeting all interest and preferential dividends payments available for the equity capital.

EAS Eurocontrol Assistance Scheme – IATA term.

Ease of handling The degree of handling facilitation with which an item/consignment may be handled either manually or mechanically. For example, palletised cargo is ideal for the fork lift truck to handle, but a 30 tonne transformer requires special equipment and arrangements to be instituted.

EASI European Association for Shipping Information.

EASIE Enhanced Air Traffic Management & Mode S Implementation – Europe Programme – IATA term.

East Asia Travel Association (EATA) Promotes travel to the area and disseminates news to members in the area.

Eastern Air Lines A National Airline and member of IATA.

Eastern bloc counties Countries embracing the former Russian/Communist bloc.

EAT Estimated or expected ship, aircraft, or consignments arrival time.

EATA Europe Asia Trades Agreement.

EATCHIP European ATC Harmonisation and Integration Programme – IATA term.

EATCOS Task Force on the Integration of the European Air Traffic Control Systems – IATA term.

EATMS European Air Traffic Management System – IATA term.

Ebb tide The receding or running out of the sea.

EBCS European Barge Carrier system.

EBI Emirates Bank International

EBTIDA Earnings before interest, taxes, depreciation and amortisation.

EBRD European Bank for Reconstruction and Development.

E Brokering Electronic Brokering.

EBU European Broadcasting Union (Eurovision).

EC European Communities/Community; Energy Corporation (PNOC) Philippines; or East coast.

ECA Economic Co-operation Administration, or Economic Commission for Africa established for the economic development of Africa.

ECAFE Economic Commission for Asia and the Far East. Established to work for economic reconstruction and development in Asia and the Far East.

E-CASys Export Customs Authority System.

ECB (European Central Bank) The ECB has legal personality. It ensures that the tasks conferred upon the Eurosystem and the ESCB are implemented either by its own activities pursuant to its Statute or through the national central banks.

ECCM East Caribbean Common Market.

ECCP European Chamber of Commerce of the Philippines.

ECCTO The European Coastal Chemical Tankers Association.

ECDC Economic Co-operation among Developing Countries (UNCTAD).

ECDIS Electronic Chart Display & Information Service. A complete electronic chart system and information service coupled with an automatic chart updating procedure. It features a vector based system.

ECE Economic Commission for Europe, based in Geneva. Its aim is to promote economic reconstruction and development in Europe. It was established by ECOSOC.

ECG European Cooperation grouping or Export Credits Guarantee.

ECGB East Coast of Great Britain.

ECGD Export Credits Guarantee Department (UK). Insurance on goods shipped against non-payment by reason of commercial and/or political risks as arranged.

ECGD Rating The rating the ECGD would evaluate for a particular risk under consideration.

Echo This option causes transmitted characters to appear on the sender's screen, enabling them to check that the data is being correctly transmitted.

Echo sounder Instrument measuring depth of water beneath a vessel.

ECI East coast of Ireland, or Export consignment identifying number.

ECLA Economic Commission for Latin America. Established by ECOSOC.

ECMA European Computer Manufacturers Association – IATA term.

ECNA East Coast North America.

ECNR European Council for Nuclear Research.

ECNs Electric Communication Networks.

ECOCOM Economic Commission for Europe – Trade and Technology Division based in Geneva.

ECOFIN See EU Council.

Ecology Science of plants and animals in relation to their environment.

E Commerce Electronic Commerce.

Econometric models A system of simultaneous equation for forecasting based on mutual dependency among the variables used.

Econometrics The adoption of a mathematical/statistical technique to an economic problem usually with the assistance of electronic data processing resources.

Economic and Financial Committee A consultative Community body set up at the start of Stage Three, when the Monetary Committee was dissolved. The Member States, the European Commission and the ECB each appoint no more than two members of the Committee. Article 109c (2) of the Treaty contains a list of the tasks of the Economic and Financial Committee, including the review of the economic and financial situation of the Member States and of the Community.

Economic and Monetary Union (EMU) The Treaty describes the process of achieving Economic and Monetary Union in the European Union in three stages. Stage One of EMU started in July 1990 and ended on 31st December 1993; it was mainly characterised by the dismantling of all internal barriers to the free movement of capital within the European Union. Stage Two of EMU began on 1st January 1994. It provided, inter alia, for the establishment of the European Monetary Institute (EMI), the prohibition of monetary financing of and privileged access to financial institutions for the public sector and the avoidance of excessive deficits. Stage Three started on 1st January 1999 in accordance with the decision pursuant to Article 109j (4) of the Treaty, with the transfer of monetary competence to the Eurosystem and the introduction of the euro.

Economic and Social Council United Nations Organisation of 144 members represent employers, trade unions, consumers and other interested groups. Expresses opinion on Commission proposals. An important UN organisation based in New York. It deals with worldwide economic and social problems, and co-ordinates international statistics.

Economic Commission for Asia and the Far East Established to work for economic reconstruction and development in Asia and the Far East.

Economic Commission for Europe Based in Geneva, its aim is to promote economic reconstruction and development in Europe. It was established by ECOSOC.

Economic Commission for Latin America Based in Santiago it is the Economic Commission for South and Central America.

Economic Community of West African states (ECOWAS) An economic group of 20 West African States.

Economic criteria A basis on which a situation is progressed/evaluated for consideration. This may include one or more of the following: taxation, unemployment, inflation, exchange rates, prices, income, distribution of wealth and so on.

Economic cycle The various stages through which a region or country economy progresses. It will commence with an underdeveloped economy to a fully developed economy. This will embrace depression, growth, inflation, decline and so on. See economic reports and economic indicators entries.

Economic growth The steady process of increasing productive capacity of the economy, and hence of increasing national income.

Economic indicators Various statistics available to measure a particular economic performance. For example, to measure the UK economy one can take index numbers and rates of change in various areas for the principal indicators. This includes industrial, and manufacturing production (broken down by sectors), engineering orders, retail sales, unemployment, exports, imports, the current account, the money supply, bank lending, earnings, wholesale and retail prices and sterling exchange rate variation.

Economic infrastructure The economic infrastructure of a country embrace all the resources which contribute to a nations economic performance. See economic indicators entry.

Economic reform The process of introducing into an economy/country/trading bloc the most recent economic strategies and thereby displacing the out of date policies.

Economic reports An economic report usually on a country or region basis giving an up-to-date assessment of the country/region under review indicating the size and main categories of imports and exports; the country/region potential as a market for exporters; the major development plans; details of overseas import and exchange control regulations; the more usual methods of payment for imports and so on. Such reports are available from Banks, Trade Associations, Chambers of Commerce and used by potential exporters.

Economics A social science studying how people attempt to accommodate scarcity to their wants and how these attempts interact through exchange. Acknowledging such a definition, it must be recognised that economics does not produce a clear line of division between economics and allied subjects. In so doing it overlaps with many other subjects such as politics, psychology, ethics, marketing together with other subject areas of wider dimension such as international trade, monetarism and so on.

Economic sanctions Measures taken by the International Community or individual governments on a bilateral or multilateral basis which penalises/legalises restrictions/constraints on the free movement of trade. Such sanctions are of a economic nature and may embrace commercial aid, inwards investment.

Economic situation The economic situation in a particular country/region/trading bloc and measured/evaluated by the various economic indicators available. These include employment level, inflation, money supply and so on. See economic indicators and reports entries.

Economic speed The optimum speed of the transport unit which will produce the best possible financial result for the transport carrier/owner.

Economic strategy The policy adopted for example by a Company, Institution or Government relative to its economic policies. This will embrace one or more of the following areas: inflation, price, income, unemployment, monetarism, distribution of wealth, taxation, exchange control, import/export policies and so on.

Economic Trends The process of analysing/evaluating one or more economic constituents. For example, such trends would materialise when evaluating over a specific period monthly figures relative to one or more of the following: inflation, prices, income, employment, exchange rates, balance of trade and so on. See also Economic Indicators.

Economic viability The degree of profitability in a particular scheme, project or other specific situation based on optimum use of economic resources.

Economic zone An area designated for economic development to improve wealth and employment in the zone with focus on industrial and service sectors. Often sponsored by governments.

Economies of scale These exist when expansion of the scale of a firm or industry causes total production cost embracing both internal and external factors to increase less than proportionately with output.

Economy class A class of passenger accommodation primarily found on airline services offering economy fare tariffs at discounted rates.

Economy fare A discount fare from the basic passenger tariff. It may have for example a 40% discount thereby offering a €12 economy fare compared to the €20 normal basic passenger tariff. It may be operative on any transport mode especially air, rail, and coach, and it is restricted to certain specified services and days on which the economy class passenger may travel.

Economy flight A flight on which economy fares/tickets are accepted. Usually such flights are operated at less popular periods or times of the day.

Economy hotel See hotel classification.

ECOSOC Economic and Social Council. United Nations Organisation of 144 members representing employers, trade unions, consumers, and other interested groups. Expresses opinion on Commission proposals. An important UN organisation.

ECOWAS Economic Community of West African states.

ECP Electronic Commerce Project.

ECPD Export cargo packing declaration.

ECPS Environment & Consumer Protection Service (European Union).

ECR Engine Control Room or efficient consumer response – a logistic term.

ECS Electronic Chart System. Several types are currently under manufacture but without an acceptable chart correction method. The ECDIS when fully developed would expect to gain worldwide approval from such organisations as IMO.

ECSA European Community Shipowners' Associations or East Coast South America.

ECSAs European Credit Sector Associations.

EC Sales List A VAT form number VAT 101 which must be completed quarterly by all traders who conduct sales to other member EC States. It details the VAT number of the buyer, the member State and the total value of the goods sold according to the customer's VAT number.

ECSC European Coal and Steel Community.

ECSI Export Cargo Shipping Instruction. Shipping instructions document from Shipper to Carrier.

ECT Europe Combined Terminals; or European Co-ordination Team – IATA term.

ECTAB Electronic Chart Table – optional accessory to the Kelvin Hughes Integrated Navigation System.

ECU (European Currency Unit) According to Council Regulation (EC) No. 3320/94 of 20th December 1994, the ECU was a basket made up of the sum of fixed amounts of 12 out of the 15 currencies of the Member States. The value of the ECU was calculated as a weighted average of the value of its component currencies. As the official ECU it served, inter alia, as the numeraire of the ERM as a reserve asset for central banks. Official ECUs were created by the EMI through three-month swap operations against one-fifth of the US dollar and gold assets held by the 15 EU national central Banks. Private ECUs were ECU-denominated financial instruments (e.g. bank deposits or securities) based on contracts which, as a rule, made reference to the official ECU. The 'theoretical' value of the private ECU was defined on the basis of the value of the individual components of the ECU basket. However, the use of the private ECU was different from that of the official ECU and, in practice, the market value of the private ECU could diverge from its 'theoretical' basket value. Following Article 2 of Council Regulation (EC) No. 1103/97 of 17th June 1997 on certain provisions relating to the introduction of the euro, the private ECU was replaced by the euro on 1st January 1999 on a one-to-one basis. The official ECU ceased to exist with the termination of the swap mechanism at the start of Stage Three of EMU.

ECUATOREANA A National Airline and member of IATA.

ECU Bond Bond denominated in the European currency unit of the European Monetary system.

ECUK East coast of United Kingdom.

ECWA Economic Commission for Western Asia.

ED Engine designer (ships machinery).

EDB Economic Development Board (Brunei) or Economic Devt. Board (Singapore).

EDC Economic Development Committee.

EDCs European Distribution Centres.

EDCS Electronic Data Capture Service. A customs term.

EDELL Enclosed double entry lifeboat launch.

EDF European development fund.

EDH Efficient deck hand.

EDI Electronic Data Interchange. The transfer of structured data from one computer system to another.

EDIA Electronic Data Interchange Association.

EDIFACT Electronic Data Interchange for Administration, Commerce & Transport. Organisation responsible to UN ECE for the development of standard EDI messages for Administration, Commerce and Transport.

EDIGEUR Pan European EDI group for the bank industry.

EDISHIP An organisation for exchanging data between Carriers and Merchants by electronic means.

Edit The process of editing a piece of written work, film, questionnaire and in so doing represent it in a more acceptable and professional standard having regard to the purpose/objective of the piece of work. In marketing terms editorship is widely used to ensure the communication skill is pitched at a lucid professional level to ensure the targeted readership comprehends it having regard to their literacy skills/levels.

Editorial publicity The article/material written in a newspaper/journal/magazine the contents of which is the prerogative of the editorial team. It may be a house magazine in which the leading article written by the editor reflects each month's topical areas of interest to the readers but basically has a slant towards promoting the company's profile/products.

EDL Electricity Authority of Laos.

EDP Electronic Data Processing. Computer processing of Data.

EDPITAF Education Development Projects Implementing Task Force (Philippines).

EDR Equipment Damage report.

EDSP Exchange Delivery Settlement Price. The settlement price for physical delivery. An IPE definition.

EE Errors excepted.

EEA Electronic Engineering Association (UK), or European Economic Area.
EEB European Environment Bureau.
EEC European Economic Community.
EEC Member States See European Union Entry.
EECP Expanded East Coast Plan (US) – IATA term.
EEP Export enhancement programme: programme of US export subsidies given generally to compete with subsidised agricultural exports from the EU on certain export markets. WTO definition.
EES Harmonisation of European Planning and Implementation Strategies in the Eastern Region – IATA term.
EET Eastern European Time.
EEZ Exclusive Economic Zones.
EF European foundation.
Effective date The date on which the insurance under a policy begins.
Effectively closed In packing terms – liquid tight closure.
Effective (nominal/real) exchange rates In their nominal version, effective exchange rates consist of a weighted average of various bilateral exchange rates. Real effective exchange rates are nominal effective exchange rates deflated by a weighted average of foreign prices or costs relative to domestic one. They are thus measures of a country's price and cost competitiveness. The euro area nominal effective exchange rate is based on BIS calculations. The figures are a weighted average of the euro area countries' effective exchange rates. The weights are based on 1990 manufactured goods trade, capture third-market effects and refer to extra-euro area trade. The real effective exchange rate for the euro area is derived using national CPLs.
Efficient Consumer response A logistics term. It is the realisation of a simple, fast and consumer driven system, in which all links of the logistic chain work together in order to satisfy consumer needs with the lowest possible cost.
EFL External financing limit.
EFP Exchange of Future for Physical – an International Petroleum Exchange term.
EFP/EFS An EFP (EFS) is the exchange of a futures position for a physical (swap) position. With EFP (EFS) two agreeing parties switch cash (swap) and futures position through the Exchange. An IPE term.
EFTA European Free Trade Association, comprising Switzerland, Austria, Sweden, Finland, Iceland and Norway.
EFTPOS Electronic funds transfer at point of sale.
EGAT The Electricity Generating Authority of Thailand.
EGC Enhanced Group Calling. A term used with GMDSS Communications.
EG (exempli gratia) (Latin) For example.
Egyptair The National Egyptian Airline and member of IATA.
Egyptian Pound The currency of Egypt. See Appendix 'C'.
EHA Equipment Handover Agreement. Agreement acknowledging condition signed when taking over Carrier's equipment and returning it, which incorporates terms of contract under which equipment is taken over.
ehp Effective horse power.
EHS Extra high strength.
EJ Aer Lingus plc. The Republic of Ireland National Airline and member of IATA.
EIA Engineering Industries Association (UK) or Environmental Impact Assessment.
EIB The European Investment Bank.
E&IC Office Excise and Inland Customs Office.

EIC Energy Industries Council (UK) or Excise and International Trade.
Eighths basis A basis used for the calculation of unearned premiums involving the assumption that the average date of issue of all policies written during a quarter is the middle of that quarter. So at the end of the year premium arising in the first quarter is ⅛ unearned, ⅜ for the second quarter etc. An insurance term.
EIL Egyptian International Line – a shipowner.
EILAT Eilat Container Line – a shipowner.
EIR Equipment Interchange Receipt.
EITZ English Inshore Traffic Zone.
EIU Even if used – chartering term or Economist Intelligence Unit.
Ejusdem generis (Latin) Things of a like nature.
EK Emirates. An IATA member Airline.
Ekuele The currency of Equatorial Guinea. See Appendix 'C'.
EL Electrical propulsion (marine engine) builder or Greece – customs abbreviation.
El Al Israel Airlines. The Israeli National Airline and member of IATA.
Elapsed flying time Duration of actual air travel after all scheduled ground time at intermediate stops has been deducted.
Elapsed travel time Duration of an air trip including time on the ground between connecting flights.
Elastic elongation The temporary change in length of a fibre or yarn under tension, which is recovered when the tension is removed.
Elasticity An economic term. There are three major areas of elasticity as given below:
(i) Price elasticity of demand and supply
(ii) Income elasticity of demand and supply
(iii) Cross elasticity of demand and supply
They all measure the proportionate change of demand and supply to the proportionate change in price, income and finally price of product. Thus elasticity is a proportional measure of response of one variable to another. The elastic (non permanent) elongation of a unit length of an element caused by a unit load. May refer to a material or a composite structure such as a mooring line. Also in maritime terms.
Elasticity of Supply An economic term measuring the responsiveness of supply to a change in price. It is calculated by dividing the percentage change in price with the percentage change in the quantity supplied.
Elastic of demand Where a small percentage change in price results in a proportionately larger change in the quantity demanded. (elasticity greater than one).
ELBA Emergency Locator Beacons Aircraft.
ELEC European League for Economic Co-operation or electric (ships machinery).
Electrolysis A process by which direct current passes from one electrode (the anode) through a liquid electrolyte to another electrode (the cathode) in a cell. The passage of electrons from the anode causes the pure metal only to be electrodeposited on the cathode. Because of the purity of the deposited metal, electrolysis is often used in refining.
Electrolyte It is the term commonly applied to the dilute sulphuric acid used in the ordinary lead plate storage batteries. The solution of potassium hydroxide used in some storage batteries is also called electrolyte. The term electrolyte is sometimes applied to the strong sulphuric acid which is meant for use in storage batteries after dilution.
Electromagnetic radiation The emission and propagation of electromagnetic energy from a source in the form

of electric and magnetic fields, which need no medium to support them and which travel through a vacuum at the velocity of light. This radiation encompasses the entire frequency range from y-rays to radio waves.

Electronic Bills The Bill of Lading which is transmitted electronically. See Bolero entry.

Electronic Commerce The process of sharing business information, maintaining business relationships and conducting business transactions by means of telecommunication networks or the production, advertising, sale and distribution of products via telecommunications networks. – WTO definition.

Electronic Data Capture Service An HM Customs facility available to traders which provides a single electronic gateway that can be used by businesses to send data for Intrastat and EC Sales list together with non-EC period entry schemes.

Electronic Data Interchange The process of sending a structured message between two user friendly computers. This has wide application in international trade. See EDISHIP.

Electronic Data Interchange Association Organisation responsible for the development of Electronic Data Interchange in international trade and related industries.

Electronic Data Interchange for Administration and Transport Organisation responsible to UNECE for the development of standard EDI messages for Administration, Commerce and Transport.

Electronic Data Processing Computer Processing of Data.

Electronic load list A cargo/freight load list produced electronically usually for a particular flight or sailing and required by customs for clearance purposes. See new export system.

Electronic mail Electronic mail is sending of messages, letters and other documents through an electronic system instead of using the postal service. This can be undertaken between users of a networked system, through network service providers. When mail is sent the user is informed that it has arrived as soon as they login to the system providing the service. Overall, it involves the user of friendly computers communicating with each other.

Electronic market exchange A market place where several sellers and buyers are present at the same time. An E-commerce term.

Electronic money (e-money) An electronic store of monetary value on a technical device that may be widely used for making payments to undertakings other than the issue without necessarily involving bank accounts in the transaction, but as a prepaid bearer instrument (see also multi-purpose prepaid card).

Electronic reporting A customs term whereby the importer submits statistical and fiscal details of goods to CHIEF electronically. See CHIEF entry

Electronic Ticket Delivery Network (ETDN) A proposed ticket delivery service to travellers on behalf of US travel agencies accredited by the Airlines Reporting Corp.

Electronic Trade Credits See ETC entry.

Electronic transmission The process of transmitting data electronically embracing diskettes, magnetic tape, VAN (Valued Added Networks) and facsimile

Electrowinning An electrolytic refining process in which the metal is recovered from the electrolyte rather than from an electrode.

Elephant's foot Lash end fitting used on Ro/Ro ships. Used in conjunction with a clover leaf or keyhole deck socket. Also, a colloquial name for the 'Bell Mouthed' shape Stripping Suction Pump.

Elevator See Grain elevator entry.

Elite The best members of a group selected to perform a specific assignment.

ELMA Empress Lineas Maritimas Argentinas – a shipowner.

El Niño Southern Oscillation (ENSO) A quasi-periodic occurrence when large-scale abnormal pressure and sea-surface temperature patterns become established across the tropical Pacific every few years.

Elongation Refers here to the total elongation (elastic and plastic) of a line.

ELT Emergency Locator Transmitter.

EM Electronic Magnetic Slip Couplings.

EMA European Monetary Agreement; or Eastbound Management Agreement – liner conference facility relative the Japan/Europe Freight Conference.

EMAC Engineering & Maintenance Advisory Sub-Committee – IATA term.

E Mail Electronic Mail.

E-manifest Electronic manifest of cargo or passenger carried on a transport unit. Ship – aircraft – road haulage vehicle

EMB Executive Management Board – IATA term.

Embargo An order of prohibition on trade or shipping/air/hovercraft services issued by the carrier, Government or other specified Institution.

Embarkation Process of passengers joining a ship.

EMBO European Molecular Biology Organisation.

EMCF European Monetary Co-operation Fund.

Emergency service A service to combat an emergency situation and primarily to cater for those needing medical attention to lessen/contain the emergency, and instituting measures to restore the situation to normality as soon as practical.

Emerging markets Countries with an economic industrial growth well above the norm such as Malaysia, Thailand, Brazil, India and Mexico. In financial terms the financial markets of developing economies.

EMI (European Monetary Institute) The EMI was a temporary institution established at the start of Stage Two of EMU (on 1st January 1994). The EMI had no responsibility for the conduct of monetary policy in the European Union, which remained the preserve of the national authorities. The two main tasks of the EMI were: (1) to strengthen central bank co-operation and monetary policy co-ordination; and (2) to make the preparations required for the establishment of the European System of Central Banks (ESCB) for the conduct of the single monetary policy and for the creation of a single currency in the third stage. It went into liquidation on 1st June 1998.

Emissivity The ratio of the emissive power of a surface at a given temperature to that of a black body at the same temperature and the same surroundings.

Empathy The strategy/technique adopted by progressive companies/entities which identifies the problems/expectations/needs of the buyer and potential buyers and their rationale in thinking. In so doing the seller responds to the buyer's needs on an ongoing basis. Accordingly, the seller reciprocates to the buyer's strategy and develops a partnership which aids business development to the mutual benefit of both parties. Empathy is basically market research led on a continuous basis and has regard especially to competition and the changing market environment.

Empty tank certificate Document confirming the tanker vessel has empty tanks immediately upon completion of discharging. The certificate is required by Charterers or Receivers.

EMs Export Management Companies.

EMS East Malaysia Shipping Ent. (Pte) Ltd. – a shipowner; Emergency Medical Service or Emergency Schedule.

EMS (European Monetary System) Established in 1979 in accordance with the Resolution of the European Council on the establishment of the EMS and related matters of 5th December 1978. The Agreement of 13th March 1979 between the central banks of the Member States of the European Economic Community laid down the operating procedures for the EMS. The objective was to create closer monetary policy co-operation between Community countries, leading to a zone of monetary stability in Europe. The main components of the EMS were the ECU; the exchange rate and intervention mechanism (ERM); and various credit mechanism. This mechanism ceased to exist with the start of Stage Three of EMU.

EMS Number Emergency schedule number. It is applicable to the shipment by sea of dangerous classified cargo and the comprehensive emergency schedules.

EMU See Economic and Monetary Union.

Emulation Software or hardware which acts in the same way as another system.

EMV Expected monetary value.

Emy Emergency.

EN Engine (Marine) builder.

En 29,000 The European standard for quality management systems which accords and recognises BS 5750. See BS 5750 and ISO 9000 entries.

ENANA Angolan – CAA term.

ENC An Electronic Navigational Chart held in a machine-readable form.

Enclo(s) Enclosure(s).

Enclosed dock A dock which is designed so that it may be separated by mechanical means – such as caissons, gates and locks – as required from the sea, river, or lake. The required height of water in the enclosed area is often maintained by pumping machinery.

Enclosed port A port which is designed so that it may be separated by mechanical means – such as caissons, gates and locks – as required from the sea, river, or lake. The required height of water in the enclosed area is often maintained by pumping machinery.

Encryption Method by which data is made unintelligible to protect its confidentiality.

End Consumer A company in whose hands the identity of the fabricated metal is finally lost in a more complex product, e.g. a refrigerator, motor car or transformer. Some of these companies also have fabricating subsidiaries. See Fabricator entry.

End frame Assembly at either end of a container consisting of two top and two bottom corner fittings, two corner posts and a bottom transverse member,.

End of line goods The remaining merchandise of a particular product. Usually it has been phased out of production due to its obsolescence or it has been superseded by more advanced/modern design and/or technical specification.

Endorsement An amendment to be fixed to an insurance policy and thereafter becoming an integral part of the policy; the payee's signature on the reverse of a cheque or bill; or the legal transfer of a title of a document by signature usually but not necessarily on the reverse such as a Bill of Lading made out 'to order'.

Endothermic A process which is accompanied by the absorption of heat.

End-use goods Goods admitted into the UK at a favourable rate of duty provided they are put to a prescribed use under customs control.

End user The person/company who uses/consumes the product/service provided by the vendor. In effect the final buyer in the supply chain.

End use relief Relief from duty and VAT (where applicable) to goods being imported into a specified country or region. The nature of the goods gaining such relief would be laid down by government. See inwards processing relief entry.

End user profile The profile of the person/company using the product/service.

End-use Trader A trader in the UK holding an authorisation from the Commissioners of Customs and Excise allowing him to import or receive end-use goods.

ENEA European Nuclear Energy Agency.

ENERGY The basic unit of energy is the Watt-second, also called a Joule. For most purposes this is inconveniently small, and the Watt-hour is used
1000 Wh = kilowatt hour
1000 kWh = 1 megawatt hour
1000 Mwh = 1 gigawatt hour
1000 Gwh = 1 terawatt hour.

ENF Enforcement.

Engine casing Plate surrounding deck opening to engine-room in a vessel.

Engine-room boy A post in the Marine Engineer's department on a vessel.

Engine transplant The process of removing one engine and installing another in its place.

English breakfast Hearty breakfast served in the UK and Ireland; usually including fruit juice, cereal, eggs, bacon, toast and beverage.

English Jurisdiction Clause This clause appears in the London market MAR policy form, and applies to all claims except where a foreign jurisdiction clause is attached, by agreement with the underwriters, to the policy. The clause provides that claims under the policy are subject to English Jurisdiction in the event that action is taken in a court of law against the underwriters.

English Law and Practice Clause This clause appears in all Institute clauses published for attachment to the MAR form of policy. It provides that when one applies law and practice to the interpretation of the policy conditions decisions must be based on English law and practice, irrespective of the country of jurisdiction.

ENG(S) Engine(s).

Enhanced Remote Transit Sheds This scheme was developed by HM Customs to provide local clearance facilities of third country imports for groupage operators and other specialised freight forwarders with access to secure premise.

Enhanced vessel traffic management system A modern system that basically tracks vessels and other transit resources, prepares schedules efficiently and makes all the operation-related information available to everyone who requires such information. It achieves maximum safety and efficiency during canal transits and provides a method for the integrated management of traffic and resources. This facility is found on the Panama canal.

Enhancement In computer terms an alteration to an existing computer system, usually involving the provision of improved facilities.

Enquiry Link The system that allows account holders in the UK RTGS Processor to interrogate balance and other information and to perform certain other functions; supported by the SWIFTNet Network.

En route (French) On the way.

ENSO El Niño Southern Oscillation. See separate entry.

Entered in Process of reporting a ship's arrival to HM Customs.

Entered ship Ship which has been entered into a P & I Club Association for insurance. An insurance definition.

Enterprise A company/entity providing/manufacturing a product or service.

Enterprise zones Areas/regions identified by a government or trading blocs as in need of financial injection with a view to their development to raise living standards, generate employment, develop infrastructure, improve the wealth of the region, support research/development of technology and so on.

Entertainment on board The provision of shipboard entertainment embracing gaming machines, film shows, live entertainment and so on. It is usually a cruise vessel.

Entreport A place/port of transhipment.

Enthalpy Enthalpy is a thermodynamic measure of the total heat content of a liquid or vapour at a given temperature and is expressed in energy per unit mass (k Joules Per 1 kg) from absolute zero. Therefore, for a liquid/vapour mixture, it will be seen that it is the sum of the enthalpy of the liquid plus the latent heat of vaporisation.

Entity An object of the real world which is relevant to an information system, e.g. customer, invoice, goods.

Entity relationship diagram A diagram showing the inter-relationship of entities and their attributes in a system.

Entrepreneur Businessman: The person responsible for the creation and organisation of a business, who also bears the risks of failure or enjoys the profits of success.

Entropy Entropy of a liquid/gas system remains constant if no heat enters or leaves while it alters its volume or does work but increases or decreases should a small amount of heat enter or leave. Its value is determined by dividing the intrinsic energy of the material by its absolute temperature. The intrinsic energy is the product of specific heat at constant volume multiplied by a change in temperature. Entropy is expressed in heat content per mass per unit of temperature. In the SI system its units are therefore Joule/kg/K. It should be noted that in a reversible process in which there is no heat rejection or absorption, the change of entropy is zero. Hence, entropy is the measure of a system's thermal energy which is not available for conversion into mechanical work. Many calculations using enthalpy or entropy require only a knowledge of the difference in enthalpy or entropy at normal operating temperatures. Accordingly, to simplify calculations, many different enthalpy or entropy tables have been produced which have different baselines. Care should be taken when using such tables as they do not provide absolute values.

Entry Declaration to Customs of goods for importation or exportation. An entry held on CHIEF IES consists of the declaration and control information (e.g. status, route). The declaration can be amended creating a new version of the entry until it is finalised.

Entry inwards The process of a ship on completion of an international voyage being processed through customs on arrival at the port. The reporting of the vessel's arrival in port by the master at the Custom House. Permission to commence discharging is obtained.

Entry outwards The process of a ship, at a start of an international voyage being processed through customs prior to her departure from the port. Permission to commence loading is obtained.

Entry processing units A computerised customs import and export entry declaration process operated by HM Customs and the Customs Handling of import and export freight. It enables exporters and agents to input the following pre-shipment and post shipment information to CHIEF through the on line computer network; full pre-shipment and post shipment information; full pre-shipment declaration; (on value pre-shipment advice, non statistical pre-shipment advice, and Simplified Declaration procedure post shipment declaration). For import cargoes, it permits import entries to allow VAT paid on the traders imports under the government contractor, registered consignee and bulk entry procedures.

Entry requirements The official documentation required by a country to allow a foreigner/non resident to enter.

Entry tax or fee See head tax.

Environment The general conditions obtaining in any activity. This applies especially in marketing, social, physical, technical and psychological. In marketing terms the environment must be studied closely to evaluate socio-economic factors, culture, buyer's behaviour, infrastructure and so on.

Environment (software) The machine, operating system and interface provisions on which software is designed to operate, e.g. 80368-MS-DOS-WINDOWS.

EO Executive Officer.

eohp Except otherwise herein provided

EONIA Euro Overnight Index Average: the weighted average, to two decimal places, of all overnight unsecured lending transactions in the interbank market, initiated by the EURIBOR contributing panel banks; calculated by the ECB.

EOQC European Organisation for Quality Control. The UK member body is the industrial division of the Institute of Quality Assurance (IQA).

EP European Parliament; or European plan.

EPA European Productivity Agency.

EPC Electric Power Corporation (Burma), or Employment Policy Committee.

EPCA European Petrochemical Association.

EPCO Entry Processing Central operations – customs term.

EPF Education Projects Fund (BC/ODA); or Employees Provident Fund (Malaysia).

Epicyclic Gear An epicyclic gear consists of two concentric gearwheels; an inner wheel known as the sun wheel, and an outer ring with internal teeth. Between them run a number of planet wheels in a planet carrier. If the carrier is restrained and the inner and outer wheels rotate, the gear is called a star gear: if the outer ring is restrained and the planet carrier rotates, the gear is called a planetary gear.

EPIRB Emergency Position Indicating Radio Beacons.

EPO European Patent Organisation.

EPOS Electronic Point of Sale. A sales check out point in a super/hypermarket.

EPP Entry Processing Points.

E-procurement The process of obtaining goods/service over the internet.

EPU European Payments Union; Economic Planning Unit (Malaysia); Economic Planning Unit (Brunei) or Entry Processing Units. Customs and Excise offices where import entries are presented for processing and payment of duties, levies and VAT etc. They also authorise release of imported goods. It is linked to CHIEF.

EPZ Export Processing Zone.

EPZA Export Processing Zone Authority (Philippines).

EQUASIS European Quality Shipping Information system. An EU maritime data base.

Equatorial country Country situated on the equator.

Equipment Damage Report Written statement concerning damage to equipment based on physical inspection.

Equipment Handover Agreement Agreement signed when taking over carriers equipment acknowledging conditions thereof.

Equipment handover charge A charge payable by the merchant moving, for example, the container under merchant haulage rules to cover the cost borne by the MTO/carrier for lifting the container from the place of rest at the container yard onto the merchant's transport (including owned/rented chassis) and vice versa, together with associated documentation.

Equipment Interchange Receipt Physical inspection and transfer receipt.

Equilibrium A situation in which the market forces are perfectly in balance. A change in the equilibrium may arise through an autonomous exogenous influence. In a equilibrium situation there is no tendency to change.

Equilibrium price The price at which the quantity of goods or service demanded in a given time period equals the quantity supplied

Equity Ordinary shares which usually take all the risk and are entitled to the balance of profits after meeting prior charges. General term for stocks and shares.

Equity Finance A method, for example, of ship finance whereby the investor is entitled to share in the risk and rewards of the business without any timescale. To liquidate the investment the holder must sell the shares.

Equity-related Bonds Bonds convertible at the holder's option into other securities of the issuing entity, usually equity capital.

Equivalence Transfer of relief status to goods equivalent to the imported goods for which relief is authorised. Overall, a method of IPR which allows the compensating products to be obtained from equivalent free circulation goods. A customs term.

ER Engine room or Extended Range.

ERC European Registry of Commerce.

ERDC Energy Research and Development Centre (PNOC) (Philippines).

ERDF European Regional Development Fund.

Ergonomics The study of the relation of a worker and the environment in which he/she work to provide the capacity for optimum efficiency. Elements of this study will relate anatomical, physiological and psychological knowledge to the machine design and the workplace.

Ergophobia The morbid dislike of work.

Erload The code name of a BIMCO approved charter party clause relative to expected to load and entitled 'Baltic Conference Expected Ready to load clause 1957'.

ERL Export Reception List

ERM Enterprise Report Management – a computer logistics term.

ERM (exchange rate mechanism) The exchange rate and intervention mechanism of the EMS defined the exchange rate of participating currencies in terms of a central rate vis-a-vis the ECU. These central rates were used to establish a grid of bilateral exchange rates between participating currencies. Exchange rates were allowed to fluctuate around bilateral central rates within the ERM fluctuation margins. In August 1993 the decision was taken to widen the fluctuation margins to ±15%. Pursuant to a bilateral agreement between Germany and the Netherlands, fluctuation margins between the Deutsche Mark and the Dutch guilder were maintained at ±2.25%. Adjustments of central rates were subject to mutual agreement between all countries participating in the ERM (see also realignment). This mechanism ceased to exist with the start of Stage Three of EMU.

ERP European Recovery Programme.

Erratum, errata (Latin) An error in printing.

Errors and omissions insurance A policy that covers damages resulting from an agent's mistakes or omissions.

ERS European rail shuttle.

ERTAT Expressway & Rapid Transit Authority of Thailand.

ERTS Enhanced Remote Transit Sheds – a customs term.

ES Spain – Customs abbreviation or DHL International EC – IATA member airline.

ESA European Space Agency.

ESA 95 European System of Accounts 1995.

ESB Eastern Seaboard (Thailand).

ESC European National Shippers Council; Economic and Social Committee (of EEC) or Portuguese Escudo – a major currency.

Escalation clause A clause allowing for the adjustment of the insured value in a ship builders risks policy or in a contract permitting a price increase usually based on a price index. A desirable feature of long-term charters permitting the rate to be increased periodically to allow for increases in operating costs.

Escalator A moving stairway found, for example, in a passenger airport.

Escalator clause A clause allowing for the adjustment of the insured value in a ship builders risks policy.

ESCAP Economic and Social Commission for Asia and the Pacific based in Bangkok. Founded by ECOSOC and reorganised in 1974.

ESCB (European System of Central Banks) The ESCB is composed of the ECB and the national central banks of all 15 Member States, i.e. it includes, in addition to the members of the Eurosystem, the national central banks of the Member States which did not adopt the euro at the start of Stage Three of EMU. The ESCB is governed by the Governing Council and the Executive Board of the ECB. A third decision-making body of the ECB is the General Council.

Escort See tour manager.

Escorted tour Pre-arranged foreign or domestic tour, usually for a group with escort service and often local guides; a sightseeing programme conducted by a guide.

Escrow Funds placed by a travel agency or supplier in the custody of a bank or other financial institution until the fulfilment of certain conditions, such as the completion of a travel contract.

Escrow clause A clause found in a charter party which insist upon an irrevocable credit of a good bank of international repute for the full period of the charter hire.

Escudo The currency of Guinea-Bissau, Chile and before Chile. See Appendix 'C'.

ESD Echo sounding device – ships navigation aid.

ESDP Eastern Seaboard Development Project (of Thailand).

ESD System Emergency Shut Down system.

ESL Eastern Shipping Lines – a shipowner; or European Community Sales List. A mandatory requirement for traders to submit to customs each quarter details of sales to other EC States indicating the VAT number of the buyer and Member State.

ESOMAR European Society for Opinion Surveys and Market Research.

ESPO European Sea Ports Organisation.

Esprit de corps (French) Respect for a collective body by its members.

ESQ Extra special quality steel cable – ships equipment.

ESRO European Space Research Organisation.

Essentials of a high quality transport and logistics service See high quality transport and logistics services.

Essovoy Code name for oil charter party.
Esterel The computerised reservation and national distribution system of France; Partner of Amadeus.
est Estimated.
EST Environmentally-sound technology – A WTO term.
EST&P EST and products – A WTO term.
Estimated time of arrival (ETA) Estimated time of arrival of a particular sailing, flight, train, consignment, etc.
Estimated time of departure See ETD entry.
Estuary Mouth of a river.
Estuarial service Shipping services operating within a river estuary and environs.
ET Ethiopian Airlines Corporation – an IATA member Airline.
ETA Estimated time of arrival of a particular sailing, flight, train, consignment, etc.
ETAN European Technology assessment Network.
ETC Estimated time of completion; explosion of the total contents or Export Trading Company.
Ect. (et cetera) (Latin) And the rest and so on.
ETCs Electronic Trade Credits. A proposal developed by the EDI Banking Interest section to facilitate an EDI alternative to Documentary Credits. An Electronic Data Input facility. See EDI entry.
ETD Estimated time of departure of a particular sailing, flight, train, consignment, etc. or Express Transportation Organisation of Thailand.
etf E-trade facilitation.
Ethical advertising The process of adopting an intrinsic honest presentation advertising strategy by a company/entity. Essentially the advertisement must be completely honest and not ambiguous to mislead or misinform deliberately.
Ethics The customers expectation and experiences of the product/service as identified in the brand image of the company. See culture entry.
Ethiopian airlines The Ethiopian National Airline and member of IATA.
Ethnocentric The strategy/process of a company or entrepreneur providing a product/service for the home/domestic market. It is the opposite to polycentric. A marketing term. See polycentric entry.
Ethnocentrism The process of sending expatriates to newly entered overseas markets.
ETOP European trade promotion organisation.
E Trade facilitation The process of conducting trade electronically. Overall, it involves Electronic Commerce and the transmission of messages/documents electronically involving the internet.
E Transport E Transport is a market place where buyers and sellers of transportation services can conduct transactions electronically. It is designed to connect all trading partners including shippers, carriers and transportation intermediaries and ports. Users of transportation services can check tariff rates for shipping specific commodities from origin to destination.
ETS Estimated time of sailing; Emergency Towing System legally required for all tankers above 20,000 dwts or Electronics Trading Standard.
Et Seq. (et sequens) (Latin) And the following.
ETUC European Trade Union Confederation.
ETV Emergency Towing Vessel.
EU Empresa Ecuatoriana de Aviaciôn SA (ECUATORIANA) An IATA member Airline or European Union; it embraces 25 countries.
EUA European Unit of Account.
EUC European Currency Units
EU Council A body made up of representatives of the governments of the Member States, normally the Ministers responsible for the matters under consideration (therefore often referred to as the Council of Ministers). The EU Council meeting in the composition of the Ministers of Finance and Economy is often referred to as the ECOFIN Council. In addition, the EU Council may meet in the composition of the Heads of State or Government. See also European Council.
eUCP Supplement to the uniform. Customs and Practice for Documentary Credits for electronic Presentation. See UCP entry.
EUI European Insurance Committee.
EUR 1 A form used to obtain preferential rates of duty in EU.
EUR/ANZ Europe/Australia/New Zealand container service.
EURATOM European Atomic Energy Community.
EUR/€ euro
EUR/EAF Europe/East Africa/Red Sea container service.
EUR/FET Europe/Far East container service.
EUR/GUL Europe/Middle East/South Asia container service.
EURIBOR Euro Interbank Offered Rate: the average rate at which euro interbank unsecured term deposits are offered by one (EURIBOR panel) prime bank to another prime bank; displayed to three decimal places and published at 11.00 central European time (CET); sponsored by the EBF and the ACI.
EUR/MED Europe/East Mediterranean container service.
Euro The name of the European currency adopted by the European Council at its meeting in Madrid on 15 and 16 December 1995 and used instead of the generic term ECU employed in the Treaty.
Euro area The area encompassing those Member States in which the euro has been adopted as the single currency in accordance with the Treaty and in which single monetary policy is conducted under the responsibility of the relevant decision-making bodies of the ECB. The euro area comprises Belgium, Germany, Spain, France, Ireland, Italy, Luxembourg, the Netherlands, Austria, Portugal, Greece and Finland. See Euro Zone entry.
Euro Banking Association See EBA entry.
Euro bond A bearer security usually issued in a currency other than that of the country of issue and sold internationally. Hence, bonds sold in a market outside that of the domestic currency.
Eurobonds The provision of longer term funds from the Eurocurrency market by way of bonds, usually in bearer form. The shipping industry, because it includes relatively few large corporate entities, has had limited access to this market.
Eurobrand A product or service which has been designed/formulated specially for the European Union market and is sold throughout it without any radical modification. See CE entry.
EURO CHEMIC European Company for the Chemical Processing of irradiated fuels.
Euro cheque A cheque with a Euro denomination.
Euroclear The clearing system for Eurobonds, based in Brussels where Eurobonds are physically exchanged and stored.
Euro commercial paper The 'managed' issues of promissory notes made, or bills of exchange drawn, by the borrower; usually guaranteed by the borrower's parent company.
Euro Control European organisation for the safety of air navigation.
Euro Co-op European Community of Consumers Co-operatives.

Eurocurrency Deposits of currency, at first dollars but now several major currencies, held by persons not living in the country whose currency it is; the deposits are maintained in a third country. London is the main market for Eurocurrencies other than Eurosterling. The expansion of the Eurodollar market in the late 1960s and early 1970s was an important factor in the funding of the rapid growth of medium-term mortgage lending to the shipping industry.

Euro currency countries A total of 36 countries trade in the euro currencies embracing: Andorra, Austria, Azores, Balearic Isles, Belgium, Canary Islands, Ceuta, Corsica, Crete, Finland, Formentera, France, French Antilles, French Guyana, Germany, Greece, Guadaloupe, Ibiza, Ireland, Italy, Luxembourg, Madeira, Majorca, Martinique, Mayotte, Melilla, Minorca, Miquelon, Monaco, Netherlands, Portugal, Reunion Isles, San Marino, Spain, St. Pierre and Vatican City. See Eurozone entry.

Euro-dollar USA currency (dollars) held by persons or organisations resident outside the USA.

Euro Fer Association of European Steel Producers.

Euro invoice An invoice payable in Euros relative to an export sales invoice.

Eurom European Federation of Precision and Optical Engineering Industries.

Euromarkets Markets in deposits loans denominated in foreign currencies.

Euronet Telecommunications network within EU to provide users with access to over 100 data banks containing scientific, technical and socio-economic data.

Euronia Euro Overnight Index Average. The weighted average rate, to four decimal places, of all unsecured euro overnight cash transactions brokered in London; calculated by the WMB.

European Article Numbering Association An international body responsible for administering the European Article Numbering system. It has affiliates in may countries such as CCCG in Germany, DCC in Japan and ANA in the United Kingdom

European Civil Aviation Conference (ECAC) An association of European aviation officials concerned with facilitating European air commerce.

European Commission The institution of the European Community which ensures the application of the provisions of the Treaty, takes initiatives for Community policies, proposes Community legislation and exercises powers in specific areas. In the area of economic policy, the Commission recommends broad guidelines for economic policies in the Community and reports to the EU Council on economic developments and policies. It monitors public finances in the framework of multilateral surveillance and submits reports to the Council. It consists of 20 members and includes two nationals from Germany, Spain, France, Italy and the United Kingdom, and one from each of the other Member States. Eurostat is the Directorate General of the Commission responsible for the production of Community statistics through the collection and systematic processing of data, produced mainly by the national authorities, within the framework of comprehensive five-yearly Community statistical programmes.

European Community member states Countries which are members of the former European Community include Belgium, Denmark, France, Germany, Greece, Republic of Ireland, Italy, Luxembourg, Netherlands, Spain, Portugal and United Kingdom. See EU entry.

European Bank for Reconstruction and Development Situated in London and responsible especially for funding approved East European projects.

European co-operation grouping A legal framework permitting companies within EEC, engaged in similar activities, to co-operate in production, manufacture and sales on a non commercial and non profit making basis.

European Council Provides the European Union with the necessary impetus for its development and defines the general political guidelines thereof. It brings together the Heads of State or Government of the Member States and the President of the European Commission. See also EU Council.

European Currency Unit A composite currency hitherto formed from the currencies of the European Union and a financial unit used for European Union accounting. See also ECU and Euro entries.

European Economic Area Formed on 1st January, 1994 it was the world's largest free trade market and embraced all the EC and three EFTA members.

European Economic Community Customs and Economic Union between Belgium, Denmark, Germany, France, Greece, Holland, Italy, Luxembourg, Republic of Ireland, Spain, Portugal and the United Kingdom. Now superseded by European Union. See European Union entry.

European foundation Based in Paris, its aim is to improve understanding between peoples within EEC.

European Free Trade Association Based in Geneva, its membership consists of Austria, Norway, Portugal, Sweden and Switzerland. Finland and Iceland are associate members.

European Monetary Agreement, Paris This Agreement, in force since 1958, replaced the older European Payments Union which became obsolete with the advent of convertibility on 29th December, 1958. Of prime importance is the Agreement's codified set of rules for monetary co-operation, particularly a multilateral System of Settlements. The EMA also operates a Fund with resources totalling $607.5 million which may be used to alleviate temporary balance of payments difficulties of the member countries. The Fund and the System of Settlements are operated under the authority of the OECD by a Board of Management and by the BIS.

European Monetary System An agreement concluded in 1979 by member countries of the European Community. Its main objective is to limit fluctuations in the rates of exchange between members' currencies. In 1955, Ireland, Holland, Belgium, Germany, France, Denmark, Portugal and Spain were members. All currencies operate within the 'wide band' range of 15 per cent of agreed central rates against other members of the mechanism except Germany and Holland which operate within the 'Narrow band' range of 2.25 per cent. It was discontinued when the euro was introduced. See Euro entry.

European Nuclear Energy Agency Based in Paris it is established under OECD for the co-ordination of the nuclear energy programmes of the member countries and the co-ordination of legislation concerning utilisation of atomic energy and for general co-ordination in this field.

European Options Market An options market which operates on the Amsterdam stock market. 'Puts', 'calls' and 'straddles' form the basic material of the market.

European Parliament It consist of 726 representatives of the citizens of the Member States. It is a part of the legislative process, though with different prerogatives according to the procedures through which EU law is to be enacted. In the framework of EMU, the Parliament has mainly consultative powers. However, the Treaty establishes certain procedures for the democratic accountability of the ECB to the Parliament (presentation of the annual report, general debate on the monetary policy, hearing before the competent parliamentary committees). This has been enlarged following the increase of EU membership from 15 to 25 countries in 2004.

European plan (EP) Hotel rate that pays for room only, meals not included.

European Productivity Agency, Paris In charge of the problems of increasing the productivity of the OECD countries.

European Quality Alliance Group formed by SAS, Austrian Airlines, Swissair and Finnair to increase the strength of European airlines against the open skies policy beginning in 1993.

European Registry of Commerce An organisation with a headquarters in Brussels and offices throughout the EEC whose objective is to provide for its members an integrated business information system. In so doing its prime aim is to promote the distribution of products know how and information between companies within the European Union.

European Seaways Company operating passenger ferries in the Italy (Brindisi) – Greece (Igoumenitsa) trade.

European style exercise An equity or index option that maybe exercised only on the expiry day.

European Style Option One that can only be exercised on a given date.

European System of Integrated Economic Accounts (ESA, second edition, 1979) This is a system of uniform statistical definitions and classifications aimed at a coherent quantitative description of the economies of the Member States. The ESA is the Community's version of the United Nations' revised System of National Accounts. The respective definitions are, inter alia, the basis for calculating the fiscal convergence criteria laid down in the Treaty. The European System of Accounts (ESA 95) was the new system of national accounts of the EU and will be implemented as from April, 1999, replacing the European System of Integrated Economic Accounts.

European Travel Commission (ETC) Co-operative agency sponsored by European nations for the promotion of tourism.

European Union Introduced on 1st January, 1995 Accession Treaty at Corfu European Council, June, 1994. Full membership 2004 – 25 Member States include Belgium, Denmark, France, Greece, Ireland, Italy, Luxembourg, Netherlands, United Kingdom, Germany, Portugal, Spain, Austria, Finland, Sweden, Estonia, Latvia, Lithuania, Poland, Czech Republic, Slovakia, Hungary, Malta Slovenia and Cyprus. Total population 500 million.

EUROPROs A European membership organisation featuring eighteen sister trade organisations in SITPRO. It provides a framework for the individual facilitation organisations to collaborate with the common good of simple trade procedures.

Eurostar A high speed rail passenger only service connecting London (Waterloo) directly with Paris and Brussels. A brand name.

Eurostat See European Commission.

Eurosystem Comprises the ECB and the national central banks of the Member States which have adopted the euro in Stage Three of EMU (see also euro area). There are currently 12 national central banks in the Eurosystem. The Eurosystem is governed by the Governing Council and the Executive Board of the ECB.

Euro terminal Purpose built British Rail freight terminals situated in the UK to serve the European Rail Freight Market consequent on the opening of the Channel Tunnel.

Euro-Tunnel A 31 mile route subterranean rail tunnel between Folkestone and Frethun near Calais. It links the British Rail and French railways systems thereby enabling the users of British Rail network direct access/use of the very extensive Continental rail network. It comprises of three tunnels two of which convey trains and one a service tunnel. The journey time is 35 minutes and conveys passengers, motorists, cars, coaches and freight vehicles. Express passenger and freight services operate through the tunnel linking major industrial centres in UK with those on the Continent. It was opened in 1994. Also termed Channel Tunnel and Fixed Link.

Euro zone The 36 countries which have the euro as its national currency. Overall, there are twelve EU currency countries trading in euros.

EUR/SAF Europe/Southern Africa container service.

EUR/USA Europe/North America container service.

Eustasy The worldwide global changes in sea level caused by changes in the water volume of the oceans due to the formation and melting of ice sheets, variations in the amount of freshwater in lakes, inland sea and underground reserves temperature of the water, or any changes in the volume of ocean basins induced by tectonic movements.

Eustatic A global change in sea level.

Even if used A chartering term. It represents time spent carrying out cargo work in 'excepted periods' not to count as laytime. Hence, such time not to count even if used.

Evening peak A peak demand materialising during the evening period. This may apply for example to transport such as rail and bus services during the period 1600/1900 hours; or television/radio peak evening respective viewing and listening

Event Marketing The process of promoting a particular event such as a car rally.

EVERETT Everett Orient Line – a shipowner.

EVERGREEN Evergreen Line (Conventional Services) – a shipowner.

Evergreen Marine Corp Ltd A major liner shipowner and container operator and registered in Taiwan – Province of China.

EVHA European Port Data Processing.

Evidence accounts A counter trade technique. It arises where a multinational company with a local manufacturing subsidiary in a developing country may be required to ensure counter purchased exports of equivalent value to its subsidiary's imports of material and equipment. Since it is not practical to balance this kind of trade item by item, the firm maintains an 'evidence account' debiting its own imports and crediting the exports it has arranged over a period.

EVO Algemene Verladers en Eigen Vervoer Organisatic. Dutch Association of Companies with own means of Transport; or Excise Verification Officers.

EVTMS Enhanced vessel traffic management system.

EWA Europische Wahrungsabkommen (European Monetary Agreement).

EWOS European Workshop for open System. An EDI facility.

Ex Examined; exchange; executed; excluding; out of; without; departing from, as in ex Sydney.

Ex ante (Latin) Forecast, based on expected results.

Ex ante, ex post Before and after a measure is applied.

Ex cathedra (Latin) With full authority.

EXCEPTED The days specified do not count as laytime even if loading or discharging is carried out on them. A BIMCO term.

Exception Part of a policy disclaiming liability from certain periods or events.

Exception clause See Exemption clause entry.

Exceptional items Items of an abnormal size and incidence which are derived from the ordinary activities of the business and therefore require separate disclosure.

Excess The term 'excess' was used for many years in the marine market to denote an amount that has to be exceeded before a partial loss claim attached to the policy. Once the excess was passed the amount of the claim above the amount of the excess was paid. The term has gradually disappeared in the marine market, to be replaced by the term 'deductible', which has the same effect on partial loss claims.

Excess baggage That part of baggage which is in excess of the portion of baggage which may be carried free of charge.

Excess baggage charge A charge raised for the carriage of excess baggage.

Excess Fares The process of paying the difference between the price of the ticket the passenger holds and the appropriate fare for the journey the passenger is making. For example it may be the difference between the First and Standard class fare.

Excess insurance The process whereby the insured bears a particular sum relative to any claim submitted. It may be the first €1,000 or €5,000 depending on the insurance policy terms.

Excess liabilities This embraces the marine insurance to cover the excess amount of liability for general average contributions, salvage charges, sue and labour charges and collision liability where the full amount is not covered by a hull policy.

Excess of line re-insurance A re-insurance to cover that part of the original underwriter's acceptance which is in excess of his retained line.

Excess of loss insurance An insurance where the insurer's liability only attaches when a loss exceeds a certain figure and then only for liability in excess of that figure.

Excess of loss re-insurance A re-insurance to cover that part of a loss paid by the re-insured which is in excess of an agreed premium.

Excess Point Term used in excess of loss re-insurance to determine the point at which the re-insurer comes on risk.

Excess stock That portion of stock on hand which is over and above the desired stock level.

Excess Value Insurance This term was commonly used in the marine market in reference to insurances effected additionally to the policy covering the hull and machinery of a ship. These could apply to insurable interest in shipowners' liabilities, or additional insurances covering hull interests (e.g. disbursements). Today, the term 'excess liabilities' is used for the former and 'increased value' for the latter.

Exchange control A system imposed by government for controlling the movement of currencies across national boundaries e.g. by restricting the amount of currency that can be held by a domestic resident, or the amount of domestic local currency that can be exported either for travel or investment. The Government may control exchange rate movements by varying the relation between supply and demand thereby keeping fluctuations to a minimum. Governments may also control trade through exchange control regulations/policies documentation.

Exchange control officer A senior managerial position in a Government Central bank, or Government Finance department which executes Government policy/regulations/control of capital/monies entering and/or leaving the country. This will involve overseas foreign loans and foreign investments and foreign exchange repatriation. Additionally it will involve overseas loans and overseas investments.

Exchange Control Regulations Government regulations of a particular state regarding the movement of currencies across national boundaries. See Exchange Control entry.

Exchange cross rates The value of one unit of currency in terms of other quoted currencies. This may include, for example, the US Dollar, Sterling, Swiss Franc, currency rate expressed against the euro, Japanese Yen, Canadian Dollar and so on.

Exchange for Future for Physical The exchange of a futures position for a physical position. In such a situation two agreeing parties swap cash and futures position through the Exchange – an International Petroleum Exchange term.

Exchange of Contracts The situation where both parties reach a stage in their negotiations that they wish to exchange contracts for ultimate agreement.

Exchange order Document issued by a carrier or other travel-related entity requesting provision of specified services or the issuance of a ticket to the person named in the document.

Exchange rate The level of one national currency expressed in terms of another. It maybe a fixed rate of exchange determined by the National Government involved, or a floating currency broadly determined by market conditions. Other factors determining the floating currency level include level of inflation in the national country, interest rates in a country of a major world currency such as USA; any dramatic change in depreciation or appreciation terms of a major currency; any agreement/membership of a particular group of countries currencies such as Eurozone; National Government policy towards influencing the exchange rate level; any period of extended market confidence; capital outflow or inflow from the country involved; international confidence in the National Government policies regarding economic, fiscal, industrial and so on; any special situation such as war, industrial dispute such as dock strike which will influence trade performance/results; seasonal fluctuations such as at particular times of the year the level of balance of payments is in deficit due to outflow of funds; the overall trend of the balance of payments performance; any policies decreed by the World Bank and IMF; world oil prices; and finally the paramount consideration of the international market confidence in the currency with no fear of any dramatic devaluation/deterioration in its value.

Exchange rate exposure The process of a currency being exposed to market pressures which may result in a variable currency movement.

Exchange rate fluctuation The process of one currency expressed in terms of another fluctuating/varying as the values of currencies alter. Fluctuations are due to Central Bank policy, or Economic and Political factors. These include rate of inflation; international interest rates; industrial relations; balance of payments; level of money supply; political unrest; trade agreement with other countries and so on.

Exchange restrictions Restriction whereby an investor's access to foreign exchange, and therefore his ability to import, is limited. A WTO term.

Exchange Risk Exchange risk arises virtually in every international payment when the exchange rates between one currency and another change/vary between the date a price is agreed between the supplier/exporter and the buyer/importer, and the date payment is made. Such an exchange risk must be faced by both the exporter and importer. To lessen such risk various options exist including the use of a third party currency; forward exchange contract and so on.

Exchange Risk Guarantee Scheme A guarantee to protect borrowers of loans from the European Investment Bank and the European Economic Coal and Steel Community against any adverse changes in exchange rates.

Exchange stabilisation The official operations in the foreign exchange markets designed to control exchange rate movements by varying the relation between demand and supply. Such operations are usually directed at keeping fluctuations to a minimum although sometimes a government may deliberately seek to raise or lower the exchange rate of its currency.

Excise Tax levied by Customs on certain classes of goods produced in UK country e.g. whisky.

Excise duties Money collected by Customs in duties in certain classes of goods manufactured, licences issued and transactions taking place within a country.

Excise Warehouse A warehouse approved for the deposit of goods subject to excise duty but without payment of duty.

EXCL Excluding.

EXCLUDED The days specified do not count as laytime even if loading or discharging is carried out on them. A BIMCO term.

Exclusion A clause in a policy excluding certain losses e.g. war perils, frustration, nuclear weapons, dangerous drugs, earthquakes, etc. An insurance definition.

Exclusion clause A clause found in a contract which is designed to exclude, or cancel out, civil liability (liability to damages) which one party to the contract would otherwise be subject to.

Exclusive Agency Agreement An agreement whereby an agent (in the strict legal sense) is given a territory in which he must sell his principal's goods. The agreement is usually a reciprocal affair whereby he gets the sole right to act as agent within the territory.

Exclusive licence A licence granted to a particular beneficiary and in so doing it is the only one awarded/granted in a particular Region/Country thereby recognising its exclusivity.

Excursion A journey, usually short, made with the intention of returning to the starting point. See excursion fares entry.

Excursion Fares A passenger tariff at discounted rate designed to develop optional travel and improve transport capacity utilisation at less busy times and increase cash flow. It may be, for example, a day, afternoon or half day excursion. Such excursions may be by rail, coach, ferry, canal or air.

Excursion Limits The minimum and maximum allowable separation between the offtake vessel and the installation to preserve mooring hawser and/or the transfer hose. A maritime definition.

Exd. Examined.

Exeat (Latin) Permission for temporary absence.

Executable Files Files with extensions of .EXE, .COM, and .BAT that can be executed from PAM or the MS-DOS command prompt.

Executable Files Menu The PAM menu that displays all the files that can be run from PAM. They have extensions of .EXE, .COM or BAT.

Execution The process of carrying out the prescribed task.

Executive class A class of passenger accommodation primarily found on airline services which offers a premium type of service: ideal for businessmen, executives, etc.

Executive Lounge Accommodation provided for passengers holding Executive class tickets or equivalent. In some cases they are available on payment of a supplement. They may be found on an airline, train, ship or coach – the latter being termed Executive coach. Also accommodation found at an airport, seaport, railway station for passengers waiting to join their transport mode. Ideal for businessmen, executives, etc.

Executor de son tort (Latin) An executor by wrongdoing or intermeddling.

Exemption An authorisation by an appropriate authority or other circumstance providing relief from any provision, statutory obligation, regulations or other situations such as found in a contract.

Exemption Clause A clause found in a contract which is designed to exclude, or cancel out, civil liability (liability to damages) which one party to the contract would otherwise be subject to.

Exercise The procedure by which the option buyer (holder) takes up the rights to the contract and is delivered a long (call) or short (put) futures position by the seller (writer). An IPE definition.

Exercise notice A formal notification that the holder of a call/put equity or index option wishes to buy/seller underlying security at the exercise price.

Exercise price The price at which an option contract gives the holder the right to buy or sell the underlying security. Sometimes it is called the strike price of the option. These are fixed by LIFFE and are set at intervals so that they exercise prices both above and below the current share price; or the price at which an option holder has the right to buy or sell the underlying commodity – an International Petroleum Exchange term.

Ex Factory Delivery cargo term of similar roles to ex works, ex warehouse.

Ex gratia (Latin) As a matter of favour.

Exhaustion The principle that once a product has been sold on a market, the intellectual property owner no longer has any rights over it. (A debate among WTO member governments is whether this applies to products put on the market under compulsory licences.) Countries' laws vary as to whether the right continues to be exhausted if the product is imported from one market into another, which affects the owner's rights over trade in the protected product. See also parallel imports. A WTO term.

Exhibition centre A place where Companies exhibit their products/services to the market place usually under the aegis of a Trade Fair, or some other commercial promotion.

Exhibition Forwarding An Agent/Freight Forwarder specialising in the International Forwarding/despatch of exhibition merchandise. It maybe for a Trade Fair or British week overseas. The merchandise could include stand material, display goods, publicity material and so on.

Ex Hypothesi (Latin) As is self-evident.

EXIMBANK The Export-Import Bank of Washington. A Bank established for the development of American Exports.

Ex mill Delivery cargo term of similar role of ex works, ex factory.

Ex nudo pacto non oritur actio (Latin) Out of a bare promise no action arises.

Ex officio (Latin) By virtue of office.

EXORS Executors.

Exotic A complex type of product e.g. exotic swap, exotic option. Distinguishes it from most basic types ('vanilla') of swap and option. An international financial term.

Exotic currencies Currencies in which there is little trading on the exchange market. Most of the currencies of the underdeveloped world would fall within this category.

Exotics Foreign Exchange dealer's term for currencies in which only a small market exists such as Kenya shilling.

Expanding market A market which is becoming larger.

Expansion trunking Continuous trunking across the top of a vessel's tanks to allow for expansion of the oil cargo.

Ex Parte (Latin) A proceeding by one party in the absence of the other.

Expatriate A national working overseas in another Country/State.

Expelling Charge An explosive charge designed to eject the projectile from the parent article without damage. A term associated with dangerous classified transported goods.

Expendable pallet See disposable pallet entry.

Expenditure Budget The process of formulating forecast and objectives relative to expenditure during a specified period for a particular area of the business activity or overall Company results.

Expense ratio The proportion that the expenses, other than claims and reinsurance premiums, of the insurer bear to his income from premiums. An insurance term.

Experience The past claims record of an assured or re-assured – a marine insurance term. It can also relate to a case history or general analysis of a specified shipping matter.

Expert system An application which enables users to benefit from the knowledge of human experts. The knowledge is built into the system as a set of rules which may be updated with use.

Expiry The point at which the option is either exercised or abandoned and therefore is no longer traded. An IPE definition.

Expiry date The date on which it expires. It may be a letter of credit, licence and so on.

Expiry month Options terminology for delivery month. An IPE term.

Ex plantation Delivery cargo term of similar role to ex works, ex factory, ex mill.

Explode Used to indicate those explosive effects capable of endangering life and property through blast, fragments and missiles. A term associated with dangerous classified transported goods.

Explosion-Proof/Flameproof Enclosure An enclosure which will withstand an internal ignition of a flammable gas and which will prevent the transmission of any flame able to ignite a flammable gas which may be present in the surrounding atmosphere.

Explosive article An article containing one or more explosive substances.

Explosives deflagrating A deflagrating explosive is an explosive which reacts by deflagration rather than detonation when used in its normal manner. Propellants belong to this type. A term associated with dangerous classified transported goods.

Explosives detonating A detonating explosive is an explosive which reacts by detonation rather than deflagration when used in its normal manner. A term associated with dangerous classified transported goods.

Explosives initiating Explosive substances which, even in very small quantities detonate on contact with flame on mild or low impact or as a result of friction; they are able to transmit detonation to other explosives close to them. A term associated with dangerous classified transported goods.

Explosives primary An explosion shall be deemed to be a primary explosion if its sensitivity is such that it requires similar handling to generally accepted primary explosives such as mercury fulminate, lead oxide and lead styphnate and to other sensitive explosives such as percussion cap compositions. A term associated with dangerous classified transported goods.

Explosives secondary Relatively insensitive substances (or mixtures) usually caused to function by primary explosives with or without augmenting charges. A term used in dangerous classified transported goods.

Explosives substance A solid or liquid substance (or a mixture of substances) which is in itself capable by chemical reaction of producing gas at such a temperature and pressure and at such a speed as to cause damage to the surroundings. A term associated with dangerous classified transported goods.

Export The process of selling/despatching a product/service to another country thereby crossing an international boundary.

Export Agent Agent responsible for promoting/selling his principals' goods overseas. Also see Commercial Agent entry.

Export berth A berth at a seaport handling/accommodating export cargo.

Export Business Plan Executive Summary; Business History; Market Research; Marketing Decisions; Legal Decisions; Manufacturing and operations; Personnel Strategies; Financial decisions and Implementation schedule. See successful export strategy.

Export Cargo Shipping Instruction A document prepared by the exporter identifying full details of the consignment and all the shipment/payment/customs arrangements including Inco Terms 2000. It accompanies the cargo at the time of collection and used by the carrier as the shippers instructions. Shipping instructions from Exporter to Carrier.

Export clearance The process of presenting goods to customs at the seaport, airport, inland clearance depot or exporters' premises in the country of their origin and subsequently being cleared by Customs Officers. The goods maybe subject to the re-exporting process or outwards processing relief. Once the goods have been cleared the goods will be released by Custom Officers to enable the consignment to commence its international transit.

Export Clerk A clerical position in a Company/entity engaged on export work which includes customs, distribution, rates, booking the consignment, documentation, export finance, insurance and so on. The clerk maybe in a Freight Forwarders, or Agents office, Transport Company, retail/manufacturing company and so on.

Export clinic A form of export consultancy service whereby exporters with specific problems seek advice from an experienced professional export consultant.

Export Clubs Informal associations formed on a regional/area/city basis with representatives of local freight forwarders, banks, insurance companies, container operators, and particularly local manufacturers engaged in international trade, with a view to develop overseas markets/trade and acting as a forum for international trade discussion.

Export Consignment Identifier A customs facility consisting of the customs registered number and up to thirteen alpha/numeric digits of unique identifying reference.

Export control The exchange control regulations imposed by a government regulating/controlling the acceptance of foreign exchange as payment for goods sold overseas.

Export Co-ordinator A post within a Company/Entity found in the Export Department. The duties would embrace the co-ordination of all the aspects of the Export Sales contract from the time the contract is secured until ultimate payment is made. This will include insurance, transportation, finance, packing, documentation, production schedule, progress chasing to ensure the goods arrive in the destination country on the date specified in the export sales contract and in accordance with the agreed specification.

Export credit Many countries have now established export credit agencies to encourage the export of goods and services. These credit facilities are offered to the exporter at fixed and preferential interest rates, and enable credit finance to be made available to importers in other countries. Most national credit agencies offer the rates of interest and length of credit terms agreed by the members of OECD which regularly reviews these export credit terms. An example is ECGD; also of significance to the shipping industry because of its dramatic impact on the financing of new ship construction in the 1960s, 1970s and 1980s. The financing of large capital equipment orders for export is supported by government guarantees and/or insurance in most industrialised countries. Competition between shipbuilding countries in the provision of officially guaranteed and subsidised credit packages became intense in the 1960s, when the Japanese shipbuilders were offering 80% dollar credit over 8-10 years at 5.5% The Organisation for Economic Co-operation and Development was obliged to intervene in 1969, initiating the so-called 'gentlemen's agreement' on shipbuilding export credit terms. This currently stipulates a maximum of 80% of the contract price repayable in no more than 17 semi-annual instalments. The minimum interest rate laid down by OECD varies by market conditions. The credit terms were under review in 2003.

Export Credit Insurance The process of insuring goods shipped against non payment by reason of commercial and/or political risk as arranged.

Exports Credits Guarantee Insurance on goods shipped against non-payment by reason of commercial and/or political risks as arranged.

Export declaration The process of an exporter declaring to customs the fullest details of the consignment being exported. See customs entries.

Export Declaration US Customs Service Form No. 75-25V.

Export Director A most senior directorate managerial position responsible for the overall policy and development of a company/manufacturer business overseas on a profitable basis. The company organisational structure will vary by the nature of the business but the Export Director will have an Export Marketing Manager as a deputy with an Export Sales Manager and Export Shipping Manager responding to him.

Export documentation All the documentation associated with the processing through Customs of an export consignment involving Bills of Lading, Export licence, certificate of origin and so on.

Export documentation procedures The code of practice to process export documentation embracing the financial, insurance, shipping, customs, etc., aspects of the international consignment.

Export Duty Revenue collected by Customs at time of exportation from duties imposed on various classes of goods.

Export enhancement programme Programme of US export subsidies given generally to compete with subsidised agricultural exports from the EU on certain export markets. A WTO term.

Exporter The entity/person/seller of merchandise to an overseas market; or for customs purposes this is considered to be the person on whose behalf the export declaration is made and who is the owner of the goods or has a similar right of disposal over them at the time the declaration is accepted. Where the ownership or a similar right of disposal over the goods belongs to a person established outside the Community pursuant to the contract on which the export is based, the exporter is considered to be the contracting party established in the Community.

Exporter Credit The process of the overseas supplier/manufacturer/exporter/seller allowing credit terms to the importer/buyer and obtains finance from a bank or export credit agency. See also supplier credit entry.

Exporters acceptance credit Credit opened by an exporter with his own bank and entitles the exporter to draw bills on his own banker.

Export Facilitation organisation A wide range of organisations of differing activities which exist to aid and facilitate the development of overseas markets for both existing and potential exporters.

Export Factoring Export Factoring is a service which relieves the seller of the responsibility of handling the sales ledger for defined areas of overseas operations. The service includes the despatch of statements, collection of monies, and provision of credit facilities for the overseas buyer. It also allows the seller/exporter to draw in advance an agreed percentage – usually between 70 and 80 per cent of those debts which are outstanding and as yet unpaid by the buyer. The advantages of using a factoring service include; the saving in time in running a sales ledger; the saving on cost of postage, telephones and facsimile and overseas bank charges; the involvement with only one debtor – the factoring company; the availability of credit facilities for the buyer on 100 per cent of the value of approved sales; the removal of responsibility for collecting overdue accounts or making provision for bad debts by the seller; the ability to predict the seller cash flow with confidence; the facility of the seller to draw cash in advance of payments being received from the buyer, and finally the ability to obtain cash discounts from suppliers using money generated from the advance payment facility. Each overseas buyer is screened.

Export Finance The financial resources required to finance an export sales contract. See Export Houses.

Export Finance House A company which finances exports under its own policy. Overall, a provider of medium to long-term export credit.

Export finance options The range of international trade funding options available. These include open account, advanced payment, mail payment, international bankers draft, Bill of Exchange, counter trade, factoring, forfeiting, etc,

Export Houses Responsible as an export merchant for buying goods outright and selling them on their own account; acting as an export department or agent on behalf of a client; or acting for an overseas buyer.

Exporting The process of despatching/selling goods to an overseas market.

Export/Import Rate (Index) A statistical measure to determine the relationship of a country exports with its imports as detailed below:-

Exports €131.69 Mill Imports €199.07 Mill = 0.66

Any figure above 100 confirms the country has a trade surplus whilst below this figure indicates the country has a trade deficit with the volume of imports in excess of exports as found in the example with a figure of 0.66.

Export invoice An invoice issued by the seller (exporter) detailing the date, name of the consignee, quantity/description of goods, marks and measurement of packages, cost of the goods, packing, carriage, freight, postage, insurance premiums, etc.

Export led company A company/entity where the core of the business is export driven.

Export led economic recovery The process of a particular economy of a region/country improving its economic performance by selling more goods overseas thereby increasing the level of employment and wealth creation in the exporting country.

Export led economy A country which is dedicated to focus attention on selling/manufacturing to overseas markets thereby increasing wealth and employment with high living standards in the exporter's country. Overall the volume of exports exceeds the level of imports thereby creating a favourable trade balance.

Export led industries Industries whose core business is export driven.

Export Licence A document issued by a Government authorising export of restricted goods/items within a specified time. Overall, it is a permit to export certain categories of goods covered by export controls, e.g. firearms, some computer parts, antiques and skins of endangered species.

Export Licence Expiry date The expiry date of a specific export licence.

Export Licence number The number of a specific export licence.

Export Licensing Controls Orders issued from time to time by HM Government-Department of Trade under the Export of Goods (Control) order made under the Import, Export and Customs Powers (Defence) Act 1930. One was issued in 1993 by DTI regarding a code of practice on export licensing control compliance.

Export Licensing Unity Advises exporters on whether or not they need a licence to export their goods.

Export logistics The process of adopting a logistics strategy for exporting goods. See outbound logistics entry.

Export Management Companies A method of indirect export trading whereby the export management companies work very closely with the export manufacturer and often in so doing become their export departments. When a full service is provided, the export management company becomes responsible for promoting export sales, processing all documentation, transportation and financial settlement arrangements including when necessary credit to the overseas buyer. It is customary for credit to be on a non recourse basis.

Export manager An individual in one company or a firm acting as an export department for one or more manufacturers.

Export Marketing It is the process of identifying or anticipating – usually through market research/intelligence – a need for a product/service and responding to that need on an acceptable basis in a particular country other than the one in which the entrepreneur/entity is situated. Overall, it is basically a management function which crosses international boundaries and involves different culture, infrastructure, life style, buyers' behaviour and so on. This involves product/service modification/adaptation and continuous strategic management to develop the importers' market.

Export Marketing Manager The person in an Export Company responsible for Marketing/Selling products overseas.

Export market malaise strategies Fourteen points identifying errors of judgement/mistakes made by exporters to develop a viable export strategy. This focuses particularly on first time exporters. These embrace: no viable export plan; absence of market research; wrong market selection; too many markets selected; wrong overseas representative/partner; absence of adequate support for overseas partner; unwillingness to modify products; failure to produce multi-lingual/promotional material; lack of ongoing commitment to the export business; failure to understand culture/protocol/language of the buyer/importer; going it alone – need to examine potential strategic partnerships; absence of customer relations/after sales department; failure to be computer literate thereby using latest IT resources/skills; absence of continuous empathy with the buyer and no negotiating plan/culture briefing.

Export market research The process of conducting market research into an overseas market such as a region/area/country for a particular product/service or range of products/services.

Export Merchants Export Merchants in the UK, for example, buy goods direct from an exporter. The transaction is almost the same as selling in the home market as the export merchants are responsible for promoting the exporters goods to overseas buyers.

Export Office An office/department usually in a manufacturing company which deals with all the aspects to process the export consignment.

Export order check list A check list or progress sheet used by an exporter to process an export order within his Company from the time the order is received, until it is despatched and ultimate payment made for the goods.

Export orientated industry An industry whose core business is export driven.

Export packers A company/entity/person dealing with the packing of goods for international delivery.

Export-performance measure Requirement that a certain quantity of production must be exported.

Export plan Process of devising a viable export plan. See export market malaise strategies, and strategy entries.

Export price The priced of an exported product/service. See Incoterms 2000.

Export processing zone A special type of foreign trade zone within which certain exemptions from duties, taxes and regulations are granted as an inducement to export oriented manufacturing. Customarily a manufacturer within the zone may import equipment and raw materials duty free for goods that are ultimately exported as finished.

Export regulations The statutory regulations laid down by government regarding the export of merchandise from a country involving the product to cross international boundaries. Some export regulations imposed by Governments restrict/regulate certain goods that may only be exported under licence. The classes of goods subject to these regulations usually include arms or military equipment, antiques, live animals and so on. Such regulations may also apply worldwide. See also Export Licensing Controls entry.

Export restitution Name for export subsidy in the EC given under the EC's Common Agricultural Policy to help exports by offsetting the difference between the internal EC price and lower world price. A WTO term.

External trade

Export selling The process of selling a product/service to a customer which has been manufactured/produced in the exporters' country. Overall, it broadly involves the transfer of the marketing mix from the domestic market to the overseas territory representing the country of the importer.

Export seminar A group of students delegates who meet under the direction of a tutor to have a systematic study relative to an export topic.

Export sequence list A list showing the order in which each container for export is to be loaded into the vessel.

Export shed Covered accommodation usually situated at an airport, seaport or agents premise provided for the handling/processing for shipment of export cargoes/consignments awaiting exportation.

Export Strategy The process of devising a viable export strategy. See successful export strategy, export market malaise strategies and strategy.

Export subsidy The process of subsidising the export product/service. It may be an off charge to the product/service produced for the domestic market.

Export System (Customs) See New Export System and CHIEF.

Export workshop A discussion group focusing on a specific export topic and usually comprises personnel from different companies. The group is led by a skilled export professional and may comprise of the evaluation of a comprehensive case study involving the formulation of a derived strategy.

Export zones A defined area which has core businesses which are export driven.

Ex post facto (Latin) Subsequent to a main event.

Express Agency This type of agency is created through the Agent receiving from his principal definite/precise instructions to do a specific assignment(s)/task.

Express Bill of Lading A sea waybill.

Express Freight A carrier which offers a high speed service for the conveyance of freight and may charge a premium freight rate.

Express operators In air transport terms, operators who convey small parcels up to 32 kilos on a predetermined schedule at a delivered tariff and conveyed by one operator throughout.

Express parcels A small parcel usually up to 32 kilos per package conveyed on a fast transit on a predetermined delivery schedule at a predetermined delivered tariff and generally provided by one operator for the whole transport process. In the European market express delivery is required within a time span ranging from less than one day to an absolute maximum of three days depending on the rates charged, the distance involved and the accessibility of the origin point and destination point.

Express services Those transport services which are classified as a high speed transport system networks and usually attract a premium rate/fare.

Express telegraphic money transfer This is the quickest method of sending money abroad. It costs more than most other methods as the importer/buyers bank has to instruct an overseas bank, by cable/telex/facsimile to pay a stated amount of money to the suppliers/exporters/sellers account.

Express terms of a contract The terms actually agreed between the parties.

Express Warranty A warranty that is specified in the policy document, or is attached thereto. A warranty that applies to the policy, but is not expressed therein (e.g. warranty of legality), is termed an /implied warranty'.

Expunged Ship removed from Classification Society due to non compliance with their rules.

ExQ Ex Quay. Cargo delivery term.

Ex Quay Cargo delivery term whereby under delivered ex quay the seller delivers when the goods are placed at the disposal of the buyer not cleared for import on the quay (wharf) at the named port of destination. The seller has to bear the costs and risks involved in bringing the goods to the named port of destination and discharging the goods on the quay (wharf). Hence, the buyers' obligation is to clear the goods for import and to pay for all formalities, duties, taxes and other charges upon import. See Incoterms 2000 and DES.

EX R/I Excess of Loss Reinsurance.

ExS Ex Ship. Cargo delivery term.

Exs. Expenses.

Ex Ship Cargo delivery term whereby the Exporter (seller) pays sea freight and insurance, loading costs at port of despatch, and any other costs, and bears the risks up to port of discharge. The buyer (importer) meets all the cargo discharge charges at port of destination including customs duty and any other charges incurred thereafter and likewise associated risk. See Inco term 2000 and DES.

Ex silentio (Latin) By the absence of contrary evidence.

Extended family A family unit embracing the husband and wife with their children plus the grand parents.

Extended port A seaport which embraces not only all the transhipment facilities/resources and related activities such as agents, brokers, ship repair facilities, but also an industrial zone where imported cargo is processed, bagged, assembled and so on. Extended ports exist in Singapore and Rotterdam.

Extended Protest Detailed statement made by the master of a vessel concerning an accident which has become the subject of a court case. See noting protest entry.

Extended use tests It is used primarily in industrial marketing to test the long-term quality of a product.

Extension Tour at additional cost to buyers of a tour or cruise; may be taken before, during or after the basic programme; or the second part of the name of an MS-DOS file. The extension may contain no more than three characters and must be separated from the filename by a period. A computer definition.

Extension forks Type of fork lift truck. A fork lift truck equipped to handle large wooden cases and other cubed cargoes.

Extension strategy A strategy to extend the life of a product.

External Command An MS-DOS command that resides in a file with a file type extension of .EXE or .COM, or .COM, or .BAT. To execute an external command, the external command file must be on the disc in the active drive.

External Disc Drive A disc drive that is connected to the computer by cables and is not installed inside the computer.

External Marketing Audit In marketing terms the process of conducting a comprehensive periodic, systematic and independent review of a company's marketing activities/performance in the market place having an especial regard to competition, strategies, opportunities, economic, political, legal, social/economic and other factors. The audit purpose is to improve performance.

External packaging The outer layer of packing on a consignment.

External procedure The form of full Community Transit procedure applicable to non community goods.

External trade The process of conducting trade which crosses international boundaries.

External trade Guarantee The process of providing insurance cover to UK manufacturers, merchants, and confirming houses to cover goods from a supplying country to the buyer's country not first imported into the UK. A facility available from ECGD.

External Trade Indicators Various statistics available to measure a particular overseas performance. For example, to measure the UK overseas performance one can take various relevant index numbers and rates of exchange. These include export and import volume, visible balance, current balance, oil balance, terms of trade and exchange reserves.

Extinguishing Water Water used to put out a fire.

Extra Charges Expenses incurred in connection with a claim under a policy.

Extract Summary or copy of something written, e.g. used in connection with the log book.

Extracts, Aromatic Liquid or Flavouring These consist of liquid extracts used for flavouring foods or beverages. A term associated with dangerous classified transported goods.

Extraneous risks Risks which are normally excluded under a cargo policy but which may be included for an extra premium.

Extraordinary Heavy Weather Bad weather conditions relative to what might be expected in that area and time.

Extra section Additional transportation equipment, such as a plane, bus or train, used or scheduled.

Extreme breadth The extreme breadth is the maximum breadth to the outside of the ships structure and in paddle steamers includes the paddle boxes.

Ex turpi causa non oritur action (Latin) No action is permissible that is based on an illegal or immoral consideration.

EXW Ex Works. An Incoterm 2000 – terms of sale – applicable to all modes of transport. See separate entry.

Ex Works This term means maximum involvement by the buyer in arrangements for the conveyance of the consignment to the specified destination. The exporter merely makes the goods available by an agreed date at his factory or warehouse. The seller minimises his obligations whilst the buyer obtains the goods at the lowest possible price, by arranging to use his national shipping line or airline and by securing insurance cover in his own country. This eliminates the need to fund such provisions using hard currency and thereby improves the importer's trade balance. This practice is much on the increase. The seller's obligations cease when the buyer accepts the goods at the factory or warehouse. It is usual for the buyer to appoint an agent in the seller's country to look after all the collection, transportation, insurance and documentation arrangements, possibly in consultation with the national shipping line or airline.

A cargo delivery term where the sellers (exporter's) only responsibility is to make the goods available at his premises (i.e. works or factory) not cleared for export. The buyer (importer) bears the full costs and risks involved in bringing the goods from there to the desired destination. The term thus represents the minimum obligation for the seller (exporter) including exports sold free of any insurance and freight (transport) charges; the seller's (exporter's) responsibilities is to make the goods available at the specified premises i.e. factory, warehouse; and the seller (exporter) is not responsible for loading the goods on the vehicle unless otherwise agreed, and only provides packing – if any – to enable the importer (buyer) to take delivery of the goods. The buyer (importer) bears the full cost and risk from the point of acceptance at the specified premises until they reach their destination thereby particularly embracing insurance and freight charges. These include to take delivery of the cargo and to pay for the goods in accordance with the contract of sale terms: to fund any pre-shipment inspection expense; to bear all the costs and risks of the goods from the time they have been placed at his disposal by the seller in accordance with sales contract terms; to fund any customs duties and taxes arising through exportation, to bear additional costs incurred and related risks inherent through the failure of the buyer to give instructions about the place of delivery within the prescribed period; to fund all costs in obtaining the documents required for the purpose of importation and exportation and for passing through the countries of transit. This term should not be used where according to regulations of the country of export the buyer cannot obtain directly or indirectly the export licence. In such a situation the FCA term should be used. An Incoterm 2000.

Ex Works – Export packing A cargo delivery term identical to Ex Works except that the seller (exporter) is required to pack the goods to a standard/specification adequate for the international transit. Usually under the Ex Works term the seller's (exporter's) obligation is merely to pack the goods to enable the buyer (importer) to collect the goods at the specified premise. See Ex Works entry.

EY Société Nouvelle Europe Aero Service – an IATA member Airline.

Eye Fitting connected to the ship's structure to which a tensioner or lash is attached. It may be called lashing eye or 'D' ring.

Eye bolt Bolt with an eye at the end for a hook.

Eyrir The currency of Iceland. See Appendix 'C'.

EZs Enterprise zones.

f Forward.
F Fresh water; Fahrenheit; forward; or forecastle.
fa Free alongside.
FA Safair (Pty.) Ltd. – an IATA associate member Airline.
FAA Federal Aviation Administration (USA term).
faa Free of all average (usually in hull policies).
Federal Aviation Administration US organisation.
FAB Forwarder Air Waybill.
Fabrication A term used to distinguish manufacturing operations for components as compared to assembly operations.
Fabricator A company which transforms refined metal (and sometimes scrap as well) in to semi-fabricated products, (e.g. wire, cable, tubes, strip, rods)) for sale to an end-consumer. Also referred to as semi-fabricator.
FAC Facilitation advisory Committee – IATA term.
fac Fast as can (loading or discharging) – chartering term or Forwarding Agents commission.
Faccop Fast as can – custom of the port.
Face Amount on the Bill The actual amount stated on the Bill of Exchange itself. In the event that a bill is 'discounted', i.e. sold prior to its maturity date, then interest charges/commission are deducted, the sum of which is dependent upon the period to maturity, currency, prevailing market rates and standing of the drawee. Consequently the amount received from discounting will be less than the face value on the bill. Upon maturity of the bill, i.e. the payment date, the holder will receive the (full) face amount on the bill.
Face to face selling The process of the seller meeting the buyer with a view to develop/negotiate/conclude a sale of a product/service.
Facilitation (of trade procedures) The systematic rationalisation of trade procedures and accompanying documentation.
Facilities Management The process of a company contracting out specific activities of its business and often identified as non core business. It may involve the management of its accommodation, services, supplies, security, catering, information technology. The Facilities Management Company takes complete control of the activity/activities thereby using all their professionalism and economies of scale and in so doing charges a fee to its principal.
Facility letter The process of the Bank issuing a Facility Letter to the exporter (seller) setting out the terms under which the Bank will make advances. Subsequently the exporter (seller) is required to signify acceptance of these terms by signing and returning to the Bank a copy of the Facility Letter. Finance is made available upon presentation by the exporter to the Bank of bills or notes accompanied by invoices and documents evidencing shipment of copies thereof.
Fac/oblig Facultative/obligatory. A reinsurance term for a contract where the reassured can select which risks he cedes to the re-insurer, but the re-insurer is obliged to accept all cession made. The term may be applied, also, to a broker's cover which entitles the broker to place business, which comes within the scope of the cover, outside the cover where it is in his client's best interests so to do.
FAC R/I Facultative Reinsurance.
Facsimile A reproduction copy of an illustration/commentary/manuscript.

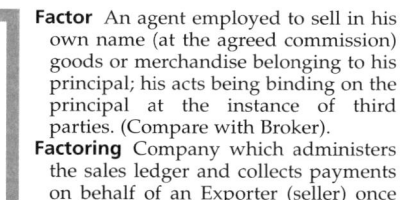

Factor An agent employed to sell in his own name (at the agreed commission) goods or merchandise belonging to his principal; his acts being binding on the principal at the instance of third parties. (Compare with Broker).
Factoring Company which administers the sales ledger and collects payments on behalf of an Exporter (seller) once the goods have been shipped.
Factors An indirect export trading technique whereby for a specified fee, the factors take over the sales ledgers (home or export or both) of companies and normally guarantee payment. See also factoring entry.
Factors of production Human and non human resources used in the production of economic goods and services. Traditionally classified as Land, Labour, Capital and Enterprise with some disagreement concerning the latter's inclusion. An economy's productive resources.
Facultative The right of option. The right of an underwriter to decide whether or not to accept a risk. Declarations under a cargo open cover are obligatory rather than facultative. Basically, a contract setting out how facultative re-insurance shall be handled by an insurer and a re-insurer.
Facultative placing Hull or cargo insurance on a single voyage from port to port.
Facultative re-insurance The re-insurance of risks that the original insurer may elect whether or not to offer for being re-insurance, the re-insurer being free to accept or reject the offer or re-insurance of a single risk negotiated and placed individually.
FAD Forms of approved documents.
FAI Féderation Aéonautique Internationale – IATA term.
Fail-safe braking System which, in the event of brake failure, automatically applies the road vehicle brakes. Normally, this involves the use of air pressure to hold the brakes off against a strong spring pressure.
Fairlead A guide for a mooring line which enables the line to be passed through a ship's bulwark or other barrier, or to change direction through a congested area without snagging or fouling.
Fairway A navigable channel for vessels often the regular or prescribed channel/route a vessel will follow in order to avoid dangerous/hazardous circumstances.
Fait accompli (French) An accomplished fact or a thing already done.
FAK Freight all kinds. Term used to show that the freight rate charge is not based on the individual commodity but freight all kinds. Uniform airline charging scale applying to a number of commodities, as opposed to SCR (Specific Commodity Rate) applying to one commodity only.
FAL Facilitation – IATA and ICAO terms.
Falconara clause The code name of a BIMCO approved charter party clause which states 'should weather or sea conditions prevent safe loading at Falconara, the vessel is either to be loaded at Ancona, or to await safe loading conditions at Falconara, in which case waiting time is to count as lay time'.
Falkland Islands pound The currency of Falkland Islands. See Appendix 'C'.
Falkland Islands pound invoice An invoice payable in Falkland Islands pounds relative to an export sales invoice.
FALPRO UNCTAD Committee on Trade Procedures and Facilitation based in Geneva.

Falsa demonstratio non nocet (Latin) A false description does not vitiate. This refers to a simple recommendation or 'puff'. The false description must not be of such a kind as to amount to fraud.

FAMA Federal Agricultural Marketing Authority (Malaysia).

Familiarisation trip A complimentary or reduced-rate trip or tour offered to travel agents and other segments of the industry to acquaint them with a destination, service or facility; usually called a 'fam' trip.

Family brand A product/service marketed under a brand name and targeted to the family environment clientele.

Family fare A type of passenger fare which includes two adults and two children.

Family life cycle The nine stages of a typical family life cycle embracing bachelor stage, newly married couples, full nest I – youngest child under six; full nest II – youngest child six or over; full nest III – old married couples with dependent children, empty nest I – old married couples, no children living with them, head of household in labour force; empty nest II – head of household retired; solitary survivor; in labour force and solitary survivor, retired. A marketing term.

Family plan Discounts offered to a family group by hotels, resorts and other suppliers.

'Fam' trip See familiarisation trip.

fan jet See turbofan jet.

FANS Future Air Navigation Systems. It is also known as ICAO, CNS, IATM system. See Global Navcom entry.

Fantail The stern overhang of a ship.

Fantainer An ISO container which is identical to the general purpose container with similar internal dimensions with the added ability to be easily converted into a Fantainer. Located high in the left-hand container door is a special hatch which has been closed and TIR sealed which allows the use of the container as a General Purpose Unit. The hatch is also fitted with an electric extraction fan which must have an electric supply when it is operated as a Fantainer. The aim is to remove any respiratory heat developed by the cargo and balance the internal temperature of the container plus the cargo with that of the varying ambients outside in order to prevent the formation of condensation.

FAO Food and Agricultural Organisation of the United Nations based in Rome.

FAQ Free alongside quay; free at quay or frequently asked question – cyberspace term.

f.a.q. Fair average quality.

FAPP Federation of Associations of Periodical Publishers in the EEC.

FAR Federal Air Regulation (USA term).

Fare The amount charged by the carrier for the conveyance of a passenger in a specified class and his/her allowable (if any) free baggage.

Fares data base The complete range of fares/passenger tariffs offered by a transport operator which are fed into a computer. It may be rail, coach, ship, airline and so on.

FAS Free alongside ship. See separate entry.

FAS (named port of shipment) Free Alongside Steamer (named port of shipment). Cargo delivery term whereby the seller (exporter) is responsible for expenses of delivering goods alongside the specified ship at a nominated port including export clearance and related costs and documentation, whilst buyer (importer) accepts all costs and risks from this point onwards including freight, etc. The insurance provision may be undertaken either by the buyer or seller. An Incoterm 2000. See separate entry.

FASA Federation of ASEAN Shipowners Association.

FASC Federation of ASEAN Shippers' Council.

Fashion Transport Transport of clothing and/or garments including shoes, belts and handbags in a dedicated means of transport.

Fast as can A chartering term indicating the vessel to load and/or discharge as fast as she can, laytime being indefinite. Sometimes this term is combined with the 'custom of the port'. See also indefinite laytime entry.

Fast Developing country A country/state whose economy is fast growing such as Malaysia, China, India and Taiwan.

Fast Ferry A vessel conveying passengers, motorists, trucks, with a speed of 40 knots and found in estuarial and short sea trades. It displaced the slower conventional ferry of 20 knots.

Fast ice Sea ice which forms and remains fast along the coast where it is attached to the shore, to an ice wall, or to an ice front, or between shoals or grounded icebergs. Vertical fluctuations may be observed during changes of sea level. Fast ice may be formed in situ from sea water or by freezing of floating ice of any age to the shore, and it may extend a few metres or several hundred kilometres from the coast. Fast ice more than one year old may be prefixed with the appropriate age category: old, second-year, or multi-year. If it is thicker than about 2m above sea level, it is called an ice shelf.

Fast ice edge The demarcation, at any given time, between fast ice and open water.

Fast Lane On a motorway system the outside lane which permits the fastest road traffic flow; or at a customs point a lane for road vehicular traffic conveying imported merchandise, often involving goods of a perishable nature.

Fast moving consumer goods A wide range of consumer goods which have a continuous buying cycle in a mass market and usually of relative low value. The main products are foodstuffs, clothing, footwear, furnishings and so on. The marketeer in order to be successful in the fast moving consumer goods market must undertake continuous market research and develop the strategy of empathy with the consumer in a competitive environment.

FASTs Flexible anti-smuggling teams – a customs terms.

Fathom Measurement for assessing sea depth on basis that one fathom equals 6 feet or 1,8288 metres.

Fathomer Instrument measuring depth of water beneath vessel.

Fatigue The tendency of a material to weaken or fail during alternate tension-tension or tension-compression cycles. In cordage, particularly at loads well below the breaking strength, this degradation is often caused by internal abrasion of the fibres and yarns but may also be caused by fibre damage due to compression. Some fibres develop cracks or splits that cause failure, especially at relatively high loads.

Faults diagnosis clinics Group discussions designed to train managers to identify areas of weakness in their companies and related strategies.

FAVC Fast as vessel can discharge. A chartering term.

FB Ferry boat.

FBD Freeboard.

FBH Free on board harbour.

FBL The FIATA Bill of Lading. It is a negotiable FIATA combined Transport Bill with negotiable status. It has been developed by the International Federation of Forwarding Agents Associations and is under the ICC Rules Uniform Customs and Practice for Documentary Credits.

FBS Franco Belgian Services – a shipowner.

FC Fibreglass covers or full cost.

FCA Free carrier (named place) cargo delivery term. See separate entry.
FCAA Federal Civil Aviation Authority – IATA term.
FCAR Free of claim for accident reported. No claim is payable under the insurance if it arises from an accident which is known to have occurred before acceptance of the risk. A marine insurance term.
FCB Frequency Co-ordinating Body (ICAO) term.
FCC Fully cellular containership; French Chamber of Commerce or Federal Comminications Commission (USA term).
FCL Full container load.
FCL delivery A container movement which commits the carrier to deliver the goods to the importers specified delivery point. See also FCL Port to (or Pier)/FCL Depot entry.
FCL Door/LCL Depot A container movement where, for example, a haulier, who was the sub-contractor of the Carrier, took an empty container to a shippers premises for the shipper to pack. Subsequently he would take the loaded container back to the depot for integration into the combined transport operator's combined transport system for the trade. At the importing end, the loaded container would then be unpacked at the combined transport operator's depot by the sub-contractor of the Carrier, who would effect delivery at the depot to the Consignee (or his agent) for carriage to the Consignees premises. The combined transport document in this instance would cover movement between Shipper premises and the importing Depot and, subject to the combined transport document terms and conditions; the combined transport document would accept responsibility for the goods from the time of receipt of the loaded container at the Shipper's premises until the time of delivery of the unpacked goods to the consignee's haulier at the importing depot. It may be also termed FCL House/LCL Depot.
FCL/FCL A container term whereby goods are conveyed from the exporter's premises and conveyed to the consignee's premises. The exporter/forwarder packs the container whilst the consignees unpacks the goods. Containers are presented to the carriers loading berth/terminal with the cargo already packed and are for ultimate delivery beyond the carriers discharging berth or terminal as a full load. Also termed House to House.
FCL House/LCL Depot See also FCL Door/LCL depot entry.
FCL/LCL Full Container load/less than container load. Container movement whereby the packing is charged to the account of the merchant and the stripping is to the carrier.
FCL Pier/FCL Depot See entry FCL Port/FCL Depot.
FCL Port (or Pier)/FCL Depot A term found in combined transport operation whereby the carrier receives from the shipper at a ship's side a container packed by the shipper and delivers same to consignee at importing depot for consignee to take away to owner's premises for unpacking.
FCO Foreign and Commonwealth Office or Forward Currency Options.
Fco. Franco (free to named point) – cargo delivery term.
FCPA Foreign Corrupt Practices Act.
FCPS A computerised cargo processing system.
FCR Forwarders Certificate of receipt. (FIATA document) or Floating crane.
FC & S Free of capture and seizure – insurers not responsible if the ship or goods are seized by a foreign power.

fcsrcc Free of capture, seizure and riots and civil commotions.
FCT Freight Forwarders combined transport bill of lading.
FCTFSTF Flight Crew Training & Flight Simulator Task Force – IATA term.
FD Franco Domicile – cargo delivery term; Free delivery or Free discharge.
fd Free discharge; free docks; free despatch; free delivery or forced draught.
FDA The Food and Drug Administration (Thailand).
F & D Freight and demurrage.
FDC Fast and developing country.
FD & D Freight deadfreight & demurrage or Freight Demurrage and Defence (Insurance).
FD & D cover Freight, Demurrage and Defence (Insurance) cover.
FDFM Flight Data and Flow Management Group – IATA term.
FDI Foreign Direct Investment.
FDISK A program that prepares a hard disc so that more than one operating system can reside on it.
FDPS Flight Data Processing System – IATA term.
FDR Flight Data Recorder – an IATA term.
Feasibility study The process of researching/studying/evaluating a particular situation/remit with a view to identifying all the relevant considerations and forming an analytical conclusion/recommendation.
FEATS Future European Air Traffic Management System – IATA term.
FEC Federal Express Corporation – an IATA Airline member or Forward Exchange Contract.
Federal Aviation Administration (FAA) An agency of the US Department of Transportation that regulates civil aviation and certifies the airworthiness of aircraft.
Federal funds rate The nearest US equivalent to the UK domestic sterling overnight interbank rate; the market is a very important money market, providing a main source of banks' short-term funds; federal funds lent or sold in the market are deposits in excess of the US banks' reserve requirements.
Federal Maritime Administration (FMA) A unit of the US Department of Transportation that administers federal programs affecting waterborne commerce, the merchant marine and the American shipbuilding industry.
Federal Maritime Commission (FMC) Independent US government regulatory agency with authority over international passenger and cargo shipping.
Federation Internationale des Associations de Transitaires et Assimilés International Federation of Forwarding Agents Associations based in Zurich.
FEDIS FIATA Electronic Data Interchange Systems Co-operative Society.
Fee-based pricing A method of pricing agency services based on the suppliers' net price, plus a mark-up that covers the cost of delivering the service and a profit.
Feedback The flow of information back into the control system so that actual performance can be compared with planned performance. Positive feedback results in increased output whereas negative feedback results in reduced output.
Feeder See Feeder vessel entry.
Feeders Wooden funnel filled with grain located in top of hold to counteract and absorb movement of grain during voyage.
Feeder port A seaport which provides maritime services linking in with major ports and thereby acts as a feeder service. See also hub and spoke system entry.
Feeder Services The use of smaller vessels to shuttle, usually, for example, from near Continent to UK with

Feeder vessel

goods when it is not a feasible or economic proposition for a vessel to call at a UK port due to its routeing or the amount of goods involved. Similar arrangements apply to air freight services and to other shipping services, such as in Far East, with the maritime trunk deep sea container service relying on smaller tonnage as a feeder service. Middle/Far East hub ports include Singapore, Colombo, Hong Kong and Dubai. More recently in oil tonnage the VLCC/ULCC relying on the smaller oil tanker such as a product carrier to convey the oil from VLCC/ULCC destination seaport to a tank farm maritime terminal. See hub and spoke entry.

Feeder vessel A vessel feeding into a trunk route liner service such as smaller container ships feeding into the deep sea liner container services, such as feeder services serving hub ports of Singapore, Hong Kong, Rotterdam and Colombo. See Feeder Services entry.

Feel of the market An impression of how a market is developing, particularly relevant to the financial and commercial aspects of a business. For example, the impression a sale and purchase shipbroker may give is the market is hardening regarding second-hand tonnage purchase thereby suggesting scope for any lower price being secured in negotiation.

FEFC Far Eastern Freight Conference.

FELCRA Federal Land Consolidation and Rehabilitation Authority (Malaysia).

FELDA Federal Land Development Authority (Malaysia).

Fen The currency of China. See Appendix 'C'.

Fender An appliance made of timber and/or rope or other materials used to prevent damage to the hull of a vessel especially during mooring and unmooring operations.

FENEDEX Federation for the Netherlands Export.

FENEX Federation of Dutch Freight Forwarders.

Fen invoice An invoice payable in Chinese Fen relative to an export sales contract.

FERD Foreign Economic Relations Department (Burma).

FERIT Far East Regional Investigation Team.

Ferry A passenger vessel engaged in estuarial or short sea trade services – the latter which may vary from 11/2 to 36 hours duration. The ferry conveys passengers, accompanied cars, coaches, motor cycles, cycles and road haulage vehicles. The longer ferry crossing would have cabins, whilst most would have restaurants and lounge accommodation. See Fast Ferry entry.

Ferry berth/ramp A berth situated at a seaport which accommodates/handles ferry vessels conveying vehicular cargo. Vehicles are transhipped from/to the ferry over the shore based ferry ramp.

Ferry flight A non revenue flight for positioning an empty aircraft.

Ferry Line Manager A senior managerial position responsible for the overall Management and policy of a particular shipping service/route/trade involved in ferry type vessels capable of conveying passengers/cars/coaches/road haulage vehicles/containers and so on.

Ferry wagon The brand name of a fleet of high capacity purpose built wagons operating between UK and the European rail network routed through the Channel Tunnel. Overall capacity exceeds 60 tonnes.

Ferticon Code name of Chamber of Shipping approved charter party used in the Fertiliser trades.

Fertisovbill Code name of BIMCO approved bill of lading used with shipments under Fertisoy charter party.

Fertisoy Code name of BIMCO approved Soviet Fertilizer voyager charter party.

Fertivoy 88 Code name of BIMCO approved North American Fertilizer voyage charter party introduced in 1978 and amended in 1988.

FESCO Fesco Lines – a shipowner.

FESPA Far East and South Pacific Association (of airport executives).

FET/GUL/SAS Far East/Gulf/south Asia container service.

FEU Forty foot equivalent unit. Container term.

FEWAC Far East/West Africa conference.

FF Tower Air Inc – An IATA Member Airline.

FFA Freight Forwarders agreement.

ffa Free from alongside – seller pays for lighterage if incurred at port of discharge or free foreign agency.

FFCR Freight Forwarders Certificate of Receipt.

FFE Freight Forward Europe – A group of major European Freight Forwarders who are focusing attention on developing improved transport services and improving services to shippers.

FFI For further instructions.

FFN Freight Forwarders Network – a computer term.

FG Ariana Afghan Airlines Co. Ltd. A National Airline and member of IATA.

FGA Free of general average or foreign general average as distinct from York Antwerp rules.

FGMDSS Future Global Maritime Distress and Safety system.

fh First half of month.

FHEx Fridays and holidays excepted. A chartering term indicating Fridays and holidays not to count as laydays – applies to non-christian countries.

FHInc Fridays and holidays included – a chartering term. Fridays and holidays to count as laydays.

FI Icelandair. The National Airline of Iceland and member of IATA or Finland – a customs abbreviation.

Fi Free in.

fia Full interest admitted.

FIA Federation Internationale de l'Automobile (French).

FIAC Foreign Investments Assistance Centre (BOI Philippines).

Fias Free in and stowed.

FIATA Fédération International des Associations de Transitaires et Assimilés. International Federation of Forwarding Agents Associations based in Zurich.

FIATA Combined Transport Bill of Lading A bill of lading approved by FIATA and used particularly in the container market. It extends the carrier's liability to cover the inland sectors in the country of export and to inland destination after discharge from the vessel. Accordingly, it features the place of receipt of cargo and likewise place of delivery together with the ports of loading and discharge.

FIATA Forwarders Certificate of Receipt A shipping document issued by the freight forwarder to the forwarders principals under the aegis of FIATA.

FIATA MTBL FIATA multimodal transport bill of lading.

FIATA Multi Modal transport bill of lading The FIATA multimodal Transport Bill of Lading is a negotiable document of title in line with the International Chamber of Commerce Uniform Rules for such documents.

FIATA NVOCC Bond A freight forwarding industry bond for use in the USA market. The bond is only available to BIFA trading members.

fib Free into barge or free into bunkers.

FIB Financial Investigation Bureau.

FIBC Flexible Intermediate and Bulk Container.

Fibre A long, fine, very flexible structure that may be woven, braided, stranded or twisted into a variety of fabrics, twine, cordage or rope.

Fibreboard A board wholly or mainly of cellulose fibrous materials, but chiefly and basically wood fibres, pulped and manufactured into sheets of various qualities, thicknesses, strengths and weights.

Fibreboard can A small cylindrical inner receptacle with the wall made of fibreboard and ends of fibreboard, metal, plastics or other suitable material.

Fibreboard cartons The most widely used form of outer packing is found in the fibreboard box or carton. Many such fibreboard cartons are bonded to a pallet to ease handling and stowage. The carton is usually a single wall, but to provide extra strength double wall and triple wall are available.

FIC Foodstuff Industries Corporation (Burma); Foreign Investment Committee (Malaysia); free insurance and carriage or Flight Information Centre.

FICS Fellow of the Institute of Chartered Shipbrokers.

FICs Forum Island Countries.

FID International Federation for Documentation.

FIDC Fishery Industry Development Council (Philippines).

Fidelity guarantee Insurance against loss arising from dishonesty of persons holding positions of trust.

Fiduciary A person who holds something in trust for another.

Field Part of a record structure for storing a particular data item (attribute). The area allocated on a screen or form design for a particular data item.

Field code Field identifier expressed in code. A data processing facility.

Field identifier Title specifying the nature of data in a data field. A data processing facility.

Field questionnaire A questionnaire which relies on face to face interviews to obtain the market research data from the respondent. See postal questionnaire and questionnaire entries.

Field reading Field identifier in plain language, full or abbreviated. A data processing facility.

Field research The process of conducting research in the market place involving face to face interviewing, postal questionnaire, telephone interviewing and so on.

Field Sales Manager A senior Sales Manager responsible for a number of Sales Managers/representatives engaged in the day to day direct selling technique with customers.

Field trials The process of conducting/testing a product/service in the market. It may be a new container type and the prototype would be subjected to trial/test transits to evaluate shippers reaction. See test transit and marketing entries.

FIEx Fellow of the Institute of Export.

FIFO First in first out.

Fifth Freedom The privilege for an airline registered in one state and en route to or from that state to take on revenue passengers, mail and freight in a second state and put them down in a third state. One of the six freedoms of the air.

Fifth special survey The fifth special hull and machinery survey of a ship which arises when the vessel is about 25 years old and decreed by the Classification Society. A critical time of a ship's life which may involve the scrapping decision being activated.

Fifth wheel Circular or wheel shaped device at the rear of a tractor unit that engages the underside of the trailer and locks into the kingpin with a spring lock device.

Fifth-wheel coupling Connecting device on articulated road vehicles comprising a support plate mounted on the tractive unit with a jaw and locking device to accept and secure the towing kingpin, and on the trailer a turntable with a kingpin projecting downwards which locates in the jaw on the tractive unit plate.

Fih Free in harbour.

Fiji dollar The currency of Fiji Islands. See Appendix 'C'.

Fiji dollar invoice An invoice payable in Fijian dollars relative to an export sales contract.

Fil The currency of Iraq, Jordan and a number of other countries. See Appendix 'C'.

File A collection of related information that is stored on a disc. Files typically have a name or number by which the information can be referenced and accessed.

Filename The first part of the name of an MS-DOS file. May contain no more than eight characters.

File protection A facility offered on most LANs to enable users to set rights to their files and sub-directories for other users, e.g. read, copy and write. These rights may also be set in most systems by adjusting the attributes of the file itself, search i.e. read only, hidden. See 'Local area network'.

File server The computer which contains the network software for a LAN and often the applications software accessible to the stations using the network. See 'Local area network'.

Fill The completed execution of an order. An IPE definition.

Filler The currency of Hungary. See Appendix 'C'.

Filler invoice An invoice payable in the Hungarian currency of Filler relative to an export sales contract.

Filler traffic Goods conveyed by the carrier on the basis they are shipped only at the last minute prior to the ship, aircraft or other transport mode departing and subject to capacity being available. Such traffic in the passenger sector is called 'standby'. Both 'filler' and 'standby' traffic sometimes attract a lower rate.

Fillet weld The fillet weld is used where plates are connected at an angle to an adjacent plate. The strength of fillet welds is inferior to that of butt welds.

FILO Free in Liner Out.

FIME Fellow of the Institute of Marine Engineers.

FIN Fin type stabilizer.

Final 'check in' The process at a passenger airport whereby passengers tickets are finally examined prior to proceeding to board the aircraft. In so doing the passenger ticket would be retained by the airline staff and the boarding ticket details checked with the passenger flight list. See also airport lounge entry.

Final dividend The last distribution for a company's trading period usually a year. Dividends may be paid twice yearly with the final one at the year's end. The dividend is recommended by the directors but authorised by the shareholders at the company's Annual general Meeting.

Finance The provision of money at a time it is wanted.

Finance House An indirect trading technique involving the provision for exporters with non-recourse finance whilst allowing agreed credit terms to foreign buyers. The credit terms vary from short to long-term. Overall, the Finance House undertakes the financial and administrative burden of arranging export finance.

Finance lease A form of lease with an option to purchase the ship. The lease payments may be considered as payment for the cost of the vessel. The lessor, whose main role is as financier, has little involvement with the asset beyond ownership. All operating responsibilities fall on the lessee who, in the event of early termination, must fully compensate the lessor. Also available for marine and cargo handling equipment. A shipping definition.

Financial Investigation Bureau An organisation which is part of the ICC and embraces the following functions: suspected frauds and fake financial instruments in circulation; reports on piracy incidents throughout the world, information on shipping and container frauds, and alerts on counterfeit pharmaceuticals.

Financially driven company
A company driven in terms of its strategy/management/development by financial considerations especially profit, share value, and financial performance.

Financial planning The process of evaluating the company/entity financial needs and determining priorities and resources provision over a specified period on an optimum basis. It embraces income, expenditure and investment.

Financial policy The financial policy of a particular company as in a particular set of circumstances. It may be, for example, in any cost analysis undertaken, ten per cent must be added on for administration expenses. It can also extend to credit control, depreciation, investment criteria and so on.

Fin Co Final Course.

Finnair Oy The Finnish National Airline and a member of IATA.

Finnish Shippers Council The organisation in Finland representing the interest of Finnish shippers in the development of international trade.

Finnres The computerised reservation and national distribution system of Finland; partner of Amadeus.

FINPRO Finnish Committee in Trade Procedures.

FIO Free in and out. The charterer, shipper, or consignee pays for the cost of loading and discharge of cargo.

FIO Charter Type of charter party whereby charterer pays for cost of loading and discharging cargo.

FIO & stowed Free in and out – charterer pays for cost of loading and discharging cargo including stowing.

FIO & trimmed Free in and out – charterer pays for cost of loading and discharging cargo including trimming.

FIOS Free in out stowed.

FIOST Free in, out, stowed and trimmed – the cargo is to be loaded, stowed, trimmed and discharged free of expense to the shipowner. A chartering term.

FIOT Free in, out and trimmed.

Firavv First available vessel.

Fire Safety Systems Code See FSS code.

Fire wire A wire rigged to the waterline over the off-berth side of a ship to facilitate towing off in an emergency.

Firm offer A definite commitment to undertake business by the party making it, and in so doing it must contain the essential features of the proposed contract and have usually a time limit for its acceptance.

Firm offer (of a cargo) The cargo offer is made of a definite nature and its acceptance subject to a time limit such as 48 hours. The prospective carrier (shipowner, chartering agent or shipbroker) is under obligation to accept, decline, or make a counter offer, within the time limit.

Firm offer (of a ship) The ship offer made is of a definite nature and its acceptance subject to a time limit such as seven days. The prospective charterer is under obligation to accept, decline, or make a counter offer within the time limit.

Firm quotation If a company wishing to deal in foreign currency contracts a bank to ask for a rate, the bank may quote its rate 'for information only' or the quotation may be firm: a firm rate implies that the bank would be prepared to deal at this rate immediately. Alternatively, a transport company or commercial entity in quoting a rate may indicate it is a 'firm quotation'. It implies the rate is firm and not subject to any further negotiation/haggling/bargaining. See firm offer.

First-ashore line A line (usually fibre) put ashore first to help in hauling the ship into berth.

First carrier The carrier who actually performs the first leg of the contracted transit. Usually associated with IATA air freight consignments.

First class A superior class of accommodation usually attracting a premium fare in a transport mode especially found on rail and airline services. It is superior to second standard class accommodation

First class hotel See hotel classification.

First Freedom The privilege to fly across the territory of another state without landing. One of the six freedoms of the air.

First in first out The method whereby the goods which have been longest in stock (first in) are used, delivered (sold) and/or consumed first (first out).

First Line Management The first line level of management below the Board of Directors in a Company.

First Mate The First Officer of the deck department on a vessel – ranking next to the Master – and responsible primarily for navigation of the ship.

First mortgage A mortgage which is not subject to any prior mortgage. This is the primary form of security in shipping finance.

First notice day The first day in which notice may be given or received of the intention to deliver actual commodities against futures market positions. These vary with each commodity and exchange. A BIFFEX term.

First Open Water A shipping term applying to a port area indicating the date/period when ice melts sufficiently for ships to commence trading at the start of a new season such as found in the St. Lawrence Seaway.

First sitting The earliest dining time; usually applies to cruises.

First surplus re-insurance/first surplus treaty The name given to an ordinary surplus treaty; the surplus must be allotted to that treaty first and in priority to any other re-insurers. Sometimes treaties are arranged as second surplus treaties, and these would receive a share of the surplus only after the first surplus treaty had received the full amount to which it is entitled.

First Three A term on a ship requiring that subsequent amendments be subject to the three leader agreement. A marine insurance term.

First world country A country with a very high GDP per capita. Such countries are termed fully developed countries and represent the richest countries in the world such as USA, Japan, UK, Germany and France.

First-year ice Sea ice of not more than one Winter's growth, developing from young ice; thickness from 30cm to 2cm may be sub-divided into thin first-year ice/white ice, medium first-year ice and thick first-year ice.

FIS Freight, insurance and shipping charges or Flight Information Service.

Fiscal Agency An agency dealing with financial matters on behalf of his Principal including invoicing, monitoring and reconciling payments into clients bank accounts, credit control and so on.

Fiscal barriers An international trade term in which goods, services and capital cannot circulate freely between countries due to distortions of competition and financial controls.

Fiscal deficit trade balance A financial deficit on a country trade balance. See fiscal policy and balance of trade.

Fiscal policy A policy executed by a government or Funding Bloc. This embraces levels of taxation, money supply, interest rates, exchange control and all areas which have a significant financial overtone in a country/Trading Bloc. It embraces both the business and population sectors.

Fiscal sanctions Measures taken by the International/Community or individual governments on a bi-lateral or multi-lateral basis which penalises/legalises restrictions/constraints on the free movement of trade. Such sanctions are of a fiscal nature and may embrace financial aid, soft loans.

Fiscal surplus trade balance A financial surplus on a country trade balance. See fiscal policy and balance of trade.

FISH Fishing boat.

Fish Dock The quayside/berth(s) at a port at which vessels/trawlers discharge their catch of fish. Alternatively, it may be a facility at a railway or road haulage depot permitting railway wagons or road haulage vehicles to be loaded or discharged of fish.

Fishtailing The name given to an offtake vessel's movement when the bow remains in approximately the same position, relative to the stern of the installation, while the stern of the offtake vessel moves to one side, so that the two vessels no longer lie in a straight line.

FSS Code The International Code for Fire Safety Systems (FSS Code) was adopted by the Maritime Safety Committee (MSC) at its seventy-third session (December 2000) by resolution MSC.98(73) in order to provide international standards for the fire safety systems and equipment required by chapter II-2 of the 1974 SOLAS Convention. The Code is made mandatory under SOLAS by amendments to the Convention adopted by the MSC at the same session (resolution MSC.99(73)). The amendments came into force on 1st July 2002.

FIT See foreign independent tour; has also loosely come to mean any independent tour arranged by travel agents, whether domestic or foreign; in hotel usage can mean full individual tariff (non discounted).

Fit operator/wholesaler A specialist in designing and operating FITs for travel agents to sell.

FITPRO Czech Republic Committee/Organisation in Trade Procedures based in Prague.

Fitting out The process of equipping a vessel so that she is in a proper seaworthy condition for the voyage.

Fitting out crane See crane fitting out entry.

Five major dry bulk seaborne cargoes These include iron ore, coal, grain, phosphate rock and bauxite and alumina.

fiw Cargo delivery term – Free into wagon. This term is similar to Free on rail/truck.

Fixed cost Those costs which remain at a constant level irrespective of whether the service is operational or immobile such as depreciation. Sometimes they are termed unavoidable cost.

Fixed crane See crane fixed entry.

Fixed exchange rate An exchange rate which is fixed officially by a government.

Fixed head A rigid end of a rod that fits directly into a corner casting.

Fixed height load carrying truck A truck carrying its load on a non-elevating platform.

Fixed income Market The market for trading bonds and preferred stock.

Fixed income sales Includes sales of our liquidity products or products with a view on interest rates, foreign exchange, and/or credit risk.

Fixed interest A transaction where the money charged for a loan is agreed in advance and does not change over the period of the loan.

Fixed interest security A security which pays a guaranteed interest rate usually maturing in a specified period.

Fixed lashing plate A lashing plate welded directly onto the hatch cover.

Fixed Link See Euro Tunnel entry.

Fixed link span A facility provided at a ramp berth accommodating a vehicular ferry/Ro-Ro vessel linking the sea-ward end of the ramp with the stern or bow of the vessel over which vehicles are transhipped (driven onto and from the vehicle deck). It is a permanent installation but if necessary can be moved to other installations.

Fixed Objects Harbours, piers, etc, with which a ship might come into contact.

Fixed platform truck Fixed height load carrying truck.

Fixed Rate Bonds Bonds with a fixed rate of interest.

Fixed rate loan A loan granted which has a fixed rate of interest throughout its term.

Fixed route A route which does not vary and usually regarded as the custom of the trade.

Fixing Chartering a vessel.

Fixture Conclusion of shipbroker's negotiations to charter a ship.

Fixture rate The rate at which the chartered vessel is hired per day or month or other mutually acceptable specified period.

FJ Air Pacific Ltd. A National Airline and member of IATA.

Flag Indicate the national authority with which the ship is officially legally registered.

Flag carrier An airline of one national registry whose government gives it partial or total monopoly over international routes.

Flag discrimination Wide variety of acts and pressures exerted by government to direct cargoes to ships of their own flag regardless of the commercial considerations which normally govern the routeing of cargoes.

Flag government See Flag State.

Flagging out The process of a shipowner transferring the registration of vessels from the National flag, to one of a Flag of Convenience such as from Italy to Liberia. Hence, this occurs when companies transfer to an alternative flag in order to overcome a perceived disadvantage of remaining on the original registry.

Flagrante delicto (Latin) In the act of committing the crime.

Flags of convenience Process of Shipowner registering ships with countries that operate national fleets tax free or virtually tax free such as Liberia or Panama. Also referred to as open regtistries.

Flags of convenience registries These include the following countries: Antigua and Barbuda, Bahamas, Belize, Bermuda, Burma, Canary Islands (Spain), Caymon Islands, Cook Islands, Cyprus, Gibraltar, Honduras, Lebanon, Liberia, Malta, Marshall Islands, Mauritius, Netherlands Antilles, Panama, St Vincent, Sri Lanka, Tuvalu and Vanuatu. In order to respond to the decline of traditional national shares of tonnage, some nations have introduced a second register such as Norway (NIS), Denmark (DIS), Germany (GIS) UK (Isle of Man) France (Kerguelen), Belgian (Luxembourg) and Portugal (Madeira).

Flag of registration Flag of the country in which a ship is registered.

Flag State The State/Maritime Government in which the ship is registered and holds the legal jurisdiction as regards operation of the ship whether at home or abroad.

Flag state control The appointed registered state/country in which the vessel is registered and mandatory all the maritime regulations imposed on the state fleet. The vessel will fly the national flag and respect all the conventions to which the state has adopted such as those initiated by the IMO. This extends to ship safety, dangerous cargo, ISM code and Port State Control.

Flake down Laying a rope in long bights on the deck with each bight clear of the adjacent one, so that it can be paid out quickly and free from turns.

Flame screen A device incorporating corrosion-resistant wire meshes. It is used for preventing the inward passage of sparks (or, for a short period of time, the passage of flame), yet permitting the outward passage of gas.

Flammable Capable of being ignited.

Flammable range The range of gas concentrations in air between which the mixture is flammable. This describes the range of concentrations between the LFL (Lower Flammable Limit) and the UFL (Upper Flammable Limit). Mixtures within this range are capable of being ignited.

Flap An extension normally hinged to the free end of a ramp, to give transition between running surfaces.

Flares Articles containing pyrotechnic substances which are designed to illuminate, identify, signal or warn. The term includes: flares, aerial and flares, surface.

Flash point Is defined as the lowest temperature at which flammable vapour is given off a liquid in a test vessel in sufficient concentration to be ignited in air when exposed momentarily to a source of ignition. This does not mean the temperature at which a liquid ignites spontaneously.

Flash powder Pyrotechnic substance which, when ignited, produces an intense light.

Flat See flat rack entry.

Flat bed A road trailer with a flat cargo-carrying surface.

Flat bed trailer A wheeled trailer or a semi trailer with a flat cargo carrying surface or deck and without any superstructure.

Flat car A railway wagon with a surface but with no side wall for carrying containerised cargo.

Flat container An open-sided container designed with corner posts for structural supports. Also termed flat rack.

Flat Line Re-insurance A re-insurance which covers losses on the original policy up to a fixed amount, any excess being borne by the re-insured.

Flat organisation An organisational structure which has a lateral formation and all executives/managers are on the same level with no vertical structure

Flat pack Garments/furniture packed in cardboard boxes.

Flat rack A flat rack is a container which has a base frame and two endwalls only, alternatively four individual cornerposts. The advantage of the flat rack is that cargo can be loaded from the top or from the side. The flat rack is mainly used for heavy machinery, steel pipes, etc. Also termed flat container.

Flat rack containers Such containers are designed to facilitate the carriage of cargo in excess of the dimensions available in either general purpose or open top containers. Overall, they consist of a flat bed with fixed ends, the external dimensions conforming in all respects to the ISO requirements.

Flat rate A basic rate which applies to a product or service. It may be a rate which is available on a bus, coach, or rail service for any journey made irrespective of distance, or a freight rate of $40 per ton irrespective of the commodity classification shipped in a particular trade/service. The criteria under which the flat rate operates will vary by circumstances. See also 'Freight all kinds' entry.

Flats An ISO container of 20ft (6.10m) or 40ft (12.20m) length designed in the form of a platform. It is ideal for a wide range of products including machinery, vehicles and so on.

Flat trailer In appearance it is similar to a rigid flat vehicle but the unit is simply a wheeled flat bed with a frontal upright to give stability to the load. After loading, the cargo is sheeted for protection. This type of trailer offers three main advantages: ease of loading; total payload is enhanced to 25,000 kilos compared with 23,500 kilos with the tilt trailer; and the trailers can be loaded 'two a top', a third flat which does the carrying and thus return the two empty trailers. This form of piggyback reduces freight costs especially with trade flow imbalances.

Flat Wrapped A method of packaging especially ideal for furniture such as chairs/tables which when unwrapped is assembled by the customer.

Flaw A narrow separation zone between pack ice and fast ice where the pieces of ice are in a chaotic state; it forms when pack ice shears under the effect of a strong wind or current along the fast ice boundary (cf. shearing).

Fleet angle The angle between the mooring line and a plane perpendicular to the axis of the winch drum.

Fleet Manager The person in control of a fleet of vessels, aircraft, road vehicles, etc. and often, in respect of Maritime Transport, the shipping Company's Registered Manager.

Fleet operation management The process of managing a fleet with particular focus on the operational aspects. This usually has regard to ship specification, manning levels, port charges, voyage cost, market forecasts, alternative options, market development/opportunities, competition and cost/revenue analysis. See Ship Management.

Fleet planning The process of planning the operational use of a fleet on an optimum basis. Factors influencing maritime fleet planning decision making process include, market forecast in terms of annual demand; new contracts; seasonal demand – peak and trough months; size of vessel acceptable at each port of call; round trip (voyage) times; warehouse accommodation resources and cost of additional storage needs to meet seasonal demand; vessel sizes available for the fleet and carrying capacities; number of days each vessel available annually and monthly for each vessel having regard to survey commitments, etc; larger vessels more economical but limited to a fewer ports; smaller ships give more operational flexibility; need to plan schedules/routes to realise optimum use of existing fleet as cost incurred on vessel whether working or not; evaluate option of planning fleet to meet average, peak, or trough demands as defined in market forecast; evaluating whether to plan for annual growth of demand; and finally evaluation of the cost/revenue production results of the plan with the profit motive/aim paramount. It also can embrace investment policy/evaluation. A similar criteria would apply to fleet planning of aircraft, international road haulage vehicles and so on.

Fleet policy Policy covering a number of motor vehicles (usually more then ten) or of ships, railway rolling stock or aircraft. It involves their deployment, future replacement specification, maintenance arrangements, utilisation and so on.

Fleet rationalisation The process of reducing the size of the fleet in terms of the number of units such as the number of vessels may be reduced from twenty to ten. A similar criteria can be applied to any transport fleet such as aircraft, road vehicles, etc.

Fleet utilisation The overall utilisation of a specific fleet of railway wagons, ships, road haulage vehicles, etc., usually measured in the total number of loaded ton miles generated by the transport fleet in a particular period.

Flexitank Polythene bag to allow bulk liquids and powders to be carried in GPs.

FLG Flag

FLIC Forwarders Local Import Control/Clearance.

Flight The operation of an aircraft from take off to landing.

Flight Band Details of flights available within a particular time period on a specified day.

Flight coupon The portion of the passenger ticket and baggage check or excess baggage ticket that indicates particular places between which the coupon is valid for carriage. An IATA definition.

Flight crew member A licensed crew member charged with duties essential to the operation of an aircraft during flight time.

Flight deck The cockpit of an aircraft where all the aircrew are situated other than stewards/hostesses.

Flight information Details of flights of particular airlines.

Flight manifest Details of air freight conveyed on a particular flight giving full details of each consignment including Air Waybill number.

Flight member A licensed crew member charged with duties essential to the operation of an aircraft during flight time.

Flight number The numerical designation of a flight – such data usually inserted in the Air Waybill.

Flight path The route allocated to a particular aircraft in aerospace by an air traffic control network.

Flight stage The operation of an aircraft from take off to landing.

Flight strategy The options available/decided in regard to a particular flight of general airline policy. It may include 'in flight' entertainment with or without payment; 'in flight' meals/drinks with or without payment and so on. It can apply to a whole range of items.

Flight time The actual journey time of a particular flight.

Flight time information A visual display found at an airport passenger terminal giving up-to-date information on individual flight number, destination, departure time and boarding gate number.

Flip flops Wedge-shaped pieces of metal set at the ends of two adjacent cells in an athwartship direction, where the cells are too close to enable the guidance system to be flared out to assist the entry of the containers.

FLIR Forward looking infra-red visual enhancement.

FLIREC Flight Recorders Study Group – ICAO term.

Floating Berths A pontoon type of berth connected usually by a causeway or link span to the shore.

Floating crane A floating crane usually found in a port, handling heavy lifts and indivisible loads. See also crane floating entry.

Floating currency A floating currency is one whose level is determined by the market forces found in international trade and the exchange rate is not preset by government. In consequence it is not subject to devaluation or revaluation.

Floating dock A floating structure which accommodates – under dry dock conditions – the repair and maintenance to the under-water parts of the ship's hull including periodical surveys. Ships requiring to use the floating dock are floated in and the dock is sealed. Subsequently the water is pumped out to afford dry dock conditions.

Floating exchange rates Currency rate which varies according to world distortions and not subject to exchange control.

Floating, open and annual marine policies A floating, open, or annual marine policy is a policy which describes the insurance in general terms and leaves the name of the ship or ships or other insurance property and other particulars to be defined by subsequent declaration. The subsequent declaration or declarations may be made by endorsement on the policy or in other customary manners.

Floating policy A cargo policy with a sum insured sufficient to accommodate a large number of shipments for a single assured. Subject to a limit on any one vessel, each shipment is covered automatically as it goes forward. The sum insured is reduced by each declaration until the amount has been exhausted. A deposit premium paid at inception is adjusted on expiry of the policy to provide the correct premium. A certificate is issued for each shipment.

Floating rate The basis on which interest is usually calculated in respect of medium-term shipping loans. The rate payable is set at a fixed margin above the relevant inter-bank offered rate, generally LIBOR. The frequency with which the rate is adjusted depends on the term of the reference LIBOR rate, but is usually at six-monthly intervals.

Floating rate bond A bond on which the interest rate is not fixed for the entire life of the bond, but is adjusted periodically on the basis of predetermined criteria.

Floating rate loans A loan where the rate varies periodically through its life.

Floating rate note A capital loan stock bearing interest that will be determined at regular intervals by a formula based upon prevailing short-term money market rates. An international finance term.

Floating rate security A security whose interest rate moves in a fixed relationship to other rates such as bank rates.

Floating stock The amount of goods in a pipeline, the sum of loading stock, goods in transit and receiving stock.

Floating warehouse The concept whereby the cargo shipment operates on a door to door basis throughout on dedicated multimodal service with no delays at terminal transhipment points. It is logistically driven.

Float time The period it takes for an importer's payments to reach the exporter's bank account – a period of up to three weeks is not uncommon costing about £75 in interest charges on a £10,000 transfer.

Floodable length Length of vessel which may be flooded without sinking below the margin line.

Floor bearing The maximum loading in kilogrammes per square metre that the floor or deck of a container/road trailer, wagon, container, ship, or aircraft is capable of withstanding.

Floor Broker One who has authority to trade in the 'ring'. A BIFFEX term.

Floor loading The static and dynamic loads imposed on the floor by the payload and the wheels of handling equipment when used.

Floor member One who has authority to trade in the 'ring' – a BIFFEX term.

Floor price The lowest price a producer will offer a particular product/service in the market place.

Floors Transverse vertical plates in double bottom of a vessel.

Floor trader One who has authority to trade in the 'ring' – a BIFFEX term.

FLOPAC Flight Operations Advisory Sub-Committee – IATA term.

Flotation The launching of a company and particularly of a new capital issue. It usually involves the listing on a recognised stock exchange of one or more classes of a

Flotsam

company's securities in conjunction with the raising of additional or start-up capital. In the United States a flotation is referred to as an initial public offering or IPO.

Flotsam Cargo cast or lost overboard and recoverable by reason if its remaining afloat.

Flow chart A plan outlining the processing sequence of a particular assignment or job from start to finish. It represents a graphical representation of the logic, flow and control of a computer program (program flow chart) or a model of the movement and interaction of documents, data and storage in a computer system (system flow chart).

Flow control A term used to describe a specific production control system. It is necessary to ensure that the data receiver is not overloaded with incoming data and is able to tell the transmitter to wait while it catches up with the transmission. One common method used is to enable the computer to issue stop and start commands, typically X-off and X-on.

Flow lines The direction of flow in which the pallets have been positioned and stowed – shown on the ship's plan by an arrow to facilitate discharge.

Flow of materials The flow of materials and components which goes to and through the factory for the product process.

Flowing off Quantity of cargo just sufficient to cover the floor of the hold.

FLT Fork Lift Truck.

FLTx Fork lift truck.

Flush deck Upper deck without side to side erections (on a ship).

Fluctuating currency An exchanged rate of a specific currency which is subject to a considerable degree of variation over a particular period due to market forces and other circumstances. A variation in excess of 2 1/2% or more over a three month period can be regarded as a fluctuating currency.

Fluidised bed combustion Combustion of a fairly finely ground fuel such as coal (although oil and gas can also be used) in a bed of inert particles such as a mixture of sand, ash and limestone. By blowing air into the bed from below, the particles and fuel are held in suspension: the bed behaves and looks like a boiling liquid. Combustion is very thorough and the heat transfer rate is high.

Flush deck Upper deck that extends the whole length of vessel and which has no poop, bridge or forecastle erection extending from side to side of the vessel.

Flush deck socket A deck socket that does not project above the ship's structure.

Fly cruise The process of cruise passengers flying to and from the port to join/disembark from the cruise vessel thereby avoiding usually a long overland journey.

Flyer A single sheet of paper which features the promotion of a particular product or service. A sales promotion aid and marketing term.

Flying Bridge Uppermost deck – like a platform (on a ship).

Flying sales executives Sales executives who visit an overseas market for the first time, conclude a sale and do not return to the market/client. It can be associated with dumping.

Flyover In transport terms a bridge permitting the transport unit to travel above one or more streams of traffic thereby avoiding any conflicting movements at the one level. It may be a railway or roadway flyover.

Fly tour A tour which may be a coach tour whereby the passenger joins the coach by flying to a convenient nearby airport from which the coach tour of several days starts.

FM Foam monitors; Facilities Management; Frequency modulation or Shanghai Airlines. An IATA Airline member.

FMB Federal Maritime Board (USA).

FMC Federal Maritime Commission. The United States Federal Authority governing sea transport.

fmcg Fast moving consumer goods.

FMD Finance Management Division – part of HMCE.

FMEA Failure modes and effects analysis.

FMET Originally a group of ships but applied to groups of other vehicles and in the US to a group of insurance companies in one ownership.

FMF Food Manufacturers Federation.

FMM Federation of Malaysian Manufacturers.

FMNSG EUR Flight Management and Navigation Systems Group – IATA term.

FMS Flight Management System – IATA term.

Fms Fathoms (timber).

FN Spanish register of Shipping – the Spanish Ship Classification Society; or fee numeral – indicates the numeral which is used to determine the fees for Special Survey and other hull classification surveys raised by Lloyd's Register of Shipping.

fo For orders; or free out terms i.e. charterer paying for cost of discharging cargo.

FO Free overside – seller pays for lighterage if incurred at port of discharge or Fuel oil.

FOA Foreign Operations Administration. Based in Washington it administers the USA foreign air programme or free on aircraft – Cargo delivery term.

FOB Cargo delivery term – Free on board – exporter (seller) responsible for meeting all charges and risk of damage or loss of cargo until the goods pass the ships' rail onto the vessel at which point importer (buyer) responsible for all costs and risks thereafter. This includes sea freight, insurance and expenses incurred in discharging the cargo including customs duty etc. This term may be used only for sea or inland waterway transport. An Incoterm 2000. See separate entry.

FOB Aeroport (French) FOB Airport – cargo delivery term.

FOB Airport Cargo delivery term. It is based on the same main principle as the ordinary FOB term. The seller (exporter) fulfils his obligations by delivering the goods to the air carrier at the specified airport of departure. The risk of loss or damage to the goods is transferred from the seller (exporter) to the buyer (importer) when the goods have been so delivered. An Incoterm 2000.

FOBAS Fuel Oil Bunker Analysis and Advisory Service.

FOC Flags of convenience.

foc Free of charge.

Focus groups Group discussion which focuses/identifies on a particular topic/area/situation for discussion. It may be the prototype design of a new car model. A marketing term.

fod Free of damage.

FOIL Forum of insurers lawyers.

Folding lashing plate A lashing plate that folds down horizontally when not in use.

Follow the Fortunes An agreement in a re-insurance contract whereby the re-insurer agrees to follow the decisions of the original insurer. Particularly important where 'ex gratia' payments are made by the original insurer, which would not be recoverable from re-insurers in the absence of this agreement. A marine insurance term.

FONASB General Agency Agreement An agreement which has been formulated by the Federation of National Associations of Shipbrokers and Agents involving the terms under which a Ship's Agent is appointed to represent the shipowner's interest in a port. It may arise whilst the ship is under charter or for liner services. Two types of agreements exist approved by FONASBA and BIMCO: the general agency agreement for liner services and the standard liner agency agreement.

FONASBA Federation of National Associations of Shipbrokers and Agents.

Food security Concept which discourages opening the domestic market to foreign agricultural products on the principle that a country must be as self-sufficient as possible for its basic dietary needs. A WTO term.

Foot print The area of the tyre measured in square inches or centimetres which actually comes into contact with the surface on which it is operating under a given load.

foq Free on quay.

Foq – Foq Free on quay to free on quay. A term found particularly in freight rates quotation in sea transport whereby the rate quoted is quay to quay between specified ports. It includes port handling charges from quay to ship (departure port) and ship to quay (arrival port) plus sea freight rate.

FOR Cargo delivery term. Free on rail – exporter meets all charges until cargo placed on railway wagon from which point importer bears all costs. See FCA entry.

Fore and aft From the bow to the stern throughout the vessel's whole length.

Fore and aft stowage Stowage from the bow to the stern (lengthwise) as opposed to stowage athwartships.

Forecasting The prediction of relevant future factors affecting an entity and its environment as a basis for formulation or reassessment of objectives and strategies and as a means to facilitate the preparation of planning decisions.

Force majeure A clause limiting responsibilities of the charterers, shippers, and receivers of cargo. Under an international contract of sale, the clause should contain/list events which would not qualify for frustration, such as strikes or other industrial relations problems, but which would be the cause for contract postponement rather than termination. For example, the clause in a metal supply contract which allows the seller not to deliver or the buyer not to take delivery of the metal concentrate or scrap under the contract because of events beyond his control. There is no force majeure clause in an LME contract. Customers affected by a declaration of force majeure by a producer or refiner can always turn to the LME as a source of metal, and suppliers can deliver their metal to the LME if their customers declare force majeure.

Forebody The part of a ship which is forward of midships.

Forecastle Forward part of a vessel where stores, ropes and anchor chains are located.

Foreclosure The taking of possession by a mortgagee of a mortgaged asset, where repayment has not been made.

Forefoot The lower part in the stem of a ship that curves to meet the keel.

Foreign Bond A bond issued overseas. Issues sold in the market of a single country on behalf of a non-resident borrower and denominated in the currency of the country of issue. An international banking term.

Foreign Buying Houses Certain large American, Canadian and Japanese retail stores have buying agencies in the United Kingdom and in the EU through which they buy the bulk of their goods. These buying houses can provide the exporter with a ready and simple means of selling his goods abroad. Likewise, Foreign Buying Houses exist in many countries throughout the world.

Foreign currency A foreign currency exists when the currency involved is not the national currency. For example, in UK it is sterling and all other currencies would be regarded as foreign currencies. See Appendix 'C'. It is becoming more popular for suppliers (exporters) sellers to request payment for their goods in a specific currency for a variety of reasons including: some goods are traded internationally in one particular currency; a supplier/exporter/seller may traditionally prefer to price a contract in one particular currency; suppliers/exporters/sellers are aware of the fluctuating nature of rates of exchange for some trading currencies and may wish payment in a currency that they expect to remain stable or rise; a supplier may wish to receive payment in his own country's currency to avoid exchange risk; by using a particular currency, a supplier may be able to gain an advantage over competitors unwilling to do likewise; or the cost of borrowing may be cheaper in one foreign currency than in another. See Euro zone.

Foreign currency inflow The movement of currency into a country.

Foreign currency loan The process of an importer raising finance in a foreign currency by taking a loan in the same currency as that to be paid to an overseas supplier. See currency borrowing.

Foreign currency outflow The movement of currency out of a country.

Foreign debt The amount of debt outstanding by a country/state overseas. See XGS and international debt entries.

Foreign debt/XGDP (%) See international debt entry

Foreign direct investment See inwards investment entry.

Foreign exchange Transactions involving purchases and sales of bank deposits in different currencies – e.g. purchase of sterling against a sale of Japanese yen. Closely related to, but distinct from, Euro-currency transactions, which are loans and deposits rather than purchases or sales. Overall, a market for buying and selling currencies either spot (deal done today) or forward (deal done today for delivery at a future date).

Foreign exchange advisory service A service/facility offered by some banks to advise customers on foreign exchange rates, economic and political conditions and on the management of exchange risk through a specialised advisory service.

Foreign exchange aid The process of a government/entity/international agency granting foreign exchange aid to an overseas territory. See donor aid entry.

Foreign exchange broker A broker operating on the foreign exchange market on behalf of a customer – possibly a financial institution, a bank or a government – and charging a fee for the service.

Foreign exchange market The foreign exchange market involves different national currencies being bought and sold by dealers on behalf of traders, travellers, and investors. Participants in this market include banks, their customers, foreign exchange broking firms, and occasionally central banks. Most dealing is done by telephone, E-commerce and facsimile with worldwide connections.

Foreign exchange tables The range of quotations in various currencies that were made during the course of the previous day's business on a foreign exchange market.

Foreign flag A foreign going registered vessel.

Foreign going passenger ship Ship operative in a foreign trade with a passenger certificate in excess of twelve passengers.

Foreign going ship Ship (UK) operative outside the following limits; the coast of United Kingdom, the Channel Islands and the Isle of Man, and the Continent of Europe between Rivers Elbe and Brest inclusive.

Foreign investors Entrepreneur/companies who are resident outside the country and choose to invest in a particular country which involves such funds crossing international borders.

Foreign Jurisdiction Clause A clause agreed by underwriters and attached to the policy, that provides for the policy to be subject to legal jurisdiction of the country named in the clause. This will, normally, take precedence over the English jurisdiction clause in the MAR policy form; but it is customary for the latter to be deleted in such cases to avoid confusion. It does not take precedence over the English law and practice clause in the Institute clauses.

Foreign key An attribute, which is not a primary or secondary key, on which the records of an entity are indexed to enable a relationship to another entity.

Foreign Nationals Persons who are not part of the indigenous population and were born overseas with a foreign passport and birth certificate.

Foreign owned container A container registered and owned in an overseas/foreign country.

Foreign portfolio investment Investment in bonds, shares and currencies of other countries.

Foreign registered ship A vessel which is registered in a country/state outside the national state and which flies a foreign flag.

Foreign Trade Enterprises See Foreign Trade Organisations entry

Foreign Trade office A government department dealing with Foreign Trade business.

Foreign trade Organisations Usually a quasi government organisation under the direct control of the Ministry of Foreign Trade, its responsibility is to import the goods required by various sectors of a particular economy of a specified country and to export specific goods. Such a situation is found in centrally controlled economies. Also termed Foreign Trade Enterprises.

Foreign visitor A visitor to a country who is not permanent resident of the host country and is subject to the immigration controls.

Forepeak The watertight compartment at the extreme forward end of the ship.

Forepeak Tank Situated below Deck Store and Chain locker and used for sea water ballast.

Forestay A wire support to the foreside of a mast.

Forest products These include logs, kiln-dried timber, telegraph poles, paper, wood pulp and board materials such as plywood, fibreboard and veneers.

Forest products terminal A terminal usually at a seaport handling timber products.

FOREX Foreign Exchange.

Forex dealer A foreign exchange dealer.

Forfaiting A method of international trade finance involving the discounting of bank-guaranteed overseas trade bills or notes. It has no recourse to the exporter as he surrenders or relinquishes his rights in return for cash payment from the forfaiter. In such circumstances the exporter agrees to surrender his rights to claims for payment on goods or services which have been delivered to the buyer. Any type of trade debt can be forfaited (i.e. surrendered for cash) but the most common is trade payee; bills of exchange accepted by the buyer, or promissory notes from the buyer. A forfaiter can provide finance for any type and duration of trade transaction but usually credit is provided for the export of capital goods which require finance for periods of between three and seven years.

For further instructions Used in place of cargo delivery instructions where final destination uncertain at time of shipment such as FFI Birmingham.

Forint The currency of Hungary. See Appendix 'C'.

Forint invoice An invoice payable in forints currency relative to a Hungarian export sales contract.

Fork lift truck A three or four wheeled mechanically or electrically operated truck designed for lifting, carrying and stowing cargo which can be loaded either into pallets or carried by the truck's forks or attachments. An item of cargo handling equipment.

Fork pockets Recess sometimes provided in the sides of a container for the entry of the forks of fork lift trucks.

Form (compared with a document) Data carrier that is designed to carry visible records of data entries.

Formal Management Development The systematic development of managers at all levels to meet present and future company needs.

Format In 'paper' applications the physical appearance of printed matter, based on the type face, binding, margins, etc. In data processing and transmission application a predetermined arrangement of characters, fields, lines, etc.; in a message (also called formatting).

Form field A space on a data collection or data entry form, into which data is entered as required.

Form M Release document which permits mutual termination of the agreement between Master and seaman.

Form O Cotton charter (freight paid on steamers net register tonnage).

Forms design sheet An application of a layout chart intended as an aid for the placing of rules and other preprinted matter in the designing of forms, containing margin indicators and a network of lines indication the location of printed rules.

fort Full out rye terms (grain trade).

FOR/T Full out rye terms (grain trade).

Fortuitous By chance, accidental – see force majeure.

Fortuitous loss Unforeseen and unexpected loss that occurs as a result of chance.

FORTUNE Fortune Lines – a shipowner.

Forty foot equivalent unit Unit of measurement equivalent to one forty foot shipping container.

Forward At, near or towards the bow or front of a vessel or an aircraft; or the purchase of sale of metal for delivery at a specified future date. Hence, Forward Price, Forward Contract – See separate entries.

Forward contract A legally binding contract between a bank and the customer where the customer agrees to receive or deliver foreign currency from or to the bank. The bank undertakes to sell or buy from the customer a specified amount of currency at a specified rate on a fixed date (fixed forward contract) or at the customer's option, within a specified period (option forward contract). No cash is exchanged at the time a contract is taken out, but both parties commit themselves to a currency exchange at the agreed rate on the maturity date or the option take up date; or a contract to buy or sell a commodity at some time in the future – a BIFFEX definition.

Forward cover The arrangement of a forward foreign exchange contract to protect a buyer or seller of foreign currency at a future date from unexpected adverse fluctuations in the exchange rate between now and that date.

Forwarder A Freight Forwarder.

Forwarder's bill of lading A bill of lading issued by a forwarding agent.

Forwarders Certificate of Receipt A shipping document issued by the freight forwarder to the forwarders principals.

Forwarders certificate of shipment A document issued by a freight forwarder certifying that the goods have been shipped on a named vessel or service.

Forwarders delivery order A document issued by a freight forwarder authorising the entitled party to deliver the goods to a party other than the consignee shown on the consignment note.

Forwarders Local Import Control A customs facility operative in the UK permitting approved Forwarders to receive goods as unit loads – e.g. in secure containers or vehicles, or specialised aircraft containers, or on pallets, roll on roll off vehicles, or railway wagons to have their goods cleared for importation at their depots provided that certain specified conditions are satisfied. See period entry traders and SPIC.

Forwarders receipt A document issued by a freight forwarder which provides evidence of receipt of the goods.

Forward exchange This is an agreement to purchase and sell foreign currency for delivery at some future date. The forward exchange rate is the rate fixed under the agreement for future delivery.

Forward Exchange contract A currency exchange contract with a bank whereby a rate is fixed at once for a purchase or sale of one currency for another which is to be completed at some fixed agreed future date. It may be a forward 'fixed' where the specific date is agreed for performance to take place, or forward 'option' where the actual date within the period is chosen by the customer.

Forward Exchange Rates The difference between the 'spot' and 'forward' rates for a foreign currency against sterling is called a 'premium' if the forward currency is more expensive than 'spot', and a 'discount' if it is cheaper. In practice the premium or discount on an annual percentage basis approximates to the difference between the interest rates obtainable on sterling and the foreign currency. Hence if the spot Swiss Franc rate against sterling is 3.50 Francs and the six month forward rate is 3.3250 Francs, the premium on Swiss Francs is 10 per cent per cent per annum calculated as follows.

$$\text{Annual Premium} = \frac{12 \text{ months}}{\text{Number of months Forward}} = \frac{\text{Spot Rate} - \text{Forward Rate}}{\text{Spot Rate}} \times 100$$

$$\frac{12}{6} \times \frac{3.50 - 3.3250}{3.50} \times 100 = 10 \text{ per cent per annum}$$

Forward – forward currency deal A forward currency deal arises for example when dollars are purchased three months forward while at the same time there is an agreement to sell the same amount of dollars six months forward. An international trade banking term.

Forwarding Agent Person responsible for exporting/importing cargo arrangements on clients behalf. Often referred to as Freight Forwarder.

Forwarding Agent's bill of lading A bill of lading issued by a forwarding agent.

Forwarding Agent's Receipt A forwarding agent arranges the transport of the goods and will issue a receipt stating that he has taken charge of the goods for delivery to the importer. The forwarder is often the agent of the importer, and exporters should ensure that the details on the receipt are exactly as required by the credit before relinquishing control of the goods.

Forwarding Charge Charges paid or to be paid for preliminary surface or air transport to the airport of departure by a forwarder, but not by a carrier under an Air Waybill (air cargo).

Forwarding Clerk A clerical position engaged in the forwarding arrangements of merchandise in Freight Forwarder/Receiver/Agents/Manufacturing Company. Duties would involve transportation arrangements, documentation, customs, booking or cargo space/reservations and so on.

Forwarding department Shipowners' organisation responsible for cargo exports and customs clearance.

Forwarding instruction A document issued to a freight forwarder giving instructions to the forwarder shipowner for the forwarding of goods described therein.

Forward margin The forward margin is the difference between the current or spot price of the currency and the price at some specified future date. An international trade banking term.

Forward market charter A charter arranged to commence at a future date usually forward several months.

Forward planning The process within a company or particular set of circumstances to devise a plan for future action. It may be, for example, an investment plan over five years or a marketing plan over a similar period. All personnel within the company involved in the plan formulation together with their subordinates work within the plan to achieve its objectives/execution.

Forward rate A type of exchange rate whereby a forward contract determines an interest rate to be paid or received on obligation beginning at a start date sometime in the future. See also other two types: exchange rate spot and swap.

Fos Free on steamer or ship – cargo delivery term.

fot Free on truck (rail) – cargo delivery term. The exporter (seller) meets all charges and risk and damage until cargo placed on truck from which point importer (buyer) bears all risks and costs embracing rail/sea freight and insurance. Also referred to as free on wagon. See FCA entry.

Foul bill A bill of lading that has been claused to show that the goods were not received in a sound condition.

Foul Bill of Lading Bill of Lading which has been claused by the shipowner as the goods described thereon do not conform to what is offered for shipment i.e. package missing, damaged, stained, inadequately packed.

Foundation A raised deck socket that is fabricated rather than cast.

Four berth cabin Passenger sleeping accommodation involving a cabin on a ship with four berths and ideal for a family of four, husband, wife and two children of ten and twelve years of age.

Four Ps Product Price Promotion and Place. Collectively known as the marketing mix. See Marketing Mix entry.

Four shipping markets These embrace freight, new ships/build, second-hand tonnage and ships for demolition/recycling.

Four TEU (4 TEU) Truck A thirty metre articulated vehicle embracing a truck with two trailers (ten axles overall) which conveys four twenty foot ISO containers.

Fourth carrier The carrier who actually performs the fourth leg of the contracted transit. Usually associated with IATA Air Freight consignments.

Fourth Freedom The privilege to take on in another state revenue passengers, mail and freight destined for the state airline registration. One of the six freedoms of the air.

Fourth freedom flight Where cargo is carried by an airline from a foreign country to the country in which it is based.

Four way entry pallet A pallet, usually a flat one, which permits entry by a fork lift truck on all four sides.

Four way pallet A pallet whose bearers permit the entry of forks or fingers from all four directions.

FOW Free on wagon (rail) – cargo delivery term. The exporter (seller) meets all charges and risks and damage until cargo placed on wagon from which point importer (buyer) bears all risks and costs embracing rail/sea freight and insurance – also referred to as Free on truck – see FCA entry: first open water – date at which ice bound port becomes navigable; or free on wharf.

fp Free piston (machinery).

FP Floating (open) marine policy; forward perpendicular or fixed pitch (propeller).

FPA Free of Particular Average. A term used in relation to cover under the SG policy (being phased out).

FP-°C Flash point in degrees Centigrade.

FP-° Flash point in degrees Fahrenheit.

FPIC Food processing industries club.

FPSO Floating Production, Storage and Offloading (Vessel).

FPT Fore peak tank.

FPV/S Floating Production Vessel/System.

FPZs Free Port Zones.

F/R Freight release.

FR Flatrack (container); Ryanair Ltd. – an IATA member Airline; France – a customs abbreviation or Freight.

Fracturing devices, explosive Used for oil wells, without detonator. Articles consisting of a charge of detonating explosive without means of initiation. They are used to fracture the rock around a drill shaft to assist the flow of crude oil from the rock.

FRAs Forward rate agreements.

Frame height The height of the top of the road transport chassis frame from the ground when the vehicle is standing level.

Frame(s) The transverse vertical member of hull structure that stiffens steel plating in a vessel. It forms the frame or ribs of the ship's hull to which is affixed the steel plating.

Framework agreement Draft agreement on the rules and principles for trade in services. A WTO term.

Franc invoice An invoice payable in francs relative to an export sales contract of a specified country such as Swiss Franc.

Franchise In an English policy the term 'franchise' has been used to define a percentage expressed in the policy (e.g. the 'memorandum' in the SG policy form) which was applied to particular average losses. If the loss fell below this percentage of the amount insured, no claim attached to the policy. If the loss attained the franchise percentage, the claim was paid in full. Franchise system is not generally used in the London market today and is used only in policies which cover freight at risk of the carrier. An insurance term; or in marketing terms a retail outlet of which there are three types. Basically, a franchise is a contract between a manufacturer, wholesaler, or service company called the franchiser and independent entrepreneur who purchase the right to own and operate the franchise. The franchiser will receive an initial fee, royalty on sales, lease fees for equipment or profit share. It may be a manufacturer sponsored retailer franchise as found in the car industry dealer retailing automobiles subject to certain conditions imposed by the manufacturer; the manufacturer sponsored wholesale franchise system as exists in soft drinks industry; and thirdly the service sponsored company retailer as found in fast food and car hire business. Overall, a facility granted by the Principal to a retailer to sell on the Principal's behalf a particular product(s)/service(s) using the Principal's name. An example is McDonald's. In such a situation the retailer must conform precisely to the code of practice stipulated by the Principal.

Franchise deductible Deductible commonly found in marine insurance contracts in which the insurer has no liability if the loss is under a certain amount, but once this amount is exceeded, the entire loss is paid in full.

Franco (French) Sender undertakes to pay all carriage charges and in addition all supplementary charges. Extended form of CIF contract. Term found in CIM documentation.

Franco Bord (port d'embarguement convenu) (French) Free on board (named port of shipment) – a cargo delivery term.

Franco de douane (French) Sender undertakes to pay all monies collected by Customs authorities from railway, in addition to the supplementary and other charges which the railway raises for Customs Clearance. Cargo delivery term. It is found in CIM documentation.

Franco de port (French) Sender undertakes to pay only carriages charges. Cargo delivery term. It is found in CIM documentation.

Franco de tous frais (French) Sender undertakes to pay all charges of every kind i.e. carriage charges, supplementary charges, Customs duties and other charges. Cargo delivery term. It is found in CIM documentation.

Franco de tous frais a l'exception de (exact description of charges sender does not undertake to pay) (French) Sender undertakes to pay all carriage charges except those indicated. Cargo delivery term. It is found in CIM documentation.

Franco domicile (French) Extended form of CIF contract – free to address of overseas buyer.

Franco le long du navire (French) Free alongside ship – cargo delivery term.

Franco pour (French) Sender undertakes to pay fixed amount in carriage charges. It is found in CIM documentation. A cargo delivery term.

Franco Transporteur (French) Free carrier – cargo delivery term. See FCA entry.

Franco Wagon (French) Free on truck/rail – a cargo delivery term. See FCA entry.

Franco y comprise (French) Sender undertakes to pay additional charges. Cargo delivery term. It is found in CIM documentation.

Franc Poincare Unit of value in which the limitation of the carrier's liability is sometimes expressed.

fr & cc Free of riots and civil commotions.

Fraud prevention Measures taken to combat fraud. See maritime fraud entry.

FRC Fast rescue craft.

Free Airport See Free Trade Zone entry.

Free Alongside Ship (named port of shipment) – (FAS) The obligations of the seller are realised when the goods have been placed alongside the ship in the quay or on lighterage at a specified port of shipment including export clearance and related cost and documentation. At this stage and thereafter the buyer has to bear all the costs including freight and risk of loss of or damage to the goods. An Incoterm 2000.

The salient features of FAS for the seller include: to supply the goods in accordance with the contract of sale terms; to arrange delivery of the cargo by the date or within the agreed period alongside the specified vessel at the loading berth and port as named by the buyer; to undertake all the customs formalities necessary for the export of the goods and related costs and documentation including governmental authorisation necessary to export the goods; to bear all costs and risks of the goods until they have been effectively delivered alongside the specified vessel – this includes the funding of any formalities necessary to fulfil the order to deliver the goods alongside the ship; to provide and pay for the customary packing of the goods unless it is the custom of the trade to ship the cargo unpacked; to render to the buyer on request and at the buyer's expense the Certificate of Origin; to assist the buyer on request and at the buyer's expense to obtain any documents issued in the country of origin or shipment, including Bill of Lading and/or consular documents required for importation of the goods into the destination country and their passage through transit countries; to provide the buyer upon request with the necessary information for procuring insurance.

The buyer's responsibilities include: to give the seller prompt notice of the name of the vessel, loading berth and delivery dates; to bear all the expenses and risks of the goods from when they have been effectively delivered alongside the vessel as specified; to fund any additional cost and to accept the risk in the event of the vessel not arriving on time or the shipowner being unable to accept the cargo; in the event of the buyer failing to notify the seller of the name of the vessel, and port of shipment within the prescribed period, the buyer would bear all consequential costs and risks from the expiry date of the notification period, subject to the goods being duly appropriated by the seller to the contract; to meet all costs to obtain the Certificate of Origin and any documents necessary for the importation of the goods, including the Bill of Lading and consular documents. The buyer must pay all costs and charges incurred in obtaining the documents or equivalent electronic messages relative to the import of the goods. The cost of pre-shipment cargo inspection is borne by the buyer unless when mandated by the authorities of the country of export. This criteria applies to other Incoterms 2000 where applicable. This term should only be used for sea or inland waterway transport.

Free baggage allowance Baggage which may be carried without payment of a charge in addition to the fare.

Freeboard The measured distance vertically amidships between the main load line of a ship and the deck line (free board deck of ship – main deck in a single tween deck, and shelter deck in a shelter deck ship). The amount of free board assigned is such that the ship, when loaded down to her load line, possesses sufficient reserve buoyancy in the portion of the hull and the erections above the waterline to ensure a satisfactory margin of safety.

Free Carrier (named place) FCA A cargo delivery term designed to meet the requirements of modern transport irrespective of the mode of transport, particularly multi-modal transport. Free carrier means the (exporter) seller delivers the goods, cleared for export, to the carrier nominated by the (importer) buyer at the named placed. If delivery occurs at the (exporter's) seller's premises, the (exporter) seller is responsible for loading. If delivery occurs at any other place, the (exporter) seller is not responsible for unloading. The carrier means any person by whom or in whose name a contract of carriage by road, rail, sea or combination of transport modes has been made. The exporter (seller) must supply and deliver the goods into the charge of the carrier or another person nominated by the importer (buyer) or chosen by the exporter (seller) at the named place on the date or within the period agreed for delivery. The exporter (seller) must provide any customary packing and carry out where applicable all customs formalities necessary for the export of the goods. The importer (buyer) must bear all costs and risks of the goods from the time they have been delivered to the carrier in accordance with the contract. An Incoterm 2000.

Free circulation Goods imported from outside the Community are in free circulation within a Community's country when (i) all the import formalities have been met, and (ii) all import duties have been paid and have not been wholly or partly refunded. Goods originating in the Community are also in free circulation. A customs term.

Free delivery Cargo delivery term. Shipper is bound to deliver the goods at a specified place free of all costs.

Free Discharge A chartering term indicating the ship is to be discharged free of expense to the shipowner.

Free domicile A cargo delivery term involving an arrangement between the carrier and shipper to deliver the goods free of charge to the consignee and whereby the shipper agrees to pay all the charges incurred. Also referred to as 'Free house'.

Freedoms of the air The concept that international air commerce should operate under six basic freedoms in any country, allowing an aircraft or airline the right to :
1. Fly over another country's territory without landing;
2. Land for technical or other non-traffic purposes; 3. Disembark traffic originating in the carrier's home country; 4. Pick up traffic destined for the carrier's country; 5. Carry traffic from one foreign country to another foreign country; and 6. Carry traffic to the carrier's home country and beyond to another foreign country.

Free enterprise economy This exists where there is the minimum of government control and trade, and industry is largely run by private enterprise as found in the USA.

Free enterprise market Free enterprise markets exist in countries which operate a free market economy based on the forces of supply and demand. Besides applying these general principles, some countries have individual characteristics in respect of distribution, pricing and design, but demand is generally influenced by competitiveness and the ability to meet the buyers requirements. Customs, traditions and climatic conditions should always be to taken into account. Changes in economic policy can sometimes affect demand for specific goods.

Free flow cargo A free flowing cargo such as grain, coal, etc.

Free house A cargo delivery term involving an arrangement between the carrier and shipper to deliver the goods free of charge to the consignee and whereby the shipper agrees to pay all charges incurred. Also referred to as 'Free domicile'.

Free house unclear Delivered at a certain destination without payment of certain duties or incurred costs.

Free in & out Charter party term indicating shippers/ consignees or charterers responsible for bearing cost of discharge and loading cargo.

Free in clause Clause in the bill of lading which means that charges on shipment are for account of the shipper.

Freeing port An opening to allow water from the deck of a ship to flow over the side.

Free in liner out Transport condition denoting that the freight rate is inclusive of the sea carriage and the cost of discharging the latter as per the custom of the port. It excludes the cost of loading and if appropriate stowage and lashing.

Free in, out, and stowed Charterer, shipper or consignee pays for cost of loading and discharging cargo including stowing. A charter party term.

Free in, out, and trimmer Charterer, shipper or consignee pays for cost of loading and discharging cargo including trimming. A charter party term.

Free in, out, stowed and trimmed Charterer, shipper or consignee pays for the cost of loading and discharging cargo including stowing and trimming. A charter party term.

Freelift The distance the forks of a fork lift truck can rise without the overall collapsed height of the mast increasing.

Freely negotiable credit A credit under which drafts or documents may be negotiated by any bank. An international trade term.

Free market A market in which the market forces of supply and demand operate.

Free of address Charter party containing no Address Commission.

Free of despatch A chartering term indicating no despatch is payable even if laytime is served.

Free of term A chartering term indicating the vessel laydays period will commence to count from the time the written notice of readiness is given by the Master, and not wait until the ship docks at the loading or discharging berth.

Free on Board (named port of shipment) Under this cargo delivery term, the seller places the goods on board a vessel – past the ships rail – at the specified port and bears all the resultant costs and risks up to this point. The risks of loss/damage passes to the buyer (importer) when the goods pass the ships rail and all the resultant costs including the sea freight and insurance. Packing costs are borne by the seller (exporter). When goods are carried on liner terms the freight will ordinarily include loading and unloading costs all borne by the buyer.

The principal seller's obligations found in a FOB sales contract include to supply the goods in accordance with the contract of sale; to deliver the cargo on the named vessel at the specified port of shipment within the agreed period or on the agreed date and in so doing promptly inform the buyer; to provide at his expense any export licence or other governmental authorisation necessary for the export of goods; to bear all cost and risks of the goods until such time as the cargo has effectively passed over the ship's rail at the named port; to provide at his expense the customary packing of the goods unless it is the custom of the trade to ship the cargo unpacked; to pay the costs of any cargo scrutiny prior to the delivery of the cargo; to supply at seller's expense the requisite documentation as proof of delivery of the goods alongside the named vessel; to provide the buyer on request and at buyer's expense with the Certificate of Origin; to supply the buyer on request and at buyer's expense every assistance to obtain Bills of Lading/Sea Waybill and other documentation issued in the country of shipment or origin necessary for the importation process both in transit countries and the destination country. The seller must provide the buyer upon request with the necessary information for procuring insurance.

The buyer's responsibilities are extensive. These include: to arrange at his own expense and risk pre-shipment cargo inspection arrangements except when such inspection is mandated by the authorities of the country of export; to bear all costs and risks of the cargo from the time it has passed the ship's rail at port of shipment and to pay the price as specified in the sales contract; to bear all the costs and risks emerging from the failure of the shipowner to fulfil the contracted pre-shipment arrangements (such as the cargo being shut out) – this is subject to the seller making the cargo available at the loading berth in accordance with the sales contract; in the event of the buyer failing to give pre-shipment details to the seller within the prescribed period all additional costs and risks to be borne by the buyer; to pay all costs to the seller to obtain Bills of Lading, Certificate of Origin, consular documents and any other documentation required to process the cargo through importation both in transit countries and in the country of destination. This term may only be used for sea or inland waterway transport. When the ship's rail serves no practical purpose, such as roll on/roll off or container traffic the FCA term should be used. When goods are carried on FIO (Free in and out), loading and unloading operations are free to the carrier and duty of the shipper. The costs and responsibility of loading may be divided between the buyer and seller according to the custom of the port. Where the seller and buyer have agreed to communicate electronically, the documents may be transmitted by an equivalent electronic data interchange (EDI) message. The seller bears all loading costs by adding term FOB stowed. An Incoterm 2000.

Free on Board – Airport (FOB – Airport) Under this cargo delivery term the seller fulfils his obligations by delivering the goods to the specified airline/agent at the agreed airport or other place designated by the buyer, and to conclude on behalf of the buyer the air freight arrangements from and to the specified airports. Additionally, the exporter provides the requisite export documentation and at sellers expense provides the customary packing. The risk of loss or damage to the goods is transferred from the seller to the buyer when the goods have been so delivered and bears all the costs of the goods from the time they have been so delivered including the air freight

Free on Rail – (normal departure point) A railway cargo delivery term in which the exporter (seller) at his risk must supply, deliver and load the goods into the specified railway wagon for full load and for part loads at the requisite railway goods depot. This also involves supply of any certificate of origin and provision of customary packing. The importer (buyer) must bear all the costs and risks of the goods from the time they have been loaded in the railway wagon or placed in the custody of the Railway Company. See FCA entry.

Free on Truck – (normal departure point) A railway cargo delivery term in which the exporter (seller) at his risk must supply, deliver and load the goods into the specified railway wagon for full load and for part loads at the requisite railway goods depot. This also involves supply of any certificate of origin and provision of customary packing. The importer (buyer) must bear all the costs and risks of the goods from the time they have been loaded in the railway wagon or placed in the custody of the Railway Company. See FCA entry.

Free on Wagon – (normal departure point) A railway cargo delivery term in which the exporter (seller) at his risk must supply, deliver and load the goods into the specified railway wagon for full load and for part loads at the requisite railway goods depot. This also involves supply of any certificate of origin and provision of customary packing. The importer (buyer) must bear all the costs and risks of the goods from the time they have been loaded in the railway wagon or placed in the custody of the Railway Company. See FCA entry.

Freight/Carriage and Insurance Paid to (CIP)

Free on Wharf Confirmation that the cargo will be brought alongside the vessel free of expense to the ship owner.

Free overside A cargo delivery term whereby seller pays charges for discharging into lighter or on to pier or wharf.

Free phone A telecommunication facility whereby telephone users can by using a specific dialling code obtain direct communication with the specific client at no charge to the caller. The entity providing the free phone facility to it potential customers funds the cost of the call. A useful promotional resource as more clients tend to do their business from their residence using the telephone network. It may be airline bookings, insurance, special offers/promotions, etc.

Free port An enclave treated as being outside the customs territory of the host state and inside which goods can be manufactured, processed and stored without payment of customs duties and subsequently exported. Customs duties and internal taxes are payable only if the goods pass from the zone into the home market. Also sometimes termed a free trade zone.

Free sale A system of reserving facilities whereby the sales agent or the CRS equipment does not need to obtain availability on a booking-by-booking basis from the principal.

Free pratique Permission granted by local medical authorities noting that the vessel has a clean Bill of Health so that people may embark and disembark.

Free-rider A casual term used to infer that a country which does not make any trade concessions but which, nonetheless, profits from tariff cuts and concessions made by other countries negotiating under the most-favoured nation principle of WTO. A WTO term.

Free sale agreement A bilateral or multilateral agreement between airline carriers which permits immediate confirmation of space in accordance with terms of the agreement without the necessity of maintaining space availability information. An IATA definition.

Free Sea Carrier A cargo delivery term. It involves the container operator and RO/Ro traffic conveyed by road trailer and shipped on container vessels or vehicular ferries. It is based on the same principle as FOB except that the seller fulfils his obligations when he delivers the goods into the custody of the carrier (container operator/road transport trailer operator) at the named point. See FCA entry.

Free trade The free flow of goods and services across national frontiers is not inhibited by government interference in the ordinary competitive process of trade, either by tariffs, or in any other way. It is based on the economic theory that trade would be carried on best, resources would be most efficiently allocated and total prosperity greatest; if all producers were allowed to concentrate on making what they do best; they would then be able to sell their products in free and open competition. Moreover, as within national boundaries, it does not make economic sense for each individual to try to satisfy all his/her needs by his/her own efforts, so all participants in international trade stand to gain from the greater efficiency that comes from specialisation.

Free-trade area: (Article XXIV) Group of countries which have removed duties and other trade barriers to almost all the trade between them. But, unlike a customs union, each member maintains its own commercial policy, including its own external tariff towards non-members of the group. A WTO term.

Free Trade Zone A defined area where trade is based upon the unrestricted international exchange of goods with customs tariffs used only as a source of revenue and not as an impediment to trade development.

Free zone Free zones and warehouses are part of the customs territory of the Community in which goods which originate in the Community, or goods imported from a non-Community country which have been put into free circulation in the Community, or goods which have been manufactured in the Community wholly or partly from materials or ports imported from a non-Community country provided that the imported materials or parts are in free circulation. Non-Community goods may be released for free circulation; undergo certain forms of handling; maybe placed under certain customs procedures e.g. inward processing, temporary importation etc.; or maybe destroyed. Goods are in temporary storage from the time they are presented to customs until they are entered or otherwise dealt with.

Free zone goods Goods within the designated free zone area which have met the necessary requirements to get free zone status.

Free Zone Manager A commercial Manager responsible for the management, development and promotion of a free zone.

Freight For insurance purposes, the term 'freight' relates to the remuneration received or receivable by a carrier for the carriage of goods, and can include the profit he derives from the carriage of his own goods. Where freight is paid in advance by the shipper, on a non-returnable basis, the insurable interest is vested in the shipper and is normally embraced within the insured value of the goods. Where freight is paid on 'outturn' of the goods at the destination port, the insurable interest is vested in the carrier (shipowner or charterer). In shipping practice, the term freight may be used to define the goods carried in transit. Thus, personnel involved in arrangements for such transit may be termed 'freight forwarders'. It is commonplace to hear goods referred to as 'freight' in regard to rail carriage, air carriage, etc. An insurance definition; freight is the reward payable to the carrier for the carriage and arrival of goods in a mercantile or recognised condition, ready to be delivered to the merchant; or cargo other than mail.

Freight account Invoice rendered by a shipping line or shipowner to a shipper charterer or transport operator showing the full amount of freight payable and the method by which this amount is arrived at, depending on the terms of carriage. The account includes any applicable surcharges and commission.

Freight all kinds A uniform rate or tariff applicable irrespective of nature of commodity goods. 'Freight all kinds' rates are charged per unit or container of a given kind and not on the weight or volume of its contents.

Freight allocation agency An agency which specialises in allocating individual shipments/consignments to specified sailing based on a allocation criteria laid down by Government and usually favouring the national flag rather than foreign tonnage. It is practiced in developing countries especially in South America and Africa. It also may feature rate fixing.

Freight Broker The Carrier's Agent.

Freight/Carriage and Insurance Paid to (CIP) Cargo delivery term. Under such term the seller pays the freight for the carriage of the goods to the named destination. However, the risk of loss or damage to the goods, as well as of any cost increases, is transferred from the seller (exporter) to the buyer (importer) when the goods have been delivered into the custody of the first carrier and not at the ship's rail. It can be used for all modes of transport.

Freight Collect Freight and charges to be paid by the consignee.

Freight container A container loaded/conveying merchandise.

Freight contingency The insurable interest of a consignee who has paid freight on goods when delivered over the ship's side, but where the goods are still subject to peril until they arrive at the final destination.

Freight desk An 'online' transportation service provider dedicated to transportation intermediaries and the shippers served by the intermediaries. It is not an online auction but deals with managing logistics processes from the standpoint of freight forwarders, customs brokers, NVOCC and their customers.

Freighter A vessel conveying cargo.

Freight Forwarder An entity/company responsible for undertaking export/import cargo arrangements on clients/shippers behalf at a seaport, airport and so on. At a seaport it would include collection of freight; collection and issuing of Bills of Lading; notification of arrival and loading of goods; customs, import, and export documentation; payment of duties and levies as necessary; issuing of landing accounts and certificates of shipment; arranging sorting of cargo, cold storage, warehousing, transport to destination including near Continent; cargo or damage surveys; Lloyd's Agents surveys and so on. The merchant's agent. Also termed Forwarding Agent.

Freight Forwarders Certificate of Receipt The forwarding agent's own 'through document' for goods (receipt).

Freight Futures contract A binding agreement to buy or sell – usually a large quantity of freight – for a definite future date at a specified price, a standardised quantity of a product or rate of fixed quality and conditions of settlement.

Freight Futures market The freight futures market enables the client to buy or sell large quantities of freight at a definite future date at a specified price, a standardised quantity of a product, or rate of fixed quality and conditions of settlement. Overall, it is financial risk management tool and benefits include: the ability to obtain price insurance or protection; improvement of cash flow and the management of capital; inventory protection; flexibility in the timing of purchases and sales and transparent pricing system. See also Baltic International Freight Futures Exchange entry.

Freight manifest Inventory of cargo shipped on board a specified vessel, aircraft or other transport unit.

Freight market That segment of the market involved in the freight business. It may be a national or international consignment which can be further segregated into bulk, or general merchandise; liquid or dry cargo and so on.

Freight or Carriage and Insurance Paid to (CIP) A cargo delivery term whereby the seller (exporter) pays the freight and insurance for the carriage of the goods to the named destination and clear the goods for export. However, the risk of loss or damage to the goods, as well as any cost increases, is transferred from the exporter (seller) to the importer (buyer) when the goods have been delivered into the custody of the contracted carrier and not at the ship's rail. The exporter (seller) must deliver the goods into the custody of the first carrier and bear all the expense for the contract of carriage of the goods to the agreed point at the place of destination. Also bear all the packing costs and provide the importer (buyer) with the commercial invoice. The importer (buyer) will receive the goods at the agreed point at the place of destination and bear all the costs and charges incurred in respect of the goods in the course of their transit including unloading charges and customs duties. The buyer must pay the costs of any pre-shipment inspection except when such inspection is mandated by the country of export. When the seller and buyer have agreed to communicate electronically, the documents may be transmitted by an equivalent electronic data interchange message. The term may be used irrespective of transport mode including multimodal transport. An Incoterm 2000.

Freight policy Marine policy underwriting cargo freight.

Freight prepaid Freight and charges to be paid by the consignee.

Freight rates Charges for the use of a vessel (e.g. for carrying cargo) See Rates entry.

Freight revenue Income from the carriage/shipment of cargo.

Freight sales representative A Company/Entity representative engaged in the selling of freight services of a transport operator which may be shipping/rail/air, etc.

Freight station facilities Facilities provided usually at a seaport or dry port to process/handle maritime cargo. It would include storage, warehouses, stuffing/unstuffing containers, cargo handling equipment, security accommodation and facilities for specialist cargoes such as hazardous materials/substances. It would handle containers and road haulage units (trucks/swap bodies). Permanent on site customs facilities would not be available usually.

Freight Strategy The process of devising/formulating a policy relative to a particular company/entity freight rates. This will include the level of rate; the scale of any rebate and the criteria applicable thereto which may be based on volume, revenue, units conveyed per day/week/month; the general viability of the rate thereby having regard to costs and associated income; the level of competition; the general loadability of the transport unit; any agreement within a liner conference; level of exchange rates and so on.

Freight tariff The freight tariff published by a liner conference, agent, shipowner, carrier and so on.

Freight tax A tax raised on freight imposed by government. Some countries consider the ship's registration flag in connection with tax exemption agreements.

Freight ton In weight terms 2240lbs English long ton; 2000lbs. American long ton; 1000Kg metric tonnes, or in cubic measurement volume terms 40 cubic feet Imperial or one cubic metric tonne – See Appendix 'D'.

Freight tonne The weight or measurement of cargo in metric in which the freight is calculated. A weight tonne equals 1000 kilos and a measurement tonne equals one cubic metre (m$\geq$) – See Appendix 'D'.

Freight Transport Association A trade organisation whose aim is to safeguard the interest of and provide services for trade and industry as operators and users of all forms of freight transport in UK.

Freight Waiver Clause The underwriter insuring hull and machinery of a ship is entitled, on payment of a constructive total loss, to take over the ship, arrange to deliver the cargo and retain any freight earned thereby. The 'freight waiver clause' in the hull policy conditions waives the right to earned freight. Prior to publication of the 1983 hull clauses this clause was commonly termed the 'freight abandonment' clause.

French Chamber of Commerce in Great Britain The organisation based in London representing France whose basic role is to develop, promote and facilitate international trade between UK and France and vice versa.

French Shippers Council The organisation in France representing the interest of French shippers in the development of international trade.

Frequent flyer programme A plan offered by airlines to award bonuses, such as free or upgraded passage, to fare-paying customers who fly specified minimum mileage; concept extended to hotels and car rental firms through frequent stay and frequent rental plans.

Fret port-paye (French) Freight, carriage and insurance – cargo delivery term.

Fret port-paye jusqu'a (French) Freight, carriage paid to – cargo delivery term.

FRGHT Freight

Frictional unemployment People moving in and out of jobs willingly.

Fridays and holidays excepted A chartering term indicating Fridays and holidays not to count as laydays. Fridays apply to non Christian countries.

Fridays and holidays included A chartering term indicating Fridays and holidays count as laydays. Fridays apply to only Christian countries.

Friendly end user A client – the buyer/consumer which is receptive to dialogue and co-operation in relation to the vendor and the product/service bought. See end user.

FRMG Framing (of a ship).

FRNs Floating rate notes. A bond issue where the interest is at a floating rate.

frns Fathoms (timber).

frof Fire risk on freight.

From the ground up A statement of an original insurer's experienced of a class of business offered for re-insurance is said to be from the ground up (FGU) when it shows the number and distribution by amount of all claims however small even though re-insurance is large claims only.

Front end approach The process of using the technical and sales resources in a business environment situation.

Front end costs The commission, fees or other payments that are taken at the outset of a 'loan', as, for example, discounting. The front end charges for capital issues are very considerable and, in calculating the total cost, a borrower should be aware of the additional cost of being short of such disbursements at the outset when compared with the cost of interest payments that are payable after the loan period and not before. An international banking term.

Front end finance The loan finance provided for the buyer; the financing of this element, usually by means of a Euro-currency loan, is called 'front end finance'.

Frontier Any place where the goods are still to be notified formally to Customs by placing them under a nominated Customs procedure.

Fronting Accepting liability for a direct insurance with the intention of re-insuring the whole risk at the original net rate, usually less an overriding commission. A marine insurance term.

Fronting insurance company An insurer who accepts direct insurance in its own name and re-insures all or most of it with a re-insurer.

Front office The area which deals with the customer embracing enquiries and reception such as office or counter in the lobby of a hotel for sale of rooms, guest registration, key and mail service and keeping guest accounts.

Front office automation Computer system used to make bookings, particularly CRSs.

Frost Heave The pressure exerted by the earth when expanding as a result of ice formations. It is a situation which can arise as a result of the low temperature effects from a storage tank being transmitted to the ground beneath.

FROTA FROTA Line – a shipowner.

FRS Forward relay station.

frt Freight

Frt fwd Freight forward.

Frt ppd Freight prepaid.

Frt ton Freight tonne.

Fruit carrier Vessel built to carry fruit often with refrigerated accommodation.

Fruit seaport A seaport provided with dedicated facilities for the handling of fruit.

Fruit terminal A terminal handling fruit. It may be a berth at a seaport; railway terminal, road depot and so on.

Frustration In legal terms a totally unforeseen event which has the effect of making the contract either physically impossible or such a different contract that the parties must be taken to have impliedly agreed that they would not go through with the contract under those circumstances.

Frustration of Adventure A circumstance whereby a ship or goods cannot reach the contemplated destination but remain undamaged and are not lost to the owner. This peril is normally excluded from policies covering war risks and strike risks.

FSA Filipino Shipowners' Association or Financial Services Authority.

FSC Finish Shippers Council; French Shippers Council or Foreign Sales Corporation.

FSD Final Supplementary Declaration – a customs abbreviation.

FSF Flight Safety Foundation (USA Term).

FSI Flag State Implementation. An IMO term. The FSI is set up to improve the performance of governments and is a forum where flag and port states can meet and find solutions to issues relating to implementation of safer ships and cleaner seas – less maritime pollution.

FSM Free Surface Movement.

FSMA Financial Services and Markets Act (2000): provides a single legal framework for the FSA, and sets the statutory objectives and principles of good regulation, for its operations in UK.

FSPC Foreign Shipping Policy Committee.

FSQC Flag State Quality Control.

FSU Floating Storage Unit. A vessel, of similar hull-form to an FPSO, but without oil processing capacity; the processing being accomplished at nearby platforms; or Former Soviet Union.

FSUs Freight Status updates.

FSW Friction stir welding.

FS W/R Indicates the specified vessel is equipped with flexible steel wire rope.

F/T Freight tonne.

ft Fuel terms or feet.

FTA Freight Transport Association; or Free Trade Agreements as for example between EFTA countries and EU.

FTAA Free Trade Agreement of the Americas.

FTB Fire tube boiler.

FTBS Fire tube boiler survey.

FTI Federation of Thai Industries.

FTL Full truck load.

FT-SE 100 Index (popularly known as 'Footsie') Nickname for the Financial Times Stock Exchange index reflecting the share price movements of the 100 largest companies; introduced on 3rd January 1984 at a level of 1,000. It is calculated by the London Stock Exchange. The index is an arithmetically weighted real time index constructed of 100 of the largest UK blue chip quoted companies representing roughly two-thirds of the total capitalisation of the UK equity market. Weighting is on the basis of capitalisation with the highest capitalised companies given a correspondingly higher weighting. The index is calculated by computer every minute throughout the working day. Overall, the index makes it possible to hedge against or profit from increases and decreases in the overall level of UK equity prices. It is a real time indicator of UK stock market direction

FTZ Free Trade Zone.

FU Air Littoral – an IATA member Airline.

Fuelcon The code name of a standard Marine Fuels Purchasing contract whereby the seller delivers to the buyer the marine fuel as quantified to the nominated vessel at the port or place or delivery if specified. It was developed by BIMCO.

Fuel oil Heavy petroleum distillates or petroleum residues or blends of those used in furnaces for the production of heat or power.

Fuel Oil Bunker Analysis and Advisory Service The process of evaluating/analysing/determining the ideal quality fuel oil for a ship/industrial plant. A service provided by Ship Classification Societies.

Fuel Oil Tank Farm A farm consisting of a number of tanks accommodating fuel oil and usually adjacent to a Maritime terminal embracing a jetty berthing oil tankers. Alternatively, it may be situated inland and road and/or rail served. It may be crude or refined oil.

Fuel surcharge A technique whereby a surcharge is raised by the shipowner on the basic freight rate to reflect increased bunker costs incurred due to upward price of fuel oil which may be attributable to currency variations or simply an increase in the basic bunker rate. The bunkering surcharge may be later consolidated with the basic rate etc. It is often called a bunkering surcharge.

FUHFUNG Fuh Fung Navigation Agency Ltd., Taiwan – a shipowner.

FUJIWARA Fujiwara Line – a shipowner.

Full air brakes Braking system dependent solely on compressed air for its operation. The compressed air is derived from an engine driven compressor and is stored in tanks mounted on the vehicle chassis.

Full and down A vessel is full and down when her cubic capacity available for cargo is completely full and the vessel is down to her plimsoll line.

Full container load A container stuffed or stripped under responsibility and for account of the shipper or the consignee; or for operational purposes a container to which no cargo can be added during the time it is transported under full container load conditions, and the container is stuffed or stripped under the responsibility and for the account of the shipper or consignee.

Full cost pricing A pricing policy based on assigning all costs – fixed and variable – to the company's products/services to determine the selling price.

Full free fork lift truck A truck on which the mast height never exceeds the fork carriage height even when the mast is fully extended. The type used for its ability to work in low spaces between decks on board ship.

Full load A container, truck, railway wagon or other transport unit which is filled to capacity with cargo.

Full pension (FP) A hotel rate, especially in Europe, that includes three meals a day; also called American plan.

Full Reach & Burden Charter party term indicating space available for cargo including deck capacity.

Full signed line When an insurer accepts a line on a slip this is called his written line. It frequently occurs that the broker over-places the insurance, whereby the total of the lines he receives exceeds the amount of the insurance required. This is accepted market practice but it means that when the insurance is finally closed (or signed) each insurer's written line must be reduced in proportion so that the total of the lines agrees with exactly the amount of insurance required. The line so arrived at is called the 'full signed line'. An insurer may use the expression 'full signed line manner' as both terms mean the same because ultimately the full written line will be reduced to the full signed line when the re-insurance is closed. In cargo insurance, a further reason for reducing the written line to a lower amount when signing occurs, is when the value, which is subsequently shipped, is less than the value contemplated by the insurance.

Full trailer A truck trailer constructed in such a way that its own weight and that of the cargo rests upon its own wheels, instead of being supported by e.g. a tractor.

Full truck load An indication for a truck transporting cargo directly from supplier to receiver.

Fully appointed Travel agent approved by the principal airline, cruise and railroad conferences for selling their travel products and services.

Fully Cellular Container ship A vessel specially designed to carry containers with cellguides underdeck and necessary fittings and equipment on deck.

Fully cellular services A group of shipping services which is a completely dedicated containerised fleet with all vessels of cellular design. See also cellular container ship entry.

Fully refrigerated gas carrier A purpose built tanker capable of conveying gas in a refrigerated condition at a controlled temperature.

Functional organisation A company whose organisation structure is based on functions such as marketing, finance, personnel, production, research and development and so on.

Funding The acquisition of liabilities to match, cover or balance the particular asset or assets for which they are required.

Fund Managers Personnel engaged in Managing the investment portfolio of institutional investors. See Institutional investors entry.

Funds held by company under re-insurance treaties The deposit or retention by a company out of funds due to a re-insurer as a guarantee that a re-insurer will meet its loss and other obligations.

Fuse/fuze Although these two words have a common origin (French fusee, fusil) and are sometimes considered to be different spellings, it is useful to maintain the convention that 'fuse' refers to a cord-like igniting device whereas 'fuze' refers to a device used in ammunition which incorporates mechanical, electrical, chemical or hydrostatic components to initiate a train by deflagration or detonation.

Fuse, igniter An article (tubular, metal clad) consisting of a metal tube with a core of deflagrating explosive.

Fuse, instantaneous, non-detonating Article consisting of cotton yarns impregnated with meal powder (quickmatch). It burns with an external flame and is used in ignition trains for fireworks, etc.

Fuse, safety Article consisting of a core of fine grained black powder surrounded by a flexible woven fabric with one or more protective outer coverings. When ignited it burns at a predetermined rate without any explosive effect.

Futures An agreement to buy or sell a standardised asset at a fixed future date at a price agreed today. Futures involve a contract to buy or sell a specified amount of a commodity or financial instrument on an agreed date and at an agreed price. Many of the products structured and traded by investment banks will involve futures at some level or contracts for purchase and sale of commodities for delivery in the future, with the freight market. A BIFFEX term.

Futures contract A contract by which the seller undertakes to deliver and the buyer undertakes to take delivery of a stipulated quantity of a standard commodity at a future date; the date and related price being fixed at the time the contract is entered into. The procedure enables dealers to safeguard their position against the risk of price fluctuations. These may be commodities of various kinds or financial assets such as bills and bonds. There is a futures contract in the FTSE 100 index; or a standardised legally binding agreement to sell or buy a specified asset at a fixed time in the future. At the time of the trade, buyer and seller agree to the price that will be paid for the asset when it is delivered in the future. A LIFFE term.

Fuzes Articles designed to start a detonation or a deflagration in ammunition. They incorporate mechanical, electrical, chemical or hydrostatic components and generally protective features. The term includes: Fuzes, detonating; Fuzes, detonating, with protective features; and Fuzes, igniting.

FV Pulkovo Aviation Enterprise (State Unitary Aviation Enterprise PULKOVO) An IATA Airline member.
FVP Full value procurement.
FW Fresh water.
FWA Fresh Water allowance.
FWBs Full Air Waybills.
FWC Full loaded weight and capacity (container)
Fwd Forward.
FWD Fresh water draft.
Fwdr Forwarder.
FWL Full written line – an insurance term.
FWR FIATA Warehouse Receipt
FX Federal Express (Federal Express Corporation) – IATA member Airline or Foreign exchange.
Fx Foreign Exchange.
FYI For your information – cyberspace term

G

G Galliot – a small Dutch sailing vessel; grain capacity in cubic feet or Gross mass or weight – an IATA definition.

g Gramme – See Appendix 'D'; or general purpose freight container.

G3 The third generation of vessel type. It may be for example the Atlantic Container Line Ro-Ro/container ships.

G7 Group of seven leading industrial countries: Canada, France, Germany, Italy, Japan, United Kingdom, United States.

G8 A group of eight developed industrialised countries. USA, Canada, Japan, UK, Italy, Germany, France and Russia.

G9 A group of nine developed industrialised countries. This group included G8 members plus China.

G15 Group of 15 developing countries acting as the main political organ for the Non-Aligned Movement.

G 77 Group of developing countries set up in 1964 at the end of the first UNCTAD (originally 77, but now more than 130 countries).

GA Garuda Indonesia Airways – A national airline and member of IATA; general authorisation; general average or general arrangement plan of a ship.

GAA General Agency Agreement. Contract between NSA and American shipowners involving State owned vessels, particularly those from the reserve fleet.

GAAKINDO Indonesian Association of Vehicle Assemblers.

GAAP Generally accepted accounting principles.

Gabriel The computerised reservation and national distribution system of Luxembourg; partner of Amadeus.

GAC General Average certificate.

G/A con General Average contribution.

G/A dep General Average deposit.

GAFTA Grain and Free Trade Association.

GA in full An agreement in a cargo insurance, whereby underwriters do not reduce a claim for general average in event of under insurance.

Gale A wind between a strong breeze and a storm of force 8 and above.

Galileo Austria, Galileo Netherlands, Galileo Portugal The computerised reservation and national distribution systems of their respective countries; partners of Galileo.

Galley The kitchen of a vessel.

Gallium A silvery-white metal with a melting point of 29.7°C (85.5°F) which may be under-cooled to almost 0°C (32°F) without solidifying. It has the property of penetrating very rapidly grain boundaries of aluminium alloys and other metals and causing embrittlement. Term associated with dangerous classified cargo.

GALWIN Galwin maritime Ltd. – a shipowner.

GAMMA Federation of Metal and Engineeering Works (Indonesia).

Gal Global area network.

Gang A number of workmen acting together especially for loading and/or discharging operations of a vessel in combination with the necessary gear.

Gangway A piece of port equipment linking the ship with the shore or (quay) enabling passengers/crew personnel to embark or disembark.

GANTRYLINE Gantry Oriental Line H.K. Ltd. – shipowner.

Gantry Aerial crane operative on parallel rails.

Gantry crane See crane gantry entry.

Gap The portion of an airline itinerary involving transportation by means other than an IATA carrier.

Gap analysis The process of examining/evaluating a particular commodity range/circumstance in the market and identifying any gaps which are currently not available and for which a market need arises. For example, a particular airline may identify the need for a family fare structure or specific flights to increase carryings/aircraft loadability.

GAPBESI Indonesian Federation of Iron and Steel Manufacturers.

Gap in the market A situation whereby in a particular market environment a need exists for a particular product/service which is currently not available. See Gap Analysis.

GAPKI Indonesian Palm Oil Producers Federation.

GAPKINDO Indonesian Rubber Producers Federation.

Garbage dues Port charges levied against vessels for the removal of garbage by special garbage boats coming alongside the vessel.

Garboard Strake Strake of bottom shell plating adjacent to the keel plate of a ship.

GARP Global Atmospheric Research Programme – a WTO term.

Garuda Indonesian Airways Indonesian State Airline and member of IATA.

GARWARE Garware Shipping Corporation Ltd. – a shipowner.

Gas Codes The Gas Codes are the Codes of construction and equipment of ships carrying liquefied gases in bulk. These standards are published by IMO.

Gas cylinder A rigid cylindrical metal receptacle designed as a portable pressure vessel for the storage and transport of gases under pressure.

Gas-Dangerous Space or Zone A space or zone (defined by the Gas Codes) within a ship's cargo area which is designated as likely to contain flammable vapour and which is not equipped with approved arrangements to ensure that its atmosphere is maintained in a safe condition at all times.

Gas detection sensor Device whose resistance varies in the presence of gas.

Gas/diesel oil Petroleum distillate having a distillation range intermediate between kerosene and light lubricating oil.

Gas drips Hydrocarbon. It is the liquid that condenses on compression of Pintsch Gas or the condensate from gas mains. It consists principally of a mixture of benzene and unsaturated hydrocarbon. It is very combustible and has a low flash point.

Gas-free Certificate A gas-free certificate is most often issued by an independent chemist to show that a tank has been tested, using approved testing instruments, and is certified to contain 21 per cent oxygen by volume and sufficiently free from toxic, chemical and hydrocarbon gases for a specified purpose such as tank entry and hot work. In particular circumstances, such a certificate may be issued when a tank has been suitably inerted and is considered safe for surrounding hot work.

Gas-free Condition Gas-free condition describes the full gas-freeing process carried out in order to achieve a safe atmosphere. It therefore includes two distinct operations: Inerting and Aeration.

Gas-Freeing The removal of toxic, and/or flammable gas from a tank or enclosed space with inert gas followed by the introduction of fresh air.

Gas Generator Assemblies (for aircraft escape slides) It consists of a steel cylinder containing a charge of a non-flammable, non-toxic gas (usually Chlorodifluoromethane) under pressure and a slow burning solid propellant cartridge (safety type) in a specially designed breech block. The assembly is installed in certain types of aircraft to provide a means of generating a supply of high pressure, low temperature gas to the aspirators that inflate emergency escape slides. Complete assemblies need to be transported for positioning as spare parts and thus comprise a non-flammable compressed gas and an explosive power device.

Gas-Safe Space A space on a ship not designated as a gas-dangerous space.

Gassing-up Gassing-up means replacing an inert atmosphere in a tank with the vapour from the next cargo to a suitable level to allow cooling down and loading.

Gastankwaybill Code name of BIMCO approved non-negotiable gas tank waybill for use in the LPG trade.

Gastime Code name of BIMCO approved time charter party for vessels carrying Liquefied gas.

Gas turbine A form of ship propulsion. A turbine in which the working medium is very hot gas. Air is heated by the combustion of fuel in the turbine itself: this fuel may be oil or a flammable gas. Fluidised bed combustion (qv) enables solid fuels to be burnt in gas turbines.

Gasvoy Gas Voyage Charter Party for the carriage of liquid gas (except LNG).

Gate keeper See gateman entry.

Gateman A senior executive/manager sometimes termed a project manager whose sole task is to approve projects/strategies/business ventures within a company and in so doing ensures each submission acceptance/approval follows a strict criteria. It may be a return on capital, level of profitability and the method of calculation, participation/sales in specific markets and so on. Overall, all such business activities are channelled through the gateman to ensure consistency of strategy/policy.

Gate number The embarkation or disembarkation access point to/from aircraft at a passenger airport terminal.

Gateway The facility on certain view data services which allows users to connect through to computers owned by 'third parties' thus giving access to a wider range of information and to processing and booking opportunities, arrival or departure city, airport or area for a flight or tour or a point at which cargo is interchanged between carriers or modes of transport.

Gateway Automation A computer facility which provides integration of data processing and communication within the computer network. In the shipping/transport field the network could embrace forwarding, shipping, shipping lines and agents, stevedores, hauliers, brokers, terminals, customs clearance, warehousing, container control and depot, maintenance and repair.

Gateway Container Corp A major container leasing company.

Gateway country A country access which permits overland geographical and/or political entry to neighbouring border countries and beyond. For example, Tanzania permits overland access to the neighbouring border countries and beyond into central Africa.

Gather Half the centering effect produced by any, device to assist containers to be handled. Overall, it is half the total off-line measurement of two flippers or two flip-flops or two cathedral arches.

Gator tail tensioner A tensioner primarily used with chains.

GATS The WTO's General Agreement on Trade in Services.

GATT General Agreement on Tariffs and Trade, which has been superseded as an international organisation by the WTO. An updated General Agreement is now one of the WTO's agreements.

GATT 1947 The old (pre-1994) version of the GATT.

GATT 1994 The new version of the General Agreement, incorporated into the WTO, which governs trade in goods.

GATT Valuation Code This is usual for the purpose of placing a value on goods at the time of importation to assess the import tariff duty payable. It operates on the basis of the 'transaction value' which is based on the price actually paid or payable for the goods. The price is only adjusted by items added or subtracted for the goods. The price is only adjusted by items added or subtracted to give an equivalent price at the first point to entry into the EC.

GAYA Gays Ocean Shipping Pte. Ltd. – a shipowner.

GBL Government Bill of Lading.

GBN GBN Line – a shipowner.

GBTA Guild of Business Travel Agents.

GC Grain cubic (thousands of cubic metres); gyro compass – ship's navigational aid; general cargo; Gas carrier; Global Company or Great Circle.

GCA Gold clause agreement.

GCC Gulf Co-operation Council.

GCIC German Chamber of Industry and Commerce.

GCMs General Circulation Models.

GCN Gencon (charter party code name).

GCR General Commodity Rate or general cargo rate.

G cr Gantry cranes.

GD Transportes Aereos Ejecutivos SA de CV (TAESA) – an IATA member Airline; or Good.

GDH Goods on Hand.

GDP Gross Domestic Product.

GDP Ratio/Trade See Trade/GDP ratio entry.

GDS Global Distribution Systems.

GE TransAsia Airways Corporation – an IATA member Airline.

Gearing Gearing is a term used to describe the relationship between shareholders' capital plus reserves and either prior charge capital or borrowings, or both as follows:-
(i) Prior charge capital – total capital in issue plus reserves.
(ii) Total borrowings – total capital in issue plus reserves.
(iii) Prior charge capital plus borrowings – total capital in issue plus reserves.
(iv) Prior charge capital x 100 – capital employed.
 Overall, it is the relationship between debenture and loan capital preference capital and equity capital. High gearing means that prior charge capital is large in relation to the equity (ordinary) capital and low geared if the reverse situation applies. Gearing is referred to as 'leverage' in the US.

Gearless vessel A ship which is not equipped with her own crane(s) or derrick(s).

GEF Global Environment Facility.

Gemini Canadian computerised reservations system owned by Air Canada, Canadian Airlines International and Gemini International.

Gen Generator.

GEN CAR General Cargo.

Gencon Code name of Baltic and International Maritime Council Uniform General Charter (as revised 1922, 1976 and 1994).

General Agent An Agent which has authority to do a whole series of acts for his Principal. The Principal is

General Agreement on Tariffs and Trade

bound by any act the agent does which is within the agent's ostensible authority, which is the authority he appears to have to a reasonable third party. A legal definition.

General Agreement on Tariffs and Trade It was established in 1948 and today over 80% of world trade is conducted under GATT/WTO. It has 144 member countries and over 30 associate members. Basically, it is a body outlining the rules governing the conduct of world trade. It has a permanent secretariat situated in Geneva which provides a forum for trade negotiation, administration of trade agreements and dispute settlement. It has now been succeeded by the World Trade Organisation.

General Account Office The US agency responsible for, among other matters, travel regulations for federal employees.

General Average A deliberate sacrifice or expenditure incurred for the common safety of the adventure. Those interests that benefits from the sacrifice or expenditure contribute rateably towards the loss.

General Average Act Any extraordinary sacrifice or expenditure voluntarily and reasonably made or incurred in time of peril for the purpose of preserving the property imperilled in a common maritime adventure.

General average bond The general average bond is given in order to secure the release of goods. It is a contract entered into with the consignees whereby in consideration of the ship master agreeing to discharge the cargo without exercising his lien, the consignees agree to pay any general average contribution chargeable on the cargo.

General Average Contribution The proportion payable by one of the parties involved in a general average act to make good the loss suffered in that act.

General Average Deposit
A deposit paid by a consignee in return for delivery of the goods where such goods are subject to general average contribution. This may be replaced by an underwriter's guarantee.

General Average Disbursements Expenses paid by the shipowner as part of a general average act. Such expenses are recovered by the shipowner from the general average fund. Unlike GA sacrifice, there is no direct liability on hull underwriters to reimburse the assured for GA disbursements. The disbursements will be included in the final GA adjustment, and incorporated in the GA contribution, underwriters paying this insofar as it is recoverable under the policy.

General Average Fund The total arrived at by adding together general average expenditure and the value of property sacrificed in a general act, plus costs of its adjustment.

General Average Guarantee An undertaking by a financial house or an underwriter to pay the contribution due towards a general average fund.

General Average Statement A statement identifying in detail the contribution of cash interest in the general average in proportion to its value.

General Cargo Variety of consumer type cargoes excluding valuable cargoes usually shipped on liner cargo or air freight services including groupage/consolidated consignments or any consignment other than one containing valuable cargo and charged at general cargo rates. It is cargo, not homogeneous in bulk, which consists of individual units or packages (parcels).

General cargo berth A berth at a seaport handling general cargo.

General Cargo Rate The rate for the carriage of cargo other than a class rate or specific commodity rate as found for example in air freight tariffs.

General Cargo Ship Single or multi-deck general dry cargo vessel with facilities for loading/discharging cargo.

General Circulation Models A computational model or representation of the Earth's climate used to forecast changes in climate or weather.

General Commodity rate The rate established for general merchandise cargo.

General Exclusions Clause A clause in the Institute Cargo Clauses 1982, which specifies risks that are excluded, irrespective of the risks covered elsewhere in the wording.

General Obligations Obligations which should be applied to all service sectors at the entry into force of the agreement (e.g. transparency, most-favoured-nation treatment). A WTO term.

General sales agent (GSA) Agency or other travel entity appointed by an airline or other supplier as its sales agent in a specific country or territory.

General system of preferences International agreement, negotiated under UNCTAD, whereby developed countries accord temporary and non-reciprocal duty preferences to imports from developing countries. Each importing country sets its own system, including product coverage, and the volume of imports affected. Such consignment/commodities must be accompanied by the appropriate General System Preference Certificate and accordingly are granted duty free entry provided the quota for the particular country of origin is not exceeded.

Generalised system of tariffs and preferences A system emerging from UNCTAD whereby developed countries grant preferential tariff treatment to certain imported goods originating in developing countries. See general system of preferences entry.

General System Preference Certificate The document which accompanies cargo from developing countries granted duty free entry. See general system of preferences entry.

General merchandise Variety of consumer type cargoes usually shipped on liner cargo services.

General purpose berth A general purpose berth at a seaport which deals with a wide variety of cargoes/shipments. These may include general cargoes, palletised consignments, loose cargoes, awkward shaped and heavy cargoes, containers, unitised consignments, conventional cargoes and so on.

General trader Tramp vessel.

General Trading conditions The trading conditions/legislation operative in a particular market for a specific product(s)/service(s).

General transire Customs warranty of inward cargo clearance.

Generic A collective term which describes a type or classification of a product or service. A marketing definition.

Generic label products A product which bears labels of a product/service which have come to be adopted as the general description for a product such as Hoover.

Generic products Merchandise sold at very competitive prices which are unbranded.

Generic term Brand names that have come to be adopted as the general description for a product such as Hoover.

Geneva round The inaugural round of GATT multilateral trade negotiations took place primarily in Geneva in 1947. Later in 1955/56 the fourth round negotiations were conducted in Geneva which was followed by the fifth 1961/62, the sixth 1963/67 and the seventh in 1973/79. The fourth called the Dillon round followed the establishment of the EC, whilst the fifth round was termed the Kennedy round involving anti dumping code and non tariff barriers. The sixth round was termed the Tokyo round focused on disciplinary codes especially in the area of non tariff barriers. Negotiations for all the rounds were held on a global basis. See WTO entry.

Genorecon Code name of Charter party used in world ore trades.

Gen set Motor generator set as power source for thermal containers.

Gentime A BIMCO approved general time charter party introduced in 1999.

Genvoy The code name of a BIMCO approved charter party clause entitled 'Baltic Conference Addendum 1946 (revised 1978)'. It requires all Bills of Lading under this Charter Party to contain general paramount general average, New Jason, and both to blame collision clauses.

Genwait The code name of a BIMCO approved charter party clause relative to a 'ready berth' and entitled 'Baltic Conference General Waiting for Berth clause 1968'.

Genwaybill Code name of BIMCO approved non-negotiable general sea waybill for use in the Short Sea dry cargo trade.

Geocentric It is the strategy/process of a company or entrepreneur providing a product or service to meet the needs of the overseas territory/market/culture on the basis that there are similarities and differences in world markets, and these can be reconciled and integrated. Overall, it follows a strategy of the value added benefit that the product/service offers to the end user and how this can be achieved through limited adaptation and minimum cost to the Exporter. A marketing term which is the synthesis of the ethnocentric and polycentric strategies.

Geographical indications Place names (or words associated with a place) used to identify products (for example, 'Champagne', 'Tequila' or Roquefort') which have a particular quality, reputation or other characteristic because they come from that place. A WTO definition.

German Chamber of Industry and Commerce The organisation based in London representing Germany whose basic role is to develop, promote and facilitate international trade between UK and Germany and vice versa.

Germancon North Code name of BIMCO approved charter party for German coal shipments.

Germanischer Lloyd A German Ship Classification Society.

German Shippers Council The organisation in Germany representing the interest of German shippers in the development of international trade.

GES Gold Exchange of Singapore.

GESAMP Group of Experts on the Scientific Aspects of Marine Pollution.

GE SeaCo The name of the merged container fleets of Genstar and Sea Container.

GF Government form (of charter party) or Gulf Air Company GSC – IATA member airline.

gfa Good fair average.

GH Ghana Airways (Ghana Airways Limited) – IATA member Airline.

GHA Greenwich Hour Angle.

Ghana Airways Corporation The national airline of Ghana and member of IATA.

Ghana Airways The Ghanaian national airline and member of IATA.

GHL Good Harvest Line – shipowner.
GHz Giga hertz.
GI Air Guinea. A national airline and member of IATA.
GIB Gibraltar.
Gibraltar Pound The currency of Gibraltar. See Appendix 'C'.
GIC General Industries Corporation (Burma).
GIDI Guidelines for data exchange and Interbridge. It is published by United Nations Economic Commission for Europe.
GIE Groupement d'interest economique (French). A form of business enterprise created by legislation in France to provide a legal framework for international company co-operation.
Giffen goods Goods which do not obey the 'law of demand'. The quantity demanded of Giffen goods rises as their price rises.
Gifts A common practice in international business whereby the visitor/vendor will tender a gift to his/her client/buyer/host as an expression of good will. Usually the gift represents the culture of the donor country. In some countries the process of offering gifts is illegal.
Gift voucher A marketing initiative designed to generate more sales. It takes two forms – one of which is in the coupon which enables the consumer to have a discount on the next purchase, or the voucher which is purchased over the counter and later exchanged to conclude a purchase at an accredited retail outlet. The latter may be a gift voucher for books, or one issued by leading retailers such as M & S, Boots, John Lewis and so on.
Gilts or gilt edged UK Government and similar stocks. Abbreviation for gilt-edged stocks, these bonds are issued by the Government to raise money. They are so called to reflect the low risk of default. Gilts are sold by the Bank of England on behalf of the Government by a variety of means: (a) Sale by tender – bids are invited at prices above a specified minimum. Successful bidders pay a common price and any stock not allotted is taken up by the Bank or sold 'on tap' (tap stock) – i.e. sold into the secondary market as and when demand appears at acceptable prices: (b) Sale by auction – bids are invited but no minimum price is set. Successful bidders pay the prices they bid.
GINSI Indonesian Importers Association.
Girders Continuous members running fore and aft under deck as a support.
GIT Goods in transit.
Giving of value The definition of negotiation. An international banking term. It applies to transactions of any tenor (sight or usance). The issuing bank must reimburse the bank which negotiates (gives value for) drafts and/or documents drawn under its documentary credit.
GL Germanischer Lloyd. German Ship Classification Society.
GLA General Lighthouse Authority.
GLAs General Lighthouse Authorities.
Glasnost The break up of the Soviet Union in 1991 involving the subsequent formulation CIS Commonwealth of Independent States.
Glass reinforced cladding Material used in ISO container construction.
Glass wool Insulation material derived from molten glass.
Global advertising The process of advertising a product/service worldwide. This involves crossing cultures. See cross cultural analysis and advertising entries.
Global branding A product/service which has a brand image and spans global markets. See also brand image entry.

Global carrier A carrier which provides a worldwide service such as British Airways and Evergreen round the world service.

Global coherence Adopting the structures of international economic consultation and co-operation to reflect the new global realities that business is already living with today.

Global company It is the process of a global company organisation targeting the selection and development of market opportunities and building up the infrastructure essential for that objective to be realised: such a process being conducted on a global basis.

Global Distriport A distriport operating on a global scale. This involves the provision of a purpose built area for the storage/processing and distribution of cargo imported through the port on a worldwide basis. See distripark and distriport entries. It provides value added services.

Global economy The worldwide economy.

Global expansion A worldwide economic expansion.

Global financial information services The process of a vendor providing a global financial information service to potential clients who require such data to support international investment decisions.

Global homogeneity A feature of globalisation where products/services have broadly the same specification subject to the value added input through culture and design. See globalisation entry.

Globalisation Globalisation of markets and trade results in the provision of a product or service which can be sold virtually in any market of the world providing the economic infrastructure and culture can support it. The key to it is the design and specification of the product or service and the added value it provides to the user or consumer.

Globalisation of production Process of a company/entity or other specified situation manufacturing/producing on a worldwide basis.

Globalisation World Bank The integration of trade, finance, people and ideas in one market place.

Global Maritime Distress and Safety System A system of radio equipment and methods of communication for ships, as regulated by SOLAS.

Global market A market which is worldwide.

Global marketing This is the process of examining markets globally – on a worldwide basis – identifying needs in specific markets and endeavouring to respond to those needs with an acceptable product/service on a profitable basis.

Global Marketing Management This is the process of focusing company resources and objectives on a global market opportunity basis.

Global marketing plan A marketing plan which spans global markets. See also marketing plan entry.

Global Navcom An annual conference conducted by IATA which attracts over 500 representatives from governments, airlines, manufacturers and service providers to discuss/evaluate the implementation of future air navigation systems involving some eleven international organisations. The first conference was held in 1993 at Seattle.

Global network A worldwide network. It may be an airline, shipping company and so on.

Global quota Explicit limits set by one country on the value or quantity of goods which may be imported through its borders during a given period on a global basis.

Global player A company which conducts its international trade on a global or worldwide basis.

Global Positioning system A satellite navigation method of fixing position either on land, at sea or in the air. See also international Maritime Satellite Organisation (INMARSAT) entry.

Global product A product which has a global market.

Global recession A worldwide economic recession.

Global sourcing The process of utilising purchasing potential on a worldwide scale. Hence, instead of the requirements of one country or plant being considered in sourcing terms, all plants in all countries within the firm have their requirements co-ordinated across the globe. Global sourcing also extends outside the company and other suppliers.

Global strategy A strategy which extends to all markets worldwide as distinct from a national strategy which would apply to a particular country and usually the home market. See also nationalism.

Global supply chain A supply chain which has a global network and is trans-national involving several modes of transport linking the supplier with the buyer. See supply chain entry.

Global Transportation System A transport system which has a worldwide network. It may be an airline or shipping containerised network

Global village Facilities available in major seaports such as Rotterdam and Singapore which provide resources to process/assemble/distribute merchandise. It involves the provision of purpose built warehouses and often designated as a free trade zone.

GLONASS Soviet Global Orbiting Navigation Satellite System – IATA term.

GM Geometric mean – a statistical term; metacentric height; or Trek Airways (Pty) Ltd. d.b.a. Flitestar – an IATA member Airline.

GMB Good merchantable brand (metal).

GMC Gerner-Mathisen Chartering AS Norway – shipowner.

GMDSS Global Maritime Distress and Safety System. See INMARSAT entry.

GML Goldrich Maritime Line – a shipowner.

GMP Garbage Management Plan.

GMRA Global Master Repurchase Agreement.

GMT Greenwich Mean Time.

GN Air Gabon – A national airline and member of IATA or guidance note.

GNP Gross national product.

GNS Group of negotiations on services in the Uruguay Round. A WTO term.

GNSS Global Navigation Satellite Systems – IATA term.

GNX Global Network Xchange – a computer logistics term.

GOB Good ordinary brand. A grade of zinc traded on the LME until December 1985; now replaced by High Grade(HG).

Go Cargo Go cargo is a website – based auction for transport services. It is independent, without any affiliation with shippers or Service providers. Through its internet website, it provides access to shippers of all sizes to receive multiple competing bids from service providers in an auction format. By viewing and evaluating alternative options that match their needs, shippers are able to select services.

GOH Goods on Hand.

Going concern concept The assumption that the enterprise will continue in operational existence for the foreseeable future.

Gold bullion standard System where standard money consists of convertible paper money.

Gold clause agreement An agreement relating to the interpretation of Carriage of Goods by Sea Act 1924 and the Hague Rules between certain insurers, cargo interest and British shipowners. This was abandoned in May 1988 and no longer in operation.

Gold Container A major container leasing company.

GOLDEN Golden Line – a shipowner.

Golden hello The process of a company/entity offering a new employee recruited from another company a separate payment as an inducement to join the new company.

Gold standard An international monetary system existing whereby all currencies are valued in terms of gold, balance of payments deficits or surpluses paid in gold, and all money supplies of all participating States is paid in gold. Hence, a particular country through its central bank is obliged to give gold in exchange, for any of its currency presented to it. This form of system existed widely up to about 1919. An example is found in country 'A' with a trade surplus receiving payment from country 'B' in gold. This would produce two results: the supply of money would fall in country 'A' thereby reducing demand and consumption of country 'B' exports. Secondly country 'B' money supply would increase, and thirdly it would result in increased demand and high prices resulting in an increased demand for imports. The trade surplus in country 'B' account would be reduced by a fall in country 'B' exports to country 'A' and a rise in country 'A' import to country 'B'.

GOLDSTAR Gold Star Line – a shipowner.

Gondola flat A container with sides and/or ends composed of bars, grilles, mesh or entirely open, with or without a roof.

Goods called forward The process of merchandise/consignment(s) being called forward from the shipper/exporter to be despatched on a specified sailing/flight on a particular date.

Good faith A basic principle of insurance. The assured and his broker must disclose and truly represent every material circumstance to the underwriter before acceptance of risk. A breach of 'good faith' entitles the underwriter to avoid the contract. See also 'Utmost Good Faith'.

Good housekeeping The process within a company/entity of optimising the use of resources thereby achieving the lowest level of expenditure compatible with the company needs. It usually applies to administration at all levels.

Goods and Services Tax An indirect form of taxation operative in New Zealand and similar in application to VAT.

Goods control certificate Document issued by a competent body evidencing the quality of goods described therein, in accordance with national or international standards or conforming to legislation in the importing country or as specified in the contract.

Goods flow The direction and path of the movement of goods and sequence of placement of those goods in a supply chain.

Goods receipt A document issued by a port, warehouse, shed or terminal operator acknowledging receipt of goods specified herein on conditions stated or referred to in the document.

Goods received note A document prepared by a recipient to record receipt of goods either (i) because an advice note has not been received, or (ii) to create a common format for use within the recipient's system.

Good Till Cancelled Order An order to buy or sell which is valid every market day during rings and kerbs (or during the entire day if specified), until cancelled by execution or client's instructions. A BIFFEX term.

Goodwill The difference between the net asset value of the company as defined by the accounts, and the value assigned to the total organisation as a result of an exchange. When a business is purchased, any amount paid in excess of its net worth represents the value placed on the goodwill. Overall, it is the intangible benefit arising from the commercial connections and reputation of a business.

Goose-neck Detachable device that connects the elevating fifth wheel of a tractor unit with a Ro/Ro trailer.

Goose-neck tunnel Recess at one end (commonly the front end) of the container designed to accommodate the raised portion of a goose-necked chassis.

Gopher A menu-based information retrieval system – a cyberspace term.

GOS Gulf of Suez.

GOSAERONAVIGACIA State Commission for Airspace Usage and ATC (CIS term).

Gourde The currency of Haiti. See Appendix 'C'.

GO (US) General Order. Issued by US Customs as notice of intention to seize goods.

Governing Body Organisation set up by governments or other accredited authorities such as Trade Associations to manage/control its members or specified companies within laid down regulations/conditions.

Governing Council One of the governing bodies of the ECB. It comprises all the members of the Executive Board of the ECB and the governors of the national central banks of the Member States which have adopted the euro.

Government intervention The process of government intervening in a particular area. It may be international trade, fiscal matters and so on. See also interventionist policy entry.

Government use For patents: when the government itself uses or authorises other persons to use the rights over a patented product or process, for government purposes, without the permission of the patent owner. See also compulsory licensing. WTO definition.

GP General Purpose (rating) or General Purpose container.

GPC General Policy Committee (of the Chamber of Shipping).

GPEI Indonesian Exporters Association.

GPI Gross Premium Income.

GPNCO Great Pacific Navigation Co. Ltd. – a shipowner.

GPO The Government Pharmaceutical Organisation (Thailand).

GPS Global positioning system.

GPWS Ground Proximity Warning System – an airport security facility.

GR Gear manufacturers; Grain or Greece (Hellas).

Grab A unit of cargo handling gear, consisting of two quarter circle metal parts which can be brought together to make a close fit. The method of using a grab is to hurl or lower it quickly in an open position, into the bulk and then to pull the two parts together, so that a full load of cargo is enclosed. The grab is primarily used for bulk cargo transhipment and some designs are magnetic, of a circular flat magnetic design, to which the cargo is attached/clings.

Grab bucket Steel bucket attached to a crane consisting of two hinged parts which are open when lowered but closed when weight is taken for hoisting.

GRACYAS CAR/SAM Aeronautical Fixed Service Regional Planning Group – IATA term.

Grain A major dry cargo bulk trade shipments embracing wheat, maize, barley, oats, rye, sorghum and soya beans.

Grain capacity Total ship's cubic capacity for grain cargoes.

Grain certificate Document giving quantity of grain shipped and vessel's draft and freeboard. A certificate to show that the regulations have been complied with when carrying a grain cargo.

Graincon The standard grain voyage charter party devised by BIMCO and introduced in 2003. It displaced the Norgrain charter party.

Grain elevator A cargo transhipment facility which may be bucket elevators or pneumatic suction which sucks the grain out of the ship's hold.

Grain loading Grain carrying ships under requirements of chapter VI of the SOLAS convention 1974 must carry the relevant approved loading manuals and plans for the authorities at grain loading ports to permit the ship to be loaded.

Grain terminal A berth where grain is transhipped into granaries.

Grainvoy North American grain charter party for Continental trade.

Grainvoybill Code name of BIMCO approved Bill of Lading for shipments on the Grainvoy Charter Party.

GRANAD Granard Line – a shipowner.

Granary Bulk grain storage warehouse.

GRAND Grand River Navigational Line – a shipowner.

Grand Alliance Major merchandise maritime trade alliance whose members are P & O Nedlloyd, NYK line, Hapag Lloyd, OOCL, Malaysia International.

Grande Vitesse (French) A fast international freight train. A UIC term.

Grandfather clause Provision in the Protocol of Provisional Application in the WTO that allows signatories to maintain certain domestic legislation, which was in effect when they joined the WTO, even though it may be inconsistent with WTO rules. A WTO term.

GRANTER The seller of an option – a BIFFEX definition.

Graving-dock Dock in which ships may be repaired or built, the water being pumped out as required.

GRD Geared.

Greater Oasis A consortium of five Japanese lines with whom P & O container trades in a wider consortium known as the 'Greater Oasis' in the Japan/Arabian Gulf/Japan trade with calls at waypoints in between.

Green card Internationally recognised document, issued to policy holders motoring abroad, which indicates that they have been granted an extension of their UK motor insurance cover.

Green channel A custom's facility where motorists, passengers and trucks/road hauliers proceed through the green channel if they have no dutiable goods to declare and are within the prescribed limits. It is usually situated at an airport, seaport or a frontier point.

GREENFIELD Greenfield Navigation Co. Ltd. – a shipowner.

Greenhouse effect An atmospheric process in which the concentration of atmospheric trace gases (greenhouse gases) affects the amount of radiation that escapes directly into space from the lower atmosphere. Short-wave solar radiation can pass through the clear atmosphere relatively unimpeded. But long-wave terrestrial radiation, emitted by the warm surface of the Earth, is partially absorbed and then re-emitted by certain trace gases.

Greenhouse gases The trace gases which contribute to the greenhouse effect. The main greenhouse gases are not the major constituents of the atmosphere – nitrogen and oxygen – but water vapour (the biggest contributor), carbon dioxide, methane, nitrous oxide, and (in recent years) chlorofluorocarbons. Increases in concentrations of the latter four gases have been linked to human activity.

Green pound The rate of exchange fixed by EU regulation between the £ sterling and the unit of account for converting sums of money into and out of units of accounts for CAP purposes.

Green seas Waves that board the ship's deck due to severe pitching or rolling.

Grenades Hand or rifle: articles which are designed to be thrown by hand or to be projected by a rifle. The term includes: grenades, hand or rifle, with bursting charge; and grenades, practice, hand or rifle. The term excludes grenades, smoke which are listed under 'Ammunition, smoke'. Term associated with dangerous classified cargo.

GREPECAS CAR/SAM Regional Planning & Implementation Group (ICAO) – IATA term.

Grey area measures Trade actions, the legality of which is not clearly addressed under the WTO. Voluntary restraint arrangements are grey area measures. A WTO term.

Grey lists A medium to high risk. See black and white list entries.

Grey market Goods bought in a lower priced market and sold in a higher priced market for less than authorised by distributor; or a market and part of a national economy which operates outside the officially regulated structure of business activities.

GRI General rate increase.

GRIAC Regional (SAM) Group on Interception of Civil Aircraft – IATA term.

Gross In financial terms the interest payable before deduction of income tax or the 'gross' amount of dividends after adding the tax credit to the actual distribution.

Gross combination weight (GCW) The total weight of an articulated road vehicle with its load, fuel and driver.

Gross Domestic Product The summation in money value terms of goods and services produced through economic activity by a nation. Such data is produced for the UK by the Central Statistical Office every three months. It also supplies details of trends of real disposable income (average living standards), of personal savings, and of profits after deducting North Sea Oil operations.

Gross earned premiums Premiums received by or due to an insurer, without deduction of the cost of any re-insurance, but adjusted to take account of the differences between the unexpired risk reserves at the beginning and end respectively of the period concerned.

Gross Excess re-insurance policies The excess of loss re-insurance policies affected by parties to the pooling agreement between, for example, the International Group of Protection and Indemnity Associations or any addendum variation or replacement to such agreement.

Gross figure The total numerical/finance data before any prescribed deductions. See net figure.

Gross form of Charter Charter party whereby shipowner pays for the cost of loading and discharging the cargo.

Gross income The total income or revenue before deduction of any expenses.

Grossing up Conversion of a net figure into a gross figure.

Gross line The share of insurance accepted by an underwriter before a deduction is made for any re-insurance by him.

Gross National Product The Gross National Product at market prices for any country is equal to the sum of the following: consumption expenditure (personal and public), gross domestic capital formation (private and public); exports of goods and services less imports of goods and services; income received from abroad less income paid abroad.

Gross net premium The gross premium for a marine insurance, before deduction of brokerage and discounts but less gross returns of premium.

Gross premium The total premium before deduction of brokerage or discounts.

Gross pricing The aggregate of all the constituents of a tariff/rate/charge for a passenger/consignment on a particular transit/service. Alternatively, it can be the maximum price charged for a product. A significant aspect of gross pricing is that it includes commission and usually maximum mark up for profit.

Gross profit (i) The difference (mark-up) between the cost price of an item and its selling price or (ii) total sales revenue less the total cost of goods sold.

Gross rate The aggregate of all the constituents of a rate/charge for a consignment on a particular transit/service. For example, the gross rate from a container freight station to destination port would include inland transport costs from the container freight station to the departure port; loading charges at departure port/customs clearance/port charges; sea freight departure port to destination port; unloading charges at the port of destination, port charges; agents commission and so on.

Gross receipts pool This arises under a liner conference system whereby each shipowner bears all his operating/investment cost and pools all the gross revenue. Accordingly, the total pooled gross revenue from each member shipowner will be distributed amongst member shipowners on the agreed percentage by the members. A major advantage of this type of pool is that it encourages/favours the low cost efficient shipowner

Gross registered tonnage In broad terms all the vessels 'closed in' spaces expressed in volume terms on the basis of one hundred cubic feet equals one gross registered ton,. A measure of the total space of a vessel in terms of 100 cubic feet (equivalent tons) including mid-deck, between deck, and the closed-in spaces above the upper deck, less certain exemptions. The GRT of most of the world's ships is recorded in Lloyd's Register.

Gross revenue The total revenue derived from a transport undertaking traffic activities during a particular period and/or specified service(s)/activity(ies).

Gross sales The total sales income without any deduction relative expenditure incurred in the selling process.

Gross tare weight The net tare weight plus weight of loose internal fittings (this weight is variable) of a transport unit e.g. lorry/truck, container, etc.

Gross terms A chartering term whereby the shipowner under a voyage charter party is responsible to employ and pay for the stevedores at either the loading port or the discharging port or both. Often under such an arrangement it is quite common to have gross terms at the loading port and net terms at the discharging port, or more commonly expressed gross load free discharge.

Gross tonnage See gross registered tonnage entry.

Gross train weight (GTW) The total weight of a drawbar road vehicle combination including its load, fuel and driver; or for rail transport, locomotive and rolling stock and passenger/freight traffic.

Gross vehicle weight (GVW) The total weight of a rigid road vehicle and its load including the weight of fuel, driver and passenger, if carried.

Gross weight The total weight of the package as presented for transport. The weight of the consignment including all the packing, blocking, etc. This includes any pallet affixed to the consignment for the throughout transit. Likewise for a container, it is the overall weight of the container, including the merchandise and any loose internal fittings.

Gross weight of container Total weight of container including all the cargo.

Gross written premiums Premiums received by or due to an insurer without deduction of the cost of any re-insurance or any adjustment for the fact that some of the income has to be reserved for the unexpired element of the policy.

Groszy The currency of Poland. See Appendix 'C'.

Ground arrangements See land arrangements.

Ground operator See land operator or contractor.

Ground slots A container term which describes the area allocated and marked out for containers which are stacked at a terminal.

Groupage Process of despatching numerous packages for various consignees as one grouped consignment under Agent sponsorship to a common destination. It is also called consolidation. It may be in a container, international road haulage vehicle, a railway wagon or air freight. Consolidation of several LCL consignments into a container.

Groupage Agent One who consolidates consignments to offer to a carrier as for example full loads in a container, railway wagon, road vehicle/trailer, or other convenient transport unit. It may be an internal or international consignment. Hence, one who consolidates LCL consignments to offer to a Carrier as an FCL.

Groupage Bill of Lading Consignment usually sponsored by Freight Forwarder containing numerous packages for various consignees to common destination.

Groupage centre The place where groupage takes place.

Groupage depot A depot where parcels of cargo are grouped and packed into containers. In so doing it involves the process of despatching numerous packages for various consignees under one grouped consignment under Agent sponsorship to a common destination in a container.

Groupage operator Transport/freight forwarder/agent engaged in the development and carriage of export/import groupage goods. See also groupage/consolidation entries.

Groupage rate A rate devised by an Agent/Freight Forwarder for consignments despatched under groupage arrangements involving a surface international transit by road container or railway wagon. It usually includes both the collection and delivery charge, customs clearance, plus the surface rate element. It may be an internal or international consignment.

Group interview In market research terms, the process of the skilled professional marketeer researcher conducting an interview with a group of which the respondents are usually pre-selected. The market researcher would conduct the interview from a brief to reflect the object/aim of the market research survey. An example could be a pre-selected group of ten men and women all small car owners of differing manufacturer to determine their preferences in terms of colour, car design, speed, parking, fuel efficiency, comfort, maintenance and so on.

Group of 7 Group of seven developed industrialised countries who participate in the annual world economic summits. The group includes Japan, USA, UK, France, Germany, Italy and Canada. See G8 and G9 entries.

Group of 77 A United Nations expression originally used to refer to the 77 developing countries attending the first UNCTAD Conference in 1964 who formed themselves into a group to establish a common bargaining front with the industrialised nations. Today the group has now over one hundred LDCs which are now usually referred to as developing countries.

Group of Negotiations on Goods (GNG) Set up during the Punta del Este meeting in 1986, this body has jurisdiction over the 14 groups on trade in goods. A WTO term.

Group of Negotiations on Services (GNS) A single body covering the mandate on trade in services which formed Part II of the Declaration. A WTO term.

Group travel Passenger travel in groups usually at discounted rates.

Grovertime The code name of BIMCO approved charter party clause relative to overtime and entitled 'Baltic Conference US Grain overtime clause 1962'. It is worded 'all overtime at loading and discharging ports shall be for account of Charterers unless ordered by the Master, except overtime payable to vessel's officers and crew which is always to be for owners' account.

GRP Glass reinforced plastic (lifeboats).

GRS Gears.

GRT Gross registered ton – ships' cubic measurement based on 100 cubic feet equals one GRT.

GRULA (Grupo Latino Americano) Informal group of Latin American countries of WTO.

GSA GSA Line – a shipowner or general sales agent.

GSC German Shippers Council.

GSEL Great South East Lines Pte. Ltd. – a shipowner.

GSM Global system for mobile communication.

gsm Good sound merchantable.

GSP Generalised System of Preferences – programmes by developed countries granting preferential tariffs to imports from developing countries. A WTO definition.

GSPB Good safe for berth.

GSSL Ports of Genoa, Savona, Spezia or Leghorn.

GSSLNCV Ports of Genoa, Savona, Spezia, Leghorn, Naples, Civetta or Vecchia.

GST Goods and Services Tax.

GT Gas turbine(ship) or gross tons.

GTC A good till cancelled order is valid for execution at any time in the future until either executed by the broker or cancelled by the client – a BIFFEX term.

GT-E Gas Turbo Electric Ship.

GTEE Guarantee.

GTS Global Telecommunications System.

GTZ German Agency for Technical Co-operation.

GU Aviateca, (aviateca, SA) – IATA member Airline.

GUA General underwriters agreement.

Guadalajara convention The 1961 convention which amended the Hague Protocol of 1955 relative to the unification of carriers responsibilities for the international carriage of goods by air. See also Warsaw convention entry.

Guar Guaranteed.

Guarani The currency of Paraguay. See Appendix 'C'.

Guarantee A bond, usually issued by a bank in warranty of its customer's performance, e.g. under a contract. Banks frequently require corporate guarantees from parent companies in respect of borrowings by subsidiaries and associates, as well as cross guarantees from a number of group companies. Shipowners equally frequently may rely on bank guarantees, for example as part of the security offered for new building finance. Whereas most guarantees imply collateral responsibility with the principal debtor, bank guarantees are really indemnities, imposing a direct responsibility on the guarantor irrespective of what happens to the principal debtor. They are accordingly costly to obtain.

Guaranteed payment reservation A hotel reservation secured by the guest's agreement, usually by credit card, to pay for the room even if it is not used.

Guaranteed tour One guaranteed to operate unless cancelled before an established cut off date.

Guaranty A contract to see performed what another has undertaken.

Guardian A legal guardian is a person acting in lieu of parents in the event of death or legal incapacity of parents.

Gudgeon Boss on rudder post to take rudder pintles about which rudder turns.

Guest history Personal profile of client's previous stays with a hotel.

Guest night See bed night.

Guided tour A local sightseeing trip conducted by a guide.

Guild of Business Travel Agents British association of travel agents concerned with business travel.

Guild of European Business Travel Agents European group of business-travel agents.

GULF Gulf Shipping Line Singapore – A shipowner.

GUL/SAS Gulf/South Asia feeder service.

Gun powder See Black Powder.

Gunwale Junction of the upper deck with the steel plating.

GUSAS Europe, Gulf South Atlantic Service.

Gusset Three folds on each side of a bag to make a pleat, which allows the bag to take up a rectangular cross section when filled.

Gutta-percha A tough hard material of similar origin to natural rubber resistant to attack by many acids and other chemicals.

Guy Rope used to guide and steady load while hoisting.

Guyanan Dollar The currency of Guyana. See Appendix 'C'.

GUYANAS Europe, Surinam, Guyana service.

GV Grande vitesse – a fast international freight train.

GVS General valuation statement – a Customs term.

GWP Gross world product.

Gypsy Small drum attached to winch or windlass.

GZ Ships righting lever.

H Hoy – a small vessel usually rigged as a sloop or hull (of a ship).
h Hopper-tainer. A form of container or hour
Ha Hatchway(s) or hectare.
HA or D Havre, Antwerp or Dunkirk (ports).
Habeas corpus (Latin) A writ to appear, Lit. You must have the body.
HACT Hong Kong Air cargo Terminals.
Hadley cell The basic vertical circulation pattern in the tropics where moist warm air rises near the equator and spreads out north and south and descends at around 29-30° N and S.
Hague Protocol An amendment to the Warsaw Convention signed at the Hague, September, 28th, 1955.
Hague Rules A complete code of rules for the carriage of goods by sea. Under such rules the carrier has certain defined responsibilities. The rules established minimum obligations, maximum immunities and the limit of the shipowner's liability. They were agreed in 1922 at an International Convention. The rules applied to most Bills of Lading, and any similar document of title. The Hague Rules were amended by the Brussels' protocol of 1968 and are known as the Hague-Visby rules. However, it should be noted not all countries have ratified the Hague-Visby rules and in such situations the Hague Rules still apply. See also Hague-Visby rules entry.
Hague Visby Rules The Hague-Visby rules arose through the amendment by the Brussels Protocol in 1968 to the Hague Rules and were given effect in the United Kingdom by the Carriage of Goods by Sea Act 1971. The reason for the amendment was that the existing Hague rules, under modern trading conditions, were considered to be too favourable to the carrier/shipowner, a view taken particularly by the developing countries. Some countries apply the rules to both inward and outward voyages. The rules contain a number of provisions such as how it defines the carrier's liability; it permits the carrier and the ship to be discharged from all liability in respect of goods unless legal proceedings were started within one year of delivery of the cargo, or the date when they should have been delivered. Furthermore, the rules indicate that the one year limit should not embrace a third person provided that proceedings are commenced within the time – allowed by the law of the Court seized of the case – for bringing an action against such third person which must never be less than three months.

The 1968 Convention provided that the Visby amendment come into effect after the convention has been ratified by ten nations, five of whom have minimum gross registered tonnage of one million tons. These were satisfied in June 1979 and by 1st January 2003 thirty eight States apply these rules. The Visby amendment is designed to all Bills of Lading where (a) the port of shipment is in a ratifying nation, (b) the place of issue of the Bills of Lading is in a ratifying nation and (c) the Bills of Lading applies Hague Visby rules contractually. See Hamburg Rules entry.
Halalas The currency of Saudi Arabia. See Appendix 'C'.
Halalas invoice An invoice payable in the Saudi Arabian Halalas relative to an export sales invoice.
Half Height Container with height of four feet.
Half height (container) An ISO container of standard length of open top with or without soft or hard cover whose height is either 4 feet (1.22m) or 4 feet 3 inches (1.30m).
Half-pension See demi-pension.
Half-roundtrip One-way portion of a roundtrip fare; the two halves may vary on a charter depending on the season; often scheduled excursion fares may be advertised on a half-roundtrip basis.
Half-tilt container A container with larger part of sides, or sides and roof, covered by tarpaulin or similar material.
Halo effect The tendency to over estimate the value of a product/service as a result of a favourable image.
Hamburg Index An index published by the Hamburg Shipbrokers Association on a monthly basis to provide a market analysis of all containership charter tonnage available in the free market.
Hamburg Rules The Hamburg Rules have been in existence since the late 1970s but until recently have not provoked great interest amongst the leading maritime nations.

Under the auspices of the United Nations, a Diplomatic Conference took place in Hamburg from 6th to 31st March, 1978, to consider the UNCITRAL Draft Convention on the Carriage of Goods by Sea which had also been approved by UNCTAD. At the end of the Conference, the delegates duly adopted a 'Convention on the Carriage of Goods by Sea' – in future to be referred to as the 'Hamburg Rules'. These Rules represent a radical departure from the tried and tested formula prescribed by the Hague Rules as refined by the Visby amendments, and form an innovative approach to the question of carriers' liability.

The major differences between the Hague Rules as interpreted and the Hamburg Rules are as follows:
(1) Period of Responsibility: The carrier is no longer able to restrict his liability to the period from 'tackle to tackle'. Instead he is liable from the time he takes over the goods at the port of loading until delivery at the port of discharge.
(2) The catalogue of exceptions in Article 4.2, of the Hague Rules is no longer available to the carrier. In particular he is no longer exonerated from liability arising from error in navigation. Instead the carrier is subject to a general rule of presumed fault under which he is liable unless he can show that 'he, his servants or agents took all measures that could reasonably be required'. Save that with respect to liability for fire the burden of proof rests on the claimant.
(3) The dual system for calculating the limits of liability has been retained, that is to say liability is limited by reference to package or weight whichever is the higher. Generally, with the exception of liability for delay, the liability of the carrier for loss resulting from loss of or damage to goods is limited to an amount equivalent to 835 units of account per package or 2.5 units of account per kgm. of gross weight. The unit of account referred to is the Special Drawing Rights as defined by the International Monetary Fund. Special units of account exist where states are not members of the IMF.
(4) The time bar period has been extended from one year to two years.
(5) Provisions have been introduced which impose liability on the carrier for delay in delivery.

The Hamburg Rules cover all contracts for the carriage by sea other than charter parties. Hence, it applies to Waybills, Consignment Notes, Bills of Lading. Also it covers shipments of live animals and deck cargoes, and to both imports and exports to/from a signatory nation.

The Hamburg Rules came into force twelve months after they have been adopted by twenty countries. These were obtained in November, 1991 and came into force in November 1992. Present signatories January 2003 total 31 nations. Eleven signatories are in land locked countries and many are in the African trades.

Hamburg Sud A major liner and container operator and registered in Germany.

Hand barrow Manual cargo/baggage handling, operated by wheeled equipment.

Hand carry A service provided by most Airlines offering a courier service whereby the courier will accompany the 'extra special' express consignment to the consignee's address.

Handing-over Ceremony Official passing of ship from builders to owners.

Handling instructions Details of how the cargo is to be handled.

Handling services The service concerning the physical handling of cargo.

Hand pallet transporter Cargo handling equipment designed to move pallets in confined spaces.

Hands on The process of actually undertaking a piece of work rather than merely discussing/evaluating such as undertaking the repairs to a car rather then evaluating how it should be done.

Handy A dry bulk carrier with a capacity range of 20,000-34,999 dwt.

Handy max A dry bulk carrier with a capacity range of 35,000-49,999 dwt.

Hangertainer A hangertainer is a dry freight ISO container equipped with removable beams in the upper part. Hangertainers are used for carriage of garments on hangers.

Hanjin Major merchandise maritime trade alliance whose members are Hanjin, Cho Yang, DSR-Senator.

Hanjin/DSR-Senator A major liner shipowner and container operator and registered in the Republic of Korea/Germany.

HANKAM Indonesian Ministry of Defence.

Hanseatic Shipping Co Ltd. Cyprus A ship management company based in Cyprus.

Hapag-Lloyd Container Line Gmbh (HMM) A major liner shipowner and liner container operator and registered in Germany.

Happy hour A one hour period allocated in a retail outlet usually a bar/restaurant when the drinks are offered at very discounted prices.

Harbour See Port entry.

Harbour dues Charges levied on vessels in port/harbour. These can include Lighterage dues, Pilotage, Towage, Anchorage dues, Port dues, Pier dues, Wharfage dues and so on.

Harbour Maintenance Tax The Water Resources Development Act, 1986 created a Harbour Maintenance Tax imposed on shipments through the USA Federally maintained harbours and channels and is payable by importers as part of the customs entry process.

Harbour master The person in charge of all the operational/navigational aspects of a harbour.

Hard Arm An articulated metal arm used at terminal jetties to connect shore pipelines to the ship's manifold.

Hard-copy Paper output from a computer system usually the original copy.

Hard currency A currency which is consistently appreciating or stable in terms of other currencies.

Hard data Data obtained at first hand and completely factual. It may emerge from field research involving face to face interviews, analysis of sales invoices, or evaluation of government statistics.

Hardening market In general business terms an upward movement of the market tariff and less flexibility in any concession by the seller in negotiation. This can also apply to insurance whereby a more selective attitude than previously adopted by underwriters is followed when accepting a risk. Likewise in transport the operator is less inclined to give a rates concession to his Agent when in negotiation.

Hard market A market in which it is difficult to obtain insurance. This is demonstrated by high premium rates.

Hard news Factual news of a topical nature.

Hard selling A very aggressive and persistent method of selling.

Hard-top container A closed container with roof that opens or lifts off.

Hardware In transport terms the transport unit and related infrastructure embracing ship, aircraft, cranes, rail track, berth, vehicle, wagon, etc. In computer terms the actual equipment used in the data processing and computer technology.

Harmonisation The progressive introduction within EU to harmonise tariff regulations and to remove tariff barriers. It also applies to other 'Free Trade areas' worldwide; or cutting tariffs on certain products to the lowest level prevailing among a number of participants. A WTO term.

Harmonisation Conference A group of shipowners operating within a common trade but each operating between different ports having varying freight rates reflecting the differing voyage distance, who agree to a policy of having identical percentage rates increase – at their annual review – such as 10% and thereby maintain their individual competitive situation with each operator's tariff differential remaining unchanged.

Harmonisation of excise duties The process in particular countries/free trade areas and so on, of harmonising – operating on a parity basis – the customs/excise duties imposed on certain classes of goods manufactured, licences issued, and transactions taking place within a country.

Harmonised Index of Consumer Pries (HICP) Protocol No. 6 on the convergence criteria referred to in Article 109j (1) of the Treaty requires priced convergence to be measured by means of the consumer price index on a comparable basis, taking into account differences in national definitions. Although current consumer price statistics in the Member States are largely based on similar principles, there are considerable differences of detail and these affect the comparability of the national results. In order to fulfil the Treaty requirement, the European Commission (Eurostat), in close liaison with the national statistical institutes and the EMI, carried out conceptual work on the harmonisation of consumer price statistics. The Harmonised Index of Consumer Prices is the outcome of these efforts. An ECB term.

Harmonised long-term interest rates Protocol No. 6 on the convergence criteria referred to in Article 109j (1) of the Treaty requires interest rate convergence to be measured by means of interest rates on long-term government bonds or comparable government securities, taking into account differences in national definitions. In order to fulfil the Treaty requirement, the EMI carried out conceptual work on the harmonisation of long-term interest rate statistics and regularly collected the data from national central banks on behalf of the European Commission (Eurostat), a task which has been taken over by the ECB. An ECB term.

Harmonised system An international nomenclature developed by the World Customs Organisation, which is arranged in six digit codes allowing all participating countries to classify traded goods on a common basis. Beyond the six digit level, countries are free to introduce national distinctions for tariffs and many other purposes. Overall, it covers some 15,000 descriptions of the products or groups of products most commonly produced and traded globally. It represents a custom's tariff classification embracing 15,000 headings, set out in 97 chapters broken down into sections, headings and sub-headings. The tariff is not used for intra EC trade. It has been adopted by 85% of the world's trading nations. It is designed for customs services but can also be used for statistics, transport purposes, export, import and manufacturing. Generally described as the Harmonised Commodity Description and Coding System. See Custom's tariff entry.
Harmonising formula Method for cutting tariffs whereby high rates are reduced more than low rates. A WTO term.
HARP Helicopter airworthiness review panel.
HASCo Shanghai Hai Hua Shipping Co – Chinese Shipowner.
Hatch beam Removal steel beam – one of several laid across the hatch (way) of a ship, on which the hatch cover rests when the hatch(way) is closed.
Hatch board Wooden board used to cover the hatchway of a ship.
Hatch coaming The steel wall around the hatch to accommodate the hatch cover, whilst at the same time to lift the hatch above the deck level to protect it against water entering the hatch and damaging the cargo
Hatch covering The fixture/fittings including beams which covers a hatch on a vessel gaining access to a ship's hold.
Hatches The openings in the deck(s) of general cargo and bulk cargo vessels to allow loading and discharge of cargo.
Hatchman A person who supervises the cranage of cargo to and from the ship's hold.
Hatchway The cargo access to the ship's hold.
Haulage The movement of goods by road between two specified points.
Haulier A road transport/truck operator involved in the conveyance of general merchandise or specialised bulk goods.
Hauliers receipt A document which gives a full description of the cargo and enables the carrier to have a signed document by the consignee or agent accepting the cargo or container, the contents of which will be discharged by the consignee or agent at the named destination. Also called Release Note.
Hav Haversine.
Havana Charter A conference held in Havana in 1948 which formed the foundation of the International Trade Organisation which was ultimately transacted in GATT/WTO as the Havana Charter but was not ratified by the US congress.
HAWB House Air Waybill.
Hawk A person/company which adopts an aggressive and non conciliatory attitude to a business situation.
Hawse Part of a ship's bows with holes from where the anchor chains can pass through.
Hawse Pipe Housing place of anchor and leading of cable from windlass through spurling pipe to chain locker.
Hawser Synthetic or natural fibre rope or wire rope used for mooring, warping and towing.
Hazardous cargo Cargo classified as dangerous. See ADR, RID, IATA, IMO entries.

Hazchem Chemical Industries' labelling scheme for vehicles carrying hazardous chemicals.
Hazchem code A code of practice for dealing with emergency situations involving chemical and other hazardous loads conveyed by road.
HAZOP Hazard and operability studies comprise a systematic review of process and engineering design in order to identify possible plant upsets, potential hazards and their consequences.
Hazzard The aspect of a risk likely to increase the severity of a claim – an insurance definition.
HBL Hydrostatically balanced loading crude oil cargo safety system.
H/C High Cube – containers 9feet 6 inches high instead of usual 8 feet 6 inches.
HC Heating coils (on a ship); hydraulic coupling or hours clause – an insurance term.
HCC Harbour Co-ordination Centre or High Cube Container.
HCMM Honourable Company of Master Mariners.
HD Half despatch.
HDLC High Level Data Link Control – IATA term.
hdlg Handling.
HDOP Horizontal dilution of precision – an expression that reflects the continual movement of satellites and the effects on the crossing angles of the range circles of GPS navigation.
HE His/Her Excellency.
Header bar Beam or bar (usually above the end doors of an open top container) which may be swung to one side or removed to improve access.
Header board A vertically mounted board to provide front wall protection against shifting cargo and used primarily on road haulage platform trailers.
Head hunter A company specialising in recruiting personnel for a particular post on behalf of a company/entity. Usually the respondents are sought through National/International/Region press plus the personnel already listed on recruitment agency portfolios. Most recruiting agencies specialise in a particular segment of the market such as banking, accountancy, export, marketing, engineering, etc. The posts available are usually in the middle upper and most senior management categories.
Heading Printed or typed text to indicate the data statement to be entered in columns, etc.
Headline point An Air freight tariff (IATA) relative to the city entry found at the top of a column, table, or page below which are shown rates applicable from/to such a city.
Head lines Mooring lines leading ashore from the fore end or forecastle of a ship, often at an angle of about 45 degrees to the fore and aft line.
Head sea The sea that rolls or hits directly against the movement of the ship.
Heads of agreement An agreement prepared based on the salient contents or brief description of each clause. This would include for example preamble, arbitration, duration of agreement, termination clause and so on. Ultimately a complete agreement is drawn up, for consideration and ultimate agreement by the parties to the agreement.
Head tax A fee collected from a visitor or passenger upon his entry into or departure from an airport or hotel; also see passenger facility charge.
Health and safety regulations Such regulations form part of the legal environment and are very wide ranging embracing conditions at the work place. These regulations apply particularly to foodstuffs, including

Health certificate

packaging and handling, and a wide range of consumer goods. These include electronics, fire resistant properties and safety in use on general consumer product design. Industrial product designs are very much focused on safety and quality standards and the use of specific materials with minimum constituents. Customs controls on importation impose stringent inspection disciplines on a wide range of goods to ensure compliance with legal standards and require appropriate documentation. See Health Certificate entry.

Health certificate Issued by a government to confirm that the agricultural and animal products which are being exported comply with the relevant legislation in the exporter's country. It is issued at the importer's request to comply with the country's health regulations and confirms the product was in a good condition at the time of inspection prior to shipment and fit for human consumption. It confirms that the Food Hygiene Regulations have been complied with. Overall, a document that may be needed when exporting agricultural products.

Heated container A container built with insulated walls, doors, floor and roof, fitted or capable of being fitted with a heating appliance which can raise and maintain the temperature inside the container at a required level.

Heating coil Heating coil apparatus usually found in individual tanks of oil tankers.

Heaving line A very light line that is thrown between the ship and the berth and is used to draw the messenger line ashore.

Heavycon Code name of BIMCO approved standard contract for heavy and voluminous cargoes.

Heavyconbill Code name of BIMCO approved bill of lading used with shipments under the Heavycon contract.

Heavyconreceipt Code name of BIMCO approved non-negotiable cargo receipt for use with the Heavycon contract.

Heavy grain Wheat, rye and maize – a chartering term.

Heavy industries Industries involved in the production/manufacture of products with a high weight and low volumetric ratios such as steel, coal, minerals and so on.

Heavy Lift A unit of cargo which cannot be lifted by the normal ship's lifting gear.

Heavy lift cargo See heavy lifts entry.

Heavy lifting beam Ships cargo handling equipment designed for heavy lift transhipments.

Heavy lift mast crane A crane capable or lifts up to 550 tons. It is usually on board a semi-submersible vessel when engaged in 'off shore' construction work.

Heavy lifts Single commodities exceeding normal loading equipment design capacities and requiring special equipment rigging techniques for handling.

Heavy lift ship Vessel designed to convey unusually heavy – often bulk – cargoes.

Heavy weather Severe weather giving rise to the possibility of damage to a cargo at sea. If a ship encounters heavy weather, her master may 'note protest' on arrival at her next port of call, which may be a necessity in order to avoid liability for damage to cargo.

HE Cls B Heat coils in bunkers (on a ship).

He Cls C Heating coils in cargo tanks (on a ship).

Hedge The action taken to reduce liability to market price fluctuations of an asset. The establishment of a position in one market which is equal and opposite to that held in another market. The reduction of risk – an insurance definition.

Hedge Funds Specialist funds that take positions in equities, foreign exchange and interest rates. Typically highly leveraged.

Hedging The establishment of an opposite position on a futures market from that held and priced in the physical commodity. Without hedging, the physical position would be at risk to price fluctuations. Hedging is a method of safeguarding against losses on foreign exchange by matching any debt in a currency with an equal and opposite claim, preferably with the same maturity date. Hedging can be achieved by buying or selling currency through forward cover or by foreign currency borrowing. Overall, it is the use of market mechanisms by a trader or operator of a business to obtain protection against loss through price fluctuations. In shipping, for example, an owner, operator or charterer may use the BIFFEX freight futures market to limit his exposure to fluctuations in spot dry cargo freight rates. Protection against fluctuations in interest rates, currencies and bunker prices may also be obtained through the respective forward swaps markets.

Heel The amount of liquid cargo retained in a cargo tank at the end of discharge. It is used to maintain the cargo tanks cooled down during ballast voyages by re-circulating through the sprayers. On LPG ships such cooling down is carried out through the re-liquefaction plant and on LNG ships by using the spray pumps.

Held covered A provisional acceptance of a risk, subject to confirmation that cover is needed at a later date. Where applicable to an existing insurance a reasonable additional premium is payable if the risk held covered comes into effect. A marine term whereby the insured's interest remains covered in the event of a circumstance arising which would, without prior agreement, cancel coverage.

Held or direct covered A condition in a marine insurance policy requiring that the insured voyage be direct from one place to another. If the voyage is delayed en route an additional premium is charged.

Helipad A landing point for helicopters. It may be on top of flat roof building, on a ship, a field, a large garden or a public place.

Heliport A terminal accommodating helicopters and thereby being an interchange point linking helicopter services with other connecting forms of transport.

Hellenic Chamber of Shipping Founded in 1936, and based in Piraeus, it represents all ships flying the Greek flag. It is an official adviser to the Greek government.

Helm Ship's steering wheel.

Help display A display that contains information to help the user use the menu from which the Help display was accessed. A computer term.

HENGLEONG Heng Leong Shipping Co. Pte. Ltd. – a shipowner.

HERMES Handling European Railway Message Exchange System.

Heterogeneous cargo Variety of cargoes.

Heterogeneous market A market which is comprised of various segments/groups and thereby having differing customs, tastes, religion, attitudes and behaviour. See heterogeneous population.

Heterogeneous population A populace in a particular country/region which is made up of various segments/ethnic groups. Overall, the populace have differing customs, tastes, religion, protocol, attitudes and nationalities. An example is New York and London. See homogeneous market.

Heuristic The process of solving problems by evaluating each step in the process, searching for satisfactory solutions rather than optional solutions. It comprises a form of problem solving where the results are determined by experienced or intuition instead of by optimalisation.

HF High Frequency (3 MHz and 30 MHz) or Hapag Lloyd (Hapag Lloyd Flug GmbH) – IATA member Airline.
HFLINE HF Line, Panama – a shipowner.
hgt Height.
HH Ports between and inclusive of Le Havre – Hamburg range.
H/H Half height (of an ISO container). An open top container only 4 feet 3 inches high.
hhd Hogshead
HHDWS Heavy handy deadweight scrap (iron).
HIC Heavy Industries Corporation (Burma).
HICP Harmonised Index of Consumer Prices.
Hidden Agenda List of items not disclosed which may represent the ultimate objective of a particular strategy which may be political, economic, social, legal, financial, technical, etc.
HIGH The highest price traded that day. IPE term.
High Cube Merchandise which has a high volume to weight ratio and therefore is likely to be bulky cargo. Examples include bales of cotton, straw, hay and so on. Such cargoes have a high stowage factor.
High cube container A container of 8 feet 6 inches (2.59m) high or more, 2.44m wide and length of 12.19m.
High cube dry freight container A high cube dry freight container is a 40 foot dry freight container where the height had been increased by 1 foot to 9 foot 6 inches. High cube containers are used for general cargo.
High cube reefer container A high cube reefer container is a 40 foot reefer container where the height has been increased by 1 foot to 9 foot 6 inches.
High cube swap body An intermodal transport unit which takes the form of a container and has a high cube factor to accommodate cargo of a high volume to weight ratio. Hence, it is likely to be bulky cargo. See high cube and swap body.
High density cargo Cargo which has a high stowage factor and therefore rather bulky with a low weight but high cube relationship such as cotton, hay, etc.
High density freight Freight/general merchandise which has a high stowage factor and therefore rather bulky with a low weight but high cube relationship such as cotton, hay, etc. An important feature when loading/stowing any transport unit.
High dollar A high value dollar currency exchange rate.
High euro A high value euro currency exchange rate.
High inventory A long list of products/items.
High level of service A high quality service.
High leverage A high level of exposure of securities to market risk in the capital structure of a company. See leverage entry.
Highlight To move the pointer to a label so that it appears brighter than the surrounding labels.
High loader Vehicle with a loading platform able to load/unload cargo units to and from an aircraft or other specified unit such as a loading bay or rack.
Highly regulated economy An economy where a very large number of government regulations exist relative to the conduct of business and economy. This embraces labour law, inwards investment, fiscal measures, company law, social benefits, property technology and taxation. It is especially found in some highly developed economies as, for example, Switzerland and EU.
High margins See high mark up entry.
High mark up A product/service bearing/featuring a high profit margin in the final selling price to the consumer/end user. Also termed high margins.
High-modulus polyethylene (HMPE) A manufactured fibre based on Ultra High Molecular Weight Polyethylene (UHMWPE).
High profile A person/company/product/service at the top end of the market position and often regarded a market leader. Usually it is associated with excessive market exposure generating much media attention
High quality transport and logistics service The essentials embrace infrastructure and technologies; security and safety; facilitation; legal aspects; market access; and exogenous facters such as distance or trade volumes.
High risk A market which lacks confidence in terms of its long-term future.
High risk business A business which lacks confidence in terms of its long-term future and is very vulnerable in regard to a change in economic conditions and infrastructure.
High risk product A product very vulnerable to competition and a change in economic conditions and infrastructure.
High seas Maritime areas that are outside the jurisdiction of any state.
High sterling A high value sterling currency exchange rate.
High Speed Craft Code 2000 See HSC Code (2000).
High stowage factor Cargo which has an excessive bulk to a low weight relationship, e.g. hay, straw.
High tech A product/service which offers/provides an advanced technical system reflecting the latest generation of technology. It may be computerised container transhipment as obtains in the Port of Singapore or advanced Air Traffic Control system. See also EDI.
High tech company A company specialising in handling/manufacturing/supplying high tech products involving the lastest generation. It is usually computer orientated.
High technology, high value added, low volume product A product which has a high technology specification with a high value added, but low in volumetric terms. An example is computer software. Such cargoes are ideal for Air Freight movement.
High technology industries Industries which are classified as high technology such as electronics, computers and medical technology.
High tech sunset industry See high tech industry and sunset industry entries.
High tech sunrise tech company See high tech industry and sunrise industry entries.
High value added products A product of high value to which further improvements and may be functions have been added such as computer software. See value added benefit.
High value cargo Merchandise of a high intrinsic value which usually attracts an ad valorem rate such as art treasures.
High value to low weight ratio A product which has a high value but low in weight terms. An example is diamonds. Such cargoes are ideal for Air Freight movement internationally.
High volume low margins market A market place where the volume of sales is high and the profit margin very low.
High yen A high value yen currency exchange rate.
High yield A high-risk bond with a credit rating that is below investment grade. An international banking definition.
Himalaya clause A clause found in a Bill of Lading or Charter Party which defines the rights and immunities of all servants and agents of the carrier/shipowner. Hence, a shipowner who wishes to protect his Master,

crew or independent contractors would insert in the charter party or bill of lading the Himalaya clause. This would exempt such persons from liability for any loss, damage, or delay of whatsoever kind arising resulting directly or indirectly from any act neglect or default on his part while acting in the course of or in connection with his employment and so on.

Hinterland That part of the industrial area in the environs of a seaport.

HIPC Heavily indebted poor counties.

Hire and reward operator A road haulage operator who promotes/advertise his road transport services – which may be national or international – for the purpose of conveying merchandise for which he is remunerated.

Hire car A rental car.

Hire purchase A contract for the hire of an asset which contains a provision giving the hirer an option to acquire legal title to the asset upon the fulfilment of certain conditions stated in the contract. In shipping, generally used in the term bareboat hire purchase, where the chartered vessel becomes the charterer's property after a stipulated number of payments.

Historical cost The actual cost of acquiring assets on goods and services.

Historical Trade A trade/traffic flow which has existed for many years.

Historic cost accounting The traditional method of accounting for profits and other balance sheet figures making no allowance for inflation as in Current Cost Accounting. Stock is valued at its original cost, not its current replacement cost. Fixed assets are entered at original cost, minus a depreciation figure based on that cost.

HIT Humber international terminal.

Hitchment Cargo An amount of goods which is added to an original consignment as the owner and the destination are the same as those of the original consignment.

HKCL Hong Kong Container Line – a shipowner.

HKD$ Hong Kong dollar – a major currency.

HKIL Hong Kong Islands Line – a shipowner.

HKSAR Hong Kong Special Administrative Region.

HKSOA Hong Kong Shipowners Association

H/L Heavy lift.

HLCS Heavy lift/crane ship.

HLDC Heavy deck cargo carrier.

h/lift Heavy lift.

HLO Helicopter Landing Officer.

HM Hard Market.

H & M Hull and machinery.

HM Air Seychelles Ltd. An IATA member Airline.

HMC Her/His Majesty's Customs.

HMC & E Her/His Majesty's Customs & Excise.

HMCG Her Majesty's Coast Guard.

HMG Her/His Majesty's Government.

HMPE High-modulus Polyethylene – see separate entry.

HMS Her/His Majesty's Ship.

HNS Hazardous & Noxious Substances. An international convention which was agreed at a diplomatic conference in 1996.

HO Antinea Airlines- IATA member Airline.

Ho Holds (of a vessel).

Hockle A knot-like twisting of individual strands of a twisted rope.

HOEGH Hoegh Lines – a shipowner.

HOEHOE Hoe Hoe Shipping Co (Pte) Ltd. – a shipowner.

H of C House of Commons.

H of L House of Lords.

Hogging Longitudinal bending stress of a vessel causing fore and aft portions to drop lower than amidships portion in a seaway which could be due to excessive weight of cargo at the ends (bow and stern). Overall the longitudinal stress causes centre area of ship to be comparatively unsupported.

Hogshead Small container – 521/2 gallons capacity – used for shipment of liquid or greasy cargoes.

Hoist Vertically operated (wagon or other transport unity) platform used particularly for bulk cargo transhipment such as coal.

Hoistable decks A vessel equipped with hoistable decks which allow the headroom to be varied for different cargoes.

Hold The interior of a vessel below the decks where cargo is stowed.

Holder The person who buys an option contract to open a position. An IPE definition.

Holding Company A company is a holding company of another if, but only if, that other is its subsidiary.

Hold space The space enclosed by the ship's structure in which a cargo containment system is situated.

Holiday A day other than the normal weekly day(s) of rest, or part hereof, when by local law or practice the relevant work during what would otherwise be ordinary working hours is not normally carried out. A BIMCO term.

Holiday and Leisure insurance Insurance provision to cover the insurers holiday or leisure activity such as cancellation expenses, medical and emergency travel expenses, personal accident, baggage and money, personal liability, etc.

Holiday package market The travel/leisure market embracing the package tour incorporating travel, holiday accommodation and sometimes optional excursions retailed at an inclusive price.

Home foreign Line of business in excess relating to general insurance business written in the UK (normally only in the London Market) in respect of risks exposed Overseas and British Overseas domiciled companies. An insurance term.

Home market The total market found in the producer's country involved in the manufacturer's process of supporting a particular product(s)/service(s).

Home port The port of registration of a vessel.

Home trade passenger ship A passenger vessel operative in the home trade. See home trade ship entry.

Home trade ship A vessel operative within the limits of United Kingdom, Channel Islands, Isle of Man and Continent between Rivers Elbe and Brest inclusive. A definition relative to vessels registered in UK. A similar criteria of a vessel plying limits applies to ships registered in other maritime countries.

Homogeneous cargo Similar cargoes.

Homogeneous market A situation in the market place involving an area/region/country whereby the market components are 'all the same' thereby constituting a mass market situation with no separate/segmented market demand. Accordingly, the customer profile/taste/fashion is all the same. Such a situation exists in Japan and China. See also homogeneous population.

Homogeneous population A populace in a particular country/region which is broadly the same and overall has similar tastes, habits, attitudes, behaviour patterns and so on. An example is Japan and China. See also heterogeneous population entry.

Homogeneous Risks Insurance of items which are of a similar nature.

Homologation The process of obtaining product approval to enable such a product to be imported into a specified country. It is usually of a technical nature. Such approval is required by the host country government.

Honeycomb slip A standard agreement form for attachment to the broker's slip.

Honeycomb slip endorsement A standard agreement form for attachment to the broker's slip. Pre-printed boxes provide space for the initials of underwriters subscribing to the agreement and help the broker to ensure that all the necessary agreements have been obtained. It is used in all cases where an agreement requires the initials of six or more underwriters.

Hong Kong dollar The currency of Hong Kong. See Appendix 'C'.

Hong Kong dollar invoice An invoice payable in the Hong Kong dollar relative to an export sales invoice.

Honour To pay or accept a bill of exchange. An international trade term found in banking.

Honour Policy One which is not acceptable as evidence in a court of law such as a PPI policy.

Hook Fitting to connect lash to tensioner or corner casting, in container securing practice.

Hook cycle The time in minutes and seconds between the hooking on of two consecutive loads.

Hook damage Damage to a cargo caused by stevedores' hooks.

Hooking back The practice of engaging the hook(s) of a sling in the upper terminal fitting so as to form a basket hitch.

Hopper Large container used to accommodate bulk cargoes being fed into railway wagon or road vehicle below.

Hopper dredger A dredger with a hopper facility thereby permitting the silt etc. to be fed into the hopper on board the dredger. See dredging and hopper entries.

Hopper tanks Tanks surrounding the ceiling of a hold used as ballast trimming accessories.

Hopping class Class hopping is the process of a shipowner changing from one classification to another on a frequent basis to gain some particular advantage.

Hor. Horizontal

Horizontal integrated Company This involves a company for example manufacturer/processing a product and sourcing all its components/materials from outside the company and subsequently selling the finished product on the open market.

Horizontal market place Its focus is to facilitate relations between parties which operate in different value chains. See horizontal integrated company.

Horizontal mergers The process of companies merging to facilitate their competitive and financial strength. See vertical mergers and horizontal integrated company.

Horizontal organisation A horizontal organisation structure found in an export office dealing with the processing of an export consignment. In so doing each clerk has a total involvement in the processing of an export consignment involving order processing, credit control, production/assembly, packing, transportation and insurance, invoicing, documentation and so on. Usually it is based on one market or group of markets such as Group market one for NAFTA; Group market two for EU and so on. The organisation encourages expertise in a group of markets and thus identification with the processing of the export consignment(s) in such markets. See also vertical organisation entry.

Horizontal thinking The process of developing a strategy within the company as, for example, the supply chain. See lateral thinking.

Horse power A standard usually adopted for measuring mechanical powers.

Hospitality Hotel room used for entertaining, usually a parlour or function room.

Hospitality suite A hotel suite, parlour or studio engaged for the entertainment of those attending a meeting, conference or convention.

Host A representative of a tour operator destination or other tour principal who provides escort service at the destination: a tour manager, as against representatives who provide only information or greeting services.

Host country The country importing merchandise or accepting tourist, businessmen and so on.

Host system Usually a CRS that gives users access to other vendors and systems through the same equipment.

Hostel An inexpensive, usually supervised lodging primarily for young people.

Hostelry An accommodation that provides lodging and/or food.

Hotel classifications Designations used throughout the world, whether the rating is made professionally or promotionally. In Europe, the general system is to rate hotels from 5-star-deluxe-to 1-star-budget or economy. There is no universally accepted system in the US.

Deluxe or luxury – a top-grade hotel; all rooms with private bath and highest standards maintained throughout.

First class – a medium-to-upper range hotel, with all rooms having private bath.

Moderate class – some rooms with private bath and most standard public rooms services.

Second class – a budget operation; very possibly no private bath and very probably limited services and amenities; also called economy or tourist class.

Hotelier A hotel owner, manager or keeper.

Hotel package A package offered by a hotel, sometimes consisting only of room and breakfast, and sometimes, especially at a resort hotel, of room meals, transportation, use of sports facilities and other services.

Hotel register The permanent record kept by all hotels of the arrival and departure of guests, each of whom must sign in on arrival.

Hotel rep A representative offering hotel reservations to wholesalers, agents and the public; some also offer marketing and other services.

Hotel Sales Management Association (HSMA) A professional trade society of executives responsible for marketing hotels.

Hot money The situation when funds flow into a country to take advantage of the favourable rates of interest in that country. In so doing they strengthen the exchange rate of the recipient country and likewise improve the Balance of Payments situation. An example is sterling improving its exchange rate against the US dollar and the Swiss Franc, Yen, Euro currencies through rising interest rates and currencies flow into the UK.

Hours clause The period of time, usually 72 to 168 hours, during which all individual losses arising out of and directly occasioned by one catastrophic event are covered under a catastrophe excess of loss treaty. An insurance term.

Hours purposes Time allowed by charterer for the dual operation of loading and discharging cargo.

House Terminology for combined transport movement indicating movement starts/finishes at customer's premises. See House to House FCL/FCL entry.

House air waybill (HAWB) Document covering conditions of goods transported in a forwarder's consolidation.

House Bill of Lading A certificate of shipment of a specified consignment and usually issued by a Freight

Forwarder in association with a groupage or consolidated international consignment.

House bills See House bill of lading entry.

House/Depot The conveyance of a consignment/merchandise from the shippers premises and terminating at the Container depot or Container Freight station in the importing area.

Household effects Household furnishings and associated accessories. Usually associated with the shipment of household removals in a container.

House magazine A magazine featuring a particular company's activities and personalities issued usually free to staff and sometimes available to selective customers. It especially highlights Company policy, financial results and aims to maintain a high morale level amongst the work force and improve the Company's public image in the market place.

House order An order for a member's proprietary account. An insurance term.

House to House Indicates goods are conveyed from the exporter's premises and conveyed to the consignee's premises and thereby providing a combined transport movement.

House to House – FCL/FCL A container term whereby goods are conveyed from the exporter's premises and conveyed to the consignee's premises. The exporter/forwarder packs the container whilst the consignee unpacks the goods. Containers are presented to the carriers loading berth/terminal with the cargo already packed and are for ultimate delivery beyond the carriers discharging berth or terminal as a full load. Also termed FCL/FCL.

House to House Transport See door to door entry.

House to Pier A container term whereby goods packed by the exporter and delivered to the carriers loading berth or terminal as a full load. At destination the cargo is unpacked by the carrier at his discharging terminal or depot for the consignee's collection. Also termed FCL/FCL.

Hovercraft A passenger carrying craft which operates on a cushion of air above the water and has a service speed of up to 50 knots. Some can also convey cars.

Hoverport A terminal accommodating Hovercraft and thereby being an interchange point linking Hovercraft services with other connecting forms of transport.

Hoy A small vessel usually rigged as a sloop or a species of a lighter; houses on a deck of a ship or horizontal engine.

HP Horse power or high pressure.

HPH Hutchison Port Holdings – see separate entry.

HPS Hazardous Polluting Substances.

HPTB High pressure Turbine (on a ship).

Hq Headquarters.

HR Hellenic Register of Shipping – the Greek Classification Society or Hahn Air Lines GmbH – IATA member Airline.

HRD Human Resource Development.

HRM Human Relations Management or Human Resource Management.

HRN House Recovery Net.

HRS Hellenic register of shipping.

HRU Hydrostatic Release Unit.

HS High strength; heating surface; Harmonised System – often referred to as the Harmonised Commodity Description and Coding System – a custom's facility or Air North Regional – an IATA member Airline.

HSC Health & Safety Committee/Executive or High Speed Craft (code).

HSC Code (2000) The International Code of Safety for High-Speed Craft, 2000 (2000 HSC Code) is a successor to the International Code of Safety for High-Speed Craft that was adopted in 1994. The 1994 HSC Code applies to high-speed craft that are involved in international voyages and for which the keels are laid after 1st January, 1996. The 2000 HSC Code applies to craft for which the keels are laid, or which are at a similar stage of construction, on or after 1st July, 2002. The application of the 1994 HSC Code is mandatory under chapter X of the SOLAS Convention; the 2000 HSC Code is similarly mandatory from 1st July, 2002.

HSCO Hungarian Shipping Co. Ltd. – a shipowner.

HSDT Hopper side tanks (on a ship).

HSE Health and Safety Executive (of the Chamber of Shipping).

HSE-MS Health, Safety and Environment Management Systems and management procedures.

HSRB Hydrographic Strategic Review Board.

HSS Heavy grain, soyas, sorghum – a chartering term; or high speed sea service, an example of which is found in the high speed catamaran operating between Holyhead and Dun Laoghaire conveying passengers, cars and road haulage vehicles with a cruising speed of 40 knots.

HSSC Harmonised System of Survey and Certification

ht Half-tilt Containers.

HT Half-time survey.

HTML Hyper Text Mark up – the programming language of the web – a cyberspace term.

Htr Heater.

HT High Tensile Steel.

HU Hungary – customs abbreviation or Hainan Airlines (Hainan Airlines Co. Ltd) – an IATA member Airline.

Hub The control transhipment point in a transport structure serving a number of consignees and/or consignors by means of spokes. The stretches between the hubs are referred to as trunks. Also a central warehouse, used for distributing to a complete country, group of countries or large region. May distribute to other, secondary warehouses rather than direct to ships or factories.

Hub airport An airport which acts as a hub of a transport network and relies on feeder services which may be rail, road, sometimes sea transport and air. Usually it involves a large volume of transhipment traffic involving passengers/cargo transferring to other airlines for onward destinations.

Hub and spoke (containerisation) The massive growth in the current decade has been due to the restructuring of the container deep sea network which has been driven by the hub and spoke system. This operates on the basis that the largest hub ports like Singapore, Hong Kong and Rotterdam rely on the smaller spoke ports to feed containers into the mega container service tonnage. The VLCS and ULCS operate between the major seaports and the feeder container vessels distribute containers to and from the hub ports. See VLCS, ULCS and feeder vessel.

Hub and spoke system An example exists in international distribution whereby packages are collected by road (parcel vans) and delivered unsorted to a central point called the 'hub'. Handling and sorting takes place at the hub taking advantage of automated handling and sorting procedures involving bar coding to scan in and scan out the individual packages. At this stage the hub movement involving a dedicated truck or aircraft convey the packages to an overseas hub centre where the goods are sorted and distributed usually by express road services to the consignee. It is usually an intermodal system road/air/road, road/air/rail. In the airline industry, a system of routeing of passengers through a central airport, permitting a carrier to operate fewer flights on less travelled routes. See hub ports and hub airports.

Hub port A seaport which acts as a hub of a transport network and relies on feeder services which may be road, rail, inland waterways or feeder shipping services. In so doing the system is often termed the hub and spoke with the spoke being the feeder service. It is particularly evident with containerisation and examples include Rotterdam and Singapore ports.

Hub reduction System in which the half shafts in the rear axle drive the wheel hubs through a reduction gear rather than direct, in order to reduce torsional strain on the shafts and the differential when running on poor ground conditions. It is found in road vehicles.

Huckepack Carriage The conveyance of road vehicles and trailers on railway wagons. See piggyback entry.

Hulk An old vessel used for storage purposes.

Hull Ship chassis.

Hull Clauses Insurance policy wordings in respect of loss or damage to the vessel itself.

Hull insurance A classification on risks referring to a hull of a ship or the fabric of a hovercraft or a plane, including its machinery and equipment. Marine hull insurance commonly includes indemnity for .75 of collision liability. Overall class of ocean marine insurance that covers physical damage to the ship or vessel insured. Typically written on an 'all-risks' basis. Also physical damage insurance on aircraft – similar to collision insurance in an automobile policy.

Hull to debt ratio Indicates the market value of a ship in relationship to the debt outstanding on the vessel such as the mortgage. In the event of ratio falling below a predetermined level, the bank may require additional security from the borrower.

Human Relations Management The process of dealing with at managerial level the employees relations in a company especially dealing with the unions, industrial legislation and industrial disputes.

Human Resource Development The development of personal resources within a company and thereby aiding each employee's career development/potential. This is especially facilitated through training and career planning within the company.

Human resources Manpower resources.

Hurricane A tropical storm marked by extremely low barometric pressure and circular winds with a velocity of 75 miles an hour or more. The name given primarily to tropical cyclones in the West Indies and Gulf of Mexico.

Hutchison Port Holdings A global container terminal operator based in Hong Kong (China).

Husbanding The task of looking after a vessel's non cargo related operations as instructed by the master or ship's owner. See also ship's husband entry.

Husbandry Agent A shipowner's agent.

HV High visibility or Transavia Holland B.V. d.b.a. Transavia Airlines – an IATA member Airline.

HVF High Viscosity fuel oil.

HW High Water.

HWB Hot water boiler(s) (on a ship).

HWH Hot water heater.

HWM High water mark.

HWOST High water ordinary spring tide.

HYC Hydraulic Coupling (of a ship).

Hydrate Inhibitors An additive to certain liquefied gases capable of reducing the temperature at which hydrates begin to form. Typical hydrate inhibitors are methanol, ethanol and isopropyl alcohol.

Hydrates The compounds formed by the interaction of water and hydrocarbons at certain pressures and temperatures. They are crystalline substances.

Hydraulically operated gangway A piece of port equipment linking the ship with the shore (quay) enabling passengers/crew personnel to embark or disembark and which is hydraulically operated/manoeuvred into position.

Hydrocarbon gas, compressed It consists of hydrocarbon gas under high pressure, but not in the liquid condition.

Hydrocarbon gas, liquefied It consist of hydrocarbon gas from natural gas or from distillations of petroleum which are liquefied by pressure.

Hydrocarbons Toxic and highly explosive gas remains following crude oil cargo discharge.

Hydrocharter Charter party code name used for nitrate cargoes from Norway.

Hydrofoil A passenger carrying craft which is propelled by water jet with fully submerged foils and an automatic control system to provide a jet smooth ride at 40 knots even in rough water.

Hydrometer Instrument to determine the relative density of liquids.

Hydroneck Detachable front end of a semi-road trailer, with built-in hydraulic bed raising and lowering operation.

Hygroscopic Substance A substance which is capable, under the right conditions, of absorbing water vapour from the surrounding atmosphere.

Hypermarket A jumbo sized self-service retail outlet with a sales area in excess of 2,500 square metres. It may be one store dealing with a large range of consumer goods, or a number of shops offering differing products. A jumbo store, combines supermarket, discount and warehouse retailing. These are found in Europe, USA and Far East.

Hypochlorite solution These are water solutions containing a soluble hypochlorite. The concentration of the solutions vary over a wide range. The solutions are alkaline and corrosive but are not flammable.

Hypothecation The pledging of a ship or cargo for financial advances made. In maritime law, the charging of a ship's cargo, or the ship herself.

Hypothesis testing (modelling) Using computer models such as spreadsheets to test possible situations e.g. modelling financial breakeven points for a business like an airline or hotel, using 'what if' queries such as changing fuel consumption, etc.

HYD PRO UN Hydraulic Propulsion units (on a ship).

Hyundai Merchant marine Co. Ltd. A major liner/operator and registered in Republic of Korea.

Hz Hertz.

I

I Inboard or insulated vans.
i Insulated containers.
ia IATA containers.
IA Iraqi Airways. The Iraqian National airline and member of IATA.
IAARC International Administrative Aeronautical Radio Conference – ITU term.
IAC Inspection and Agency Corporation (Burma).
IACA International Air Carrier Association – IATA term.
IACF International Air Cargo Forum.
IACS International Association of Classification Societies.
IACSC Improved Airworthiness Communications Steering Committee – IATA term.
IADB (or IDB) Inter-American Development Bank, Washington. Established by 19 Latin-American states (excluding Cuba) and the USA to foster economic development of the Latin-American states through utilisation of non-American financial resources, and to administer certain funds.
IAEA International Atomic Energy Agency.
IAEC International Association of Environmental Co-ordinators (Brussels).
IAF Inflation Adjustment Factor.
IAFS International Association of Flag States.
IAG Infrastructure Action Group – IATA term.
IAIS International Association of Insurance Supervisors.
IAL International Aeradio Ltd. (UK).
IALA International Association of Lighthouse Authorities.
IAMA International Aeronautical and Maritime Search and Rescue Manual.
IAOPA International Aircraft Owner's and Pilots Association.
IAPA International Airline Passengers Association.
IAPH International Association of Ports and Harbours.
IAPIP International Association for the Protection of Industrial Property.
IARA International Agricultural Research Institute (UN/IBRD).
IAS Institute of Aerospace Sciences (USA term).
IASA International Air/Shipping Association.
IASTA International Air Services Transit Agreement – IATA term.
IATA International Air Transport Association.
IATA Air Waybill An air freight consignment note used on International Air Transport Association services.
IATA Cargo Agent An agent recognised and approved by IATA, which is appointed by a carrier and authorised by the respective carrier to receive shipments, execute his carrier's Air Waybills and to collect charges.
IATA Clearing House An institution founded by IATA in 1947 which settles claims and accounts between IATA member airlines.
IATA low density rule This changed from 6,000 cubic cm per kilo to 5,000 cubic cm per kilo.
IATA Member An Airline which is a member of IATA.
IATAN International Airlines Travel Agent Network (wholly owned Subsidiary of IATA).
IATA rate A specified or constructed Air Freight rate devised in accordance with IATA regulations.
IATA Standard Interline Traffic Agreement The facility amongst IATA members totalling 150 and some 70 non member airlines permitting inter availability of passenger tickets and air waybills.
IATF International Airline Training Fund.
IATP International Airline's Technical Pool – an IATA term.

IAWPR International Association of Water Pollution research.
IB In bond; invoice book; or IBERIA (Líneas Aéeas de España SA) – an IATA Airline Member.
IBA Indonesian-British Association.
IBAP Intervention Board for Agricultural Produce. UK organisation responsible for collecting export CAP charges and paying refunds; also responsible for CAP import accounting and issuing all CAP licences.
IBC Intermediate Bulk Container or International Broadcasting Co. Ltd.
IBCC International Bureau of Chambers of Commerce.
IBC Code International Bulk Chemical Code – Lloyd's Register of Shipping term.
IBEC International Basic Economic Co-operation. American organisation for the investigation of investment possibilities in the developing countries.
Iberian Airways The Spanish National airline and member of IATA.
IBIA International Bunker Industry Association.
Ibid (short for ibidem) (Latin)
In the same place i.e. the place previously mentioned.
Ibidem (Latin) In the same place i.e. the place previously mentioned.
IBM International Business Machines Ltd.
IBM Code International Safety Management Code – Lloyd's Register of Shipping term.
IBNR Incurred but not enough reported.
IBO Institutional Buy Out – a corporate vendor buying the business.
IBRA Insurance Brokers (Registration) Act 1977.
IBRC Insurance Brokers Registration Council.
IBRD International Bank for Reconstruction and Development.
IBRS International Business Reply Service.
IBS International Business Systems.
IBTS Industry Baggage Tracing System – IATA term.
ic Inland container.
IC Indian Airlines. The Indian National airline and member of IATA; Integrated Circuit or Import Certificate.
ICA International Co-operation Administration.
ICAA International civil airports Association.
ICAC International Cotton Advisory Committee, Washington, exchange providing information about cotton.
I-Ca/cM Import Customs Account Manager.
ICAHG International Collaboration Ad Hoc Group (re communications satellites) – IATA term.
ICAO International Civil Aviation Organisation. A specialised agency of the United Nations, with headquarters in Montreal. Its task is to promote general development of civil aviation (e.g. aircraft design and operation, safety procedures, contractual agreements).
ICAs International Commodity Agreements (UNCTAD).
I-CASys Import Customs Authority System.
ICB International Container Bureau based in Paris.
IC & C Invoice cost and charges.
ICC International Chamber of Commerce. A Paris-based international forum that aims to facilitate trade; Institute Cargo Clauses. There are basic sets of these clauses (A, B and C). The A clauses cover 'all risks' subject to specified exclusions. The B and C clauses cover specified 'risks' subject to specified exclusions; Integrated Computer Control; Interline Communications Committee (ATA/IATA) or (US) Interstate Commerce Commission. The US governmental body to regulate interstate trade.

ICCAIA International Co-ordinating Council of Aerospace Industries Association – IATA Term.
ICCGB Italian Chamber of Commerce in Great Britain; or Indian Chamber of Commerce in Great Britain.
ICCH International Commodities Clearing House. See separate entry.
ICCICA Interim Co-ordinating Committee for International Commodity Arrangements.
ICCL International Council of Cruise Lines.
ICCO International Council of Containership Operators based in London.
ICC Rules of Arbitration and conciliation The rules of the Arbitration of the International Chamber of Commerce operative from 1st January, 1988 and embracing 35 articles including definitions; multiple parties; challenge of arbitrators; place of arbitration; new claims and notification. They are found in ICC publication No 581.
ICD Inland clearance/container depot: placed where cargo (primarily containerised) cleared through customs. Overall, a place approved by HM Customs and Excise to which goods imported in containers/vehicles may be removed for entry, examination and clearance, and at which goods intended for export in container/vehicles may be made available for export control before being moved to a place of export.
ICDV Import Certificate and Delivery Verification.
ICE Institution of Civil Engineering.
ICE A Ice strengthened class – ship classification term.
Iceberg principle Psychological term suggesting that most wants and desires are hidden.
Ice Breaker A vessel specially strengthened and constructed to break up ice in order to open up a navigable channel for other ships to use. There are two main types: Polar and Baltic; or process of initiating discussion amongst strangers.
Ice class ships See ice class tanker entry.
Ice class tanker A tanker vessel specially strengthened and constructed and classified to operate/trade in maritime designated ice zones.
Ice clause A clause found in a Charter Party whose purpose is to prevent a situation in which the shipowners and masters are left with no choice but to attempt to proceed to the contractual destination irrespective of ice conditions.
Ice dues Dues raised by a port authority on ships using a port. An example arises involving all ships calling at Danish ports during the period December, 12th to March 31st.
Icelandair The National airline of Iceland and member of IATA.
Icelandic Crown (Krona) The currency of Iceland.
Iceloadcon The code name of a BIMCO approved charter party clause relative to 'ice' and entitled 'Baltic Conference Ice Clause, 1938'.
ICE LRZ etc. Ice strengthening class with previous or existing society – ship classification term.
ICEM Inter-Governmental Committee for European Migration.
ICEMA Nv Tot Keuring Van Elecgtrotechnische Materialen. A national standards body – See ISO.
ICES International Council for the Exploration of the Sea.
ICET IATA Capacity Enhancement Team (IATA term Europe).
ICETT Industrial Council for Educational and Training Technology (UK).
ICFC Industrial & Commercial Finance Corporation.
ICFTU International Confederation of free Trade Unions.
ICHCA International Cargo Handling Co-ordination Association.

ICIA International Credit Insurance Association.
ICIE International Centre for Industry and the Environment.
ICITO Interim Commission of the International Trade Organisation.
ICJ International Court of Justice (based in the Hague).
ICL International Computers Ltd.
ICM International Consignment Message.
ICN Intrastat Combined Nomenclature – a customs term. See also Intrastat.
ICO International Coffee Organisation.
ICOM International Council of Museums which is based in Paris.
ICOTAS International Committee on the Organisation of Traffic at Sea.
ICPEMC International Commission for Protection against Environmental Mutagens and Carcinogens (UN).
ICPL International Committee of Passenger Lines.
ICR Intercooled and recuperated.
ICRC International Committee of the Red Cross.
ICS International Chamber of Shipping or Institute of Chartered Shipbrokers.
ICSU International Council of Scientific Unions – IATA term.
ICU Implementation Co-ordination Unit (Malaysia).
ICVA International Council of Voluntary Agencies.
ICWP Inter-conference working party.
ID Import Duty or induced draft.
IDA International Development Association. Specialised agency affiliated to IBRD or Import Duty Act.
IDB Inter American Development Bank. See IADB entry or Integrated Data Bank.
ID Card Identity card. A card identifying the holder, usually featuring the holder's name, address and photograph.
IDD International Direct Debits.
IDE Inland depot equipment.
IDEC Interline Data Exchange Centre.
Idem (Latin) The same. As before mentioned.
Identity card A card identifying the holder, usually featuring the holder's name, address and photograph.
IDF Inter-Departmental Flexibility (Ratings).
IDG Instructions for Despatch of Goods form.
Idle Time A period when a product/services/resources is not used.
Idle tonnage Ships 'laid up' due to lack of employment.
IDM International Direct Marketing.
IDPs International Driving Permits.
IDRC International Development Research Centre.
I & E Office Imports & Exports Office.
IE Solomon Airlines. A National airline and member of IATA; or Republic of Ireland – customs abbreviation.
i.e. (id est) (Latin) That is.
IEA International Energy Agency examining world wide energy problems; or International Economic Association based in Paris dealing with economic science.
IEAT Industrial Estates Authority of Thailand.
IEC International Electro Technical Committee.
I – EDI Interactive Electronic Data Interchange – an EDI research organisation of relevance to businesses which need fast response times.
IEE Institute of Electrical Engineers.
IEEIE Institution of Electrical and Electronics Incorporated Engineers.
IEM Industrial Environmental Management Programme (Thailand).
IES (CHIEF) Import Export System – UK custom abbreviation.
IFA Irish Farmers Association or International federation of Airworthiness.

IFAD International Fund for Agricultural Development.
IFALPA International Federation of Air Line Pilot Associations – IATA term.
IFAP International Federation of Agricultural Producers.
IFAPA International Foundation of Airline Passenger Associations – IATA Term.
IFATCA International Federation of Air Traffic Controllers Associations – IATA term.
IFBP In-Flight Broadcast Procedure – IATA term.
IFC International Finance Corporation. Specialised agency affiliated to IBRD.
IFCT Industrial Finance Corporation of Thailand.
IFIA International Federation of Inspection.
IFO Intermediate fuel oil.
IFPMA International Federation of Pharmaceutical Manufacturers.
IFR Instrument Flying Rating.
IFRB International Frequency Registration Board – ITU term.
IFTMFR International Forwarding and Transport message.
IG Meridiana (Meridiana S.p.A.) – an IATA Airline.
IGC Code International Liquefied Gas Carrier Code.
Igloo A bottomless shell made of fibreglass, metal or other suitable material; its shape conforms to aircraft cargo compartment contours. It covers the maximum usable area of an aircraft pallet to which it is secured during flight.
Igniters In general, any device, chemical, electrical, or mechanical used to ignite something.
Ignorantia juris neminem excusat (Latin) Ignorance of the law is no excuse.
IGS Inert Gas System.
IH Falcon Aviation AB – an IATA associate Airline Member.
IHO International Hydrographic Organisation.
ihp Indicated horse power.
II Illegal immigrant; or Business Air Ltd. – an IATA Airline Member.
IIAP Insurance Institute for Asia and the Pacific.
IIC Institute of Insurance Consultants.
IICL Institute of International Container Lessors.
IIEC International Institute for Energy Conservation.
IIESWG Interline Information Exchange Standards Working Group – IATA term.
III (Triple Eye Group) Implications of Infrastructure Investment or Insurance Information Institute.
IIL The Insurance Institute of London
IITA International Institute for Tropical Agriculture (IARI).
IITC Insurance Industry Training council
IIWG International Industry Working Group (IATA) – IATA term.
IKAPI Indonesian Publishers Association.
I/L Import licence.
ILA International Longshoremen's Association (based in New York) or International Law Association.
Illegal immigrant A person who has entered a country illegally.
Illiquidity The inability to service debt and redeem or reschedule liabilities when they mature and the inability to exchange other assets for cash.
ILO International Labour Office.
ILOA Industrial Life Offices Association.
ILOC Irrevocable Letter of Credit.
ILRAD International Laboratory for Research on Animal Diseases (IARI).
ILS Instrument landing system – an air navigation system which is being replaced by the microwave landing system or Inventory Locator Service.

ILU Institute of London Underwriters.
ILWU International Longshoremen's and Warehousemen's Union (based in USA).
IMA Industry Monetary Affairs – IATA term.
Image See Brand Image entry.
Image based advertising The projection of the company/entity image in advertising terms. See image projection entry.
Image projection The projection of the company/entity image in the market place. This can be greatly facilitated by skilful advertising/press relations techniques and market acceptance of the Company logo. A good public image of the Company in the market place favours development of its products and aids sales.
IMB International Maritime Bureau.
Imbalanced Working A transport term involving the operation of transport units between two points inasmuch that in one direction a greater number of services operate in comparison with the return direction. Hence, there may be six services north bound and only five south bound
Imbalance of trade An international trade term which arises when trade between two countries is such that the value of exports exceeds substantially the value of imports or vice versa.
IMC International Maritime Committee (based in Antwerp); or International Materials Conference. The latter based in Washington aims to investigate the possibilities of increased production of raw materials which are scarce and secure their effective distribution; Information Management Committee – IATA term or Internal Management Committee (IATA).
IMD Information Management Division.
IMDG International Maritime Dangerous Goods (Code).
IMDGC International Maritime Dangerous Goods Code.
IMDG classification The International Maritime Dangerous Goods code classification spanning nine classes.
IMDG Code International Maritime Dangerous Goods Code. The IMO recommendations for the carriage of dangerous goods by sea.
IME Institute of Marine Engineers.
I Mech E Institute of Mechanical Engineers.
IMF International Monetary Fund.
IMFA International Maritime Forwarding Association.
IMG International movement of goods.
IMGS International Medical Guide for Ships.
IMIF International Maritime Industries Forum.
Immediate rebate A rebate system whereby the shipper is granted an immediate rebate/discount on his freight rate invoice at the time it is billed, settled. Hence, an €100 invoice subject to an immediate rebate of ten per cent would involve a €90 payment.
IMO International Maritime Organisation or International Money Order.
IMO resolution A890 See principles of safe manning.
IMO – Vega data base Developed jointly by IMO and DNV it embraces the fullest details on ship specification including year of build, ship type, ship size, cargo, trade area and flag. It also includes historical data including regulations which have been superseded – such data very essential in port State control.
IMP Information Message Procedures – Computer term (IATA term).
IMPA International Maritime Pilots Association or International Marine Purchasing Association.
Implied Agency A situation/circumstance whereby a person receives implied authority to act for another.
Implied condition A condition which does not appear in the policy but which is understood to be incorporated

therein and is equally binding as though it were expressed in the policy. Breach of an implied condition by the assured entitles the underwriter to avoid the contract from inception – insurance definition.

Implied term of a contract A term which is not actually agreed to but which the law deems has been agreed to. It may for example arise by Statute, by the Courts, or from previous dealings.

Implied Warranty A warranty that is not expressed in a policy, but which is implied to be therein, by law (e.g. warranty of seaworthiness in a voyage policy, warranty of legality in all policies). Overall, it is understood by both parties to be incorporated in the contract. An implied warranty must be strictly complied with and in the event of a breach of warranty the insurer is discharged from liability as from the breach of the date, but the insurer may waive the breach or the breach may be waived by Statute.

Import berth A berth at a seaport handling/accommodating imported cargo.

Import clearance The process of presenting goods to customs at the seaport, airport, inland clearance depot or importer's premises in the destination country and subsequently cleared by Customs Officers. This may involve payment of import duty. The goods may be subject to the inwards processing relief process. Once the goods have been cleared the goods will be released by Customs Officers and handed over to the consignee or other specified party and placed in free circulation.

Import clerk A clerical position in a company/entity engaged on import work which includes, customs, documentation, distribution, insurance, financial settlement and so on. The clerk may be in a Freight Forwarders or Agents/Receivers Office, or retail manufacturing company and so on.

Import clinic A form of import consultancy service whereby importers with specific problems seek advice from an experienced professional import consultant.

Import country Entry of goods from an overseas territory or relative to the European Community from a third country.

Import cover An insurance cover for an imported consignment.

Import Documentation supervisor A senior clerical position in a Manufacturing/Freight Forwarders/Agents office responsible for the general supervision of an office – or a section of an office – involved in the processing of import documentation for an international consignment. In particular it will involve customs clearance, onward distribution, finance settlement and so on. The paramount factor will be the need to have absolute accuracy of the completion and presentation of the documents which are likely to be computerised. The role/responsibility of the job will vary by Company.

Import duties Customs duties and charts having an effect equivalent to customs duties payable in the importation of goods, and agricultural levies and other import charges introduced under the common agricultural policy or under the specific arrangements applicable to certain goods resulting in the processing of agricultural products.

Import duty Revenue collected by customs at the time of importation from duties imposed on various classes of goods. The term 'import duty' does not include excise duty (see also customs duty).

Importer The entity/person/buyer/trader of merchandise from an overseas market.

Import facilitation organisation A wide range of organisations of differing activities which exist to aid and facilitate the development of imports for both existing and potential importers such as the British Importers Confederation.

Import Groupage Operator A transport/freight forwarder/agent involved in the development and carriage of imported groupage merchandise. See also groupage/consolidation entries.

Importing The process of buying goods from an overseas market.

Import led economy A country where the volume of imports exceeds the level of exports thereby creating a trade deficit.

Import Licences Import licences are broadly of two kinds: (i) An Open General Import Licence which allows anyone to import any of the goods to which it applies. Importers of goods covered by an individual Open General Import Licence do not have to present an import licence to customs officials at the port of entry; and (ii) Individual import licences which must be obtained for any goods not covered by an open general licence. These are issued to a named importer and there are two types. See individual import licences entries.

Import loan A loan provided to an importer to fund an overseas transaction embracing the procurement of goods. It enables the buyer sometimes to negotiate better terms with the supplier and permits the importer to structure an effective borrowing schedule and match expected cash flows.

Import logistics The process of adopting a logistics strategy for importing products. See inbound logistics entry.

Import logistics and outbound distribution A logistic term. Its objective is to manage an inbound supply of products to an exacting schedule and distribute the finished products to prescribed destinations. It may be foodstuffs, car components, white furniture, household furniture or industrial components.

Import malaise strategy See procurement malaise strategies.

Import Manager A senior managerial position in a company/entity which may be engaged in the retail business, manufacturing and so on. The prime task is to ensure most satisfactory arrangements are concluded and undertaken regarding the import of products. This will include documentation, finance, insurance, customs, distribution, and all the elements involved in the processing of an import consignment from the time the sale has been agreed until the goods arrive and financial settlement undertaken. The job specification will vary by company/entity and circumstances.

Import penetration A situation whereby imported products are eroding the domestic market products such as Australian wine being bought in preference to locally produced French wine.

Import plan The process of formulating a viable procurement plan to buy goods/services in overseas markets. See successful procurement strategy.

Import price The price of an imported product/service. See Incoterms 2000.

Import quota A quota system introduced by a government limiting the quantity of goods of a specific nature which are permitted to be imported. Such a policy is designed to protect the home industry and restrict the level of imports thereby reducing the external trade balance.

Import regulations The statutory regulations imposed by government regarding the import of merchandise from a country involving the product crossing an international boundary to enter the country of importation. See import tariff.

Import Sales Executive A managerial position in a company/entity engaged in Freight Forwarding/ Agents, Transport operation – particularly air Freight. The prime task is to develop/secure more import business for the company/entity. The job requires not only a good knowledge and experience of international selling promotion techniques, but also comprehension of import operations including the commercial and operations aspects.

Import seminar A group of students/delegates/trainees who meet under the direction of a tutor to have systematic study relative to an import topic.

Import shed Covered accommodation usually situated at an airport or seaport provided for the handling/ processing through customs of imported cargoes/ consignments for subsequent distribution.

Import strategy The process of devising a viable procurement strategy. See successful procurement strategy, procurement malaise strategies and strategy.

Import substitution A policy adopted by a country whereby it is government/business policy to develop domestic production/manufacture of a product rather than rely on the imported product.

Import tariff A government statutory tariff that is raised on imported goods and often termed customs tariff. See tariffs entry.

Import tariff duty A customs duty raised on specified imported goods. See customs duties entry.

Import workshop A discussion group focusing on a specific import topic comprising personnel usually from different companies. The group is led by a skilled import professional and may undertake the evaluation of a comprehensive case study involving the formulation of a derived study.

Impounding The process of pumping water to the required height into an enclosed dock.

IMRO Investment Management Regulatory Organisation.

IMRS Industrial Market Research Services (Thailand).

IMS Institute of Manpower Studies.

IMSL IMSL Shipping Limited (Hong Kong) – a shipowner.

IMT Immediate money transfer.

IMTA International Maritime Transit Association (now renamed Interferry).

IMTEG Working Group on EUR ILS/MLS Transition (EANPG term of IMO).

IMV Internal Motor Vehicle.

In Insulated or refrigerated capacity in cubic feet (of container).

IN Indonesian Register of Shipping – the Indonesian ship classification society or Macedonian Airlines – MAT – IATA member Airline.

In a deliverable state A legal term indicating the goods must be in such a condition to be able to be handed over. Hence, an engine weighing 100 tons was sold, but was not in a deliverable state as it was concreted to the floor and cost €450 to dismantle it.

Inaugural Institution of a new airline route or equipment, often with free tickets for travel agents, along with free accommodations.

Inaugural cruise The first cruise offered to the market for a particular vessel which may be from Southampton to a range of Mediterranean ports.

Inaugural flight The operation of (i) an air service over an entirely new route; or (ii) the operation of an air service over an existing route with an extension at either the origin and/or destination points; or (iii) the operation of an air service to a new intermediate point; or (iv) the operation of a new type of aircraft different from that previously operated by the carrier on the route. An IATA definition.

Inaugural sailing The operation of (i) a shipping service over an entirely new route-trade; or (ii) the operation of a shipping service on an existing route with an extension at either the origin and/or destination points; or (iii) the operation of a shipping service to a new intermediate point; or (iv) the operation of a new type of vessel different from that previously operated by the carrier on the route.

Inaugural schedule The operation of (i) a service for a particular transport mode over an entirely new route; or (ii) the operation of a service for a particular transport mode over an existing route with an extension at either the origin and/or destination points; or (iii) the operation of a service for a particular transport mode to a new intermediate point; or (iv) the operation of a new type of transport unit for a particular transport mode from that previously operated by the carrier on the route.

Inaugural transit The operation of (i) a transit for a particular consignment/goods involving a specified transport mode over an entirely new route; or (ii) the operation of a transit for a particular consignment(s)/ goods involving a specified transport mode over an existing route with an extension at either the origin and/or destination points; or (iii) the operation of a transit for a particular consignment/goods involving a specified transport mode to a new intermediate point; or (iv) the operation of a transit for a particular consignment/goods involving a new type of transport unit for a particular transport mode from that previously operated by the carrier on the route.

In Bond Goods liable to Customs Duty placed in a Bonded warehouse under Customs surveillance or in transit under Customs seal.

Inbound agent or operator A ground operator specialising in serving incoming visitors, particularly those from foreign countries.

Inbound logistics The process of adopting a logistic strategy for importing products. This involves receiving goods.

In-built acceptability Usually associated with a product(s) or service(s) whereby the user/customer expects it to have specific qualities in terms of its quality, durability, performance, reliability and so on.

In camera (Latin) In secret.

Ince Insurance.

Incendive spark A spark of sufficient temperature and energy to ignite a flammable gas mixed with the right proportion of air.

Incentive Commission A form of commission which is devised to maximise sales results by offering a commission geared to a graduating level of sales. For example, it may be 'X' per cent on €10,000 and 'X + Y' per cent on €15,000 gross turn-over per month.

Incentive contracting Incentive contracting objective is to motivate the contractor to reduce contract performance. Overall, the technique is to offer him the chance to earn increasing profit by sharing with him any savings in relation to an agreed cost, but to reduce his profit, if costs exceed the target.

Incentive marketing A marketing initiative designed to stimulate sales. It may involved participation in competition with generous prizes, vouchers/coupons providing good discounts on the buyer's next purchases: three products for the price of two sales strategy; extended warranty/guarantee; and small gifts. Such strategies are growing in the retail markets of developed and developing countries.

Incentive travel A marketing initiate designed to stimulate travel.

Inception date The date when a risk commences – insurance term.

Inchmaree Clause A clause in the hull policy extending the perils to include negligence of master and crew and other additional perils.

Incident The generic term incident is used to include accidents and near-misses. An insurance definition.

Incidental charges Those charges which do not form part of the bulk of a billed account such as telephone, fax, etc. An example arises through the rendering of the total transit cost of an international consignment where the bulk of the cost may arise through transportation charges, customs, packaging, whilst the incidental charges are those incurred in sending facsimiles, e-mail, making telephone calls and any documentation charges.

Incidental non-marine Non-marine insurance written by a marine underwriter as an adjunct to his marine insurances account.

Incl Including.

Inclad Including address.

Included angle The angle subtended by the two legs of a two leg sling or by the diagonally opposite legs of a four leg sling provided the two legs in question are symmetrical about a vertical centre line.

Inclusive resort A property that prices itself to include the room, meals and amenities in a single package rate; also called all-inclusive resort.

Inclusive tour excursion (ITX) British and Continental usage for inclusive tour fare.

Inclusive tours A product retailed by the Travel Trade such as a shipowner, airline, travel agent, etc., which includes not only the travel arrangements but also the hotel accommodation and sometimes optional excursions all quoted in the inclusive price.

Income bonds Securities, the interest on which is payable only out of profits.

Income Budget The process of formulating forecast and objectives relative to income during a specified future period for a particular area of the business activity or overall company results.

Income effect An income effect takes account of the fact that a change in price of a good you buy changes the purchasing power of your income (i.e. on real income). An increase in real income, brought about by a fall in price of a good, will usually result in an increase in the quantity of the good purchased. If, however, an increase in real income tends to decrease the quantity bought, then the good is an inferior good.

Incompatible Describing dangerous goods which, if mixed, would be liable to cause a dangerous evolution of heat or gas or produce a corrosive substance.

Inconvertibility The failure of the buyer's country to transfer foreign exchange as a result of economic, financial or political difficulties. Trade finance term.

INCO Terms 2000 International rules drawn up by the International Chamber of Commerce relative the delivery trade terms of international consignments and as specified in the export sales contract which is usually evidence in the export invoice. Overall, there are thirteen terms and include EXW, FCA, FAS, FOB, CPT, CIP, CFR, CIF, DAF, DES, DEQ, DDU and DDP. See separate entries

Increased Value Clause (Cargo) A clause in a cargo policy which provides that, where both an underlying policy and an increase value policy cover the same goods for the same transit, the insured values expressed in both policies shall be aggregated; so that the sum insured by each policy shall be deemed to be part of the whole. The effect is that, where both policies settle a claim, any recoveries in respect of that claim are shared between the underwriters subscribing both.

In curia (Latin) In open court.

Incurred but not reported Amount set aside for claims that are anticipated to have occurred but which have not yet been advised to the Company. The accounting code definition states that as a reserve it also represents any general reserve for inflation. For direct business written within the UK it normally constitutes the average number of claims which are reported late X the average claims costs. London Market Business, however, normally represents the difference between known case reserves and the ultimate loss ratios determined by the Company. An insurance term

Ind Independent.

Indefinite laytime This arises where the laytime depends upon the custom of the port at which the cargo operations are being performed, or on the speed at which the ship can load or discharge or both.

Indemnity Compensation for loss/damage or injury. The insured who has sustained a loss is restored to the same financial position (as far as possible) that he enjoyed immediately before the loss. Hence, an insurance principle by which the policy holder shall be put in the same financial position after a loss as he was in immediately before it, or a document usually countersigned by a bank indemnifying the shipowner/ carrier against all consequences arising from certain acts e.g. the issue of a duplicate set of bills of lading or the release of cargo without the production of bills of lading.

Indent Agent Also called a Buying agent.

Indent House An indirect trading technique whereby numerous large retail organisations in any country establish offices in other countries in order collectively to purchase from manufacturers in those territories. The extension of the system relates to the establishment of independent houses who buy on behalf of well known foreign 'department stores'. An ideal outlet for manufacturers of consumer products and often referred to as Buying Houses.

Independent contractor An agent who arranges travel for clients; may sell a supplier's products directly or through an agreement with an agency.

Independent hotel One not affiliated with a chain or group.

Independent Lines Shipping companies which operate outside shipping conferences and regarded as non-conference lines.

Independent wire rope core (IWRC) A type of construction of wire rope.

In depth discussion A lengthy discussion by a group of people on a single topic featuring all the constituent elements of the subject area. This may involve the merits of increasing the process of a range of products in a competitive market discussed by the sales force/ Agents/Distributors and market research personnel. The discussion can extend to a group of items.

In depth study The process of evaluating/studying a particular situation/area/problem in depth involving usually some degree of research and enquiry. For example, the reason for the failure of a new product launch may result in the company/entity initiating an in-depth study/enquiry embracing all the aspects including design, price, specification, late delivery causes, faulty equipment, inadequate after sales services, distribution arrangements, unsuccessful advertising and so on.

INDIAFRICA Indiafrica Line – a shipowner.
Indian Airways The national Indian airline and member of IATA.
Indian Chamber of Commerce for Great Britain The organisation based in London representing India whose basic role is to develop, promote and facilitate international trade between Great Britain and India.
Indian Register of Shipping The Indian register of Shipping Classification Society and based in Bombay, India.
Indian Rupee The currency of India. See Appendix 'C'.
Indian Rupee invoice An invoice payable in Indian rupees relative to an export sales invoice.
Indifference curve map A map of an individual's utility function showing a consumer is indifferent, between any collection of the two goods represented by a point on a single curve. With any point on a higher curve being preferred to any point on a lower curve.
Indigenous administration An administration which is comprised of local nationals of the country and were born in the country in which they are now resident and work.
Indigenous capital Capital which is subscribed from within the country/region/market place and thereby originates from that country. This must be compared with inwards investment involving capital coming from outside the country and in effect is 'imported capital'.
Indigenous Company A company which is a local or home market company and has arisen through local organic growth in terms of capital, labour, product and so on. It contains no inwards investment.
Indigenous labour force A labour force which forms the 'nationals' of the country and were born in the country in which they are now resident and work. This compares with an expatriate labour force involving nationals working overseas in another country/state.
Indigenous market A market which is basically the home market which responds to a locally produced product.
Indigenous production Production undertaken from within the country/region/market place and thereby originates from that country.
Indirect air carrier A charter tour operator, an agent or other operator who may contract for charter space from an airline for resale.
Indirect damage Damage caused by an insured peril but not proximately caused thereby. An insurance definition.
Indirect exporting The process of exporting a product/service through a third party into an overseas territory. Examples include Export Houses, Distributors, Agents and so on. Hence, the export of goods is not handled by the original supplier.
Indirect export trading (sales abroad) In international trade export terms, the process of a manufacturer wishing to sell his goods/services abroad and in so doing not become directly involved in contacting overseas buyers and associate product distribution arrangements but to rely on the use of external and shipping departments. Basically there are seven differing forms of assistance available including: export management companies; using merchants who buy and sell as principals in their own right; manufacturers' export agents; buying or indent houses; and confirming houses. See also separate entries.
Indirect marketing The process of selling a product/service through an intermediary and not direct to the end user. The intermediary may be an Agent, Distributor, Export House, franchises and so on.
Indirect port related job A hybrid post at a seaport such as a maritime lawyer, shipbroker and so on.

Indirect procurement The process of the buyer outsourcing its procurement through a third party.
Indirect route Any route between two points other than the direct route
Individual import licences There are two types of individual import licence; (i) an open individual licence which allows the import of a specified commodity with restriction on the quantity or value. An open individual licence is usually valid for twelve months but can be valid for longer or shorter periods in specific cases; and (ii) a specific individual licence which allows the import of a stated quantity or value of a specified commodity from a specified source. A specific licence is valid only for a specified time.
Individual Licence Licence issued for one particular import/export consignment. See individual import licences entries.
Indivisible load Consignment shipped as one complete unit e.g. transformer as distinct from it being knocked down into various units for shipment.
Indoor Sales Manager A managerial post responsible for the Administration of the sales force especially the processing of enquires and correspondence matters. In particular to process quotations and associated matters up to the point of sale. When the sale has been made, subsequent matters arising are usually processed by the After Sales Manager. See After Sales Manager entry.
INDO/SIN Indonesia/Singapore feeder (shipping service-fully containerised).
INDORIENT PT Indonesian Oriental Line – a shipowner.
Indorsee The person to whom a bill of exchange, promissory note, bill of lading etc. is assigned by endorsement, giving him the right to sue thereon.
Indorsement The process of a person indorsing a document. For example, it may be a Bill of Exchange involving the following types of indorsement: indorsement in blank involving a simple signature of the indorser; special indorsement where the indorser does not name an indorsee; conditional indorsement and restrictive indorsement.
INDPRO Indian Committee/Organisation on Trade Procedures and Facilitation based in New Delhi.
Industrial action The process of a person/work force going on strike, working to rule refraining from doing overtime etc. due to some grievance with the Management of the company/entity or some other specific reason.
Industrial centre An area region where industry is concentrated.
Industrial consumption goods
Products consumed in the manufacturing process including coal, gas, oil, chemicals, lubricants, stationery, timber, office supplies, metals and other raw materials.
Industrial co-operation The process of two parties agreeing to co-operate in industrial issues. It may be governments, companies or economic/trading blocs. Overall it may involve industrial research, collaborative agreements, exchange of personnel. An example is a counter trade technique. It arises when a plant, factory or licence is sold to country B. Long-term finance is secured in Country A. Products connected with the technology transferred are accepted over an agreed period – usually a number of years – by country A. The sale of these goods provides the means for repayment of the long-term finance.
Industrialised country A country whose economy is based primarily on industry involving the manufacture of goods.

Industrial demand The market demand for an industrial product or service. It may be the local factory wishing to avail themselves of a local firm industrial floor cleaning service, or the road haulage company wishing to purchase more lorries.

Industrial design Ornamental or aesthetic aspect of a functional article.

Industrial dispute A confrontation between Management and Union(s) representing the work force resulting in a strike, work to rule, ban on overtime and so on. It jeopardises the productive use of labour resources, inflates cost, reduces production, and overall places at risk the company product/service in the market place. Above all the financial well being of the Company may be undermined and existing orders/contracts placed at risk.

Industrial durables Products provided for the industrial market and classified as replaceable items through age/specification change/new technology such as industrial plant, computers, security equipment, vehicles and office equipment.

Industrial market economy An industrial market driven economy focusing on manufactured goods. See industrialised country.

Industrial Marketing The process of promoting/marketing those manufactured goods provided for the industrial sector, both at home and overseas.

Industrial marketing research The process of conducting market research into industrial products with a view to improving marketing performance. This can embrace industrial consumption goods, industrial durable goods, industrial services and so on.

Industrial products Products which are manufactured and often of high value.

Industrial relations The relationship between Management and Unions – the latter representing the labour force – on the conduct, management and development of the company/entity especially in the role of negotiations and consultation relative to pay and employment conditions, of the labour force.

Industrial Relations Director A most senior Managerial appointment at Directorate level responsible to the company/entity board for all aspects of Industrial Relations policy. This involves primarily the task of formulating and executing Board policy in the field of Industrial relations involving the Union(s) representing the work force in the negotiations/consultation with Board Management. This involves conditions of employment, productivity, annual wage reviews, manning levels and so on. Also called Human Relations Director.

Industrial research The process of undertaking research into industrial products with a view to their development. This may involve a research chemist conducting research into areas of chemistry in laboratory conditions.

Industrial restructuring The process of modernising the industrial base in a country/company and thereby introducing modern technology and work practices.

Industrial services Services provided in the market place for the industrial user such as consultancy, cleaning, maintenance, security and employment agency.

Industrial spirits Refined petroleum fractions with boiling ranges up to 200°C dependent on the use to which they are put – e.g. seed extraction, rubber solvents, perfume, etc.

Industrial targeting When a government establishes a long-term programme of direct and indirect support to a particular industry, usually export-oriented (e.g. semiconductors). A WTO term.

INERC Interim New En-Route Centre – IATA term.

Inert Gas A gas, such as nitrogen, or a mixture of non-flammable gases containing insufficient oxygen to support combustion.

Inert gas system A system of preventing any explosion in the cargo tanks of a tanker by replacing the cargo, as it is pumped out, by an inert gas, often the exhaust of the ship's engine. Gas-freeing must be carried out subsequently if workers have to enter the empty tanks. Overall, use of flue uptake from ship's exhaust gases to provide oxygen-free neutraliser to cargo tanks.

Inertia reel A safety line connected to a lashing worker to prevent falls from container stacks.

Inerting The introduction of inert gas into an aerated tank with the object of attaining an inert condition suited to a safe gassing-up operation, or the introduction on inert gas into a tank after cargo discharge and warming-up with the object of: (a) reducing existing vapour content to a level below which combustion cannot be supported if aeration takes place; (b) reducing existing vapour content to a level suited to gassing up prior to the next cargo and (c) reducing existing vapour content to a level stipulated by local authorities if a special gas-free certificate for hot work is required. Overall, the act of avoiding an explosive atmosphere by replacing the oxygen in an oil tank by an inert gas.

In extenso (Latin) At full length.

In extremis (Latin) At the point of death.

INF Information.

Infant industry An industry in the earliest stages of its development especially in terms of product manufacturing and development whose aim is to secure an adequate viable market share which may be in an environment of keen competition from overseas or at home. In the case of overseas competition tariff barriers are introduced by some nations to protect the infant industry thereby aiding its early market development and support.

Inferior goods A good for which demand falls when its consumers' incomes rise.

Infini Japan-based computerised reservations system for agents owned by All Nippon and Abacus; has ties to Worldspan.

Inflatable dunnage Flexible bags positioned within the stow and inflated so that movement of cargo might be prevented. The air filled bags are placed between units of cargo to prevent damage of goods in transit.

Inflatable life-rafts Life saving apparatus

Inflation The process of constantly rising prices resulting in diminishing purchasing power of a given nominal sum of money. This may be due to excessive wage awards, lower exchange rates causing imports to cost more, increased prices, available money supply, poor productivity in high labour cost intensive industries resulting in higher prices, overall lack of investment and so on. The extent to which the foregoing items and others will depend on an individual economy and government strategy.

In flight catering The provision of catering resources during the course of a flight to passengers.

In flight entertainment The provision of entertainment resources during the course of a flight to passengers. This includes films, magazines, pre-recorded music and so on.

In flight facilities The provision of those passenger facilities found on aircraft and available during the course of the flight. These include video and audio facilities, meals, facilities for nursing mothers, duty free products and gifts and so on.

In flight film The process of an aircraft in flight providing films as a form of entertainment to passengers during their journey. The airline may or may not charge the passengers for such entertainment.

In flight information The process of the airline aircraft crew giving information about the flight to passengers. This would include aircraft speed, altitude, expected time of arrival, meals service, duty free facilities, in flight entertainment and so on.

In flight magazine A magazine published/produced specially for 'in flight' passengers. Usually produced on an Airline basis it will feature strongly tourist, etc, information about the National Airline country; in flight duty free tariff, and other items of interest to the airline passenger. It is available free and issued monthly/quarterly, etc.

In flight meals The process of an aircraft in flight providing meals to passengers during their journey.

In flight movie The process of an aircraft in flight providing films as a form of entertainment.

In flight music The process of an aircraft in flight providing music as a form of entertainment to passengers during their journey.

In flight research The process of conducting market research involving interviewing passengers or asking respondents to complete a questionnaire during the flight.

In flight service
Provision of passenger facilities available during the flight on a passenger aircraft such as films, meals, other entertainment, etc.

Info Information.

Information desk A place/counter/office in an airport/seaport/railway/station/hotel where information is disseminated to clients/passengers/guests.

Information flow A diagram which shows, for example, types of information and its flows between functions within an organisation, e.g. customer orders received by a sales function which are passed to an accounts function for processing.

Information system A computer/logistic term. It contains accessible information available to users of the system. It may be financial, economic, company profile, technical, sailing schedule and supply chain information. The quality of the data depends on three factors availability of information, accuracy of information and effectiveness of communication.

Informative advertising Advertising designed to educate and inform rather than openly persuade the potential/existing clients in the market place.

Infra dig. (infra dignitatem) Beneath one's dignity.

Infra (Latin) Below, beneath, after.

Infrastructure In general terms the transport, communications, ports, airports commercial and technology obtaining in a country. In marketing terms the infrastructure of a company is its production network. In the travel industry, the network of highways, water supply, airport, port facilities, lodging, restaurants and all the elements needed to support tourism.

Inherent nature The natural tendency of certain cargoes to suffer from inevitable loss; usually engendered by delay in transit. An example would be fruit which deteriorates naturally during delay. Unless the policy provides otherwise, inherent nature of the subject matter insured is excluded by the MIA, 1906. See inherent vice entry.

Inherent vice A defect or inherent quality of the goods or their packing which of itself may contribute to their deterioration, injury, wastage or final destruction without any negligence or other contributing causes. Examples of those properties of certain goods which lead to their arrival in damaged condition without accident or negligence include unprotected steel will 'weather', bales of rubber stick together, copra is almost invariably infested. It is always excluded by the insurers of the cargo because of its inevitable nature.

In house A product or facility confined to an internal resource within a company or organisation. For example 'In House' magazine would be a publication featuring events and personalities within the Company.

In house documentation Documentation systems within a company.

In house documentation completion Documents completed and filled in relative to an international consignment by the exporter.

In house systems Computer systems within a company.

In house training The process of providing training within a company resources usually under the aegis of a training manager.

Initial boiling point The temperature at which the liquid under test boils first.

Initial commitments Trade liberalising commitments in services which participants are prepared to make early on. WTO term.

Initial margin The initial outlay of money required to open a futures position. The size of this deposit varies from commodity to commodity, but is usually a fixed amount, representing a small percentage (around 10%) of the value of the contract. Overall, the returned collateral required to establish a futures or options position – an International Petroleum Exchange term.

Injunction A legal term involving a court order to a defendant not to do something he has contracted to do. It may be prohibitory (ordering the defendant not to do something) or mandatory (ordering the defendant to do something).

Injuria sine damno (Latin) A breach of a legal right not resulting in damage.

INKINDO Indonesian Consultants Association

Inland clearance depot Customs cargo clearance depot.

Inland marine insurance A broad type of insurance, generally covering articles that may be transported from one place to another as well as bridges, tunnels and other instrumentalities of transportation. It includes goods in transit (generally excepting trans-ocean) as well as numerous 'floater' policies such as personal effects, personal property, jewellery, furs, fine art and others.

Inland marine risk Inland waterway (river, canal etc.) risks but in practice the term is loosely applied to certain non-marine risks placed in the marine market. A marine insurance term.

Inland rail depot Depot where international rail-borne cargo cleared through customs.

Inland Waterway All water areas available for navigation that lies inland of the inland waterway boundary. The inland waterway boundary is taken to be the most seaward point of any estuary which it would be reasonable to bridge or tunnel and in UK conditions this is defined as 'where the width of water surface area is both less than 3 kms at low water, and less than 5 kms, at high water (springs)'. Inland waterways are of two types: Locked Canals/River Navigations and Tidal River Navigations/Estuaries. See separate entries.

Inland Waterways Bill of Lading Transport document made out to a named person, to order or to bearer, signed by the carrier and handed to the sender after receipt of the goods.

Inland Waterway Vessel A vessel which does not normally proceed seaward of the inland waterway boundary. This includes vessels which do not require load lines and legally are allowed to trade anywhere within the Partially Smooth Water Area.

INMARSAT The International Maritime Satellite Organisation.

INN Denotes an international non-proprietary name approved by the World Health Organisation (WHO).

Inner Harbour The innermost landward part of a harbour which is the area which affords the maximum protection from the elements and includes the facilities accommodating the ships berths/quayside, etc.

Inner packing The packing protecting the goods and regarded as the first layer of packing.

Innkeepers' lien Legal right of an innkeeper in some countries to keep the property of a guest for unpaid charges.

INNM Denotes a modified international non-proprietary name approved by the World Health Organisation (WHO).

Innovative financing These embrace elements of non-traditional structures employed, for example, in ship financing transactions with the objective of easing constraints of the financing transactions with the objective of easing constraints of the borrower. It can be structured by using some characteristics of a traditional source of finance such as loans and a non-traditional one such as a lease.

Innovative management techniques A form of management which is unique/novel to the business environment. It may be driven by efficiency, market exposure, technology or financial considerations.

Innovative Strategies A form of strategy which is unique/novel to the market place. See innovative marketing entry.

Innovative marketing A form of marketing which is unique/novel to the market place. It is designed to stimulate sales and provides favourable exposure of the product/service.

INP Input.

In pari delict ((Latin) Of equal terms.

Inpatriate A person who is employed by a company originating from another country.

In personam (Latin) Valid only against a definite person.

In-plant agency A travel agent's sales outlet located on the premises of a company and doing business primarily for that company only; also called a customer-premises agency location.

Input The process of putting information into a system such as data input for a computer.

Input devices These include: keyboard, mouse, digitiser, joystick, bar code reader, MICR, OMR, OCR, voice (speech recognition), scanner, sensor devices, data logger. A computer term.

Input Tax The VAT charged on the goods and services purchased for a business operative within the EU.

INRA International Research associates (Thailand).

In rem (Latin) Valid against all the world.

'Inroads' A commercial term indicating that a particular policy is making headway. An example is found in an advertising campaign whereby increased sales are improving the Company market share and thereby making 'inroads' into the competitors' sales as the latter market share is declining.

INS Integrated navigation system.

Ins Insurance or insulated (refrigerated space on a ship).

INSA International Shipowners Association. Based in Gdynia, it is an Association of shipowning companies from eleven countries primarily in South Western Europe, or Indonesian National Shipowners' Association.

INSC Indonesian National Shippers' Council.

Insert In Marketing terms the process of placing/inserting in a journal/magazine, newspaper etc., advertising merchandise. It may be a brochure insert in a Journal.

Inshore Traffic Zone An area specified between the landward boundary of a maritime traffic separation scheme and the adjacent coast earmarked for coastal traffic.

Inside cabin A cabin/passenger berth usually on a cruise liner with no port hole or picture window.

Inside cross lash A lashing arrangement where the lashings cross but are independent of and do not cross the lashings of adjacent container stacks.

Insider dealing Trading in shares with the benefit of 'inside knowledge' of the company, so as to gain unfair advantage over others. An illegal practice in most countries.

Insolation (from incoming solar radiation) The solar radiation received at any particular area of the Earth's surface, which varies from region to region depending on latitude and weather.

Insourcing The process of buying goods/services such as car components for a local/national car assembly plant. See out sourcing.

In specie (Latin) In its own form.

Inspection Tax A tax raised by open registries/flags of convenience when vessels are inspected by surveyors.

Inspector A managerial post which, for example, in the transport sector is one of a staff supervisor.

INSPIRES The Indian Ship Position & Information Reporting System.

INSPR Inspector.

INSROP International Northern Sea Route Programme.

INST Installed.

Institute Cargo Clauses Standard insurance conditions, published by the Institute of London Underwriters, for policies covering goods in transit overseas.

Institute Cargo Clauses (B) and (C) These clauses specify that loss or damage is only recoverable from insurers in those situations where the loss or damage is reasonably attributable to fire or explosion, vessel being stranded, grounded, sunk, or capsized, overturning or derailment of land conveyance; collision of carrying vessel or vehicle with external objects (other than water); discharge of cargo at port of distress. These cargo clauses were introduced in January 1982 under the approval of the Institute of London Underwriters and replaced the 'old' clauses 'All risks', 'With average' and 'Free of Particular Average' found in the cargo SG policy adopted by Lloyd's in 1779. The new clauses and replacement of the SG policy form were mandatory from March, 1983. See MAR Policy Entry.

Institute Clause A standard clause published by the Institute of London Underwriters.

Institute of Certified Travel Agents (ICTA) A professional organisation in the US concerned with developing and administering educational programmes for travel agents; see Certified Travel Counsellor.

Institute of Export The trade association for personnel involved in exporting especially Export Executives/Managers/Directors.

Institute of London Underwriters An organisation representing the interests of member insurance companies. Although the majority of members are British companies, membership is available to foreign companies. The Institute maintains a close liaison with Lloyd's marine market, and provides facilities for a number of joint committees to operate. Overall, a bureau of insurers transacting marine, aviation and transport insurance in London.

Institute of Marine Engineers Based in London, it is the professional organisation for personnel seeking to become Marine Engineers.

Institute of Travel Managers UK organisation of passenger traffic managers.

Institute of Warranty Limits It defines the maritime geographical area a ship may operate within marine insurance cover.

Institute Time Clauses Standard insurance conditions, published by the Institute of London Underwriters, for policies covering ships for a period of time.

Institute Warranties A set of express warranties, published by the Institute of London Underwriters, for use in hull and machinery policies. The set comprises five locality warranties and one trade warranty; the latter relating to the carriage of Indian coal as cargo. In event of breach of any of the warranties, the assured is held covered by the breach of warranty clause in the Institute hull clauses (see Breach of Warranty Clause).

Institutional investors Mutual funds, banks, insurance companies, pension funds and others that buy and sell stocks and bonds in large volumes.

In store promotion A form of promotion in a retail outlet. It may be a sales video, demonstration of a product or special offers with supporting sales assistance.

INSTRASTAT The EU system for collecting intra community trade statistics.

Instructions for despatch of goods form A document which is presented by the Shipper to the Airline giving the fullest details of the consignment for shipment, consignor, consignee, flight data, etc. This data is used by the Airline to complete the Air Waybill.

Insulated container A refrigerated container particularly suitable for the conveyance of perishable cargoes such as meat/dairy products/fruit/drugs/photographic items.

Insulated tank container Container for holding one or more thermal insulated tanks for liquids.

Insulation Flang An insulating device inserted between metallic flanges, bolts and washers to prevent electrical continuity between pipelines, sections of pipelines, hose strings and loading arms or other equipment.

Insurable interest It is illegal for anyone to insure without an insurable interest or a reasonable expectation of acquiring such interest. One has such interest when his relationship to property at risk may expose him to loss or liability or where he stands to gain by the safety of such property. Hence, the legal right to insure. For a contract of insurance to be valid the (potential) policyholder must have an interest in the insured item to the extent that its loss, death, damage or destruction would cause him loss. This is called insurable interest and must exist as follows: Marine insurance – at the time of the loss. Life assurance – at the time the policy is taken out. Other insurance – at the time the policy is taken out and also at the time of the loss.

Insurable risk An insurable risk exists when one's relationship to the property at risk may expose one to loss or liability or where one stands to gain by the safety of such property. It is illegal for anyone to insure without an insurable interest, or a reasonable expectation of acquiring such interest.

Insurable value The value of the insurable interest which the insured had in the insured occurrence or event. It is the amount to be paid out by the insurer (assuming full insurance) in the event of total loss or destruction of the item insured.

Insurance In legal terms a contract between one person called the insured (or assured) and another called the underwriter or insurer, whereby in return for a premium the underwriter agrees to indemnify the insured against losses covered by the insurance. An exporter will be concerned with a number of types of insurance such as credit insurance and transport insurance.

Insurance agent A representative of an insurer, whether an employee or an independent contractor, who negotiates, effects and sometime services contracts of insurance.

Insurance Broker One who advises persons in their insurance needs and negotiates insurances on their behalf with insurers, exercising professional care and skill in so doing. A registered broker sometimes acts as an insurance agent.

Insurance certificate Proof that an insurance contract for a particular shipment has been concluded.

Insurance contract A contract whereby one party, the insurer, in return for a consideration, the premium, undertakes to indemnify the other party, the insured, against loss upon the happening of a specified event that is contrary to the interest of the insured.

Insured value The insured value of the commodity specified in the marine insurance policy is formulated on the basis to customarily allow a further 10% to be added to the estimated insurable value to arrive at the insured value.

INT Intermediate survey. See separate entry.

Int Interest.

Intangible asset Any asset which does not have a physical identity e.g. good will.

Intankbill 78 Code name of BIMCO approved bill of lading used with shipments under the Tanker voyage charter party.

INTASAFCON III International Tanker Safety Conference.

Intascale International Tanker Nominal Freight Scale A schedule of nominal rates for movements of tankers between world ports. Publication ceased in 1969 and succeeded by Worldscale. See Worldscale.

Int. Co. Initial Course.

Integral A refrigerated container with the machinery built in.

Integrated Carrier Forwarder which uses own aircraft, whether owned or leased, rather than scheduled airlines.

Integrated global community The fusing together of individual countries to produce an integrated global society with common ideology, complete transparency and mission statement but retaining individual cultures.

Integrated oil companies These are the groups with a significant interest in more than one of the functions in the industry, namely exploration and production, refining, transportation and marketing. A fully integrated company is involved in all functions.

Integrated Producer A producer of metal who owns mines, smelters and refineries and sometimes also fabricating plants.

Integrated Tug-Barge This is a high capacity combination of a sea-going barge pushed by a tug, with a rigid, or articulated, connection between the two vessels. The system is widely used in coastwide shipping in North America.

Integration The process of fusing together a number of related compatible parts/elements to produce in transport terms an acceptable marketable service or product. An example arises where in a conurbation area complementary road (bus service) and rail (underground rail services) are available in the common market place. In so doing, each form of transport service is optimising its resources compatible with market demand.

Integration programme The phasing out of MFA restrictions in four stages starting on 1st January, 1995 and ending on 1st January, 2005. A WTO term.

Integrators A large transport operator who either owns or controls all the segments of the transport chain on a door to door basis thereby placing one carrier as liable for the throughout transit. It is particularly evident, in the international/global air express markets involving Fedex, TNT, United Parcels Service and DHL.

Intellectual behaviours The study of intellectual behaviour embracing problem analysis, creativity and judgement. See inter personal behaviours entry.

Intellectual property Ownership of ideas, including literary and artistic works (protected by copyright), inventions (protected by patents), signs for distinguishing goods of an enterprise (protected by trademarks), and other elements of industrial property (see Paris convention). A situation where property rights exist and title can be established. Examples include trade marks, patents, registered brand names, copyright and registered designs on which very stringent legislation exists both nationally and internationally.

Intellectual Property Rights Ownership of ideas, including literary and artistic works (protected by copyright), inventions (protected by patents), signs for distinguishing goods of an enterprise (protected by trademarks) and other elements of industrial property. A WTO definition.

INTELSAT International Satellite Communications.

INTER Intermaritime Ltd. – a shipowner.

Interaction Forces causing tendency for a ship to pull towards/away from other ships/objects when passing too closely, and at too high a speed.

Inter alia (Latin) Amongst other things.

Inter-American Development Bank, Washington Established by 19 Latin-American states (excluding Cuba) and the USA, to foster economic development of the Latin American states through utilisation of non-American financial resources and to administer certain funds.

Inter-American Travel Congress Annual professional meeting on travel and tourism in the Western Hemisphere.

Inter availability In transport terms the facility of a ticket holder to travel by an alternative route which may involve another Company service. It usually arises in an emergency situation.

Interbank funds transfer system (IFTS) A funds transfer system in which most (or all) participants are credit institutions. An ECB term.

INTERBANK (IBRD) See International Bank for Reconstruction and Development entry.

Interbank market The market in which banks lend to one another on a wholesale, unsecured basis.

Interbank Time 80 Code name of BIMCO approved International Association of Independent Tanker Owners Tanker time charter party.

Interbarrier space The space between a primary and a secondary barrier of a cargo containment system, whether or not completely or partially occupied by insulation or other material.

Intercargo It was set up in 1980 to represent shipowners, managers and operators in the dry bulk sector. Overall, it is a forum for the shipping community to discuss issues of mutual interest to act as a platform for the proper development of the dry cargo sector. It works closely on common issues with maritime associations such as BIMCO, ICS and Intertanko.

Interchange Reciprocal exchange of e.g. information between two or more parties.

Interchange container A container allowing the transfer between different types of aircraft or vessels usable on different types of aircraft or vessel whether owned by the same or different companies.

Interchange flight A flight that gives passengers/cargo the benefit of a through service and is operated by two or more carriers from the boarding point to the deplaning point using the same aircraft. An IATA definition.

Intercoa '80' Code name of tanker charter party involving the contract of affreightment. It was formulated by the International Association of Independent Tanker Owners.

Inter coastal Calling ports along the coast.

Interconsec 76 Code name of consecutive voyages charter party for tanker shipments. It was formulated by the International Association of Independent Tanker Owners.

Inter Container An organisation formed in 1967 by eleven European Railway Companies as a means of co-ordinating European railway's activities in the containerisation field, and of developing the technique a type of rail product. Overall, some 29 European Railway Companies and Interfrigo are members.

Intercontinental scheduled service A scheduled transport service operating between two continents such as Europe and North America.

Inter depot transfer The process of transferring by road or rail goods from one depot to another.

Interest payments Interest on loans and dividends from investment.

Interest Policy One underwriting specified subject matter.

Interest rate parity When the level of interest rates from two or more banking sectors are the same. For example, interest rates remain at parity throughout the Euro zone as the European Central Bank sets the interest rate applicable in all States forming the Euro zone.

Interest rate swap An arbitrage transaction between issuers who have access to different money and capital markets. See Arbitrage entry.

Inter-ethnic wealth distribution The distribution of wealth amongst various ethnic groups in a country such as Malaysia embracing Indian, Chinese and Malaya.

Interface A common boundary e.g. the boundary between two systems or two devices e.g. computer produced master and photo-copying machine, or facsimile transmitter. In the travel agency industry, a direct link between an airline reservations system and an agency's computer.

Interface limit line That line which defines the safe distance between the seaward end of the ship ramp landing area and the outer face or the shore ramp.

INTERFRIGO International Railway Company for Refrigerated Transport.

INTERFUND (IMF) International Monetary Fund. Based in Washington, it encourages monetary co-operation, establishes international standards for exchange policy, promotes stable exchange rates among member nations, and makes short term loans and standing credits to members in temporary payment difficulties.

Inter-Governmental Committee for European Migration Based in Geneva, the Organisation assists in the emigration of surplus population of Europe and the transport of displaced persons of European origin.

Intergovernmental Transfers An international financial term indicating governments borrow from and lend to each other in the same way as private individuals and companies.

Interim Co-ordinating Committee for International Commodity Arrangements Based in New York it deals with International Commodity arrangements.

Interim dividend The process of a company granting to shareholders an interim dividend at the half year/six month point, followed by the final dividend at its year end. It is declared by the directors, after they have considered the results of the first six months' trading. See Final dividend entry.

Interim Receipt A receipt given by a carrier pending execution of an Air Waybill (air cargo).

Interline Co-operation among air carriers that permits travellers to fly on two or more lines on the same trip using a single ticket and to check luggage at point of origin for the connecting flight or flights. Mutual agreement between airlines to link their route network.

Interline agreement A contract between two or more Airline carriers to expedite exchange of traffic between the parties to the agreement.

Interline carrier An Airline carrier with whom the carrier has an interline agreement.

Interline connection The transfer of passenger baggage or cargo between flights of different airlines.

Interliner An airline employee travelling on a carrier not his own.

Interline rep An airline salesman who deals with other airlines.

Interlinking mechanism One of the components of the TARGET system. The term is used to designate the infrastructures and the procedures which link domestic RTGS systems in order to process cross-border payments within TARGET. An ECB term.

Inter-market The period between the close of the morning kerb and the opening of the afternoon ring when LME members conduct inter-office dealings by telephone. This period is important because the opening of Comex (qv) takes place during this period.

Intermediary An agent or broker through whom a transaction is arranged between parties. An Insurance term.

Intermediate bulk container A portable receptacle of capacity between $0.45m^3$ and $3.0m^3$ and designed for mechanical handling.

Intermediate container A container with a capacity of up to one tonne. Such containers tend to be privately owned by the shipper.

Intermediate customers A person/company/entity which purchases a product(s)/service(s) for resale who may be a dealer or distributor.

Intermediate survey Part of a special survey carried out at two year intervals on a ship.

Intermodal A transport system which permits the interchange of transport units and unit load devices between different but compatible modes of transport. Examples exist in regard to containers, swap bodies, and covers sea, rail, road and air. Overall, the intermodal transport network permits through dedicated integrated services to be provided.

Intermodal Container A container allowing carriage by different modes of transport i.e. rail, road, sea, and air, etc.

Intermodal hub A facility offering two forms of transport as the hub of a transport network with feeder services integrated with each. An example arises with the seaport and airport of Singapore which are virtually contiguous to each other. The airport and seaport operate on the hub and spoke system. See the hub and spoke entry.

Intermodal packaging A packaging specification for a commodity being despatched under combined transport operation arrangements.

Intermodal transport chain The elements/legs through which a consignment travels and is basically the ensemble of operations from the point of origin to the point of destination involving two or more modes of transport. It extends not only to the carriers involved but also to the forwarding agent, consignor, consignee, packers, handlers and so on.

Intermodal transport law The provision of legislation defining the carriers conditions or liability embracing different modes of transport i.e. road, rail, sea and air involving the provision of a through connecting service usually of an international nature.

Intermodal transport system The provision of through connecting transport service(s) involving different modes of transport, i.e. road, rail, sea, and air, etc.

Internal Command An MS-DOS command that resides within the COMMAND.COM file. A computer term.

Internal disc drive A disc drive that is installed inside the computer.

Internal flight A flight undertaken within the confines/boundaries of a country and therefore not subject to customs and immigration needs controls. See domestic flight entry.

Internal frontier A frontier common to two member states.

Internal marketing The process of a company/entity marketing its products/services within the company. It may be to other employees or clients of other departments, subsidiaries.

Internal marketing audit In marketing terms the process of conducting a comprehensive periodic, systematic independent review of a company's in-house marketing performance with a view to seeking improvements. See Marketing Audit.

Internal motor vehicle A motor vehicle unit termed a tug used for towing trailers within a terminal such as Container Freight Station, Container depot and so on.

Internal support Encompasses any measure which acts to maintain producer prices at levels above those prevailing in international trade; direct payments to producers, including deficiency payments, and input and marketing cost reduction measures available only for agricultural production. A WTO term.

International Agencies Organisations which are non political and have an international focus many of which are under the aegis of the United Nations. These include WHO, IMO, ICC, BIMCO, UNCTAD, etc.

International Agency An Agency with an international focus such as IMO, WHO, UNCTAD, UNCITRAL.

International Aid and Loan bulletin A bulletin issued by a bank or other service which gives brief details of financial aid and loans which have been arranged for the development of projects around the world and which are likely to be of interest to potential exporters.

International Air Carrier Association (IACA) An airline association based in Brussels that primarily promotes the interests of charter airlines.

International Airline Passengers' Association UK organisation concerned with the rights of travellers.

International Airlines Travel Agent Network (IATAN) Wholly owned subsidiary of the International Air Transport Association, responsible for accrediting US travel agents member airlines.

International Air Transport Association (IATA) World trade association of international airlines, proposing rates, conditions of service, safety standards and other elements, and appointing and regulating travel agents who deal in international ticketing and likewise air freight agents appointed on an accredited IATA basis.

International Association for the Protection of Industrial Property Based in Zurich it is a private organisation established in 1897 which promotes the international protection of patents and trademarks. Its members are mostly West European nations, Argentina, Australia, Canada, Colombia, Israel, Japan and the United States.

International Association of Amusement Parks and Attractions (IAAPA) Trade group dealing with promotion of theme parks, zoos, amusement parks, resorts and other facilities.

International Association of Classification Societies An association representing the world's major Classification Societies. Its main objectives are to promote the highest standards in ship safety and the prevention of marine pollution. It was formed in 1968 and based in London. Over 90% of the world's merchant fleet in terms of tonnage is covered by the eleven member standards (classification societies) for hull structure and essential engineering systems, which are updated, applied and monitored on a continuous basis. The major classification societies include Bureau Veritas, Lloyd's Register of Shipping, Germanischer Lloyd and so on.

International Association of Convention and Visitors Bureau (IACVB) An industry association.

International Association of Tour Managers (IATM) A professional society of tour escorts.

International Atomic Energy Agency Based in Vienna, the Agency was established in 1956 to promote the peaceful uses of atomic energy and to administer research programmes. It is a special but independent, agency of the UN. Its membership includes the United States and the CIS, all of Western and Eastern Europe except Ireland, most Latin American Republics and many Asian and African nations.

International bankers draft A cheque drawn by one bank upon another. This method of payment is particularly suitable for non-priority, low value international payments or for those which are to be accompanied by documentation such as invoices.

International Bank for reconstruction and Development The IBRD was established with the IMF in 1945 and because it provides long-term loans in convertible currencies to governments and government institutions for economic development projects, it is widely known as the World Bank. Although it is loosely affiliated with the UN, the Bank's administration and Budget are independent. Total capital resources are in excess of $50 billion. Its membership consists of most of the nations. It is based in Washington.

International Banking Banking transactions conducted involving banks in more than one country.

International Basic Economic Co-operation An American organisation concerned with investigating investment possibilities in developing countries.

International boundaries The boundary which separates one country from another and normally is manned by Customs posts.

International Bulk Chemical Code This provides safety standards for the design, construction, equipment and operation of ships, carrying dangerous chemicals. An additional code, the BCH code is applicable to existing ships before 1st July, 1986. A certificate of fitness in accordance with the provisions of IBC or BCH code is mandatory under the terms of either 983 amendments to SOLAS 1974 or MARPOL 73/78. For National Flag Administrations not signatory to SOLAS 1974, a statement of compliance would be issued by the classification society acting on behalf of the shipowner.

International Bureau of Expositions An organisation that sanctions expositions worldwide.

International Business International Business embraces the movement of goods, services, capital and personnel; transfers of technology, information or data; supervision of employees and so on. It is a field of Management which deals with business activities which cross National boundaries.

International Business environment The environment in which an international company operates – does business – overseas. This involves the economic, political, technical, legal and social aspects of a country as portrayed in the country analysis. See country report entry.

International Business Manager A manager usually at senior level responsible for identifying and developing international businesses which cross national boundaries. See international business entry.

International Business Strategy It is the strategic management processes by which companies/governments evaluate their changing international business environment and respond/shape an appropriate organisational response that involves the crossing of international borders and their cultures.

International Buying The process of purchasing goods/services which cross international boundaries.

International Cargo Handling Co-ordination association Organisation whose aim is to facilitate improved handling techniques in the world transport system.

International carriage The conveyance of traffic in circumstances/situations whereby the place of departure and the place of landing/arrival/destination are situated in different States/Countries. It may be passengers/cargo/parcels, etc.

International cash management The process of managing cash payments across international borders. It is complex as it involves purchases in both world trading currencies and non-convertible currencies. It is a risk area and requires special management skills, hedging techniques or other alternatives.

International Certification Document required in USA for any load with a gross weight exceeding 29,000 lbs before the load is taken on the highway required under the International Safe Container Act

International Chamber of Commerce Based in Paris, its role is to introduce – with its members agreement – standard rules for commercial transactions, settlement of disputes (arbitration), and deals with general business and economic problems. Within such a framework it facilitates the development of international trade under conditions of reliability and business/market confidence.

International Civil Aviation Organisation Based in Montreal, it deals with the problems of international civil aviation and seeks to establish international standard terms and regulations in Civil Aviation.

International Clearing House A clearing house where participating countries/companies permit their accounts to be settled regularly such as monthly.

International Commodities Clearing House Limited Situated in London the role of the Clearing House is to provide for the registration, clearing and settlement of contracts and the collection of deposits and margins. It also acts as an intermediary whenever commodities are tendered in due performance of purchase and sales. Thus the clearing house acts as a banker to the Baltic Exchange and provides a guarantee of performance to its Clearing members who can be confident that contracts will be fulfilled.

International competitiveness The ability of a company/product/service to compete successfully in the international market place.

International Confederation of Free Trade Unions
Since 1949, when the ICFTU was created, the 137 member labour federations and unions have consulted and collaborated with each other in opposition to the Communist labour union federation. It is based in Brussels.

International Conference on Training and Certification of Seafarers A conference convened by IMO in London 1978 which adopted for seafarers revised requirements relating to watch keeping, training certification and continued proficiency. See STCW.

International Congress and Convention Association (ICCA) European trade organisation of convention organisers.

International Consignment Message The standard message structure used to transfer information about the international movement of goods by traders registered under the International Trade Prototypes arrangements.

International Control Centre (ICC) An operator assistance interface between the national and international networks.

International Convention on Arrest of Ships 1999 To take effect it requires ten contracting parties.

International Convention on Load Lines 1966 and its 1988 Protocol See Load Lines

International Convention on Maritime Liens and Mortgages 1993 Not yet in force – requires ten contracting parties – presently five.

International Convention on the Control of Harmful Anti-Fouling Systems on Ships, 2001 (AFS 2001) The International Conference on the Control of Harmful Anti-Fouling Systems for Ships, 2001, was held in London from 1st to 5th October 2001. The Conference adopted the International Convention on the Control of Harmful Anti-Fouling Systems on Ships, 2001 (the AFS Convention), together with four Conference resolutions, relating to the early and effective application of the AFS Convention, approval and test methodologies for anti-fouling systems on ships and the promotion of technical co-operation.

International Convention on the limitation of liability for Maritime Claims A 1976 limitation convention enacted in the UK by the Merchant Shipping Act, 1979.

International Conventions International Conventions set out the minimum terms and conditions to all contracting states which are signatories to it. The International Conventions are normally under the auspices of the United Nations and cover a range of topics/subjects

International Co-operation Administration The American Government organ which administers US aid programmes abroad, economic as well as military. (ICA's regional office for Europe, see AID). It is based in Washington.

International Cotton Advisory Committee Based in Washington it deals with exchange of information about cotton.

International Council for the Exploration of the Sea (Conseil international pour l'Exploration de la Mer) Based in Charlottenlund, Denmark, it initiates and co-ordinates the scientific exploration of the sea in the Eastern Regions of the North Atlantic.

International Court of Justice Based in The Hague, the Court was established in 1945 as an affiliate of the UN, thus superseding the Permanent Court of International Justice that had existed from 1920 under the auspices of the League of Nations. Its membership consists of all members of the UN plus Switzerland. The Court had jurisdiction over private and governmental international litigation, but compliance with its findings is not binding on the member nations. Nevertheless, a large body of international law has evolved from its decisions and its overall contribution to international business has been enormous.

International credit clubs A facility available to help exporters of substantial items of capital equipment, so that overseas buyers can obtain instalment credit finance quickly and cheaply.

International Credit Union An organisation or association of finance houses or banks in different countries such as in Europe. The finance houses or banks have reciprocal arrangements for providing instalment finance. When a buyer in one country wants to pay for imported goods by instalments, the exporter can approach a member of the credit union in his own country which will then arrange for the finance to be provided through a credit union member in the importer's country. The exporter receives immediate repayment without recourse to himself. The buyer obtains instalment finance.

International dateline The line at 180 longitude where, by international agreement, each day begins; eastbound travellers gain a day when they cross the line, westbound lose a day.

International debt Often referred to as foreign debt. The debt incurred by a country or group of countries with an International Bank such as IMF and World Bank. This is particularly onerous with less developed countries – termed the third world debt – whereby in many situations a high percentage of their exports earnings fund the servicing of the nation's debt. Examples of the key variables to measure debt are 'debt service/exports (%) and 'foreign debt/GDP (%)'. The former may be 16.3% and the latter 9.4%.

International Development Association Based in Washington, this institution lends long-term funds in convertible currencies which can be repaid in inconvertible currencies. Such 'soft' loans are extended to governments or to government institutions of developing countries. IDA was established in 1959 as an affiliate of the IBRD, but is separately financed; its total resources are in excess of $1,000 million.

International Direct Debits An electronic method by BACS of collecting regular payments from abroad.

International Direct Marketing The process of the exporter mailing directly to potential clients with the business reply facility or conducting the tele sales technique over the telephone network.

International Donor Agency An international organisation providing funds to an overseas territory to facilitate usually economic and social development. The funding areas targeted would be far ranging and include infrastructure, medicine, education, tourism and so on. Such funds are accredited and involve soft loans. An example includes the World Bank and the International Monetary Fund.

International Electrotechnical Commission Founded in 1906 it is the electrotechnical counterpart of ISO. It comprises of national electrotechnical committees of 43 countries. Standards or reports are approved if not more than 20% of National Committees cast a negative vote. Approximately 1,800 current standards or reports have been produced. Over 80 technical committees and over 120 sub-committees exist. It is governed by IEC Council and its Committee of Action.

IEC special committees:
 ACET – Advisory Committee on Electronics and Telecommunications.

ACOS – Advisory Committee on safety.

CISPR – International Special Committee on radio Interference.

ITCG – Information Technology Co-ordinating Group.

International Environment The environment in which international trade is conducted. This will vary on a country by country basis and embrace the economic, political, social, technical and legal aspects. It will also encompass the infrastructure of the country such as transport, banking, culture, social, trade facilitation, technology and economic/trade factors.

International Fax Directories A fax directory containing fax numbers of clients/companies situated in range of countries.

International Federation of Agricultural Producers Based in Paris, it is an International Federation of Agricultural Organisations.

International Federation of Air Line Pilots Associations Worldwide association of national pilots' groups.

International Federation of Tour Operators (IFTO) Organisation of European tour operators promoting, among other matters, safety measures for hoteliers and tour packagers.

International Federation of Women's Travel Organisations (IFWTO) Professional group of women travel executives.

International Finance Corporation Based in Washington it provides long-term loans in convertible currencies to private companies in less developed countries.

International Financial Management The definitive information service for all those involved in formulating and implementing financial policy internationally. This includes international expansion policies; international mergers and take-overs; international sources of external funds; financing of international trade; financing long-term projects and acquiring major assets; profit, cash and financial control in international groups; foreign exchange interest rates and commodity risks; and factors influencing the investment decision.

International Futures Exchange of Bermuda A Computerised trading opportunities facility in the futures contracts markets which provides an alternative to the existing conventional futures markets. It is based in New York and used by shipowners and charterers and uses the Baltic Freight Index.

International Hotel Association A group of hotel associations from different countries dedicated to promoting and teaching top professionalism in the industry.

International Labour Organisation Based in Geneva and established in 1919, the ILO is now closely affiliated with the UN. It studies labour problems throughout the world, promotes the improvement of working conditions and publishes monthly statistics on wages. The membership includes most UN members, Germany and Switzerland.

International Liquefied Gas Carrier Code This requires that the design, constructional features and equipment of new ships minimise the risk to the ship, its crew and the environment. There are additional gas carrier codes applicable to existing ships built before 1st July, 1986. A certificate of fitness is mandatory under the terms of the 1983 amendments to SOLAS 1974. For national flag administrations not signatory to SOLAS, 1974, a statement of compliance would be issued by the classification society in accordance with the shipowner's request.

International liquidity Assets which can be used to settle international payments imbalances.

International load line certificate A document/certificate issued by a ship classification society on behalf of a Maritime Government/National Administration. It is issued under the International Convention on Load Line 1966 and required by any vessel engaged on international voyages except warships, ships of less than 24 metres in length, pleasure yachts not engaged in trade and fishing vessels. An initial survey ensures that all arrangements, materials and scantlings fully comply with the Convention before the ship goes into service. Freeboards are computed, marked on the hull and verified by a surveyor. Periodical surveys are required at least every five years together with annual inspections on the anniversary of load line certification

International Management The Management structure of a company involved in International Business crossing international boundaries.

International Maritime Dangerous Goods Code A code devised by the International Maritime Organisation representing the classification of hazardous cargo into some nine classes and identifying the maritime shipment of such cargoes under an international legal framework.

International Maritime Organisation A specialised agency of the UN: its role is to facilitate co-operation amongst governments on technical matters affecting international shipping such as navigation, safety, dangerous cargo shipment code, security and pollution control.

International Maritime Satellite Organisation (INMARSAT) Inmarsat is an internationally owned co-operative which provides mobile communications worldwide. It was established in 1979 to serve the maritime community, and has 75 member countries. The service that the INMARSAT satellite can now support include direct-dial telephone, telex, facsimile, electronic mail and data communications for maritime applications; flight deck voice and data, automatic position and status reporting and direct-dial passenger telephone for aircraft; and two way data communications positioning reporting, electronic mail and fleet management for land transport and emergency communications.

International Marketing The process of evaluating/researching a series of markets, identifying their needs and responding to those needs on an acceptable and co-ordinated basis. It is a management function and essentially market research led with the entrepreneur building up a number of markets which cross international boundaries and serving them. The product core is usually retained and the modifications are consumer/market research led to develop the acceptable strategy of empathy. Long-term it is profit led. See successful export strategy.

International Marketing Strategy It is the strategic and operational marketing issues arising in the management of an entity/international operations to achieve its business and marketing objectives. The firm operating within international markets develops its own international marketing strategies and implements them in the context of a complex and changing environment and opportunity. See successful export strategy and successful procurement strategy.

International Monetary Fund The IMF encourages monetary co-operation, establishes international standards for exchange policy, promotes stable exchange rates among member nations, and makes short-term advances and standby credits to members in temporary payments difficulties. Its resources come mainly from subscription of its members. Total assets are in excess of $35 billion. The IMF was established in 1945 and based in Washington.

International money draft A method of international payment whereby the supplier/exporter/seller arranges through a bank to make payment through an international money draft payable at an overseas bank which may be in US dollars or sterling and despatched direct to the supplier by registration airmail.

International money transfer A method of international payment whereby the importer/buyer can arrange an international money transfer through a bank which will instruct an overseas bank, by airmail, to make payment to the supplier.

International multimodal transport The carriage of goods by at least two different modes of transport on the basis of a multimodal transport contract from a place in one country at which the goods are taken in charge by the multimodal transport operator to a place designated for delivery situated in a different country.

International office of Episootics Deals with international standards concerning animal health. A WTO term.

International organisation A company whose organisation is responsible for all the international operations of the business. The nature of the organisation structure will vary, but at headquarters it would be responsible for policy and global strategic planning especially marketing and finance for international operations.

International Organisation for standardisation Founded in 1947, it comprises of national standards bodies of over 132 countries (87 member bodies and 36 correspondent members). More than 13,000 ISO Standards have been published, based on approval by 75% of member bodies. More than 2,700 technical bodies for the preparation of international standards exist (160 technical committees, 600 sub-committees and some 1,350 working groups), governed by ISO Council.

ISO	– Council committees.
CERTICO	– Committee on certification.
COPOLCO	– Committee on consumer policy.
DEVCO	– Development Committee. Aimed at the needs of developing countries.
EXCO	– Executive Committee. Includes responsibility for finance.
INFCO	– Committee on information.
PLACO	– Planning Committee.
REMCO	– Committee on reference materials.
STACO	– Committee on standardisation principles.

The National Standard bodies that make up the ISO membership include: ANSI, API, ASME, BSI, CEN, CSA, DIN, DEMKO, JIS, ICEMA, MITI, NATA, SAE and SEMKO. See separate entries.

International Organisation of Employers Based in Brussels, the IOE was established in 1949 to promote wider communication and better relation among employers. The members are from all the West European countries (except Spain), the United States and 40 other nations of the free world.

International payment cycle It focuses on the time/period between placing an order for goods and receipt of payment. This may be a lengthy process in international trade depending on payment method: open account, bills for collection, documentary credits and advance payment. See separate entries including payment cycle.

International Petroleum Exchange of London It was set up in 1980 and introduced in the following year its initial contract trading in gas oil futures. Subsequently further energy contracts have expanded into a complex which includes Brent Crude Oil, futures and options; Gasoline futures and gas oil options. It is a non-profit making organisation limited by guarantee, set up and owned by its Floor Member Companies. It is located in London.

International Press Institute An association of editors all over the free world, established to protect the freedom of the press. It is based in Zurich.

International Purchasing The process of buying goods and services overseas which cross international boundaries.

International Purchasing strategy See successful procurement strategy and procurement malaise strategy.

International Registers International registers exist with the specific aim of offering shipowners internationally competitive terms and a means of earning revenue for the flag state. Examples include Liberia, Panama and Cyprus. The terms and conditions offered by international registers vary considerably from the very professional and enforceable international conventions to less vigilant registers. Also termed open registers and flags of convenience.

International Road Transport Union The International Road Transport Union is an international federation of national associations founded in 1948 in Geneva. Its main objectives are to contribute to the development and the prosperity of national and international road transport in all countries and to defend the interests of road transport for 'hire or reward', and on 'own account' operations.

International Safety Guide for Oil Tankers and Terminals It features the safe transportation of crude oil and petroleum products and promotes the concept of global best practice. It is continuously updated.

International Safety Management Code A scheme/code which demonstrates a shipping company's commitment to the safety of its vessels, cargo, passengers, and crew, and to the protection of the environment in compliance with the ISM Code. It is usually administered by a Ship Classification Society such as Lloyd's Register of Shipping.

International Sales Director The position at Director level responsible for a company/entity international sales strategy and budget performance. Such international sales strategy embraces goods and services sold in a country/state crossing international boundaries.

International Seafarers Code It represents nearly all aspects of the conditions of work and life of merchant seafarers as developed/sponsored by the International Labour Organisation. See ILO entry.

International Ship and Port Security Code Amendments to SOLAS chapter XI embrace the International Ship and Port Security Code. It became operative in 2004 under a certification regime. The aim of the code is to expedite clearance of ships, crews, passengers and cargoes.

International Ship Managers Association It represents over 16 countries controlling over 2,300 ships and obtains a commitment by members of quality assured management system which meets the requirements of the Code of Ship Management Standards and to submit these quality assured systems to audit by an independent body.

International Shipping Federation Based in London, it is an international association of shipowners organisations dealing with employer's problems of international shipping.

International (ship) registry A ship's registry which attracts not only national tonnage but also foreign tonnage. It may also attract flagged out ships owned by nationals. See Lloyd's Register of Shipping.

International Ship Supplier Association A ship stores supply source.

International Ship Security Certificate Document issued by Classification Society confirming marine management compliance with International Ship and Port Security Code.

International standard industrial classification A numerical classification of all economic activities issued by the UN.

International Standards Organisation A worldwide federation of national standards institutes which embraces the ISO container.

International Statistical Institute Based in The Hague, the institute provides for international development and co-ordination of statistical methods and procedures.

International Student Identity Card Issued by the Council on International Student Exchange to qualified students to help them secure special travel rates, low-cost accommodations and other benefits.

International Telecommunications Union Based in Geneva, its aim is to maintain and extend the co-operation between the nations in respect of an improvement and rational utilisation of the means of telecommunication. One of the tasks of the organisation is to allocate most rationally radio wavelengths to the various countries.

International Tin Council Based in London, the Council administers the International Tin Agreement (ITA), the object of which is to stabilise the international tin market.

International Tourism Bourse Largest international travel trade show held each year in Berlin.

International Trade The activities, practices, and formalities involved in collecting, presenting, communication and processing data/goods required for international trade.

International Trade Centre Originally established by the old GATT and is now operated jointly by the WTO and the UN, the latter acting through UNCTAD. Focal point for technical co-operation on trade promotion of developing countries.

International Trade Facilitation See trade facilitation entry.

International Trade Law Legislation relative to International Sale of Goods, Law of Carriage, Law of Insurance and so on.

International Trade Procedures Working Group A group of experts in trade facilitation and international trade procedures relating to administration, commerce and transport, working under the mandate of UN/CEFACT, which aims to harmonise, simplify, identify and align public and private sector practices, procedures and information flows relating to international trade transactions both in goods and related services.

International Trading certificate An international trading certificate is a transferable instrument issued by a country to its overseas buyer. It bears the authorisation of the Central Bank, or appropriate monetary authority, granting the holder the irrevocable right to hard currency for goods or services sold to the issuing country. The International Trading certificate aim is to stimulate exports (from developing countries); obviate the need for a Western exporter to be directly involved in a purchase from his buyers country; turn counter-trade into a documentary form of business; provide an increased level of payment security to the exporter; and multi-lateralise international trade.

International Transport and Information System An organisation whose aim is to improve within the Port of Rotterdam the development, realisation and exploitation of a communication and information network for computers.

International Transport Workers Federation Organisation based in London which is a free trade union body established to defend and further internationally the economic and social interests of transport workers of all kinds and their trade unions. It focuses attention on seafarers, dockers, inland navigation workers and fishermen, and collaborates with the International Confederation of Free Trade Unions. See blue certificate entry.

International Travel Industry Exposition (ITX) Annual event in the US including a conference and exhibition representing tourist bureaux, tour operators, cruise lines and other travel industry elements; aimed at the trade and the public.

International Tribunal for the Law of Sea An international maritime tribunal based in Hamburg.

International Trucking The process of operating an international road haulage network which crosses international boundaries.

International video conferencing The process of conducting a meeting/conference amongst delegates/personnel situated in various countries via a video satellite. See video conferences.

Internet communication The process of communicating – sending messages/documents between two or more parties on the internet system.

Internet trading The process of conducting business on the internet involving the buyer and seller or their representatives.

Inter nos (Latin) Between ourselves.

Inter-Parliamentary Union Based in Geneva, it establishes contact between Parliaments of individual countries.

Interpersonal behaviours The study of interpersonal behaviours embracing impact, persuasiveness, team work, written communication, oral communications, listening, planning and organising, delegation, leadership and environmental awareness. See Intellectual Behaviours entry.

Interpool group A major container leasing company.

Interport removals The interport removal facility enables the importer to move uncleared goods from the place of import to another approved port or airport for customs clearance. The procedures are similar to those which operate for removal to an ICD. A customs term.

Inter praesentes (Latin) Between those present.

In terroreom (Latin) By way of threat.

Inter se (Latin) As between themselves.

Interstate Commerce Commission (ICC) An independent US agency created by Congress to regulate surface transportation by common carriers such as railroads and bus companies.

INTERTANKO The International Association of Independent Tanker Owners. It is based in Oslo and has represented the interest of independent tanker owners around the world since 1970 and is committed to upholding the principles of safe transport, cleaner seas and free competition.

Intertanktime '80' Code name of tanker voyage or time charter party. It was formulated by the International Association of Independent Tanker Owners.

Intertankvoy '76' Code name of tanker voyage or time charter party. It was formulated by the International Association of Independent Tanker Owners.

Intertropical Convergence Zone (ITCZ) A narrow low-latitude zone in which air masses originating in the northern and southern hemispheres converge and generally produce cloudy, showery weather. Over the Atlantic and Pacific it is the boundary between the north-east and south-east trade winds. The mean position is somewhat north of the equator but over the continents the range of motion is considerable.

Interval service A transport service which operates at regular intervals throughout the timetable period. For example, it may be a bus service every hour leaving at ten minutes past the hour between the time bands 07.10 to 22.10 hours daily. The great advantage of such a service is that it aids marketing of the product as the market finds it easy to memorise the service pattern.

Interventionist policy The process of a Government becoming directly involved in the conduct of its international trade. It may be raising import duty, extending the import licence system and so on.

Interventionist strategy The strategy adopted by a government or economic/trading bloc in the conduct of its international trade. See interventionist policy entry.

Intervention Price The price at which national intervention agencies buy commodities offered to them. It is a form of guaranteed price to the producer.

Intervention Stocks Agricultural products under CAP which have been brought off the market as part of the Community market support arrangements.

INTEX International Futures Exchange of Bermuda.

In the field The market place embracing the overall environment/place where sales/business is undertaken/conducted.

In-the-money A call/put equity or index option where the exercise price is below/above the current market price of the underlying security i.e. it has intrinsic value. It applies to traded options where the buyer purchases a call option that has a strike price beneath current market price or has a strike price above the current market price for a put option. A BIFFEX term. Also an option which has intrinsic value – an insurance definition. See also Out the Money entry.

Intinerary The route followed by a particular consignment in a transit. It may be road, rail, air or sea or intermodal.

INTIS International Transport and Information System.

In toto (Latin) On the whole; completely; root and branch.

Intra Asian trade Trade originating within Asia.

Intra business EC In this category all internal organisational activity can be included, usually performed on intranets, that involve exchange of goods, services or information. Activities can range from selling corporate products to employees to online training and cost reduction activities.

Intra Community cargo Cargo which originates within the European Community and is subsequently transported inside the European Community market.

Intra community trade statistics Statistical data relative to trade between member states of the European Community.

Intra Community transit A consignment which is despatched within the European Community and therefore subject to its customs and transport regulations.

Intra EC Trade Statistics Statistical data relative to trade between member States of the European Community.

Intra European cargoes Cargoes originating within Europe.

Intra European Transport Transport services operating within Europe.

In transitu (Latin) In transit.

INTRASTAT The system for collecting statistics of trade between EC Member States.

Intrastat Combined Nomenclature The commodity codes in the Customs and Excise tariff.

Intra trade Trade within a group of countries such as trade amongst the EC states can be termed intra EC trade. Other EU countries exceeding £150,000 p.a. (valid 2002) must be analysed by commodity code (the UK Tariff, based on the Harmonised System) in a Supplementary Statistical Declaration.

In transit The status of goods or persons between the outwards customs clearance and inwards customs clearance.

Intra vires As permitted by the Memorandum and Articles of Association of a company, or by appropriate board resolution.

Intra vires (Latin) Within one's powers.

Intrinsically safe Equipment, instrumentation or wiring is deemed to be intrinsically safe if it is incapable or releasing sufficient electrical or thermal energy under normal conditions or specified fault conditions to cause ignition of a specific hazardous atmosphere in its most easily ignited concentration.

Intrinsic value A call put equity or index option has intrinsic value of the exercise price if the option is below/above the current market price of the underlying security. The intrinsic value is then the difference between the exercise price and the price of the underlying security.

Introductory offer The process of launching a product/service and in so doing offering a market incentive to use it. It may be a special low price; free participation in a competition; gifts/souvenirs; press launch for first sailing/flight and so on.

INTTRA One stop shop internet service embracing five mega container shipping lines: Maersk Sealand. P & O Nedlloyd; Hamburg Sod; Mediterranean Shipping Company and CMA CGM.

Inv Invoice.

Inventory A record of goods received, stored and delivered.

Inventory Management The process of managing/controlling products/articles/equipment on hire or lease by the lessee to ensure to their most productive use and the lessor is complying to the terms of the lease/hire contract.

Investment Gross fixed capital formation as defined in the European System of Integrated Economic Accounts. An ECB term.

Investment bank A bank proving long-term fixed capital for industry, in exchange for which it takes over shares in the companies so financed. It may also perform a number of merchant banking functions, such as merger making, providing corporate finance and investment advice and broking.

Investment buying The process of buying a product for ultimate resale. Examples for an international buyer embrace wine, antiques, machinery, services and installations including commercial and residential property.

Investment centre A profit centre in which inputs are analysed in terms of expenses and outputs are measured in terms of revenues, and in which assets are also examined; the excess of revenue over expenditure then being related to assets employed.

Investment criteria The code of practice adopted to process an investment scheme from the time of its conception to authorisation of the project. In particular it involves the factors to consider and techniques to use in the evaluation of the scheme.

Investment income Interest and dividends arising from the investing of premium monies pending claims being made.

Investment objectives The scale of objectives of an investment programme/project(s). This may be improved quality of service in a transport service; lower manning cost in an industrial plant; much improved production output in a mining venture and so on.

Investment overseas The process of buying property/equity etc. in markets which cross international boundaries.

Investment rate of return The rate of return accredited to a particular investment. It is usually expressed as a percentage of the total investment.

Investment review The process of critically reviewing an investment programme/project(s) with a view to determining their adequacy/necessity/scale of financial return in the light of the new situation presented. For example, a trade recession may require a rethink on the investment programme to ensure production capacity matches market demand and there is not an excessive over capacity situation thereby rendering new plant being idle on completion.

Investment trust A limited company set up to invest in the shares of other limited companies and so spread the risk; its shares are quoted on the Stock Exchange like any other listed company.

Invisible exports Income received by a country from its services rendered to foreigners and services received from foreigners embracing banking, shipping, insurance, air transport, tourism and interest on foreign investments.

Invitation to tender An invitation by a buyer for various sellers to quote their prices for specific goods and contract terms.

Invitation to Treat An invitation to make an offer – a legal term.

Invoice A document prepared by a supplier showing the description, quantities, price and value of goods delivered, or services performed. To the supplier this is a sales invoice; to the purchaser the same document is a purchase invoice. The invoice may also state terms of payment. See also commercial invoice, customs invoice, consular invoice and proforma invoice entries.

Invoices in a third currency Contract quotations in a third currency, i.e. in a currency that is foreign to both the exporter and the importer, are particularly common when consideration is being given to the purchase of capital goods. An example arises: when the goods are exported from the United Kingdom to Italy and payable in Swiss Francs. Also, when trade is conducted with a country whose currency is regarded as 'exotic' it is usual to invoice in an internationally traded currency such as the US dollar.

In these cases, both the importer and the exporter will face the problems that are associated with invoicing in a foreign currency. Some countries' exchange control regulations, however, may preclude the use of a third currency

Invoice in the currency of the overseas buyer An invoice payable in the currency of the overseas buyer's country. It may be the buyer in Switzerland paying for the merchandise in Swiss Francs exported from United Kingdom.

Invoice in the currency of the overseas seller An invoice payable in the currency of the overseas seller's country. It may be the seller in United Kingdom invoicing/selling the goods in sterling to the Italian buyer.

Inward cargo Cargo in process of being imported from a country.

Inward charges Pilotage and other expenses incurred in entering a port.

Inward Clearing Bill Document issued by Customs to Master on completion of all cargoes discharge formalities.

Inward dues Charges/dues raised by a Port Authority on vessels entering a port and usually based on a ship NRT.

Inward Freight Department Shipowners organisation responsible for cargo imports and clearance.

Inward looking Company A company/entity with a tendency to focus on the domestic local market for development rather than have an international focus crossing international boundaries and encountering a new business environment/infrastructure and culture. See outward looking Company entry.

Inward processing The importation of goods for process and export from the European Union in the form of compensating products.

Inwards investment Capital subscribed from outside the country/region/market place and thereby is in effect 'imported capital'.

Inwards mission A marketing technique to facilitate an exporter develop/promote his product/service. Under such a project, groups of companies invite overseas businessmen and journalists – who can influence exports – to visit the exporter's country to see their products at first hand. This usually involves product demonstrations and factory visits. It is usually arranged by the exporters trade association and the number of delegates tends to range from ten to twenty persons.

Inwards processing relief The process of importing products from outside the EU and subsequently exported to non EU countries. In so doing the goods benefit from customs import duty relief. This is available either by the suspension or drawback methods.

Inwards trade strategy A strategy directed towards developing trade within a region/economic bloc/customs union.

In warehouse Sale point in LME contracts.

IN WRITING Any visibly expressed form of reproducing words; the medium of transmission shall include electronic communications such as radio communications and telecommunications. A BIMCO term.

IO Information Officer; Investigation Officer or Immigration Officer.

IOB Insurance Ombudsman Bureau.

IOC Inter-Governmental Oceanographic Commission (based in Paris).

IOE International Organisation of Employers. Based in Brussels it role is to provide wider communications and better relations among employers.

IOI International Oil Insurers.

IOPC International Oil Pollution Compensation Fund.

IOPC Fund The International Oil Pollution Compensation Fund (Based in London).

IOPP International Oil Pollution Prevention (certificate).

IOTTSG International Oil Tanker and Terminal Safety Group.

IP Institute of Petroleum.

IPA Institute of Practitioners in Advertising.

IPACG Informal Pacific ATS Co-ordinating Group (JCAB/FAA).

IPC Industry Policy Committee – IATA term; or BIA Investment Protection Committee.

IPD Industrial Planning Department (Burma).

IPDS Inmarsat Packet Data Services.

IPE Institute of Petroleum Exchange.

IPF Intaken piled fathom.

IPG Inter-Governmental Preparatory Group on a Convention for International Multimodal Transport.

IPI International Press Institute (Based in Zurich).

IPLC International Product Life Cycle.
IPMS Integrated Platform Management System.
IPO International payment order or Initial Public offerings.
IPR Inward Processing Relief. The system of duty relief for goods imported from non-EC countries for process and re-export. An EU definition.
IPRs Intellectual Property Rights.
Ipso Facto (Latin) By the very fact itself.
IPTC International Press Telecommunications Committee – IATA term.
IPU Inter-Parliamentary Union. Based in Geneva, its role is to establish contact between parliaments of individual countries.
IQ Augsburg Airways (Augsburg airways GmbH) – IATA Member Airline.
IR Indian Register of Shipping – the Indian Ship Classification Society; Iran Airways, the Airline of the Islamic Republic of Iran – the Iranian National airline and member of IATA or Issued and Renewed – an insurance term.
Iran Airways The Iranian National airline and member of IATA.
IRAQI Iraqi Line – a shipowner.
Iraqi Airways The Iraqian National airline and member of IATA.
IRC International Revenue Code.
IRD Inland Rail Depot. Similar to inland clearance depots (ICD) but handles only rail traffic, either in train-ferry wagons or containers.
IRIS Industry Rates Information System – IATA term.
IRM Institute of Risk Management.
IRN Import Release Note or iron.
Iron ore carriers Vessels specially built for the shipment of iron ore in complete/bulk shiploads.
IRR Internal Rate of return.
Irrevocable Cannot be revoked or cancelled without the agreement of all parties to the transaction. See Irrevocable Letter of Credit entry.
Irrevocable credit A letter of credit which cannot be cancelled or amended without the agreement of both parties. See irrevocable letter of credit entry.
Irrevocable Letter of Credit An irrevocable letter of credit cannot be altered or cancelled without the agreement of yourself and your supplier. It therefore offers both you and your supplier a high degree of security.
In order to obtain maximum protection against the risk of non-payment your supplier might ask you to instruct your bank to have the credit confirmed by a bank in his country. This means that the overseas bank will pay him even if events occur to prevent you or your bank from making payment. However, when a credit is issued by a well established international bank it is unusual for a supplier to request that his buyer obtains additional confirmation which would of course add to the cost of the goods.
The following are various types of irrevocable letter of credit:-
(i) Confirmed. A 'Confirmed' irrevocable letter of credit is an irrevocable credit to which the advising bank (at the request of the issuing bank) has added its confirmation that payment will be made. The advising bank thus confirms that it will honour drawings which conform to the terms of the credit.
(ii) Unconfirmed. If the irrevocable letter of credit is 'Unconfirmed' the advising bank merely informs the exporter of the terms and conditions of the credit without adding its own undertaking to pay or accept under the terms of the credit.
(iii) Transferable. A 'Transferable' letter of credit is one under which the exporter has the right to issue instructions to the paying bank (or to the negotiating bank) to make the credit in whole or in part to one or more third parties, providing partial shipments are not prohibited.
(iv) Others. There are other forms of credit – for example 'Back to Back', 'Red Clause' and 'Revolving' – but they are not widely used. See separate entries.
Irrevocable transferred credit An irrevocable credit in which the named applicants are not the buyers of the goods but are usually intermediaries in the transaction.
IRRI International Rice Research Institute.
IRS Indian Register of Shipping – Indian Ship Classification Society.
IRSG International Rubber Study Group. Based in London it investigates and convenes meetings regarding production, processing and utilisation of rubber.
IRU International Road Transport Union. Based in Geneva, its role is to develop national and international road transport or International Recruitment Unit (OD).
IS Independent Spherical Aluminium Tank (on a ship); Industrial systems or Iceland – customs abbreviation.
ISA International Sugar Agreement; Indonesian Sawmill Association; Iron and Steel Authority (Philippines) or International Seabed Authority.
ISASI International society of Air safety Investigators – IATA term
ISBN International Standard Book Number or Integrated Satellite Business Network.
ISC International Sugar Council (based in London); Italian Shippers Council or Israeli Shippers Council.
ISD Information Systems Directorate – part of HMCE.
ISDN Integrated Services Digital Network.
ISF International Shipping Federation.
ISGOTT International Safety Guide for Oil Tankers and Terminals, produced by ICS, OCIMF and IAPH.
iShip iShip provides an Internet-based shipping service that allows shippers and carriers to make shipping transactions on line. Its focus is mainly on small-sized shipments. It allows users of shipping services to access and compare rates and services that are provided by major integrated carriers, such as UPS, Fed Ex, Airborne, US Postal Services, Yellow Freight System. It provides answers to basic shipping questions such as the cost of shipment by alternative carriers, delivery times, by weight and destination, etc.
ISI International Statistical Institute. Based in The Hague it deals with the international development and co-ordination of statistical methods and procedures.
ISIC International Standards Industrial Classification – a numerical classification of all economic activities issued by UN.
ISIN International Securities Identification Number: a standardised identification (e.g.GB0031790826) of securities and other financial instruments within a uniform system.
ISIS International Shipping Information Service or IATA Statistical Information system.
ISLE Isle Shipping Line – a shipowner.
ISLWG Working Group on International Shipping Legislation (UNCTAD).
ISM International safety Management.
ISMA International Ship Managers Association.
ISMA Code A voluntary code of ship management standard based on a combination of ISO and IMO declarations on quality control and ship management. It embraces all the former ISO 9002 and ISM requirements of the new ISO 9000.

ISM Code International Safety Management Code.
ISNAR International Service for National Agricultural research (IARI).
ISO International Standards Organisation; International Sugar Organisation or International safety organisation.
ISO 9000 The international standards organisation accreditation for quality management systems which accords with and recognises BS 5750. See also BS 5750 and EN 29000 entries.
ISO 9002 An accredited quality standard.
ISO base The base of a stacker tailored to fit the ISO hole.
Isocyanates This includes a number of chemical products used in the manufacture of plastic foams, synthetic rubber, etc. Some are sufficiently toxic or lachrymatory to need classification as poisonous substances, particularly iso-cyanates in pure form. Others may need to be classified as flammable liquids, dependent on their characteristics. A number may be non-hazardous in transportation.
ISO hole The shape of the corner casting holes in top and bottom of containers as specified by the ISO.
Isothermal Descriptive of a process undergone by an ideal gas when it passes through pressure or volume variations without a change of temperature.
ISP Internet Service provider.
ISPACG Informal south Pacific ATS Co-ordinating Group (FAA/ACNZ/CAAA).
ISP/ASP Internet Service Provider/Active Server Pages.
ISPS Code (and December 2002 amendments to SOLAS) The International Ship and Port Facility security Code (ISPS Code) was adopted by a Conference of Contracting Governments to the International Convention of the Safety of Life at Sea (SOLAS), 1974, convened in London, December, 2002. The aims, primarily to establish an international frame work for co-operation between Contracting Governments, Government Agencies, Local Administrations, and the Shipping and Port Industries to detect security threats and take preventive measures against security incidents affecting ships or port facilities used in international trade and to establish relevant roles and responsibilities at the national and international level. These objectives are to be achieved by the designation of appropriate personnel on each ship, in each port facility and each ship owning company to make assessments and to put into effect the security plans that will be approved for each ship and port facility. The Conference also adopted several related resolutions and amendments to chapters V and XI (now divided into chapters X1-1 and X1-2) of the 1974 SOLAS Convention, as amended. Under the new chapter X1-2, which provides the umbrella regulations, the ISPS Code became mandatory on 1st July 2004. The Code is divided into two parts. Part A presents mandatory requirements, part B recommendatory guidance regarding the provisions of chapter X1-2 of the Convention and part A of the Code.
Israel Shippers Council The organisation in Israel representing the interest of Israeli shippers in the development of international trade.
ISRO International Securities & Regulatory Organisation.
ISS International Social Service (based in Geneva).
ISSA International Ship Suppliers Association ship store supply source.
Issue price The price at which stock or shares are issued to the public. This may or may not be the same as the nominal or par value of the shares.
Issuing bank The bank that opens the Credit and which extends its legally binding undertaking to the Seller on behalf of the Importer/Buyer. Otherwise known as the Opening Bank.

Issuing carrier The carrier who issues the Air Waybill or ticket. A term associated with IATA traffic and especially relevant when the issuing carrier is not the airline which carries the traffic and/or a second carrier is involved to complete the final leg of the flight.
Issuing house A merchant bank, stockbroking firm or other institution which arranges and sponsors new issues of capital stocks and shares and arranges their under writing.
ISU International Salvage Union
it Insulated tank containers.
IT Air Inter (Lignes Aériennes Intérieures), an IATA Member Airline; information technology; independent tank; inclusive tour; or Italy – customs abbreviation.
ITA Institut du Transport AÈrien; International Trade Administration; or information Technology Agreement – See separate entry.
Italia di Navigazione SPA A major liner operator.
Italian Chamber of Commerce in Great Britain The organisation based in London representing Italy whose basic role is to develop, promote, and facilitate international trade between Great Britain and Italy.
Italian Shippers Council The organisation in Italy representing the interest of Italian shippers in the development of international trade.
ITB Integrated tug barge or Insurance Technical Bureau.
ITC International Tin Council (based in London); independent tank centre (of a ship); International Trading certificate; International Telecommunication Convention; International tonnage certificate or International Trade Centre – See separate entry.
ITCB International Textiles and Clothing Bureau – Geneva-based group of some 20 developing country exporters of textiles and clothing.
ITCZ Inter topical Convergence Zone.
ITF International Transport Workers Federation (based in London).
ITI Customs Convention on the International Transit of Goods.
ITIC International Transport Intermediaries Club Ltd.
IT number Number assigned to an inclusive tour containing certain specified elements that make it eligible for commission overrides from the airlines.
ITO International Trade Organisation. The 23 countries which originally signed the GATT in 1948 were at the time engaged in drawing up the charter for a proposed International Trade Organisation which would have been a United Nations specialised agency. Plans for the ITO had to be abandoned when it became clear that its charter would not be ratified, and the GATT, which is based largely on parts of the draft ITO charter, was left as the only international instrument laying down trade rules accepted by nations responsible for most of the world's trade. A WTO term.
ITOA Independent Tanker's Owners Association.
ITOPF International Tanker Owners Pollution Federation Ltd. which provides advice on all aspects of preparing for and responding to oil spills from tankers.
ITP Independent Television Publication or International Trade Prototypes.
ITPWG International Trade Procedures Working Group. See separate entry.
ITSC International Transit Switching Centres.
ITU International Telecommunications Union. Based in Geneva, its role is to maintain and develop co-operation between nations in respect of an improvement and rational utilisation of the means of telecommunications.
ITW Independent tank wing (of a ship).
ITX Independent tank common (of a ship) or inclusive

tour excursion.
ITXs Independent Trading Exchanges – a computer logistics term.
IUA International Underwriting Association (of London).
IUAI International Union of Aviation Insurers.
IUMI International Union of Marine Insurance.
IUOTO International Union of Official Travel Organisations. Based in Geneva, its role is to develop international travel.
IUR International Union of Railways. Based in Paris, its role it to co-ordinate and improve conditions regarding construction and operation of railways in international traffic.
iv Increased value or invoice value.
IVHS Intelligent vehicle highway systems.
IVIS Instantaneous Vertical Speed Indicator.
IW AOM – Minerve S.A. d.b.a. AOM French Airlines – an IATA Member Airline.
IWA International Wheat Agreement. Based in London, it is an agreement aiming at securing supplies of wheat for the importing countries, and markets for the exporting countries, at reasonable prices.
IWC International Wheat Council (based in London) or International Whaling Commission (FAO).
IWL Institute of Warranty Limits which defines maritime area ship may operate within marine insurance cover.
IW International Wool Secretariat. Based in London, its role is to develop co-operation between the wool producing countries.
IWSG International Wool Study Group. Based in London, it is an information bureau for wool.
IWTC Inland Water Transport Corporation (Burma).
IY YEMENIA Yemen Airways – an IATA Member Airline.
IZ Arkia Israeli Airlines – an IATA Member Airline.

J2 Azerbaijan Airlines (Azerbaijan Hava Yollari) – IATA Member Airline.
JA Air Bosna – IATA Member Airline.
JAA Joint Aviation Authority – IATA term.
Jack-knife Situation which occurs when the driving wheels on the tractive unit of an articulated road vehicle lock under heavy braking causing the trailer, which still has forward motion, to push the rear of the tractive unit out of line and into a skid. This eventually results in the tractive unit swinging round and closing on the trailer.
JAFZ Jabal Ali Free Zone.
JAFZA Jabal Ali Free Zone Authority.
Jamahiriya Libyan Arab Airline The Libyan National airline and member of IATA.
Jamaican Dollar The currency of Jamaica. See Appendix 'C'.
J & WO Jettisoning and washing overboard – marine insurance policy optional cover.
Japan Air Lines The Japanese National airline and member of IATA.
Japanese Chamber of Commerce and Industry in the United Kingdom The organisation based in London representing Japan whose basic role is to develop, promote and facilitate international trade between UK and Japan and vice versa.
Japan External Trade Organisation An organisation based in Japan with offices throughout the world. Its objective is to facilitate the development/growth of Japanese imports.
JAPANISCON Japan/West Africa (Nigeria/Senegal Range) Freight Conference.
JAPANLINE Japan Line – a shipowner.
JAPWACCON Japan-West Africa (Angola/Cameroon Range) Freight Conference.
Jason Clause A clause in a contract of affreightment relating to liability of the shipowner under the US Harter Act in disputes concerning general average. See New Jason Clause.
JAS REP Japanese Ship Reporting System.
JASTPR Japanese Committee/Organisation on Trade Procedures and Facilitation based in Tokyo.
JASTPRWY Japanese Committee/Organisation on development of electronic data processing techniques for International Trade based on Tokyo.
JCCC Joint Customs Consultative Committee.
JCCIUK Japanese Chamber of Commerce and Industry in the United Kingdom.
JCS Joint declaration of interest.
J curve A statistical term whereby by way of an example following the depreciation or devaluation of a currency a country's current account may initially worsen and thereafter gradually improve. In graphic presentation terms, it follows the profile of the letter 'J'.
JD Japan Air System (Japan Air System Co. Ltd.) – an IATA Airline Member.
JDI Joint declaration of interest.
JE Manx Airlines Ltd. – an IATA Airline Member.
JECO Jeco Shipping Line – a shipowner.
JEFC Japan/Europe Freight Conference.
Jerque note Inwards clearance certificate confirming ships cargo discharged and related customs formalities completed.
Jerricans These are metal or plastic packagings of rectangular or polygonal cross-section.

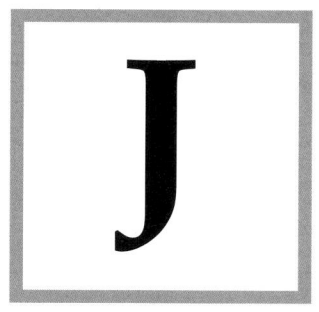

Jet foil A passenger carrying craft which is basically a ship whose weight is supported by foils which produce lift by virtue of their shape and forward velocity through the water. Two types exist: those with surface piercing foils and those with fully submerged foils. Overall, it is an advanced design high speed hydrofoil with a speed of 60 knots and capacity of 150/400 passengers.
Jet lag Physical and mental fatigue experienced by many airline passengers following completion of a long flight and changes in time zones.
Jet perforating guns An oil well, without detonator. Article consisting of a steel tube or metallic strip into which are inserted shaped charges connected by detonating cord, without means of ignition.
JETRO Japanese External Trade Organisation.
Jetsam Goods thrown overboard to lighten vessel and afterwards washed ashore.
Jet stream Strong winds in the upper troposphere whose course is related to the major weather systems in the lower atmosphere and which tend to define the movement of these systems.
Jettison Voluntary act of throwing overboard cargo, stores, etc., in time of ship's peril.
Jetty A mole or breakwater running out into the sea to protect a harbour or coastline. It may be used as a landing pier.
Jetway A loading and unloading bridge giving passengers protected entry and exit from an aircraft; the term is a registered trademark.
JHC Joint Hull Committee.
Jiao The currency of China. See Appendix 'C' and Yuan.
Jiao invoice An invoice payable in Jiao currency and relative to an export sales contract. See Yuan.
Jib The projecting arm of a crane, or attachment connected to the top of a crane boom.
JICA Japan International Co-operation Agency.
Jiffy bag The bags are an envelope with bubble wrap padding inside and ideal for small consignments such as medical supplies, promotional items and computer discs. Overall there are 67 different types.
JIS Japanese Industrial Standard. The National Standards Body in Japan. See ISO.
JIT Just in time.
Jitney A small bus or motorcar that serves a route, usually on a flexible schedule.
JJ TAM Linhas Aereas – IATA member Airline.
JK Spanair – an IATA Airline Member.
JL Japan Airlines Co. Ltd. The Japanese National airline and member of IATA.
JLAURITZEN J Lauritzen Line – a shipowner.
JLCD Joint Liaison Committee on Documents.
JM Air Jamaica Ltd. – an IATA Airline Member.
JMC Joint Maritime Commission. Based in Geneva under the aegis of the ILO, it deals with labour conditions at sea and is composed of representatives of shipowners and seamen's organisations of leading maritime countries.
JMDPC Japan Maritime Disaster Prevention Centre.
JMSDC Joint Merchant Shipping Defence Committee (of the Chamber of Shipping).
JMT Japan Maritime Transport Law.
JO JALways Co. Ltd. – IATA Member Airline.
JOA Joint Oceanographic Assembly.
JOCE Journal Officiel des Communautes Europeenes.

Jobber A Stock Exchange Member whose Firm acts as a dealer on securities. In reality the Jobber is a stall holder in the market unlike Brokers, who are Agents for their clients and are paid by commission. Jobbers deal as principals buying and selling shares at their own risk.

Jobbing The process of trading in and out of positions on the same day – a BIFFEX term.

Job creating expansion A policy which may be economic, legal, political, technical or social generating new jobs.

Job specification A specification of the duties and responsibilities of a particular post within a company. It may also specify the grade or salary range, and how it fits in with the organisational structure of the Company. It will also specify to whom the post is responsible e.g. Sales manager to Sales Director, and who the post controls/supervises eg. all Sales Offices in the Arab States.

Joint Account An account operated by two people usually with equal responsibility such as husband and wife joint current bank account.

Joint Cargo Committee A committee composed of representatives from Lloyd's and London company market, to examine matters of interest to cargo insurers and to make recommendations or action to be taken by the market.

Joint charge A tariff which embraces two carriers offering a through service under a combined rate quotation.

Joint consultation The process of consulting between two parties usually found in the realm of industrial relations. This involves Union and Management, and could arise in an investment programme involving changes ultimately in work practices and work environment. At some stage later, the actual wage scales and code of work practice would be negotiated when the scheme is more advanced. Consultation also arises when the Management inform the Union of new commercial initiatives to develop the business.

Joint fare Fare applying from the point of origin to the destination through one or more intermediate points for travel on more than one airline; see through fare.

Joint Hull Committee Similar to the Joint Committee, but in respect of hull interests.

Joint Hull Understandings Agreement between London hull underwriters to provide uniformity in their consideration regarding hull insurance conditions and rating structures.

Joint Maritime Commission Based in Geneva under the aegis of the ILO, it deals with labour conditions at sea and is composed of representatives of shipowners and seamen's organisations of leading maritime countries.

Joint Negotiation The process of negotiating between two parties usually found in the realm of industrial relations. This involves Unions and Management negotiating on pay; conditions of employment; company re-organisation; manning levels in a factory, redundancy and so on. Joint negotiation may be on a national level such as pay and conditions throughout the company, or on local level involving factory re-organisation of work force and procedure in the interest of cost efficiency.

Joint notice of change Form containing information about an agency's status and about proposed new owner or owners; submitted with a change of ownership application to conferences to continue appointments under new ownership.

Joint operations Two or more carriers operating a service using one aircraft, with one carrier administering and one carrier controlling the reservations.

Joint service A service/timetable operated by two different carriers under agreed terms involving the allocation of specific schedules to each operator and revenue apportionment. It may involve, for example, two shipowners of differing nationality one operating forty per cent of the sailings and the other sixty per cent.

Joint Traffic Conferences Conference of IATA members from two or all three IATA Traffic Conference Areas.

Joint venture The process of two trading companies from different countries – at least one of them being local – forming an agreement to manufacture/produce goods on a joint basis. Moreover, the local firm provides expertise in the prospective market and the multinational firm provides management and marketing expertise. Usually it is undertaken to strengthen the competitive advantage of both companies in the market place and does involve a transfer of technology. For example, a Japanese Motor Manufacture may form a joint venture with a UK Motor Manufacture with a view to Japan supplying the engine and UK the chassis with the production/assembly of the car in the UK. Often the local Company has the major share of the equity and director portfolio thereby retaining the control of the business. The OECD definition is an equity joint venture which implies the sharing of assets, risk and profits, and participation in the ownership (e.g. equity) of a particular enterprise or investment project by more than one firm or economic group; or a project undertaken by two or more persons/entities jointly together with a view to profit, normally in connection with a single operation; or a project undertaken by two or more person/entities joining together with a view to profit, normally in connection with a single operation.

Jolanda A set of rails fitted into a container either as a single or double pair on which pallets may be placed and easily transported by a single jacking device which allows loads of goods to be shifted in or out of the unit with the minimum effort.

Jordan Dinar The currency of Jordan. See Appendix 'C'.

Jordan Invoice An invoice payable in Jordanian dinars relative to an export sales invoice.

Journal A record of financial transactions, such as transfers between accounts and correction of bookkeeping errors; or a magazine/news bulletin which is published every two months.

Journey cycle See call cycle entry.

Journey Planner A visual display as found on London Underground or printed presentation as found in a book/brochure/leaflet giving information whereby the respondents can plan their rail/road/air, etc., journey.

JP Adria Airways – an IATA airline or Japan.

JPPCC Joint Public-Private Consultative Committee.

JQ Trans-Jamaican Airlines Ltd – an IATA Airline Member.

JR Jugoslav Register – the former Yugoslavian Ship Classification and Survey Society or Aero California (Aero California, SA de CV) – IATA Member Airline.

JS Joint Support (ICAO term) or Air Koryo – IATA Member Airline.

JSA Japanese Shipowners Association.

JSC Japan Shippers Council.

JSEA Japan Ship Exporters Association.

JSP Jackson Structured Programming. A method of using structure diagrams for analysing data and programs.

JTIDS Joint Tactical Information Distribution system – IATA Term.

JU JAT (Jugoslovenski Aerotransport) – IATA Member Airline.

Juggernaut A road vehicle of 32 tonnes and over gross weight.

JUGOLINIJA Jugolinija East – Rijeka – a shipowner.

Jugoslav register The former Yugoslavian Ship Classification and Survey Society.

Jumbo A wide bodied aircraft (jet).

JUMBO The Consortium of Australia/West Pacific Lines. This includes Malaysia International Shipping Corporation and Southern Shipping Lines.

Jumbo derrick Ship's derrick with high lifting capacity provided for heavy lifts.

JUMBO Ferry The largest ferry of its class usually in excess of 40,000 GRT and capable of conveying passengers, cars, road haulage vehicles and coaches.

Jumboisation The process of converting/enlarging a small vessel to one of jumbo proportions. It could involve increasing the length, beam, draught and capacity of the vessel.

Junior motorman A junior post in the Marine Engineer's department on a vessel and formerly designated as Engine Room boy.

Junior seaman A junior post in the Deck department on a vessel and formerly designated as a deck boy.

Junior suite A large room with a partition separating the bed and sitting areas.

Jurisdiction clause A clause in a bill of lading or charter party which stipulates that any dispute between the parties arising from the contract be resolved in a court of law as opposed to arbitration. It also specifies which country has jurisdiction that is, the authority to administer justice.

Jurisprudence Judicial decisions used for explanation and meaning of law.

Jury Mast A temporary mast.

Jus disponendi (Latin) Law of disposal

Just in time (distribution) A distribution system whereby the goods arrive at their specified destination at a specific time. It involves the whole supply chain from the ordering of raw materials and components for manufacturing to delivery of the finished goods to the sales organisation or retailer. Overall, it closely involves inventory control and the effective use of electronic data interchange. The system eliminates or reduces stock and requires guaranteed quality levels and elimination of waste to be effective.

JV Joint Venture.

JVX Joint Services Advanced Vertical Lift Aircraft.

JWP Joint Working Party.

JY Jersey European Airways – an IATA Airline Member.

JZ Skyways AB – an IATA Airline Member.

K6 Khalifa Airways – IATA Member Airline.

K Knots, or Ketch – a small coasting vessel.

KA Hong Kong Dragon Airlines Ltd. (DRAGONAIR) – An IATA Member Airline.

KADIN Indonesian Chamber of Commerce & Industry.

KAIUNION Kai Union Shipping Pte. Ltd. – a shipowner.

Kanban A method which during storage uses standard units or lot sizes with a single card attached to each. It is a pull system in which work centres signal with a card that they wish to withdraw parts from the supplying party

Kangaroo The carriage of unaccompanied road vehicles and trailer on wheels on rail flat cars.

Kangaroo crane Crane evolved on the principle that in the crane cycle the time spent in slewing is wasted. If lifts a grab load of bulk cargo and by luffing up the full extent, deposits the load into a hopper (or pouch) built onto the structure of the crane facing the quay. Thus there are two motions of the crane instead of the conventional luffing, slewing and lowering/raising.

KANNIVELL Kannivell Lines – a shipowner.

KANSAI Kansai Steamship Co. Ltd. – a shipowner.

KARLANDER Karlander (Australia) Pty. Ltd. – a shipowner.

Kassaschiene A privately placed interest bearing certificate of debt – traditionally medium term – that are issued by domestic borrowers. It is issued mainly by Swiss public sector entities and large industrial concerns as an instrument of medium-term financing.

Kassenobligation The listed bearer stock issued by the German government agencies or Girobanks which differs from most Euro market paper in that its interest is subject to withholding tax.

Katabatic wind A wind created when very cold air forms in upland areas and becomes sufficiently dense to drain downhill: when part of larger scale weather patterns (e.g. the Bora or Mistral in the Mediterranean, or around Antarctica) these offshore winds can be a major hazard to shipping.

KC Kilogram per square centimetre.

KD Kendell Airlines – an IATA Airline Member or Kuwait Dinar – a major currency. See Appendix 'C'.

kDWT Deadweight in thousands of tonnes.

KE Korean Air – an IATA Airline Member.

Keg A small cask.

Kelvin waves Gravity-inertia waves which occur in both the atmosphere and the oceans, where either the effect of the Coriolis Force is negligible (i.e. close to the equator) or where this force is balanced by the pressure gradient. The most important examples are in the equatorial stratosphere and in the thermocline of the equatorial Atlantic and Pacific close to the equator (in both cases the waves propagate eastwards relative to the Earth).

Kempenaar ship A vessel navigable on the navigable inland waterways of Netherlands and Belgium with a length of 63 metres and beam of 7 metres. The vessel has a capacity of 32 TEUs and 48 TEUs when stacked three high. The ship also operates on parts of the French and German inland waterway networks. See pallet ship.

Kennedy Round The sixth round of the Gatt multilateral trade negotiations conducted during the period 1963-1967.

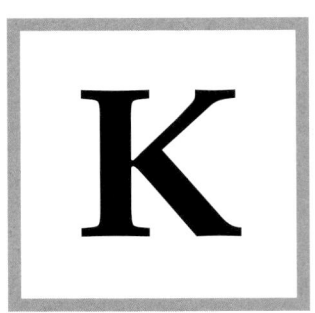

KEN PRO Kenyan Committee/Organisation of Trade Procedures and facilitation based in Nairobi

Kenya Airways The Kenyan National airline and member of IATA.

Kenya Shilling The currency of Kenya. See Appendix 'C'.

Kenya Shilling invoice An invoice payable in Kenyan shillings relative to an export sales contract.

Kerbside weight The weight of a motor vehicle with no person in it, a full supply of fuel in the tank, adequate other liquids and no load except normal loose tools and equipment.

Kerb trading This is a period of fifteen to twenty five minutes at the end of each session (am and pm) where all metals are traded simultaneously around the ring. Kerb trading is often expressed simply as 'on the kerb' or 'kerb'. The term comes from the early history of commodity futures markets when trading after hours was literally conducted on the kerb of the street. Trading which continues after the market's official close normally for a specified length of time – a BIFFEX term.

Key One of the attributes of an entity on which an index has been created or a relation has been set. Primary keys are the key attributes in a table, secondary keys are used to sort records with the same primary attribute value, foreign keys are the attributes in a table which provide the facility for relationships to be set with primary keys in the parent table. A computer term.

Key account A major account in a sales portfolio.

Key currency Currencies kept as reserve by IMF members in addition to their own currency and gold reserves. Key currencies are less subject to large exchange rate fluctuations.

Key facts The salient points in any discussion/memorandum/decision making process and crucial in any ultimate evaluation including the realisation of any objective. This applies to transport, marketing, international trade and so on.

Key Financial Variables (in a country analysis) A range of international financial data to measure a nations economic/political/financial performance and usually found in a country report as published by leading Banks, such as Barclays. The key variables include growth (%); inflation (%); Govt Balance/GDP (%); Trade Balance (USD bn); Current Account (USD bn); Current Account'GDP (%); Import cover (months); International reserves (USD bn); Foreign Debt/GDP (%); Debt Service/Exports (%); Int. Reserves/Short Term debt (%) and Exchange rate.

Keyhole socket A deck fitting with a keyhole-shaped hole used for lashing ISO containers.

Key markets Markets which in the view of the entrepreneur offer the best prospects both long and short term and usually of low political risk. Overall, such markets are central to the marketing strategy of the company/entity.

Keynesian A distinguished Economist J. M. Keynes (1883-1946) who primarily laid the foundation for today's macro economics. In particular he favoured liquidity preference; schedule; multiplier policies; and many other economic measures.

Key Officers The key officers of a ship are the master, the chief officer (second in command) and the chief engineer.

Key Pallet The first pallet which has to be discharged to enable an entry to be made in the stow.

KF Koff. A two masted sailing vessel; or Air Botnia (Oy Air Botnia Ab) – IATA Airline member.
KFTA Korean Foreign Trade Association.
Kg Kilogram.
KG Kirgizia – customs abbreviation; or measure distance from the ships keel to the ships centre of gravity.
Kg(s) Kilogram(s).
Khoums The currency of Mauritania. See Appendix 'C'.
Khoums invoice An invoice payable in Khoums currency relative to an export sales contract.
KHz Kilo Hertz.
KI Biro Klasifikasi Indonesia – the ship Classification Society of Indonesia.
Kicker An equity interest in the residual value of an asset, accepted by a bank as additional security or a sweetner in an otherwise marginal loan proposition.
Kickstart The immediate activation of any process/operation. It maybe short lived in terms of the duration of the operational period. Usually kickstart tends to be unscheduled and caused by a particular situation emerging which pressurises the urgent need to commence operations.
KID Key industry.
KIEGWAN Kie Gwan shipping (Pte) Ltd – a shipowner.
Kilogram (1000g) 2.2046223 lbs – unit of weight in metric system.
Kilometre Major measurement of distance throughout the world except the US; equivalent to 0.62 statute mile.
Kina The currency of Papua New Guinea. See Appendix 'C'.
Kina invoice An invoice payable in Kina currency relative to an export sales contract.
King pin The coupling pin, welded or bolted in the centre of the front underside of a semi-trailer chassis, which couples to the fifth wheel of the towing tractor or dolly converter.
Kingpin position The location of the Kingpin, which determines the load distribution of the articulated outfit, its overall length and the amount of swing clearance on a road vehicle.
King post A name sometimes given to a Samson Post or small mast on board a ship and used for supporting derricks on either side of the deck at the position of the hatchways. It is also the term used for the mast round which a 'Scotch derrick' or other types of slewing cranes pivot.
King or Queens enemies Enemies of the shipowners sovereign – a common law exception.
King room A hotel room with a king-size bed.
Kings warehouse Where goods seized by the Customs are stored.
Kip The currency of Peoples Democratic Republic of Laos. See Appendix 'C'.
Kip invoice An invoice payable in Kips currency of Laos relative to an export sales contract.
KL KLM (Royal Dutch). The National airline of Holland.
'K' Line A major liner shipowner and container operator and registered in Japan.
KLM KLM Royal Dutch Airlines. The Dutch National airline and member of IATA.
KLSE Kuala Lumpur Stock Exchange (Malaysia).
KM Air Malta Company Ltd. The Maltese national airline and member of IATA; or Measured distance from the ships Keel to the ships Metecentre.
Km Kilometre.
Kmer's Syndicate Premium Limit The limit prescribed by or on behalf of a member of a syndicate on the amount of insurance business allocatable to a year of account which is to be underwritten on his behalf through that syndicate (such limit being expressed as the maximum permissible amount of his member's syndicate premium income allocatable to that year of account); or where a limit lower than that referred to above is prescribed by or under the authority of the Council of the Committee, that lower limit.
Km/h Kilometres per hours.
kmph Kilometres per hours.
Kn Knot(s).
Knocked down condition Goods (e.g. vehicles) dismantled for transit.
Knock for Knock Agreement whereby each insurer pays for the damage to its policy holder regardless of who was to blame, providing the policy covers own damage. An insurance term.
Knock-out axle System which enables the in-line axle at the rear of a low-loading road trailer to be detached from the trailer. Wheels and axle maybe detached from the trailer complete with suspension units to enable the loading bed to be lowered to the floor for ease of loading.
Knot A ship's unit of speed – a nautical mile (i.e. generally 6080 feet) – 1.151 statute miles or 1853 metres per hour.
Known Loss A loss known to one or both parties when a broker and underwriter are negotiating a placing (see also FCAR).
Known outstanding losses Where claim notifications have been received for potential loss – insurance term.
Known shipper A definition devised by the Department of Transport and in the wake of the Lockerbie disaster and the subsequent legislation of the 'Aviation and Maritime Security Act 1990' which reads as follows:- 'a customer who has established contractual arrangement with the airline having met the requirements for establishing such an arrangement and whose procedures the airlines judge to be secure or whose procedures have been checked by the airline and found to be secure'.
KNUTSEN Knutsen Line – a shipowner.
KNV Koninktiijke Nederlandse Verenigingvan Transport Ondernemingen (Royal Netherlands Association of Transport Enterprises).
KNVR Koninktiijke Vereniging van Nederlandse Reders (Royal Dutch Ship-owners Association).
Kobo The currency of the Federation of Nigeria. See Appendix 'C' and Naira.
Kobo invoice An invoice payable in Kobo currency relative to an export sales contract.
Kommandittselskap A type of Norwegian limited partnership, which has been widely used in the financing of ship purchase. At lease one participant in the partnership (the komplementar) must be fully liable (although the komplementar may itself be a limited liability company), while at lease one, usually many more, has limited liability. A limited liability partner is known as a kommandittist. The attractions of the K/S partnership to a Norwegian investor paying an exceptionally high marginal rate of tax are that he achieves the same tax benefits as would apply to direct ownership of an asset, which having only limited liability.
Kopecks The currency of CIS. See Appendix 'C'.
Kopecks invoice An invoice payable in Kopecks relative to an export sales contract.
Korean Register Korean Ship Classification Society.
Korean Register of Shipping The Korean Ship Classification Society based in Deajeon, Korea.
Korins Korins Maritime Services Pte. Ltd – a shipowner.
KOSTPRO Korean Committee for Simplification of International Trade Procedures.

KQ Kenya Airways Ltd. The Kenyan National airline and member of IATA.
KR Korean Register of Shipping – Korea Ship Classification Society.
Krona The currency of Iceland and Sweden. See Appendix 'C'.
Krona invoice An invoice payable in Krona relative to an export sales contract.
Krone The currency of Norway. See Appendix 'C'.
Krone invoice An invoice payable in Krone relative to an export sales contract.
KSA Korea Shipbuilders Association.
KSC Korea Shipping Corporation – a shipowner; or Korean Shippers Council.
Kts Knots
KU Kuwait Airways Corporation. The Kuwait National airline and member of IATA.
Kurus The currency of Turkey. See Appendix 'C'.
Kurus invoice An invoice payable in Kurus currency relative to an export sales contract.
Kuwait Airways Corporation The Kuwait National Airline and member of IATA.
Kuwaiti Dinar The currency of Kuwait. See appendix 'C'.
Kuwaiti Dinar invoice An invoice payable in Kuwaiti Dinar currency relative to an export sales contract.
kV Kiolvolt

KVR Kamer van Koophandel en Fabrieken (Dutch Chamber of Commerce).
Kw Kilowatt(s).
Kwacha The currency of Malawi and Zambia. See Appendix 'C'.
Kwacha invoice An invoice payable in Kwacha currency relative to an export sales contract.
Kwanza The currency of Angola. See Appendix 'C'.
Kwanza invoice An invoice payable in Kwanza currency relative to an export sales contract.
kwh Kilowatt-hour.
Kyat The currency of Burma. See also Appendix 'C'.
Kyat invoice An invoice payable in Kyat relative to an export sales contract.
Kyosei Kyosei Line – a shipowner.
Kyoto Kyoto convention. This represents work undertaken by the WCO for the simplification and harmonisation of customs procedures. A significant development is a change in emphasis from consignment based controls at the frontier to control inland, using traders' records. See local clearance procedure and WCO entries.
Kyowa Kyowa Shipping Co Ltd – a shipowner.
KZ Kazakhstan customs abbreviation; or Nippon Cargo Airlines (NCA) – an IATA Airline Member.

L4 Lauda Air S.p.A. – an IATA Airline member.
L Liquid capacity in cubic feet (of a tanker vessel); or length.
LA Lifting appliance; or LA – Lan Chile S.A. – IATA airline member.
L/A Lloyds Agent; landing account; or letter of authority.
LA (LAN) Chile airline. The Chilean National airline and member of IATA.
LAB Live Animals Board – IATA term.
Label A slip inscribed and affixed to a package for identification or description.
Labour One of the primary factors of production. It embodies all human economic effort, both physical and mental contributing to the production of wealth, in the creation of utility. Economic term.
Labour cost Production/servicing cost which represents the labour input.
Labour intensive Production/servicing process which has a very high labour input and usually limited capital investment involvement. Teaching and nursing in the community are both labour intensive.
Labour migration The process of country labour force emigrating.
Labour Union Regulations Rules in respect of the requirement for crew members to be represented by trade unions (with consequent effect on wages and conditions).
LAC Lights Advisory Committee (of the Chamber of Shipping).
LACAC Latin American Civil Aviation Commission.
Lacquer base or lacquer chips, dry It may consist of a colloided solid mixture of nitrocellulose, pigment, gums, and a plasticizer. Those containing nitrocellulose are highly flammable.
LAFTA The Latin American Free Trade Association – now termed LAIA.
Lagan Cargo jettisoned and later buoyed for recovery.
Laggards The last group (16%) of consumers in a market to purchase a specific new product – a marketing term.
LAIA Latin American Integration Association. (Formerly LAFTA).
Laid up A vessel is laid-up when moored in a harbour etc., for want of employment.
Laid-up tonnage Vessels which are laid-up when moored in a harbour etc. for want of employment.
Laissez faire The policy of Government not to interfere with industry and thereby leave it to its own devices allowing the market forces to play their role in framing industry future.
Laker A vessel specially constructed for navigation in the waterways of the Great Lakes and canal systems of North America, carrying mainly ore, grain and timber.
Lake SS Periodical survey of ship's classed for Great Lake service in North America.
Lamcon Code name of charter party used in iron ore trade (area).
LAN Chile airline. The Chilean National airline and member of IATA; or Local Area Network.
Lame duck The term used to describe a company/entity with a very poor financial rating and maybe in financial difficulties.
Lanai A room with a balcony or patio overlooking water or a garden, usually in a resort hotel.
LANBY Large automatic navigational buoy.
Lancashire flat A type of flat with head-board at one end.

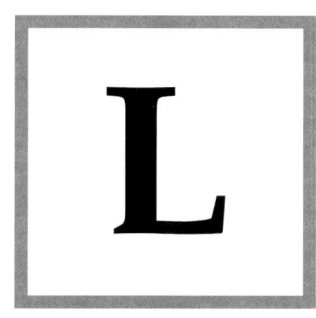

Landbridge The provision of through international, dedicated multi modal transport service operation on a regular basis. It maybe the Far East/European service involving sea/rail/sea with the containers being rail conveyed between the West and East coast ports; the Trans Siberian railway linking in with ocean transport services; and the international road haulage service embracing two separate voyages with a road transit intervening thereby giving a sea/road/sea transit with the road leg linking the maritime transits. Overall, such services offer a through rate, single carriers multi modal document and guaranteed schedules/transit times.
Landbridge rate A freight rate embracing two maritime tariffs and surface transport rate. It maybe a consignment from a European port to Japan and involving a voyage across the Atlantic, a rail journey across North America and the final journey to Japan by sea.
Landed prices The price/value of a product(s) at the time of importation into a specific country. Usually it is a reflection on the market.
Landing account The charges raised at a particular port to cover the handling of the goods overside from the vessel to the quay/lighterage or other specified area. The items featured in the landing account will vary by port.
Landing and delivering cargo The process of landing/discharging cargo from a ship onto the quay or lighterage and subsequently delivering/despatching the cargo to its specified destination – after all the documentation and customs formalities have been undertaken.
Landing card A card which is usually issued to each passenger on board a vessel in exchange for a ticket. On disembarkation the passenger surrenders the ticket thereby ensuring all passengers have purchased a ticket.
Landing charges Expenses incurred in the discharge of cargo at a port. Such expenses maybe included in the sea freight or such as under the CIF where the charges are payable by the importer/buyer/receiver.
Landing gear Legs supporting a semi-road trailer when uncoupled from a tractive unit.
Landing Officer HM Customs officer responsible for control of imported/exported cargo.
Landing order A document authorising the dock company or wharfinger to receive goods from a ship.
Landing stage A shore based installation, usually situated on a river estuary or lake, where passengers embark and disembark from/to a steamer/ferry boat.
Land locked A country which has no seaboard.
Land locked country A country with no seaboard such as Nepal and Switzerland thereby relying on overland transport for distribution of merchandise to and from a seaport situated in a third country.
Land locked port A port which is situated inland and has to rely on rivers/canals on access to the sea board.
Land operator A firm that provides sightseeing, tickets and other services as a destination.
Lane length Total maximum linear lane length in metres available for stowage of vehicles/wagons on a Ro/Ro vessel.
Language audit The process of evaluation the language needs in a particular situation relative to the objectives. It may be French, German or Arabic for a particular company exporting in such markets involving medical products.

Lapse rate A measure of the temperature profile of the atmosphere with height – the fall of temperature in a unit height.

Laree The currency of Republic of Maldives. See Appendix 'C'.

Large-value payments Payments, generally of very large amounts, which are mainly exchanged between banks or between participants in the financial markets and usually require urgent and timely settlement.

LASH Lighter aboard ship.

Lash end fitting A fitting for connecting a lash to either corner casting or tensioner, in container securing practice.

Lasher A dockside labourer responsible for fitting lashing gear. Also termed Rigger.

Lashing Container or cargo securing member effective in tension – usually rod, wire or chain; in general the securing of cargo.

Lashing cage A man-carrying cage lifted by a container gantry crane from which lashing operations can be accomplished.

Lashing eye A fitting connected to the ship structure to which a tensioner or lash is attached. Also termed 'D' ring.

Lashing plate A flat plate projecting above deck or hatch cover to which a tensioner is connected.

Lashing point Deck fitting to which a lashing is attached.

Lashing pot A flush or low profile deck fitting to which a lash can be attached.

Lashing ring A fitting connected to the ship's structure to which a tensioner or lash is attached. Also termed 'D' ring.

Last cargo A chartering term whereby the shipowners warrant that the three cargoes last carried by the vessel – usually a tanker – prior to the commencement of loading of the cargo, shall have been clean unleaded products.

Last carrier Transport operator over whose route(s) the last section of carriage under the consignment note is performed. It usually involves a combined transport operation or single transport mode such as road, rail, or air involving connecting services. For example, the participating carrier over whose air routes the last section of the carriage under the air Waybill or ticket is undertaken/performed. An IATA definition

Last in, first out price A method of pricing material issues using the last purchase price first.

Last-seat or room availability An agent's ability to book the last seat on a flight or room in a hotel; for travel agent, this access is an issue when a vendor cannot or will not release through a CRT operation by another supplier.

Last trading day The day on which trading ceases for a particular month.

Lat Latitude.

LAT Lowest Astronomical Tide.

LATAM/CAR Latin American/Caribbean (Region – IATA).

LATCC London Air Traffic Control Centre.

Late charge Charges appearing on a guest's credit card for services such as restaurant and telephone that do not appear on the bill at checkout.

Late majority The fourth group (34%) of consumers in a market to purchase a specific new product after innovators, early adopters, and the majority. They have usually below-average income and social prestige.

Late night trading ('Late') The period of unofficial LME trading from the close of the afternoon kerb to the close of the comparable US markets.

Latent defect A fault in insured property (e.g. machinery or ship's hull) which exists before the insurance commences and comes to light during the policy period. A defect not obvious from cursory inspection.

Latent head of vaporisation The amount of heat absorbed or emitted during the change of state from a liquid to vapour or vice versa without change of temperature

Latent heat The heat required to cause a change in state of a substance from solid to liquid (latent heat of fusion) or from liquid to vapour (Latent heat of vaporisation). These phase changes occur without change of temperature at the melting point and boiling point, respectively.

Latent market A quasi dormant market which becomes energised when a certain pattern/series of events emerges.

Lateral and front moving track High lift stacking track capable of stacking and retrieving loads ahead and on either or both sides of the driving direction.

Lateral cargo mobility The lateral cargo mobility coefficient records the number of different cargo units that the vessel can convey.

Lateral diversification The process of a Company/entity embarking on a new area of the business sector to which their core business is not related. An example would be a shoe manufacturer entering the ceramics field.

Lateral thinking The process of developing a strategy outside the company. See horizontal thinking.

Late show Passenger holding a reservation who arrives at the check-in desk after the designated time.

Latex Liquid rubber – milky juice or sap of plants.

LATF Lloyds American Trust Fund.

Latin American Integration Association – formerly LAFTA Its aims are the gradual elimination of all types of duties and restrictions (with special treatment for agriculture) that affect the import of goods originating in the territory of any member, the co-ordination of industrial and agricultural development within the area and the harmonisation (as far as possible) of their import and export systems, in respect of capital goods and services originating outside the area. Free trade was achieved within the area by 1980 and it now includes eleven countries embracing Argentina, Bolivia, Brazil, Chile, Colombia, Ecuador, Mexico, Paraguay, Peru, Uruguay and Venezuela.

Latticed sided container An open or closed container with at least one side consisting of elements with openings between them.

LAUA Lloyd's Aviation Underwriting Association.

Launch The process of launching/placing the vessel for the first time after construction into the water; an open deck or half decked small boat with or without an engine; or the process of placing on the market a new product for sale.

Launch campaign A marketing campaign to launch a product or service.

Lawful goods Cargo or merchandise which is considered to be legal in accordance with the laws of a country.

Law of comparative advantage Whenever opportunity costs differ among countries, specialisation of each country in producing those commodities in which it has comparative advantages will make it possible to increase production of all commodities relative to the quantities available if each country attempted to be self-sufficient.

Law of demand The quantity of a good or service demanded in a given time period increases (decreases) as its price falls (rises). Economic term.

Law of Diminishing Returns (Law of Variable Proportions – Law of Proportionate Returns) The hypothesis is that if successive increments of one factor of production are employed in conjunction with a constant amount of all factors, these increments will, after a point yield successively smaller and smaller additions to total output. A law stating that the marginal physical product of a variable input declines as more of it is employed with a given quantity of other fixed inputs. Economic term.

Law of supply The quantity of a good or service supplied in a given time period increases (decreases) as its price increase (decreases). Economic term.

Law of the flag The applicable maritime law relative to the Nationality of the ship's flag.

Laydays The days agreed as found in a charter party between the parties during which the shipowner will make and keep the ship available for loading and discharge. They maybe separated into days for loading and days for discharge, or if agreed a total taken for the two operations when they are known as 'reversible laydays'.

Layout chart A sheet provided with scales and other indication conforming to the characteristics of the majority of character printed machines in general office and data processing use.

Layout key A pro forma document used for indicating space reserved for certain statements appearing in documents in an integrated system.

Laytime The period of time agreed as found in a charter party for loading and discharging the vessel. It is fixed on the basis of loading/discharging rate, e.g. 100 tons per day per workable hatch. In the event of the laytime becoming expired, the cancelling date clause takes effect. Overall, it is the period of time agreed between the parties during which the owner will make and keep the vessel available for loading or discharging without payment additional to the freight.

Laytime saved Despatch to be paid by the shipowner to the charterer for the laytime covered in accordance with the terms of the charter party. A chartering term.

Lay up The temporary withdrawal of tonnage (laid up) by shipowners from freight/passenger earning activity. This occurs during periods of depressed freight rates/passenger tariffs when trading is no longer commercially viable.

Lay up berth A berth situated at a seaport where vessels temporarily out of service/without employment are moored usually for an indefinite period.

Lay up of a vessel The temporary cessation of trading of a vessel by the shipowner.

Lay up return The premium charged on a hull time policy is based on the vessel navigating and being exposed to full maritime hazards during the whole currency of the policy. If the ship is laid up in protected waters for an agreed amount of time the risk is considerably reduced and therefore the insured will receive a return of part of the premium. Insurance definition.

Lazaretto A place where persons are kept during quarantine; or where freight is accommodated for the purpose of fumigating it.

LB Lloyd Aéreo Boliviano S.A. (LAB) – an IATA Member Airline.

LBC Local Baggage Committee – IATA term.

LBO Leverage buyout. See separate entry.

LBP Length between perpendiculars (of a ship's hull).

lbs Pounds

LC London clause; label clause; lay can; or Loganair Ltd. – an IATA Airline member.

L/C Letter of credit. The document in which the terms of a Documentary Credit transaction are set out.

LCA Life Cycle Analysis.

LCCI London Chamber of Commerce & Industry.

LCD Liquid crystal display – Electronic display screen widely used in various navigation instruments.

LCE The London Commodity Exchange (where BIFFEX is traded); or Local Currency Exchange.

LCG Longitudinal centre of gravity.

LCH The London Clearing House (a company owned by the major clearing banks which guarantees against counterparty default on BIFFEX).

LCL Less than (full) container load. A parcel of goods too small to fill a container which is grouped by the Carrier at a CFS with other compatible goods for the same destination.

LCL Door/LCL Depot Under this term used in container movement the carrier collects the cargo from the shipper and takes it to the container depot for groupage assembling/consolidation into a container. At the destination the carrier delivers the container to the consignee at the importing depot.

LCL/FCL A container term. The exporter/forwarder delivers the shipment to the carrier's depot or container freight station as 'break bulk'. The carrier packs the container at the container freight station/depot whilst at the destination destuffing of the container is performed by the consignee. Also termed Pier to House.

LCL/LCL A container term. The cargo is packed for shipment and unpacked at destination by the ocean carrier. Both operations are undertaken at the carrier's premises. Hence, the inland movement prior to and subsequent to the sea carriage is effected by the merchant or his agent. Also termed Pier to Pier.

LCL Service Charge Charge made by ocean carrier for packing LCL cargo and sometimes described as a packing charge. Also describes unpacking of inbound containers with mixed LCL cargoes.

LCM Lateral cargo mobility.

LCP Local clearance procedure – customs term.

L/D Loading/Discharging

LDCs Less Developed Countries.

ldg. Loading.

Ldg & Dly Landing & delivery.

LDMC Livestock Development and Marketing Corporation (Burma).

lds. Loads.

ldt Light displacement tons.

ldwt Light deadweight tonnage – the unladen weight of a ship.

Lead The direction a mooring line takes up whilst being handled or when made fast.

Leadage The cost of transporting coal from colliery to place of shipment.

Leader An underwriter whose judgement is so respected by other underwriters that they will follow his lead in accepting a risk. His syndicate or company will be the first on the slip. A marine insurance term.

Leading Underwriter Agreement A provision on a broker's slip whereby subsequent agreements (other than those which materially alter the risk) are acceptable to the underwriters on the slip when initialled solely by the leading underwriter.

Lead manager The principal issuing house or sponsor of a new issue or IPO.

Lead slip To lead a slip is to be the first, e.g. syndicate or company on the slip. A marine insurance term.

Lead time The period required for example to prepare for the introduction of a new facility/scheme/plan/product etc. It includes time for order preparation, queuing, receiving, inspection, and transport.

Lead time delay The time delay incurred relative to the

introduction of a new facility/scheme/plan/product etc. For example the preparation time for a particular scheme originally was six months which has now been extended to nine months – a three month lead time delay.

League International for Creditors An association of independent collectors situated throughout the world whose task is to collect monies from the importer (buyer) in accordance with the terms of the export sale contract.

League of Arab States A group of 33 Arab States.

Lease A contract between a lessor and a lessee for the hire of a specific asset. It is a means of financing the use of an asset. The lessee has full use of this asset but he is not the owner, although in many countries the law may allow him to become the owner on termination of the lease. Under a leasing agreement a leasing company will purchase the capital equipment for the lessee and rent it to the lessee for a period of two years or more. Items suitable for leasing are manufacturing plants, computers, commercial vehicles etc. There are many different types of leasing contract, the two most well known being finance leases and operation leases. The former recovers the leasing company's investment in the equipment during the term of the leased contract, whilst in the latter the leasing company will probably lease the equipment to several different users during the life of the equipment. Leasing advantages include: capital is freed for other uses; lessee does not have to wait until sufficient funds are available to purchase the equipment; rental terms are flexible and can be adjusted to suit individual needs; costs are usually fixed for a known period; and finally in some countries tax concessions are available.

Lease back A situation in which the owner of a product which may be a ship, property, plant and machinery etc., sells it to a person or institution, and then leases it back again for an agreed period of time and rental as specified in the agreement. Such a scheme has tax benefits and can raise cash for the owner in times of cash flow problems within the Company. A common method of raising liquid funds without losing control of an asset.

Leased-space operations Where a carrier leases seats or cargo capacity on a service to another carrier; also known as shared operations.

Leasing The hiring of an asset for the duration of its economic life. It is a method of financing which has been widely used in the shipping industry, and in many other capital intensive industries. Tax leases are based upon the ability of the lessor to make use of the capital allowances associated with the acquisition of expensive capital equipment. These tax benefits are then passed on to the lessee via an adjustment to the level of rental payments. Leveraged leases in the United States take advantage of the tax depreciation obtainable on the asset acquisition cost, although the majority of this cost is met through borrowings. Again, the benefits of these tax savings are reflected in the effective interest cost element of the rental payments.

Lebanese Pound The currency of Lebanon. See Appendix 'C'.

LEFO Land end for orders.

Left-locking twistlock A twistlock which is locked when the handle is pointing to the left.

Leg Legalisation; or the portion of a flight/sailing between two consecutive scheduled airport stops/seaports.

Legacy currencies The former national currencies of the countries participating in EMU.

Legal barriers In trade terms barriers which prohibit/restrict the free flow of trade on a bilateral or multilateral basis.

Legalised invoice An export invoice which has been certified/authenticated by an accredited Chamber of Commerce or Consulate of the country to where the goods are being exported.

Leisure led A market situation which is leisure led in terms of its influence and development.

Lek The Albanian currency. See Appendix 'C'.

Lempira The currency of Republic of Honduras. See Appendix 'C'.

Len Lengthened.

Lending Selling metal on a nearby date and simultaneously buying it back on a forward date, thereby extending a long position. See long entry.

Length BP (between perpendiculars) The distance on the summer waterline from the fore side of the stern to the after side of the rudder post, or to the centre of the rudder stock if there is no rudder post.

Length from bow to centre manifold The distance from the bow to the centre of the rudder stock if there is no rudder post.

Length overall (LOA) The extreme length of a ship.

Leone The currency of Sierra Leone. See Appendix 'C'.

LES Land earth stations.

Le Shuttle The 'brand name' given to the dedicated driver, motorist, coach operator and freight vehicle 'on rail' services between the road/rail terminal portals at Folkestone and Calais involving Euro tunnel. See Le Shuttle Freight service.

Le Shuttle Freight Service Eurotunnel's rail freight service carrying freight lorries and their drivers. See Le Shuttle.

LESO Land earth station operators.

Less developed countries Such countries are regarded as at an early stage of industrial development with an industrial base primarily serving the domestic market. Overall, the country would be agriculturally driven in economic terms. Such countries are now more usually referred to as developing countries.

Lessee Holder of, tenant of house etc., under lease.

Lessor Person who lets a lease.

Less than container load A container which is filled with consignments of cargo for more than one consignee or from more than one shipper. Overall, it is a parcel of cargo too small to fill a container which is grouped by the carrier at a container freight station/depot or container base with other compatible cargo for the same area/country/region.

Letter of assignment A document with which the assignor assigns his/her rights to a third party.

Letter of credit A Documentary Letter of Credit is an instrument by which a bank undertakes to pay the seller (exporter) for his goods providing he complies with the conditions laid down in the credit. Whilst ensuring that the exporter receives payment for his goods it also gives the importer (buyer) the confirmation that he will not have to part with his money until he is presented with documents evidencing that the stipulated conditions have been fulfilled. The credit specifies when payment is to be made. Usually this is either: when the documents are presented to the paying bank or at some future date, usually within 180 days or receipt of the documents by the paying bank. Such a method of payment is common in export sales contracts. Overall, it is a letter issued by a Bank authorising credit to a correspondent.

Letter of Hypothecation Bankers document outlining conditions under which an international transaction will be executed on exporters behalf, the latter which have given certain pledges to his banker.

Letter of Indemnity Request to a company's registrar to issue a replacement stock or share certificate when the original has been lost, destroyed, or stolen. In it the holder undertakes to indemnify the company for any loss incurred as a result of issuing a duplicate; most companies require this undertaking to be countersigned by a bank or insurance company. Alternatively, it is a document indemnifying the shipowner or agent from any consequences, risk or claims which may arise through 'clean' Bill of Lading being irregularly issued. Also, a guarantee signed by shippers to avoid a clause being put into a Bill of Lading.

Letter of introduction A letter written usually by a Bank addressed to Group Offices or banking correspondents in the centres which the traveller (businessman/businesswoman) intends to visit and request that assistance be given by way of information and advice. It is usual to indicate the purpose of the bank customer's journey and can be linked to a Trade Enquiry. The letter is written at the request of the Bank customer and forms an important aid available to customers or their representatives travelling abroad on business. An International Banking definition.

Letter of Renunciation Form attached to an allotment letter which is filled in should the original holder wish to pass his entitlement to someone else. A Stock Exchange term.

Leu The currency of Romania. See Appendix 'C'.

Leutwiler report A report published in March 1985 by a group of seven eminent persons and entitled 'Trade Policies for a Better Future – Proposals for Action'. It was chaired by Dr Fritz Leutwiler. The report was initiated and submitted to GATT.

Lev The currency of Bulgaria. See Appendix 'C'.

Level The grade of Marine Terminal staff for which a particular competence is most appropriate. This could be Supervisor or Operator for example.

Levelling pedestal A stacking fitting that permits bridge fittings to be used with containers of different heights.

Level of service The quality of service provided by an entity/Company in the market place.

Level playing field A term to describe the situation where all the conditions are broadly the same in a particular area. An example arises where in the EU there is no restriction on the flow of goods amongst the twenty five States and on that basis each State competes on an equal basis.

Level Seas An internet chartering system.

Leverage The degree of exposure of securities to market risks; the capital structure of a company may be increased by issues of loan stock as well as equity, and the risk relationship between the two may be described as the leverage.

Leveraged buyout (LBO) An acquisition of a company by an investor group in a transaction funded principally with loans intended to be repaid out of asset disposals. Where management is involved, such a transaction is referred to as a management buyout (MBO). The relative transparency of shipping company values and the high levels of gearing common in the industry have made LBOs and MBOs quite rare in shipping, although there are signs of increasing activity of this type.

Leverage (gearing) In view of the low option premium needed to establish a market position in the underlying security, any given change in the share price can produce a much larger percentage change in the value of the option than in the value of underlying shares. A LIFFE term.

Lever tensioner A tensioner primarily used with chains. See also chain tensioner entry.

Lex mercatoria (Latin) The low 'merchant'. The usages and customs of merchants ratified by the Courts.

Lex non cogit ad impossibilia (Latin) The law does not compel one to do impossible things.

Lex scripta (Latin) The written or statue law.

LG Luxair – an IATA Airline Member.

L(gas) Capacity in cubic metres of liquefied gas cargo in a tank.

LGC Laboratory of the Government Chemist, or liquefied gas carrier.

LGSS London General Shipowners Society.

LH Deutsche Lufthansa A.G. (LUFTHANSA). The National airline of Germany and member of IATA.

Lh Last half of the month.

LHA Local Hour Angle.

LHAR London, Hull, Antwerp or Rotterdam (ports).

LHX Light helicopter family, experimental.

L/I Letter of Indemnity. Sometimes used to allow consignee to take delivery of goods without surrendering B/L which has been delayed or become lost.

Liability insurance Insurance to cover the legal liability of the assured to the extent of such liability but subject to any limitations expressed in the policy.

LIANHUAT Lian Huat Shipping Co. (Pte) Ltd. – a shipowner.

LIAN SOON Lian Soon Shipping & Trading Co. (Pte) Ltd. – a shipowner.

LIARS Lloyd's Instantaneous Accounting Record System.

LIBA Lloyd's Insurance Brokers Association.

Libelling (a vessel) American term for arresting a vessel.

Liberian dollar The currency of Liberia. See Appendix 'C'.

Liberalisation of trade A situation whereby a particular country conducts overseas trade with other countries with no trade restrictions in terms of quotas, tariff barriers, licensing and so on.

Liberalisation of trade barriers The withdrawal of trade barriers in a specific country/trading bloc thereby removing impediments to trade. See trade barriers entry.

Liberalised trade A situation where the trading position between two countries is not constrained and no impediments exist to discourage/develop trade.

Liberty ship A vessel built during the period 1942-45 in USA on a mass production basis.

LIBOR London Inter-bank offered rate; the key rate at which banks will lend to each other in the Eurocurrency or inter-bank markets. It is often used as a guide for interest rates on other loans. LIBOR will vary according to market conditions and will of course depend upon the loan period as well as the currency in question; it maybe found that at the same time, for the same currency and for the same period, the quotation of a LIBOR figure by one bank in London and another in London will differ slightly; this would be expected if one bank were already 'long in a currency' for that period, having just taken in a matched deposit, and the other bank's position were different.

Libyan Dinar The currency of Libya. See Appendix 'C'.

Libyan Dinar invoice An invoice payable in Libyan Dinars currency relative to an export sales invoice.

LIC League International for Creditors. An association of independent collector's situated throughout the world whose task is to collect monies from the importer (buyer) in accordance with the terms of the export sale contract.

Licensable A consignment requiring an import or export licence.

Licensing

Licensing A situation whereby a licence is granted to an overseas company to manufacture the products on a royalty basis using either for example the UK manufacturer's brand name or the name of the licensee.

Licensing requirements Trade barriers or regulators requiring an import licence for specific products imported into a country.

Lido The area around the swimming pool on a cruise ship; a fashionable beach resort.

Lien Retention of property until outstanding debt is discharged.

Lifeboat A boat carried by a ship for use in an emergency or land based at a lifeboat station.

Life cycle See Product life cycle.

Life Cycle Analysis The process of analysing the product life cycle of a service or product involving all its stages of growth and ultimate decline. See product life cycle entry.

Life cycle cost All costs associated with feasibility studies, research development, design, production, support training and operating costs generated by the acquisition, use and replacement of physical resources.

Life extension programme Ship management extending the life of a vessel through the survey programme.

Life saving apparatus Shipboard equipment provided for passengers/crew use in an emergency such as abandoning ship.

Life-seeing Activity or orientation of a tourist seeking to observe and meet residents of the destination to experience how they live.

Life span In business terms the commercial life of an asset/product.

LIFFE London International Financial Futures and Options Exchange.

LIFO Last in first out (price), or Liner in free out.

Lifting on/Lifting off A cranage charge raised by the carrier, upon receipt at depot or container yard of FCL on exporter's or forwarder's haulage (merchant haulage). This term is also used when import FCL's are placed at carriers facility on to importers or forwarders haulier.

Lift on/lift off type of vessel A ship whose cargo transhipment arrangements are cranes, using either the ship derricks or shore based lifting apparatus, such as a cellular containership.

Lift unit frame A system in use with ro/ro transport of handling numerous containers at one time and other cargo to achieve high handling rates.

Light bill A customer's receipt for the payment of light dues.

Light cargo Merchandise which fills cargo holds to capacity but does not bring vessel down to her load line.

Light dues Monies collected by the UK Customs on behalf of Trinity House for the maintenance of lighthouses and buoys. Dues are levied on vessels according to their net registered tonnage.

Lightening or lightering The process of transferring cargo from a tanker to another ship. This often involves the operation of reducing the draft of a VLCC by discharging part-crude oil cargo into a smaller tanker.

Lighter A dumb barge without its own mechanical means of propulsion. See dumb barge entry.

Lighter aboard ship An ocean going barge carrier with a capacity of 44,000 dwts, and capacity of 73 barges each of about 400 tons of cargo capacity. The barges are transhipped from the mother vessel on arrival at the port area to be towed along the various inland waterways thereby providing a door to door service. The vessel can be converted to carry up to 1400 TEUs.

Lighterage The provision of barge resources. Also the price (dues) paid for loading or unloading ships by lighters or barges.

Lighterage berth A berth specially designed for the loading/discharge of cargo to/from lighterage.

Lightering A procedure for discharging part of a vessel's cargo into other vessels, so that it may enter a port which could not take it when fully laden.

Lighters fuse Articles of various design actuated by friction, percussion or electricity and used to ignite safety fuse.

Light grain Barley and oats – a chartering term.

Light industries Industries involved in the production/manufacture of products with a low weight but high volumetric ratio such as furniture, white furniture, electrical goods especially components, and footwear.

Light on demand A situation/circumstances whereby the demand is rather below expectations/predictions such as 100 seats occupied on an airliner which is some ten below the predicted figure of 110.

Lightweight In maritime terms it is the weight of the vessel as built, including boiler water, lubricating oil and the cooling system water. See displacement light entry.

Lilangeni The currency of Swaziland. See Appendix 'C'.

Lilangenic invoice An invoice payable in Lilangeni currency of Swaziland relative to an export sales invoice.

Limit A restriction set on an order to buy and sell, specifying a minimum selling or maximum buying price. When giving a limit the client should say for how long it is to be kept in force, e.g. goods for the (current) account, goods till cancelled etc., or a figure which has been agreed by Bank management up to which a customer is able to go overdrawn on the customer's account, or in some futures markets (but not on the LME) there is a limit to the rise or fall in the price of a commodity during a given period of time. Once this limit is reached, trading is suspended until the time period expires, when trading can resume. The suspension relates only to forward contracts, the spot month continuing to trade.

Limit any one bottom The maximum amount that can be declared under open cover for shipment on any one vessel featured in the insurance cover. Bottom refers to ship's bottom in this contract.

Limit any one location The maximum amount of cargo at underwriter's risk. Under an open cover or floating policy, in one location. The Institute location clause applied the limit solely to accumulation risk prior to shipment but a special location clause also published by the Institute of London Underwriters applies the limit to any one location throughout transit.

Limitation of liability Maximum sum of money payable by a carrier to a shipper or bill of lading holder for any damage or loss to the cargo for which the carrier is liable under the contract of carriage. The basis of the limitation may be per piece or package or per tonne or per container according to the particular contract. The amount of the limitation is determined by agreement of the two parties or by law.

Limited Authorisation Authorisation valid for a fixed period of time by the Commissioners of Customs & Excise, and issued to traders not handling end-use goods on a continuing basis.

Limited partnership A partnership consisting of one or more limited partners whose liability is limited to the amounts they have invested and one or more general partners who are fully liable for the partnership's debts.

Limited postponed accounting An imports accounting system relative to VAT which grants limited postponed accounting for those companies necessarily importing parts and raw materials for incorporation into or processing into export consignments.

Limited risk An option buyers maximum risk is the purchase cost of the options, e.g. the amount of premium paid. A LIFFE term.

Limited slip differential Design of differential which limits the extent to which one driving wheel on an axle is allowed to spin in relation to the opposing wheel, when it cannot obtain grip on slippery surfaces.

Limited terms Insurance conditions that provide limited cover; being less than full cover (examples – FPA, TLO).

Limit order An order to buy or sell a 'futures contract' within the freight market at an exact specified price – a BIFFEX term.

Limit up/down A restriction on price movement administered by the Exchange (currently 50 points on BIFFEX).

LIMNET (London Insurance Market Network) This links computers of more that 1,000 participating firms and other organisations who combine to form the London Insurance Market, providing them with a unique data interchange facility.

Limousine service At hotels, the transportation provided to guests, especially between the airport and the property.

Line The amount or percentage in a broker's slip or policy which establishes the extent of the underwriter's liability. The line is written by the underwriter on the slip when he accepts the risk. The signed line is the underwriter's proportion of the risk as shown in the policy.

Linear-cut approach Method for cutting all tariffs across-the-board by an agreed percentage.

Line featuring A marketing term whereby the advertising and sales campaign features the complete product line available in the market place. A retail marketing term.

Line haul The actual transporting of items between terminals as distinguished from pick up delivery and other terminal services.

Line haul carriers See line haul entry.

Line management A structural organisation whereby the line of command/responsibility passes from a common function/responsibility area. An example arises whereby the Divisional Marketing Manager would respond to the Chief Marketing Manager at headquarters and have no direct responsibility to the Divisional Manager whose role maybe more of a co-ordinator.

Line of credit A loan to buyers to enable them to pay in cash term exports of capital, semi capital goods, and associated services; such lines maybe in respect of a specific project or for unconnected contracts with overseas purchasers. Overall, a bank credit facility available to a customer on condition that it is used for agreed purposes and as long as the customer's account remains at the lending bank.

Line pruning The process of dropping/eliminating a product(s) from the product line available in the market place. A retail marketing term.

Liner A vessel habitually employed on a regular schedule and loading and discharging at specified ports for passengers and/or cargo on given routes. It maybe under Liner Conference conditions.

Liner Broker A shipbroker who finds the cargo for vessels which regularly trade as common carriers on specific scheduled services.

Liner cargo A wide range of commodities conveyed on liner cargo services embracing both industrial and consumer products such as timber, steel, fertilizers, textiles, coffee/tea, oils and fats, wheeled vehicles and so on.

Liner conference Voluntary organisation whereby a number of shipowners – often of different nationality – offer their regular services to a series of ports on a given sea route on conditions agreed by the members.

Liner conference code A liner conference code 40/40/20 adopted at the UNCTAD V Manila conference 1979, where the participating trading national shipping lines convey each 40% of the cargo and third flag carriers the remainder.

Liner conference lines Shipping Companies which operate within a specific liner conference system.

Liner consortia A number of shipowners providing liner cargo service(s) operating on a consortium basis. This is particularly evident in the container shipping market.

Liner freight indices A liner freight index published monthly by the German Ministry of Transport. It is devised in three parts, overall index; homebound index; and outbound index. It measures/records the liner freight movement in each of the three categories monthly. See tanker freight index.

Liner in free out The qualification to a freight rate denoting that it is inclusive of the sea carriage and cost of loading. It excludes the cost of discharging which is payable by the shipper or receiver. There maybe a lay time and demurrage arrangement at the port of discharging since the carrier has no control over the discharging.

Liner rate Cargo rate based on measurement, weight, value or linear footage.

Liner Shipping Company A shipowner offering regular sailings for cargo and/or passengers involving specified ports.

Liner terms An inclusive freight rate usually found on cargo liner services which embraces not only the sea freight, but also loading and unloading costs for a particular consignment. This varies widely from country to country and within countries, from port to port: in some ports, the freight excludes all cargo handling costs, while in others the cost of handling between the hold and the ship's rail or quay is included. Under a voyage charter the shipowner is paid freight which includes the cost of loading, stowing and discharging the cargo.

Liner time Code name of BIMCO uniform time charter Spanish edition.

Liner trade route A liner cargo trade which is usually containerised offering regular all the year round services. The three major liner trades are Transpacific: Asia/United States and United States/Asia; Europe Asia/Asia-Europe; and Transatlantic: United States/Europe and Europe/United States.

Liner voyage simulation A computer facility designed to assist liner management in the decision making process. Overall, the broad aim is to assist the management of a shipping company to assess the optimum fleet of vessels for a defined trade or trades, or to investigate the results of various service strategies before firm decisions are made in the actual service.

Line slip An arrangement entered into between underwriters and brokers whereby, for a given type of risk, the broker needs to approach only the leading and second underwriter who will accept or reject each risk on behalf of all the underwriters concerned in their agreed proportions. This is an administrative convenience where a broker is placing a large number of similar risks with the same group of underwriters. Insurance term.

Line stamp A rubber stamp used by a Lloyd's underwriter. It incorporates the syndicate's pseudonym and number and is impressed on the broker's slip by the underwriter who inserts his line and reference.

Line stretching A retail marketing term whereby the product line extends into new areas.

Line way bill The code name of a BIMCO approved seaway bill. It is for port to port shipments but can also be used for through transport where, unlike the multimodal liability regime, the carrier is responsible only when in control of the goods. At other times the carrier acts as the shipper's agent when, for example, entering contracts for other forms of transportation and during pre and onward carriage.

Linkage The inter-connections between international trade and other factors, including Third World indebtedness, foreign investments, interest rates and currency misalignments. A WTO definition.

Links between securities settlement systems The procedure between two securities settlement systems for the cross-border transfer of securities through a book-entry process (i.e. without physical transfer).

Link span A bridge ramp on which the vessel's ramp can rest and thus connects a vessel with the quay or shore and allows for the working of ro/ro trailers and cargo. In so doing the vehicles are transhipped (driven to and from the vehicle deck) over the stern or bow of the vessel via the link span.

LINX London Internet Exchange. A not-for-profit partnership of Internet Service Providers, providing a physical interconnection for the exchange of internet traffic.

Liquefied Gas A liquid which has a saturated vapour pressure exceeding 2.8 bar absolute at 37.8° C and certain other substances specified in the Gas Codes.

Liquefied natural gas carrier A specially constructed ship designed to carry natural gas at low temperatures or under pressure.

Liquidation The winding up of a company, in which assets are sold, liabilities settled as far as possible and any remaining cash is returned to the owners; or the closing out of a long or short position – a BIFFEX term.

Liquid bulk cargo Merchandise conveyed in bulk and of a liquid nature/specification such as oil, wine, chemicals and so on.

Liquid dangerous goods Unless otherwise provided, dangerous goods with a melting point or initial melting point of 20° C or lower at a pressure of 101.3 kPa must be considered as liquids. A viscous substance for which the specific melting point cannot be measured is considered a liquid if it is so determined when subjected to ASTM D 4359-90 test or to the test for determining fluidity (penetrometer test) prescribed in Appendix A.3 of the United Nations publication ECE/TRANS/100 (Vol.1) (ADR) with the medication that the penetrometer must confirm to ISO Standard 2137:1985 and that the test must be used for viscous substances of any class.

Liquid gas tanker Vessel constructed and arranged for the carriage of liquefied gases either in integral tanks or independent tanks under pressure or refrigerated.

Liquidity The ability to service debt and redeem or re-schedule liabilities when they mature and the ability to exchange other assets for cash. In the case of shipping companies, this should entail the maintenance of substantial balances of cash and short-term investments, since the relationship between current income and short-term payables may quickly become adverse.

Liquid petroleum gas carrier Ship designed to carry liquid petroleum gas, such as butane or propane, by means of tanks within the holds. The gas is kept in liquid form by pressure and refrigeration.

LIRMA London Insurance and Reinsurance Market Association.

Lisbon Agreement Treaty, administered by WIPO, for the protection of geographical indications and their international registration. WTO term.

LISCR Liberian International Ship and Corporate Registry.

Lisente The currency of Lesotho. See Appendix 'C'.

Lisente invoice An invoice payable in Lesotho Lisente relative to an export sales invoice.

Listed company A company that has obtained permission for its shares to be admitted to the London Stock Exchange's Official List.

Listing The obtaining of a quotation on a stock market for loan stock or equity which maybe traded on the Stock Exchanges.

List price The price published by the supplier/retailer to the market and without any discounts.

Liter (litre) Volume measurement used in most of the world except the US; equal to 1.057 quarts (liquid).

Lithium battery or lithium cells A battery is two or more cells which are electrically connected together by a permanent means. A cell is a single encased electromechanical unit which exhibits a voltage differential across its two terminals.

Lithium silicon It is a so-called alloy of metallic lithium and silicon used for industrial purposes. The material is somewhat combustible and reactive with water. The containers must be maintained in a condition to prevent entrance of moisture.

Litramar Litramar Line-Spain – a shipowner.

Livestock The number of cattle, sheep, pigs and/or poultry as featured, for example, in a statistical table.

Live storage Material held in storage where movement is allowed to take place especially useful where stock rotation is necessary on a first in – first out principle.

LJ Sierra National Airlines (Sierra National Airlines Co Ltd) – IATA member Airline.

LK Air Luxor (Air Luxor SA) – IATA airline member.

Lkg & bkg Leakage and breakage.

LL Load line.

Lla Long length additional. See separate entry.

LLC Land locked country.

LLDCs Least-developed countries.

LLMC International Convention of Limitation of Liability for Maritime claims. A 1976 limitation convention enacted in the UK by the Merchant Shipping Act 1979.

Lloyd's A & CP Anchors and cables tested by Lloyd's ship surveyor.

Lloyd's Agent Persons appointed by the Corporation of Lloyd's and stationed in most world ports. One of their many functions is to safeguard Lloyd's interests and report movements and losses of ships. This involves reporting local developments and shipping movements to Lloyd's, also involved in surveys of damaged ships.

Lloyd's broker A firm or individual that is authorised to place insurance at Lloyd's and who in doing so represents the insured. There are over 250 firms of Lloyd's brokers, and they must pass solvency tests and comply with Lloyd's regulations. With limited exceptions all business in and out of Lloyd's must pass through Lloyds's brokers.

Lloyd's List Shipping paper issued each weekday.

Lloyd's Loading List Shipping paper issued each weekday detailing vessels 'in port' for cargo transhipment and other ship movements

Lloyd's members Individuals who personally insure risks in whose names Lloyd's policies are issued and who pledge their potential wealth to pay losses arising.

Lloyd's of London An association of private underwriters based in London which provides an international market for the insurance of marine risks. It is world renowned both for its marine insurance business and many other aspects of its worldwide shipping intelligence network. It is known as the Corporation of Lloyd's comprising underwriters and brokers.

Lloyd's Open Form Standard salvage agreement involving 'No cure no pay'.

Lloyd's policy signing office A central service provided by the Corporation of Lloyd's. Its main functions are to: validate and number transactions, sign policies on behalf of underwriters, take down entries, process syndicate reinsurances, produce daily tabulations of transactions, produce periodic settlement statements, provide special schemes (as required), provide statistics, manuals, circulars, advise etc., operate LCA.

Lloyd's Register of Shipping An independent non-profit-making society. It undertakes surveys and classification of vessels; also produces various annual publications. See Lloyd's Register Quality Assurance and International (Ship) Register.

Lloyd's Register Quality Assurance Part of Lloyd's Register the ship classification society – it is involved in the audit and certification of quality, environmental and health and safety management systems. It provides a management system certification and features strongly in the ISM Code Certification involving Total Quality Management.

Lloyd's RMC Ships refrigerated cargo installation classified with Lloyd's Register of Shipping.

Lloyd's syndicate A group of underwriting members of Lloyd's who are bound by the signature of an active underwriter on their behalf. When an insurance is accepted on behalf of the syndicate, each member is liable for his stated fraction only of the insurance, not for the fractions of other members. His personal liability is unlimited.

Lloyd Triestino A major container operator.

LLWSAS Low Level Wind Shear altering System – IATA term.

LM ALM (Antillean Airlines) – an IATA Member Airline.

LMA Lloyd's Manoeuvring Assessment; Lloyd's Marine Association or Loan Market Association.

LMAA London Maritime Arbitrators Association.

LMC Ships machinery classified with Lloyd's Register of Shipping.

LME London Metal Exchange.

LMEVFS London Metal Exchange Vendor Feed System.

LMIS Lloyd's Maritime Information Service.

LMXL London Market Excess of Loss.

LN Libyan Arab Airlines – an IATA Member Airline.

LNC Lloyd's Navigational Certificate.

LNC(A) Lloyd's Navigational Certificate for Periodic one man watch.

LNG Liquefied natural gas carrier, the principal constituent of which is methane. A specially constructed ship designed to carry natural gas at low temperatures or under pressure.

Lo Low loaders.

LO Polskie Linie Lotnicze (LOT). The Polish National airline and member of IATA.

LOA Length overall (of ships hull).

Load To copy information from a location, usually a disc. into the computer's memory. A computer definition. Also in transport terms, weight, volume, quantity of cargo on a transport unit ship, road haulage unit, railway wagon, barge, or aircraft.

Loadability The extent to which the merchandise having regard to its weight/measurement characteristics utilises adequately the transport provided bearing in mind the cube measurement and weight constraints.

Load bulge The process of a package bulging. It maybe due to over stuffing; compression; or the merchandise settling.

Load centre A concept in liner ship operation whereby shipowners concentrate their services on one port and rely on feeder services. This is especially relevant to container schedules. See hub and spoke.

Load centre port See load centre entry.

Loaded vessel A ship where cargo has been put on board.

Loader The person or persons responsible for the physical loading of the goods to the ship, aircraft or TIR vehicle.

Load factor The quantity of cargo or passengers conveyed on the transport unit e.g. ship, coach, wagon, on a particular service or group of services and formulated by dividing the total carryings into the available capacity and expressing it as a percentage e.g., 4000 tons carried on a vessel of 10,000 tons capacity has a load factor of 40%.

Loading Agent An Agent situated at a port involved in the process of concluding all the arrangements relative to the loading of merchandise/cargo on a specified ship/sailing.

Loading allocation The process of the transport operator allocating space on the transport unit to individual clients. It maybe a ferry vessel of 200 car capacity whereby 100 car spaces are allocated to Agency bookings and the residue to bookings made by the motorist direct with the shipowner's reservation office.

Loading arms Oil transfer units between ship and shore for discharge and loading; maybe articulated all-metal arms (hard arms) or a combination of metal rams and hoses.

Loading bank See loading platform entry.

Loading broker Person who acts on behalf of liner company at a port.

Loading dates Dates during which cargo accepted for loading on the ship on specified sailing.

Loading dock A facility within a warehouse to load and unload road vehicle or railway wagons.

Loading gauge The maximum height and width above rail level to which all locomotives, vehicles, containers and conveyed goods must conform. The loading gauge on British Rail/Network Rail is slightly more restricted than that of the mainland European railways.

Loading list A Commercial document which may be used instead of continuation sheets to supplement SAD CT documents when more than one item is being shipped.

Loading overside Process of loading overside over ship's rail into the vessel's hold or deck from a barge/lighter moored alongside the ship.

Loading platform A flat surface sometimes termed a loading bank to facilitate loading usually alongside a warehouse.

Loading Port Port at which a vessel was loaded with cargo.

Load Line The Load Line, sometimes called the Plimsoll Line, or 'marks', indicates the depth in the water down to which a ship may be loaded; the position of these marks is governed by international convention.

Load Line Certificate Ship's document containing port of registry, gross registered tonnage, ship's name, freeboard and load line.

Load Lines Hull markings of a ship indicating loading states in various conditions. See Load Line zones.

Load Lines – International Convention on Load Lines 1966 and its 1988 Protocol The International Convention on Load Lines, 1966 has been accepted by many States since it was adopted in 1966 and entered into force in July 1968. The Convention was modified by a Protocol in 1988; other States have accepted the Convention as modified by this 1988 Protocol, which entered into force in February 2000.

Load Line zones

Load Line zones Such zones are areas of the world divided by international convention into geographical regions. Some are described as tropical, others as summer and the remainder as winter. A number of zones are permanently described as tropical, summer, or winter, whilst others alter during various seasons of the year. The different drafts allotted to voyages in each condition are in accordance with lines (marks) painted on the amidship portion of the ship's hull in positions located by the international regulations dependent upon her strength, construction and ultimate classification.

Load list A list of goods intended for export also providing their destination.

Loadmaster A load calculator designed for a vessel approved by a classification bureau for the calculation of the vessel's stability.

Loadmate Ship board computer calculations of shear force, bending moment, and trim and stability.

Load on top system The mixture of oil and water which results from cleaning tanks is pumped into a special slop tank where it is allowed to separate. When it is clean enough, the water at the bottom of the tank can be pumped overboard and fresh oil loaded on top of the oil which remains. This technique applies to oil tanker tonnage.

Load plan A plan depicting location of freight/merchandise/parcels which is stowed/accommodated in a transport unit. It maybe a container, railway wagon, road vehicle, ship or aircraft.

Load port The seaport where a specified vessel loads her cargo.

Load sensing device Device which senses the degree of braking effort in relation to the rolling resistance or the road wheels of a vehicle and which automatically controls the braking effort to prevent wheels locking and skidding on the road surface.

Load sheet A document showing the weight of the transport unit, the weight of the load, the description and distribution of the load, and giving the overall balance of the transport unit when required. It maybe a road vehicle, railway wagon or aircraft.

LOADSTAR Loadstar International Shipping Co. Inc. (Manilla) – a shipowner.

Load transfer System by which the weight on an individual axle bogie under braking, heavy loading or on uneven ground is transferred to retain an even distribution of load.

Loan agreement The process of negotiation/concluding an agreement between two parties – one of which involves a Bank/Finance House – for a particular sum of money under specific conditions including repayment terms.

Loan default Failure to undertake the loan repayment as scheduled.

Loan monitoring Prime function to ensure the borrower is in compliance with all the financial and non financial terms and conditions of the loan agreement. This will involve monitoring the borrower's position in respect of any payment and/or technical defaults, particularly in areas of adhering to covenants and maintaining security at adequate levels.

Loan payments The issue of a loan in one country on behalf of a borrower in another gives rise to a payment across the exchanges from the country of the lender to that of the borrower which will have an adverse impact on the exchange rate of the lending country's currency and cause that of the borrowing country to appreciate. An International Finance term.

Loan rate The price per unit at which the government for example makes loans to farmers enabling them to hold their crops for later sale.

LOB Line of Business – insurance term.

LOC Letter of Compliance.

Local area network The cable interconnection of items of computing equipment over a small local area such as a single building site. Such systems enable the sharing of data, software and equipment resources.

Local assembly Goods assembled in the country of import. See Free Trade zone entry.

Local buying office A buying office situated in the exporter's country representing the interest of the buyer. Many marketing organisations in foreign countries buy so much from a particular overseas market that they find it financially advantageous to maintain an overseas buying office. It is usual for the buying office to look after affreightment, insurance and finance.

Local charge A charge which applies for carriage over the lines of a single carrier. An IATA definition.

Local clearance procedure A customs facility operative in the UK and in many overseas countries permitting approved exporters to declare at their own or other nominated inland premises the goods for export. For exports of excise goods only, a warehouse keeper maybe approved to operate a local clearance procedure. When the goods are presented and ready for export, the exporters will submit an electronic notification to CHIEF. Once CHIEF has accepted the declaration, permission to progress is given to move the goods to the frontier. Where Community Transit is required to cover the movement of the goods from the local clearance premises, then they must not be removed until the Community Transit documents have been authenticated. By using the scheme, the exporter benefits from the speedier handling at the office of export because any examination of the goods will normally take place at the exporters approved inland premises. See simplified clearance procedure.

Local content measure Requirement that the investor purchase a certain amount of local materials for incorporation in the investor's product. A WTO definition.

Local currency The national currency of a particular country, e.g. Japan – Yen.

Local customs office The customs office responsible for controlling a participating trader's 'account' with the relevant customs authority.

Local import control A customs facility operative in the UK and in many overseas countries permitting approved importers who regularly receive/import goods as unit loads – e.g. in secure containers or vehicles or specialised aircraft containers or on pallets, roll on/roll off vehicles, or railway wagons – to have their goods cleared on importation at their own premises provided certain specified conditions are satisfied. See local clearance procedure.

Localised marketing strategies This usually arises in Global Marketing Management whereby the marketeer will have regard to the local needs of individual markets and endeavour to respond to those needs especially in the area of produce specification/acceptance and the media plan.

Local officer The local Customs and Excise Officer who is responsible for dealing with the CFSP authorised trader. A customs definition

Local operator See land operator.

Location clause Used in cargo open covers and floating policies, this limits underwriters' liability in any one location prior to shipment.

Locator The top part of stacking fitting that fits into the bottom hole of ISO corner casting and usually termed cone.

Lock An enclosure with gates at either end for lifting or lowering vessels from one level to another. See also locked canals/river navigations entry.

Lock box See negotiable cheque entry.

Lock boxes A type of payment by cheque. The buyer sends the cheque to the seller's account at a bank in the buyer's country and the cheque is credited the day of receipt. Any additional paperwork is sent to the seller and the account information is made available to the seller by electronic mail or courier.

Locked canals/river navigations These include totally man-made canals and river navigations improved by major engineering works, including locks.

Lock-chamber The portion of the lock between the front and rear gates in which the vessel is accommodated during the period of change in water level.

Locked train gear A gear construction enabling the mechanical forces to be equally shared between several gear-wheels in mesh, thus increasing the total power which can be transmitted by the gear.

Lock free access A canal route/network which is free of any locks.

Locking A locking device associated with containers. See also twistlock and pinlock entries.

Lockmatic stacker A patented automatic stacking fitting for containers.

Loco Price of goods includes the cost of packing and conveyance to the place named.

LO Code Location code – computer term. It maybe an airport, seaport etc.

Lodicator A computer used to simulate the loading and discharge of a vessel, and allows for fuel consumption during the voyage.

LOF'95 Lloyd's Open Form 1995. A standard form of salvage agreement.

Log book Ship's official diary signed daily by the Master.

Log carrier Ship designed to carry logs, usually geared and having long wide hatchways.

Log extracts Details obtained from the written records of the voyage from the ship's log book.

Logical operation The Boolean operations such as AND, OR etc.

Logistics The process/mechanism of moving/distributing goods in a cost effective and efficient manner, and the related organisation and technology required to achieve this objective which can conveniently be summarised as the ability to get the right product at the right place at the right time. It basically involves the 'just in time' concept and embraces an integrated and high quality package of services in which emphasis is focused on the care of the cargo and responsibility for the most efficient co-ordination and management of the transport process. Hence, it is the time related positioning of resources, ensuring that material, people, operational capacity and information are in the right place at the right time in the right quantity and at the right quality and cost. Overall, the process of managing all activities required to strategically move all raw material, parts and finished stock from suppliers premises to a manufacturing conversion enterprise to finished product store and finally to the customer. Effectively it combines the roles of Material Management and Distribution Management into an overall function. See 'just in time' and supply chain management entries.

Logistics chain The process of moving/distributing goods through all the elements of the logistic chain originating from the supply source and moving to the manufacture/assembling/processing stage and terminating at the distribution point. The length of chain elements will depend on the product and the market environment. See also logistics entry.

Logistics hub The hub of a logistics network as found in Hong Kong seaport and airport which attracts a substantial volume of mainland China exports. It is facilitated by Hong Kong free port status.

Logistics literate A person/company/nation competent in logistic skills.

Logo A Company symbol, badge or name style usually associated with the development of its Brand image. See Brand Marketing entry.

LOI Letter of Indemnity.

L(oil) Capacity in cubic metres of oil cargo in a tank.

LO/LO Lift on/Lift off vessels. Usually cellular vessels which are loaded or unloaded with containers by means of a gantry crane, over the ship's side. Also load on/load off

Lome Convention The aim of the Lome Convention first signed in February 1975 is to assist the development of the economies of the developing world. The Fourth Convention is subject to mid term review in 1995 and came up for renewal in 2000. Its members include the African, Caribbean and Pacific countries.

London Clearing House It was founded in 1888 with the task of clearing coffee and sugar trades. By November 1993 it cleared some 150 million contracts a year registered on four London Exchanges embracing: International Petroleum Exchange; London Commodity Exchange, London International Financial Futures and Options Exchange, and London Metal Exchange. It is owned by six major Banks, Barclays, Lloyds, Midland, National Westminster, Royal Bank of Scotland and Standard Chartered. Its major role is to act as the central counterparty for trades executed on the four exchanges.

London Commodity Exchange See London Clearing House entry.

London Insurance Market Network An electronic data transfer system embracing a network of computers. This system is designed to reduce the flow of paperwork and improve the flow of information within the London insurance market.

London International Financial Futures and Options Exchange It was created by the merger in March 1992 of the London International Financial Futures Exchange and the London Traded Option Market. It is based in London and is Europe's leading exchange and the third largest exchange in the world devoted to financial futures and options. See London Clearing House.

London landed terms Goods sold when at the docks or in bonded warehouse – buyer has to collect them.

London market excess Excess loss reinsurance on business written entirely in the London market.

London Metal Exchange A market place in London where members buy and sell by specification such metals as copper, lead, tin and zinc. See London Clearing House entry.

London Tanker Brokers Panel Ltd The panel formed of six Directors of Shipbroker Companies form the Management Committee of World scale Association (London) Ltd together with Worldscale (NYC) Inc of New York. The Committee are jointly responsible for the worldscale tanker nominal freight rate schedule. See worldscale entry.

LONG Longitudinal; or the starting of a transaction by purchase of a futures contract within the freight market. A bought open position. An open purchased futures position embracing an open purchased futures position.

A BIFFEX term; a bull position or description of a holder of a particular stock or share; or the position established by the opening purchase of a contract. IPE term.

Longer-term refinancing operation A regular open market operation executed by the Eurosystem in the form of a reverse transaction. Longer-term refinancing operations are executed through monthly standard tenders and have a maturity of three months.

Long haul The longer distance transit which usually involves ocean transport and long distance air freight. Additionally it can involve rail land bridge services such as the trans-Siberian railway and rail container services linking the North American West and East coast ports. A transit more than 2500 km.

Long haul airline An airline which offers services involving flight times in excess of 3 1/2 /4 hours.

Long in a currency A term found in the foreign exchange market whereby if a dealer is 'long in a currency', it means that holdings exceed liabilities for the currency.

Longitudinal bulk head The bulk head that runs fore and aft instead of transversely on a vessel.

Longitudinal framing Hull framing that runs fore and aft instead of transversely of a vessel.

Long lead time A situation whereby the date of its conception of a scheme/project to the actual introduction date extends from six months upwards.

Long length additional Extra charge applied by liner conferences and shipping lines on cargo exceeding a length specified in their tariff, often 40 feet or 12 metres.

Long liquidation The closing of a long position – a BIFFEX term.

Long room Customs House office where documentation of goods declared are received and checked and where ships are reported inwards and cleared outwards at customs.

Long run A period of time long enough for all inputs to be varied; there are no fixed costs. A firm can vary the quantity of all its productive resources.

Longs Government and similar stocks with repayment dates (lives) of more than fifteen years.

Longshoreman A person who loads/discharges cargo from a ship at a seaport in USA.

Long Tail A term used to describe a risk that may have claims arising long after the risk has ceased to exist. So that he can close the underwriting account for the year it is often necessary for an underwriter to arrange reinsurance protection to cover claims which may arise after the account has been closed.

Long tail claims Claims notified or settled normally a long time after the expiry of a period of insurance.

Long term A policy exceeding 12 months. A Marine Insurance term.

Long term budget A long term plan usually prepared in monetary terms.

Long term commitment An undertaking extending to beyond seven years.

Long term forecast A market forecast which extends beyond three years. Such forecast could extend to five, seven or ten years.

Long term planning The process within a Company or particular set of circumstances to devise a plan for future action which may be five, seven, or in exceptional cases rather longer up to ten years. It could include manpower resources, investment, marketing, research and development and so on. All personnel within the Company's sector of management would be under obligation to work within the plan to ensure its objectives/execution is realised. It is likely the plan will be reviewed at regular intervals such as every two years to update any elements of it in the light of changed circumstances, thereby ensuring the plan is capable of realistic execution.

Long ton 2240 lbs.

Loose cargo Merchandise despatched on an individual consignment basis and not in a unitised form as a container or pallet.

LOP Line of Position.

LORD Lord Steamship Co SA – a shipowner.

Lorenz distribution A strategy in which the majority of product/service buyers emerge from relatively few clients. An example is the life boat industry.

Lorry-mounted crane Hydraulic crane mounted on a vehicle and driven by a power take-off which powers the hydraulic pump. Such cranes are normally controlled by the driver from within, or from the rear of the vehicle cab.

Lorry reception area A place where road vehicles are received prior to their shipment at a seaport. When the vehicle has been processed through the reception area the vehicle is placed in a parking/assembly area immediately prior to shipment of the vehicle on a specified sailing/ship/date.

Loss The data on which the incident giving rise to a claim occurred; an event giving rise to a claim under an insurance; a claim; or the disappearance of the subject matter of insurance through theft or some other cause as opposed to its survival in a damaged state. Insurance definition.

Loss adjuster An independent and highly trained claims expert, who acts as a consultant to insurers in assessing the true extent and value of any loss which has resulted in a claim being made against them. Although paid a fee by the insurer, a member of the Chartered Institute of Loss Adjusters is required to act with the claimant's legitimate interests in mind.

Loss damage waiver (LDW) Protection offered by a car rental company against responsibility for damage to the rental car resulting from loss, theft, vandalism or collision.

Losses occurring basis A provision in a reinsurance treaty whereby the reinsurance applies to all losses occurring during the treaty period even if the risk attached before that period. The reinsurance does not apply to claims occurring after the reinsurance period even though the underlying insurance began to run during that period. Professional Indemnity Insurance is usually underwritten on this basis, although 'claims occurring' are usually advised and accepted on a potential claim basis. Insurance term.

Loss leader A product/service offered for sale in a competitive market place at a low price with little or no mark up. Such a strategy is designed to encourage other consumers into for example, the retail outlet to buy other products. Another example is to offer very cheap flights or hotel accommodation to develop long term business.

Loss of specie A change in the nature of character of insured goods so that they are no longer the thing insured.

Loss ratio The proportion of claims paid or payable to premiums earned.

Loss reserve The setting up of a reserve for outstanding losses: where it is agreed that the reassured shall retain a proportion of the net premiums – for a given period to provide a fund to settle gross claims without recourse to the reinsurer. Sometimes called cash flow reserve; or a fund set aside to pay claims outstanding at the end of a period of account. Insurance term.

Loss sharing rule (or loss sharing agreement) An agreement between participants in a transfer system or a clearing house arrangement regarding the allocation of any loss arising when one or more participants fail to fulfil their obligation; the arrangement stipulates how the loss will be shared among the parties concerned in the event that the agreement is activated.

Loss sustained cover A type of reinsurance treaty covering only those losses which are sustained during the term of the treaty.

LOST Law of the Sea Treaty

Lot The minimum amount of a commodity in which one may deal. On the LME a lot is often called a warrant or a contract.

LOT Polish Airlines. The Polish National airline and member of IATA; letter on tape, or contract – see contract entry.

Louvre Agreement An agreement formulated in 1987 intended to stabilise the dollar and enhance economic co-operation between the Group of seven major currencies including UK, Germany and France.

Low The lowest price traded that day.

Low context cultures Cultures in which most information is conveyed overtly and explicitly rather than through cultural nuances.

Low cube Merchandise which has a low volume to weight ratio. Examples include steel rails, steel plates and so on. Such cargoes have a low stowage factor.

Low density cargo Cargo which has a low stowage factor and thereby has a low cube but high weight relationship as found in iron ore, steel rails etc.

Low dollar A low value dollar currency exchange rate.

Lower deck The deck below the main deck on an aircraft or ship.

Lower deck container A unit load device shaped to fit into the lower deck of a high capacity aircraft.

Lower flammable limit (LFL) The concentration of a hydrocarbon gas in air below which there is insufficient hydrocarbon to support combustion.

Low inventory A short list of products/items.

Low level of service A poor quality service.

Low leverage A low level of exposure of securities to market risk in the capital structure of a company. See leverage entry.

Low-loader General term covering any trailer designed to provide a low-loading facility for heavy plant and equipment.

Low margin See low mark up entry.

Low margins The low level of profit margin featured in the selling price of a product or service.

Low mark up A product/service bearing/featuring a low profit margin in the final selling price to the consumer/end user. Also termed low margins.

Low pressure selling See safe selling.

Low profile A person/company/product/service which tends to be at the lower end of the market position and often regarded as a laggard. Usually it is associated with very limited media attention.

Low profile coasters Sea-going ships which possess special features, such as a low sea draught, to maximise their capabilities for navigating inland waterways. They are sometimes known as 'Sea-going Barges'.

Low profile deck socket A deck socket that projects above the ship's structure, but not so high as to restrict the use of vehicles.

Low specific activity (LSA) material Radioactive material which by its nature has a limited specific activity or radioactive material for which limits of estimated average activity apply.

Low sterling A low value sterling currency exchange rate.

Low stowage factor Cargo which has low bulk to high weight relationship, e.g. steel rails.

Low toxicity alpha emitter (Radioactive Material Only) Natural uranium; depleted uranium; natural thorium; uranium-235 or uranium-238; thorium-232; thorium-228 and thorium-230 when contained in ores or physical or chemical concentrates; and radionuclides with a half-life of less than 10 days.

Low value added products The process of adding value to a low value product such as food stuffs. See value added benefit.

Low value but high weight ratio A product which has a low value but high weight ratio. An example is coal. Such cargoes are ideal for sea transport movement internationally in bulk.

Low value goods When low value goods are exported from the UK under the New Export System a special customs procedure code must be adopted to identify the goods to CHIEF. A customs term. See CHIEF and New Export System.

Low volume high margins market A market place where the volume of sales is low and the profit margin is high.

Low yen A low value yen currency exchange rate.

Loyalty contract See loyalty rebate.

Loyalty factor See customer loyalty.

Loyalty rebate A rebate offered by a carrier usually a shipowner under the deferred rebate system whereby provided the shipper remains loyal to the liner conference and does not ship outside the conference, the shipowner will grant a rebate on the freight rate.

LP Low pressure; or Lan Peru (Lan Peru SA) – IATA associate member Airline.

LPDR Lao Peoples Democratic Republic.

LPG Liquid petroleum gas carrier. This group of products includes propane and butane which can be shipped separately or as a mixture. LPGs may be refinery by-products or may be produced in conjunction with crude oil or natural gas.

LPH Landing platform helicopters.

LPN National Padi and Rice Authority (Malaysia).

LPSO Lloyd's Policy Signing Office.

LPTB Low pressure turbine (on a ship).

LQT Liverpool quay terms.

Lr Lugger.

LR LACSA (Lineas Aéreas Costarricenses SA), IATA member Airline; Loss Ratio – see separate entry; Lloyd's Register – British Ship Classification Society; or long range.

LS Light Ship.

Ls Lump sum (freight).

LSA Life Saving Appliance, or Apparatus.

LSC Live stock carrier.

Lsd Last safe date (Marine Insurance).

LSD charges Landing, storage and delivery charges.

LSO Loading Sequence Accelerated.

LST Loading Sequence Normal.

LT LTU – Lufttransport-Unternehmen GmbH & Co – an IATA Member Airline; long tons; or long term.

LTB London Tourist Board.

LTFP Long Term Freight Policy.

LTL Less than truck load.

LTS Lay time saved – despatch to be paid for lay time saved.

Lt V Light vessel.

LU Luxembourg – Customs abbreviation.

LUA Liverpool Underwriters Association; or Lloyd's Underwriter Association.

LUAA Lloyd's Underwriting Agents' Association.

LUAMC Leading underwriters' agreement for marine cargo business.

LUAMH Leading underwriters' agreement for marine hull business.
LUBS Lubricants.
LUCKDRAGON Lucky Dragon Line – a shipowner.
LUCO Lloyd's Underwriters Claims Office.
LUF Lift unit frame.
Luffing crane It replaced derricks as supports for cargo pipes during loading/discharging. See crane luffing entry.
Lufthansa The National airline of Germany and member of IATA.
Luggage Articles, effects and other personal property of a passenger as are necessary or appropriate for wear, use, comfort or convenience in connection with his/her trip, journey. Unless otherwise specified it includes both checked and unchecked baggage. Usually termed baggage. An IATA definition.
Lumber North American term for sawn or split timber.
Lump sum allowance A fixed amount which an institution may deduct in the calculation of it reserve requirement within the minimum reserve framework of the Eurosystem.
Lump sum charter Vessel or part of it chartered for a voyage for a fixed sum.
Lump Sum Freight Remuneration paid to shipowner for charter of a ship or portion of it irrespective of quantity of cargo conveyed up to a certain limit.
LUNCO Lloyd's Underwriters Non-Marine Claims Office used to denote Million.
LUS London Upper Section – IATA term.
LU'S Lu's Brother Co SA – a shipowner
LUT Local Users Terminal Communication receiver terminal employed with GMDSS communications.
LV Albanian Airlines (Albanian Airlines MAK SH.p.k.). IATA airline member; or low value.
LVG Lauro/Viceroy/Global Joint Service – a shipowner.
LVO Local VAT office.
Lw Lower (hold) (of a ship).
lw Low water.
Lwei The Angolian currency – see Appendix 'C'.
Lwei invoice An invoice payable in Angolian Lwei relative to an export sales invoice.
LWOST Low water ordinary spring tide.
LWP Lashing Work Planner.
Lwr Lower.
LWT (Ships) Lightweight tonnage.
LWUA Local Water Utilities Administration (Philippines).
LX SWISS (Swiss International Air Lines dba SWISS) – IATA member Airline.
LY El Al Israel Airlines Ltd. The Israeli National airline and member of IATA.
Lykes A major container operator and shipowner.
Lying off A vessel at anchor or moored in an estuary/port environs. Usually such a vessel maybe waiting a berth or taking shelter from a storm.
LZ Balkan Bulgarian Airlines – an IATA Airline Member.

M4 Avioimpex (Avioimpex A.D.p.o.) (limted) – IATA Airline Member.
m Midship or metre.
M Motorship machinery or motor.
m³ Cubic metre.
MA MALEV – Hungarian Airlines Public Ltd. Co. (MALEV plc) – an IATA Airline Member.
M/A Mediterranean/Adriatic.
M & A Mergers and acquisitions.
MAASA Malaysian Shipowners' Association.
Maastricht Treaty A European Community treaty signed in 1991 in Maastricht embracing the single currency adoption by 1999, removal of Social Chapter, promise of more EC funds for Spain and Portugal, creation of common foreign and security policy, increased role of European Parliament, and other measures including training, education, trans European networks, health, industry, consumer protection and culture. Basically, it was the enacting powers to create the single European State.
MABEFTA Malaysian British Economic, Friendship & Trade Association.
MAC Medical advisory Committee IATA term.
MACH Modular automated container handling.
Machinery Ship's engine.
Machinery damage additional deductible clause An Institute clause available for use in a policy covering hull and machinery that provides for an additional deductible to be applied to claims for damage to machinery which is attributable to negligence of master, officers or crew.
Machinery Damage Co-ins. Clause A hull clause applying a deductible of 10% to machinery etc., damage claims, attributable in part or whole to negligence of the master, officers or crew.
MAC-NELS Mac-Nels Agencies Pte. Ltd. – a shipowner.
Mac pak See pal box entry.
Macroeconomics That part of economics which is primarily concerned with the evaluation of the relationship between broad economic aggregates. This includes National income, aggregate savings and consumers expenditure; investment, aggregate employment, the quantity of money, the average price level and the balance of payments.
Macro environment Factors which consist of the larger societal forces embracing demographics, economics, natural forces, technological environment, politics and culture. All these factors influence product/service development. See micro environment entry.
Macro marketing That part of marketing which is primarily concerned with the evaluation of the relationship between marketing strategy and a country's broad national needs. For example, a country with a serious trade deficit emerging from an import led economy would view with disfavour a Multi National Company strongly promoting its products which resulted in an accelerated decline of a major home industry with substantial decline in the labour force.
MAD Magnetic anomaly detector.
Made goo The sums paid to a general average fund to make good losses incurred by the general average act.
MADJU Madju Sdn. Bhd. – shipowner.
Madrid Agreement Treaty, administered by WIPO, for the repression of false or deceptive indications of source on goods.
Maersk Sealand A mega container shipping line and registered in Denmark.

Mafi trailer See roll trailer entry.
Maglev (Magnetic levitation) New concept in high-speed surface travel where vehicles would float off the ground following a guide-way.
Magnesium scrap These are borings, clippings, shavings, sheets, turnings or scalpings from machining operations or cuttings from thin magnesium metal sheets. The scrap can be ignited by external flame and burns intensely and persistently. It does not heat spontaneously. The scrap may have a bright metal lustre or may be dull and sometimes have a painted surface.
Magnetic levitation train A train which has no wheels but floats along elevated track on a magnetic cushion as operates between Tokyo-Osaka. It travels up to 250 mph.
MAIB Marine Accident Investigation Branch.
Maiden Voyage The first voyage of a newly built vessel (new building) after she has passed her sea trials.
Mailbox Refers to the requirement of the TRIPS Agreement applying to WTO members which do not yet provide product patent protection for pharmaceuticals and for agricultural chemicals. Since 1st January 1995, when the WTO agreements entered into force, these countries have to establish a means by which applications of patents for these products can be filed. (An additional requirement says they must also put in place a system for granting 'exclusive marketing rights' for the products whose patent applications have been filed.) A WTO definition.
Mail box A mail box is a repository for electronic mail (e-mail) into which messages are delivered for an address and from which messages are accessed by the owners of the mail box.
Mailing list A list of classified names and addresses representing a particular portfolio. It may be all International Banks in London and other major cities in the UK. Alternatively, the mailing list may represent the entrepreneur clients during the past five years such as the local garage. Overall, mailings are used to promote the entity products/services and keep in touch with the Company client base. Mailing lists can be bought in by an entrepreneur.
Mail merging Combining a master file with a secondary file containing variable data such as names and addresses, to produce multiple documents – each of which contains the same master information but is addressed to a different addressee. A computer term.
Mail order business A type of business whereby the distributor promotes his products through a postal catalogue despatched to individuals who respond by postal order form. Ultimately the goods are despatched by the distributor through the postal network. It can involve a mail shot technique, newspapers/journal advertisements involving a postal coupon, or local agent visiting people's homes. Such a business avoids the middleman as the product is promoted and sold direct between distributor and client.
Mail shot The process of despatching by letter mail to the recipient details of a particular product(s)/service(s) on a promotion basis. It may be a new shipping service, new rates structure, launching new car model and so on.
Mail transfer A debtor's instruction to his bank to request its correspondent bank in the exporter's country to pay the specific amount to the exporter.
Main Maintenance.

Mainland Europe The continent of Europe which is enveloped by a seaboard and excludes Great Britain and Republic of Ireland.

Main Menu The PAM menu that contains the names of the applications that have been Added to PAM. An Added application can be run directly from PAM when its programs are on the disc in the appropriate drive. A computer term.

Main port Port which handles a significant proportion of a country's seaborne trade. It is normally one which can accommodate a large number of ships and which has a wide range of facilities.

Main processor unit This includes: CPU, mother board, controller boards (e.g. video, disc), special processors (e.g. maths), input and output ports (serial, parallel etc). A computer term.

Main refinancing operation A regular open market operation executed by the Eurosystem in the form of a reverse transaction. Main refinancing operations are conducted through weekly standard tenders and have a maturity of two weeks. An ECB term.

MAINS Maritime information system – a PSA facility.

Maintenance contract A contract devised to ensure a product is maintained to a prescribed level. See maintenance schedule entry.

Maintenance period The period over which compliance with reserve requirements is calculated. The maintenance period for Eurosystem minimum reserves is one month, starting on the 24th calendar day of each month, and ending on the 23rd calendar day of the following month. An ECB term.

Maintenance schedule The schedule devised to maintain a product to a reasonably adequate standard on a cost efficiency basis, and having regard to statutory provisions and safety standards.

Major currency A currency which features strongly in financing of international trade. Overall, there are some 28 major world currencies including US dollar, UK sterling, Swiss Franc, Euro and so on.

Major foreign exchange markets Major financial centre dealing with the buying and selling of currencies on behalf of their customers. Major foreign exchange markets exist in London, New York, Singapore, Hong Kong, Tokyo, Zurich, Frankfurt and Paris.

MAJUTERNAK National Live-stock Development Authority (Malaysia).

Make port To reach the port.

Makuta The currency of the Republic of Zaire. See Appendix 'C'.

Makuta invoice An invoice payable in Makuta currency relative to an export sales contract.

MAL Malaysia Ringgitt – a major currency.

Mala Fide (Latin) In bad faith.

Malaysia International Shipping Corporation (misc) A major liner operator.

Malaysian Ringgitt The currency of Malaysia. See Appendix 'C'.

Malaysian Ringgitt invoice An invoice payable in Malaysian Ringgitts currency relative to an export sales contract.

MALD Maldives islands.

MALDIVE Maldive Lines – a shipowner.

Maldivian Rupee The currency of the Republic of Maldives. See Appendix 'C'.

Maldivian Rupees invoice An invoice payable in Maldivian Rupees currency relative to an export sales contract.

Malicious Acts Clause A clause which excludes claims on a hull policy arising from detonation of an explosive or any weapon of wear used by persons acting maliciously or from a political motive.

Malicious Damage Deliberate and intentional damage to property.

Malicious Damage Clause A clause published by the Institute of London underwriters for use in a cargo policy that is subject to the Institute Cargo Clauses (1982) B or C. It adds the risks of malicious acts, vandalism and sabotage to the cargo policy.

Mali Franc The currency of the Republic of Mali. See Appendix 'C'.

Mali Franc invoice An invoice payable in Mali Francs currency relative to an export sales contract.

Maltese Pound The currency of Malta. See Appendix 'C'.

Maltese Pound invoice An invoice payable in Maltese pounds currency relative to an export sales contract.

MAN Metropolitan Area Network.

Manage Applications Menu The PAM menu that allows the user to select from five application management functions: Add, Delete, Modify, Reorder and Auto Start. A computer term.

Managed currency A situation whereby a government controls or influences through its strategies/regulations/policies on its national currency.

Managed floating exchange rates The process whereby through an exchange control mechanism the level of exchange rates is managed by government/agency, etc. It may take the form of an agreement between two companies/countries.

Managed loans A circumstance whereby syndicated credits extended by banks to rescheduling countries in support of economic and financial adjustment programmes agreed between the debtor country and International Monetary Fund. An International Banking term.

Managed rates The process of a government controlling or influencing through its strategies/regulations/policies its national currency exchange rate levels.

Managed Trade The process of two countries in a bilateral agreement to regulate/control trade with a view to achieving a balanced trade performance.

Management A business process involving the continuous planning of operations and the control of resources in order to achieve an aim (usually profit).

Management aid A facility to aid management in the conduct of its entity on a cost efficiency basis and/or to facilitate the decision making process. This may involve computerisation, management information, work study, budgetary control and so on .

Management by objectives The process of a company sector of a business, or other specified area, executing the technique of management by working to a plan of objectives. This is usually allied to a timescale and regularly reviewed to monitor progress and institute any remedial measures if the plan objective is at risk, such as not meeting the target date of completion.

Management Company A company, which provides specialised services for the management of ships which it does not own itself. See third party ship management.

Management contracts In a local economy to acquire advanced technology and management expertise without the permanence of direct foreign investment.

Management culture This embraces the manner and protocol in which the company conducts its business overseas and also the company structure; in particular the decision maker and the influencers and the overall decision making process. It also reflects the company business plan objectives and constraints in dealing with any particular country, company or product overseas.

Management development The process of developing management resources. This is undertaken on an individual basis within a company having regard to age, experience, education, character, existing job performance, overall ability, qualifications, potential, job opportunity within the Company or Group and so on.

Management fee Payment by the shipowners to a ship management company for the management of a ship. Payment may be on a monthly or annual basis. An example is the Shipman 98 Standard Ship Management BIMCO agreement.

Management floating The process of a currency value being influenced by its government intervention with a view to levelling out short/stabilising run fluctuations in the market. See also floating currency and clean floating entries.

Management information The provision of statistical/financial data for management use, interpretation and action thereon. It may for example include monthly sales results; credit control such as outstanding debts and the clients involved; factory production; output, unit cost calculated weekly; or a daily punctuality statement on rail services. It is a management aid to help run the business and requires to be regularly reviewed regarding its adequacy. Computers are frequently employed for such data production to ensure it is currently produced, as out-of-date information tends to be only of historical interest. Management information is often aligned to budgetary control techniques/data.

Management Plan Organisational structure of a Company/Entity featuring various layer(s) of management and managerial positions.

Management Structure The way a particular Company or organisation structures its Management organisation. It may be, for example, Regional, Functional, Line or Divisional Management systems. A very wide range of management structures exist and ideally each would be so formulated to maximise the Company efficiency and potential in the market place.

Managerial qualifier Agency staff person who meets the requirements for managerial or sales promotion experience established by the airline conferences.

Managing change The ability of a company/human resources to be able to manage effectively change within a company/business/marketing environment. This usually involves a change in the direction a company is focused and also culture management change. See management culture.

Managing Director The person responsible for the overall day to day running of a Company and closely involved in Company policy formulation together with the execution of such policies and accordance with Company policy. This is an important post in a Company and usually the deputy to the Chairman of the Board. A good Managing Director can very much favourably influence the development of the Company business on a viable basis.

Managing Lloyd's Agent See Agent – Lloyd's Underwriting Agent entry.

Managing owner's Person or company appointed by the shipowner to be responsible for ship operation, manning levels, maintenance of vessels and so on. Usually the Managing owner is paid on the basis of an annual management fee and this may extend to commission or freight earnings. The Managing owner may be responsible for several vessels and has overall responsibility to ensure the vessel and her crew comply with all the statutory regulations based on the National registration of the vessel/fleet.

Managing trade receivable The efficient collection of accounts payable. This embraces trade finance and significantly reduces the cost of bad debt.

Mandatory application A legal requirement imposed by Government which must be carried out – without demur/deviation.

Mandatory provisions A requirement for such provisions to be carried out/undertaken without deviation or demur.

Mandel shackle A special shackle used to connect a wire mooring line to a synthetic tail.

Manfrs Manufacturers.

Manifest Inventory of cargo and stores on board a ship/aircraft, or list of passengers which is usually termed passenger list. See load list

Manifest, cargo A list of cargo to be despatched by a particular flight or sailing.

Manifest clerk A clerical position which deals with the ships manifest and is responsible for its compilation.

Manifest of cargo A summary of cargo loaded on a ship. This would include its weight/measurement, freight details and total amount – whether prepaid or payable at the destination, port departure and destination; nature of cargo and so on. Such a manifest is usually termed a freight manifest.

Manifold Point on a tanker where the cargo is loaded/discharged from/to the jetty Chiksans.

Manning level The level/number of personnel/staff required to man/operate an industrial or transport unit. In transport terms this may be a train, aircraft, ship and so on. It will particularly reflect statutory obligations, safety considerations, transport mode and specification. It is usually subject to negotiations/agreement between Management and Unions. A further code of considerations/criteria would emerge regarding the manning levels of a manufacturing factory plant and especially negotiations/agreement between Management and Unions, together with the degree of automation.

Manning Requirements Rules detailing the size and quality of crew required. See Principles of safe manning.

Manning scales The composition of the personnel involved in the crewing of a particular transport unit such as a vessel or aircraft. For example, it may be ten officers and twenty-five seamen on a ship. The manning scale will detail the rank of the personnel involved such as Chief Engineer, First Engineer, Second Engineer and so on. See Principles of safe manning.

Manoeuvring basin An area of water for turning vessels usually found in a seaport/harbour and usually termed a turning basin.

Man overboard A control element fitted to most GPS units which allows the watch officer to obtain an immediate fix in an emergency; e.g. as in man overboard.

Manpower planning The process within a company or particular set of circumstances to devise a plan for future action which may be of three, five or even ten years' duration. It will take account of all aspects of the Company business/policy with a view to ensure manpower resources are adequate particularly in terms of staff calibre to meet business opportunities throughout the duration of the plan on a viable basis. Furthermore, this may involve a reasonable spread of age groups; new technology; good calibre personnel at all levels; adequate training for new opportunities and so on. It will be reviewed, for example at two year intervals to ensure the adequacy/realisation of the plan. All concerned with the plan must work towards its realisation/objectives.

MANRES Management of Natural Resources and Environment for Sustainable Development (Thailand).

Manufacturer's agent Agent with rights to promote and sell a manufacturer's products in one or more markets. See Manufacturers' Export Agents.

Manufacturers' Export Agents An indirect export trading technique whereby agents usually undertake the exporting activities of several complementary manufacturers in one particular branch of industry. Usually they specialise in particular market areas – such as kitchen furniture – and negotiate directly or in the names of clients. The Agent can either help a new manufacturer to enter an overseas market, or help large established exporters expand their markets. Payment is usually by commission or profit participation.

Manufacture under contract An entity/company which has entrusted a third party to manufacture its products under contract. It may be a Company in the UK who choose to have its component electrical products manufactured under contract in Taiwan.

Manufacturing based economy An economy based primarily on producing/manufacturing goods/products as distinct from one for example based on agricultural products.

Manufacturing measure Requirement that an investor manufacture a specific product or a limitation on what the investor can manufacture. A WTO definition.

Manufacturing Resource Planning (MRP2) An extension of Materials Requirement Planning where the whole company operates as one system co-ordinating the activities of Manufacturing, Finance, Marketing, Engineering, purchasing.

MAP See modified American plan.

MARA Council of Trust for the Indigenous People (Malaysia).

MARAD Maritime Administration (USA).

MARDEC Malaysia Rubber Development Corporation.

MARDI Malaysian Agricultural Research and Development Institute.

MAREP Marine Reporting System.

Margin The sum required as collateral from the writer of an option. It may be lodged as cash or in the form of certain specified securities. See also marginal cost/revenue entries. Also termed profit margin.

Marginal The additional unit produced and consumed. It is the margin, a unitary change increase or decrease in an economic aggregate.

Marginal cost The additional cost of producing one more unit of output. It can be calculated, by dividing the 'change in output' into the 'change in total cost' as given below:-

$$\text{Marginal cost} = \frac{\text{Change in Total Cost}}{\text{Change in output}}$$

Marginal cost pricing The process of setting a price level which covers variable cost and makes a contribution to fixed cost and profits. See also fixed and variable cost entries.

Marginal lending facility A standing facility of the Eurosystem which counterparties may use to receive overnight credit against a pre-specified rate. An ECB term.

Marginal Product The increase in output which results from increasing the quantity of an input by one unit with the quantities of all other inputs remaining constant.

Marginal Revenue The revenue gained from selling one extra unit of output.

Margin (maintenance) A request for additional funds to maintain an original deposit level. A BIFFEX term.

Margin (Original) The initial outlay required to open a futures position within the freight market. A BIFFEX Term.

Marine diesel oil A heavier type of gas oil suitable for heavy industrial and marine compression-ignition engines.

Marine Engineer Person responsible for maintaining ships machinery and associated equipment.

Marine insurance The process of providing insurance cover for a vessel (hull and machinery) and cargo shipped.

Marine insurance frauds The deliberate scuttling of an over insured vessel. Such an example arises with an old vessel very much over insured with a shipload of cargo and sent to sea by the shipowner with a view to scuttling the vessel to make an insurance claim on the ship and her cargo.

Marine insurance Policy Certificate A document confirming the cargo value as declared by the shipper has been insured for the maritime transit.

Marine Liability Legal responsibility for loss or damage caused as a result of certain common marine activities.

Marine oil berth A purpose built berth at a seaport handling marine oil. The berth is usually situated in an isolated part of the port for safety and environmental reasons. Oil is transhipped by pipe line. Modern storage tanks are provided. In major ports blending, refining, storage and distribution facilities are provided.

Marine Piracy Illegal acts of violence or detention by the crew or passengers of a private ship against another ship committed on the High seas outside the jurisdiction of any State.

Marine Risk analyst A managerial position responsible for systematic analysing/evaluating marine risk relative to cargoes/ships.

Marine risk management A marine management activity responsible for the prevention of and response to accidents involving vessels and related maritime activities.

Marine Supt. Shore based person responsible for navigation aspects and deck officers in ship management.

Marine surveyor A duly qualified person who examines ships or any parts thereof to ascertain their condition on behalf of owners, underwriters and so on. Also called ship surveyor. See classification surveyor.

Marine Terminal Manager The person, or management team, having overall responsibility for the safe and efficient operation of a Marine Terminal.

Maritime Aspects relating to the sea and shipping.

Maritime canal Canal running into a sea and capable of accepting ocean going ships.

Maritime dues Charges/dues raised by a Port Authority. An example arises in Uruguay which is based on a percentage of the pilotage dues, fees for entry and dispatch of vessels, declaration of inflammables, etc.

Maritime Fraud Any species of fraud arising out of commercial dealings transacted in and intended to be carried out in a marine environment. Overall, it is the process of obtaining money, or services, or property in the goods, or a pecuniary advantage by one or more parties to a transaction from the other party or parties, by unjust or illegal means.

Maritime fraud prevention A wide range of methods can be adopted to prevent Maritime fraud including: safeguard of buyers through education programmes; a central registry of Bill of Lading; secure Bills of Lading; Bank action; licensing of carriers; better information services; agency investigation; registration of ships; monitoring of ships movements; licensing of shipping agents and action against pirates embracing the establishment of an expanded jurisdiction involving a new convention.

Maritime insurance aids The funding of the fleet insurance partially or wholly by the state. A form of flag discrimination.

Maritime Law Legislation relating to and focusing on the shipping environment.

Maritime Lien The claim a master and crew have on the vessel for the payment of wages due. The terms may be applied also, to the rights of a salvor in regard to property that is subject to a salvage award; also, to the rights of a carrier in regard to cargo that is liable for a general average contribution.

Maritime Terminal A port, quay, jetty, berth, accommodating vessel.

Maritime trade credit management Credit risk is a potential situation that a company will fail to meet its obligations in accordance with agreed terms. The goal of credit risk management is to maximise the control or risk of return by maintaining credit risk exposure within acceptable parameters. It arises in trade finance, foreign exchange transactions and guarantees.

Maritime Transport Shipping services operating in a maritime environment.

Mark A general term used for a navigational mark such as a buoy, structure, or topographical identity which may be used to fix a vessel's position or a symbol on packages used for identification or indication of ownership purposes.

Market See market place and marketing entries.

Marketability The degree of investment demand for a particular asset offered at a given price. A banking definition.

Marketable amount The amount of stock or number of shares in which a jobber quoting a price would reasonably be expected to deal. A BIFFEX term.

Market access The process of having the facility of an exporter to gain access to an overseas market.

Market Access Negotiations Includes commitments to eliminate or reduce tariff rates and non-tariff measures applicable to trade in goods. A WTO term.

Market aim The objective of a company/person to secure a specified market usually as detailed in the Marketing and Sales plan.

Market analysis The process of analysing a market of all its components parts for a particular product(s) or service(s).

Market Analyst Person who specialises in analysing markets or a specified market of all its components parts of a particular product(s) or service(s).

Market awareness The strategy of a company to ensure the range of products/services reflect the needs of consumer/buyer in the market and more especially the Company is conscious of the need to ensure the market place is aware of its products/services the Company is promoting.

Market base The overall structure of a market foundation for a particular product or service. For example, the market base for children's tricycles is between the two to eight years age group.

Market behaviour The behaviour or reaction of a market to a particular product, circumstance or general trend over a specified period of time. It would embrace all the market elements, viz economic, political, technical, etc. This may involve a change in price, product design change and so on.

Market Capacity The maximum amount an insurance market can absorb as liability to its policy holders while maintaining a proper solvency margin.

Market capitalisation The total value of a company's issued securities at their current market prices. It includes all different types of security issued by the company.

Market concentration The strategy of focusing/concentrating on a number/cluster of markets rather than one individual market place. An example is the Scandinavian market.

Market confidence The level/degree of confidence found in the market place relative to a particular product(s) or service(s). For example, market confidence was seriously impaired when oil prices rose sharply in 1973 and consumers looked to other means of energy sources, whilst at the same time conserving their consumption of oil. Conversely, consumer market confidence strongly supports any future car specification which would offer very substantial improvements in petrol/diesel consumption

Market coverage The extent to which advertising reaches target audiences (readership/viewers/listeners). Television has a wide market coverage whereas a technical journal has a very limited specialised readership. The term can also be applied to the area covered by the sales team/retail outlets/distributors etc. For example a Company may have a market coverage in seventeen countries.

Market creation The process of creating a market for a particular product(s)/service(s) on an innovative basis.

Market Dealing The process of dealing/negotiating in the market place.

Market demand The specified demand for a product(s) or service(s) in a particular Market.

Market Development The expansion/growth/development of a specific market. This may be facilitated by the product specification and other factors. It is usually related to an entity evaluation of a market for its product(s) and how best to develop the market to its own advantage.

Market driven company A company driven in terms of its strategy/management/development by marketing considerations especially responding to the needs of the customer and increasing market share. See consumer led marketing policy entry.

Market driven entry strategy The process of entering a market for cogent marketing reasons. See market driven company entry.

Market economy An economy which is market driven and allows the full interplay of an open free market with no government interventionist policy/strategy and freedom of choice by the consumer in a competitive environment.

Marketeer A person involved professionally in marketing.

Market entry The process of entering a specific market(s) for a particular product(s) and/or service(s).

Market entry option The options available to an exporter to enter a particular overseas market. These include joint venture, direct selling, Export Houses, licence agreement, Agent and so on. It also involves the criteria for market selection and analysis of the impact of environmental variables.

Market Entry options strategy The strategy a Company will adopt to enter a market. This will involve a number of factors including size of Company and resources available; risk in the market; political/economic situation in market place; market forecast and level of profitability including return on investment; level of competition; product life cycle and nature of product/service. See also market entry option entry.

Market environment

Market environment A broad profile or composition of the market in which the product or service is retailed. It may be a consumer market with good growth potential and a large number of the populace on high salaries producing a very favourable disposal income level and very receptive to modern technology. See micro and macro environment entries.

Market Evaluation An evaluation of a particular market for a product(s) or service(s).

Market expansion The process of a market expanding. It may be, for example, a ten per cent growth in house purchase, new car sales, tourism and so on.

Market exposure A situation whereby a product/service has been promoted in the market place through the media thereby exposing all its features to potential clients/buyers/competitors.

Market extension strategy The process of entering new markets such as moving from a domestic market to an export market.

Market foothold To establish a business/product/service in a particular market/region/country.

Market forces In economic terms the interplay of the forces of supply and demand in a particular market place for a specified product(s) or service(s).

Market forecast The process of forecasting/predicting a level of market demand for a particular product(s)/ service(s) over a specified period of time. It may be during the next three years – recorded on a monthly basis – a particular product such as a new shipping service.

Market haulage Exporter/importer/forwarder attends to haulage of LCL or FCL cargoes to or from the carriers premises/terminal/depot.

Market holding A pricing strategy technique of attempting to set a price that will hold a company share of a market of a specific commodity/service.

Market if Touched A term usually used by chartists. It is an instruction to buy or sell at market if a certain level above or below the current price is reached.

Market indicators Evidence emerging through research and statistical data often government sourced indicating a particular trend. It may be lower employment, increased credit demand, lower inflation, improving GDP, higher car sales, rising birth rate and so on. All such data may be useful to the entrepreneur to develop his/her business depending on the product/service involved.

Market infrastructure The framework/network/fabric of the market place embracing transportation, commercial network, social/economic aspect, technology and so on.

Marketing The process of evaluating/researching a particular market, identifying a need and responding to that need on an acceptable basis to the consumer/buyer usually in a competitive environment. It is a management function which is market research driven and motivated ultimately by profit.

Marketing and Sales Plan See marketing plan entry.

Marketing asset The process of marketing areas of the business where the Company has a competitive advantage such as brand names, retail locations, technology, patents, copyright and so on.

Marketing audit The process of a Company or entrepreneur conducting a comprehensive periodic, systematic, impartial, independent review of a company's environment, strategies, objectives and activities to identify opportunities and problem areas with a view to devising a plan to improve company marketing performance. It would embrace not only the 'in house' Company internal marketing audit, but also the lateral marketing audit embracing economic, political and social/economic features.

Marketing Board A Board responsible for the promotion of a product(s) and usually found in the environment of representing a large number of producers. This applies especially in Agriculture, embracing milk, tomatoes, potatoes and so on.

Marketing Budget The process of devising a budget covering a one year period or other specified time span involving the marketing operations/effort of a particular company, etc. It may be 1.5 million Euros embracing all the marketing department personnel, brochures, advertising, trade exhibitions, sales conferences, sponsorships and so on.

Marketing Channel Traces the movement of ownership or title of a product from the producer to the user or ultimate consumer, including all middlemen.

Marketing Communications The system/network established to communicate with the market. See Market Information system entry.

Marketing conference A conference involved in the promotion of a product(s) or service(s).

Marketing cost The cost incurred in researching the potential markets and promoting products in suitable attractive forms and at acceptable prices.

Marketing cost variance The difference between the budgeted costs of marketing (including selling and distribution costs), and the actual marketing costs incurred in a specified period.

Marketing culture An attitude/focus adopted by Government/Company which focus primarily on the marketing evaluation in the strategy/management/ development of the business. See market driven company

Marketing data The assembly/collection of data relative to an entity marketing operation. It may be also termed marketing information.

Marketing Director The person at Director level within a Company responsible for all the Marketing policy of the business. He is responsible to the Managing Director and his duties include market research, advertising, selling, pricing policy, public relations, appointment of agents, development of the business and so on. The basis of his policy and action plan is found in the marketing and sales plan produced annually.

Marketing Information System A system developed within a Company to provide management – especially decision makers – with all the data they require to take decisions embracing Market intelligence, Channel of Distribution, costs, sales, competitors, market trends, market profile, pricing and so on.

Marketing Manager Person responsible for promoting (marketing) his company's business or a specified part of it.

Marketing Mix Usually understood to represent the four 'P's embracing product, price, promotion and place (channels of distribution). Overall it is the fusion of the product, price, promotion and place and their constituents which forms the basis of the marketing plan in the most cost effective manner acceptable operating in a competitive market environment. The manufacturing sector has four P's as earlier described, but the service sector has seven P's embracing product, price, promotion, place, people, process (logistics) and physical aspects. A similar criteria of fusion obtains.

Marketing Mix Strategy The strategic and operational marketing issues emerging in the fusion of the marketing mix embracing product, price, promotion, place for the four Ps, and for the seven Ps, product, price, promotion, people, process (logistics) and physical aspects. See marketing mix entry.

Marketing Mix Transfer The transfer of the marketing mix from one country to another as found in a domestic driven market company into an export driven market company. See marketing mix entry.

Marketing Officer Person responsible for promoting (Marketing) a particular product, range of product's or service(s). He/she usually reports to the Marketing Manager.

Marketing Plan The provision of a plan outlining the marketing techniques and programme covering a specified period to promote a specified product(s) or service(s). Also termed Marketing and Sales plan . Overall, planning, which focuses on a particular product or market and details the resources, strategies and programmes for achieving the products objectives in the market.

Marketing Planning System A technique for making a time schedule for the implementation of a project.

Marketing Policy The marketing policy of a particular company or in a particular set of circumstances. It maybe for example that only thirty per cent of the advertising budget will be allocated to commercial television promotion, ten per cent to sponsorship, thirty per cent to press campaign and so on. Such a marketing policy can be reviewed annually in the light of market sensitivity.

Marketing Profile The composition of the market promotion.

Marketing Research Research to develop appropriate marketing strategies for a product.

Marketing Strategy The strategic and operational marketing issues arising in the management of an entity operation to achieve its business and marketing objectives in a competitive target market. It embraces decisions in expenditure, composition of the marketing mix, allocation of marketing resources and empathy. Overall it is market research led.

Market Intelligence Information/data obtained relative to the constituents of a particular market relative to a particular product(s)/service(s).

Market Intervention Within the European Union, intervention refers to, for example, public purchasing of farm produce in order to support its market price at previously determined levels.

Market Leader The product(s)/services which leads the market in total sales/product design/general durability and other constituent parts.

Market Led Economy An economy which responds to market opportunities and provides the product/service at an acceptable price and specification thereby displaying the strategy of empathy.

Market Life Cycle The process of plotting the stage of a product/service in its life cycle in the market place.

Market makers Like former stock jobbers, these are firms which buy and sell shares on their own behalf making a profit out of the difference between the buying and selling prices. Market makers also exist in the foreign exchange and gilt-edged markets where they quote prices for currencies and gilts respectively.

Market mechanism The economic process where supply and demand in a free market determine product/service price and production levels.

Market objective The aim/objective of a company/ person to secure a specified market usually as detailed in the Marketing and Sales plan. This may involve the increase of market share from twenty to thirty per cent.

Market opportunity The ability to identify a market opportunity with a view to obtaining a sale.

Market order An order to buy or sell a futures contract at the first obtainable price or prices on the market for tonnage involved.

Market orientated business sectors An entity organisational structure which is so geared towards maximising its marketing objectives. In so doing the organisational structure gives prime emphasis to marketing at all levels and the need to improve market support, market share, and overall Company viability.

Market penetration The extent to which a product/ service penetrates a market and thereby improves its market share. See price penetration.

Market penetration pricing See price penetration and market penetration.

Market performer A stock that analysts believe will rise and fall in line with the market as a whole. An outperformer is a stock which is believed will rise by more than the market, and conversely an underperformer is believed will rise by less.

Market position The entrepreneur/entity position/ranking in the market place in a particular country/region or globally having regard to sales volume relative to other suppliers. To be top of the sales volume and thereby become the market leader is a very prestigious position to hold and has significant competitive market advantage.

Market potential The growth/decline factor evaluation in a specific market relative to a nominated service/ product. Such evaluation is realised through market research.

Market place The place where goods or services are sold. It may be a street, town, country, region and so on.

Market price Price which commodity would reach if sold on the open market.

Market pricing A process of devising a passenger or freight tariff at a level which will maximise revenue and traffic volume, and is correlated to the potential market demand, sensitivity, and optimum utilisation of shipping/aircraft capacity. A similar situation obtains with the inclusive tour operator offering lower tariffs embracing travel and hotel accommodation at the less busy periods of the year.

Market profile The composition of a specified market.

Market prospects An analysis of a particular market giving details of its market prospects.

Market rate Rate charged by brokers, discount houses, joint stock banks and other market members for discounting first class bills. Alternatively it may be current rate/price of a product/service obtaining in the market place. In so doing, it is the rate the market will sustain.

Market rate interest rates See market rate.

Market report A report produced on a particular market, or group of markets in various countries/regions. It may be an evaluation of market prospects for one or more products, or a general market situation report. It may be specially assigned or provided by a Bank, Consul, Marketing Board, or undertaken by the Exporting Company. It can embrace all ingredients of the market, including economic, political, technical, competition, and so on.

Market research The systematic gathering, recording, and analysing of data about problems/opportunities relating to the marketing of goods and services. A business function concerned with the collecting and supplying of information to manufacturers, wholesalers and retailers about consumer trends and preferences.

Market research briefing The process of holding a meeting to inform participants of the strategy/methodology to conduct the market research programme.

Market research debriefing The process of holding a meeting to inform participants of the outcome/result of the market research conducted.

Market research led A marketing strategy adopted by a firm/entity which bases its marketing decision on the research data obtained – usually continuously – from the market place and in house data obtained through sales and resultant trends. This embraces pricing, design, after sales, product range and so on.

Market resistance The general reluctance of a market to accept a service or product on offer. For example, a particular country may tend to be reluctant to accept a new hair style as it goes against national custom.

Market risk premium The extra return required from a share to compensate for its risk compared with the average risk of the market.

Market segment A specific portion of a market. Overall, it is a subdivision of a market representing a distinct section of customers to be reached with a distinct marketing strategy.

Market selection overseas The process of the seller endeavouring to find a suitable market overseas. It involves the following salient considerations: the size of the prospective market and nature of products/service; the potential demand for the product; the major competitors and their activities; the present and future economic conditions in the country; the local tastes, language and traditions; the level of import duties and government restrictions; the methods of trading and distribution; the acceptable forms of packaging and advertising; cost/income evaluation; company resources; potential buyers, and credit rating.

Market sensitivity The general sensitivity or reaction of a market to a particular product, circumstance, or general trend over a specified period of time. It would embrace all the market elements vis economic, political, technical etc. It could arise through product design changes, variation in price, and so on.

Market servicing operation The process of servicing a market involving distribution, control, cost, advertising, market mix and so on.

Market Share One entity's sales of a product or service in a specified market expressed as a percentage of total sales by all entities offering that product or service.

Market share strategies The process of a Company/Entity adopting strategies to improve its market share which may involve price reduction increased advertising and so on.

Market situation The situation in the market place at any particular time which may constitute high unemployment, limited consumer spending potential, low inflation, etc.

Market size The total size of the market in a particular area/region/country. It maybe motor vehicle licences, TV licence holder or demographic data.

Market skimming A pricing strategy technique whose basic aim is a deliberate attempt to reach a segment of the market that is willing to pay a premium price for a product because it has high value to them. See price skimming.

Market slump A decline in a segment of the market relative to a particular commodity, a range of commodities, or a general decline in the market.

Market spreading The process of a product/service serving a number of overseas markets in different territories.

Market stability The stability of a competitive market place showing little signs of any significant movement/variation in the profile of the market.

Market support The market response to a service or product sale. Hence the provision of a new local bus service in a conurbation has attained favourable loadings thereby proving the service has good market support; or the provision of market analysis, market research and numeracy data relative to the evaluation of a product/service.

Market trends The trends emerging in a particular market over a specified period of time relating to a particular product or circumstance, and can embrace all the component parts of the market, or particular elements of it.

Marks A term used in shipping practice to refer to the loadline marks on the ship's hull. The term may be used also in regard to a bill of lading which shows the identification and destination markings on packages shipped.

Marks and numbers Markings distinctly displayed on goods being shipped, or on their packaging, for ease of identification. These include the port or place of destination and a package number, if there is more than one.

Mark to market Evaluating an open position on the basis of the current market price, usually to assess the need for a margin call (qv). Hence the process whereby options and future positions are devalued daily using current prices to determine profit/loss and therefore variation margin – an International Petroleum Exchange term.

Mark up A term used in retailing, wholesaling and factoring to indicate the addition to the cost price of goods to produce a required selling price, often expressed as a percentage. It is often termed profit margin.

MARPOL International Convention for the Prevention of Pollution from Ships (as amended). The principal body of rules, framed by IMO to control pollution of the marine environment. The Convention contains six Annexes, respectively, covering pollution by oil, noxious liquid substances in bulk, harmful substances carried in packaged forms, sewage, garbage, and air pollution.

MAR POL 73/78 The International Convention for the prevention of pollution from ships 1973 as modified by the protocol of 1978.

Mar policy A cargo insurance policy introduced on 1st January 1982, following UNCTAD's report on Marine Insurance and Institute of London Underwriters recommendation. Its use became obligatory on 1st April 1983. Overall the new policy merely acts as a vehicle for Institute of London Underwriters which constitute the terms and conditions of cover. All that is recorded on the MAR policies is the full name of the assured, the place from which the insurance attaches, the overseas vessel/conveyance, ports of loading and discharge, final destination, insured value and currency of cover and details of the subject-matter insured and its packing. Spaces exist for indication of the ILU clauses incorporated within the policy and details are given of the procedure in the event of loss or damage for which underwriters maybe liable.

Mar Policy Form A simplified form of marine insurance policy that was introduced in the Lloyd's marine insurance market to replace the SG policy form. A similar form of policy was introduced by the Institute of London Underwriters, for use by member companies, at the same time. The new policy form was introduced in 1982 for cargo insurance and in 1983 for hull insurances.

Marshall-Lerner Condition The sum of the import and export elasticities of demand must be greater than one for a devaluation to improve the Balance of Payment on current account.

MARVS This is the abbreviation for the Maximum Allowable Relief Value Setting on a ship's cargo tank – as stated on the ship's Certificate of Fitness.

MAS Malaysian Airlines System; or Monetary Authority of Singapore.

Masking The techniques used during a one run documentation system to ensure the suppression of any details appearing on the master which are not required on a particular form aligned to it.

Maslow theory A theory of human motivation developed by A H Maslow who identified that peoples desires/needs/aspirations can be formulated into a hierarchy of needs embracing physiological, safety, social esteem and self actualisation.

MASPS Minimum Aircraft System Performance Specification – IATA term.

Mass market A large homogeneous market for consumer products/services.

Mass media Basically the media range which reaches a mass audience such as radio, television and newspapers/journals/magazines. Overall the media represents channels of communication within the market place.

MASSOP Management structures of shipowners and ship operators – an EU project.

Mass Transit Railway A passenger rail transport system operative in a major city offering a frequent service in the peak periods. Examples include the Paris metro, London Underground, Hong Kong Mass Transit Railway Corporation and so on.

Mass Transit system A high capacity transport system usually found in a city centre or conurbation area. Examples are the underground railway systems in Paris, London, New York, Rome and so on. In the wider sense it may involve the high capacity trainload or ship usually fully integrated with another transport mode when the situation so demands.

Master Person in sole charge of a ship.

Master Air Waybill An Air Waybill covering a consolidated shipment showing the consolidator as the shipper.

Master at arms Principal police officer on a ship – a petty officer.

Master budget A budget which is prepared from, and summarises, the functional budgets. It may be also called a Summary Budget.

Master document A document having an imprint (usually typed or from a computer terminal) from which individual documents are to be derived by reprographic or electronic means.

Master owner A ship's master who owns the vessel which he operates. It is often found in the coasting trade whereby the Master will trade the vessel under a charter party or other form of contract to enable the ship to be regularly employed in a particular trade.

Master pallet A pallet which can be lifted by a hook and is not used for loose cargo but for made up pallet loads. One means of conveying a pallet with cargo between ship and shore. Also termed slave pallet.

Master production schedule A timed plan in which the complex of activities, available personnel, working hours (management), policy and goals are visualised and given concrete form in such a way that beginning, intermediate and end period give a clear view of the process. Overall an authoritative statement of how many end items are to be produced and when.

Master's certificate Certificate of Master's competency.

Masters Portage Bill The ships account presented at the end of the voyage in the case of a tramp vessel or the end of the month for liner ships to the shipowner. It includes all the income and expenditure during the month or the voyage.

Master Unique Consignment Reference A reference allocated by the authorised trader used to link several customs declarations of unique consignment reference. See unique consignment reference.

Mast step The foundation on which a mast is erected.

MAT Material; or Moving Annual Total – a statistical term.

Matches, safety (book, card, or strike-on-box) These are matches intended to be struck on a prepared surface. Term associated with dangerous cargo classification.

Matching system The system operated for the matching and confirmation of Exchange contracts.

Mate, The Chief Officer in charge of deck department and second-in-command of ship.

Material circumstance Any circumstances which would influence the judgement of a prudent underwriter in determining whether to accept a risk and the amount of premium to charge.

Material culture The standard of living or level of economic development which a society has achieved with emphasis on quality products/service and the range and choice available extolling high professional standards.

Material representation A statement made to the underwriter before acceptance of risk which is material to his decision in accepting and rating the risk.

Material requirements planning An inventory and purchasing planning system that integrates product components, lead times and deadlines. It determines gross and net requirements (discrete period demands for each item of inventory) in order to generate information needed for correct inventory order action.

Materials handling The activities of loading, unloading, placing and manipulating material and of in-process movement.

Materials management Materials management involves bringing together the activities concerned with the goods flow under one managerial function. It is the structured organisational solution for providing integral supervision of the flow of goods in, through and out of a company whilst making optimal use of the available means of production and co-ordination of all the related activities.

Materials transfer note A document which records the transfer of materials from one store to another, from one cost centre to another, or from one cost unit to another, usually in both quantity and value.

Mates receipt Document issued to the shipper for ships cargo loaded from lighterage and later exchanged for bill of Lading. A receipt signed by the mate to say the cargo has been received on board in good order and condition.

Matrix analysis Techniques (of which the Boston Matrix is the most famous) of analysing a firm's portfolio of products.

Maturity The date on which a loan, bill or other debt instrument falls due for repayment.

Maturity date The date by which the amount in the bill of exchange must be paid to the seller.

Maturity yield The return on a security held to maturity at a given price.

Mauritanore Code name of charter party for Mauritanian ore trade.

Mauritius Rupee The currency of Mauritius. See Appendix 'C'.

Mauritius Rupee invoice An invoice payable in Mauritius Rupees relative to an export sales invoice.

MAWB Master Air Waybill.
Max Maximum.
Maximum gross weight, permissible (Applicable to ULD only) In respect to unit load device the maximum gross weight permissible means the sum of the tare weight and maximum net weight permissible of the unit load device.
Maximum net mass The maximum net mass of contents in a single packaging or maximum combined mass of inner packagings and the contents thereof expressed in kilograms. Term associated with dangerous cargo classification.
Maximum net weight, permissible (Applicable to ULD only) In respect to unit load device, the maximum net weight permissible means the manufacturer's declared rating for the unit load device, i.e. the maximum weight of contents which a particular unit load device is capable of carrying.
Maximum normal operating pressure (Radioactive Material Only) The maximum pressure above atmospheric pressure at mean sea-level that would develop in the containment system in a period of one year under the conditions of temperature and solar radiation, corresponding to environmental conditions of transport in the absence of venting, external cooling by an ancillary system or operational controls during transport.
Maximum transport temperature The maximum temperature likely to be encountered by a substance in transport.
Mb/d Million barrels per day (of crude oil/natural gas liquid).
MBI Management Buy Ins – a group of managers outside the company or subsidiary using venture capital to fund their purchase.
MBL Minimum breaking load.
MBM Multi-buoy mooring.
MBO Management by objective or Management Buy Out (of a company).
Mbps Million bits per second.
MBS Machinery (ships) classified according to British Corporation rules.
mCi millicurie.
MC Metalling clause; mobile cranes; or marginal cost.
MCA Management Consultants Association; or Maritime and Coastguard Agency (UK).
MCAs Monetary compensatory Amounts. Charges and payments applied to imports and exports to prevent currency fluctuations from distorting agricultural trade. A customs term.
MCD Miscellaneous Cash Deposit.
Mchts Merchants.
MCHY Machinery.
Mchy aft Machinery (ships) aft.
Mchy fwd Machinery (ships) forward.
MCM Mine counter measures.
MCO Miscellaneous Charges Order – IATA term.
MCP Maritime cargo processing. A customs term.
MCr Mobile crane.
MCSC Movement Control Sub-Committee – IATA term.
MCT Moment to change trim; or Minimum connecting time – IATA term.
M/D Months after date.
MD Managing Director; Moldavia – Customs abbreviation; or Air Madagascar – an IATA airline member.
MDC Marina Di Carrara.
MDD Meteorological Data Dissemination System.
M'dise Merchandise.
MDL Measurement Devices Ltd.
MDO Marine diesel oil.

m dwt Million tons deadweight.
ME Middle East Airlines Airliban (MEA). The National Middle East Airlines and member of IATA; or Main Engine.
MEA The Metropolitan Electricity Authority (Thailand); or multilateral environmental agreement – WTO term.
MEAF Middle East and Eastern Africa (Office – ICAO term).
Mean low water The average height of the low water determined from records taken in some predetermined period.
Mean range of tide The range between the mean high water and mean low water as for example: MHW MSL +2.08m.
Mean sea level The average height of the sea surface, based on hourly observation of the tide height on the open coast, or in adjacent waters that have free access to the sea. In the United States, MSL is defined as the average height of the sea surface for all stages of the tide over a nineteen-year period.
Measurement tonne Freight rate cubic assessment measurement based on forty cubic feet equals one ton (2240 lbs) or 1,000 kilogrammes equals 35½ cubic feet (one cubic metre). A criteria used to calculate a freight rate W/M.
Measure of indemnity The extent of liability of the insurer for loss.
Measures of natural wealth See natural wealth measures.
Measures with a temporary effect Comprise all non-cyclical effects on fiscal variables which:
 (1) reduce (or increase) the general government deficit or gross debt (deficit ratio, debt ratio) in a specified period only ("one-off" effects), or
 (2) improve (or worsen) the budgetary situation in a specified period at the expense (or to the benefit) of future budgetary situations ("self-reversing") effects.
 A term associated with Eurozone.
MEBO Management Employee buyout.
MEC Marine Evacuation Chute.
Mechanical handling equipment Cargo handling equipment used to handle goods in store, and in transit. It maybe powered or non-powered thereby relying on manual operation. This includes pallet trucks, fork lift trucks, cranes, conveyors and so on.
Mechanically ventilated container A closed container equipped with means of forced air ventilation.
Mechanic's lien The legal enforceable claim which a person who has performed work or provided materials is permitted to make against title to the property or as a preferential person in the event the estate or business is liquidated.
Mechanisation of cargo handling The technique of handling cargo – often in a unitised form – with mechanical equipment such as fork lift trucks and pallets.
MEDAFRICA Med-Africa Lines – a shipowner.
MED/ANZ Mediterranean/Australia/New Zealand container service.
MED CLUB The Consortium of Cie Maritime Des Chargeurs Reunis (CMCR). This includes Compania Naviera Marasia, SA Flotta Lauto; Lloyd Triestino; Mitsui-OSK Line; NYK Line; and Hapag Lloyd.
Medcon Code name of Chamber of Shipping approved charter party for coal trade.
Media Television, radio including commercial radio, and the press.
Media Budget The marketing budget allocated to media or Television advertising for a particular period involving a product(s)/service(s).

Media pack The composition/ingredients of the data provided by a newspaper/journal/TV company relative to its readership/viewing audience profile, advertising rates, special advertising features and so on. All of such data will interest the potential client wishing to advertise a product or service.

Media plan A marketing or advertising plan featuring promotion of a designated product or service on TV/Radio, and the press on specified dates.

Media research The process of researching the media relative to its readership, audience and circulation data. Areas evaluated would be cost effectiveness, characteristics of the media involved, qualitative and quantitative factors.

Media schedule A marketing or advertising plan featuring promotion of a designated product or service on TV, Radio and the press on specified dates.

Mediterranean Seaboard Countries situated on the Mediterranean seaboard such as Egypt, Morocco, Italy, Greece, Turkey and so on.

Mediterranean Shipping Co SA (MSC) A major liner shipowner and container operator.

Meditore Code name of Chamber of Shipping approved charter party for ore trade.

Medium-term loan Usually considered to be loan with a term of 3 to 10 years. Most shipping loans fall into this category.

Medium-term notes A flexible facility to issue notes of varying maturity, varying currency and either fixed or floating – all within one set of legal documentation.

Medivac Medical evacuation.

MED/USE Mediterranean/North America container service.

MEG Middle East Gulf.

Mega carrier A very large transport operator such as British Airways.

Mega container shipping line A container shipping line with a fleet capacity in excess of 250,000 TEUs.

Mega cruise liner A jumbo sized cruise liner with a passenger certificate in excess of 2500 passengers.

Mega port A major seaport such as Singapore, Rotterdam and New York.

Mega warehouse A very large warehouse and may be described as a jumbo warehouse.

MEIC Myanma Export Import Corporation (Burma).

MEISHIN Meishin Line – a shipowner.

Melbourne-Darwin rail link Australia's new North South rail link opened in January 2004. See The Ghan entry.

Member Lloyd's Agent See Agent – Lloyd's Underwriting.

Member state Member of the EC.

Member states Members/countries of the European Union or any other Economic bloc/Customs Union.

Member state of departure The Community country where the goods are presented to validate the carnet. A customs definition.

Member state of temporary use The Community country where the goods are temporarily used. A custom definition.

Membrane type gas tanker The most modern liquefied Natural Gas carrier afloat which offers cost savings in shipbuilding and operation, high stability and safety in manoeuvring and easy maintenance and repair.

Memorandum of agreement Sale and purchase of ship form. See Sale form 1993 entry.

Memorandum of association Legal document giving information about the nature of a limited company, vis-a-vis outsiders – e.g. its name and objectives.

MENAS Middle East Navigation Aids Service.

Men rea (Latin) A guilty state of mind.

MEP Member of the European Parliament.

Mep Mean effective pressure.

MEPC Marine Environment Protection Committee.

MEPME State Warehouse for Foreign Missions (Laos).

Merchandise In marketing terms brochure, point of sale display/materials to aid sales; or goods/cargo/freight.

Merchandiser A facility dispensing brochures/sales literature. It may be a person or a rack/holder accommodating brochures/sales literature in a Travel Agency.

Merchandising All the activities directed towards selling products/services at the point of sale embracing packaging, display, pricing, special promotion, sales videos and so on.

Merchant A method of indirect export trading where merchants buy and sell as principals in their own right. The nature of the merchants business makes it possible for them to assess the marketability of products and prospects for successful selling. So far as the exporter is concerned all credit risk is removed. The merchant usually pays the exporter against purchase of the goods. For example, a metal merchant in contrast, as distinct from a producer's agent or broker, often acts as a principal, buying metal or concentrate from producers and others and selling it on to others. He will often hold metal on his own account while waiting for a buyer.

Merchant bank In United Kingdom terminology, merchant banks are that group of financial institutions which grew out of the business activities of a number of foreign merchants who settled in London in the eighteenth century. Their international trading led them into various types of financing, including the raising of long-term loans for foreign governments and major enterprises. Most of them developed expertise in the financing of shipping. Today the UK merchant banks are a diverse group, the largest of them are able to offer a wide range of investment banking, corporate banking and fund management services, while the smallest are effectively private banks. In the United States, merchant banking has come to mean the investment of the bank's own funds in transactions – often as a form of bridging finance – with a view to realising substantial profits, usually in the short term.

Merchant haulage Inland transport of shipping containers provided and paid by the shipper or receiver of goods rather than by the ocean carrier.

Merchantable quality Goods so described are reasonably fit for the normal purpose for which goods of that sort are usually bought – a legal definition.

Merchant inspired carrier haulage Carrier haulage by a carrier, which is nominated by the shipper or receiver of the goods, but paid by the carrier.

Merchant Navy and Airline Officers Association The UK ships officers union.

Merchant Shipping (certification of Deck and Marine Engineer Officers) Regulations, 1977 The regulations governing the certification of Deck Officers and Marine Engineer Officers in the UK.

Merchant Shipping (Radio Installations) Regulations 1992 The UK Merchant Shipping regulations determining the number of radio officers required for different vessels and the need for a valid certificate of competency.

Merchant Shipping (Registration etc) Act 1993 UK Shipping legislation re-establishing the British Register and the facility to register foreign vessels under bareboat charter to British companies.

Merchant Shipping (Repatriation) Regulations 1979 The UK legislation governing the repatriation of seafarers. It superseded the Merchant Shipping (Repatriation) Regulations, 1972.

MERCOSUR Mercado Comun del Sur – Trade agreement which entered into effect on 29 November 1992 between Argentina, Brazil, Paraguay and Uruguay. It has 190 million consumers and internal trade barriers were eliminated in 1994. It has a common external tariff.

Merge-in-transit A hybrid of cross docking. The process of tracking several containers of freight moving through a cross docking facility upon mode of entry. It forms a pipeline, a regulated and continuous flow of product which can be managed in the absence of warehouses and inventory cost incurrence. By utilisation of merge-in-transit, inventory becomes a just-in-time function allowing inventory costs to be minimised while dependability of product delivery is maximised as it maintains its allotted schedule. See cross docking entry.

Merger A specific type of takeover which is subject to varying definitions in different jurisdictions, but which usually involves two companies of approximate equal size coming together in such a way that one does not dominate the other but that both sets of shareholders continue to have a substantial equity interest in the new group. The distinction between a merger and an acquisition is particularly important in terms of accounting treatment of the deal since mergers permit an aggregation of individual statements without any consequent generation of goodwill. It unifies marketing and finance. See joint venture.

Mersail Code name of Chamber of shipping approved charter party for coal trade.

MERSAR Merchant Vessel Search and Rescue Manual.

Merseycon Code name of Chamber of Shipping approved charter party for Mersey district coal trade.

MES Marine Evacuation System.

Messages A grouping of segments arranged according to both the design guidelines and the syntax which completely represents a specific transaction such as an invoice. An EDI definition.

Message transfer A series of information exchanges about the international movement of goods by traders registered under International Trade Prototypes arrangements. Each exchange is individually denoted by "MTn".

Messenger lines A light line attached to the end of a main mooring line and used to assist in heaving the mooring to the shore or to another ship.

MET Meteorology.

met Measurement.

Metacentric height The distance between the metacentre and the centre of gravity, which is also known as the GM. This changes as the weight in the ship is raised or lowered. When a cargo is loaded, the aim is to give the ship the metacentric height which gives her the easiest and safest movement at sea.

METAG MET Advisory Group (of EANPG).

Metal can A light receptacle of tin plate or other metal requiring an outer packaging for conveyance.

Metal catalyst, finely divided, activated, or spent This is metal in an extremely finely divided form. It must be shipped wet in moisture-tight and airproof packagings. If exposed to air, the metal may become hot and may even ignite. A term associated with dangerous cargo classification.

Metal fatigue Weakening of metal structure due to repeated stress and flexing.

Methane carrier Ship designed to carry methane (natural gas) in liquid form. This state is maintained by refrigeration.

Methodology The sequence of constituents by which a task is undertaken. For example the process of framing a picture involves a code of procedure or methodology.

Methods of payment The options available to effect payment of the goods under an export sales contract which require the intervention of international banking facilities. Overall, the most common payment options include payment in advance; open account; Bill of Exchange (at sight or terms); documentary sight bill (document on payment); documentary term bill (documents on acceptance); counter trade; and documentary letter of credit. See separate entries.

Methylacetylene and propadiene mixture, stabilised A flammable gas mixture that is reasonably stable at ordinary temperatures. While this is an acetylene derivative, the gas is not shipped dissolved in liquid and the cylinders do not require an absorbent filler. A term associated with dangerous cargo classification.

Methyl bromide Fumigant used to kill infestation in various commodities. It may on occasion be used to fumigate in the container.

Methyl bromide fumigation An effective way of treating timber for export.

Me-too firm Firm entering a market, after the introduction phase, with a product similar to the existing products and thus shifting to competition on a price basis.

Metric ton Freight rate weight or measurement based on one metric tonne equals 1,000 kilogrammes or one cubic metre. 1000 kilogrammes equals 2204.6 lbs.

MeV Mega electron volt.

Mexicana The National Mexican airline and member of IATA.

Mezzanine finance A type of debt generally provided to middle market or non-investment grade companies. A method, for example, of funding a vessel which contains elements of both debt and equity. Capital ranking after senior debt but before ordinary equity. It may take many forms – secured and unsecured loan stock, preferred shares, subordinated notes, second-mortgage loans, junk bonds. In some cases, such financing will include the option to convert into ordinary equity. Mezzanine finance is naturally less secure than senior financing, and consequently the interest cost tends to be appreciably higher. Its use in shipping finance is expected to increase as a means of bridging the gap between first mortgage loans and owner's equity, but financiers have found it difficult to develop mezzanine instruments suited to the peculiar characteristics of the shipping industry.

MF Medium Frequency (300 KHz to 3MHz); or Xiamen Airlines – IATA member Airline.

MFA Multifibre Arrangement (1974-94) under which countries whose markets are disrupted by increased imports of textiles and clothing from another country were able to negotiate quota restrictions. WTO term; or Ministry of Foreign Affairs (Thailand).

MFAG Medical First Aid Guide (for use with accidents involving dangerous goods).

MFIs Monetary Financial Institutions.

MFN Most-favoured-nation treatment (GATT Article 1, GATS Article 2 and TRIPS Article 4), the principle of not discriminating between one's trading partners. WTO term.

Mfrs Manufacturers.

MFTB Myanma Foreign Trade Bank (Burma).

MG Metra – centric height.

MGN Marine guidance notice.

MGR Manager.

Mgt Million gross tons.

MH Main hatch; merchants haulage; or Malaysia Airline

System Berhad – an IATA Airline Member.
MHE Mechanical Handling Equipment.
MHHW Mean high high water.
MHLW Mean high low water.
MHS Message handling system.
MHWI Mean high water interval.
MHWS Mean high water springs.
MHz Mega hertz.
MIA Manchester International Airport.
MIB Merchandise in Baggage.
MIBOC Marketing of Investments Board Organising Committee.
MICC Malaysian International Chamber of Commerce.
Micro bridge A through movement in which cargo moves, for example, between an inland US point and a port via rail or truck connecting with a steamship line for movement from or to a foreign port. The ocean carrier accepts full responsibility for the entire movement on a single through bill of lading.
Micro-computer A system of hardware and software comprising: main processor unit, keyboard, VDU, auxiliary storage and possibly other peripheral units together with an operating system.
Micro economics That part of economics which is concerned with the evaluation of the individual decision units such as the consumer, households and firms. In particular the way in which their decisions interrelate to determine relative prices of goods, and factors of production and the quantities of these which will be bought and sold.
Micro environment Factors which consist of the smaller societal forces embracing competitors, customer markets, marketing channels, suppliers and the Company itself. All these factors influence product/service development. See macro environment.
Microfiche A type of information retrieval system where documents are photographed then reduced in size and contained in a card. When required, the card is placed in a microfiche reader and the photographs are magnified and projected onto a screen to enable reading.
Microfilm As with microfiches, except the reduced photographs are held on a roll of film instead of a card.
Micro marketing That part of marketing which is concerned with the smaller company and its strategy and the tactics/mechanism to achieve it.
Micrometre (μm) 10-6 metres.
Microsecond One millionth of a second.
MID Middle East (region – IATA term); or middle.
MIDA Malaysian Industrial Development Authority.
MIDA's Maritime Industrial Development Areas.
Mid Atlantic rates Air freight rates applicable between the Mid Atlantic Area and points in IATA Area 2.
Mid-deck tanker A new oil tanker ship design developed by Marine Safety System, Inc of USA. It is configured without double sides and the cargo sides divided into pairs, port and starboard, by two longitudinal bulk heads that form the sides of centrally located ballast tanks.
Middle East Airlines National Middle East Airlines and member of IATA.
Middle range of the market The mid point value of a product price range, for example, in a market range of one hundred clients involving a particular product, there may be three prices: bottom, middle and top. The middle range of the market would feature the middle price which may represent forty per cent of the market share with the bottom price covering thirty-five per cent and the top range twenty-five per cent.

Middle rate The average of the buying and selling rates for a given currency. An International Banking term.
MIDF Malaysian Industrial Development Finance.
Midlock A patented automatic stacking fitting used for containers.
MIFT Manchester International Freight Terminal.
MIGAS Directorate General of Oil & Gas (Indonesia).
Mileage proration The proration on the basis of the applicable local mileages (air cargo).
Milliard A thousand million or an American billion.
Milliemes The currency of Egypt and Sudan. See Appendix 'C'.
Milliemes invoice An invoice payable in Milliemes currency relative to an export sales contract.
Millisecond One thousandth of a second.
Mils The currency of Cyprus and Malta. See Appendix 'C'.
Mils invoice An invoice payable in Mils currency relative to an export sales contract.
MIN Marine information notice.
Min Minimum.
Min B/L Minimum Bill of Lading.
Mines Articles normally consisting of metal or composition receptacles and a bursting charge. They are designed to be operated by the passage of ships, vehicles or personnel. The term includes "Bangalore torpedoes". A term associated with dangerous cargo classification.
Minibar A small refrigerator unit in a hotel room stocked with alcoholic and soft drinks as well as snacks; guests are usually charged for each item taken.
Mini bridge A service offering, for example, a through movement of cargo between Europe and the Pacific Coast of the US or between the Far East and the Atlantic or Gulf Coast of the US. A Bridge of unit train movement across the US is substituted for the all-waters route through the Panama Canal. The ocean carrier accepts full responsibility for the entire movement on a single through bill of lading. See land bridge and micro bridge.
Mini bridge services A dedicated service linking maritime routes with an overland scheduled rail service. The rail service is relatively short.
Minimal data content Term used to describe the concept of limiting the size of message by including the smallest amount of information consistent with the defined purpose of the message.
Minimal breaking load The minimum breaking load of a mooring line as declared by the manufacturer for a new line. It does not include allowance for splicing or for wear and tear.
Minimum charge The lowest amount which will be charged for the conveyance of a specified consignment between two points.
Minimum connecting time The time allowed between the arrival of one scheduled flight and the departure of a connecting flight, established separately for every commercial airport.
Minimum crew certificate See principles of safe manning.
Minimum inventory The planned minimum allowable inventory for an independent demand item.
Minimum land packages The minimum tour, in terms of cost and ingredients, that must be purchased to qualify a passenger for a certain tour or bulk air fares.
Minimum longitudinal strength standard An ICAS requirement which specifies a technical basis for the determination of a fundamental component in the assessment of hull strength on a vessel.
Minimum/maximum stay requirements Condition of sale of airline tickets and other service providers that require travellers to stay at the destination for a certain period.

Minimum price fluctuation The minimum movement that a futures price can move. Otherwise known as a tick. IPE term.
Minoan Lines Company operating passenger services in the Greek maritime trades.
Mint par Fixed ratio between two currencies.
MIP Marine Insurance Policy.
MIPRO Manufactured Imports Promotion Organisation – a Japanese organisation.
MIRANS Modular integrated radar and navigation system.
MIRDC Metals Industry Research and Development Centre (Philippines).
MIS Marketing Information System or Management Information System.
MISC Malaysian International Shipping Corporation – a shipowner.
Miscellaneous charges order A document issued by an airline carrier or its agent requesting issue of an appropriate Passenger Ticket and Baggage check, or revision of services to the person named in such a document.
Miscible The ability of a liquid (or gas) to dissolve uniformly in another liquid (or gas). Miscibility depends on the chemical nature of the substances involved and in some cases, liquids may only be partially miscible. Liquids which do not mix at all are said to be immiscible. A term associated with dangerous cargo classification.
Misconnection A passenger who due to late arrival or non-operation of his/her original delivery flight/sailing etc, arrives at the inter line point by his/her original delivery flight/sailing etc, an alternative flight/sailing etc, or surface transportation too late to board/join/embark his/her original receiving flight/sailing.
Misdating The practice of ante dating or post dating a document. If the date the goods are loaded is before or after the commencement or expiry date on a letter of credit, the carrier and/or his Agent maybe asked to predate the bill of lading. Hence the practice of misdating. Thus any attempt by the shipowner to persuade the carrier to issue a bill of lading showing a date other than the date on which completion of loading occurred, should be strongly resisted. Often the value of the cargo or price to be paid will be governed by the issue of the bill of lading. A letter of credit example.
Misdescription Incorrect information concerning a ship given by a shipowner to a charterer or concerning cargo given by a charterer or shipper to a shipowner or shipping line. This may give rise to a claim for extra cost or damages or, in some cases, cancellation of the contract of carriage.
Mis-matched maturity When the maturities of the funding cover and the loan or other asset do not coincide.
Misrepresentation A mis-statement of fact made by the assured or his broker to the underwriter, before acceptance of the risk, which misleads the underwriter in assessing the risk. If the representation is material and amounts to misrepresentation, it is a breach of good faith.
Missing flight An overdue flight.
Missing ship A ship is deemed to be 'missing' when, following extensive inquiries, she is officially posted as 'missing' at Lloyd's. She is then considered to be an 'actual total loss' and policy claims for both hull and cargo are settled on that basis.
Missing vessel An overdue vessel.
Mission A business visit to an overseas market. See outward and inward mission entries.
Mission statement The stated objectives of an entity; business, marketing, media or sales plans.

MIT Market if touched.
MITI Ministry of International Trade and Industry. The Japanese National Standards body. See ISO.
Mitsui OSK Lines Ltd (MOL) A major liner shipowner and container operator.
Mixed cargo A consignment which contains a number of different commodities which do not qualify for the same rate and conditions. An IATA definition.
Mixed consignment A consignment of different commodities, articles or goods whether packed or tied together or contained in separate packages.
Mixed economy A situation where there are some State run industries, but many industries and most trading activities are in the hands of private firms as found in the UK.
MJ LAPA (lineas AÈreas Privadas Argentinas) – IATA member Airline.
MK Air Mauritius. The National airline of Mauritius and member of IATA.
mL Millilitre.
ML Motor launch.
mlc The metres liquid column; it is a unit of pressure used in some cargo pumping operations.
MLA Maritime Law Association.
MLC Market Life Cycle.
MLLW Mean low low water.
MLR Minimum Lending Rate – formerly Bank rate; or Medium long range.
MLS Microwave landing system – an air navigation system.
MLW Mean low water.
MLW MSL 1.49m – Mean range of tide 3.57M.
mm Millimetre.
MM Multimodal.
m/m Month on month.
MMC Monopolies and Mergers Commission; or Metro Manila Commission.
MMO Multi modal operator.
mm 2/s Square millimetre per second = unit of kinematic viscosity.
MMSA Mercantile Marine Services Association.
MMSI Maritime mobile service identity number.
MMTC Metro Manila Transit Corporation.
MN Commercial Airways (Pty) Ltd (COMAIR) – an IATA Member Airline; Machinery numeral – used for ships classified under Lloyds' Register of Shipping or British Corporation to calculate fees for periodic engine survey; or Merchant Navy.
MNAOA Merchant Navy and Airline Officers' Association.
MNC Multinational Company.
MND Ministry of National Development (Singapore).
MNDO Merchant Navy Discipline Organisation.
MNE Merchant Navy Establishment; or Multinational Enterprise.
MNEA Merchant Navy Establishment Administration.
MNEMONIC The abbreviation of a Member's company name or individual trader used on a trading badge. IPE term.
MNE's Multinational Enterprises.
MNI Member of the Nautical Institute.
MNOPF Merchant Navy Officers' Pension Fund.
'M' Notices Merchant Shipping Notices.
MNPS Minimum Navigation Performance Specifications – IATA term.
MNPSA Minimum Navigation Performance Specifications Airspace – IATA term.
MNR Ministry of National Resources (Philippines).
MNTB Merchant Navy Training Board (UK).

MO Money order; or medical officer.
MOA Ministry of Agriculture (Philippines); or Memorandum of Association.
MOB Ministry of Budget (Philippines); or man overboard.
Mobile cranes A crane which is mobile and does not operate from a fixed platform/site. Such a crane is vehicular and operates within the operators requirements in a seaport, warehouse, depot and so on. Lifting capacity varies up to 20 tonnes.
Mobile lift frames These machines resemble the straddle carrier, but without its high-stacking ability, and its capacity for extended travel at reasonable speeds. There are a number of configurations and different degrees of elaboration, but in all cases a mobile steel structure is positioned astride the container, which is lifted on the top or bottom four twist locks. Container lengths of 20 ft (6.10m), to 40 ft (12.20 m) can be handled, and alignment over the load is made less critical by incorporating chains or side shift in the hoist.
MOC Myanma Oil Corporation (Burma).
MOD Ministry of Defence.
Modalities Provisions defining the manner in which an agreement will be implemented. A WTO definition.
Model (computer) A software representation of a real situation or system which can be used for analysis of its operation. A simplified version of a process. Examples of models include: financial budgets with variable costs and profits, journey planning between geographical points using roads available, queues at checkout desks and the number of people waiting, traffic lights controlled by numbers of vehicles and pedestrians, producing a three-dimensional model of a building to investigate environmental effects on nearby surroundings, pilot simulation etc. A computer definition.
Modem Short for modulator-demodulator, a device that converts signals from one form to another for use with difference types of equipments, such as from electronic pulses into audio tones for phone transmission; needed by automated agencies to allow in-house CRTs to communicate with outside airline computers and by agents who use PCs to communicate with any off-premises computer.
Mode of operation (software) The way in which a software package enables user interaction, e.g. control key combinations, menu, object selection, function key control and the use of peripheral devices. A computer definition.
Modes of communication See 'transmission modes'.
Modes of delivery How international trade in services is supplied and consumed. Mode 1: cross border supply; mode 2: consumption abroad; mode 3: foreign commercial presence; and mode 4: movement of natural persons. WTO term.
Modified American plan (MAP) A hotel room rate that includes breakfast and one other meal; also known as demi-pension or half-pension.
Modified re-buy The process of re-ordering. A requirement which has been subjected to changes such as in design of a product.
MODU Mobile Offshore Drilling Unit.
MODU Code The Code for the Construction and Equipment of Mobile Offshore Drilling Units, 1989 (1989 MODU Code) was adopted by Assembly resolution A.649(16) and concerns MODUs built since 1 May 1991. The 1989 MODU Code superseded the 1979 MODU Code adopted by Assembly resolution A.414(XI). The Maritime Safety Committee (MSC) adopted amendments to the 1989 MODU Code in May 1991 and decided that, to maintain compatibility with SOLAS, the amendments should be become effective on 1 February 1992, Further amendments were adopted in May 1994, to introduce the harmonised system of survey and certification (HSSC) into the Code, provide guidelines for vessels with dynamic positioning systems and introduce provisions for helicopter facilities. The Committee decided that the amendments introducing the HSSC should become effective on the same date as the 1988 SOLAS and Load Line Protocols relating to the HSSC (i.e. 3 February 2000), and that those providing guide-lines for vessels with dynamic positioning systems and provisions for helicopter facilities should become effective on 1 July 1994.
Modulation Method of using radio frequency carrier waves to transmit audio frequency signals.
Modus operandi (Latin) Plan of working.
MOE Ministry of Energy (Philippines).
MOF Ministry of Finance (Philippines).
MOH Medical Officer of Health.
MOI Ministry of Industry; Ministry of Interior (Thailand).
Moisture proof Material which resists penetration by water.
MOL Mitsui OSK Lines. A major container operator.
Molar Volume The volume occupied by one molecular mass in grams (g mole) under specific conditions. For an ideal gas at standard temperature and pressure it is 0.0224 m^3/g mole.
Mole The storage yard between two docks; or the mass that is numerically equal to the molecular mass. It is most frequently expressed as the gram molecular mass (g mole) but may also be expressed in other mass units, such as the kg mole. At the same pressure and temperature the volume of one mole is the same for all ideal gases. It is practical to assume that petroleum gases are ideal gases.
Mole fraction The number of moles of any component in a mixture divided by the total number of moles in the mixture.
Mollier Diagram A graphic method of representing the heat quantities contained in, and the conditions of, a liquefied gas (or refrigerant) at different temperatures.
MOLOO More or less in owners option.
Monetarism The technique of devising monetarist policies involving interest rates, money supply and so on.
Monetarist A person following/adopting monetarist policies.
Monetary aggregates A monetary aggregate can be defined as the sum of currency in circulation plus outstanding amounts of certain liabilities of financial institutions that have a high degree of "moneyness" (or liquidity in a broad sense).
Monetary and financial institutions In a UK context, monetary and financial institutions are broadly equivalent to banks and building societies.
Monetary compensatory amounts Charges and payments applied to prevent currency fluctuations from distorting agricultural trade given that community agricultural support prices are converted into national currencies at fixed representative rates. An EEC/CAP facility.
Monetary indicators A range of indicators to measure monetarist policies performance such as broad money supply, exchange rates, interest rates on deposits and consumer price inflation. Important indicators when selecting countries in which to conduct business.
Money Cash or currency of a country.
Money at call Money loaned in a form that can be withdrawn at any time.
Money Laundering The misappropriation of money such as fraudulent notes/cheques, fraudulent money transfers through the banking international system.

Money market A situation where borrowers and lenders of large sums of money are in contact with each other.

Money measurement The concept that financial accounting information relates only to those activities which can be expressed in monetary terms.

Money supply The amount of money circulating in the country, usually expressed in terms of M numbers. The higher the number, the larger or broader the measure – for example, M0 (the narrowest definition) includes notes and coins in circulation and commercial bank operational deposits at the Bank of England; M4 comprises the notes and coins in circulation, plus all private sector sterling bank and building society deposits.

Money up front The process of an entrepreneur having money immediately available to support his/her offer/bid.

Mongo The currency of the People's Republic of Mongolia. See Appendix 'C'.

Mongo invoice An invoice payable in the Mongo currency relative to an export sales contract.

Monitoring a market The process of recording specific areas of movement within a particular market for a product(s)/service(s). This would include sales performance, competition, pricing, market share, economic indicators, product development and so on.

Monitoring of rates The process of recording specific actual rates alongside predicted/permitted/authorised rate levels. This usually arises under a liner conference system whereby members who quote rates below a certain minima or outside an agreed rates calculation criteria are penalised heavily by the conference by a fine and withdrawal of the rate.

Monkey Island Area above Wheelhouse housing navigating equipment and mainmast with statutory lights.

Mono hull A single hull vessel.

Monopoly A situation where a sole seller or buyer completely dominates the market and in so doing controls supply and fixes the price.

Monopsony A situation where only one buyer exists and many suppliers are available from which to choose.

Mono rail The operating of a railway network on a one rail system.

Monsoon A seasonal reversal of wind which in the summer season blows onshore, bringing with it heavy rains, and in winter blows offshore – it is of greatest meteorological importance in southern Asia. The word is believed to be derived from the Arabic work 'mausin' meaning a season.

Monthly sales The monthly sales results of a particular Company/Entity for a specified product(s) or services(s).

Montreal Convention 1999 The Montreal Convention is a unifying convention and consolidates and updates the provisions of the amended Warsaw Convention. See Warsaw Convention.

Montreal Protocol An MEA dealing with the depletion of the earth's ozone layer. WTO term.

MOPGC Malaysian Oil Palm Growers' Council.

Moor Mooring.

Mooring basin An area designated for the mooring of vessels. It is usually found in a seaport/harbour and tends to be protected from adverse weather storm like conditions.

Mooring buoy A facility to which a vessel may be moored.

Mooring restraint The capability of a mooring system to resist external forces on the ship.

MOPS Minimum Operational Performance Standards (RTCA term).

Moratorium An agreement between creditors and an insolvent debtor that payment will not be enforced for a specific period.

More or less in charterers option Option allowed to a voyage charterer to load up to a certain quantity, normally expressed as a percentage or a number of tonnes, over or under a quantity specified in the contract of carriage. This option may be sought if the charterer is not certain of the exact quantity which will be available at the time of loading.

More or less in owner's option Option allowed to a shipowner to carry up to a certain quantity, normally expressed as a percentage or number of tonnes, over or under a quantity specified in the voyage charter. This option maybe sought if the shipowner is not certain what the ship's cargo capacity will be, taking into consideration bunkers, stores and fresh water, or if he wants flexibility to adjust the ship's trim.

Morphological analysis The process of having a three dimensional analysis to identify possible market segments which are most prone to yield an opportunity to generate a profitable situation.

Mortgage A long-term loan secured on fixed assets – i.e. ship or property.

Mortgage bonds A fixed interest certificate of debt issued by officially recognised central mortgage bond institution for the purpose of long term fund raising and secured by registered mortgages.

Mortgagees' interest insurance Banks holding mortgages on vessels may take out insurance through the Lloyd's market protecting them against the eventuality that the value of the mortgage may be lost through its being subordinated to large prior liens or legal claims. Interest on mortgagees' interest insurance increased during 1990 as a result of the enactment of the Oil Pollution Act in the United States, which could open the door to unlimited liability claims against owners and operators of tankers held liable for oil pollution incidents in US waters. A specific class of mortgagees' interest insurance called 'additional perils (pollution)' has been made available to cover such risks. A form of insurance which protects the bank in the event of a policy becoming void if certain warrantees are broken by the mortgager.

Mortgage of a ship A legal registered instrument of security giving the right of possession to the mortgagee. More than one mortgage may be registered against one ship, each mortgage being ranked according to priority by time of registration. A Legally valid ship mortgage is generally regarded as an effective form of security, especially because of the ability to sell the asset anywhere in the world. On the other hand, ship valuations are subject to sharp fluctuations and a ship my ordinarily be supposed to depreciate quite rapidly, unlike most land-based property.

MOS Months.

M-OSK Mitsui OSK Lines – a shipowner.

MOSTE Ministry of Science, Technology and Environment (Thailand).

Most favoured nation A commercial treaty between two or more countries which guarantees that all parties to the agreement will automatically extend to each other any trade concession which they may subsequently grant to non member countries. All members of WTO comply to the principle but Groups like EEC & EFTA are exempt.

MOT Ministry of Tourism (Philippine); or Ministry of Transport (of various countries).

MOTC Ministry of Transport and Communications (Philippines).

Mother ship Ship which performs the main ocean leg of a voyage, being fed by smaller ships or barges. See Baco Liner.

Motif A symbol representing a particular Company and devised primarily for promotional purposes. It is a marketing aid. For example, the motif of Lloyds Bank is a 'Black Horse'. Also termed 'Logo'.

Motivational behaviours The study of motivational behaviours embracing self motivation, work standards, energy, initiative and resilience. See inter personal behaviours.

Motivational research The process of studying the psychological reasons determining/influencing human behaviour especially relative to buying situations.

MOTNE Meteorological Operational Telecommunications network for Europe (ICAO term).

Motor fuel anti-knock mixture A mixture of one or more organic lead compounds such as tetraethyl lead, triethylmethyl lead, diethyldimethyl lead, ethyltrimethyl lead, and tetramethyl lead, with one or more halogen compounds such as ethylene dibromide and ethylene dichloride.

Motorist inclusive tour A travel package available, for example, in UK for the motorist including sea travel for a car and passenger(s) plus hotel accommodation available in the Continental market.

Motorcoach A highway passenger vehicle often equipped with a restroom, air conditioning and other amenities.

Motorman Grade I A post in the Marine Engineer's department on a ship and formerly designated as fireman or greaser's.

Motorman Grade II A post in the Marine Engineer's department on a ship and formerly designated as ordinary engineer assistant.

Motor premium The insurance premium raised on a motor car.

Motor spirit Blended light petroleum distillates used as a fuel for spark-ignition internal combustion engines other than aircraft engines.

Motorway The most major trunk road providing fast road transits involving up to six lanes in each direction and free of any bottle-necks inasmuch no contra flows of traffic cross it.

MOT(S) Motor(s).

MOU Memorandum of Understanding.

Moulded breadth The moulded breadth is the greatest breadth at amidships from keel of frame to keel of frame.

Moulded draught The vertical distance from the top of the keel plate to the designed summer load waterline at a specified station.

Movable link span A facility provided at a ramp berth accommodating a vehicular ferry/Ro-Ro- vessel linking the seaward end of the ramp with the stern or bow of the vessel over which vehicles are transhipped (driven to and from the vehicle deck). It is so designed to enable/permit the link span to be moved from one berth to another or elsewhere.

Movement Certificates See SAD.

Movement inventory The inventory during a production process caused by the time required to move goods from one place to another.

Movement of factors of Production The cross-border movement of capital or labour necessary for the provision of services.

MP Multi purpose; or maritime patrol.
MPC Multi Purpose Carrier.
MPCU Marine pollution control unit.
MPD Minimum pre-marketing set of data.
MPEs Multinational production enterprises.
MPHW Ministry of Public Works & Highways (Philippines).

MPIB Malaysian Pineapple Industry Board.
MPO Management and Personnel Office.
MPP Multi purpose (type of ship).
MPR People's consultative Assembly (Indonesia).
MP's Meridional parts.
MPS Minimum Performance Specifications (EUROCAE term); or Master Production Schedule.
MPV Multi-purpose vehicle.
MR Marginal revenue; mates receipt; or market research.
M.rads Metre radians.
MRBLE Marble.
MRCC Marine Rescue Co-ordination Centre.
MRELB Malaysian Rubber Exchange and Licensing Board.
mrem/h Millirem per hour.
mr/h Milliroentgen per hour.
MRO Maintenance, repair and overhaul; maintenance repair and operational; or Main Refinancing Operation: regular OMO used by the ECB to provide 14 day funds to the banking system.
MRP Marginal revenue product.
MRR Medium range recovery.
MRRDB Malaysian Rubber Research and Development Board.
MRT Mass Rapid Transit (Singapore).
M/S Member State.
MS Machinery survey; motor ship; manual system; Egyptair – the National Egyptian airline and member of IATA; or manuscript.
m/s Metres per second; or months after sight.
MSA Merchant shipping Act; Mutual Security Agency (USA); the Multilateral Steel Agreement negotiated among twenty steel-exporting and importing participants, including the United States, the EU and Japan. A WTO term; or Marine Safety Agency linked with Department of Transport and Coastguard (UK).
MS Act(s) Merchant shipping Act(s).
msbl Missing Bill of Lading.
MSC Manpower Services commission; a major container operator and registered in Switzerland; or Maritime Safety Committee (of IMO).
msca Missing cargo.
MSI Marine safety information.
MSL Mean sea level.
MSN Merchant Shipping Notice.
MSR Mean Spring Range.
MST Multi purpose shuttle tanker.
MSTS Military Sea Transport Services (USA).
MSU Main Switching Units.
MSV Multi Support Vessel.
mt. Million tons.
MT Mail transfer – a remittance purchased by the debtor from his banker in International trade; minimum transfer; mean time; minimum temperature; motor tanker; metric tons; medium term; or message transfer.
MTA Maltese pound – a major currency.
MTBF Mean time between failures.
MTBL Multi modal transport Bill of Lading – see FIATA.
MTC Maritime Transport committee. Based in Paris under the aegis of the OECD: it deals with Maritime matters.
MTD Multimodal Transport Document.
MTEU Million TEU – see TEU entry.
MTIB Malaysian Timber Industry Board.
MTH Month.
MTL Medium term loan; or Mean Tidal level.
mtm Million ton miles.
MTN's Multilateral trade negotiations.
MTO Motor Transport Officer; or Multimodal Transport Operator.

MTO/WTO/ITO Proposals for institutional reform of the GATT include submissions dealing with the establishment of an international body variously referred to as the Multilateral Trade Organisation (MTO), the World Trade Organisation (WTO) or the ITO. A GATT term.
mtons Metric tonnes.
MTR Mass Transit Railway.
MTS Motor turbine ship.
MTSO Mobile telephone switching offices.
MTTA Machine Tool Trades Association.
MTX Mobile telephone exchange.
MTX-Trade Moving technology exchange Trade – a computer logistics term.
MU China Eastern Airlines – an IATA Member Airline.
MUCR Master Unique Consignment Reference – customs term.
MUL Manufacturing under license.
Multi-access reservation system Computerised reservations system offering travel agencies access to the computers of various carriers and other suppliers; can also be used to book space on non-participating airlines and other suppliers.
Multi accident An accident involving more than two situations such as a vessel which is moored being damaged during a period of heavy weather lasting several days, and subsequently during two periods of heavy weather separated by an interval of a number of days of fair weather.
Multi buoy moorings A facility whereby a tanker is usually moored by a combination of the ship's anchors forward and mooring buoys aft and held on a fixed heading. Also called conventional buoy moorings.
Multi category stacking The process of stacking container types of differing specification as distinct from stacking containers of all the same type.
Multi channel distribution A situation whereby a range of channels of distribution are available. See channels of distribution.
Multi country The focus on two or more countries. See multi country contract.
Multi country contract A contract involving two or more countries. This may be a capital project involving a number of companies each situated in different countries.
Multi country international marketing plans An international marketing plan involving two or more countries.
Multi country international market research An international market research programme embracing two or more countries.
Multi crew levels A ships crew with a variable crew complement usually allied to the passenger certificate level. For example, the ship in the peak summer season may have a crew complement of 100 personnel for a passenger certificate of 1600 passengers. This figure would fall to 70 personnel when the passenger certificate was 800 passengers.
Multi crew operation See multi crew levels.
Multi culture A situation/circumstance which covers a range of cultures.
Multi culture crew A ships crew embracing several ethnic groups such as European, Asian, Philipinos, Chinese and Indian.
Multi deck car terminal A car terminal at a seaport or other transport terminal where cars can be loaded/discharged at two levels.
Multi deck vessel Ship with several decks or levels, most suited to carrying general cargo.
Multi delivery The process of a vehicle/delivering various consignments at different 'off loading' points during a specific schedule.

Multi destination journey A transit/journey (voyage/flight) involving two or more destinations of the itinerary.
Multi disciplined A situation/circumstance which covers several disciplines/activities/sectors. It may be a multi disciplined business offering a range of products targeting a range of market segments
Multi double-stack trailers Terminal operation equipment found in a container terminal especially at a seaport. It has up to two times the capacity of double stack trailers involving eight 20 feet or four 40 feet loaded containers. Such equipment operates in the port of Singapore. See also double stack trailer entry.
Multi flag fleet A maritime fleet which is composed of various ships which spans various registration flags. For example, a shipowner may have vessels registered in Liberia, Panama and France.
Multi fibre arrangement (MFA) Arrangement concluded in 1974 under which countries whose markets are disrupted by increased imports of textiles and clothing from another country are able to negotiate quota restrictions. A WTO definition.
Multi functionality Idea that agriculture has many functions in addition to producing food and fibre, e.g. environmental protection, landscape preservation, rural employment, etc. See non-trade concerns entry. WTO term.
Multi job A means of operating ICL System 4 machines by which the computer can perform several tasks so quickly that it appears to be doing them simultaneously. A computer definition.
Multi jurisdictional A situation whereby litigation investigations are conducted in a number of countries which may vary between four to ten countries.
Multi King 22 A multi purpose tween deck general cargo ship with an overall length of 155.5m, breadth of 2.255 m, draught of 9.75 m, deadweight tonnage of 21,500 tonnes and speed of 15.3 knots. The vessel has a hold grain capacity of 30,340 m^3 or a hold bale capacity of 27,950 m^3; or a container capacity of 746 TEU's. Four holds are provided and cargo handling equipment includes electro-hydraulic deck cranes of two single 25 tonnes and one twin of 25 tonnes. The vessel is suitable for shipment of general cargoes, dry bulk, long steel products, grain cargoes, timber and containers. The vessel is built by British Shipbuilders.
Multi lateral agreements Agreements signed by countries from multiple regions of the world, the overall purpose being the reduction of tariffs and other barriers to world trade.
Multi lateral aid The process of giving aid to more than one recipient country/organisation/agency etc, such as the provision of financial multilateral aid direct to the development agencies such as the World Bank, regional bodies like the African, Asian and Inter-American Development Banks, UN Agencies and so on.
Multi lateral donor A donor – usually a government or trading bloc providing financial aid to more than one country/region.
Multi lateral financing Funding of a particular contract/project involving a range of sources. This happens in new tonnage build involving a mega cruise liner, or the channel Tunnel project.
Multi lateralism The framework for trade between many countries and thus the means of extracting maximum gains from the efficient use of resources and division of labour.
Multi lateral security measures A range of security measures embracing security on board and ashore. These embrace appointment of security officers on board

and ashore; training of personnel; preparation of security plans; maintaining security records on board; and development of biometric identity cards for seafarers. Overall, a response to the United States government initiative and introduced in 2002. See SST.

Multi lateral Trade Agreement An agreement amongst a number of trading nations. It may be over a specified period involving particular goods and services up to a specific overall value such as £500 million over five years, or what is more common simply in general terms to facilitate trade amongst the trading nations. The latter could involve relaxation of customs tariff barriers.

Multi lateral trade negotiations Trade negotiations conducted on a global basis involving a group of countries as exemplified in the eightieth round of WTO negotiations known as the Uruguay round. See WTO entry.

Multi layer distribution network A distribution system whereby the goods are distributed through several processes before they reach the consumer/end user as obtains in Japan. The imported product is processed by the Agent, transported to the wholesaler, bought by the Distributor and finally sold by the retailer to the end user or consumer.

Multi legged sling Cargo handling equipment comprised of a two or three legged sling with a single ring at the upper terminal for attachment to the crane hook. Four-legged slings (or quads) have two intermediate rings joining the legs to the ring for crane hook attachment.

Multi level rates Range of rates, such as rack or corporate, that may be applied to one or more room types.

Multilingual The ability to speak/communicate in more than three languages, or a document which is presented in more than three languages.

Multi market A product/service serving several markets.

Multi market product A product serving several markets such as EU, NAFTA and Far East.

Multi market strategies A marketing strategy extending to several markets.

Multi modal A transport service offering more than one transport mode. Multi modalism is the process of operating/providing a door to door/warehouse to warehouse service to the shipper embracing two or more forms of transport, and involving the merchandise being conveyed in a unitised form in the same unit for the throughout transit. It involves a scheduled/dedicated service. Forms of multi modalism include containerisation – FCL/LCL/road/sea/rail; land bridge – trailer truck – road/sea/road; land air bridge – pallet/IATA container – road/sea/air/road; swap body – road/rail/sea/road; and trailer truck – road/sea/road.

Multi Modal Transport Bill of Lading A combined transport document with negotiable status issued by the freight forwarder to the forwarders principals.

Multi Modal Transport Document This form of transport document covers at least two different forms of carriage (i.e. multimodal transport). It indicates the place at which the goods are taken in charge by the Carrier, as stipulated in the Documentary Credit (which may be different from the port, airport or place of loading) and the final destination (which may differ from the port, airport or place of discharge); and/or it contains the qualification 'intended' in relation to the vessel, port of loading and port of discharge.

Multi modal transport law The law/regulations relating to the carriers liability of a combined transport service.

Multi modal transport operator A person (or company) who, on his own behalf or through another person, concludes a multi modal transport contract involving at least two modes of transport such as road/ship, road/rail/, road/air and so on.

Multi modal transport service A scheduled coordinated transport service involving two or more transport modes.

Multi National Company The Multi National Company is an international business which is a non-governmental organisation vehicle for transferring technology, financial resources, management techniques, and marketing experience among nations at various stages of development. Overall a company with production and distribution facilities in more than one country, but which has its ownership and control based in one country (see 'holding company').

Multi national corporation A corporation that acts and thinks internationally, instead of being oriented to a single national market.

Multi national crew A ship's crew comprised of different nationalities such as UK officers and Italian, Greek and Indonesian seamen.

Multi national marketing The process of marketing a product/service across several countries thereby crossing international boundaries and cultures.

Multi national trading A method of direct export trading whereby the exporter establishes a Company overseas. In so doing, the decision to operate overseas by incorporated body in the foreign territory enables the exporter to have the choice to form a company, to buy one, or to merge. The law or the situation in the overseas territory may reduce these options

Multiple choice question A question found in a market research questionnaire presenting the respondent with several answers from which to make a selection. For example the purpose of journey question may feature work, visiting friends/relatives, leisure, education and so on.

Multiple containers Several containers covered by the same transport document.

Multiple contract finance – general purpose line of credit This is designed to provide a framework for financing purchases from a number of exporters by a number of overseas importers.

Multiple contract finance – projected line of credit This is designed to cover purchases from a number of exporters by one overseas buyer in connection with a specific project.

Multiple currency clause bonds A bond which gives the investor a choice between two or more currencies for the repayment of principal and usually also for the payment of interest.

Multiple drop A road haulage term indicating the driver has a number of consignments to deposit at varying places during the course of his daily schedule.

Multiple Travel Agent Association Organisation of British travel agency chain companies.

Multiplier An economic term indicating a measure of the effect of total National income of a unit change in some component of aggregate demand.

Multiplier effect Concept that tourist expenditures in an area generate even more expenditures and thus more money as the tourist income is spent by residents who receive it as wages or profits; can be estimated statistically.

Multiple readership A situation where several people read a newspaper, journal, magazine or periodical. It is quite common for example with technical journals to be read by up to four people especially when the issue is company bought and the journal is circulated amongst technical personnel.

Multiple table input forms Input forms which enable data (attributes) to be entered into more than one table (entity) at the same time

Multi port itineraries A sailing schedule of a vessel involving a number of ports of call during a voyage conveying for example a bulk cargo of oil.

Multi port operation The process whereby a vessel has a number of ports of call in order to discharge completely her cargo. A term frequently found in tanker operation whereby the vessel will discharge her shipment of oil at maybe up to three ports. Usually under such situations such multiple ports are considered to be one for the purpose of lay time calculations.

Multi product berth A berth at a seaport handling a range of different commodities/products.

Multi purpose A product or service which provides more than one function or purpose. It maybe a multi purpose vessel able to carry more than one type of traffic such as a ferry conveying cars, lorries, coaches, passengers etc, or a multi purpose transport operation involving road haulage, parcels, vans and coaches.

Multi purpose berth A seaport berth able to handle a wide variety of cargo commodities transhipped in varying forms, i.e. pallets, containers, break-bulk merchandise involving vessels of differing specifications.

Multi purpose offshore support vessel A very specialised vessel designed to support offshore activities such as an oil rig, diving support ship, sub-sea construction support, cable laying, repair ships, dredgers, fire floats, and geotechnical-hydrographic survey.

Multi purpose vessel A ship so designed to carry a variety of cargoes in differing forms such as containers, vehicles, unitised shipments, break-bulk etc. Examples include Combi Carrier, Vehicular Ferries, Freedom type tonnage.

Multi quote The process of giving two or more rates relative to a particular product(s) or service(s). It maybe the rate from 'A' to 'B' involving options by air and sea and/or different routes and/or operators.

Multi racial
A situation where more than one group of nationality obtains.

Multi route A situation where more than one route is involved probably on an optional basis such as the sea routes to France from Dover include Dover/Dunkerque, Dover/Calais and Dover/Boulogne.

Multi sector business trip An overseas business trip which involves visiting several countries on business.

Multi sourcing Process of an entity obtaining supplies/raw materials/component parts from two countries/companies/suppliers or more. A procurement term.

Multi stage sampling A statistical technique where by from a given group, series, etc, a sample is taken embracing all the constituents within the group, series etc.

Multi stop A container movement where more than one packing/stripping place is involved.

Multi tiered price structure A price structure containing several layers of price – for example car ferry motorist tariff may differ according to the time of year winter – summer – autumn.

Multi trade agreements A trade agreement – usually a government or trading bloc to more than one country.

Multi union plant An industrial plant such as a factory where the work force are members of differing unions reflecting their different crafts and skills.

Multi user berths A berth facility situated at a port which is used for more than one cargo type such as a bulk dry cargo, general cargo and so on.

Multi user route A route used by many transport operators.

Multi user/shared user warehouse Warehouse used for storage and distribution activities by more than one retailer and/or manufacturer. In general, will be operated by a third party storage or distribution specialist on behalf of its clients. Increasingly used as a tool to cut storage and distribution costs for users unable to justify a dedicated facility.

Multi wall A packaging term relative to the fibre board carton with a multi wall structure comprising of several layers of fibre board to afford improved protection to the goods.

Multi Year ice Old ice up to 3m or more thick which has survived at least two Summers' melt. Hummocks are smoother than in second-year ice, and the ice is almost salt-free. Colour, where bare is usually blue. Melt pattern consists of large interconnecting irregular puddles and a well-developed drainage system.

Murmapatitbill Code name of BIMCO approved Bill of Lading for shipments on the Murmapatit charter party.

Murmapatit Code name of charter party used for shipments of Apatite ore and Apatite concentrates from Murmansk.

Mutatis mutandis (Latin) After making the necessary changes.

Mutual insurers Organisations where the profits go to the policy holders – there being no shareholders.

MV Motor vessel.

M/V Motor (merchant) Vessel.

MVT Aircraft Movement Message – IATA term.

(M) with date Date when modified special survey on a vessel took place. The period for which the class is maintained follows the date.

MW Mega watt.

MWSS Metropolitan Waterworks and Sewerage system (Philippines).

MWWA The Metropolitan Water Works Authority (Thailand).

MX Compañia Mexicana de Aviación SA de CV (MEXICANA). The National Mexican airline and member of IATA.

MY Air Mali. A National airline and member of IATA; or motor yacht.

MZ Melpati Nusantara Airlines – an IATA member airline.

N New; or North – a point on the compass.
n/a Not applicable; or no account (banking).
N/A No advice.
NAA Not always afloat; or Nigerian Airports Authority.
NAABSA Not always afloat but safe aground.
NB North bound.
NAC North Atlantic (Shipping) Conference.
NACA National Air Carrier Association (US charter operators).
NACFA National Clearing and Forwarding Agency.
NACS National Association of Chinese Shipowners.
NADIN North American Data Interchange Network (USA term).
NAEGA North American Export Grain Association.
NAFEC National Aviation Facilities Experimental Centre (US).
NAFED Indonesian National Agency for Export Development.
NAFTA North American Free Trade Agreement of Canada, Mexico and the US.
Naira The currency of the Federation of Nigeria. See Appendix 'C'.
Naira invoice An invoice payable in the Federation of Nigeria naira relative to an export sales invoice.
NAITA National Association of Independent Travel Agents.
NAM North America(n).
NAMAS National Measurement Accreditation Service.
NAMCAR North American and Caribbean (Office – ICAO term).
Name An underwriting member at Lloyd's. An insurance definition.
Names A control of the interbank money markets is exercised by the individual banks limiting the total volume of business they will be prepared to conduct with every other bank on a name by name basis; or in marine insurance terms: individual underwriting members of Lloyd's of London who group into syndicates and trade in insurance with unlimited liability.
NANAC National Aircraft Noise Abatement Council (USA term).
Nanyozai '1967' Code name of charter party for shipment of logs. It was formulated by the Japanese Shipping Exchange. Under the BIMCO 'Shipwood load' slip, it is applicable to any fixture of sawnwood from Finland and Sweden to all destinations except the UK and Republic of Ireland.
NAO North Atlantic Oscillation.
NAOCC Non Aircraft Operating Common Carrier.
NAPE National Association of Port Employers.
NAPOCOR National Power Corporation (Philippines).
NAR North American route (structure) – IATA term.
NARE National Airspace Review Enhancement Advisory Committee (USA term).
NARG Navigation Aids Development Criteria and Area Navigation Working Group (EANPG term).
Narrow body An aircraft with one corridor such as a DC-9 and a Boeing 737.
Narrows A narrow channel between two sheets of water.
NARTEL North Atlantic Radio-Telephone (Committee – SITA).
NASA North Atlantic Shippers Association; or National Aeronautics and Space Administration (US).
NASC National AIS System Centre – IATA term.
NASUTRA National Sugar Trading Co (Philippines).
NAT North Atlantic – IATA term.
Nat Natural logarithm.
Natias National Testing Laboratory Accreditation Scheme.
National A person who has the citizenship of a country, either by birth or by naturalisation.
National Accounts Manager A senior Sales Manager position responsible for major Customers National Sales Accounts.
National Administrator A government organisation/department such as Ministry of Shipping/Trade/Transport.
National Air Carrier Association (NACA) US trade group of airlines flying low-cost schedule and charter services.
National Association of Cruise Only Agencies (NACOA) US trade group of travel agencies specialising in sea voyages.
National Association of Independent Travel Agents UK trade group.
National Business Travel Association US trade group representing corporate travel managers; formerly known as the National Passenger Traffic Association.
National currency The currency of a particular country such as the Swiss Franc is the currency of Switzerland.
National firms Companies that focus their business on domestic markets.
National Flag Ship A vessel registered in a particular country and flying the national flag thereby conforming with the maritime fleet legislation imposed by that country. See cabotage.
National Flag tonnage Vessels registered in a particular country and flying the national flag thereby conforming with the maritime fleet legislation imposed by that country. See cabotage.
National geodetic survey A branch of the US National Ocean Service Administration. It is responsible for the supply for GPS orbit data via the NIS bulletin board.
National Income A measure of the money value of the total flow of goods and services produced in an economy over a specified period of time.
Nationalism An international trading technique adopted by a country or state to protect its national shipping/ airlines and related infrastructure, e.g. seaport/airports through discriminatory commercial policies and with maybe similar policies applicable to its overall international trading techniques; or the culture of a country/region. This identifies the populace with the country/region heritage and social/economic factors and prejudices.
Nationalistic market A country/state/region where the indigenous population tend to purchase/prefer their national products rather than rely on imported products. Countries favouring their domestically produced goods are Japan and France.
National load centre A transport terminal which processes/distributes cargo passing through it and usually on an intermodal transport system such as the Port of Singapore.
National Park Service US federal agency within the Department of the Interior responsible for operation of the US national park system.
National quotas A situation/circumstance whereby a particular Free Trade area conducts trade on the basis of limiting the quantity of a particular commodity(ies) exported within the Trading group from each country.

National Register A national register is one that treats the shipping company in the registration of its vessels in the same way as any other business in the country.

National schedules The equivalent of tariff schedules in GATT/WTO, laying down the commitments accepted – voluntarily or through negotiation- by WTO members. A WTO term.

National Sovereignty The supreme right of nations to determine policies, free from external controls.

National Standard Shipping Note A six-part document which is completed by the supplier of the goods or the shipper/freight forwarder giving full details of the goods similar to those found on the Bill of Lading, against which it is matched before issue. It accompanies the merchandise to the port or terminal and is unacceptable for use with dangerous classified goods shipments. Copies of the National Standard shipping Note are retained by those parties handling the goods until they are finally on board the vessel from which a 'shipped' bill of lading is issued. It is used primarily for container groupage points and only goods for shipment to one port of discharge on one sailing and sometimes relating to only one Bill of Lading may be grouped on one shipping note.

National strategy A strategy which applies to a particular market usually the home market, as distinct from a global strategy which would apply worldwide. See also global strategy and nationalism.

National Subsidiary Structure A company whose organisational structure requires each Foreign subsidiary reporting directly to the parent board of the Company without any intermediate layers of management either at regional headquarters or international headquarters.

National Tour Association (NTA) Organisation of US motorcoach tour operators, wholesalers and other tour suppliers to deal with standards, training and marketing; formerly called the National Tour Brokers Association.

National Transportation Exchange A USA national business to business e-commerce trading exchange for member shippers and truck/maritime carriers.

National Transportation Safety Board (NTSB) The US agency that recommends safety standards for all modes of public transportation and investigates accidents.

National treatment The principle of giving others the same treatment as one's own nationals. GATT Article III requires that imports to be treated no less favourably than the same or similar domestically-produced goods once they have passed customs. GATS Article XVII and TRIPS Article 3 also deal with national treatment for services and intellectual property protection. WTO term.

National wealth measures The process of measuring the national wealth of a country which will embrace one or more of the following gross domestic products, GDP per capita, private consumption per capita, private consumption per capita, GDP per capita, and gross domestic product.

NAT/NAM North Atlantic/North America (Region – IATA term).

NATO North Atlantic Treaty Organisation signed in 1949 between certain European countries, Canada and the USA.

NATS National Air Traffic Services (UK).

Natural persons People, as distinct from juridical persons such as companies and organisations. WTO term.

Natural resource based products The main products considered under this heading are: fish and fish products, forestry products and non-ferrous metals and minerals. Energy products (especially coal) have been discussed also.

Nautical mile Maritime measurement of 6080 feet (1.858 kilometres) equivalent to one knot.

Nautical tables Book containing tabulated data, arithmetical, geometrical, astronomical and geographical, for use in navigation.

N aux B date New Auxiliary boilers with date fitted for vessels.

Naval Architect Person responsible for ship design.

Navigation The process of determining the correct course(s) of a ship or aircraft to enable it to arrive safely at the destination sea or airport.

Navigational aid A facility on a transport unit to aid navigation or equipment located in the path of the transport unit and to facilitate the safe movement of the ship or aircraft. It includes radar, radio telephone, buoys, maritime traffic systems etc. Navigational aids feature strongly at airports and seaports and many are computerised involving satellite systems.

Navigational limits Geographical restrictions on the areas where a vessel may travel.

Navigation lights Those lights compulsorily shown by vessels at sea in accordance with international rules.

NAVSEP Navigation and Separation (Panel – EUROCONTROL).

Navtex Manual Navtex is an international automated direct-printing service for promulgation of navigational and meteorological warnings and urgent information to ships. It has been developed to provide a low-cost, simple and automated means of receiving maritime safety information on board ships at sea and in coastal waters. The information transmitted is relevant to all sizes and types of vessel and the selective message-rejection feature ensures that every mariner can receive a safety information broadcast which is tailored to his particular needs. The Manual is intended for use by seafarers, shipowners, maritime administrations and others concerned with the preparation, broadcasting and receiving of maritime safety information.

NAWFA North Atlantic Westbound Freight Association.

NAWK National Association of Warehouse Keepers.

NB North bound.

NBCC Netherlands British Chamber of Commerce, or Nigerian British Chamber of Commerce.

NB Date New boilers with date fitted for vessels.

NBL No berth list.

Nb (nota bene) (Latin) Observe what happens, Lit note well.

NC The Nordic Council; or non continuous liner.

N/C New crop; or new charter.

NCAD 'Notice of cancellation' at anniversary date. Cover is cancelled automatically at the anniversary date of a long term insurance unless the notice is withdrawn.

NCB National Computer Board (Singapore).

NCBs National central banks.

NCC Norwegian Chamber of Commerce; or National Computer Centre (Philippines).

NCITD National Council on International Trade Documentation – a United States organisation based in New York.

NCL National Carriers Ltd; or Norwegian Caribbean Line – a Norwegian shipowner.

NCSPA North Carolina States Ports Authority.

NCTS New Community Transit System. See separate entry.

NCV No commercial value or no customs value.

ND No discount; new deck; or natural draught.

NDFCAPMQS No dead freight for charterer account provided minimum quantity supplied – chartering term.

ndp New domestic boiler(s).

NDC National Development company (MTI Philippines).

NDLV Westbound Freight agreement. Liner conference embracing westbound trade from Northern Germany, Holland and Belgium to the United States.

NDT Non destructive testing.

NE North East; or no effects.

ne Not exceeding.

NEA The National Energy Administration (Thailand).

Neap tide Tide with which high-water level is at its lowest point.

Nearby A date/contract that is relatively close to the cash/spot/expiry date.

Near market A market which is close to the suppliers (exporters) location but crosses international boundaries. To the UK exporter near markets include Spain, Germany, Belgium, France, Netherlands and so on.

NEB National Electricity Board (Malaysia); or Office of the National environment Board (Thailand).

NEC National Exhibition Centre (Birmingham); or net explosive content – dangerous cargo term.

Necore Code name of Chamber of Shipping approved charter party for Mediterranean ore trade.

NEDA National Economic Development Authority (Philippines).

Needs Analysis The process of identifying a specific requirement or undertake a particular task/work assignment.

Negative pledge A clause in a loan agreement prohibiting the borrower from pledging its assets elsewhere without the lending bank's consent.

Negligence Absence of care according to the circumstances.

Negligence and Exceptions clauses Charter Party clause detailing shipowner's exemptions from all liability for loss or damage.

Negligence Clause A clause extending the policy to cover the peril 'negligence', (e.g. Inchmaree Clause). In the absence of this clause the SG policy form does not cover negligence.

Negotiable Bill of Lading One capable of being negotiated by transfer or endorsement.

Negotiable instrument Negotiable document.

Negotiated cheque This permits the exporters to obtain the immediate value of the cheque or for a pre-set forward value date in the currency of the cheque. Lock boxes provide a fast and secure method of receiving cheque payments from overseas. Cheques are not sent to the seller but are paid directly by the buyer into the seller's bank via a PO Box number reducing the clearance cycle period.

Negotiated rates Usually a commercial rate, which is available to specific clients in return for volume of business.

Negotiating bank The bank which negotiates under a credit. It may be a bank nominated by the issuing bank or another authorised under the credit. In so doing the Bank under a documentary letter of credit purchases a bill of exchange drawn by the exporter payable at another bank. Unless the Negotiating Bank has confirmed the credit or unless otherwise indicated, such negotiations are normally made with recourse to the seller.

Negotiating plan A plan devised by an individual/entity/company which is used as a negotiating base. It is usual in the international business field to have a strategic base embracing a stated objective with a range of options and identify key negotiating points. Overall, the plan must be costed and have a legal technical and economic focus within the environment in which the plan is situated. The plan would have a timescale and reflect the culture. The foregoing constituents will depend on circumstances of both parties to the negotiations.

Negotiating position Situation of a particular client from which negotiations over a commercial deal, industrial or other specified matter, commence.

Negotiation The conversion of a foreign cheque into the domestic currency, giving credit to the customer in advance of receipt of proceeds or action by which the negotiating bank buys the documents. A Banking definition.

Nemine contradicente (Latin) or nemo con (Latin) Without opposition.

Nemo dat quod no habet (Latin) No one gives that which he has not.

Neo bulk cargo A term given to a wide range of staple products that can be united as break bulk including a wood pulp, board, paper, logs, steel, and newsprint.

NEQ Net explosive quantity – dangerous cargo term.

NERC New En-Route control Centre – IATA term.

NES Not elsewhere specified; or new export system – see separate entry.

NESDB Office of the National Economic and Social Development Board (Thailand).

Nesting Packing hollow-ware cargo (e.g. earthenware bowls) so that one item nests within another to save space. Paper or straw may be used to separate each item and avoid damage.

Nesting berth A type of berth at a port designed in the shape of a letter 'U' which enables the stern or bows of the vessel to nestle into the berth thereby facilitating stability of the ship during loading or discharging of cargo/passengers.

Net Abs Premium is payable net of all discount, including any over-rider applicable. Return premiums are usually paid on this basis.

Net asset value The amount by which the assets of a Company exceed all liabilities including loan and preference capital, divided by the number of equity shares in issue.

NETC No explosion of the total contents.

Net capital expenditure Comprises a government's final capital expenditure (i.e. gross fixed capital formation, plus net purchases of land and intangible assets, plus changes in stocks) and net capital transfers paid (i.e. investment grants, plus unrequited transfers paid by the general government to finance specific items of gross fixed capital formation by other sectors, minus capital taxes and other capital transfers received by the general government). An ECB definition.

Net daily loss A chartering term which is calculated on the basis of the gross freight income from a particular voyage minus the gross running cost divided by the number of days taken to perform the voyage. If the voyage incurs a loss with the gross freight income being less than the gross running cost, a financial loss will result. This is termed as 'net daily loss'.

Net daily surplus A chartering term which is calculated on the basis of the gross freight income from a particular voyage minus the gross running cost divided by the number of days taken to perform the voyage. If the voyage is profitable it will produce a net daily surplus.

Net exporter A company, country, seaport, airport, carrier handling/generating a greater volume/value of exports than imports.

Net external asset or liability position (or the international investment position (iip)) The statistical statement of the value and composition of the stock of an economy's financial assets or financial claims on the rest of the world, less an economy's financial liability to the rest of the world.

Net figure The total financial/numerical data less any prescribed deductions. See gross figure.

Net form charter Type of charter party whereby charterer pays for port charges and cost of loading and discharging the cargo.

Netherlands British Chamber of Commerce The organisation based in London representing Holland whose basic role is to develop, promote and facilitate international trade between UK and Holland and vice versa.

Netherlands Shippers Council The organisation in Holland representing the interest of Dutch shippers in the development of international trade.

Net importer A company, country, seaport, airport, carrier handling/generating a greater volume/value of imports than exports.

Net Line The amount, of the original line, which is retained by the re-insured (Net retained line).

Net liquid funds Cash at bank and in hand and cash equivalents, e.g. investments held as current assets, less bank overdrafts and other borrowing repayable within one year of the accounting date.

Net present value The value obtained by discounting all cash outflows and inflows attributable to a capital investment project by a chosen percentage, e.g. the entity's weighted average cost of capital.

Net pricing The tariff/sale/charge raised for a consignment/passenger on a particular service but exclusive of all ancillary charges. This could include commission payments, port handling cost, other port charges and so on.

Net profit The amount of money left after the indirect costs of a business are deducted from the gross profit.

Net quantity The mass or volume of the dangerous goods contained in a package excluding the mass or volume of any packaging material, except in the case of explosive articles and matches where the net mass is the mass of the finished article excluding packagings.

Net rate The rate/charge raised for a consignment on a particular transit/service but exclusive of all the ancillary charges. For example, the shipowner may receive a net rate of €100 for a consignment between port 'A' and port 'B' whilst charging a gross rate of €120. The €20 difference would represent the port handling costs, other port charges, and so on; or a wholesale rate to be marked up for retail sale.

Net realisable value The price at which goods in stock could be currently sold less any further costs which would be incurred to complete the sale.

Net receipts pool A pooling agreement amongst transport operators whereby each participant contributes only the profit element of his business. For example, in May a transport operator may have received €12,000 in traffic receipts, and incurred some €10,000 in cost, in which case the pool contribution would be €2,000. At the year end the net receipts in the pool are distributed proportionate to each participant pool allocation which may be 10%, 20%, 15% and so on. A major weakness of such a system is there is no incentive to maintain cost efficiency and the inefficient operator, who incurs a loss annually, can cross subsidise the efficient operator who makes a profit each year. Such a system is found in shipping and road transport operation.

Net Registered Ton A traditional unit of measurement of a ship's size which is little used now. It is derived from the gross tonnage by deducting spaces for crew accommodation, propelling machinery and fuel. Net registered tonnage, the GRT minus the spaces that are non-earning, machinery, permanent bunkers, water ballast, and crew quarters. Over the range 0-6,000 NRT there is a reasonably good correlation between NRT and DWT: DWT = 2.5 NRT.

Net revenue This represents in shipping terms the freight income accruing from a particular voyage less the related voyage expense. See daily net freight revenue (voyage).

Net settlement system (NSS): a funds transfer system The settlement operations of which are completed on a bilateral or multilateral net basis. ECB term.

Net tare weight Weight of empty container, railway wagon or trailer.

Net terms Under these terms the cargo is loaded and discharged at no cost to the shipowner. The cost of stevedores at the loading port is borne by the shipper and at the port of discharge by the receiver. The term 'net terms' is not in common use but is generally referred to as free in and free out (FIO) with exactly the same meaning.

Net tonnage This is the gross tonnage less the machinery, boiler and bunker, crew and stores spaces.

Net weight The weight of the goods excluding all packing.

Network See 'Local area network' and 'wide area network'.

Network analysis A graphical method of planning a project in a logical sequence by plotting major activities with start/finish dates and times for each activity enabling overall time required to complete project to be estimated.

Networking The process of setting up/establishing contacts within a business environment.

Network of sales outlets A number of sales outlets forming a network for a particular Company/Entity.

Network organisation An organisation in which activities and decision making are widely dispersed as found for example in a multinational industry.

Neutral unit of construction (NUC) Unit devised by IATA as a common denominator for constructing air fares whereby the local selling fare is multiplied by a pre-established rate of exchange for the country of origin.

New build Vessels which have recently been built.

New building The provision of new tonnage.

New building contract A contract to build a new vessel in a shipyard.

New community transit system The new community transit system is the European wide system to allow the electronic processing of customs documentation designed to provide better management controls; lower costs in respect of community and common transit; improve security and fraud prevention. It was introduced by HM Customs in 2003 and Customs requirements include: complete electronic operation; time out release for authorised traders; guarantee and authorisation validation; transit accompanying document will not require stamping for authorised traders; bar coding required (UN ISO format 228); list of items replaces load list, and LCP premises can be used if an authorised trader under NCTS. See new export system.

New Export System The new export system – fully computerised – was introduced in 2003 and applies to exports to non-EU countries originating from EU countries. It replaces period entry, local export control and simplifying procedures. Overall, it supersedes the community transit system. There are three messages to be submitted to the Customs' CHIEF Computer: pre-shipment advice (export declaration); goods arrived at port (arrival), and goods loaded aboard ship (departure). After each message CHIEF will issue an acknowledgement concerning clearance to proceed to the next stage in the process. Only once full clearance has been given by CHIEF, after successful completion of the first two stages, can the goods be despatched. The benefit to the exporter include: export consignments being eligible for immediate positive clearance, 24 hours a day throughout the year; increased predictability of clearance; the choice of Customs declaration procedures submitted electronically – local clearance procedure, simplified declaration procedure, Low value, Non-Statistical declarations, and the full declaration procedure; paperless trading for majority of declarations submitted; the opportunity to build a partnership with customs around risk based audit control providing both operator and Customs with a high level of system assurance; a choice of electronic routes to Customs, including via the Internet; access to links with other UK government departments and agencies, such as the CAP export refund claims; improved quality and assurance in obtaining official evidence of export for VAT and economic regimes; the provision of Designated Export Place facilities for inventory linked consolidators to clear cargo at their own approved premises. Approved DEP operators can use SDP or full electronic declarations, but not LCP; and allowing ICDs to continue under NES. Goods can be declared to Customs; and brought under their control using normal and SDP (but no LCP) procedures. See CHIEF entry.

New for old Sometimes called 'thirds', this is the deduction from the cost of hull repairs, provided in law, to allow for depreciation due to age of material or parts being replaced. In practice, hull underwriters waive thirds. When new material or parts replace damaged material or parts during repairs to a ship, underwriters are entitled to make a deduction from the claim as a result of betterment but they waive this right in practice. Average adjusters may apply the principle in general average for vessels over 15 years old. Marine insurance definition.

New ice A general term for recently formed ice, which includes frazil ice, grease ice, slush, and shuga. These types of ice are composed of ice crystals which are only weakly frozen together (if at all) and have a definite form only while they are afloat.

New Jason Clause Protective clause inserted into a charter party or bill of lading which provides that the shipowner is entitled to recover in general average even when the loss is caused by negligent navigation. The need for such a clause arises from the decision of an American court that, while American law exempted a shipowner from liability for loss or damage to cargo resulting from negligent navigation, this did not entitle the shipowner to recover in general average for such a loss.

Newly industrialised countries A country which has been newly industrialised from the third world such as Brazil, South Korea and Malaysia which strive to enter the world free trading system under WTO.

New Peso The currency of Chile. See Appendix 'C'.

New Peso invoice An invoice payable in the Chilean new peso relative to an export sales invoice.

New Taiwan Dollar The currency of Taiwan (Republic of China). See Appendix 'C'.

New Taiwan Dollar invoice An invoice payable in the Taiwan (Republic of China) new dollar relative to an export sales invoice.

New Tonnage Vessels which have recently been built.

New Uruguayan Pesos The currency of Uruguay. See Appendix 'C'.

New Uruguayan Pesos invoice An invoice payable in Uruguayan new pesos relative to an export sales invoice.

New World Alliance Major merchandised trade alliance whose members are APL, Hyundai MM, Mitsui OSK Line.

New York Convention The United Nations Convention signed in 1958 on the recognition and enforcement of foreign arbitral awards.

New York prime loan rate The US banks' equivalent of the UK base rate; it is not a market rate as such but forms the rating basis for some short-term commercial loans. The rates are not altered on a daily basis for practical loan administration reasons. However, when the trend of the cost of the underlying funds changes, NYPLR will be moved into line; loans may be based on fixed or on floating prime rates; the majority are based on the latter so that the loan will fluctuate broadly in line with market conditions; loans that are based on fixed prime will run to maturity on the basis of the prime rate ruling at the time funds are drawn down, and as such must be considered as any other fixed interest lending.

New Zealand Dollar The currency of New Zealand. See Appendix 'C'.

New Zealand Dollar invoice An invoice payable in New Zealand dollars relative to an export sales invoice.

New Zealand Tonnage Committee The Conference Committee for the Europe – New Zealand route.

New Zealand UK Chamber of Commerce The organisation based in London representing New Zealand whose basic role is to develop, promote and facilitate trade between UK and New Zealand and vice versa.

N/F No funds.

NF Air Vanuatu Ltd – an IATA airline.

NFA National Food Authority (Philippines).

NFAC National Food and Agricultural Council (Philippines).

NFAS National Federation of American Shipping (based in Washington).

NFD Non-Federated.

NFPA National Fire Protection Association (USA term).

NFTZ Non Free Trade Zone.

NFY Nylon filament yarn.

NG Nigeria customs abbreviation; or Lauda Air (Lauda Air Luftfahrt Aktiengesellschaft) – IATA member Airline.

NGLs This is the abbreviation for Natural Gas Liquids. These are the liquid components found in association with natural gas. Ethane, propane, butane, pentane and pentanes-plus are typical NGLs.

NGO Non-governmental organisation.

NGS New Guidance System – IATA term; or National Geodetic Survey.

NH All Nippon Airways Co Ltd – an IATA Member Airline.

N H crude No heat crude (oil).

NHP Nominal horse power.

NI PGA – Portugalia SA – an IATA Member Airline.

NIA National Irrigation Administration (Philippines).

NIC Newly Industrialised Country.

Niche marketing The process of identifying a small segment of the market relative to a particular product/service and directing marketing into that niche. It usually has a small volume base and features innovation in terms of the product/service which enhances its competitiveness and thereby premium pricing. To become effective niche marketing requires continuous market research.

NIDA National Institute of Development Administration (of Thailand).

NIDC National Investment and Development Corporation (Philippines).
NIEO New international economic order – UNCTAD term.
NIFs Note Issuance Facilities.
Nigerian British Chamber of Commerce The organisation based in London representing Nigeria whose basic roles is to develop, promote, and facilitate trade between UK and Nigeria.
Nilas A thin elastic crust of ice, easily bending on waves and swell under pressure, thrusting in a pattern of interlocking 'fingers' (finger rafting). Has a matte surface and is up to 10 cm in thickness. May be subdivided into nilas and light nilas.
Nil desperandum (Latin) Never despair.
Nil paid A new share offer, usually a rights issue on which no payment has been made.
NIMEXE The nomenclature for the foreign trade statistics of the EEC. It is the harmonised nomenclature that EU countries have adopted to classify goods for compiling their foreign trade statistics and is consistent with CCCN and SITC.
NINUS An Integrated Navigation System developed by Kelvin Hughes. NINAS for Nucleus Integrated Navigation System. Nucleus being a trade name for a sophisticated radar set.
Nip Ice is said to nip when it forcibly presses against a ship. A vessel so caught, though undamaged, is said to have been nipped.
Nippon coal Code name of charter party used for shipment of coal.
Nippon Kaiji Kyokai The Japanese Ship Classification Society based in Tokyo, Japan.
Nipponore Code name of charter party used for shipment of iron ore. It was formulated by Japanese Shipping Exchange.
Nippon Yusen Kaiska (NYK) Line A major container liner shipowner and registered in Japan.
NIS Norwegian International ship Register or Navigation Information Service – operated by the US.
NISA National Invasive Species Act.
NITPRO Nigerian Committee/Organisation on Trade Procedures and Facilitation based in Lagos.
Nitrating Acid Mixture A mixture of nitric and sulphuric acids used for the nitration of glycerine, cellulose or other organic substances. This acid mixture coming in contact with organic matter commonly causes fire, unless the mixture contains much water. Term associated with dangerous cargo classification.
NIWO Dutch international road transport organisation.
NJ New Jason (charter party) clause.
NK Nippon Kaiji Kyokai – Japanese Ship classification Society.
NKR Norwegian Krone – a major currency.
NL No liner; Netherlands – customs abbreviation; or Air Liberia – an IATA airline.
N/m No mark.
nm Nautical Miles.
NM Mount Cook Airlines – an IATA Member Airline.
NMA National Maritime Authority or Norwegian mapping agency.
NMAC Near Mid-Air Collisions (USA term).
NME Near miss encounters which reflect the situation of a near accident at sea. It arises after two ships pass within eight cables of each other.
NMFTA National Motor Freight Traffic Organisation.
NMOHSC National Maritime Occupational Health and Safety Committee.
nmp Net material product.

Nmt New means of transport. A term associated with HM Customs relative to the purchase of vehicle, boat, or aircraft overseas and the relevant customs import regulations.
NMT Nordic mobile telephone.
NN Air Maltinique – an IATA Member Airline.
NNI Nederlands Normalisatie Instituut (Dutch Standards Institute).
NNR Non-negotiable receipt.
NNRF Non-negotiable report of findings. See preshipment inspection
NNWB Non-negotiable Waybill.
NO Norway – customs abbreviation; or number.
NOAA National Oceanic and Atmospheric Administration – a USA organisation.
NOB National Association of professional goods transporters (the Netherlands).
No Cure – No Pay Salvage provision whereby no award is paid to a salvor if he is unsuccessful.
NOLA North lakes.
NOL/APL A major container shipowner and registered in Singapore.
No market Broker's term indicating that the risk offered is not placeable in the market(s) available to him. This might arise where the market capacity is saturated with the risk already (e.g. reinsurance on a large vessel) or in the case of a particularly hazardous risk.
Nominal effective exchange rates See effective (nominal/real) exchange rates.
Nominal price An estimate of the price for a future month date which is used to designate a closing price when no trading has taken place in that date. Also used of current price indications in similar circumstances in physical trading.
Nominated Bank The bank nominated in the documentary credit to pay, accept, negotiation or incur a deferred payment undertaking.
Nominee One who is nominated by another.
Nominee Name Name in which a security is registered that does not indicate who is the 'Beneficial Owner'.
Non-commercial cargo Cargo shipped free such as under a foreign aid programme.
Non-compliance A deviation from the Code, identified at an assessment.
Non cond Non condensing.
Non-conference lines Shipping Companies which operate outside the liner conference system.
Non-conference line vessels Vessels trading outside liner conference conditions.
Non-conference operator A shipowner which is trading outside liner conference conditions.
Non-contribution clause A clause providing that the policy does not pay any part of a loss that is covered by another policy in force on the same risk.
Non-convertible currency A currency which is not readily and freely transferable/traded on the inter-national currency exchanges. Such a currency is subject to restrictions and exchange control. See convertible currency.
Non-core business That part of the business of an entity/company which is subservient to the main thrust of the company activities and outside the main stream of the business. A railway company's main activity or core business may be passengers, but its non core business may be in train and station catering; estate management of leased land and buildings.
Non-cyclical factors Indicate influences on the government's budget balances that are not due to cyclical fluctuations (the cyclical component of the budget balance.). They can therefore result either from structural,

i.e. permanent, changes in budgetary policies or from measures with a 'temporary effect' (see also measures with a temporary effect). An ECB definition.

Non-demise charter party Charter party under which shipowner provides ship and crew, and charterer cargo.

Non-disclosure Failure of the assured or his broker to disclose a material circumstance to the underwriter before acceptance of the risk. A breach of good faith.

Non est factum (Latin) This was not my deed. Hence no defence if the person signing is negligent – a legal term relative to a contract.

Non-exclusive licence A licence granted to a particular beneficiary which is one of several granted/awarded. See exclusive licence entry.

Non-Executive director A Director of a Company whose role is to contribute an independent view to the Boards deliberations to help the Board provide the Company with effective leadership; to ensure the continuing effectiveness of the executive directors and management; and to ensure high standards of financial probity on the part of the Company.

Non-fixed radioactive contamination (Radioactive Material Only) Radioactive contamination that can be removed from a surface by wiping with a dry smear.

Non-flammable Incapable of burning.

Non-gateway country A country which has no direct entry access to other countries usually due to its geographical and/or political situation.

Non IATA member An Airline, not a member of IATA, but which may work within the terms of IATA, like Interline Partners.

Non-linearity The lack of direct proportionality of the input and output of a physical system.

Non-member States Countries situated in Europe which are not members of the European Union or any other International Trade Association.

Non-negotiable Bill of Lading Bill of Lading not capable of being negotiated.

Non-negotiable report of findings A report/document issued by SGS to confirm a satisfactory checking of documents that the goods are correct in regard to their quantity, quality and price. In due course the document is presented to the Bank to facilitate payment of the goods. It is a preshipment document.

Non-negotiable Sea Waybill A Documentary Credit may specify presentation of a Non-negotiable Sea Waybill rather than a Bill of Lading as the document of movement. It is, as its name implies, a Non-Negotiable Document which does not constitute a title to the goods, therefore title cannot be transferred to another party using this document and permits the consignor to change the consignee en route. It does not, therefore, offer the guarantees of a Bill of Lading and must be viewed differently.

Non-oil exports Products exported from a country/state excluding oil. It maybe quoted in value or tonnage.

Non-oil GDP The GDP of a country excluding its oil input. See gross national product.

Non-oil imports Products imported into a country/state excluding oil. It maybe quoted in value or tonnage.

Non-oil sectors An economy which features the non-oil segment of parts/elements of the economy which are not oil related.

Non-performing debt A debt not being repaid. See loan default.

Non-performing loan A loan on which the debt is not being repaid. See loan default.

Non-recourse finance Finance provided to a borrower without a specific liability on the borrower to repay. Repayment is assured by other means, for example by a government guarantee, a confirmed documentary credit or, in project finance, by a stream of contractual income from the project which will be channelled through the financing bank. See recourse.

Non-refundable ticket Air ticket whose dates cannot be changed and which cannot be turned in for a refund because of its low cost; sometimes elements may be changed by paying an extra fee.

Non-reusable As applied to receptacles, those which are in fact used only once for the transport of dangerous goods.

nonrev See non-revenue.

non revenue A flight on which there are no paying passengers; also, a passenger, usually an employee of the airline, who has not paid for a ticket.

Non-reversible laytime This arises where separate allowances are made for the loading or unloading and discharging operations of a particular vessel. An example arises with 'four days load/1,000 tonnes discharge', or '6,000 shinc load/2,000 tonnes shex discharge'. Once the 'statement of facts' have been received from the load port, a laytime statement can be drawn up based on the charter party laytime clauses, and demurrage/despatch calculated and settled whether or not the vessel has even reached her discharge port(s). The calculation for laytime at the loading port is entirely separate from that calculation for the discharging port; the term non-reversible being sometimes used to describe this state of affairs. Also termed normal laytime.

Non sequitur (Latin) It does not follow.

Non-sked Non-scheduled; an airline or other carrier that operates at irregular times, usually at a lower fare.

Non-smoking lounge A public passenger lounge on a ship or at a passenger terminal which prohibits smoking.

Non-stop A flight or other journey that goes from origin to destination without interruption.

Non-tariff barriers A non-tariff barrier is a measure, public or private, adopted by a country/trading bloc relative to a product or service whose objective is to discourage/frustrate international trade in such a way to create an obstacle to selling/conducting business in a particular market which crosses international barriers. It includes quotas and trade control, restrictive customs procedures, discriminatory government and private procurement strategies usually favouring the local national supply source, restrictive administrative and technical regulations, and selective monetary controls and discriminatory exchange rate policies.

Non-trade concerns Similar to multifunctionality. The preamble of the Agriculture Agreement specifies food security and environmental protection as examples. Also cited by members are rural development and employment, and poverty alleviation. WTO term.

Non-user profile The profile of a recipient of product or service who fails to use it.

Non-unionised work force A work force which does not rely on unions to represent their views and negotiate with the management on terms and conditions of employment/pay rises but employs other methods/resources.

Non-unitised freight Goods not handled in individual unit loads, e.g. raw materials and fuels.

Non-verbal message A gesture, a smile, eye contact, etc. Overall it is an expression of culture in communication.

Non-vessel operating carrier A carrier issuing Bills of Lading for carriage of goods on ships which he neither owns nor operates. It is usually a freight forwarder issuing a 'house' Bill of Lading. A carrier defined by maritime law is one offering an international cargo transport service. Also termed 'non-vessel operating (common) carrier' or non-vessel owning carrier'.

Non-vessel operating (common) carrier See non-vessel operating carrier.

Non-vessel owning carrier See non-vessel operating carrier.

Non-violation case A measure or any other situation which does not violate the provisions of the WTO but which, nonetheless, causes nullification or impairment.

nop Not otherwise provided.

NOP Notice of protest.

No pay no cure The basis of Salvage. A Salvage operation where no pay is awarded unless the salvage operation is successful.

NOR Notice of readiness. A chartering term.

Nordic Council The Council has no central office but maintains a secretariat in each capital of the member nations. The Council was established in 1952 for the mutual consideration of economic, social, cultural, legal and communications questions. Its members are Denmark (Faroe islands and Greenland), Finland (Aland Islands), Iceland, Norway and Sweden.

Nordic country These include Denmark, Iceland, Norway, Finland and Sweden. See Nordic Council.

Nordice The code name of a BIMCO approved charter party clause relative to 'ice' and entitled 'Baltic Conference Special Ice Clause 1947'.

Nordics Norway, Sweden, Finland and Iceland (informal grouping).

No rec No record: a PNR that cannot be found because it was lost or delayed during its transfer.

Norgrain Code name of BIMCO approved North American Grain voyage charter party 1973 issued by the Association of Shipbrokers and Agents (USA) Inc. See Graincon.

Normal charge A specified general cargo rate without any quantity discount (air cargo).

Normalisation The process of converting an invalid data model into a valid data model, ensuring consistency and integrity of the data model. A computer term.

Normal laytime This arises where separate allowances/periods are made in a charter party for the port of loading and discharging operations of a particular vessel. An example arises with 'four days load/1,000 tonnes discharge' or, '6,000 shinc load/2,000 tonnes shex discharge. Once the 'statement of facts' have been received from the load port, a Laytime Statement can be drawn up based on the charter party laytime clauses, and demurrage/despatch calculated and settled, whether or not the vessel has even reached her discharge port(s). The calculation for laytime at the loading port is entirely separate from that calculation for the discharging port, the term non-reversible being sometimes used to describe this state of affairs. Also termed non-reversible laytime.

Normal profit That profit which is just sufficient to induce the entrepreneur to remain in his present activity.

NORPRO Norwegian Committee/Organisation on Trade Procedures and Facilitation based in Oslo.

Norske Veritas The Norwegian Ship Classification Society.

North American Free Trade Agreement A unified market of 360 million consumers embracing Canada, Mexico and United States of America. It was inaugurated in 1993 and spans fifteen years in transitional period by which time all trade barriers will be eliminated. Overall, it is designed to promote growth and expansion, and thereby advance the competitiveness of the NAFTA market globally.

North Atlantic Oscillation An index of the circulation in the North Atlantic which is measured in terms of the difference in pressure between the Azores and Iceland. In winter this index tends to switch between a strong westerly flow with pressure low to the north and high in the south and the reverse: the former tends to produce above normal temperatures over much of the northern hemisphere, the latter the reverse.

North Atlantic rates Air freight tariffs between point in IATA Area 1 and all points in IATA 3 except the South West Pacific.

North bound The process of proceeding in a north bound direction. It may be a consignment, transport unit, passenger, transport route and so on.

Northern Range United States Atlantic ports of Norfolk Va., Newport News, Philadelphia, Baltimore, New York, Boston & Portland, Me.

Norwegian Chamber of Commerce The organisation based in London representing Norway whose basic role is to develop, promote and facilitate trade development between UK and Norway and vice versa.

Norwegian Shipowners Association Based in Oslo the Association focuses attention on the development universally applicable rules and regulations relative to the maritime industry. It has nearly 200 members which own and operate the largest merchant fleets in the world.

Norwegian Shippers Council The organisation in Norway representing the interest of Norwegian shippers in the development of international trade.

nos Not otherwise specified.

No show(s) A person, Agent or Company who reserves space on an aircraft, ship or train on a specified service, but neither uses it nor prior to dispatch cancels the reservation.

Nostro An international Banking term meaning 'our'. An example is found with the UK Bank Barclays having an (nostro) account in an American Bank situated in New York. It would be a US dollar based account. Hence, the overseas currency account of a bank with foreign bank or subsidiary.

NOTAM Notice to Airmen – IATA term.

Notary Public An official certified to take affidavits and depositions from members of the public.

NOTAS Notas Line – a shipowner.

Note A promise or obligation to pay; promissory notes, bank notes, and floating rate notes all contain the issuer's primary responsibility for payment.

Note Issuance Facilities A circumstance whereby medium-term loans are funded through the sale of short-term paper where banks guarantee the availability of funds to borrowers by purchasing any unsold notes or by providing standby credit. Variations include revolving underwriting facilities and prime under-writing facilities. An International banking term.

Notes Notes are privately placed interest-bearing certificates of debt (traditionally medium term) that are issued by foreign borrowers.

Notes with warrants A long or medium term certificate of debt to which is detached a document/warrant entitling the holder to purchase equity securities (shares or participation certificates) within a predetermined period (exercise period) at a predetermined price (exercise price).

Notice day A trading day during which notices of intention to deliver actual commodities against short positions are made. A BIFFEX term.

Notice of abandonment The initial action to be taken by an assured who wishes to claim a constructive total loss.

Notice to underwriters must be given with reasonable diligence as soon as the assured is aware of the circumstance. Its purpose is to give the underwriter the opportunity to take action to prevent or minimise the loss. Overall, a condition precedent to a constructive total loss. If the assured fails to give notice to the underwriter the loss can be treated only as a partial loss unless an actual total loss is proved. An underwriter who accepts notice admits liability for the loss. Notice is not necessary where it would not benefit the underwriter, where the underwriter waives the obligation or in the case of a reinsurance. Action taken by an underwriter to prevent or reduce the loss is not deemed to be an acceptance or abandonment.

Notice of cancellation at anniversary date Cover is cancelled automatically at the anniversary date of a long term insurance unless the notice is withdrawn.

Notice of de-registration Process of de-registering a vessel from a flag nation state registry.

Notice of readiness Notice to the charterer, shipper, receiver or other person as required by the charterer that the ship has arrived at the port or berth as the case may be and is ready to load/discharge.

Notify address Address mentioned in a bill of lading or an air waybill, to which the carrier is to notify when goods are due to arrive.

Notify party Party to whom ANF is sent. Most international trade contracts involving transport overseas specify a party who is to be advised of the arrival of the goods. This maybe the consignee himself or an agent and his name and address must appear on the transport document.

Noting protest Master's deposition before Notary public, Consul or magistrate detailing anything unusual which has occurred on a voyage and any consequential emergency measure taken by the Master.

Not negotiable Cannot be transferred to another with the same rights as belonged to the original holder.

NOTO Non-Official Trade Organisation.

Not restricted Means not subject to or restricted by these Regulations.

Not to Insure Clause A clause, in a policy, which prohibits the passing of benefit of the insurance to a bailee, or other party who has care of the insured goods (see 'Benefit of Insurance Clause').

Not to Inure Clause A clause in a cargo policy stating that the policy shall not inure to the benefit of a carrier or other bailee. The intention is to deny the right of carriers to benefit from the insurance when they claim such a right in the contracts of carriage.

Novus actus interveniens (Latin) An intervening act of a third person.

NOx Nitrogen oxides.

np Net proceeds.

NPA Nigerian Ports Authority.

NPB National Productivity Board (Singapore).

NPC National Power Corporation (Philippines): News and Periodicals Corporation (Burma); or National Petrochemical Complex (Thailand).

NPCC National Pollution Control Commission (Philippines).

NPD Norwegian Petroleum Directorate; or New product development.

NPDES National Pollutant Discharge Elimination System.

NPIS New Product Information Service – a USA resource designed to give publicity of US product available for export.

NPL National Physical Laboratory.

NPRM Notice of Proposed Rule Making (USA term).

NPSH Net Positive Suction Head. An expression used in cargo pumping calculations. It is the pressure at the pump inlet and is the combination of the liquid head plus the pressure in the vapour space.

NPV Net present value.

NR Nuclear Reactor.

nr No risk until confirmed, or net register.

nrad No risk after discharge.

NRDC National Research and Development Corporation.

NRMC Natural Resources Management Centre (MNR) (Philippines).

NRS National Registration office.

NRT Net registered tonnage – one NRT equals 100 cubic feet.

N/S Not sufficient funds (banking).

NS No sparring; or Eurowings AG – an IATA member airline.

NSA National Shipping Authority (based in Washington); or Norwegian Shipowners Association.

NSC Netherlands Shippers Council; Nigerian Shippers council; Norwegian Shippers Council; or National Steel Corporation (Philippines).

NSCSA National Shipping Company of Saudi Arabia.

NSO Nedlloyd Standard Operations; or National Statistical Office (of Thailand).

nspf Not specially provided for.

NSR Northern Sea Route.

NSSN National Standard Shipping Note.

NSTT National Sea Training Trusts.

NSWA Next shift work advice.

N/t Net terms.

Nt Night trunk – an overnight road haulage service, or other specified form of transport.

NT Nusan Tenggara – a shipowner.

NTBs Non tariff barriers to trade.

NTC National Telecommunications Commission (Philippines).

NTE National Transportation Exchange.

NTMs Non-tariff measures such as quotas, import licensing systems, sanitary regulations, prohibitions, etc.

NTNF Royal Norwegian Council for Scientific and Industrial Research.

NTT Nippon Telegraph and Telephone Corporation.

NTUC National Trade Union Council (Singapore).

Nubaltwood Baltic Wood Charter Party 1964 – code name of Chamber of Shipping approved charter party for the timber trade Baltic and Norway, to UK and Ireland.

NUC Not under command.

Nuclear Exclusion Clause A clause that is common to all insurance policies (though the wording may vary), whereby all loss of or damage to the subject matter insured arising from a nuclear weapon, or similar, is excluded from the policy cover.

Nudum pactum (Latin) An agreement devoid of legal effect.

Nuisance duties Import duties of 5% or less of value which cost more to collect than the fiscal revenue raised.

Nuisance tariff Tariff so low that it costs the government more to collect it than the revenue it generates. WTO term.

Nullification and impairment Damage to a country's benefits and expectations from its WTO membership through another country's change in its trade regime or failure to carry out its WTO obligations.

Null modem A cable used to enable two computers to communicate with each other by emulating a modem. The null modem cable is wired up so that the end connections from one to the other are: 2 connected to 3; 4 and 5 connected to 8; 6 connected to 20. A computer term.

NUMAST National Union of Marine Aviation and Shipping Transport Officers (ex MNAOA, NMSA and REOU).

Nuvoy 84 Code name of BIMCO approved universal voyage charter party 1984 (revised voyage charter party 1964) published by the Polish Chamber of Foreign Trade Gaynia.

Nuvoy bill Code name of BIMCO approved Bill of Lading for shipments on the Nuvoy Charter Party.

NV Norske Veritas – Norwegian Ship Classification Society.

NVD No value declared.

NVE Night vision equipment.

NVOC Non-vessel owning carrier; or non-vessel operation carrier; or non-vessel operation (common) carrier.

NVOCC Non-vessel owning common carrier – usually a freight forwarder issuing a House Bill of Lading.

NVO-MTO Non-vessel operating – multimodal transport operator.

NVQ National Vocational Qualification.

NW North west – a point on the compass; or Northwest Airlines Inc – an IATA Member Airline.

NWC National Wages Council (Singapore).

NWRC National Water Resources Council (Philippines).

NWTB New Water Tube Main Boiler(s).

nwtdb New water tube domestic boilers.

NX Air Macau (Air Macau Company Ltd) – IATA airline member.

Nymex New York Mercantile Exchange.

NYPE New York produce exchange (charter party).

Nype 93 Code name of BIMCO approved New York Produce Exchange Time Charter.

NYPLR New York prime loan rate.

NYSA New York Shipping Association.

NYSE New York Shipping Exchange.

NZ Air New Zealand Ltd – an IATA member airline.

NZTC New Zealand Tonnage Committee.

NZ UK CC New Zealand United Kingdom Chamber of Commerce.

O Outboard
OA Olympic Airways SA. The National airline of Greece and member of IATA; open account; or over aged.
oa Overall.
O/a On account of.
OAA Orient Airlines Association – IATA term.
OAB Operational Advisory Broadcasts – a service provided by the USCG Navigation Information Service.
OAC Oceanic Area Control – IATA term.
OACC Oceanic Area Control Centre – IATA term.
OAP 90 The US Oil Pollution Act of August 1990.
OAPEC Oil organisation of Arab Petroleum Exporting Countries.
OAS Organisation of American States.
OB Occurrence basis.
Obiter dictum (Latin) A statement by the way; an opinion expressed incidentally by a Judge, concerning a question upon which no actual decision is called for.
OBQ On board quantity.
O & D Study Origin and Destination study. Market Research term.
O & M Organisation and Methods. A generic term for a number of techniques like work study which are used in the examination of jobs in order to effect their improvement.
OAS Organisation of American States (based in Washington).
OASIS A consortium of five Japanese lines with whom Overseas Container Ltd trades in a wider consortium known as 'Greater Oasis' in the Japan-Arabian Gulf-Japan trade, which calls at 'wayports' in between; or Oceanic (and selected non-Oceanic) Area System Improvement Study – IATA term.
OATs Obligations Assimilables du Trèsor: French Government securities issued with maturities of 7 to 30 years.
OAU Organisation of African Unity (Organistion de l'Unite Africaine). Based in Addis Ababa, its role is to promote the unity and solidarity of African States, to co-ordinate and intensity their efforts in respect of economic and social improvements, and to remove all traces of colonialism from Africa.
OB Oil bearing.
Ob. (obit) (Latin) Died.
OBC Ore bulk carrier.
OBCT Ore bulk container.
Obiter dictum (Latin) Said as an aside.
Oblige Line A line written by an underwriter to help a broker in placing a risk in circumstances in which the underwriter would not, otherwise, have accepted the risk.
OBO Oil bulk ore.
OBS Oil Bunker Surcharge.
Observation car A railroad car with special provisions for sightseeing, some of them with high windows that curve into the ceiling and are called bubble cars or dome cars.
Obsolescence An indication a product is out moded and non-competitive. Hence, it is out-of-date and at the end of its product life cycle in technical/operational/fashion/design/taste terms.
OC Air Zaire The National Airline of Zaire and member of IATA; old crop; open charter; open cover; oil bearing continuous liner; old charter; ore carrier or On carriage (by road/rail/barge).

o/c Overcharge.
OCA Oceanic Control Area – IATA term.
OCAL Overseas Containers Australia Ltd.
OCAM Organisation Commerce Africain et Malgache.
OCBC Overseas Chinese Banking Corporation.
Occupancy rate Ratio expressed as a percentage of bed nights or room nights sold to the total offered for sale by a hotel.
Occurrence basis Excess of loss reinsurance by which the reinsurer's liability is determined in relation to the aggregate net retained loss sustained by the reinsured as a result of a single occurrence.
OCD Operations Compliance Directorate – part of HMCE.
OCDE Organisation de Co-operation et de Developpement Economiques (French). (Organisation for Economic Co-operation and Development).
Ocean Bill of Lading A Bill of Lading used primarily in deep sea trades.
Ocean container freight rates Deep sea container freight rates.
Ocean freight The freight rate – freight income derived from the conveyance of merchandise in deep sea operated vessels.
Oceanfront A hotel room directly facing the ocean.
Ocean going vessel A ship which plys in deep sea water. The Panama Canal authority categorise an ocean going vessel as one of 300 net tons and over.
Ocean liner operators Deep sea liner cargo shipping companies.
Ocean marine insurance Insurance for sea-going vessels, including liabilities connected with then, and their cargoes.
Ocean sea waybill A sea waybill used primarily in deep sea trades.
Ocean Shipping Reform Act 1998
US legislation – which substantially amends the Shipping Act of 1984 – strives to promote the growth and development of US exports through competitive and efficient ocean transportation and by placing a greater reliance on the marketplace. It focuses on the deregulation of common carriage in the US and seeks to ease the longstanding control of shipping conferences over their members, protect the confidentiality of private contracts between carriers and shippers, reduce government oversight of pricing and encourage carriers to offer intermodal inland transportation.
Ocean view A hotel room, usually at a side, with a view of the ocean.
Ocean Waybill A sea waybill used primarily in deep sea trades.
OCIMF Oil Companies International Marine Forum.
OCL Overseas Containers Ltd. A consortium of ship-owners operating deep sea ship container services.
OCN Out of Charge Note.
OCO One cancels other.
OCPL Overseas Containers (Pacific) Ltd.
OCR Optical Character recognition.
OCT Octuple screw (on a ship).
OCTI Office Central Transport Interionaux par Chemins de Fer.
OCT's Overseas counties and territories of member states associated with the EU.
O/D Overdraft.
o/d On demand.

ODAPS Oceanic Display and Planning System (USA term).
ODAS Ocean Data Acquisition Systems.
Odessa Odessa Ocean Line – a shipowner.
ODETTE Organisation for data exchange by tele-transmission in Europe.
ODI Overseas Development Institute (UK).
ODM Ministry of Overseas Development.
OEC Overpaid entry certificate.
OECD Organisation for Economic Co-operation and Development. Based in Paris its fourfold role amongst its members is to co-ordinate economic policy, promote and co-ordinate aid to developing countries, further liberalise international trade, and promote sound economic expansion.
OECD Industrial Production Index An Industrial Production Index produced for OECD countries. The base year of 100 is 1995.
OECF Overseas Economic Co-operation Fund (OECD).
OEEC Organisation for European Economic Co-operation.
OEI One engine inoperative.
OEM Original equipment manufacturing.
OESO Organisatie voor Economische Samen werking en ontwikkeling (Dutch term for OECD).
OF Open Full (side door containers).
OFEX The UK stock market for smaller companies.
Off-balance sheet Borrowing which does not appear on a company's balance sheet. Leasing has been the principal form of off-balance sheet financing, although accounting practice in a number of countries now requires finance leases to be included in the balance sheet as fixed assets and matching long-term liabilities. Shipping has been particularly susceptible to the attractions of off-balance sheet financing, since the capital expenditure required to purchase ships outright has often been out of proportion to the net worth of the shipowner. The chartering in of vessels over long periods, equivalent to operating leases, has been a favoured method of fleet expansion. The present value of these off-balance sheet liabilities has not infrequently become dangerously excessive, and sometimes brought about the financial collapse of shipping groups during freight market recessions.
Offer A firm offer to sell at a stated price. Hence, the price the seller asks for the commodity on offer. (See also Bid). A BIFFEX term definition; the price over which a loan may be based; a security purchased from the market; a bill discounted; or a country's proposal for further liberalisation – WTO term.
Offer document An official document from a bidder in a takeover battle sent to shareholders in target company. The offer period extends until the closing date which will be mentioned in the document.
Off hire A chartering term indicating the chartered vessel is no longer attracting a daily fixture rate. It arises when owing to poor performance such as slower speed than prescribed in the voyage charter, the voyage time is extended by 'X' number of days and excess bunker consumption is incurred. Another example is a time charter when due to an engine mishap lasting a few hours, 'off hire' arises.
Off hire days In a time charter the owner is entitled to a limited time for his vessel to be off hire until such time as she maybe repaired or dry docked. The permissible time is prearranged by the contracting parties.
Office of departure The Customs office where the carnet is validated. A customs term.
Office of entry The Customs office where the goods enter the Community country of temporary use. A customs term.
Office of exit The Customs office where the goods leave the Community country of temporary use. A customs term.
Office of transit The Customs office where goods in transit enter or leave a Community country. A customs term.
Official intervention The process of a government or central bank buying or selling its national currency to influence/stabilise its international value. Official intervention affects the level of a country's reserves of gold and foreign exchange. See also dirty floating.
Official log Ship's log.
Official No. The ship's registration number allocated to every merchant vessel and yacht.
OFF-LINE Describes an airline that sells in a market to which it does not operate. An Off-Line carrier will use another operator to link with its network; or any employee, function or facility located or performed off a carrier's regular route.
Off-line carrier Any airline other than the one whose computer is being used to obtain information and make reservations; any carrier other than the airline an agency is using to make a multi-airline booking, whether by computer or by telephone.
Off-loading The unloading of a transport unit of merchandise.
Off-peak A time of year when business is traditionally slow.
Off-route charter A flight by a scheduled airline to or from a point it is not authorised to serve on a regularly scheduled basis; limited by some governments.
Off-season Time when business is traditionally slowest; rates are often lowest then.
Offset A counter trade technique involving a condition of exporting some products, especially those embodying advanced technology, to some markets is that the exporter incorporates into his final products specified materials, components or sub-assemblies procured within the importing industry. An example and long established feature of the trade is found in defence systems and aircraft, but it's becoming more common in other sectors; or the closing out of a futures or options position by buying or selling an opposite position.
Offshore Outside the jurisdiction of a particular country – an international banking term; or contracting work carried out at sea such as drilling for oil.
Offshore account An account conducted outside the country in which the transaction took place.
Offshore business The process of a Company/person conducting business overseas – an exporter.
Offshore company A company based overseas. The motivation maybe tax advantage, lower labour and costs and various marketing advantages.
Offshore industry An industry which is located overseas. See on shore industry.
Offshore installation Any offshore structure such as a drilling rig, production platform etc, which may present a hazard to navigation.
Offshore installation manager The person in charge of an installation who is responsible for everyday planning, co-ordination, monitoring and control of operations and within oil field limits.
Offshore installation operator A company with responsibility for the management of an offshore loading installation. In the text such companies are usually revered to as installation operators.
Offshore investment The process of investing capital/resources outside the domiciliary country thereby crossing international boundaries.

Offshore loading installation A fixed or floating oil production and/or storage facility used for loading crude oil to offtake vessels – usually referred to in the text as an installation. The more complex floating versions are usually ship-shaped. They can be FPSOs (Floating Production, Storage and Offtake) vessels. Other systems can include the FSUs (Floating Storage Units), receiving the stabilised crude from elsewhere, and where offshore loading takes place at another facility, such as an SPM, via pipelines. Yet other systems include the SPAR types or articulated towers which may be moored to the sea-bed by CALM or SALM moorings.

Offshore markets Financial markets which have developed outside a country's boundaries, but which offer facilities in competition with that country's domestic markets. Examples include the Euro currency markets and investment facilities in tax havens.

Offshore multi purpose support vessel See multi purpose offshore support vessel.

Offshore product A product which is manufactured/assembled/processed outside the buyer's country and therefore classified as an imported product crossing international boundaries.

Offshore register See Ship register.

Offshore sourcing The process of relying on/using raw materials/componentised units from a country outside the domiciliary country involved in the manufacturing/processing/assembly of the product.

Offshore support vessel Vessel usually associated with servicing offshore miniature installations such as an oil rig and gas installations. This involves platform supply vessels. See separate entry.

Offshore transhipment terminals A transhipment maritime terminal situated offshore thereby avoiding the need for megaships to transit shallow sea areas and through busy narrow channels. Their location provides for fast express ocean trunk hauls as minimal deviation time from main trade lanes arise whilst ocean crossing times are reduced. Examples include Malta, Gioia Tauro, Sines, Algeciras, Colombo-Freeport, Grand Bahamas, Salalah, Halifax and Scapa Flow. They are most evident in the mega container tonnage and oil terminals.

Offtake vessel An oil tanker often specially designed or adapted for loading crude oil at offshore loading installations. These ships are sometimes, known as Shuttle Tankers.

Offtake Vessel Operator A company with overall responsibility for the management of an offtake vessel – usually referred to in the text as vessel operators.

OFISP Operational Flight Information Services Panel (ICAO term).

OFT Office of Fair Trading.

Oftel Office of telecommunications.

OGD Other government department.

OGIL Open general import licence.

OGL Open general licence – a type of import licence.

OGP International Association of Oil and Gas Producers (formerly E&P Forum); or Original Gross Premium.

OGPI Original gross premium income.

OGR Original gross rate – insurance term.

O/H Overheight: a container with goods protruding above the top of the corner post. Likewise the term can be applied to a road haulage vehicle.

OIE International office of Epizootics.

Oil Liquid maybe classified in five different ways. (a) Petroleum base producing petrol/gasoline, paraffin/kerosene, light fuel oils, lubricating oils and thick asphalt; (b) animal oils e.g. lard oil; (c) vegetable oil e.g. linseed, tung, cotton seed, palm and caster oils; (d) coal base involving distillation of coal and shale producing benzol etc, and (e) marine animal and fish oils extracted from cod, sharks, seals, whales and other fish.

Oil analyst Person specialising in the oil market especially in economic evaluation, market forecasts, price levels, etc.

Oil/bulk/ore carriers Multi-purpose bulk carriers.

Oil Companies International Marine Forum Founded in 1971 and based in London, its aim is to improve safety at sea and protect the environment. It focuses attention on the prevention of pollution from oil tankers and terminals.

Oil gas A gas made by the reaction of steam at high temperatures on gas oil or similar fractions of petroleum, or by high temperature cracking of gas oil. The gas is flammable, but is classified as a toxic gas because it contains a high proportion of Carbon Monoxide.

Oil pollution The process of an oil leakage emerging from a ship or other resource and in so doing entering the maritime environment causing the sea to be polluted.

Oil record book The oil record/log book on a particular ship. See log book.

Oil tanker Vessel designed to ship oil in bulk.

OIM Offshore Installation Manager.

OIML International Organisation for Legal Metrology Founded in 1955 to resolve the technical and administrative problems of legal metrology raised by the construction, use and checking of instruments of measurement and to facilitate co-operation between States in this field. It has 42 member states.

OIRT Organisation International de Radiodiffusion et de Television.

OJ Official Journal (of the European Union).

OJ(EC) Official Journal of the European Union.

OK Ceskoslovenske Aerolinie (CSA) – an IATA member airline.

OLB Official Log Book – ship's diary under Master's surveillance and daily signature.

OLDI On-Line Data Interchange – IATA term.

Old ice Sea ice which has survived at least one Summer's melt. Most topographic features are smoother than on first year ice. May be subdivided into second-year and multi-year ice.

Oligopolistic market structure A situation where few firms are sharing a large proportion of the industry as found in the liner trades.

Oligopoly A type of market in which there is basically a larger number of small firms which account for a large proportion of the market ingredients embracing output, employment, capital and so on. An important feature in such a market that each company must be able to operate within a high degree of independence within the market, thereby permitting the market forces of demand and supply to operate with its attendant risk by each company.

Olympic Airways A national airline and member of IATA.

OM MIAT (MIAT Mongolian Airlines) – IATA member airline.

OMA Orderly Marketing Agreements.

OMBO One man bridge operation.

Omnibus survey A survey which runs continuously and covers a range of topics. It is usually funded by Companies who have a direct interest in some of the data/survey report. An example is the British Travel Association conducting a twelve month survey of foreign visitors to the UK. Hoteliers, transport operators, caterers, leisure companies etc, would be interested in such data.

Omni Carrier An all purpose general cargo vessel engaged in deep sea liner trades capable of conveying containers, loose cargo, indivisible loads, vehicular decks, container accommodation, derricks of 36 tons and 126 tons lifting capacity, and ramps for vehicular traffic transhipment both at the stern and amidships. The vessel thus is equipped with roll on/roll off, lift on/lift off and side cargo transhipment facilities. It is often called a Combi carrier.

Omnipresence A situation which is present at the same time in all places.

OMO Offshore Mining Organisation (Thailand); or open market operation: transaction undertaken between a central bank and its money market counterparties, at the initiative of the central bank.

OMS Organisation Mondiale de la Sante (World Health Organisation).

on Official number.

'On (all) laytime saved' Despatch money shall be payable for the time from the completion of loading or discharging the expiry of the laytime excluding any periods excepted from the laytime. A BIMCO term.

On carriage The process of conveying the goods/merchandise from a specified point/terminal following completion of the major portion of the transit. This usually involves the trunk route completion and the delivery arrangements.

On cost The contribution of the cost of overheads added to the direct cost of production.

On deck Bill of Lading A Bill of Lading endorsed that the cargo has been shipped 'on deck'.

On demand bond A bond which can be called at the sole discretion of the buyer as the nature of the document is such that claims must be met without being contested. An international banking term.

On demurrage The laytime has expired, unless the charter party expressly provides to the contrary the time on demurrage will not be subject to the laytime exceptions.

One cancels other A trading instruction incorporating two orders with a built-in further instruction to cancel one if the other is executed. The order can indicate two different price levels or a combination of price and time limits.

One class A common class of accommodation found on a transport unit such as a one class ship, aircraft or train. On a train it would be second or standard class.

One off movement A transit which is unlikely to be repeated.

One run method The use of reproduction process to transfer all or part of the information recorded on a master document into one or more forms comprising an aligned series.

One stop shopping The situation where the consumer undertakes all the shopping in one retail outlet such as at the hypermarket.

One stop ports The situation where a major shipping company will opt to serve one port in an area/region rather than a series of ports, and rely on feeder services to develop trade through that port. An example of a one stop port is Singapore.

One trip sling A piece of cargo handling equipment suitable for bag or case transhipment and which is used only on one journey, e.g. to convey/tranship the goods from or to the vessel or vice versa and accompanies the cargo throughout the voyage.

One way lease A lease of a container for the forward voyage only, the container being returned to lessor at or near destination.

One way pallet A pallet intended to be discarded after a single cycle of use.

One way trip Transportation from a point of origin to a destination without provision for returning.

On going A situation/circumstance which is continuous.

On going management system The process where personnel within the Company are appointed to Managerial positions and thereby not relying on personnel being recruited/appointed to such posts from outside the Company from the open labour market.

On going project A project which is continuous.

On-line charge See on-line rate.

On-line connection The transfer of passenger baggage or cargo between flights of the same airline.

On-line enquiry A computer term involving the client endeavouring to obtain data from the vendor using the interlinked computer network. An example is an Exporter using the personal computer processing enquiry such as rates and sailing data through the Shipowner Computer network.

On-line rate An air freight rate which applies for carriage over the lines of a single carrier.

On-line reservation system Automated access to a reservations database.

On-line retrieved A computer term involving the client endeavouring to obtain data from the contracting party using the inter linked computer network. For example such data searches/enquires could involve the importer enquiring from the shipowner through the on-line computer, information of the estimated time of arrival and ship berth details.

On-line tracking A computer term involving the process of identifying through the inter linked computer network the location of a particular consignment in transit. Such an enquiry could emanate from the Freight Forwarder computer who would have direct access to the shipowner computer network to obtain the data.

On/off-line survey A charter party clause. The owners shall bear all expenses of the on-survey including loss of time, if any, and the charterers shall bear all expenses of the off-survey including loss of time, if any, at the rate of hire per day or pro rata. A BIMCO clause recommendation.

ONPI Original net premium income.

ONR Original net rate.

On sell The process of buying a product and then selling it to a third party.

On shore industry An industry which is indigenously domiciliary located in the country in which it is registered. See off shore industry.

On sold Process of selling to a third party.

Onward flight The process of a passenger continuing his/her flight such as London/Frankfurt and Frankfurt/Bangkok. The portion Frankfurt/Bangkok being the onward flight.

Onward shipment Process of a cargo shipment containing the transit such at Rotterdam/Singapore and Singapore/ Kobe.

Onus (Latin) Burden.

Onus of proof See burden of proof.

Onus probandi (Latin) Burden of proof.

O/o Order of; or Oil/ore ship.

OOC Ore oil carrier.

OOCL Orient Overseas Container Line – a shipowner.

OOG Out of Gauge: a transport term applying to the conveyance of a consignment by road or rail which requires special operational arrangements as the size of the consignment is outside the weight or dimensions limitations on the particular transport mode. For example, goods whose dimensions exceed those of the container/flat rack in which they are packed.

OOL Orient Overseas Line – a shipowner.

OOW Officer of the watch.

OP Open or floating cargo insurance policy.

OPA Oil Pollution Act.

OPA 90 Oil Pollution Act 1990. A US act with very onerous provisions covering all vessels trading to/from USA.

OPAC Operations Advisory Committee (ECAC term).

OPAS Programme for Provision of Operational Assistance (UN).

OPEC Organisation of Petroleum Exporting Countries.

Open account An agreement whereby the seller (exporter) agrees to despatch the goods to the buyer (importer) on the understanding that he will pay for the goods after he as received them, usually on a monthly basis. Although this is a simple and inexpensive way to arrange payment, involving little paperwork, it carries considerable risk since the buyer may withhold or delay payment or the transfer of funds might be delayed as a result of exchange control regulations in his country. Before agreeing open account terms the seller should be satisfied of the integrity and credit worthiness of the buyer and that local exchange control regulations permit payment to be made. There are various ways in which the buyer can send money to the seller under open account and the seller may wish to stipulate the method to be used. This includes buyer's cheque, banker's draft, or mail payment order. Overall, open account business is transacted on credit terms where no security in forms of Bills of Exchange or Promissory Notes is obtained from the buyer (importer). It is a most popular method of payment within the EU and the most common method of payment is by cheque, banker's draft, and telegraphic transfer.

Open account business Business transacted on credit terms, where no security in the form of Bills of Exchange or Promissory Notes is obtained from the buyer (i.e. on trust).

Open base A raised deck fitting with draining holes used to assist the securing of containers.

Open berth A berth in a seaport which is available to all port users/shipping companies. See closed berth.

Open bid The process of inviting anyone to tender/bid who fulfils the necessary requirements/conditions in response to an advertised request for bidders for a particular product(s)/service(s).

Open border A country/territory with no restrictions on access/entry.

Open charter Vessel can be fixed for any port with any cargo.

Open conference A liner conference with no restrictions on membership entry and does not require existing members to vote on admission of new members. See liner conference.

Open container A container with sides and or ends composed of bars, grilles, mesh or entirely open, with or without roof.

Open contract (or open position) A contract to buy or sell futures that has not been closed by an opposing futures transaction within the freight market. A BIFFEX term.

Open cover A cargo insurance agreement covering all shipments of the assured for a period of time, subject to a cancellation clause and a limit to the amount insured in any one ship. Other conditions include a classification clause.

Open deck The deck on a vessel exposed to the weather and not enclosed/covered.

Open dock A dock which is designed without any mechanical means such as a lock gate of separating its docks or quays from the sea, river, or lake in which it is situated.

Open economy See open trade entry.

Open ended flight reservation A circumstance where by a passenger reserves a seat on a specified flight on a provisional basis and in so doing has the option of confirming the reservation nearer the flight departure date or deciding to take another flight.

Open ended question A question found in a market research questionnaire which is so worded to allow the respondent complete freedom of expression. It may be 'what is your opinion of the quality of service offered by the retailer' and the respondent simply responds using their own expression. It is in complete contrast to the classified dichotomous type of question.

Opener An international trade term found in banking. It is the person usually the buyer, on whose behalf the credit is issued. Also known as the 'applicant'.

Open exchange economy An economy with no trade barriers.

Open general import licence An import licence permitting the import of a specific commodity without restrictions as to quantity or value. It is normally valid for one year and issued by the Government.

Open ice Floating ice in which the ice concentration is 4/10 to 6/10, with many leads and polynyas, and the floes are generally not in contact with one another.

Open individual licence An import licence permitting the import of a specific commodity without restrictions as to quantity or value. It is normally valid for one year and issued by the Government.

Opening Bank The bank which opens the credit on behalf of the importer (buyer). It may also be termed the issuing bank.

Opening of borders Process of relaxing international border controls which maybe in the form of permitting inwards investment or deregulating trade controls thereby permitting selected commodities to be imported into a country.

Opening of the credit When the importer (buyer) has decided to accept the exporter (seller) agreed terms of offer and signed the export sales contract, the importer (buyer) must open with his bank a documentary credit in favour of the exporter (seller).

Opening procedure The market procedure to establish the opening level of a particular contract.

Opening purchase A transaction in which the buyer becomes the holder of an option. This is sometimes described as a long position. A financial definition.

Opening sale A transaction in which the seller of an option becomes the writer. A financial definition.

Opening up new markets Process of entering a new market maybe in a different country such as China or a new product in an existing market.

Open Interest The total number of contracts remaining open in the freight market. The market will also publish details of uncovered contracts which is the minimum number of contracts that would be required to be traded to close out every open contract; this future will therefore have made allowance for the 'netting off' of open purchases against 'open purchases' against 'open sales'. A BIFFEX term.

Open jaw Round trip in which the return trip begins as a point other than the arrival point; thus, New York to Chicago with return from Detroit.

Open market access The process/opportunity to enter an overseas territory thereby crossing international boundaries without any trade barriers such as quotas, licences, import tariffs and so on.

Open market competition A market place where no constraints exist both in regard to domestic sourced goods and those imported. The most intense competition possible.

Open market operation See OMO.

Open market trading A technique adopted by a country or state to permit/allow trading with a particular nation(s) without any restriction/trade barriers. An example is found in membership of a Free Trade area which encourages/permits the concept of free trading usually on a common basis thereby ensuring no one country has any competitive advantage over another insofar as trading conditions/customs regulations etc are concerned.

Open network A system where many computers are linked together are able to share the data they hold, operating as either one unit or several. The size of the network can vary as conditions demand.

Open outcry The method of dealing employed on the LME and other futures markets. A broker announces to the entire ring the amount of a commodity he has for sale (or seeks to buy), and the required delivery date and price. When another broker responds to accept the bid or offer, the deal is fixed and that price becomes the latest traded price for the delivery date. A BIFFEX term.

Open policy One in which a maximum amount is stated and the goods insured are allowed to be shipped in varying amounts in one or more vessels up to maximum sum, but within the general terms of the policy. Shipments are advised by declarations and evidenced by the issue of certificates.

Open port A port which is designed without any mechanical means such as a lock gate of separating its docks or quays from the sea, river or lake on which it is situated.

Open position This arises where the dealer's assets and liabilities in any currency do not balance thereby leaving the dealer in an open position; or a bought or sold futures contract which has not been liquidated. A forward market position which has not been closed out. For some clearing house exchanges the net open position is an important statistic and is published daily. A BIFFEX term.

Open pricing The agreement of prices and pricing policies between Companies in an industry.

Open rate A situation in which airlines have failed to negotiate a uniform fare leaving each carrier able to charge its own fare; applies only to international market.

Open registries Process of shipowners registering ships with countries that operate national fleets tax-free or virtually tax-free. Major open-registry countries include the flags of Bahamas, Bermuda, Cyprus, Liberia, Malta, Panama and Vanuatu.

Open-sided container Doors, shutters or tarpaulin allowing one or both sides to open up completely.

Open side/open top container An open side/open top container is equipped with removable side-gates, top rails, roof bows and door header, and is covered by tarpaulin. Open side/open top containers can be used as an open top or open side container, and have the same accessibility as flatrack containers when stuffing/stripping.

Open slip A broker's slip covering a sum sufficient to cover a number of shipments which are advised as they take place.

Open storage Storage accommodation which is not covered or protected from the weather. It may be at a seaport, railway terminal, road depot, manufacturer's premises and so on.

Open systems A computer term which arises where different types of computer equipment can communicate with each other with no constraint of vendor, platform or operating system. Open systems are able to inter work with each other if they adhere to international standards for open systems communication and inter-operability.

Open systems inter connect The international standards in process of development by suppliers, user groups, and government agencies to facilitate the use and development of the open system computer network. See Open systems.

Open tendering The process of inviting anyone to tender who fulfils the necessary requirements/conditions in response to an advertised request for bidders, for a particular product(s)/service(s).

Open ticket A ticket that does not specify when a service is to be performed leaving the holder responsible for reserving at a later time.

Open top container An open top container is a container which has a removable tarpaulin cover on top instead of a permanently fixed roof as on the ordinary dry freight container. Open top containers are used when cargo is not suitable for loading through the door manually or by forklift, but can be lowered into the container by the use of a crane.

Open trade A trading nation which has no trade barriers in the conduct of trade in a global market and therefore operates on an open market basis. See managed trade.

Open trading economy See open trade entry.

Open-wall container Container without one or more side or end walls, but having at least a base, end structures and a top frame with corner castings.

Open water A large area of freely navigable water in which sea ice is present in concentrations less than 1/10. No ice of land origin is present, although the total concentration of all ice shall not exceed 1/10.

Operating alliances Operating alliances emerge when two or more shipping companies come together through merger, acquisition or simply working together and sharing common facilities. Overall they result in the concentration and centralisation of the company's facilities enabling the operating alliance to have common resources such as berth, ships agents, computerised network involving information technology, marketing resources, marketing and sales plan, shipbroking, ship management, logistics and training resources. In situations of simply working together and sharing common facilities, costs are apportioned in accordance with the company' schedule, but share capital remains unchanged.

Operating cost The cost associated with the operation of a transport unit which may be a vessel, aircraft, coach, train, road vehicle and so on. This includes crew, lubricants, survey, maintenance, fuel, insurance, administration and so on.

Operating department Shipowners organisation responsible for all aspects of ship operation.

Operating efficiency A measure of the operating efficiency/performance of a transport operator on the basis of striving to achieve the optimum performance at all times. This includes loadability, punctuality, fuel consumption, manning levels, total cost/total revenue – net loss/profit, turn round time at the terminal and so on. Operating efficiency must be measured against the circumstances obtaining, quality of service and the prescribed schedule. The factors involved will vary by circumstance, type of operation and Company/entity.

Operating expenses of a ship See voyage estimate entry.

Operating lease A lease other than a finance lease. The risk and benefits of ownership are usually the responsibility of the lessor. He is usually responsible for repairs, maintenance and insurance. He carries the risk of obsolescence and the residual value at the end of the lease is important to him.

Operating plan A plan devised in a transport entity relating to the employment/disposition of its fleet. This will include allocation of transport units by type to individual route/service/trade; provision for transport units under repair/maintenance; crew arrangements; standby transport units, and so on. It will have regard to the commercial/market requirements, and the plan's viability; or a plan devised in a manufacturing company involving a manufacturing schedule and the required resources.

Operating profit An accountancy term. It is the difference between the sale volume income and the product/service cost. See profitability ratios.

Operating schedule A timetable of a particular transport mode which may be ship, aircraft, train, road vehicle and so on. It will include the depot/departure and destination/arrival times; the stopping places en route and their arrival/departure times; the type of transport unit including capacity; the speed of transport unit; manning arrangements and so on.

Operating strategy The process of devising/formulating a policy relative to a particular company/entity ship operations which may include consideration of sailing schedules, ports of call, speed of vessel, manning levels, port turn round time, revenue, expenditure, liner conference agreement and so on. Usually the operating strategy devised would have special emphasis on the need to attain viability on the service/trade/route and have regard to all the options available particularly the need to realise optimum performance. A similar criteria could apply to any transport mode especially air transport, rail, road and so on.

Operating subsidy Government financial grant towards operating a transport service.

Operating system The software program which provides the environment in which applications programs can be used. The operating system controls the operations of handling: input, output, interrupts, storage and file management.

Operational control The process of controlling operations and services in accordance with strategic and tactical plans.

Operational productivity of the world fleet The total cargo carried and ton miles performed per deadweight ton (dwt) of the total world fleet produced on an annual basis. For example in 2001 world fleet (millions of dwt) 825.7; total cargo carried (millions of tons) 5832. Total ton miles performed (thousands of millions of ton-miles) 24,338; tons of cargo carried per dwt 7.1; and ton miles performed per dwt 29,500.

Operational research A largely mathematical problem-solving technique.

Operation time The period of time required to carry out an operation on a complete batch exclusive of set up and break down times.

Operator In transport terms a person, organisation or enterprise engaged in or offering to engage in a transport operation involving a particular mode such as aircraft, ship, road vehicle, barge, railway and so on.

Opinion poll A sample survey of the general public opinion undertaken by face to face interview or through the process of completing questionnaires which are subsequently evaluated.

OPMC Outside plant maintenance centre.

OPMET Operational Meteorological Information – IATA term.

Opportunity or Alternative cost doctrine The cost of using productive resources for a certain purpose, measured by the benefit given up by not using them in their best alternative use. It is the cost of other goods which must be sacrificed to gain more of a particular good. An economic term.

OPR Outward Processing Relief. A system of duty relief which applies to goods sent outside the EU Community for process and subsequent re-importation.

OPRC-HNS Protocol Protocol on Preparedness, Response and Co-operation to Pollution Incidents by Hazardous and Noxious substances, 2000. The 1990 conference on International Co-operation on Oil Pollution Preparedness and Response invited IMO to initiate work to develop an appropriate instrument to expand the scope of the International Convention on Oil Pollution Preparedness, Response and Co-operation, 1990, to apply, in whole or in part, to pollution incidents by hazardous substances other than oil and prepare a proposal to this end.

The Marine Environment Protection Committee adopted in principle, at its forty-second session in November 1998, a draft protocol on preparedness, response and co-operation to pollution incidents by hazardous and noxious substances. Together with the OPRC Convention, the OPRC-HNS Protocol will provide a framework for international co-operation in establishing systems for preparedness and response at the national, regional and global levels.

In accordance with Article 2(b) of the Convention on the International Maritime Organisation and resolution 10 of the above-mentioned conference, the MEPC, at its forty-second session recommended and the Council, at its eighty-second session decided to convene a conference to consider the adoption of a protocol on preparedness, response and co-operation to pollution incidents by hazardous and noxious substances.

This conference was held at IMO headquarters from 9 to 15 March 2000 and in addition to the Final Act, the Conference adopted the Protocol on Preparedness, Response and Co-operation to Pollution Incidents by hazardous and Noxious Substances, 2000. The Conference also adopted six resolutions which are contained in the Attachment to the Final Act.

OPS Operations – IATA term.

OPSP Operations Panel (ICAO term).

Ops/Rg Operational range.

Option Charter party clause permitting alternative or additional cargo to be shipped at the charterer's option, in ports of call for loading or discharging purposes; in financial terms the right on payment of a consideration to buy or not to buy (call option) a particular security at a stated price, or to sell or not to sell (put option) a particular security also at a stated price over a given period of say one, two, or three months. A double option gives the right to buy or sell. Also in financial maritime terms, options are a right to buy (a call option) or sell (put option) as a security or asset at some future time at a given price. The buyer of an option pays an amount of money up front in exchange for being granted this right by the seller. A tradable option in certified form is often called a Warrant. Most frequently used in the securities and commodities markets, options have become more commonplace in ship sale and purchase transactions in recent years, especially where period charterers are given options at intervals during the term of the charter to buy the vessel at pre-determined prices from the existing owners; or a tour extension or side trip offered at extra cost.

Option date The date upon which definite commitment, often a deposit, must be made or the space, seat or other facility will be made available for sale to others.

Option Forward Contracts If the date of payment is uncertain relative to an export sales contract, an option forward contract may be agreed between the importer and his bank, where the importer will undertake to buy the currency from the bank at some time between two specified dates. The length of the period of an option is agreed between the bank and the customer and can be measured in weeks or months. An importer who has arranged an option forward contract does not have to take delivery of all the currency at one time; several separate deliveries can be made within the option period. This facility is called 'part take up' or 'part delivery' of a forward contract.

Option port A port of call on a schedule which is optional.

Options The right, but not the obligation, to buy/sell equities, bonds, foreign exchange or interest rate contracts by a future date at a price agreed now. 'Traded options' means the options themselves can be bought and sold.

OR Owners risk – shipowners conditions of carriage; or open registries.

Orange book A publication issued by the United Nations every two years. It is prepared by the United Nations Committee of Exports on the Transport of Dangerous goods and takes the form of recommendations but has no legal status. Ultimately such recommendations filter through the international organisations responsible for each mode such as IMO, IATA, ADR, RID and ICAO.

ORCAM Originating Region Code Assignment Method – IATA term.

Ord Order.

Order A buy or sell instruction, or combination thereof. Marine insurance term.

Order Bill of Lading Bill of Lading made out to order of shipper or consignee. See straight Bill of Lading entry.

Order/call ratio The ratio between orders secured and number of calls made on a Company/entity by a sales person. A test of efficiency of sales staff.

Orderly marketing agreements Ad hoc and usually bilateral agreements to protect domestic industries. A WTO definition.

Ordinal data entry Data entry intended for identification of an individual document or an item, or for classification or sorting but not as a quantity for calculation.

Ordinary engineer assistant A seaman's post in the Marine Engineer's department on a ship.

Ordinary loss A term arising in marine insurance which describes the natural loss in weight or volume due to leakage evaporation or ordinary wear and tear of a commodity/cargo.

Ordinary return A type of travel ticket valid for three months from date of issue for a return journey. Its validity will vary by transport mode and is at full tariff.

Ordinary seaman A seaman's post in the Deck department on a ship.

Ordinary shares These are the most common form of shares. Holders receive dividends which vary in amount in accordance with the profitability of the company and decisions of the directors. The holders of the ordinary shares are the owners of the company.

Ordinary single A type of travel ticket valid usually for three days from date of issue for a single journey. Its validity will vary by transport mode and is at the full tariff.

Ordinary ticket Passenger travel tickets which are usually at the full fare level. For example, it may be an ordinary single valid for three days, or ordinary return for three months. Its validity will vary by transport mode.

Ore/Bulk/Oil Carrier A multi-purpose ship which is capable of carrying full cargoes of either ore, other bulked dry cargo or oil. This is a combination bulk carrier arranged for the carriage of either bulk dry cargoes or liquid cargoes often in the same cargo spaces but not simultaneously.

Ore carrier Ship built to convey bulk ores – a single deck dry cargo vessel.

Orecon Code name for a charter party used primarily in Scandinavia to Poland ore trade.

Oresund Bridge A four lane motorway with twin track railway bridge of 7.8km length built in 2000. It is part of an overall project completed in 2000 embracing a 4km tunnel and the construction of an artificial island – Pepper Island, linking the tunnel and motorway. The tunnel/bridge links Malmo Sweden and Copenhagen Denmark. The rail/motorist journey takes ten minutes and opens up the economy of Scandinavia. Hitherto the journey was by train, ferry and passenger ferry.

Orevoy Code name of BIMCO approved charter party used for shipments of iron ore.

Orevoy bill Code name of BIMCO approved Bill of Lading for shipments on the Orevoy Charter Party. This was superseded by Coal-Orevoy. See Coal-Orevoy.

Organic Organic substances are compounds of carbon except those that are binary compounds for example, carbon oxides, the carbides, carbon disulphide, etc; ternary compounds such as metallic cyanides, metallic carbonyls, phosgene (COC12), carbonyl sulphide (COS), etc; and the metallic carbonates, such as calcium carbonate and sodium carbonate. Some examples of organic substances are hydrocarbons, alcohols, ethers, aldehydes, ketones, organic solvents and organic acids. (See also 'Inorganic'). Term associated with dangerous classified cargo.

Organic growth The growth of a company/entity within its portfolio and without a merger or acquisition intervention.

Organisational chart A plan outlining the organisation of a particular company or department.

Organisational integration A situation in a company/entity whereby all personnel – whether in production, research, development, design, finance, selling, distribution, after sales service etc, departments are involved in a particular marketing, finance, etc process/policy formulation on a completely integrated basis.

Organisation Commerce Africaine et Malgache An economic trading bloc consisting of Central African Republic, Dahomey, Gabon, Ivory Coast, Mauritius, Niger, Rwanda, Senegal, Togo, and Upper Volta. It was introduced in 1965.

Organisation for Economic Co-operation and Development Based in Paris its fourfold role amongst its members is to co-ordinate economic policy, promote and co-ordinate aid to developing countries, further liberalise international trade, and promote sound economic expansion.

Organisation of African Unity (Organisation de l'Unite Africaine) Based in Addis Ababa, its role is to promote the unity and solidarity of African States, to co-ordinate and intensify their efforts in respect of economic and social improvements, and to remove all traces of colonialism from Africa. It has 65 member countries.

Organisation of American States (OAS) Group of 45 American States.

Organisation of Petroleum Exporting Countries Based in Vienna its membership represents the main petroleum exporting countries. Its prime objective is to control production through a quota system and thereby endeavour to sustain the maximum level of the price of oil the market will support.

Organised market A market which is structured/organised in terms of its promotion, channels of distribution, retail outlets and so on. Such a market lends itself to analysis/evaluation and market research.

Orient Overseas Container Line (OOCL) A major shipowner and container operator and registered in Hong Kong (China).

Original Bill Original Bill of Lading.

Original gross premium The premium charged by the reassured to the original assured before any discounts have been taken into account. The reinsurer usually requires knowledge of the original gross premium to ensure that he is not backing the liability of a reassured who is retaining high commissions and thus encouraging bad underwriting.

Original slip The placing slip used by a broker when negotiating the insurance with the underwriter.

Originating products Products which have been 'wholly produced' in the EC. An EU customs term.

Origin based system A customs term. It confirms the goods have been manufactured/processed/assembled at a particular place/country which determines the customs arrangements/procedures/tariff.

ORION Shipboard weather routing service provided by Oceanroutes of WNI (Weathernews) Group.

Orlop Lowest deck in a vessel.

ORS Overseas reporting system.

ORTPEACE Orient Peace Corporation SA – a shipowner.

OS Austrian Airlines. The National Austrian Airline and member of IATA; or Ordinary seaman.

O/S Offshore.

OSC On Scene Co-ordinator.

OSD Open shelter deck – type of vessel.

OSD/CSD Open shelter deck-closed shelter deck – type of vessel.

OSE Oslo Stock Exchange.

OSI Open system interconnection – IATA term.

OSO Ore – slurry – oil (ship).

OSRA Ocean Shipping Reform Act 1998 – USA legislation.

ost Ordinary spring tides.

OSV Offshore support vessel or offshore supply vessel.

OT Oil tight; or open topped container.

OTAR Overseas tariff and regulations. It provides UK exporters with information about tariffs, regulations, etc, for all overseas countries.

OTC Organisation for Trade Co-operation.

OTD Ocean Travel Development – organisation responsible for promoting passenger sea travel.

Other persons ARC concept referring to agencies that could be appointed by individual carriers outside the ARC framework but would report sales and remit through the ARC area settlement plan.

Other service information Information included in a client's airline booking record that does not require specified action by the carrier, such as a VIP designation.

OTIF Organisation of International Carriage by Rail.

OTIMBERS Ocean Timbers Line – a shipowner.

OTS Organised Track Structure – IATA term.

OU Croatia Airlines – an IATA member airline.

OUB Overseas Union Bank.

OUE Other underwriting expenses.

Ouguiya The currency of Mauritania. See Appendix 'C'.

Outbound logistics The process of adopting a logistics strategy for exporting products. This involves distribution goods.

Outer packaging The outer protection of a composite or combination packaging together with any absorbent material, cushioning and any other components necessary to contain and protect inner receptacles or inner packagings of a product.

Out of charge note A document issued by Customs confirming goods have been cleared by Customs. See Entry Processing Unit.

Out of gauge Cargo whose dimensions exceed those of a container, railway wagon, or road haulage unit in which it is accommodated and consequently requires special arrangements being instituted to convey it. Usually regulations exist detailing the maximum dimension permitted on a transport mode without resorting to special arrangements.

Out of the money A call option which has a strike above the market price of the futures contract and an out of the money put option a strike below. An option with no intrinsic value. A BIFFEX term. See also in the money entry.

Out port A small seaport which relies on feeder services and would be the spoke and the hub system. See also hub and spoke system entry.

Output devices Include: visual display unit, printer, plotter, controlled devices, speech and audio. A computer term.

Output gap The difference between the actual and potential output level of an economy as a percentage of the potential output. For monetary policy purposes, the rate of growth of potential output may be considered to be the rate of real output growth consistent with price stability in the medium term (i.e. no tendency for either inflationary or deflationary pressures to appear). This may be similar in practice to the trend rate of growth of the economy. A positive (negative) output gap means that actual output is above (below) the trend or potential level of output, and suggests the possible emergence (absence) of inflationary pressures. An ECB definition.

Output per head The output per person employed.

Outright forward purchase If a currency is purchased forward without a corresponding sale of that currency at spot, it is termed an outright forward purchase. An international trade banking term.

Out shipment Passengers/cargo refused shipment as vessel already fully to capacity.

Out sourced The process of a Company contracting out the process of sourcing products/components/materials etc for a Company.

Out sourcing The process of a Company/supplier operating through a Distribution Company from a central distribution centre overseas which undertakes all the sale administration, sales order scheduling, invoicing on clients letter head, monitoring and reconciling payments into the clients bank account, credit control and warehouse activities. All such activities are transferred from the exporter/suppliers base/premises to the overseas central distribution centre thereby out sourcing such activities. Overall such a development enables the exporter to become more competitive.

Outstanding claims losses The total of losses or claims which have been advised but at a given time are still outstanding and as such are only estimated amounts.

Outstanding claims portfolio An amount payable by a cedant to a reinsurer in consideration of the reinsurer accepting liability arising under a contract of reinsurance in respect of reinsurance claims incurred and arising prior to a fixed date.

Out-the-money A call (put) option where the exercise price is above (below) the current price of the underlying – insurance term.

Outturn report A written statement by a stevedoring company in which the condition of cargo discharged from a vessel is noted along with any discrepancies in the quantity compared with the vessel's manifest.

Outward cargo Cargo in process of being exported from a country.

Outward dues Charges/dues raised by a Port Authority on vessels leaving a port and usually based on a ship NRT.

Outward freight department Shipowners department responsible for marketing outward (export) cargo.

Outward mission The prime purpose of an outward mission involves a group of exporters usually under a Chamber of Commerce or Trade Association to visit an overseas market, either to explore and assess the market prospects of their goods or services, or to reinforce their overseas market effort.

Outward processing relief A customs facility available to exporters whereby goods may be despatched overseas (outside the European Union) for processing or repair and subsequently reimported and in so doing remain immune from any customs duty being raised, provided certain conditions are complied with and adequate preplanning is undertaken by the exporter.

Outwards investment Capital which is invested/exporter overseas.

Outwards trade strategy A strategy directed towards developing trade outside the region/economic bloc/ customs union.

Outward Trade Mission The prime purpose of an outward trade mission involves a group of exporters usually under a Chamber of Commerce or Trade Association to visit an overseas market either to explore and assess the market prospects of their goods or services, or to reinforce their overseas market effort. It may involve Government support and Ministerial participation.

Over age An additional premium charged on an open cover declaration because the carrying vessel is outside the scope of the classification clause. It may be applied, also, to additional premium charged for breach of navigational warranties (e.g. Institute Warranties) where the ship is over 15 years old.

Over age insurance Any additional insurance on the cargo required because of the age of the vessel.

OV Estonian Air – an IATA member airline.

Overall length The length, for example, of a transport unit from the foremost point to the rear most point. It may be road vehicle, railway wagon, ship and so on.

Overall width The width, for example, of a transport unit from the outermost projection on one side to the outermost projection on the other side. It may be road vehicle, railway wagon, ship and so on.

Overbooking The practice by a supplier of confirming reservations beyond capacity, either in expectation of cancellations or no shows or in error.

Over carried A carrier who carries cargo beyond the allotment distributed to him.

Overcentre tensioner A tensioner designed primarily for use with wire lashings. Also termed quick release tensioner.

Overdrive An extra high gearbox ratio which permits economical cruising speeds to be achieved at lower engine revolutions than those required to propel the vehicle in top gear.

Overhang If the stock market knows that a major investor (or group of investors such as underwriters) is eager to sell a large line of shares in a particular company, they are said to overhang the market.

Overhauls The process of maintaining annually or other prescribed period a transport unit and/or its associated equipment. It may be a road vehicle, aircraft, ship and so on.

Overheating An economic term which arises in a situation whereby the demand in an economy is growing at much too fast a pace, with a consequent outbreak of inflationary pressures and a shift into a larger deficit on the current account.

Overheight See O/H and over height cargo entries.

Overheight cargo Cargo loaded into an open-top container so that the level of the cargo rises above the normal level of topside rails, accepted by certain operators under certain conditions.

Over insured The process of affecting an insurance cover at a value in excess of the market value of the insured item/circumstance.

Over investment The over provision of investment driven resources. It maybe over tonnage or a seaport with too many berths which the trade cannot support.

Over invoicing An over priced item on an invoice.

Overland transport Transport services provided overland such as rail, or road.

Overlay A guide, usually made of transparent film with certain data field areas blanked out with opaque vinyl material and other fields left clear as transparent apertures. When used with a master document and photocopying machine a particular overlay will produce a specific form in the aligned series

Overlength cargo Cargo exceeding the standard length.

Over manning An over provision of labour resource having regard to all the circumstances to undertake a particular job/assignment.

Overnight travel The process of travelling overnight.

Overpack An enclosure used by a single shipper to contain one or more packages and to form one handling unit for convenience of handling and stowage. Dangerous goods packages contained in the overpack must be properly packed, marked, labelled and in proper condition as required by these Regulations. (A Unit Load Device is not included in this definition). Shrinkwrap or banding may be considered an overpack.

Over Pivot area The rate per kilogramme to be charged for the over pivot weight for an Air Freight consignment.

Over Pivot weight The weight in excess of the pivot weight for an Air Freight consignment.

Overplacing Placing more than 100% of the amount at risk.

Overplanning See safety stock entry.

Over provision An excess of provision of resources in relation to the market demand or other criteria. For example too many aircraft to convey declining passenger market.

Over resourced The provision of an excess quantity/ adequacy of resources available to undertake a particular assignment. See resource available entry.

Override Extra commission paid by suppliers as a sales incentive.

Overriding Commission An additional discount allowed to a reinsured who effects reinsurance at original net rates, to help cover his underwriting costs – an insurance definition; or a level of commission rate which is over and above the norm found in the market place. For example, if the norm is nine per cent special circumstances such as a large retail chain store may have

ten per cent or one per cent overriding commission. A travel trade definition; or a discount allowed to an agent or ceding insurer in addition to normal commission. In reinsurance it is commonly by way of contribution to the direct insurers overheads. In direct insurance it may be payable to an employee or agent in respect of insurance's written within his territory, even without his mediation – insurance definition.
Oversale See overbooking.
Overseas Branches A method of direct export trading where an adequate support service cannot be provided by local agents or distributors. The export company manufacturer would consider the cost and effectiveness of setting up their own organisations abroad. This is likely to be economic only in larger or long term markets.
Overseas investment cover An insurance cover facility which is available to cover losses, which are attributable to political events, on new foreign investments and loans (with repayment periods of more than three years) made to overseas enterprises.
Overseas market The total market found in a specific country/region/area overseas relative to a particular product(s)/service(s).
Overseas market visit See visiting overseas market.
Overseas seminar A seminar conducted overseas by a Company to an individual audience of potential customers. Such Company presentations provide good product/service exposure and enables potential buyers to meet face to face with the Export Sales Executive. It is usually conducted in a hotel and can feature films and refreshments/luncheon arrangements.
Overseas tariff regulations It provides UK exporters with information about tariffs, regulations, etc, for all overseas countries.
Overseas trade fairs and seminars A marketing facility whereby exporters can bring their products or services to the attention of a specific audience in an overseas territory. They are organised by Trade Associations or other representative bodies.
Overshooting A situation where an exchange rate changes/fluctuates by more than is necessary to remove an underlying payments imbalance.
Overside discharge Process of discharging cargo overside from vessel's hold or deck over ship's rail into a barge/lighter moored alongside the ship.
Overside loading Process of loading cargo overside over ship's rail into the vessel's hold or onto the deck from a barge/lighter moored alongside the vessel.
Oversized container A container longer than 40 ft (12.20m).
Overspend The process of spending/buying above the level of expectations/budget. It maybe a budget figure of $3,000 and an amount of $3,500. An overspend of $500.
Overspent The process of exceeding a budgeted/target figure in financial terms.
Overstowed The process of overloading a transport unit (cargo hold) or warehousing racking resulting in the package being damaged. It maybe bulging, splitting of bags, disfigurement of the carton etc.

Over stuffing The process of overloading a package resulting in a load bulge. An example arises in a carton of clothes.
Over subscription A financial situation where an offer of securities is over subscribed when applications exceed the amount offered.
Over-the-counter A tailor-made contract traded off-exchange.
Over tonnaging The provision of too much shipping capacity in a particular trade/route/market in relation to market demand, thereby producing a surplus of shipping capacity.
Over trading The process of executing more business in a company than working capital will prudently allow to be serviced. This results in a serious cash flow problem which primarily arises from the lag in payment by clients, subsequent to the placing of orders.
Over valuation The exchange rate of a currency which is in excess of its purchasing power parity. In consequence the country's goods will be uncompetitive in the export markets.
Overvalued currency Currency whose rate of exchange is persistently below the parity rate.
Overview The process of giving a broad/widely based evaluation of a particular circumstance/situation.
Overwidth cargo Cargo exceeding the standard width. For example, a container with goods protruding beyond the sides of the container/flat rack onto which they are packed.
O/W Overwidth. A container with goods protruding beyond the sides of the container/flat rack onto which they are packed.
Own account operator A road haulage operator who conveys solely his own merchandise and does not advertise/promote his business as a 'hire and reward' operator. For example, Marks and Spencer convey/distribute by road their products to their retail stores.
Owner In P & I club membership terms, when related to an 'entered ship' it includes shipowner, owners in partnership, owners holding separate shares in severalty, part owner, mortgagee, trustee, charterer, operator, manager or builder of such ship by or on whose behalf the same has been entered in the Association for insurance whether he be a member of the Association or not.
Owners account A chartering term which indicates that all charges/overtime so described/identified in the charter party are to be paid by the shipowners.
Owner's Broker Person who acts for the shipowner in finding a cargo for a vessel which would be chartered.
Owner's clauses Clauses specially agreed for a particular fleet, probably giving wider cover than is normally allowed.
Owners' marine superintendent Expert with overall responsibility for the care, maintenance and repair of shipowner's vessels.
Owners risk A carriers' conditions of carriage.
OWS Owners.
Oxygen analyser Instrument used to measure oxygen concentrations in percentage by volume.
Oxygen-deficient atmosphere An atmosphere containing less than 21 per cent oxygen by volume.
OZ Asiana Airlines Inc – IATA Airline member.

P Prop; paddle; platform; poop; or port side of a vessel.

PA Press Association. One of the news agencies which serve the British media; Post Assistant; Private account: Power of Attorneys; Peoples Association (Singapore); propulsion assistance; particular average; personal assistant, for example, Managing Director; personal accident; or percentage adjustment.

P/A Public address system.

PAA Philippine Automotive Association; or personal accident (assault).

PABMA Philippine Association of Battery Manufactures Inc.

PABX Private automatic branch exchange.

PAC Public Account Committee (of the Chamber of Shipping); Patents Advisory Committee; Pacific – IATA term; or Port Authority charges.

Pacific Asia Travel Association (PATA) Government and travel industry organisation promoting tourism throughout the Pacific and Indian Ocean areas.

Pacific Forum Line Established by ten member governments in 1977 as a means of providing an adequate and reliable regional shipping service, aimed at encouraging economic development in the region. It is owned by twelve South Pacific Forum Governments excluding Australia.

PACIS Port and Cargo information study.

Package The complete product of the packing operation consisting of the packaging and contents; in computer terms a generalised program written for a major application which can be easily adapted for use by many different users – for example stock control package, payroll package, and so on; or in commercial terms the formulation/conclusion of a deal involving a number of ingredients such as a package holiday involving travel, accommodation and entertainment at an 'all in' inclusive price by a Travel Agent for his clients.

Package deals A price quotation which includes a number of components. This could include an Agent who would process a consignment through customs including documentation, stevedoring, etc for an overall inclusive price. Alternatively, it may be a retailer to provide and install, and provide free maintenance for one year of a household product such as electric cooker or central heating system.

Packaged timber Timber shipped in bundles (unitised) form.

Package goods Merchandise which is despatched in a packaged condition.

Package holiday A product retailed by the Travel Trade such as shipowners, airline travel agent etc, which includes not only the travel arrangements, but also the hotel accommodation.

Package operator A Company/entity involved in travel and related resources for a particular passenger journey. The package operator may not own the travel unit of a coach or aircraft, but may charter/hire them. The package maybe providing holidays in Italy involving all the travel, accommodation, insurance, excursions and other arrangements. It is usually retailed through a Travel Agent.

Packager Organiser and usually operator of tour packages, wholesaler.

Package Tour A travel/leisure market product whereby the sponsor, who is usually the carrier, offers to the customer a tour for an inclusive price. This may be the airline, coach or rail travel, plus accommodation at an hotel. Additionally, it could include optional excursions in the area visited. Some tours are based on one centre such as Berlin, whilst others may include a number of centres called 'stop overs' and thereby usually involve a more lengthy itinerary. Travel Agents usually offer the package tour on behalf of operators and the overall inclusive price is very competitive.

Package Tour operator A person/company which provides tours embracing hotel accommodation, travel-coach/rail/air, excursions, and any equipment such as skiing equipment and training for a skiing holiday.

Packaging The assembly of one or more containers and any other components including closures, which are necessary to assure compliance with any minimum packing requirements for the envisaged transit, to ensure the goods arrive at their destination in an undamaged condition. Factors involved determining packaging needs include type of commodity; transport mode(s); packaging cost; value of consignment; size/quantity/dimensions/weight of goods; terms of sale; marketing considerations; statutory obligations and so on. Packaging regulations are particularly severe for dangerous classified merchandise.

Packaging certificate Confirmation the cargo has been correctly stowed and labelled in an ISO container.

Packaging cost The total packaging cost of a particular product or group of products. This embraces labour, materials, and design cost.

Packaging credit Advance granted by a bank in connection with shipments of storable goods guaranteed by the assignment of the payment expected later on under a documentary credit.

Packaging list In some trades the packing list document is used. This provides a list of the contents of package/consignment(s). In particular it will include the number and kind of packages, their contents, overall net and gross weight, usually in kilogrammes, the dimension(s) of the package(s) including length, width and height, and finally, the cube of the package(s). The document is often referred to as a 'packing not' and may feature the package marking.

Packaging research The process of conducting research into the effectiveness/adequacy of packaging for a particular commodity with particular emphasis on its suitability for transportation.

Packaging sift proof One which prevents the escape of powdered material.

Pack ice See Drift Ice.

Packing The art and operation by which articles or commodities are enveloped in wrapping and/or enclosed in packaging or otherwise secured. Term includes carton, sack, crate or other forms of merchandise packaging.

Packing instructions A document issued within an enterprise giving instructions on how goods are to be packed.

Pacta sunt servanda Contracts are to be kept – a Latin term.

PACWID Presidential Actions Committee on Wood Industry Development (Philippines).

PADC Philippine Aerospace Development Corporation.

Pad eye A type of lash end fitting. See lash end fitting entry

PADIS Passenger and Airport Data Interchange Standards (IATA term).

PAEC Philippine Atomic Energy Commission.

PAFMI Philippine Association of Feed Millers Inc; or Philippine Association of Flour Millers Inc.

P & I Protection & Indemnity – shipowners P & I insurance club.

Paid losses or claims The total of claims or losses which have been settled during a given period.

Pail A metal drum either cylindrical or with tapered sides usually fitted with a bale handle.

Paisa The currency of India, Bangladesh and Pakistan. See Appendix 'C'.

Pakistan International Airlines The National airlines of Pakistan and member of IATA.

P & L Profit and Loss.

PAL Philippines Airlines and member of IATA.

Pal Box A jumbo sized fibreboard box placed on a pallet which is sealed or strapped to provide a unitised transit.

Pallet Mounted steel or wooden platform of 800 x 1,200mm or 1,000 x 1,200mm designed to accommodate and facilitate cargo transhipment and through cargo movement. Overall, the load board has two decks separated by bearers or a simple deck supported by bearers constructed with a view to transport and stacking, and with the overall height reduced to a minimum compatible with handling by fork lift truck and pallet trucks.

Pallet, Aircraft Is a platform with a flat under surface and with standard dimensions used for assembling goods into unit loads equipped with mechanical handling system. It is designed to restrain loads aboard the aircraft through the use of nets, straps or other tie-down equipment. As such, it is a certified component of the aircraft loading and restraint system.

Pallet converter Superstructure which can be applied to a pallet to convert it into either a box or post pallet.

Pallet forks A pallet lifting device attached to the cargo hook consisting of two forks fixed to a triangular frame. It is so designed that the forks always remain horizontal, empty or under load.

Pallet intermodality The ability of a pallet to be transferred from one transport mode to another during an international transit such as road/air/road.

Palletised stowage The stowage of merchandise in a warehouse in a palletised condition.

Pallet net A network of webbing which is affixed to a pallet to restrain the load on the pallet.

Pallet rack A skeleton framework of fixed or adjustable design to support a number of individual pallet loads.

Pallet ship Operative in the Netherlands and Belgium inland waterway networks and parts of the French and German networks, she has a capacity of 650 euro pallets. The vessel is of similar capacity of the Kempenaar ship with a capacity of 32 TEUs and 48 TEUs when stacked three high. The pallets are automatically self-loading and unloading. The vessel was introduced in 2002. See Kempenaar ship.

Pallet-track Runners incorporated in the vehicle floor, in which hydraulically raised arms run from front to rear to enable heavy pallets to be lifted off the floor to facilitate easy movement.

Pallet truck Cargo handling equipment designed to move in confined spaces.

Pal van A type of railway freight wagon specially designed to accommodate palletised freight.

PAN Panama.

Pan African Appertaining to all the States of Africa. It may be culture, living standards, literacy, religion, economic factors and so on.

Panama Canal A waterway linking the Atlantic and Pacific Oceans of 50 miles length and forming an important world trade route.

Panama Canal tonnage Panama canal authority require special tonnage dues certificate for dues payment when a vessel passes through it.

Panama type fairlead A non-roller type fairlead mounted at the ship's side and enclosed so that mooring lines may be led to shore with equal facility either above or below the horizontal. Strictly pertains only to fairleads complying with Panama Canal Regulations, but often applied to any closed fairlead or chock.

Pan American Appertaining to all the States of North/Central and South America. It may be culture, living standards, literacy, religion, economic factors and so on.

Panamax vessel A vessel capable of passing through the Panama canal which has limitation/constraint of below 33m beam and 270m length overall. The dry bulk and tankers Panamax carrier capacity range is 50,000 – 79,999 dwt.

Panasia Pan Asia Line Ltd – a shipowner.

Pancake ice Predominantly circular pieces of ice 30m in diameter, up to 10cm in thickness, with raised rims due to the pieces striking against one another. May form on a slight swell from the grease ice, shuga or slush, or as a result of the breaking of ice rind, nilas or, under severe conditions of swell or waves, of grey ice. Sometimes forms at some depth at an interface between water bodies of different physical characteristics, then floats to the surface. Its appearance may rapidly cover wide areas of water.

Pancan The code name of BIMCO approved charter party clause relative to the Panama canal and entitled 'Baltic Conference Panama canal clause 1954'.

Panel A group of experts formed to assess and report on a situation; or in WTO terms, consisting of three experts, this independent body is established by the DSB to examine and issue recommendations on a particular dispute in the light of WTO provisions.

Pan European Appertaining to all the countries of Europe. It may be culture, living standards, literacy, religion, economic factors and so on.

Pan European Brand A brand representing a product known and accepted throughout the European Market.

Pan European media The media which has been produced for the European market. It may be TV commercials or advertising programme for the national press in each country which would be featured in the local language of the buyers market.

Pan European Product A product which has been designed for the European Market.

Pan European strategy The strategy/policy adopted in all the European States. This may arise in a Multi National Enterprise, trading blocs, and so on.

Pan European systems The network/system which obtains throughout the European States. It may be customs, trading patterns, computer/telecommunications network and so on.

Pan Euro zone customer A customer which is situated throughout the Euro zone market.

PANISLAMIC Pan Islamic Steamship Co Ltd Karachi – a shipowner.

Pan regional advertising An advertising programme which extends throughout a particular region.

PANS Procedures for Air Navigation Services (ICAO term).

Panstone Code name of Chamber of Shipping approved charter party for stone trade.

Panstop The code name of BIMCO approved charter party relative to the Panama canal and entitled 'Baltic conference Stoppage of Panama canal traffic clause 1962'.

Panting stresses It arises when the vessel is pounded in rough weather and the ship is rising and falling, and at the stern of the vessel the local stresses are increased by the racing of the propeller. The fitting of panting 'beams', 'breasthooks' and heavier plating, help to combat the 'in' and 'out' movement of the shell plating and stresses created by increased and decreased water pressure.

PANTRAC Pan American Tracing and Reservations system.

PAOHSING Poa Hsing Navigation (Panama) SA – a shipowner.

Paper It usually means commercial paper such as bills of exchange or promissory notes, but may refer to any securities. For example, Par (1) is a nominal value of a security; Par (2) is an issue price of a security if floated at its full face value; and Part (3) is when the 'spot' and 'forward' prices of a currency are the same.

Paper bag A flat paper receptacle, normally of two or more plies, i.e. separate sheets with or without a gusset.

Paper kroft A strong paper, manufactured wholly from bleached or unbleached cellulose fibres.

Paperless communication system The electronic transmission of structured data between user friendly computer systems. It is usually termed electronic Data Interchange. See Electronic Data Interchange.

Paperless trading Process of accepting electronic transmitted data for the international exchange of goods involving customs, insurance, finance, transport etc. Hence no documents are involved. See Bolero.

Paper trading Trading of crude oil or products when there is no intention and possibly no facility to enable physical delivery to take place. Paper trading is used for a range of purposes including hedging – an IPE term.

Paper – waterproof Paper which has been manufactured or treated subsequently so as to be waterproof.

PAPI Precision Approach Path Indicator – IATA term.

PAPM Philippine Association of Paint Manufacturers Inc.

Par Par or parity arises in a specific foreign currency when there is a zero discount or premium between the two currencies. In such a situation the spot and forward contract are at par or parity. An international trade banking term. See also currency at a premium and currency at a discount entries; or a share is at par when it is quoted or sold at its face value.

Parador Castle, monastery or such converted into first-class or luxury tourist accommodations by government, particularly in Spain and Puerto Rico.

Paraffin wax Includes paraffin wax which is a white crystalline hydrocarbon material of low oil content normally obtained during the refining of lubricating oil distillate, paraffin scale, slack wax, microcrystalline wax and wax emulsions used for candle manufacture, polishes, food containers, wrappings etc.

Parallel imports When a product made legally (i.e. not pirated) abroad is imported without the permission of the intellectual property right-holder (e.g. the trademark or patent owner). Some countries allow this, others do not – WTO definition.

Parallel indexing (PI) Automatic or manual radar safety device for coastal navigation and harbour entrance.

Parameters Instructions and options that are added to an MS-DOS command. Computer term.

Paras The currency of Yugoslavia. See Appendix 'C'.

Paratroop salesmen The process of a salesman visiting a country on a single visit to sell merchandise and has no subsequent follow up visits to help secure a sales contract commitment.

Parcels market That segment of the market involved in the parcels business. Often referred to as small packages, and weighing up to 15 kilogrammes. It involves the Post Office, Airlines and so on. The tariff structure, terms of carriage, and limitation on size, weight and commodity type varies by carrier.

Parcels van A road vehicle designed to convey parcels.

Parent company A company which owns the majority (51% or more) of shares in another company (its subsidiary).

Parent directory Any directory that has one or more sub-directories.

Parent sailing A sailing which forms the basis of a shipping service and around which relief or duplicate sailings are operated.

Parent service A transport service such as road, rail or air which forms the basis of the service and around which relief or duplicate services are operated.

Pareto law The theory that a small proportion has a disproportionate effect on the whole. It is more widely known as the 80/20 rule where 20 per cent of the customers are responsible for 80 per cent of the turnover and vice versa.

Pareto ratio The process of a small percentage of clients being responsible for a large percentage of orders. It may be 20/30% of the customers being responsible for 70/80% of the Company's turnover. An industrial marketing term.

Pari Passu (Latin) Equally without preference.

Paris Club Created in the late 1950s to deal with the liquidity problems then faced by a small number of debtor countries, mostly in Latin America. Today it has devised strategies on debt rescheduling to a large number of countries and enhanced under the Toronto terms adopted in December 1991.

Paris Convention Treaty, administered by WIPO, for the protection of industrial intellectual property, i.e. patents, utility models, industrial designs, trademarks and trade names, indications of source and appellations of origin. This Treaty also seeks to repress unfair competition. A WTO definition.

Paris Moll's The Paris memorandum of Understanding on Ports.

Parity Parity or par arises in a specific foreign currency when there is a zero discount or premium between the two currencies. In such a situation the spot and forward contract are at parity or par. An international trade banking term. See also currency at a premium and currency at a discount entries; or when two shipping/freight forwarders/airlines etc quote identical rates for the same traffic. This arises especially under liner conferences. See also parity on rates entry.

Parity Clause A clause which automatically adjusts the premium rates on one open cover in parity with a change in a similar (related) cover placed in another market. An insurance term.

Parity on rates A situation where two rates quoted by different companies are identical. This happens in liner conferences where parity on rates exist but competition arises on service quality and other areas of business.

Parking bays Space allocated at a terminal for vehicles.

Parlour A living room or sitting room that is part of a hotel suite and not used as a bedroom; in Europe often called a saloon or lounge.

Parlour car In the US or Canada, rail cars featuring individual swivel seats and food and liquor service.

Part cargo Goods which do not represent the entire cargo for a particular ship but whose quantity is sufficient to be carried on charter terms.

Part charter A situation in which a schedule airline transport charter passengers, using some of the seats, on a regularly scheduled flight. Also may apply to a container slot charter.

Partial loss Any loss that is not a total loss; an insurance term.

Partial pressure The individual pressure exerted by a gaseous constituent in a vapour mixture as if the other constituents were not present. This pressure cannot be measured directly but is obtained firstly by analysis of the vapour and then by calculation using Dalton's Law.

Participating carrier A carrier participating in a tariff and who therefore applies the rates, charges, routings and regulations of the tariff. In so doing it results in the carrier over whose routes one or more sections of carriage is undertaken/performed. It maybe by air, rail/sea/rail, road/sea/road and so on.

Participating trader A trader registered to use the ITP procedures. A customs term. See ITP.

Participation certificate A participation certificate is a security representing a portion of the property rights in a company but carries no right of membership.

Particular Average A fortuitous partial loss to the subject matter insured, proximately caused by an insured peril but which is not a general average loss. In a freight at risk policy the term may be applied to a claim for loss of freight following a particular average loss of goods.

Particular Charges Particular Charges – used in relation to expenses incurred in connection with a Particular Average claim. The term is being phased out and is replaced, in practice, with 'Extra' charges.

Part load Less than a full load conveyed in a transport unit. It is especially relevant to a railway wagon, container, or road vehicle. In the case of a container it is usually termed – less than container load. (LCL); or a technique developed by the Forwarding Agent involving the TIR vehicle whereby the Agent would collect a large consignment from the exporter, deliver the same vehicle without any intermediate transhipment and use the major part of the available residual space in the vehicle for groupage.

Partners People who own and run a business together and who bear unlimited liability for its debts.

Part shipment A shipment which is not transported in one lot but in two or more parts.

Party ticket A travel, etc ticket for a group.

PASG Pacific Air Traffic Services Group – IATA term.

Pass Passenger (ship).

Passage plan The maritime route of a proposed voyage of a ship.

Passenger Any person, except a member of the crew, carried or to be carried in a transport mode such as an aircraft, train, coach, ship, etc, with the consent of the carrier.

Passenger aircraft An aircraft that carries any person other than a crew member, an operators' employee in an official capacity, an authorised representative of an appropriate national authority, or a person accompanying a consignment. An IATA definition.

Passenger analysis The process of analysing passenger traffic data for a particular service, group of services, retail outlets such as a station, travel agent for a particular period. It will embrace type of ticket, revenue, number of journeys and so on.

Passenger care The process of a transport passenger operator giving particular attention to passenger welfare during the journey. This involves clean comfortable transport units; provision of refreshment facilities during the journey; entertainment facilities involving films, etc, and so on. It also involves adequate well appointed passenger lounges in passenger terminals. The extent of passenger care facilities varies and tends to become more extensive the longer the journey, especially on international travel.

Passenger control The process of a transport operator controlling/regulating the number of passengers on a particular schedule/flight(s)/sailing(s) etc. This maybe achieved by a compulsory reservation system and/or the issue of a control ticket in association with the passenger ticket.

Passenger coupon The portion of the passenger ticket and baggage check that constitutes the passenger's written evidence of the contract of carriage. An IATA definition.

Passenger Cruise Liner Vessel built to convey passengers on cruises.

Passenger dues These dues are raised on a passenger basis embarking or disembarking from the vessel. It maybe $2 per passenger and is sometimes called a port tax or passenger toll.

Passenger facility charge (PFC) A head tax allowing US airports to impose a fee to be used for federally approved airport improvements; airlines collect the tax and remit the funds to the airports.

Passenger fares The tariffs devised for passenger traffic. They may be single or return; child, juvenile, or adult; individual or group; weekly, monthly, annually, first or second class; and so on.

Passenger lift The provision of an elevator/lift linking one or more levels with another. It maybe at a railway underground station linking street level with the railway platforms; in a large retail store/shop; and so on. A large passenger lift can carry up to 100 passengers.

Passenger lounge Accommodation provided at a passenger terminal such as a railway station, airport, coach/bus station, seaport and hover port, for passengers usually waiting to join their transport mode. Alternatively, it maybe purely passenger accommodation on a vessel.

Passenger market That segment of the market involved in passenger travel. It embraces rail, coach, car, ship, canal, air and other transport modes.

Passenger miles The total number of passengers carried on a transport unit multiplied by the distance travelled by each passenger.

Passenger miles per vehicle hour The total number of passengers carried on a transport unit multiplied by the distance travelled by each passenger.

Passenger name record (PNR) The record of a booking made and stored in a computerised reservations system, including all the pertinent information, such as passenger name, flight number, travel times and dates, the airline(s) and price.

Passenger number certificate Document confirming maximum permitted number of passengers to be shipped and compliance with relative statutory obligations.

Passenger rated traffic Traffic conveyed on a service at passenger rates. This may include ordinary (foot) passengers, motorists, coach passengers and ancillary traffic such as parcels, letter mails and post office traffic.

Passenger return Inventory of passengers shipped on a vessel or conveyed on an aircraft.

Passenger safety certificate Document confirming maximum permitted number of passengers to be shipped and compliance with relative statutory obligations.

Passenger service agent Airline employee helping passengers with check-in and boarding procedures.

Passenger ship A vessel conveying more than twelve passengers is classified as a passenger vessel and accordingly is required to have a specification and survey schedule for passenger tonnage. It maybe an estuarial ferry, a cross channel ferry conveying passengers, cars, coaches, accompanied road haulage vehicles, or a cruise liner.

Passenger ship safety certificate A document/certificate issued by a ship classification society on behalf of a Maritime Government/National Administration. It is required by any passenger ship engaged on international voyage, except troopships. A passenger ship is a vessel which carries more than twelve passengers. Pleasure yachts not engaged in trade do not require a Passenger Ship Safety Certificate. The classification survey ensures compliance with the 1974 SOLAS convention embracing survey arrangements for subdivision, damage, stability, fire, safety, life saving appliances, radio equipment and navigational aids.

Passenger strategy The process of devising/formulating a policy of a particular company/entity relative to passenger aspects of the business. This includes fares, group travel, ticketing, accountancy, refunds, computerisation, passenger amenities/welfare, discounted travel and so on.

Passenger tariffs The rates devised for passenger traffic. See passenger fares.

Passenger throughput The total number of passengers passing through a station, airport, seaport during a specified period.

Passenger tolls Charges raised by a Port and Harbour Authority per passenger passing over the quay on embarkation or disembarkation.

Passenger tonnage Vessels designed/built to convey passengers.

Passenger traffic manager (PTM) See corporate travel manager.

Passenger vessel Vessel which carries more than twelve fare paying passengers whether berthed or unberthed accommodation.

Passing of risk The process of passing the responsibility for making the insurance claim against loss or damage. In the event of the goods being uninsured during their transport from seller to buyer, the risk is undoubtedly for the owner of the goods at the time of loss or damage, the owner being identified according to the transfer of property in the goods.

Passive exporter An exporter who adopts a reactive as compared with a proactive strategy. In so doing the exporter waits for a potential overseas buyer to approach him/her rather than initiate an overseas promotion of the product/service to obtain business and thereby give exposure of the entity – products/services available to the overseas territory.

Passive importer An importer who adopts a reactive as compared with a proactive strategy. Insodoing the importer waits for a potential overseas seller to approach him/her rather than research the overseas market place for a particular product/service.

Passport An official licence of permission for a person to enter or leave a country.

Passr Passenger.

Pasteboard A type of fibreboard manufactured by the combination of several layers of papers or of board by the use of a suitable adhesive.

PA system Public Address system.

PAT Port Authority of Thailand.

PATACA The currency of Macao. See Appendix 'C'.

Patent The process of registering a new/modified product with the patent office and thereby obviating the risk of the product being copied. See also intellectual property.

Patrol launch A small vessel which plys ahead of a ship to form a navigation aid and usually operative in confined or estuarial waters especially when proceeding to or from a seaport.

Pay as paid A factoring term indicating the factor pays the balance when the buyer pays.

Payback The period, usually expressed in years, which it takes the cash inflows from a capital investment project to equal the cash outflows.

Paying Agent A trader known to customs and formally set up on CHIEF who transmits importer Customs warehouse removal declarations to CHIEF. A customs definition.

Paying bank The bank which is to pay under the credit. It can be the issuing bank itself or a bank nominated by it, usually the advising/confirming bank.

Payload The revenue producing/earning capacity of a transport unit(s), e.g. aircraft, ship, road vehicle, train, etc. For example, in the case of a road vehicle, it is the amount of load that may be carried on a vehicle within the maximum permissible limit by law, after deducting the tare weight.

Payment against documents See cash against documents entry.

Payment at Sight The legal definition as per the Bills of Exchange Act 1882, Section 3 of a Bill of Exchange is: 'An unconditional order in writing addressed by one person to another, signed by the person giving it, requiring the person to whom it is addressed to pay on demand, or at a fixed or determinable future time, a sum certain in money to or to the order of, a specified person or to bearer'. Bills of exchange are widely used in international trade as a means of claiming payment by the supplier (drawer) from the buyer (drawee) and can be used as a vehicle for obtaining trade finance. A 'sight' bill calls for payment immediately upon presentation, as opposed to a 'term' bill which calls for payment upon a fixed or determinable future date. Sight Documentary Credits need not necessarily include a Bill of Exchange, in which case payment is purely against the shipping/other documentation specified in the credit.

Payment cycle The time/period between a buyer placing an order for goods and the seller's receipt of monies in respect of the execution of the order. The buyer maybe purchasing finished goods for stock or to fulfil pre-sold orders. Alternatively, the goods may be raw materials for re-manufacture or components requiring assembly. Season peaks, long transit times and lengthy credit terms all add to the demands on the international buyer.

Payment in advance The process of the exporter, paying for the goods in advance – prior to their despatch. This form of payment is not common as it means the importer is extending credit to the supplier (instead of vice versa, as is normal). It can arise for small quantities of samples where the supplier asks for payment in advance and an international money draft or an express telegraphic money transfer may be used.

Payment order Brokers' authority to act for the assured in collecting a claim.

Payment terms These are the terms by which payment is required.

Pax Travel industry abbreviation for passengers.

PB Permanent bunkers; poop and bridge; or permanent ballast.

PBA Paid by Agent.

PBF Poop, bridge and forecastle.

PBH Partial bulkhead.

PBPC Passenger and Baggage Processing Committee – IATA term.
PBR Payment by Results.
PBS Preferred Berth Scheme. A seaport term.
PBT Polybuthylene terephthalate.
Pc Platform/Crane.
PC Passenger certificate; postcard; personal computer; Private care; Per Capita; product carrier or Poly Carbonate.
P & C Private and confidential.
P/C Percent, petty cash or price current.
PCA Philippine Contractors Association; or Panama Canal Authority.
PCARD Philippine Council for Agriculture Resources, Research and Development.
PCB Polychlorinated biphenyl or printed circuit board.
PCBA Printed circuit board assembly.
PCC Pure car carrier; Panama Canal Commission; or Processing under customs control.
PCCI Philippine Chamber of Commerce and Industry.
PCH Purchase.
PCIC Paper and Chemical Industries Corporation (Burma).
PC-INP Philippine Constabulary – Integrated National Police.
PCN Personal communication network.
PCNT Panama Canal Net Ton.
PCR Performance capability rating.
PCs Participation certificates.
pcs Pieces.
PCT Polychlorinated triphenyl.
PCTC Pure Car Truck Carriers.
PC/UMS Panama Canal Universal Measurement System.
PCWBT Protective coatings in water ballast tanks (in a vessel).
PD Port dues, post dated; or property damage (third party).
Pd Paid or passed.
pd Per day.
PDCP Private Development Corp of the Philippines.
PDI Pre-delivery inspection. A task undertaken prior to taking delivery of new vehicle.
PDK Promenade deck.
PDM Physical Distribution Management.
PDNA Programme for Developing Nations' Airlines – IATA term.
PE Portfolio entry – insurance abbreviation.
PEA The Provincial Electricity Authority (Thailand).
Peace clause Provision in Article 13 of the Agriculture Agreement says agricultural subsidies committed under the agreement cannot be challenged under other WTO agreements, in particular the Subsidies Agreement and GATT. Expires at the end of 2003.
Peak Fare, rate or season when travel and tourism are traditionally at the highest level.
Peak rate A term associated with a freight rate, currency rate and so on, indicating the rate has reached its highest level having regard to market conditions.
Peak tank Small tank situated at the extreme forward end – fore peak tank or after end – after peak tank of a ship. It normally holds water ballast and is used to help to trim the ship, that is, to adjust the draughts forward and aft.
PEC Pilotage exemption certificate.
PECO Period Entry Central Office.
Pedestal Usually termed a levelling pedestal which is a stacking fitting that permits bridge fittings to be used with containers of different heights.
Pedestal roller fairlead A roller fairlead usually operating in a horizontal plane. Its purpose is to change the direction of lead of a mooring or other line on a ship's deck.
PE (Exports) Period Entry (Exports)
Peer group A group of people who are perceived as having the same characteristics as oneself. It may be education, class, status and so on. Some peer groups formulate opinions in the market place and thereby have much influence on buyers' behaviour.
PEEST Political, Economic, Environmental, Scientific and Technological.
PE(I) Period Entry (Imports).
PEKANBARU Pekanbaru Shipping Pte Ltd – a shipowner.
Pelican Hook A one piece hook assembly used to connect a tensioner to a 'D' ring or lashing plate. It is used for cargo securing.
PELNI Pelni Lines – a shipowner.
Pelorus Portable, hand operated compass, lined up with ship's gyrocompass, and used to take bearing of objects blind to the repeaters due to obstructions (i.e.: funnel).
PELT Personnel Licensing and Training (Panel – ICAO term).
PEL/TRG Personnel Licensing and Training – ICAO term.
Penalty clause A clause in an agreement which outlines/specifies the indemnity which arises consequent on a breach of the agreement. An example arises in a voyage charter which indicates the amount payable for non-performance of the charter.
Pence The currency of a number of countries throughout the world, including United Kingdom and Gibraltar. See Appendix 'C'
Pendulum service A shipping service which serves a group of seaports which may be in close proximity to each other but the sequence of serving them varies throughout the monthly sailing programme.
Penetrating the market The process of entering a market for the first time relative to the sale of goods and/or services.
Penetration pricing The strategy adopted by a Company/entity to price a product/service at a level which will attain maximum penetration of the market, thereby increasing substantially/significant sales volume and or market share. Introduction of a new product at a low price to penetrate the market.
Peniche Long flat-bottomed, usually wooden, canal boat latched and with raised cabin amidships.
Pennia The currency of Finland. See Appendix 'C'.
Pension Often a private home renting a bedroom to paying guests; often including some meals, in which case it may be called a boarding house.
Pentiolite A detonating explosive which is an intimate mixture of pentaerythrite tetranite (PETN) and trinitro-toluene (TNT). A term used in the transportation of such merchandise relative to dangerous classified cargo.
PEPOA Philippine Electric Plant Owners Association.
PER Price/earnings ratio. See separate entry.
per Through; by means of; or according to.
Per (Latin) By means of, according to.
Per capital income The average level of income earned by the working population of a particular country during a particular period which is usually one specified year.
Perceived benefit See perceived benefits to the user entry.
Perceived benefits to the user The benefits derived by the client/user of the product/service purchased. It can be also expressed as value added benefits but is not a tangible evaluation.
Perceived value The value/price of the product/service in the eyes/judgement of the buyer/consumer/industrialist.

Per cent (per centum) (Latin) In every hundred.

Percentage of depreciation The proportion of loss suffered by a cargo owner expressed as a percentage of the sound value of the cargo. This percentage maybe assessed by a surveyor and is applied to the insured value to determine a claim for partial loss.

Percentage of product price spent on its advertising/promotion The promotion cost incurred to retail a product in the market place expressed as a percentage of the selling price. For example it may cost 3% (€150) in promotional cost to sell a car at €5,000.

Perdiem (Latin) By the day.

Perfect competition Situation where a single producer acting alone can influence the price level.

Perfect monopoly Situation where monopolist can act without the possibility of any sort of direct competition.

Performance Bond An undertaking give by a bank or an insurer at the request of a seller to the effect that the guarantor will pay the buyer in the related international sales contract an agreed sum of money in the event of the seller's default. Three agreements are usually involved: between the guarantor and buyer; between the seller and buyer (the related sales contract); and between the guarantor and the seller in which the seller agrees to reimburse the guarantor if the guarantor has to pay under bond. It replaces the bid bond if the contract is awarded to you. It covers specific aspects of the contract such as shipment of goods etc.

Performance clause Clause in a time charter which stipulates that, should the ship be unable to achieve the agreed speed or should she consume too much fuel, the charterer is entitled to recover from the shipowner the cost of time lost and extra fuel, normally by means of a deduction from hire money; or a contractual clause to measure/outline the criteria of the performance of a specified product/service. See performance testing and performance monitoring.

Performance guarantees Guarantee to the buyer that the supplier will carry out the terms of the contract.

Performance letters of credit A letter of credit featuring a performance clause in the letter of credit. See performance bond entry.

Performance marketing A marketing term involving the process of a marketing/promotional technique being tested in the market place to determine its adequacy/effectiveness. It maybe a sales campaign in a specific area to encourage consumers to purchase a particular domestic floor cleaner. Also termed test marketing.

Performance monitoring The automatic calculation and recording at predetermined intervals of critical plant performance criteria such as specific fuel consumption to monitor gradual changes in efficiency occurring on a service.

Performance Standards for Shipborne Radiocommunications and Navigational Equipment The international Maritime Organisation specification for general requirements and performance standards for shipborne radiocommunications and navigational equipment.

Performance testing A packaging term involving the process of a consignment/package being tested in transit relative to the adequacy of its packaging specification.

Per hatch per day It shall mean that the laytime is to be calculated by dividing (A), the quantity of cargo by (B), the result of multiplying the agreed daily rate per hatch by the number of the vessel's hatches. Thus:

$$\text{Laytime} = \frac{\text{Quantity of cargo}}{\text{Daily Rate} \times \text{Number of Hatches}} = \text{Days}$$

Each pair of parallel twin hatches shall count as one hatch. Nevertheless, a hatch that is capable of being worked by two gangs simultaneously shall be counted as two hatches. A BIMCO term.

Peril A term used in the Marine Insurance Act (1906) to denote a hazard. The principle of proximate cause is applied to an insured peril to determine whether or not a loss is recoverable. In modern practice the term 'risk' replaces 'peril'.

Perils Events or circumstances that may cause losses.

Perils of the sea Something fortuitous and unexpected in a maritime adventure.

Period entry scheme See period entry traders.

Period entry system (importers) A customs entry system facility available to large shippers/traders who have adequate computing resources to render periodic returns electronically. Advantages include reduced entry cost, better management information, reduced stock levels, simplified customs documentation, speedier clearance, improved cash flow, deferment of duty and VAT and better customs relations resulting from quicker and more reliable services. See period entry traders.

Period entry traders Period entry allows authorised importers in UK who use computers for accounting and stock control purposes, to provide HM Customs with the required import information on a periodic post-import basis in the form of supplementary declarations on computer media. Authorised importers obtain clearance of their goods either by making a simplified declaration on a SAD at the frontier or, if they are also authorised for Local Import Control (LIC), by an input to their computer system. Under Period Entry, duty and VAT are paid using the deferment system. Therefore, the importer must ensure that the level of your deferment guarantee is sufficient to cover the liability for VAT as well as duty.

Periodical survey To maintain the assigned class the vessel has to be examined by the Classification Society's surveyors at regular periods.

Period warehousing A customs facility whereby approved traders warehouses are permitted to store goods. Such merchandise is monitored on a computer system for accounting and stock control. Receipts and deliveries are controlled at transaction level. Advantages include delays payment of duty and VAT until removed for delivery; goods may be re-exported without the payment of duty and VAT; reduces paperwork to a minimum; fully computerised system, and overall provides useful management information data.

Peripheral A hardware device, such as a disc drive, printer or plotter, that is external to, yet controlled by the computer.

Perishable cargo Foodstuffs and other products which tend to deteriorate in extended transits and require special facilities to maintain the quality of the product and need controlled temperatures. Many such goods travel in refrigerated units.

Permanent storage Storing data such as computer BIOS and other Boot programs which are usually stored on ROM. See 'Storage devices'.

Permit on deck shipment Authority to ship a cargo on deck such as dangerous, heavy or awkward sized goods. Such a phrase would feature on the Export invoice/Letter of Credit.

PERNAS National Corporation Ltd (Malaysia).

Peroxide A compound formed by the chemical combination of cargo liquid or vapour with atmospheric oxygen or oxygen from another source. In some cases these compounds may be highly reactive or unstable and

a potential hazard. A term associated with dangerous cargo shipments.
Per pro (Latin) On behalf of.
Per se (Latin) By itself.
Persona (Latin) Person in the eye of the law.
Personal effects An individual's clothing, toiletries and the like accompanying a passenger. It also embraces the shipment of personal effects/furnishings etc, in an ISO container, e.g., household removal overseas.
Personal Export Scheme A facility to tourists visiting a country whereby certain commodities can be purchased free of VAT/Custom Duty subject to the relevant regulations being adhered to and completion of any documentation.
Personal interview The process of two or more people meeting to discuss a project, market research survey report, product launch etc, or having face to face interview/negotiation as found in personal selling.
Personal lines Those types of insurance, such as auto or home insurance, for individuals or families rather than for businesses or organisations.
Personal selling The process of conducting face to face selling or telesales involving seller/buyer in the market place, designed to promote and sell a product or service. See telesales.
Persuasive advertising An advertising campaign designed to encourage a sale/purchase rather than to inform the client/market place.
PERT Project Evaluation and Review Techniques.
PERTAMINA Indonesia State Oil Corporation.
PERUMTEL Indonesian Telecommunications State Company.
"Per working hatch per day" (WHD) or "Per workable hatch per day" (WHD) It shall mean that the laytime is to be calculated by dividing (A), the quantity of cargo in the hold with the largest quantity, by (B), the result of multiplying the agreed daily rate per working or workable hatch by the number of hatches serving that hold. Thus:

$$\text{Laytime} = \frac{\text{Largest Quantity in one Hold}}{\text{Daily Rate per Hatch} \times \text{Number of Hatches serving that Hold}} = \text{Days}$$

Each pair of parallel twin hatches shall count as one hatch. Nevertheless, a hatch that is capable of being worked by two gangs simultaneously shall be counted as two hatches. A BIMCO term.
PES Period entry system. See period entry traders.
PESA Philippine Electrical Suppliers Association
Peseta The currency of Spain. See Appendix 'C'.
Pesewas The currency of Republic of Ghana. See Appendix 'C'.
Peso The currency used in some countries such as Argentina, Colombia and so on. See Appendix "C".
PEST Political Economic Social and Technological – a marketing term.
PET Pacific Engineering Trials – IATA term; or Polyethylene terephthalate.
PETCOKE Petroleum coke.
Petit dejeuner French term for breakfast.
Petite-Vitesse (French) Ordinary (non-express) international goods train.
Petrol chemical complex A petroleum and chemicals complex as found in a seaport involving storage, transhipment, berthing processing and distribution resources.
Petrol chemical currency An International currency which is largely based on its exchange rate relative to its oil production. This includes OPEC members, UK etc. Such countries have an economy which is predominantly geared in a trade balance through its oil exports.

PETRONAS National Oil Co (Malaysia).
Petty Officer (Deck) A post in the Deck Department on a vessel and formerly designated Bosun's Mate.
Petty Officer (Motorman) A post in the Marine Engineer's department on a vessel and formerly designation Donkeyman.
PE(W) Period Entry (Warehousing). See period entry traders.
PF Poop and forecastle; or Palestinian Airlines – an IATA Member Airline.
PFA Philippine Foundry Association.
PFC Processing under customs control for free circulation. Dutiable goods may be imported or released from customs warehouse without paying the duty, processed under customs control and the products put into free circulation on payment of the duty appropriate to the products rather than to the imported goods; or passenger facility charge.
Pfd Position fixing device – ships navigation aid.
PFDA Philippine Fisheries Development Authority
PFI Position finding instrument – ships navigation aid; or Private Finance Initiative.
PFL Pacific Forum Line.
PFPC Passenger Form and Procedures Committee – IATA term.
PFSO Port Facility Security Officer.
PFSP Port Facility Security Plan.
PFY Polyester filament yarn.
PG Bangkok Airways Co Ltd – IATA Member Airline.
pgke Package.
PGP Pretty good privacy – the program which encodes your e-mail so that only the intended recipient can decode and read it – a cyberspace term.
PH Polynesia Airlines Ltd – an IATA Member Airline.
PHA Port Health Authority.
PHI Permanent health insurance.
PHIC Pharmaceutical Industries Corporation (Burma).
PHIL Philippines.
PHILCITE Philippine Centre for International Trade and Exhibitions.
Philippine Airlines The National airline of Philippines and member of IATA.
Philippine Peso The currency of Republic of Philippines. See Appendix 'C'.
PHILSUCOM Philippines Sugar Commission.
Phlegmitiser A solid or liquid which is added to a substance such as explosives or organic peroxides in order to reduce its sensitivity to heat and impact thereby assisting in ensuring safety during transport. A term used in the transportation of such merchandise relative to dangerous classified cargo.
Phosphate berth A purpose built berth at a seaport handling bulk shipments of phosphate. Usually the berth is rail and/or road served and the phosphate is loaded by a conveyor system and discharged by grabs.
Physical delivery The transfer of ownership of an underlying commodity between a buyer and seller party to a futures contract following expiry. IPE term.
Physical distribution The movement and handling of goods from the point of production to the point of consumption or use. It is based on four elements – transportation, marketing, distribution and management. This involves the process of moving the goods from the manufacturer seller to the distributor/end user involving one or more carriers as distribution channels.
Physical distribution management The techniques used to distribute merchandise in the most cost effective basis practicable conducive with market conditions. This involves all transport modes especially road, rail, containerisation and includes the code of practice

Phytosanitary (Plant Health)

adopted at the factory/warehouse to ensure the optimum distribution method is employed.

Phytosanitary (Plant Health) A certificate issued by plant health authorities. Such a document is required in particular countries for planting material, forest trees and other trees and shrubs, and certain raw fruit and vegetables. In some countries the importation of certain species of plants from certain areas of the world is prohibited.

PI Sunflower Airlines Ltd – an IATA Member Airline; Primary insurer; Professional indemnity; Personal injury; premium income; or parallel indexing – see separate entry.

P & I (Protection and Indemnity club) Non-profit organisation providing insurance cover on a mutual basis, principally for third-party risks in maritime business.

PIAP Printing Industries Association of the Philippines.

Piastres The currencies of Egypt and Lebanon and the Republics of Sudan and Syria. See Appendix 'C'.

PIC Petrochemical Industries Corporation (Burma).

PICC Philippine International Convention Centre.

Pick and Pack Taking goods out of stock and packing them according to customer conditions.

Picking list A list used to collect items from stores needed to fulfil an order.

Pickings Damaged material picked off bales of cotton to make the remainder saleable.

Pick up The collection of cargo by the carrier at shipper's premises or other prescribed point.

Pick up and delivery A service concerning the collection of cargo from the premises of the consignor and the delivery to the premises of the consignee.

Pick up cost The cost of collection goods/merchandise from a specified point such as a warehouse for distribution.

Pick up service The surface carriage of 'out bound' consignments from the point of pick-up to the airport, seaport, ICD, CB, or other specified point of departure.

Pictogram A symbol featuring a particular facility used especially in publications, directional, and location sign provision. For example, a level crossing with gates would feature a locomotive and gates on road signs approaching the crossing. Also used in poster presentation of statistical data.

PIE Pan Island Expressway (Singapore).

Pier The location in a seaport at which cargo arrive or departs. A dock for unloading ships or vessels. A type of wharf running at an angle with the shore line of the body of water. Also termed a quay.

Pier/Pier The conveyance of a consignment/merchandise starting and finishing at the ships side. Hence the inland movement prior to and subsequent to the sea carriage is effected by the merchant or his agent.

Pier to House A container term. The exporter/forwarder delivers the shipment to the carriers' depot or container freight station as 'break-bulk'. The carrier packs the container at the container freight station/depot whilst at the destination de-stuffing of the container is performed by the consignee. Also termed LCL/FCL.

Pier to Pier A container term. The cargo is packed for shipment and unpacked at destination by the ocean carrier. Both operations are undertaken on the carrier premises. Hence the inland movement prior to and subsequent to the sea carriage is effected by the merchant or his agent. Also termed LCL/FCL.

Piggyback The carriage of unaccompanied road vehicles and trailers on wheels on rail flat cars; in international marketing terms the process of an exporter fresh to the business in a particular overseas country joining forces with an established exporter and thereby promoting/selling both companies products/services on a joint basis.

Piggyback promotion The situation where two products are promoted – one on the back of another such as garden furniture and greenhouses, or a product/service is promoted which is accompanied by a voucher or free gift.

Pig's ear A funnel.

PIL Pacific International Lines. A South-East Asia container line providing container services in the region.

PIL/MSC Pacific International Lines/M'sia Shipping Corporation – a shipowner.

Pilot Person who aids the Master in ship navigation in confined waters; or the person in charge of an aircraft.

Pilotage A charge raised on a vessel entering a river/port/estuary/canal where the services of a pilot are obligatory.

Pilotage slip Document detailing dues to be paid on Pilotage engaged.

Pilotage tugs Tugs operative usually in a port which facilitate the mooring or departure of vessels to/from the berth under the guidance of a pilot.

Pilot in command The pilot responsible for the operation and safety of the aircraft during flight time.

PIM International goods regulations governing conveyance of goods by rail in Europe.

PIN Personal identification number.

Pinch bar Cargo handling equipment used by the docker to facilitate moving cargo particularly in confined spaces.

Pinion Small cog wheel engaging with larger one.

Pinlock A stacking fitting locked into a corner casting by a pin that projects through a side hole of the corner casting into the cone of the stacking fitting.

Pintles Pins that hinge the rudder to the gudgeons on the rudder post.

PIOPIC Protection and Indemnity of Oil Pollution Indemnity Clause.

Pipe laying ship Vessel designed to lay pipes on the sea bed.

Pipeline inventory The amount of goods in a pipeline; the sum of loading stock; goods in transit and receiving stock.

PIRA Packaging Division of the Research Association for the Paper and Board, Printing and Packaging industries. An organisation available to exporters relative to packaging techniques and training.

Piracy In maritime terms and as defined by the United Nations Convention on the Law of Sea, adopted in 1982 Article 101. Piracy consists of any of the following acts: (i) any illegal acts of violence or detention, or any act of depredation, committed for private ends by the crew or the passengers of a private ship or a private aircraft, and directed (a) on the high seas, against another ship or aircraft, or against persons or property on board such ship or aircraft, (b) against a ship, aircraft, persons or property in a place outside the jurisdiction of any State; (ii) any act of voluntary participation in the operation of a ship or of any aircraft with knowledge of facts making it a pirate ship or aircraft, (iii) any act inciting or of internationally facilitating an act described in sub-paragraph (i) or (ii). Hence any act of violence against ships, particularly those which occur within ports or the territorial seas of States are not acts of piracy under international law, but classified as armed robbery; or in WTO terms, unauthorised copying of materials protected by intellectual property rights (such as copyright, trademarks, patents, geographic indications, etc) for commercial

purposes and unauthorised commercial dealing in copied materials.
PIR/IST Pireaus/Istanbul feeder service.
PISC Philippine International Shipping Corporation – a shipowner.
PISO Philippine Investments Systems Organisation.
Piston-engine plane Aircraft powered by internal combustion engine.
Pit Trading arena where open outcry takes place. See outcry.
PITC Philippine International Trading Corporation.
Pit card The trading card used by traders to detail all elements of trade whilst in the pit.
Pitching Motion of a ship along her fore and aft axis as she heads into a moderate/rough sea.
Pitwoodcon Code name of Chamber of Shipping approved charter party for timber (pitwood) trade.
Pivot port A port at which cargo from outlying points is routinely collected for onward carriage.
Pivot weight Minimum chargeable unit load device weight for Air Freight consignments.
Pixel The smallest element of an image which can be displayed on a video display screen.
Pixpinus Code name of Chamber of Shipping approved charter party for timber (pitch pine) trade.
PJKA Indonesian State Railways.
Pk Pakistan International Airlines Corp (PIA). The National airline of Pakistan and member of IATA.
PKD Partially knocked down – a consignment.
Pkg Packing.
Pkge Package.
Pkg instr Packing instruction.
PKN State Economic Development Corporation (Malaysia).
PKNS Selangoe State Development Corporation (Malaysia).
4PLs Fourth party logistics (companies). Overall they form clusters of logistics service providers.
P/L Partial loss.
3PL Third party logistics.
PL Empresa de Transporte Aéreo del Per' (AEROPERU) – an IATA member airline; partial loss; protective location; Poland – customs abbreviation; passenger lifts; pipe layer; public liability; personal liability; or products liability.
PLA Port of London Authority.
Place of acceptance The place where cargo is accepted for transit and carriers liability commences.
Place of delivery The place where cargo is delivered and carriers liability ceases.
Place of origin The precise location at which goods are received by the carrier or his agent.
Place of receipt Place where goods are received for transit and carrier's liability commences.
Placing An interbank deposit which may be a straight market placing; a private placing of short, medium or long term funds; or a private sale of a new issue of shares to a limited number of investors. The sponsoring investment banks, or placement agents, will generally undertake to carry out such a placing on a best efforts basis, whereas a flotation or IPO will generally be underwritten.
Plaintiff The litigant who sues another. A person claiming compensation or damage – a legal term.
Plaited rope A rope structure consisting of two pairs of strands twisted to the right and two pairs of strands twisted to the left and braided together such that pairs of strands of opposite twist alternatively overlay one-on-another.
Planetary gear See Epicyclic gear.
Planned maintenance system A planned maintenance system covers maintenance schedules based on running hours or calendar period, survey maintenance is stipulated and repair schedules to cover various breakdowns of machinery and system. A marine engineer term.
Planned selling The process of evolving and executing a sales plan. It involves strategies/tactics and a network to monitor results and respond to opportunities and threats in the market place. Usually the sales plan is market researched and budget driven with each sales person being accountable for his/her sales performance.
Planning The process of regulating and co-ordinating activities on a time basis together with the resources necessary to carry out these activities in order to achieve set objectives.
Plan Position Indicator Pumproom Radar picture appearing on Screen of cathode ray tube space (usually aft on VLCC) housing ship's cargo and ballast pumps.
Plastics A wide variety of synthetic materials used for packaging, among them are polypropylene, polyethylene or other plastic material.
Plastic solvent A name commonly used for mixtures of liquids employed for dissolving plastics, or for thinning plastic cements. In general they may contain flammable liquids such as acetone, amylacetate, or some of the alcohols or cetones. The dangerous cargo classification is determined by the flash point.
PLAT Platform.
Plate/B/L Bill of Lading used in connection with cargo from River Plate.
Plated weight Maximum gross weight designated for a vehicle or its axles by the manufacturer (i.e. manufacturer's plated weight) or the Department of Transport (Dtp plated weight).
Plate lifting clamp Cargo handling equipment used for metal plate transhipment.
Platform body A truck or trailer without ends, sides or top but with only a floor.
Platform Flats An ISO container of 20ft or 40ft size designed to carry awkward, oversize and project cargoes. The unit can be used as temporary 'tweendecks' for the carriage of large indivisible loads. Also ideal for trades where there is cargo in one direction and nothing in the other. They can be returned as one stack of seven units. See flatrack containers.
Platform supply vessel A vessel specially designed to ship/handle all the necessary equipment such as pipes, cement, tools and provisions to offshore oil and gas platforms and drilling rigs.
Platform vehicle A two axle load haulage vehicle of platform carrying unit design.
PLATO Pollution Liability Agreement Among Tanker Owners.
PLB Personal locator beacon.
PLC Public Limited Company; or product life cycle.
PLDT Philippines Long Distance Telephone Company.
Plimsoll mark Horizontal line situated amidships inscribed on ships outer plating depicting limit to which ship maybe submerged.
PLN Indonesian State Electricity Utility Undertaking.
PLR Paid loss ratio – insurance abbreviation.
PLS Please.
PL/SBT Protectively located segregated ballast tanks.
PLTC Port Liner Terms Charges.
PLTS Pallets.
Plug A lash end fitting that is inserted into the corner casting.
Plurilateral agreement An agreement covering two or more areas/disciplines/activities.
Pluralistic society A society/community with many individual aspirations and high individualism.

Plying limits Maritime area in which a ship may operate.
Plywood Laminations of wood veneers glued together with the grain of alternate plies lying at right angles.
Plywood drum A cylindrical receptacle consisting of, either a plywood sheet butt joined and held in position either by means of a plywood strip on the inside, or a metal strip on the outside, or a continuous winding of plywood veneer glued with water resistant adhesive.
PM Performance monitoring; or position mooring.
pm Premium.
PMA Port of Marseille Authority.
PML Probably/possible maximum loss – insurance abbreviation.
PMS Policy management system – insurance abbreviation.
PMT Per metric ton.
PN Promissory Note.
PNB Philippine National Bank.
Pneumatic tensioner A tensioner of turn buckle type powered by a pneumatic gun.
PNOC Philippine National Oil Co-operation.
PNPF Pilot's National Pension Fund.
PNR Philippine National Railways; or passenger name record.
PNSC Pakistan national shipping Corporation, Karachi – a shipowner.
po Postal order.
PO Port operations.
P&O Peninsular and Oriental Steam Navigation Company.
P&O Ferry master Freight Forwarder specialising in logistics and intermodal distribution throughout Europe.
P&O Ned Lloyd A mega container shipping line.
P&ONL P & O Ned Lloyd – a shipowner.
POA Place of acceptance.
POB Pilot on Board.
POC Port operator charges.
POCB Philippine Overseas Construction Board.
POD Paid on delivery; place of delivery; proof of delivery; proof of debt; or port of discharge/delivery.
PODIP Pacific Operational Development & Implement-ation Programme – IATA term.
Poincare Franc The fictitious gold franc of defined weight and finesse used to assess carriers liability in an inflation proofed manner in the Hague Visby Rules.
Point of origin The precise location at which goods are received by the carrier or his agent.
Point of purchase Place where goods are bought such as from a mail order or shop.
Point of sale Place where goods are retailed/sold.
Point to point fare A basic fare from one destination to another.
Point to point transport See door to door entry.
POL Pacific Ocean Lines – a shipowner; or port of loading.
Polar low Small, often intense depressions that form at high altitudes in cold arctic air masses.
Polcoalbill Code name of BIMCO approved Bill of Lading used for shipments under the Sovcoal charter party.
Polcoalvoy Code name for coal charter party.
Polcon Code name of charter party for coal trade from Poland.
Policy A document detailing the terms and conditions applicable to an insurance contract and constituting evidence of that contract. It is issued by an underwriter for the first period of risk. On renewal a new policy may not be issued, but the same conditions would apply and the current wording would be evidenced by the renewal receipt. A change of conditions is evidenced by an endorsement.
Policy Proof of Interest (PPI) The underwriter waives proof of insurable interest which must exist at the time of loss, but the policy is deemed void in the eyes of the law.
POLISH Polish Ocean Lines – a shipowner.
Polish Airlines The Polish National Airline and member of IATA.
Poliski Rejestr Statkow Polish Ship Classification Society based in Gdansk, Poland.
Political, Economic, Environmental, Scientific and Technological A technique used to evaluate the effects of external influences on a business.
Political infrastructure The structure of a government in a country and how the country is governed reflecting the democratic process, government and opposition parties; central government; monarch led government; or military rule. Also it will embrace all levels of government both central, regional, local, or trading/economic blocs as found in the European Union.
Political risk Risk of a political nature encompassing all political events which have an impact on any contractual relationships covered by civil law. These embrace war and other such disturbances; cancellation of an export licence and insurance; foreign currency conversion risks; transfer risk – also embracing third country transfer risk – where one country can freeze the assets and bank accounts of another country held locally; any action of a foreign government which in some way hinders the enactment of the contract including import/export restrictions; and transactions between exporters and public purchasers – default on payment by a public purchaser. An export credit insurance definition.
Poll Pollution control.
Pollution In maritime terms the process/consequences of an escape/leakage/discharge of cargo. See TOVALOP.
Pollution insurance Special insurance cover to combat claims from maritime pollution.
Pollution liability Exposure to lawsuits for injury or cleanup costs that result from pollution damage.
Polycentric The strategy process of a Company or entrepreneur providing a product/service to meet the local needs of a particular country/culture/region. It is the opposite of ethnocentric which operates on the basis the home country product has a global market without modification or adaptation. A marketing term. See ethnocentric entry.
Polymeric beads, expandable These are semi-processed products used to manufacture polymeric articles. When impregnated with a flammable gas or liquid as a blowing agent, they may evolve small quantities of flammable gas during transportation. A term associated with dangerous classified cargo.
Polymerisation The chemical union of two or more molecules of the same compound to form a larger molecule of a new compound called a polymer. By this mechanism the reaction can become self-propagating causing liquids to become more viscous and the end result may even be a solid substance. Such chemical reactions usually give off a great deal of heat. A term associated with dangerous classified cargo.
Polymerisable material Any liquid, solid, or gaseous material which, under conditions incident to transport-ation, may polymerise (combine or react with itself) so as to cause dangerous evolution of gas or heat. A term used in the transportation of such merchandise relative to dangerous classified cargo.
Polynya Any non-linear shaped opening enclosed in ice. Polynyas may contain brash ice and/or be covered with new ice, nilas or young ice. Sometimes the polynya is

limited on one side by the coast and is then called a shore polynya or by fast ice and is called a flaw polynya.

Polystyrene beads expandable These are semi processed products used to manufacture polystyrene articles. When impregnated with a flammable gas or liquid as a blowing agent, they may evolve small quantities of flammable gas during transportation. A term associated with the conveyance of dangerous classified cargo.

Polyvalent A system of ship officer training. A joint deck and engine certificate. This involves an officer could be a second engineer on one voyage and the next trip a chief officer.

Pomerene Act The Federal Bill of Lading Act operative in the USA relative to the 'straight Bill of Lading'. It applies to USA exports and not imports, and states it is not a document of title which must be surrendered to secure goods on delivery. Moreover, it indicates the document must be made out to a nominated consignee and marked 'not negotiable' and allows delivery without surrender of the documents.

Pon Pontoon.

Pontoon A flat bottomed shallow draught vessel or landing stage.

Pontoons Large sections of hatch cover.

POO Port operations officer.

Pool Arrangement whereby shipowners – usually under a liner conference system – pool their receipts which are distributed amongst members on an agreed basis.

Pool co-operation Two or more airlines co-ordinating services in a market and pooling costs and revenue; such flights are administered by the operating carrier, which also controls the reservations.

Pooling system An arrangement where by transport operators – which may be road, rail, ship or air – pool their receipts which are distributed amongst pool members on an agreed basis. For example, four airlines may operate a service between A and B and distribute the pool receipts on the basis of 10/25/30/35 as agreed.

Pool schemes Various arrangements for processing very small premium or claim entries which are entered in a pool rather than being applied to individual syndicates. An insurance definition.

Poop Raised deck at the stern of a vessel.

POP Point of purchase; or a point of presence. The location of the internet base which basically determines the internet/phone account.

POR Place of Receipt.

PORAM Palm Oil Refiners Association of Malaysia.

Port A port is a terminal and an area within which vessels load or discharge cargo whether at berths, anchorages, buoys or the like, and shall also include the usual places where vessels wait for their turn or are ordered or obliged to wait for their turn no matter the distance from that area. If the word 'PORT' is not used, but the port is (or is to be) identified by its name, this definition shall still apply. A BIMCO term. Usually it has an interface with other forms of transport and in so doing provides connecting services; or it is the left hand side of the ship/aircraft when facing forward; or the external connection point on a computer to which peripherals devices are attached – computer term.

Port access Access (entry) to a port. This maybe from the seaward end direct from an enclosed lock system, river estuary, or merely an open roadsted or from the landward side by road, rail, or inland waterway.

Port Agent An Agent at a particular port is primarily involved with looking after the needs of his Principal which is usually a shipowner. Overall, the local representative producing services to the Master and owners of a ship. In such a situation this would involve the provision of berth resources; cargo handling equipment; dock labour; transport distribution; custom clearance; victualling needs of the vessel; ship crew needs such as crew change over; documentation; liaison with port authorities; general security; provision of shipboard pilotage and tugs assistance; and so on. Overall, the Agent must ensure the prompt turn round of the Principal vessels and efficient processing of cargo including customer liaison and cargo distribution/ collection.

Portable gangway A piece of port equipment linking the ship with the shore (quay) enabling passengers' crew/personnel to embark or disembark and which can be manually handled with or without the assistance of a crane.

Portage Bill See Masters Portage Bill.

Portainer crane Quayside lifting apparatus used in the transhipment of ISO containers. It is also employed in container stacking areas.

Portal A berth accommodation Ro/Ro Vehicular ferry type vessel with the vehicles (cargo) being transhipped (driven on/driven off) over the portal ramp. It may be a single or double deck portal ramp thereby permitting loading/unloading simultaneously; or the terminals gaining access to the subterranean rail terminal between Folkstone (UK) and Frethun near Calais, France.

Portal crane See crane portal entry.

Port and Security Act 2001 USA legislation which together with the Anti-terrorism Act 2002, provides for conducting security assessments in foreign ports and refusal of entry into United States ports for vessels transiting in ports with unsatisfactory security procedures.

Port Captain A shore based appointment embracing an experienced former Ship Officer. The Port Captain is responsible for all aspects of the ISM Code embracing Ship Management and is a link between the shipboard personnel and its management within the ISM code and the shore based personnel in the shipping Company.

Port charges All those charges/tariffs raised by a port authority relevant to the conduct of its business especially in the areas of cargo – all types, passengers, craneage, tonnage dues, and so on; or a voyage charter party which specifies the vessel to discharge her cargo at a specified port and in so doing the vessel will not have officially arrived until she actually reaches the named port where cargo operations are to be performed.

Portcon The code name of a BIMCO approved charter party clause entitled 'port and dock charges'.

Port congestion charges A charge raised usually by shipping companies payable by the shipper when due to port congestion long delays occur in the loading and discharge of cargo.

Port control The organisation which controls the day to day traffic control of a port. This includes the movement of ships to and from the port; allocation of berths and dock labour; liaison with customs and railway company regarding cargo/ship clearance and rail conveyance; liaison with security embracing both landward and river policy; liaison with pilotage association; equipment provision; special arrangements such as heavy lift craneage required for heavy indivisible load(s) transhipment involving floating crane; liaison with lighterage companies and so on. The prime task is to keep cargo moving through the port in an efficient manner and to minimise port turn round time of vessels. In some ports it could be called Port Traffic Control.

Port Director A most senior directorate managerial position responsible for the overall Management of a port. This will involve not only the overall day to day management of the port, but also the general policy development of the port including investment. He would be strongly supported by a team of Chief Officers for the business, such as Marketing Officer, Harbour Master, Port Operations Officer, Chief Financial Controller and so on. The Port Director would be a member of the Board of Directors.

Port disbursements Cost/expenses incurred directly attributable to a vessel visiting a port on a voyage. These include, for example, pilotage, towage, line handling and lights dues. A chartering term.

Port dues A charge raised on a vessel entering a port usually based on ships gross registered tonnage. Also dues raised on cargo and passengers.

Port efficiency indicators Statistics/indicators provided to measure the efficiency of the port in management and competitive terms. This includes number of TEUs handled per hour per berth; throughput of each berth; timescale of cargo passing through the port; labour cost and so on.

Porter A marketeer/academic who extolled the development of introducing the powerful notion of the value chain and narrowing the gap between the strategy domain of marketing and corporate planning.

Porterage Baggage-handling service, may be included in the price of a tour.

Port facilities The range of facilities found at a port designated primarily for the shipowner. These include bunkerage, pilotage, tugs, storage facilities, police, fire fighting equipment, cargo handling equipment, range of berths, distribution facilities, chemists, lighterage, shipyard and so on.

Portfolio A bank or investors loan and investment assets; or in insurance terms (a) the totality of the business of an insurer or reinsurer, and (b) a segment of the business of an insurer or a reinsurer, e.g. fire insurance business is called his fire portfolio.

Portfolio entry The mechanics of attachment of a reinsurance treaty at inception may be arranged on varying bases – to new and renewal business or business in force, etc – all referred to as Portfolio entry.

Portfolio management The process of managing a range of products.

Portfolio premium The premium payable for the reinsurance of a portfolio of insurance or reinsurance business.

Portfolio return The return of reinsurance premium and losses to the ceding company in respect of in-force business as of a specified date upon termination of a reinsurance treaty.

Portfolio run-off The opposite of Portfolio return – permitting premiums and losses in respect of in-force business to run to their normal expiration upon termination of a reinsurance treaty.

Portfolio transfer A procedure for transferring a long term insurance contract from one reinsurer to another, including outstanding losses.

Porthole Insulated container relying on external source of cold air; or window on port/starboard side of a ship.

Port hopping The process of a vessel sailing from one port to another usually along an extensive seaboard. It is widely practised by cruise liners and particularly evident in the Mediterranean and the Caribbean.

PORIM Palm Oil Research Institute of Malaysia.

Port infrastructure All the facilities provided at a seaport embracing berths, warehouses, labour, rail/road/canal networks, district parks, ship repair, victualling, customs, immigration, passenger terminal, security, navigation aids and so on.

Port interchange The provision of interchange transport facilities linking the port with other forms of transport such as rail, road, air and lighterage facilities.

Portion control The effort by commercial food services to provide equal-size portions to all customers.

PORLA Palm Oil Registration and Licensing Authority (Malaysia).

Port layout The layout of a particular port. This will include the berth layout, standage areas; rail and road access to the port; graving-dock; work-shops; transit sheds; bonded warehouse; bonded area; dangerous classified cargo area and so on.

Port liner terms charges A circumstance whereby the shipper pays the port handling charges. See also liner terms entry.

Port Manager A senior managerial position responsible for the overall management of a port. This will embrace operations, commercial, personnel, technical, statutory obligations, finance, marketing, navigation, and so on.

Port Management The technique/process of managing a competitive seaport whose usual aim is the maximum throughput of traffic at a low unit cost yielding a reasonable level of profitability in a safe port operational environment.

PORTNET The Port of Singapore code name given to its computer network – available to port users with on time computer equipment. It has an information data base for vessels, cargo, containers and shipping details and allows real time information and electronic processing 24 hours per day. It is very comprehensive and includes customs clearance, and many other disciplines and data resources.

Port of call The specified port of call on a sailing schedule.

Port of destination The destination port to which the goods are consigned or to which the vessel is sailing.

Port of discharge The port specified in a voyage/time charter or other document where cargo is to be discharged.

Port of loading The port specified in a voyage/time charter or other document where cargo is to be loaded.

Port of refuge Port where ships can take shelter, particularly from very severe weather conditions, hostilities etc. Basically a safe haven for ships and their crew/passengers/cargo. Overall, a port not included in the vessels passage plan.

Port of Registry Home port of a vessel at which place she is registered.

Port of safety Port to which a damaged vessel is taken for repair. See also port of refuge.

Port operation The operation of a port. This includes all the operating aspects of managing a port embracing berth allocation; access to and from the port; deployment of dock labour; cargo transhipment; security and so on.

Port Operations Officer A senior managerial person who is responsible for all port operation including especially ship movements, cargo transhipment; and cargo distribution to and from the port.

Port Operators Committee A Committee operative at a sea port formed of Port Users and Operators. Its purpose is to facilitate the effective management and operation of the port on an optimum basis and having regard to the port users/shippers/needs. Members of the Port Operators Committee include major Shippers, agents, stevedores, carriers, customs, security, shipbrokers and so on.

Port/Port The conveyance of a consignment/merchandise starting and finishing at the ship's side. Hence the inland movement prior to and subsequent to the sea carriage is effected by the merchant or his agent.

Port related industries Those industries situated in the port environs including ship repairs, victualling, container servicing/repairs, lawyers, shipbrokers, agents, stevedores, police, customs and so on.

Port rotation The sequence of ports a particular ship is scheduled to call.

Portside cabin A cabin/passenger berth usually on a cruise liner with a picture window or porthole.

Port State The National registry relative to a ship which is registered in a particular seaport.

Port State control The organisation/authority that plays a key role covering inspection and targeting ships which fail to comply with ISM Code. Hence the port state control checks whether visiting foreign ships comply with the requirements of international conventions on standards of safety, pollution prevention and crew conditions. Inspection is usually undertaken by the Port State control maritime government which has a maritime agency employing professionally qualified maritime personnel such as marine surveyor, marine superintendent etc, who carry out a ship inspection, examine all the records and check all the certificates/documents are valid. This extends to the crew and compliance with the STCW code. The code of practice is found in the Paris Memorandum of Understanding on Port State Control.

Port statistics Statistical data produced relevant to a port(s). This will include cargo tonnage; number of passengers; number of ships using port; average revenue yield per cargo ton and passenger; number of workforce; and so on. The statistics produced for the local port management tend to be more orientated towards measures of efficiency within the port such as average cost of handling cargo unit related to revenue yield; berth throughput; average turn round time of ships and so on. National statistics on Ports tend to be more towards total business volume, number of ships handled, size of dock labour force, cargo valuation imports and exports, and so on.

Port surcharge A charge raised by some port authorities on the shipowner when excessive port delays are experienced. Alternatively, it can be an additional charge raised by the Port Authority for a particular facility such as the provision of extra cargo handling equipment.

Port tariffs All the rates and charges raised at a port. This includes cargo handling; ship dues; cargo dues; hire of equipment; demurrage charges and so on.

Port tax A charge/tariff raised on passengers passing through a port and usually included in the overall passenger fare.

Port throughput The volume of traffic passing through a port during a specified period.

Port to Port The conveyance of the merchandise between the two specified ports.

Port Traffic control The process of controlling/regulating/managing a port primarily in operational terms. This includes port traffic management systems to assist in the safe and efficient flow of vessels traffic; operating and information service such as acceptance of bookings for pilots, ship movements, berth occupancy etc; a navigation information service dealing with the movement and safety of ships outside port limits; and domestic communications service involving telephones, radio networks, computerisation and so on.

Port turn round time The period taken by a vessel to discharge and/or load her cargo and/or passengers at a particular port.

Port user Companies/shippers who use the facilities at a seaport. See Port Operators Committee.

Port Users Committee A committee at a port whose prime purpose is to facilitate the smooth and cost effective running of the port for the benefit not only of its users, both existing and potential, but also the overall consideration of developing trade/industry throughout the port and its environs. Members of the committee would include representatives of Port Authority, Customs, Agents, Police, Shipowners, Immigration, Stevedores, Chamber of Commerce, Shippers Council and so on.

Positioning The movement of a vehicle on land, air or sea to place if for boarding and departing; or defining the location of a product relative to others and promoting it accordingly – marketing term.

Positioning in the market The process of identifying a product or service position in the competitive market place. It may be a market leader or a laggard.

Position media The media which occupies a permanent site in the market place environment such as a poster or neon sign/advertisement.

Positive space A confirmed reservation.

POSREP Position Report.

Post After.

Postal exports Parcels or packets despatched by post containing export merchandise.

Postal questionnaire A questionnaire which is despatched by post to the respondent. See questionnaire and field questionnaire.

Postconvention and preconvention tour An extension offered as a supplement to a convention.

Post delivery finance scheme The process of arranging payments after ship delivery from the ship yard.

Post delivery loan The tenor of the loan which may be 15 years.

Posted price Originally the phrase indicated the price at which US refiners were prepared to buy crude oil, but it now means the price at which products or crude oils are available for purchase – an IPE term.

Post entry The system of lodging details of consignments after shipment in the case of exports; or after clearance in the case of imports.

Poster advertising The process of advertising on a poster which may be situated alongside a railway line or roadway, at a railway or bus station, or any other site with a favourable viewing audience.

Poster display A site displaying a poster. See poster site entry.

Poster site A site on which is displayed a poster. This may be on a hoarding, a railway station, a bus terminal, an airport, a shopping precinct, and so on. Some posters are illuminated.

Post flight information Data on a flight completed which includes details of traffic, passenger list/cargo manifest; departure/arrival time; weather conditions, and so on.

Post invoicing The process of invoicing the buyer after the goods have been despatched.

Post pallet A pallet which has a post at each corner permitting one pallet to be stacked on another without damage to goods on the pallet below caused by the weight of the pallet above.

Post purchase All the arrangements undertaken after the actual sale has been completed. An example arises relative to the servicing of a car.

Post purchase advertising The process of conducting an advertising campaign after the goods have been purchased by clients with a view to increasing repeat sales and reassuring existing customers.

Post sailing information Data on a sailing completed which includes details of traffic/passenger list/cargo manifest; weather conditions; departure/arrival time and so on.

Post shipment finance The process of arranging payment of the goods after the date of shipment which would include bank overdrafts and loans, discounting of Bills under letters of credit, clean acceptance credits, counter trade, factoring, invoice discounting, forfeiting, leasing and so on.

Post shipment inspection The process of having the international consignment being inspected at destination in the importing country by an independent surveillance company such as SGS. Subsequently the company will issue a certificate of inspection to confirm the goods have been supplied/received in accordance with the contract. The importer will send the certificate of inspection to the exporter who may be able to claim some customs duty concession.

Post socket Deck socket with a circular shaped hole.

Post stacker Stacking fitting that can only be used at the base of a stack and which fits into a post socket.

Post visit debriefing (of an overseas visit) The process of communicating/informing personnel/departments within a Company/entity the outcome of an overseas business visit. It would be conducted when the executive returned and usually takes the form of a meeting and/or a report on the visit featuring an action plan.

Pot A sunken deck socket into which a lashing is connected.

Potassium metal liquid alloy This is a metal mixture or alloy of potassium and another metal, existing as a liquid at ordinary temperatures and which flows more or less readily depending upon composition. Avoid contact with moisture since such contact may cause the mixture to become ignited and burn. A term associated with dangerous classified cargo conveyance.

Potassium sodium alloys These are mixtures of metallic sodium and potassium that are solid at ordinary temperatures. All mixtures, regardless of physical state, will react vigorously with water and may be self-igniting. The mixtures are all combustible. A term associated with dangerous classified cargo conveyance.

Potassium sulphide A reddish-coloured solid having a strong odour. It is hygroscopic and oxidizes spontaneously on contact with air. Spontaneous ignition may occur in material improperly packed. A term associated with the conveyance of dangerous classified cargo.

Potash berth A berth specially constructed to tranship potash.

Pound 'spot' and 'forward' The price of sterling for immediate delivery and prices on contracts struck for settlement one month or three months.

Powder – smokeless Any propellant explosive based on nitrocellulose. A term associated with the conveyance of dangerous classified cargo.

Power of recourse Power to take a person/company to court.

Power pack A transportable electrical power station built into a 6.10m (20ft) ISO container. It is designed for use at sea in vessels with insufficient electrical power for refrigerated containers; or on land for container depots and ports to power refrigerated containers and cranes. It can also provide power on site for temporary housing or any other purpose.

Power steering System in which the effort required to steer a vehicle is reduced by the assistance of an hydraulic ram powered by a pump driven from the vehicle's engine. As the steering wheel is turned, the ram assists by pushing the steering arm.

Power take-off Auxiliary drive attached to the vehicle gearbox to direct engine power for working ancillary equipment such as hydraulic pumps for tipping gear, lorry-mounted cranes, tail-lifts, refrigeration units and tanker discharge pumps.

Power to weight ratio The relationship of the power of the engine of a transport unit to the overall gross weight as found for example in road vehicles.

POUCC Post Office Users Co-ordination Committee.

POY Polyester pre-oriented yarn.

pp Picked ports.

PPA Philippine Ports Authority; or Philippine Posts Authority; Process Plan Association (UK); or Pharmaceutical producers association (of Thailand).

PPB Policy holders protection board.

PPC Production Planning and Control (Sub-Committee: Handbook) – IATA term.

PPD Primary Production Department (Singapore).

Ppd Pre-paid.

PPF People's Police Force (Burma).

PPFC People's Pearl and Fishery Corporation (Burma).

PPI Postage Paid Impression; or plan position indicator – see separate entry.

ppi Policy, proof of interest.

PPIA Philippine Poultry Industry Association.

PPI Policy 'PPI' means 'Policy Proof of Interest'. The underwriter waives proof of insurable interest which must exist at the time of loss, but the policy is deemed void in the eyes of the law.

PPM Process and production method – WTO term.

ppm Parts per million.

PPO Polyphenylene oxide.

PPP Profit and performance planning; or Purchasing power parity.

Purchasing power parity Economic term. A statistical adjustment for cost of living differences by replacing normal exchange rates with rates designed to equalise the prices of a standard 'basket' of goods and services. It is used to obtain the GDP per head of population on a national basis with the USA as a 100.

PPPI Indonesian Fisheries Association.

PPS Precise positioning service; or Polyphenylene sulphide.

PPSC Petroleum Products Supply Corporation (Burma).

PPSE Propose.

ppt Prompt (loading).

PR Polski Rejester Statkow – Polish Register of Shipping; Public relations; primary reinsurer; prospective rating – insurance abbreviation; position report – maritime term; pro rata; portfolio return – insurance abbreviation; or plan and report.

PR Agency An Agency specialising in public relations work.

PRC People's Republic of China.

PRD The Public Relations Department (Thailand).

Preamble The introductory clause of an agreement/contract.

Pre-booking shipping document A document used to pre-book cargo on a specified sailing. This enables the shipper to be assured that cargo space has been reserved on the particular sailing and thereby enables the exporter to plan the distribution arrangements accordingly.

Pre-carriage The carriage of goods from the place of receipt to the place of loading into the main means of transport.

Pre-CIF All or some freight and insurance costs are not included in the invoice price. See CIF entry.

Predatory pricing Dramatic price cuts designed to force a rival out of the market.

Pre-delivery loan Loan agreement negotiations prior to delivery of vessel from shipyard.

Pre-entry Process of lodging with customs for scrutiny, the appropriate documentation giving details of consignments available for shipment which maybe by sea or air. The customs pre-entry in the case of import products will be before their arrival, whilst for export consignments before shipment. See period entry traders.

Pre-existing condition A physical condition that existed before the effective date of coverage.

Preference An agreement between the Community and a non-community country allows goods originating in that country to be imported at a preferential (often nil) rate of duty. The quantity of goods allowed to be imported at preferential rates may be restricted by tariff quotas or ceilings. Customs term.

Preference certificates A customs document permitting a claim to preference import duty level for a particular trader and product.

Preference goods Goods from certain non-community countries on which importers may claim lower or nil rate of customs duties.

Preference shares Shares which give their holders some preference in receiving their dividends, which are paid before the ordinary shareholders receive theirs. Preferential shareholders are also repaid their shares before the ordinary shareholders are repaid theirs should the company go into liquidation. Most preference shares pay a fixed dividend.

Preferred bidder The bidder most favoured by the buyer. See preferred supplier

Preferred port The port most favoured by the shipper or shipowner.

Preferred supplier The supplier most favoured by the buyer and accordingly tends not to be subject to any competitive tendering to obtain the business. Usually the preferred supplier has special qualities not available elsewhere in the market place and is so designated preferred suppliers after favourable experience/relationships have been built up between the buyer and supplier.

Pre-finance goods Common Agricultural Policy goods on which an export refund is paid in advance of the actual export (e.g. when the goods are put under the customs free zone regime for export).

Pre-financing A system whereby the export refund can be paid before export takes place, the goods being under customs control either in an approved warehouse, or for processing, and to be exported within a prescribed period. (Not to be confused with advance payment).

Pre-invoicing The process of invoicing the buyer prior to the goods being despatched.

Pre-loading inspection The process of inspecting/examining the goods/merchandise prior to shipment or loading onto the vehicle.

Pre-market Trading among brokers before the LME opens for ring-dealing. Pre-market trading can be very active, influenced by the hedging operations of producers and consumers. This is because the previous day's settlement price is still the valid producer price, usually until 11.30am or midday. Consumers, therefore, are matching their sales to customers with price fixing on the producer price and then hedging the out-of-balance metal position on the LME. See back-pricing. LME term.

Premium The amount paid to an insurer or reinsurer in consideration of his acceptance of risk. The broker is responsible for the payment of a marine insurance premium; in finance terms a security stands at a premium when the price is paid up or par value; or the non returnable payment made by the buyer to the seller – a BIFFEX term; or (i) the cost of buying an option; (ii) the amount by which the price for one delivery date is greater than that for another date; (iii) the difference in price between different refined products, e.g. between ingot and billet.

Premium additions The amounts payable to the Company if the exposure under a policy is increased or the rate required by the Insurer is increased.

Premium adjustment The amount due on the adjustable Policy as in the case of employers' liability, or where a deposit premium has been paid.

Premium Brands A British Airways Brand name given to the dedicated reservation desk to handle the carriage of valuable goods, livestock and human remains.

Premium earned The proportion of the total premium of a policy which refers to a particular period normally calculated pro rata.

Premium gross The amount of premium after reinsurance, but before commission is deducted.

Premium price A price above the recognised/normal price and usually emerges when goods are in short supply and demand is exceptionally heavy. This enables the retailer to place a 'mark up' on the goods at a level above the norm, associated with such a product or service.

Premium pricing The process of pricing a product/service in the market place at a level above its norm on the basis of its quality, scarcity, specification, design, technology, excessive demand or other reasons of a commercial nature. Overall, the product/service must have usually quality and design features above the mass market product range.

Premium reserve A provision in a reinsurance contract allowing the reassured to retain an agreed proportion of the premium against payment of claims.

Premiums closed A premium having been received in full for the period of insurance.

Premiums gross The amount of the premium before any deductions.

Premiums reducing An insurance covering the premium paid by a shipowner. The amount insured reduces by one twelfth each month during the period of twelve months covered.

Premiums reinsurance outwards Premiums payable by the company other than return premiums on original risk written either in respect of contracts written by or as retrocession's or original reinsurances accepted by it, also includes excess of loss premiums or similar protective premiums payable by the company to protect its own underwriting account. Excludes insurance premiums payable by the company in respect of xc motor vehicles or property it owns or third party risks it may suffer, etc, which are included elsewhere.

Premiums renewal The premium to renew an existing policy which is expiring.

Premiums returns The amount due to the insured if a policy is reduced by rate, exposure or cancelled.

Premiums revenue Net premiums shown in the long term business revenue account, including consideration for the granting of an annuity and premiums recoverable from the Inland Revenue; or premiums booked in a specified accounting period.

Premiums unearned The pro-rata amount of premium in respect of a policy that has not expired.

Premium transfer Syndicate accounting is on an annual basis but some long term insurance contracts run for more than twelve months. Where the premium paid in the first year relates in part to subsequent years, a transfer is arranged at the end of the year to carry forward the relevant premium to the subsequent underwriting year's account. The operation becomes a book entry only.

Pre-paid Paid in advance, e.g. before a service is given or before goods are handed over.

Pre-paid charges The process of entering charges on the Air Waybill/CMR/CIM consignment note for payment by the shipper.

Pre-paid ticket advice The notification from a carrier or a travel agent in one city asking a carrier in another city to issue a ticket for prepaid transportation to a specified person.

Pre-planning The process of preparing a course of action/sequence of events to counter a particular foreseen situation/circumstance. It maybe the code of action to be adopted in the event of industrial action being taken in a particular Company resulting in a much reduced frequency of sailings.

Pre-registration Assignment of room and filling out registration information before a guest's arrival; often used for convention, meeting and tour guests and sometimes involving a lower rate.

Presentational graphics Bitmapped graphic images which are used to present information such as pictures or electronic artwork, Elements of an image such as lines cannot be edited as an entity, i.e. moved and rotated etc. Computer term.

Presentation of documents The process of presenting the requisite documents to customs for clearance of the cargo, or to a specified Bank – provided all the terms of the credit have been complied with – to effect payment of the goods.

Present of Natural persons Individuals travelling from their own country to supply services in another such as fashion models or consultants officially. WTO term.

Present value The cash equivalent now of a sum of money receivable or payable at a stated future date at a specified rate.

Pre-shipment advice An abbreviated export declaration presented to Customs at the time of export.

Pre-shipment declaration (often called pre-entry) A full statistical export declaration presented to Customs with the goods before shipment. This may be in the form of a SAD or CHIEF.

Pre-shipment finance The process of arranging payment of the goods prior to the date of shipment involving bank overdrafts and loans, and progress payments under ECGD.

Pre-shipment inspection The process of having the international consignment awaiting shipment inspected by an independent surveillance company such as SGS. Subsequently the company will issue a certificate of inspection to confirm the goods have been supplied in accordance with the contract. This involves the practice of employing specialised private companies to check shipment details of goods ordered overseas – i.e., price, quantity, quality, etc. Used by over 65 developing countries many of which are mandatory.

Pre-sling Units of cargo which have the sling fitted prior to shipment and purposely left attached to assist in discharging the key pallet in the square.

Pre-slinging The process of assembling loose bundles cargo on the quayside/berth in slings prior to the commencement of loading the cargo onto the ship.

Press Campaign The process of a Company/Entity devising a press campaign for a particular purpose. It maybe to launch a new product. The campaign plan would be devised to facilitate the product launch involving product demonstrations, sponsorships, press conferences and so on.

Press Conference See Press Launch entry.

Press Launch The process of informing the press details of a particular new product(s)/service(s) etc, a company/entity has just launched thereby enabling the market to be made aware of is existence and encourage strong market support. It is usually in the form of a Press Conference to which journalists are invited to attend representing varying newspapers/journals and so on. Alternatively the Press Conference may be called for another reason such as company restructuring, new price structure, annual report results, or any other topic which will interest the market/media/readership.

Press Officer A managerial position in a particular company/entity whose primary role is to liaise with the press on all matters of press interest, especially to represent the company/entity image/policy etc, in the most favourable way practicable in accordance with their policy. In particular, to ensure good press relations are maintained and maximum publicity is gained of company policy, promotion and successes. The Press Officer will supply editorial material, press releases, and other material to gain maximum favourable impact in the market place through the press media.

PREST Scientific and technical research policy committee.

Pre-tension Additional load applied to a mooring line by a powered winch over and above that required to remove sag from the main run of the line.

Pre-testing The process of testing a product before a full scale test.

Preventative maintenance The process of devising a maintenance schedule for a particular item of equipment and adhering to it very closely. In so doing, the schedule devised will ensure the regular maintenance will maximise the operational life of the equipment on an economical basis and thereby strive to attain the highest practicable economic standards of reliability. The salient feature will be to ensure equipment components will be regularly, and not spasmodically maintained, and in situations where the component part shows signs of deterioration beyond economical repair/servicing, it will be replaced and not as in a sporadic maintenance schedule only be detected in terms of it malaise/failure when the equipment breaks down/becomes unserviceable. The prime benefits of preventative maintenance include equipment reliability; reduced cost of maintenance overall; extended economic/operational life of equipment and its component parts, and maximum utilisation/availability of the equipment with much reduced risk of breakdowns/unserviceability.

Price The perceived value of a product/service in the eyes of the buyer in a competitive market place. A significant factor is the value added benefit which accrues to the buyer in the light of the purchase made and this will vary by culture, country, buying power and the social-economic factors in the country.

Price ceiling The maximum price the goods/service may be sold.

Price convergence The process of prices moving towards parity on a comparison basis within, for example, the Eurozone.

Price demand elasticity The degree of sensitivity of the pricing mechanism whereby a reduction in a product/

service price yield more than a proportionate increase in demand. See elasticity.

Price discrimination The process of charging different prices to different market segments.

Price/earnings ratio This represents the current share price dividend by the last published earnings (expressed as pence per share). It is used as a measure of whether a share should be considered 'expensive'. Thus, a share selling at 50p with a P/E Ratio of 10 would be considered dearer than one selling at 100p with a P/E Ratio of 5.

Price Leader The company/entity/person/product/service which leads the market in terms of price.

Price mechanism A system whereby the market forces of supply and demand are permitted to work without outside intervention thereby allowing the individual buyers and sellers to make their own decision without intervention.

Price sensitive market A market where price is the dominant factor in consumer or industrial choice. It may be a product or service.

Price sensitive product/service See price sensitive market.

Price setting process The process/rationale of determining the price of a product/service in the market place.

Price skimming The strategy adopted by a company/entity to price a product/service at a high level and thereby generate high profits – usually in the short run. Its objective is to recover development cost.

Price undertaking Undertaking by an exporter to raise the export price of the product to avoid the possibility of an anti-dumping duty. A WTO term.

Price (used as a transitive verb) To fix the price of a purchase or sale on one or more LME quotations, usually the official settlement price for the metal concerned.

Price/value The price of a product/service and the correlation of its perceived value – especially the value added benefit – as judged by existing and potential buyers and the rationale.

Price variation clause A clause found in a contract which permits either party to vary the price – usually increase it – in accordance with an agreed criteria detailed in the contract document. For example, the contract may be valid for five years and each year the price is reviewed on the basis it would be varied in accordance with the index of retail prices.

Price war The circumstance in which there exists keen competition in price in the market place as each entity/manufacturer/retailer etc, strives to improve their market share of the product/service and in so doing improve revenue production/cash flow.

Pricing cutting The process of reducing the price of a product/service in order to gain increased volume sales and abstract business from competitor(s) thereby improving potentially the vendors market share performance. Also termed pricing down.

Pricing down The process of reducing the price of a product(s)/service(s).

Pricing fixing Buying known quantity, quality and shipment date with the price being determined later – a BIFFEX term.

Pricing-in When a merchant agrees to buy a quantity of metal from a consumer, both parties will agree in advance that the price basis will be the closing price of a specific ring or rings on the LME. The buyer will therefore be pricing-in on the close of that ring, and will probably hedge his purchase by selling during the closing stages of that ring.

Pricing-out The opposite of pricing in where the seller is selling a quantity of metal at the closing price and will, therefore, hedge his sale by buying during the closing stages of the ring.

Pricing penetration The process of conducting a pricing strategy which will penetrate the market and thereby increase volume sales and/or market share. See also market penetration entry.

Pricing research The process of conducting market research to determine its acceptability and, the role the price level influences the consumers choice of the product/service.

Pricing review The process of reviewing prices usually with a view to increasing them.

Pricing strategy The process of planning within an overall objective the price formulations with regard to product qualities/perceived benefits, specification, distribution cost, consumer needs, profitability, product cost, market demand, market share/forecast, buying power/behaviour, and so on. It will accord with the mission statement and company business plan.

Pricing up The process of increasing the price of a product(s)/service(s).

Prima Facie (Latin) At the first glance.

Primage Originally small payment to a Master made by shipper for safe conveyance and custody of cargo, but nowadays termed deferred rebate and granted under different conditions.

Primary balance Government net borrowing or net lending excluding interest payments on government liabilities.

Primary barrier This is the inner surface designed to contain the cargo when the cargo containment system includes a secondary barrier.

Primary capital markets Where a newly issued security is first offered. All subsequent trading of this security is done in the secondary market.

Primary data The process of collecting raw data from the market place involving the execution/collection/analysis from a research programme. It usually involves questionnaires, face to face interviews and so on which has been obtained through field research – the process of asking respondents to complete a questionnaire face to face, or by postal application.

Primary industry An industry involved in raw materials.

Primary key The attribute or attributes used as the primary and unique index key for an entity. Computer term.

Primary products Products which make up the bulk of an article.

Primary storage Storing data and instructions in a computer's ROM and RAM. See 'Storage devices". Computer term.

Prime entry A Custom House form used when the cargo is required for immediate use on payment of duty at importation. A statement of the export goods which is based on bill of lading details.

Prime interest rate The highest or top interest rate.

Prime listening time A period on radio which attracts the largest listening audience and on commercial radio attracts the highest rates for commercial advertising.

Prime rate The highest or top rate for a particular product(s) or service(s).

Primers – cap type Metal or plastic caps containing a small amount of primary (initiator) composition that is readily ignited by impact. They serve as igniting elements in small arms cartridges. A term associated with the conveyance of dangerous classified cargo.

Prime site A location possessing all the attributes which the discerning entrepreneur would seek to secure a site at the top end of the range. It is likely to command a premium price/rent/leasehold terms and would possess strong competitive advantage and potential.

Prime time A period on television or radio which attracts the largest viewing/listening audience and on commercial television radio attracts the highest rates for commercial advertising.

Prime viewing time A period on television which attracts the largest viewing audience and on commercial television attracts the highest rates for commercial advertising.

Principles of safe manning Contained in IMO resolution A890 it specifies the minimum safe manning of a ship taking into account all relevant factors embracing size and type of ship, number, size and type of main propulsion units and auxiliaries, construction and equipment of the ship, method of maintenance used; cargo and/or passengers to be carried; frequency of ports of call, length and nature of voyages to be undertaken; trading area(s), waters and operations in which the ship is involved; extent to which training activities are conducted on board; and applicable work hour limits and/or rest requirements.

Principal A primary producer of an element of travel – such as transportation, accommodations or a cruise – that pays a travel agent a commission to sell it; or a person or organisation able to make, and obliged to fulfil, all the conditions of a contract.

Principal carrier The carrier who issues a combined transport document regardless of whether or not goods are carried on his own or a consortium member's vessel.

Principals' market A futures market where the Ring dealing members act as principals (qv) for the transactions they conclude across the ring and with their clients.

Printer An output device producing characters or graphic symbols. Common types are impact dot matrix, ink jet and laser. Resolution varies from 100 dots per inch to 1200 dpi and speed from 100 characters per minute to tens of pages per minute. Computer term.

Printer server The computer which contains the printer server software for a LAN and controls the printer queue. See "Local area network". Computer term.

Print out Data obtained from a computer in tabulated form which maybe, for example, the effect of a ten per cent increase would yield on selected products, or the total sales, by commodity within the past six months.

Prior importation A customs facility which enables replacement goods to be imported before the displaced goods for repair are exported. It is linked to the Standard Exchange System which provides duty relief on goods imported as replacement of goods exported for repair.

Priority delivery A service offered by some carriers, especially airlines, whereby the consignment will receive priority delivery at destination airport/seaport/ICD/CFS.

Priority of booking (cargo) strategy A booking system devised by a shipowner which gives priority/preference to cargo booking on a specified sailing based on, for example, rate level, contractual obligations, nature of cargo, freight rate profitability, maritime government regulations (flag discrimination) urgency of shipment, ship type/capacity availability, and liner conference regulations. See filler traffic.

Priority list The process of devising a list of products/services/items in accordance with their importance/significance – the most important being at the top of the list.

Private Finance Initiative A government strategy adopted in some countries where private sourced finance are invited to invest in government funded projects. It maybe hospitals, motorways, schools, universities and so on.

Private key Any technically appropriate form, such as a continuation of numbers and/or letters which the parties may agree for securing the authenticity and integrity of transmission under the CMI rules for electronic Bills of Lading.

Private placing A sometimes quicker method of raising unlisted medium or long-term funds than a listed public issue, since only one or a limited number of investors will be involved.

Private sector That part of the economy not under the direct government control.

Private siding A railway siding situated on private property and usually serving an industrial activity. It would be connected to the main railway network who would be responsible for operating wagon/train loads to and from the private siding onto the National railway network.

Private wide area network A network which uses privately-leased or owned lines and does not offer the general public connection facilities, although it may provide network services to subscribers. Many of the well-known network service providers offer these facilities and enable access through gateways from other networks, e.g. Internet, JANET, CompuServe.

Privatisation The transfer of ownership from the state to the private sector. The opposite process is called nationalisation.

Prix fixe French term for the price for a specific complete meal.

PRO Public Relations Officer; Poop and raised quarter deck; or portfolio run-off – insurance abbreviation.

Proactive policy A policy which leads in its initiative and does not operate on a reactive basis.

Proactive strategy A strategy formulated/adopted by a company/entity which leads in its initiatives.

Probable maximum loss The maximum amount of loss that can be expected under normal circumstances.

Problem analysis A personal service designed to help management identify problems and act upon them appropriately.

Problem solving The process/technique of solving a particular problem by problem identity and analysis, and to devise optional solutions together with a recommended course of action. All available management techniques would be employed. See Problem Analysis.

Pro bono publico (Latin) For the public good.

PROBO's Product/ore/bulk/oil carriers.

Procedures for Port State Control Since the adoption of resolution A.466(XII), Procedures for Port State Control, in 1981, IMO has adopted a number of resolutions on the subject. In 1995, in recognition of the need for a single document to facilitate the work of maritime administrations in general, and Port State Control officers in particular, a comprehensive review resulted in the adoption of resolution A.787(19). This resolution provides basic guidance to Port State Control officers on the conduct of inspections, in order to promote consistency worldwide and to harmonise the criteria for deciding on deficiencies of a ship, its equipment and its crew, as well as the application of control procedures. Developments in the period since 1995 have prompted proposals for amendments to resolution A.787(19) – for example, to take the ISM Code and the 1969 Tonnage Convention into account, to provide for suspension of inspections and to define procedures for the rectification of deficiencies and release. After consideration by the relevant subcommittees of the Maritime Safety Committee and the Marine environment Protection Committee, amendments were finally adopted by the Assembly in 1999 by resolution A.882(21).

Production possibilities curve (transformation curve)

Processing Any operation which changes the condition of imported goods. This includes both the manufacture and assembly of goods. A customs definition.

Procurement The management and provision of a purchasing and supply service. Internationally this involves purchases on behalf of ministries, companies, public organisations and international agencies. This involves ordering, competitive pricing, technical support, financing the purchases and management.

Procurement cycle The various stages and timescale to complete a procurement. See procurement.

Procurement logistics Control of the flow of materials up to the manufacturing process.

Procurement malaise strategy The following points identify errors of judgement/mistakes made by importers to develop a viable procurement strategy. This focuses on particularly first time importers/procurement operators; failure to identify company objectives in business plans; no viable procurement plan; absence of researching potential overseas supplier; no feasibility study undertaken; failure to devise adequate product specification; wrong overseas representative/partner chosen; failure to screen the supplier; absence of professional in-house experience multilingual procurement personnel; failure to do a cost analysis of procurement strategy; absence of any payment cycle analysis/strategy; failure to attend trade fairs or feature on inward trade missions to evaluate potential suppliers; no price strategy and terms; lack of any product/cost/value added benefit of competitors and their procurement strategies; absence of any product sample and/or testing; absence of computer literacy and satellite communication resources; no logistics/supply chains/international distribution plan; wrong suppliers selected; absence of adequate in-house company resources; no commitment to procurement business; failure to meet/visit suppliers; absence of any import regulation evaluation; failure to examine legal and political constraints in the suppliers market; no risk analysis; no negotiating plan; no country/supplier credit evaluator; and no central system through budgetary management. See successful procurement strategy.

Procurement market A country/place where the buyer (importer) chooses to buy the product/service crossing international boundaries.

Procurement strategy See successful procurement strategy and procurement malaise strategy.

Product Form Time Charter issued by New York Produce Exchange.

Product acceptance The acceptance by the market place of a particular product.

Product analysis Any analysis of a particular product or range of products in the market place. For example, some six colour television receivers may be available in the market place and all produced by differing manufacturers. Accordingly an analysis may be made of all six to determine which provides the best value for money. This will involve product design, specification, cost, reliability, aesthetic appeal, durability, after sales service, size of screen, range of stations and so on.

Product awareness The degree to which a product/service is known in the market place by the potential consumer/buyer.

Product brochure A brochure promoting a particular product/service.

Product carrier Small tanker used to carry refined oil products from the refinery to the consumer. In many cases, four different grades of oil can be handled simultaneously.

Product chain All phases in the transformation or production process of one product.

Product contract The contents of a product.

Product core The major product of a company and its central/core business. See core business.

Product cost The cost of a finished product built up from its cost elements.

Product creation The process of creating/originating a product for a market.

Product design The design of a particular product.

Product development The process of developing a product for a particular Market(s). This will involve product composition; design of product; compatibility of product specification with any statutory provisions; outcome of any product research including test marketing; production cost and market price correlation; competition; market demand and so on. It may be for example a new kitchen range or inclusive tour for the leisure market.

Product development cycle The chain of events leading up to the birth/initiation of a new product embracing concept, mock up, prototype, reproduction batch and full production.

Product differentiation The degree to which consumers have the ability to distinguish between different products, especially its unique selling proposition. Overall, supplying modified versions of a basic product to different market segments.

Product driven company A company driven in terms of its strategy/management/development by product specification considerations especially the perceived value. See product marketing strategy.

Product driven market entry strategy The process of entering a market by product specification considerations. See product driven company.

Product evaluation The perceived benefits in value added terms the product/service offers to the client/consumer.

Product exposure A situation whereby a product has been promoted in the market place through the media thereby exposing all its features to the potential clients/buyers/competitors.

Product extension A marketing term involving developing/extending the product/service and thereby adding value to it. Examples include the television with the remote control.

Product hire pricing The process of calculating prices so that parity exists across a range of goods/services, even though there may be cost differences between products within the range.

Product image The general customer awareness for the product in the market place and the degree of goodwill which is derived therefrom.

Production capacity The production capacity of a factory/plant/entity/company based on the availability of specific resources including labour/materials/equipment/plant, etc.

Production cost The cost of producing/manufacturing a product/service.

Production department The department of a particular entity/manufacturer/exporter which undertakes the manufacture/production of the goods/merchandise.

Production platform An offshore platform involving the production of oil subterranean/sea bed.

Production possibilities curve (transformation curve) A curve showing all the alternative quantities of output which can be produced with fixed amounts of production factors. A curve indicating how the production of one good 'A', can by reducing the output of 'A' and transferring the resources used in the production of 'B'.

Production Test A term associated with exploring for offshore oil and gas. If an exploratory well strikes a suitable rock formation containing hydrocarbons, a production test may be carried out, using a controlled release of hydrocarbons to test the volume and quality. Excess hydrocarbon is burned or 'flared' off the rig.

Productivity The relationship between the quantity of an item produced and the resources used to produce it.

Productivity of world fleet This relates to an analysis/correlation of the world fleet (dwt), total cargo carried (tons), total ton-miles performed, tons of cargo carried per dwt, and ton miles performed per dwt. Such data usually available from UNCTAD. Overall, it measures the global ship management efficiency and/or operational productivity for the world fleet. See operational productivity of world fleet.

Product knowledge The knowledge a person has of the constituents of a particular product/service. For example, a successful professional marketeer must be completely informed of all the marketing techniques available together with the product involved such as a garden furniture range and market profile to ensure the successful promotion of the product.

Product launch The process of conducting a marketing campaign to launch a product or service.

Product liability All sellers of goods, whether they be manufacturers, repairers, intermediaries, for example, importers, exporters and wholesalers, retailer of suppliers of credit, may incur liability to their customers and others for injury, illness, loss or damage arising from the supply of goods.

Product liability insurance Insurance against the liability of a producer, supplier, tester or servicer of goods for injury to third parties or loss of or damage to their property caused by a deficiency in their goods.

Product life cycle The operational/commercial life of a product or service involving its launch growth and ultimate decline. This embraces introduction/development, growth, maturity, saturation and decline.

Product management The process of managing effectively on a viable basis a product in its development, research and information systems, promotion, pricing, planning, budgetary control and so on. A marketing term.

Product-mandating measure Requirement that the investor export to certain countries or regions. A WTO term.

Product market audit The process of evaluating the range of constituents of the products in the market place.

Product marketing The process/technique of marketing a product.

Product marketing strategy The strategy adopted by a company/entity to market a specific product/service.

Product mix The constituents of the product: design, specification, taste, warranty, price, servicing, packaging and so on.

Product mix transfer The process of transferring the product mix from one market to another as experienced by a domestic focused company moving into an export driven entity. See product mix.

Product/ore/bulk/oil carriers A bulk carrier vessel capable of conveying both clean and dirty oils including crude, as well as such dry cargoes as ore, coal, grain, alumina, and cement. Additionally a number of such vessels can convey caustic soda up to the full deadweight capacity of the ship. Furthermore, the holds and hatches can accommodate containers and other unitised cargoes.

Product organisation A company whose organisational structure is based on product responsibility. Each product group has primary responsibility for planning and controlling all activities for its products which maybe on a national or worldwide basis.

Product positioning The position of a product or service in the competitive market place. It may be a market leader or merely in the top ten featuring at number eight or nine annually in terms of volume sales each year. See also market leaders and laggards.

Product, price, place and promotion The marketing mix.

Product planning The process of evolving a plan for a product(s) in terms of sales, development, launch etc.

Product range The range of products in the market place. This will include for example all the types of washing machines being retailed in the market place.

Product research The process of researching into a product relative either to its market acceptance, adequacy, market image, customer evaluation analysis and so on, or to its composition and further technological design specification developments.

Products carriers A tanker designed to carry refined products such as gas, oil, aviation fuel, kerosene etc. Segregated loading, stowage and discharge facilities are provided to enable a number or separate parcels or types of products to be conveyed on any one voyage.

Product screening The process of evaluating a product/service embracing all its constituents. A form of market research. See vessel screening.

Product spec See product specification entry.

Product specification The specification of a particular product or service presented to the buyer by the vendor. The specification will provide fullest details about the product/service.

Product standard The specification of a product. In the wider competitive context the adequacy of the product in the market place especially as a saleable item on a growth basis. This includes product specification, durability, design and after sales service.

Product strategy The process of developing a course of action relative to a product. See strategy and strategic planning.

Product tanker An oil tanker designed to convey refined oil.

Product value The value of a particular product. This is usually the market value. It may be required for insurance or customs purposes in which case a commercial invoice could be requested to substantiate the product valuation. In transport terms it may be the product valuation at so much per ton such as $300 per ton involving a 30 ton bulk load.

Professional indemnity insurance The insurance of professional people against claims by clients or third parties for damages for error, negligence etc.

Professional reins A term used to designate a company whose business is confined solely to reinsurance and the peripheral services offered by a reinsurer to its customers as opposed to direct insurers who exchange reinsurance or operate reinsurance departments as adjuncts to their basic business of primary insurance.

Profile A drawing giving an elevation of a vessel indicating the location of decks, bulkheads, crew accommodation etc; information, such as travel preferences and typical methods of payment stored in a reservations systems computer on frequent travellers and other important customers: also known as a frequent travel file; or in social/economic terms the outline of a persons/company focusing on particular areas relevant to the evaluation/enquiry being undertaken.

Profitability ratios An accountancy term whereby it enables a company to evaluate its profit levels with regard to its size, and the effectiveness of its management. It embraces gross profit, operating profit, net profit before tax, net profit after tax and net profit attributable to ordinary shareholders.

Profit centre An organisation unit which will be held responsible for its own profit and losses.

Profit commission A commission based upon profit (as defined in an agreement). Two specific cases are: (a) the commission received by a Lloyd's underwriting agent as part of its remuneration from the Names; (b) the commission received by a cedant where business ceded to a particular reinsurer has been profitable.

Profit led A company/entity which operates on a strategy of profit motivation.

Profit repatriation The process of a company/entity transferring/repatriation its profits to the parent company. For example, a Bank may have separate offices worldwide and all the profit therefrom are transferred to the headquarters registered bank in Switzerland.

Pro forma invoice An invoice issued by the seller (exporter) endorsed "Pro forma' invoice and giving full description of goods, packing specification, price of goods with period of validity, packing details/cost, and where relevant cost of freight, insurance, and terms of payment. It is usually required for the letter of credit.

Program A complete set of program statements (instructions to the computer) structured to meet a given set of processing needs.

Program generator A program which expands simple statements into program code, enabling users to write their own programs.

Programmed deliveries Where production materials can be accurately determined well in advance, standing orders for regular deliveries from suppliers are placed.

Programme trading The automatic buying and selling of shares on the instruction of a computer.

Programming language Software which enables the production of computer programs. Each program is produced as code which must be translated into machine code for execution. There are a wide variety of such languages but the basic types are procedural, declarative and object-event.

Progressing Export Consignment The technique/stages/ criteria through which an export consignment processes from time quotation given to Importer until goods finally arrive. See Appendix 'A'.

Pro hac vice (Latin) For this occasion only.

Project Agent An agent involved in the project business. See project business.

Project cargo Quantity of goods connected to the same project and often carried in different movements and from various places.

Project Evaluation and Review Techniques A methodical planning technique used especially with major marketing plans.

Project Finance Process of funding projects which constitute major capital outlays such as provision for airports, hotels, seaports and highway systems. It usually involves government aid funds and development of bank funding often involving highly innovative mixed credit financings. Overall, finance arranged to appraise, initiate and execute a major capital project. Usually provided by a syndicate of financing institutions, it will be based on repayment from contractually assured project cash flows without recourse to the project participants.

Project Forwarding The despatch/conveyance arrangements which stem from a contract award. For example, a company in country A has a contract with a consortia in country B to build a factory which involves the importation of substantial quantities of merchandise, especially technical equipment. This requires much co-ordination involving buyer/seller in terms of despatch arrangements and site construction programme. Usually a Freight Forwarder undertakes the despatch arrangements such as a power station project.

Projectiles Any object such as shell or bullet, that is projected from a cannon (mortar, howitzer, or other artillery gun) rifle, or other small arm. A term associated with the conveyance of dangerous classified merchandise.

Project Management The process of having overall managerial responsibility for co-ordinating and controlling the many activities that contribute to the fulfilment of a project, ideally starting from the time it is conceived and evaluated, and continuing through all the phases of its execution. For the concept to be applied efficiently and effectively, there must be a thorough understanding by the project manager and his team members of its purpose and aims and also of the principles, structures, systems, and techniques involved. An example is a ship conversion assignment under a project manager.

Project Participants Insolvency Cover An insurance cover facility available for major contracts – for example, those in excess of £100 million – to cover losses due to insolvency of subcontractors or consortium members.

Project planning The process of formulating a plan to execute a project. An example emerges in project forwarding whereby the appointed Freight Forwarder organises all the despatch arrangements on behalf of his principal. It may be building a hospital, power station or mass transit system. This involves a major planning and logistic operation to construct the project with all the components/materials sourced overseas with the Freight Forwarder being responsible for the planning and operation of the despatch arrangements.

Projects business An overseas market which is involved in major projects. These arise in the areas of transport – a major road project; power generation – hydro electric scheme; communications – telephone system; petrochemical – oil refinery and so on.

PRO-LINE Pro-Line – a shipowner.

Prom dk Promenade deck.

Promissory note 'An unconditional promise in writing by one person to another signed by the maker (i.e. the importer), agreeing to pay on demand or at some fixed future date, a sum certain in money to a specified person or to the bearer of the note'. It differs from a Bill of Exchange in that a promissory note is written by the buyer to the seller rather than being drawn by the seller on the buyer. A promissory note provides a commitment to pay, as does the acceptance of a Bill of Exchange.

Promotion In marketing terms, the process of making known in the market place a product or service. It may be through advertising, direct selling, trade exhibition, trade fair, publicity and so on.

Promotional fare An air or other tariff offered at discounted levels to stimulate travel in slack times or for other reasons; usually roundtrip and restricted in one way or another.

Promotional freight rate Concessionary freight rate offered to a shipper by a shipping line or liner conference to facilitate the sale of goods into a new market.

Promotional mix The constituents involved in promotional strategy adopted by a company. It usually embraces advertising, publicity/press relations, personal selling and sales promotion embracing merchandising/brochures/below the line advertising.

Promotional mix transfer The process of transferring the promotional mix from one market to another such as a domestic focused company moving into an export driven company. See promotional mix.

Promotional sheets Promotional material featuring a particular product/service with a view to increasing its sales and market awareness. It is likely to be classified as a soft sell and may be used as a back up resource of a media campaign.

Promoter The initiator and developer of a new business venture, usually involving the forming of a new company and raising of capital by means of a private placing or flotation. The promoter's recompense – known as promotion money or simply the 'promote' – may take the form of fees, shares priced at a discount to the placing or flotation price, share options or warrants, or the right to participate in future profits.

Proof of citizenship Document that establishes the nationality of a traveller to the satisfaction of a government or carrier.

Proof of delivery A signed receipt showing time and date of delivery of the consignment.

Propane Hydrocarbon containing three carbon atoms, gaseous, at normal temperature, but generally stored and transported under pressure as a liquid. Used mainly for industrial purposes, and some domestic heating and cooking. A term used in the transportation of such merchandise relative to dangerous classified cargo.

Propcon Code name of charter party for timber trade.

Propellant liquid A substance consisting of a deflagrating liquid explosive, used for propulsion.

Propellant solid A substance consisting of a deflagrating solid explosive, used for propulsion.

Propeller A means of propelling a vessel through the water.

Proper shipping name The name to be used to describe a particular article or substance in all shipping documents and notifications and, where appropriate, on packages.

Proportional or pro-rata reinsurance Includes Quota Share, First Surplus, Second Surplus, and all other sharing forms of reinsurance where under the reinsurer participates pro-rata in all losses and in all premiums.

Proportional rate A domestic or regionally specified rate for the construction of international air freight rates. An IATA term.

Proportional reinsurance Reinsurance of part of original insurance premiums and losses being shared proportionately between Reinsurer and Insurer.

Proposer Person or Company who wishes to take out insurance. An insurance definition.

ProPRR Proportional or pro-rata reinsurance.

Propulsion Means by which a vessel is propelled or driven.

Pro-rata freight Freight charges on proportion of voyage completed.

Pro rata (Latin) In proportion.

Pro-rata reinsurance Proportional reinsurance.

Proration Division of a joint rate or charge between the carriers concerned on an agreed basis for an international consignment.

Proration mileage Proration on the basis of the respective in line route mileages between the carriers concerned for an international consignment. Such proration mileage will form the basis of the rate apportionment between carriers.

Proration rate Proration on the basis of the respective on line rates. Such rates apportionment between carrier for an international consignment maybe on the basis of mileage or some other acceptable method.

Prospective rating Determination of premium in advance from past experience whether on a fixed basis or on the basis of a stated rate to be applied to a variable exposure.

Prospectus A circular containing information about the borrower and the terms and conditions of the proposed loan issue. Also a document setting forth the main features of a proposed commercial undertaking, generally a new company for which equity investors are being sought. The terms of such a document are in most countries strictly controlled by company law and stock exchange requirements.

Pro tanto (Latin) To what extent: proportionately.

Protected A guarantee by a supplier or wholesaler to pay commission to a travel agent and full refunds to clients on prepaid, confirmed bookings regardless of subsequent cancellation by the supplier; see guaranteed tour.

Protection and Indemnity Club A mutual association of shipowners to protect themselves against those risks for which they are not covered under ordinary marine insurance. In particular third party liability which is not covered by the usual hull and cargo policies obtained in the marine insurance markets at Lloyd's or elsewhere. These include personal injury to passengers and crew damage to or loss of cargo, claims arising from another ship or other object, removal of wreck and so on. Also known as P&I Clubs.

Protection and indemnity risks Certain risks have in the past not always been readily insurable in the marine insurance market and shipowners have formed mutual associations (clubs) to cover them. Protection risks include quarantine expenses, liabilities to crews, and for collision and impact damage, wreck removal expenses and liability under towage contracts, also liability for damage to other vessels apart from collision etc. Indemnity risks include shipowners' liability for cargo lost or damaged, fines for inadvertent immigration or customs offences etc.

Protectionism A technique adopted by a country or state to protect their trade which is the converse of the open market trading principles. Essentially protectionism relates to the imposition by governments of measures which restrict or distort trade. This involves measures to protect a domestic industry from competition from imports, or to give its exports an artificial advantage such as a subsidy.

Protectionism 'offshore' The process/circumstance of a particular country adopting/practising 'offshore' protectionism/relative to favouring their national fleet. Examples arise through Governments favouring their national tonnage for National based companies.

Protest A sworn statement by crew members of voyage particulars involving bad weather or accident encountered and any emergency measures taken by the Master. The protest is made before a Notary Public or Consul whenever possible – a shipping definition; or in banking terms the legal procedure noting the refusal of the drawee to accept a draft (protest for non-acceptance) or to pay (protest for non-payment). It is essential in order to preserve – recourse on the endorser – endorsement.

Protocol Original draft of diplomatic document, especially of terms of treaty agreed to in conference and signed by the parties; or it embraces the manner in which a person presents himself. It extends to dress code, mannerisms, language, body language, propriety, discretion, non verbal message, codes of behaviour when conducting negotiations and the decision making process. See culture briefing and Protocol analysis.

Protocol analysis Process of analysing a person/service protocol and what tends to influence them to formulate the protocol constituents such as language, dress, body language, greetings, entertainment etc.

Protocol of Accession The legal document that establishes the rights and obligations of a country when it joins the WTO.

Protocol of Provisional Application The Protocol, agreed at the time of the establishment of the GATT, includes the 'grandfather clause' which requires countries to apply the provisions in Part Two of the GATT (Articles III to XXIII) to the fullest extent not inconsistent with legislation existing at the time when the acceding country becomes a signatory to the GATT now designated WTO. The Protocol of Accession (see above) also includes this grandfather clause.

Protocol on Preparedness, Response and Co-operation to Pollution Incidents by Hazardous and Noxious Substances 2000 See OPRC-HNS Protocol.

Provisional rate, premium or commission Tentative figures for subsequent adjustment under a reinsurance retrospective rating plan.

Provisions The money a business sets aside to cover bad and doubtful debts. In recent years, substantial provision have been made by banks against outstanding loans to third world countries and to domestic borrowers. Financial definition.

Provisions and bunkering The process of providing at a port provisions (Food), goods and fuel on board a ship for a particular voyage.

Proviso A conditional clause in any legal document, the observance of which its validity depends.

Prow See bow.

Proximate cause Proximate cause is the original cause in an unbroken chain of events which gives rise to a claim.

Proximo (Latin) In or of next month.

PRPorC Provisional rate, premium or commission.

PRS Poliski Rejestr Statkow – Polish Ship Classification Society.

PRS(PR) Position Reference system – employed with Dynamic Position operations.

PRT Petroleum Revenue Tax.

P/S Public sale.

PS Paddle steamer; private siding (rail); Personal secretary; port and starboard thrusters; or Air Ukraine International – an IATA Member Airline.

PSA Philippine Shipbuilders Association; Product Standards Agency (Philippines); Port of Singapore Authority or Pre-Shipment Advice – a general term encompassing LCP 'notifications' and SDP 'incomplete declarations' sent to CHIEF under NES simplified procedure. A customs term.

PSAC Policy Signing and Account Centre – insurance abbreviation.

PSAI Philippine Shipowners Association Inc.

PSB Philippine Shippers' Bureau.

PSBR Public Service Borrowing Requirement.

PSC Passenger Services Conference – IATA term; Philippine Shippers Council; Price Stabilisation Council (Philippines); Public Service commission (Singapore); port state control; or Pakistan Shippers' Council.

PSDF Parcel size distribution function.

P/Sec Personal Secretary.

Pseudo-PNR Reservation or other information stored in an airline reservations system that does not include an air booking; for example a package tour, car rental or travel insurance.

PSF Polyester staple fibre.

psf Per square foot.

PSI Pounds per square inch; or preshipment inspection – the practice of employing specialised private companies to check shipment details of goods ordered overseas – i.e. price, quantity, quality, etc – WTO definition. See separate entry.

PSR Postal Services representative.

PSSCG Passenger Ship and Special Craft Group.

PSTN Public switching telephone network.

PSV Public Service Vehicle; or platform supply (ship).

PSWA Partially smooth water area.

Psychographics The strategy adopted to identify/segment markets using psychological criteria to distinguish between the different segments.

Psychological price The price that matches the perceived value of the product/service in the judgement of the customer.

PT Pallet trucks; or Portugal – customs abbreviation.

pt Private terms; or Part (of a deck).

PTA Prepaid Ticket Advice – IATA term; purified terephthalate acid; Petroleum Authority of Thailand; or Preferential Trade Area (for Eastern and Southern Africa).

ptB Part Bunkers.

PtC Part cargo tanks.

PTC Posts and Telecommunications Corporation (Burma).

PTD Post and Telegraph Department.

Pte Private Company (Singapore).

PTL Partial total loss.

PTM Passenger traffic manager.

PTON Per ton.

P to P Port to Port.

PTP State Owned Plantation (Indonesia).

PTS Polar Track Structure – IATA term.

PTT Petroleum Authority of Thailand; or Postal, Telephone and Telegraph (Administrations – various countries).

Pty Proprietary company (Singapore).

PU PLUNA – Primeras Lìneas Uruguayas de Navegaciôn Aérea – an IATA Member Airline.

PUB Public Utilities Board (Singapore).

Public awareness The degree to which a product/service is known by the public at large in the market place.

Public charter The predominant form of vacation charter travel in the US airline industry; the rules allow US and foreign carriers and tour operators to offer public charters directly or through travel agents, with or without inclusive land packages, no purchase or duration requirements, but rigid consumer protection, refund, bonding and escrow provisions.

Publicity The process of making the market place aware of an entity/product/services. This involves the media including press releases, press conferences, articles for the press in company products/services, sponsorship, videos, in house publications, open days and so on.

Public relations The deliberate, planned and sustained effort to establish and maintain mutual understanding between an organisation and its public. An Institute of Public Relations definition. See also publicity.

Public Relations Officer An executive in an entity responsible for developing, initiating and executing the public relations activities in accordance with Company strategy. See also Publicity.

Public Service Vehicle(s) A bus or coach vehicle(s).

Public service vehicle/dormobile A passenger service carrying vehicle of two axle design with a seating capacity for up to fifteen persons.

Public utilities Companies providing a public service such as gas, electricity, transport and so on.

Public wide area network A network which is generally intended to be accessible to the general public, e.g. telephone and cable TV networks which are wired up to many homes and may be used in a variety of ways.

Published charge The tariff which is specifically detailed in the carriers or agent public tariff.

Published fare The fare which is specifically detailed in the carriers or agents public tariff.

Published rate A rate, the amount of which is specifically detailed in the carrier's or agents public tariff.

Published research The process of publishing a completed research assignment which maybe for example an enquiry into the consumer profile of customers travelling by air between London and New York. This maybe of age group and sex, salary range, purpose of journey, travelling alone or with a family, type of ticket and so on.

Puddling A chartering term. See sweeping.

Pull distribution system A system to provide warehouses with new stock on request of the warehouse management.

Pullman Railroad sleeping and parlour car in the US.

Pulpboard Board manufactured in one separate thickness or by bringing two or more thicknesses of the same board or paper into a single structure.

Pumping log An oil tanker record of 24 hour pumping.

Pump room Area where cargo pumps and related equipment are located on a tanker. Protected by oil-tight bulk heads, the pump room is normally situated between cargo tanks and machinery places.

Punctuality The process of a transport unit departing and arriving on time according to the schedule.

Punctuality analysis The process on a particular route/service/period of analysing/evaluating the punctuality performance of a transport mode. This will establish the number of trains/coaches/buses, airliners arriving 'right time' and those within five minutes and at other convenient periods/intervals. The analysis will feature the reason for the lack of punctuality including weather conditions; railway signal failure; traffic congestion; traffic diversion; shortage of crews/staff; heavy traffic volume; accident, and so on.

Punctuality performance The result of a monitoring exercise to measure the effectiveness the punctuality performance of a particular transport mode covering a particular period/route(s)/service(s) and so on. See also punctuality analysis.

Puncturing The process of a package being pierced.

Punta del Este Resort in Uruguay at which the Uruguay eight round of GATT negotiations commenced.

Purchase The various means of elevating or traversing the hook – i.e. crane, derrick, gantry, sheer-leg.

Purchasing power The ability/financial resourcefulness of a market to make the desired purchases. This involves Companies, sections of the populace such as social economic categories, specific geographical areas, and so on. See also buying power entry.

Pure car carrier's A purpose built vessel designed for the conveyance of cars with a capacity of up to 6,000 cars.

Pure Car Truck Carrier's A purpose built vessel designed for the conveyance of cars, Ro/Ro vehicles with a capacity of up to 6,000 cars/Ro/Ro vehicles.

Purpose built A product such as a transport unit built for a particular specified product market rather than for a general or multi purpose market operation. Examples are found in a road milk tanker, railway cement wagon. The prime advantage of such purpose built transport units is lower operational cost and much improved payload, but in times of recession the problem arises of alternative employment for such units which usually proves very difficult to find, other than through expensive modification of the transport unit.

Purpose built tonnage A ship designed and built for a particular trade or product such as a sugar carrier, chemical carrier, and so on. Usually such vessels cannot be employed to convey any other cargo which tends to make them inflexible in their operation particularly in periods of trade recession.

Purpose built vehicle A road vehicle designed and built for a particular product conveyance such as a road oil tanker, bulk cement carrier, milk tanker and so on. The vehicle maybe of a rigid or articulated construction.

Purser Officer in charge of ships passenger administration/welfare sometimes including catering.

PUS Past us (commissions).

Push distribution system A system to provide warehouses with new stock upon decision of the supplier of the goods.

Push/pull strategy Basically it is the two options available to an entity to move its products through a distribution channel. The push strategy involves trade and sales force incentives such as cash discounts, direct mail shots, credit facilities, competitions – trade and sales force, exhibitions and demonstrations. Alternatively the pull strategy is based on consumer incentives. This includes consumer competitions, free samples available offered in store demonstrations, packaging, sponsorships, price reductions by the manufacturer, and so on. Overall, the promotion strategy involves extensive advertising and consumer promotion.

Push-tow barges A push-tow barge is a modern form of dumb barge which is pushed by a push-tug (a tow-boat in the USA). Rigid couplings exist between the push-tug and barges, as opposed to the line connections between tugs and conventional dumb barges.

Put option An option that gives the holder the right (but not the obligation) to sell a specified quantity of the underlying instrument at a fixed price, on or before a specified date. The writer has the option to take delivery of the underlying instrument if the option is exercised by the holder. A put option is bought on the expectation of a fall in prices. An International Petroleum Exchange term.

PV Pressure/vacuum relief valve (of a ship); Petite Vitesse – slow goods train; or Peru.

pvc Price variation clause. A clause found in a contract permitting the parties to vary the price at a certain date and in specified circumstances in the life of the contract. For example, the contract may permit on 1st October each year a road haulier to impose on his client a rate increase based on the level of increase found in the index or retail prices of the previous twelve months.

PVP Payment-Versus-Payment: a mechanism ensuring that final payment of one sum takes place only if the final transfer of the other sum occurs simultaneously.

PWB Permanent water ballast.

PWD Public Works Department (Singapore) (Brunei).

PWEMA Philippine Electric Wire Manufacturers Association.

(P) with date Date when progressive survey took place on a specified vessel.

PWWA The Provincial Water Works Authority (Thailand).

PX Air Niugini – an IATA Member Airline.

PY Surinam Airways – an IATA Member Airline.

PYA Prior year adjustments – insurance abbreviation.

Pyas The currency of Burma. See Appendix 'C'.

PYE Previous year experience or equivalent – insurance abbreviation.

PYN Prior year now – insurance abbreviation.

Pyrometer Instrument for measuring very high temperatures.

Pyrophoric liquid A liquid which may ignite spontaneously when exposed to air the temperature of which is 55°C (130°F) or below. A term used in the transportation of such merchandise relative to dangerous classified cargo.

Pyrotechnic substance A mixture or compound designed to produce an effect by heat, light, sound, gas, or smoke, or a combination of these, as the result of non-detonative, self sustaining, exothermic chemical reactions. A term used in the transportation of such merchandise relative to dangerous classified cargo.

Pyroxylin solution It consists of pyroxylin (nitrocellulose) or soluble cotton dissolved in amyl acetate or other organic solvents. Pyroxylin solution is used as a basis for the manufacture of lacquer, leather coating compounds, leather substitutes, cements, etc. It is generally more viscous than ordinary lacquers. A term used in the transportation of such merchandise relative to dangerous classified cargo.

PYT Payment.

PZ Transportes Aereos del Mercosur (TAM – Transportes aéreas del Mercosur) – an IATA member Airline.

Q3 Zambian Airways (Zambian Airways Ltd) – IATA associate member airline.
3Q³ China Yunnan Airlines – IATA member airline.
Q7 Sobelair SA – IATA member airline.
Q Quadruple screw (propeller); raised quarter deck; or quadruple expansion engine.
QA Qantas Airline – an Australian international airline; or Quality assured.
Qafcobill Code name of BIMCO approved bill of lading used with shipments under Qafcocharter party.
Qafcocharter Code name of BIMCO approved fertilizer voyage charter party.
Qantas Airline An Australian international airline and member of IATA.
Qatar Riyal The currency of Qatar. See Appendix 'C'.
Qatar Riyal invoice An invoice payable in Qatar Riyals currency relative to an export sale contract.
QBE Query by example (relational). A simplified way of entering SQL queries, enabling user to enter the query using a menu or keystroke sequence which is automatically converted into an SQL command. Computer term.
QBO Quasi-biennial oscillation – see separate entry.
QC Air Zaire – an IATA airline; or Quality Control.
Qco Quantity at captains' option.
QD Quadruple screw.
QE Quadruple expansion.
Qe (quod est) (Latin) Which is.
QED Quicker Export Documentation.
Qed (quod erat demonstrandum) (Latin) Which was the thing to be proved.
QF Qantas Airways Ltd – an IATA Member Airline.
QGC Quayside gantry crane.
Qindarka The Albanian currency. See Appendix 'C'.
Qindarka invoice An invoice payable in Albanian Qindarkas currency relative to an export sales contract.
QM Air Malawi Ltd – an IATA airline.
QN Quintuple screw; or Air Queensland Ltd – an IATA airline.
qn Quotation.
QO Aeromexpress (Aeromexpress SA de CV) – IATA member airline.
QPC Quasi-propulsive coefficient.
QR Qatar Airways (Qatar Airways Co (WLL)) – an IATA member Airline.
QRR Quarterly Reserve Review.
QRs Quantitative restrictions – specific limits on the quantity or value of goods that can be imported (or exported) during a specific time period. A WTO term.
QS Quota Share.
'Q' sailing A non-published sailing which will only be operated when traffic conditions warrant it, such as to cope with the seasonal traffic volume or to provide relief/duplicate sailings.
QSCS Quality system Certification Scheme.
QT Qatar.
QTE Quote.
QTEV Quadruple turbo-electric vessel.
Qty Quality.
Quad Room suitable for occupancy by four persons; or Canada, EC, Japan and the United States.
Quadruple stacker A stacking fitting connecting four corner castings – each belonging to different containers.
Qualifying Category Codes An alphabetical coding system to indicate to the data processing services the category of an entry (e.g. Premium, claim etc).

Qualitative research Often referred to as soft data, qualitative research represents the applied behavioural science data and embraces attitudes and motivations.
Qualitative survey See qualitative research.
Quality (ISO 8402) The totality of features and characteristic of a product or service that bears on its ability to satisfy stated or implied needs.
Quality assurance (ISO 8402) All those planned and systematic actions necessary to provide adequate confidence that a product or service will satisfy given requirements for quality.
Quality certificate A certificate required by the importer in some countries confirming the quality/specification of the product is in accordance with the export sale contract at the time of shipment. It is often issued by SGS.
Quality control A process of monitoring a transit(s) or service performance or critically examining a production line product on a regular or sporadic basis to establish its adequacy/standard with a view to improving the service/quality in accordance with the market need. Several production samples selected at random must be tested in accordance with the requirements and frequency described for each type of package. If the requirements are not met, the whole production lot must be rejected unless otherwise specified. Quality control is becoming increasingly important in the realm of mass production technology and transportation. This is to ensure standards are maintained and no malpractices, particularly the labour force, develop to the detriment of the product competitiveness in the market place.
Quality control (ISO 8402) The operational techniques and activities that are used to fulfil requirements for quality.
Quality of service The process of providing a transport service which is reliable, and free of any inadequacies such as poor time keeping, low quality transport units, indifferent staff and so on.
Quality Service Management Managerial personnel responsible for monitoring/controlling/maintaining a level of standard acceptable within the criteria set. This involves BS 5750 or ISO 9000.
Quality System Certificate scheme Introduced by the ICAS it is based on the requirements of ISO 9001 and sets rigorous standards to create and maintain uniformity and consistent quality of Member Societies internal operations. It covers classification of ships and offshore installations in respect of design, construction and operation, together with statutory work carried out on behalf of National Maritime Administrations. The Society Certificate of Conformity is valid for three years and subject to intermediate audits.
Quantitative research Often referred to as hard data, quantitative research represents the numeracy data. This embraces size of market, production volume, readership levels and all data which can be expressed in a tabulated form.
Quantitative restrictions (QRs) Specific limits on the quantity or value of goods that can be imported (or exported) during a specific time period. A WTO definition.
Quantitative survey See quantitative research.
Quantity For packaging tests, quantity means a certain number of random samples, or in certain cases all of each

individual type and size, out of a lot or out of a production series.

Quantity Data Entry Numerical data entry which can be used as a quantity for calculation.

Quantity discount (1) A discount allowed for buying in bulk.

Quantity discount (2) A percentage reduction of a rate based on quantity either on a specified consignment or contracted period usually covering one year.

Quantum A portion, par or amount.

Quantum meruit (Latin) As much as he has merited or earned or it is worth.

Quarantine A state of forced isolation – often to establish that something is not contaminated or infected – as for live animals being imported which may carry disease.

Quarantine dues Dues/charges raised by an airport or seaport relative to retention of livestock for a particular period to conform with the appropriate quarantine regulations.

Quarter deck The upper deck on a raised basis situated at the aft end of the vessel.

Quasi-biennial oscillation The alternation of easterly and westerly winds in the equatorial stratosphere with an interval between successive corresponding maxima of 20 to 36 months. Each new regime starts above 30 km and propagates downwards at about one kilometre a month.

Quay The location in a seaport at which cargo arrives or departs. A dock for unloading ships or vessels. A type of wharf running at an angle with the shore line of the body of water. Also termed a pier.

Quay apron Portion of a wharf or pier lying between the waterfront edge and the transit shed. Overall, it is that portion of the wharf carried on piles beyond the solid fill.

Quay dues Dues raised by a port authority on cargo handled/passing through a berth. In Hamburg, for example, it is based on two elements: tonnage dues and weight dues.

Quay fitting The technique of inserting twistlocks upside down into the base of a container before loading into the ship.

Quay to Quay The basis of which some freight rate quotations are given and in so doing it excludes any port dues, handling cost and other port charges.

Queen room A hotel room with a queen-sized bed.

Queens or Kings enemies Enemies of the shipowners sovereign – a common law exception.

Questionnaire A list of questions prepared to elicit data from the respondents which may be for example, enquiring from customers their views on a new washing powder, or new/existing shipping/airline service.

Queue system Computerised tickler file that is part of an automated reservations system and can provide such information as waitlisted clients, bookings affected by fare or schedule changes and ticket-printing reminders.

Quetzal The currency of Guatemala. See Appendix 'C'.

Quetzal invoice An invoice payable in the Quetzal currency relative to an export sales contract.

Quick assets Assets that can be converted into cash quickly.

Quick release tensioner Tensioner designed primarily for use with wire lashings. Also termed overcentre tensioner.

Quick tite A patented tensioner used in containerisation.

Quid pro quo (Latin) One thing for another.

Qui facit per alium Facit per se (or facit ipse) (Latin) He who does a thing through another does it himself.

Quintal One hundred pounds.

Quo ad hoc (Latin) To this extent, as far as this.

Quod vide (Latin) Which see.

Quoin Shaped timber wedge used to secure barrels against movement.

Quondam (Latin) Former – erstwhile.

Quota Quantitative restrictions sometimes imposed on individual supplier countries. When the quota is exhausted, no additional imports are admitted. In some situations additional imports are permitted but at a higher tariff rate. Overall, a mechanism for restricting the amount of goods imported into a country which may benefit from a preference.

Quota ceilings The maximum volume/value a particular country restricts/limits the importation of a particular product or range of products.

Quota sample A group chosen to represent the views of a larger group. A statistical sampling technique.

Quota share reinsurance A form of pro rata reinsurance. Treaty reinsurance providing that the reinsurer shall accept a certain percentage of all or certain classes of or parts of the business of the reinsured.

Quota system A situation whereby a particular country limits the volume/value of imports of a particular product or range of products.

Quotation The price of a product(s) or service(s); or a price made in securities by market makers. The term has come to imply that the security has been officially approved. Overall, the method used to express the price of a contract for trading purposes – an IPE definition.

Quotation expiration date The date as from which a quotation is no longer valid.

Quote A price estimate given to the potential consumer as he/she decides to which company a formal application will be submitted. Company may be legally bound to honour this quote in some jurisdictions and/or lines of business.

Qv (quod vide) (Latin) Which see.

QY European Air Transport – an IATA Airline Member.

QZ Zambia Airways Corporation Ltd – an IATA Airline Member.

R3 Armenian Airlines – an IATA Member Airline.
7R Red Sea Air (Red Sea Air Plc).
R22 R22 is a refrigerant gas whose full chemical name is monochlorodifluoromethane and whose chemical formula is $CHClF_2$. It is colourless, odourless and non-flammable. It is virtually non-toxic with a TLV of 1,000 ppm. Its relatively low toxicity and flammability levels render it suitable for use on gas carriers and is approved for such use under the IGC Code. Other refrigerant gases listed in the IGG Code are shown in Appendix 2, although many are now controlled with a view to being phased out under the Montreal Protocol (1987). Dangerous cargo associated term.
r Refrigerated containers.
R Refrigerated; reconditioned; riveted; reduce class rate (rate classification); retrocession – insurance term; or retrocedent – an insurance definition.
(R) Refrigeration.
R/A Refer to acceptor.
RA Regulatory amount. This is a charge applied to certain tariff subheadings in the wine sector from Spain. Customs term.
RAA Reinsurance Association of America.
R & CC Riots and Civil Commotions.
RAC Rules of the Air and Air Traffic Services (ICAO term).
Racking The deformation of the container and frame in shear, or similar cross-sectional deformation in other aspects such as ships.
Rack price An American term meaning the price quoted by refiners to Jobbers for truck loads picked up at the refinery or terminal – an IPE term.
Rack rate The regular, public rate for a hotel room.
R & D Research and Development.
RAD Raised after deck or radar.
Radar The use of radio waves or microwaves to measure the distance of objects measuring the time taken for a pulse of radiation to travel from the transmitter to the object and back to an adjacent receiver.
Radar altimeter An instrument used on satellites to measure the height of the Earth's surface, including wave heights, the profile of currents and other large scale features of the oceans.
Radiation level (Radioactive Material Only) The corresponding dose-equivalent rate expressed in millisieverts per hour (previously in millirem per hour). A term associated with dangerous cargo shipments.
Radiatively active trace gases Gases present in small quantities in the atmosphere, that absorb incoming solar radiation or outgoing infrared radiation, thus affecting the vertical temperature profile of the atmosphere. These gases include water vapour, carbon dioxide, methane, nitrous oxide, chlorofluorocarbons, and ozone.
Radioactive contents (radioactive material only) The radioactive material together with any contaminated solids, liquids and gases within the package. A term associated with dangerous cargo shipments.
Radioactive materials (radioactive material only) Any material having a specific activity greater than 70 kBq/kg (0.002 µCi/g). A term associated with dangerous cargo.
Radio data terminal See RDT.
Radio hand held terminal See RHT.
Radio log book The Radio Officer's ships diary.
Radio meter An instrument which makes quantitative measurements of the amount of electromagnetic radiation falling on it in a specified wavelength interval: widely used in satellite systems to measure the properties of the atmosphere oceans and the Earth's surface.
Radio officer Ship's officer responsible for radio transmission and maintenance.
Radio-sonde A free balloon carrying instrument which transmits measurements of temperature, pressure and humidity to ground by radiotelegraphy as it rises through the atmosphere.
RAF Royal Air Force
RAFC Regional Area Forecast centre – IATA term.
Rafting Pressure processes whereby one piece of ice overrides another. Most common in new and young ice (cf finger rafting).
Rai Unit of area (2.23 Rai = 1 acre = 4,048 square meters).
Rail air link A train schedule providing a through connecting service with a passenger airport. The railway station usually forms an integral part of the airport. Examples include London (Victoria) to Gatwick, and London (Paddington) to London (Heathrow) Airport.
Rail box operator A container operator specialising in the movement of containers by rail usually on an intermodal basis.
Rail car A railway wagon used for carrying cargo – an American term; or railway wagons which are purpose built to convey cars – automobiles.
Rail Guided tour A facility offered by a railway company whereby a particular rail excursion is advertised and at various stopping points en route – maybe two – a local guided tour is provided. This may involve a coach trip. Throughout the journey promotion, hostess/tour guides are provided to inform passengers of places of interest.
Railhead Terminal which acts as a collection and delivery point by road for rail borne traffic.
Railrover A ticket permitting unlimited rail travel in a specified area or region for one week or other stated period.
Rail tour A usually scenic rail tour with particular areas of attraction to the passenger (tourist) and undertaken on a guided tour basis. It may embrace both rail and coach journeys to produce the most attractive tour to the market.
Railway consignment note The railway consignment used for international transport of goods by rail and usually referred to as the CIM consignment note. It is not a document of title and is not negotiable.
Railway timetable A timetable giving details of train services featuring stations served en route.
Raised deck socket A deck socket projecting above the ship's structure which does not project below the deck plating.
Rake The inclination from the vertical of masts, funnel, etc.
Rake of wagons A number of railway wagons usually between three to six which are all coupled together and being shunted in a railway marshalling yard or siding.
Ram An underwater ice projection from an ice wall, ice front, iceberg, or a floe. Its formation is usually due to more intensive melting and erosion of the unsubmerged part; or Random Access Memory. Electronic, read and write memory which is volatile. It loses its contents when power is removed. See 'Storage devices'.
Ramadan A culture strictly observed whereby the ninth month of the Mohammedan year is observed as thirty day fast during the hours of daylight by all Mohammedans.

Ramp An artificial inclined path, road or track along which wheeled vehicles, cargo and trailers may pass for the purpose of changing their elevation, or vehicle transhipment access to and from a vehicular ferry at a port.

Ramp agent An airline employee who loads and unloads baggage, cargo and food supplies on an aircraft.

Ramped cargo berth A berth specially designed with a portal ramp thereby permitting the vehicles (cargo) being transhipped driven on or off the ship/ferry over the portal ramp. Alternatively the ramp may consist of a quay which slopes gently (about 1 in 12) down to a sill at the dockwater's edge on which the vessel's ramp can rest when lowered, forming in fact a continuation of the ramp for vehicles entering or leaving the vessel.

Ramp stillages Developed for the road vehicle swap-body system, the principle can also be applied to demount freight containers. For example, wheels or skates attached to the corner castings engage with inclined rail to lift or lower the box, as the carrying vehicle moves horizontally.

RAN Regional Air Navigation (Meeting: Plan) – ICAO term; or The South African rand – a major currency.

RAND South African Rand – a major currency.

Random sample A statistical sampling technique whereby each member of the group has an equal chance of selection.

Range of performance indicators A collection of efficiency statistics/indicators to measure the competitive performance of an entity which maybe a seaport, shipping company, MNE and so on.

Ranging Clause A clause in a hull policy exempting underwriters from liability under the policy (for damage received as well as for damage done) in the event that the insured ship collides with another ship while the vessels are approaching or ranging alongside each other at sea for the purpose of transferring cargo. Cover may be obtained by arrangement with the underwriters, in practice, provided they are notified, of the intention to transfer cargo in this manner, in advance. The exemption does not apply to customary transhipment in port areas involving inshore harbour craft.

RAPE Resource analysis profit evaluation.

Rapid Transit system Mode of transport providing fast disciplined controlled form of conveyance such as an electric railway service between city centres or railway underground system found in a major city.

Rappen (Centimes) The currency of Switzerland. See Appendix 'C'.

RAR Restricted articles regulation.

RASC Regional AIS System Centre – IATA term.

RASP Radar Applications Specialist Panel (EUROCONTROL) – IATA term.

RASRO Rescue at Sea Resettlement Offers.

RASU Rangoon Arts & Science University (now Rangoon University).

Rate A tariff devised by a transport operator or his agent/retailer to convey merchandise/passengers, which maybe based on weight, volume, or value. Overall the amount charged by the carrier(s) for the carriage of a unit of goods and is the current rate which the Carrier, in the publication it normally uses to publish rates holds out to the, appropriate segment of the public, as being applicable for carriage of a unit or weight (or volume) and/or value of goods.

Rate cabotage A rate applicable to transportation between the territories of the same State.

Rate card The current rate for advertising in a particular newspaper/journal etc, which maybe in mono or colour. It maybe a full page; half page or other specification.

Rate class A rate applicable to a specifically designated class of goods in or between specified areas.

Rate construction The formulation of a rate as for example in accordance with an IATA Traffic Conference rate construction resolution.

Rate cutting The process of cutting rates below the published tariff level in order to secure business for a company. This runs the risk of generating a rates war among other Companies in the same type of business.

Rate desk Department of an airline that calculates fare constructions.

Rate dilution The process of reducing the level of rate/tariff usually arising in a competitive environment or through a volume rate negotiation involving a guarantee of traffic from a shipper in response to a lower rate.

Rate flexibility A rates policy of a shipowner(s)/agent(s) whereby the rate would be negotiated and subject to market forces. It is likely to have regard to competition; cargo volume, carriers cost, nature of commodity and so on.

Rate general cargo A rate for the carriage of cargo other than a class rate or specific commodity rate.

Rate incentive A freight rate structure so designed to encourage volume business from a particular client, such as a shipper, agent and so on. In so doing when a particular volume level of business is achieved, an improved rebate is offered or some other financial incentive.

Rate, joint A rate which applies for carriage over the lines of two or more carriers and which is published as a single amount.

Rate, normal The full under-45-kilogram rate established for general cargo. An IATA definition.

Rate of Exchange Price of one currency in terms of another.

Rate of loading The number of tonnes of cargo loaded per day into a ship. Such provision is often included in the terms of a voyage charter.

Rate of return The ratio between net profit and the capital employed by the business.

Rate of turnover The number of times the company's average stock is sold in a 12 month period.

Rate, proportional A rate which is used in combination with other rates to establish a through rate.

Rate proration Proration on the basis of the applicable local rates (air cargo).

Rate, published A rate, the amount of which is specifically set forth in the publication the carrier normally uses to establish such rates.

Rates below cost A tariff/rate which is below the cost price of the service/product.

Rates, combination of An amount which is obtained by combining two or more rates.

Rates clerk A person engaged in the compilation and quotation of rates.

Rate, sectional The rate established and used by a scheduled air carrier(s) (including any local or joint rate) for a section of a through route.

Rates formulation The process of devising a rate which maybe merchandise, parcels, or passenger tariff. Factors determining rates formulation include competition, transport cost, nature of commodity, quality of service, volume of traffic, statutory regulations, viability of traffic, general load ability and so on.

Rates Officer A Managerial post responsible for overall rates quotation and policy in accordance with Company policy.

Rate(s) of yield The amount of compensating product obtained from processing imported goods. Customs term.

Rate, specific commodity A rate applicable to carriage of specifically designated commodities.

Rate, specified A rate specifically set forth in an IATA Cargo Tariff Co-ordinating Conference Resolution. An IATA definition.

Rates review The process of reviewing the level and structure of a rates system for merchandise with a view to determining their adequacy and general viability. This may coincide with the annual rates increase – to keep pace with inflation – and to make any adjustments to reflect market changes, rebate levels, increased competition, improved distribution system, above the normal increased cost, general level of profitability or loss and so on.

Rates strategy The policy/strategy adopted/followed by a transport operator/freight forwarder/agent relative to the level of rates quoted to a shipper/client. It maybe that no rebate/discounts are quoted below the published public tariff on a particular route due to a very strong market demand and oversubscribed shipping capacity coupled with the fact that the great majority of the shipments are small consignments.

Rate structure The composition of a tariff structure for merchandise. It maybe based on commodity classification, a 'freight all kinds' basis, an ad valorem basis, a weight/measurement basis and so on. Some transport operators such as Air Freight have a wide variety structured rates system whilst a freight forwarder may operate a weight/measurement system. A parcels operator, however, may have a simple system based on weight and aligned with distance on a zonal basis with limitations being imposed on maximum parcel weight and dimensions.

Rate, through The total rate from point of departure to point of destination.

Rate trends The movement over a specific period of a rate or group of rates. For example, the merchandise rates during a five year period may have increased by 4% each year while costs have risen 6%. Such a shortfall out of alignment with cost and no compensating increase in merchandise volume conveyed, produces a loss situation and needs to be corrected to bring the business back into a profitable situation. This may prove difficult if the rates are depressed due to intense competition and the prevalence of rate wars.

Rate wars The process of competitors within the market place for a particular product(s) or service(s) undertaking a policy of continuing rate/tariff cutting/reductions to secure the business and improve their market share.

Rathole Access spaces cut into horizontal tank structures.

Rating The assessment of the quality of an issue by an established rating agency.

Rating guidelines Common practice in respect of the influence of rating factors on premiums.

Ratio decidendi (Latin) The basic thought or principle underlying a decision.

Rationalisation The process based on economic criteria or some other evaluation by which resources are reduced to become more correlated to their use and market demand on a viable basis. This may involve for example, rationalisation in ports of call to improve productive ship utilisation; rationalisation of station stops thereby placing more reliance on bus and car to feed into the inter city rail station; rationalisation in the parcels depots in a given area to provide one large depot. Overall, the process of eliminating by objective evaluation items/activities/products/service which are regarded as unnecessary and surplus to the situation.

Ratio of advertising/promotional expenditure This represents the allocated/budgeted expenditure identified for advertising related to promotional cost/expenditure by a particular company/entity for a specific service(s)/product(s) in a specified period such as one year.

RATS Restricted Articles Terminal systems – an IATA term.

RB Rubber bearing; Syrian Arab Airlines – An IATA Member Airline; or risk basis.

RBA Royal Brunei Airlines.

RBMR Royal Brunei Malay Regiment.

RBS Radio base station.

RBT Rebuilt.

r/c Return cargo.

RC Reversing gear clutch.

RCAG Remote Communication Air/Ground – IATA term.

RCB Registro Cubano de Buques – Cubian Ship Classification Society.

RCC Rescue co-ordination centre; or Royal Caribbean Cruises.

RCCA Route capacity control airline.

RCC & S Riots, Civil Commotions and Strikes.

Rcd Received.

RCDC Rangoon City Development Committee.

RCDS Raster chart display system.

RCG Regional Co-ordinating Group – IATA term.

RCH Radio Channel.

RCL Royal Caribbean Line – a Norwegian shipowner; Regional Container Line – a shipowner; or Rubber continuous liner.

RCM Statement of Basic Operational Requirements (OR), Planning Criteria (PC) and Methods of Application (MA) – IATA term.

RCU Rate construction unit. A fictitious basic currency unit used to construct unpublished rates, regardless of local currency; or Remote concentration unit.

RCV Receiver.

RD Refer to drawer; or running days – chartering term.

rd Running days.

R/D Refer to drawer.

R & D Research and development.

RDBTN Round bottom (of a ship).

RDC Running down clause – charter party or marine insurance policy clause.

RDEC European research and development committee.

RDF Radio Direction Finder – ship's navigation aid.

R dk Raised deck.

RDLY Redelivery.

RDR Radar – ships navigational aid.

RDS Real time data service.

RDT Radio data terminal – associated with container berths; or Radio data terminal transfer.

re With reference to.

Reachable on Arrival or Always Accessible In such a situation the charterer undertakes that an available loading or discharging berth be provided to the vessel on her arrival at the port which she can reach safely without delay in the absence of an abnormal occurrence. A BIMCO term.

Reactive policy A policy which responds to competitors initiatives and changes in the market place rather than initiating such policies and becoming a market leader.

Reactive strategy A strategy adopted by a Company/entity which responds to changes/new initiatives in the market place.

Readership The number of persons reading a particular newspaper, magazine or other news/information media. Thus a particular newspaper may have a circulation of one million daily with each newspaper having a readership of 1.5 persons thereby giving a total readership of 1.5 million persons.

Ready berth The code name of BIMCO approved charter party clause relative to a 'ready berth' and entitled 'Baltic Conference Ready Berth clause 1938'.

Real effective exchange rates See effective (nominal/real) exchange rates.

Realignment A change in the central parity of a currency participating in an exchange rate system with a fixed but adjustable peg. In the former ERM a realignment consisted of a change in the ECU central rate and the bilateral central rates of one or more of the participating currencies. An ECB term.

Real terms Economic term. Figures adjusted to exclude the effect of inflation.

Rear-ramp loading trailer Low-loading trailer with drop-down tailboard usually used for carrying wheeled plant.

REART Restricted articles – an IATA term.

Rebagging The process of bagging cargo/merchandise having previously been bagged but emptied for a variety of reasons. It maybe for customs examination, processing the commodity, and so on.

Rebate An allowance made as a discount in an account/rate. For example, a particular freight rate which is usually based on volume such as ten per cent on €100 yields a €10 rebate. It maybe based on the condition the client despatches 10,000 tons of traffic during a particular year. Likewise an account paid within one calendar month may attract a 2½% rebate/discount.

Rebateable A rate which offers a discount/rebate on the basic tariff. This may be subject to various conditions. For example, a rate of €100 per ton may be subject to a discount/rebate of thirty per cent if the traffic volume exceeds 100,000 tons annually in which case the net rate would be €70 per ton.

Rebating Giving any valuable consideration (commission) to a prospect or insured as an inducement to buy – insurance term.

Re-buy The process of repeat ordering of an item(s) previously ordered.

REC Receive.

rec Rectangular.

Recd Received.

Receipt A written acknowledgement of anything, e.g. merchandise, document received.

Receivable portfolios See trade receivable portfolios evaluation.

Received for shipment Bill of Lading It confirms the shipping company has received the goods in custody for shipment. The received for shipment Bill of Lading can be converted to a 'shipped bill' by an endorsement from the shipping company when the goods have been loaded aboard the ship. The received for shipment Bill of Lading is widely used with container shipments and inland clearance depots with the container being transported by road or rail to the port for shipment.

Receiver Name of the Agent, company, consignee, person receiving the consignment; or the person appointed to 'receive' and administer the rents and profits, or other monies, accruing to an estate or business undertaking which is administered or wound-up under the supervision of the court. Official Receivers are officials permanently employed to act in that capacity in bankruptcy proceedings or the winding-up of joint-stock companies.

Receiver of the Wreck An official who is responsible for all wreckage that is salved on the coastline, or which is found at sea and is brought to a British port. Usually a Senior Customs Officer.

Receivers Agent One who represents a principal and in so doing looks after the interest of the importer/receiver at the port of destination relative to the processing and facilitation of clearance of the cargo through customs at a specified port. In so doing the Receivers Agent may undertake the clearance of the cargo through customs or engage a Clearance Agent to do this work. The paramount consideration is to ensure the receivers agents merchandise is promptly discharged/processed through customs in an efficient cost effective manner following arrival of the ship in the port or its environs.

Receiving airline A carrier that will transport a passenger after arrival at an interline point.

Receiving and loading cargo The process of receiving/accepting cargo at a port and subsequently loading the merchandise on the vessel after all the documentation and customs formalities have been undertaken.

Receiving carrier The carrier who receives/accepts the consignment from a carrier, Agent, or shipper for onward transportation.

Receiving date Date from which cargo accepted for shipment for specified sailing.

Receptacle A containment vessel, including closures, for receiving and holding substances or articles.

Reception area A facility where clients/passengers/consignments are received and usually 'checked in' by a particular entity/Company. This maybe a motorist having his ticket/reservations checked at a port before embarking on a particular ferry service; a passenger checking in at an Airport passenger terminal prior to joining a specific flight; or entrance area to a Company office where all visitors/clients are 'checked in' by the receptionist.

Reception depot A form of warehouse accepting cargo for subsequent despatch.

Reception point A facility where clients are received and usually 'checked in' by a particular entity/company. See also reception area entry.

Reciprocal Trading Trading between nations or companies using the exchange of goods of similar value for payment rather than currency or cash. Overall, bilateral or multilateral trading based on reciprocity entry.

Reciprocity The notion that trade negotiations should be mutually advantageous to all parties through the reciprocal exchange of concession on a multilateral basis. It is used to describe the notion that a country should be entitled to a reciprocal trade advantage and even balances with each of its trading partners on a bilateral basis – a view not supported by WTO. Overall, the principle governing WTO negotiations under which governments extend concessions in the expectation that they will receive comparable concessions. Also the exchange of reinsurances between two reinsurer's – insurance definition.

Re-circulating ball steering Steering design where worm gearing is used. The nut which moves up and down the worm runs on ball bearings to reduce friction and provide smoother steering effort.

Reclaim area An area at an airport/seaport where passenger baggage is placed following the discharge/transhipment from a ship/aircraft. In so doing the passenger collects his/her baggage and accompanies it when presenting it to customs for baggage examination and customs clearance.

Reclaiming The retrieval by mechanical means from a stockpile of ores or other bulk materials; or the process of a passenger collecting his/her baggage from the reclaim area following discharge/transhipment from a ship/aircraft. See reclaim area.

Recmd Recommissioned.

Recognition The measure of the effectiveness of a media/campaign plan activity in terms of its advertising impact and measure of recall by consumers. This involves testing the recall of the advertisements by consumers.

Reconditioned packagings These include metal drums that are cleaned to original material of construction, with all former contents, internal and external corrosion and all external coatings and labels removed; restored to original shape and contour, with chimes (if any) straightened and sealed, and all non-integral gaskets replaced; and inspected after cleaning but before painting, with rejection of packagings with visible pitting, significant reduction in material thickness, metal fatigue, damaged threads or closures, or with other significant defects. A term associated with dangerous cargo shipments by air.

Reconditioning All activities connected with testing and/or adjusting the packaging of a product in such manner that it can be presented to the customer in the requested form.

Reconfirmation Notice of intent to use a reservation, required by some airlines.

Reconsign Redirect consignment thereby involving a change in the original point of destination to a new one in accordance with the shipper's most recent instructions.

Recourse The right to claim against the joint and several guarantors (endorsers, drawers) of a Bill of Exchange or a cheque on payment made. A banking term.

Recourse agreement The process of an agreement being signed by an exporter – for example in favour of ECGD – undertaking to reimburse ECGD any sums paid by them under the bank guarantee.

Recourse, with In a credit transaction. If a buyer fails to pay in accordance with the agreement, then the bank (or other party) which has paid and failed to recover the payment can claim the payments from the exporter.

Recourse, without In a credit transaction. Once the exporter fulfils their part of the contract and receives payment, it is not liable to pay back the money to the bank (or other party) if the buyer fails to pay.

Recovery Amount recovered from a third party responsible for a loss on which a claim has been paid. Overall, money received by an insurer in respect of a loss, thus reducing the loss, by way of subrogation, salvage or reinsurance.

Recovery vehicle A road vehicle which recovers usually abandoned or immobilised vehicles. It is so designed/equipped to permit the vehicle to be loaded onto the recovery vehicle by means of a ramp at the tail board.

Recreational vehicle (RV) A van or other vehicle designed for camping or other recreational use.

Recruitment Manager A Managerial post in a Company/entity responsible for recruitment of employees.

Recto (Latin) The right hand page.

Recycling In financial terms internationally, the process of transferring the investible surpluses of certain oil producing countries to borrowers in other countries, particularly those with a current account deficit; or the process of scrapping vessels.

RED Reduction.

Red/amber/green/lights These refer to the three categories of subsidies: 'red' for prohibited subsidies; 'amber' for actionable subsidies (may be subject to countervailing action); and 'green' for the generally-permitted subsidies. A WTO term.

Red & Green system Process of passengers customs examination whereby those with nothing to declare go through the green barrier and the remainder proceed through the red channel.

Red channel A customs facility where motorists, passengers and truck/road hauliers proceed through the red channel if they have dutiable goods to declare and are outside/above the prescribed limits. It is usually situated at an airport, seaport or frontier point.

Red clause Clause contained in a documentary credit authorising the paying bank to make advances on an unsecured basis to the beneficiary even before shipment of the goods. A rarely used type of letter of credit which allows the beneficiary to draw advances up to a set limit. This advance will be deducted from the funds which become due when the documents are presented.

Red Clause Credits Red Clause Documentary Credits, sometimes known as Anticipatory Credits, make a percentage of the Credit value available to the Seller as a pre-shipment advance without the prior condition of presenting the documents called for under the Credit. The term originated in the past, when the clause enabling the Seller to receive a pre-shipment advance was printed in red. This type of Credit primarily issued to cover shipments of wool from Australia is rarely used today. Red Clause Credits fall into three categories – Clean, Unsecured and Secured (also known as Green Clause). Banks understand the meaning and intent of the Red Clause and use their own wording and format to suit the customer's requirements.

Redelivery The return of a consignment to the party who originally delivered it to the carrier.

Redelivery clause A time charter party usually contains a provision setting out conditions for re-delivery of the vessel at the end of the charter period. The charterer is obliged to re-deliver the vessel in the same good order and condition, except for ordinary wear and tear.

Redemption The cancellation of a security by payment; redemption may be mandatory on a certain date, optional by the borrower after a certain date or conditional upon certain described and defined circumstances; or in marketing terms, it is the process of redeeming or trading in vouchers, coupons, or trading stamps for a specified product or other benefit.

Redemption date The date on which a security is due to be redeemed by the issue at its full face value. The year is included in the title of the security; the actual redemption date is that on which the last interest payment is due.

Redemption yield The flat yield plus an annualised proportion of the profit which will accrue on the purchase of a fixed interest stock at a price below its redemption price. Alternatively, if the market price is more than the redemption price, the flat yield less the annualised loss on repayment.

Redeployment The process of utilising equipment or labour elsewhere. Examples arise in the labour sector when a particular workshop closes at a factory and staff are redeployed elsewhere within the plant. Likewise, due to a road depot closure the vehicles maintenance equipment and staff are all transferred and redeployed in another depot some thirty miles away.

Red eye Transportation late at night or overnight.

Red Line Clause A clause in a cargo policy or certificate which sets out the procedure the consignee must adopt in the event that he does not receive the goods from the carrier, or when he receives damaged goods.

Re-domination of securities The denomination of a security is the currency in which the par value of the security is expressed (in most cases, the face value of a certificate). Re-denomination refers to a procedure through which the original denomination of a security, issued in national currency, is charged into euro at the irrevocably fixed conversion rate. An ECB term.

REDS Registered Excise dealers and shippers. A customs term.

Reduced fares Those fares which offer a reduction on the basic full fare. For example, a 20% discount on a fare of €8 would produce a tariff of €6.40, which maybe an excursion fare.

Reduced lead time A situation whereby the date of the concept of a scheme/product relative to the actual original introduction date is reduced such as from ten months to eight months.

Redundancy An engineering term indicating permanently fitted back-up equipment, to be used in the event of failure of the primary unit. Also the process of a person/employee becoming surplus to the employers' needs.

REEFER Reefer Lines Pte Ltd – a shipowner.

Reefer Specialised dry cargo vessel with 80% or more insulated cargo space or; a refrigerated ISO container.

Reefer cargo Cargo requiring temperature control.

Reefer container A thermal container with refrigerating appliances (mechanical compressor unit, absorption unit etc) to control the temperature of the cargo. A reefer container is capable of maintaining the internal temperature within the range of +20°C to -22°C (+68°F to -8°F).

Reefer trade An international shipping trade dealing with conveying refrigerated cargoes.

Re-engineering the business process The process of applying electronic trading EDI techniques to business practices with a view especially to improve upon them and introduce new methods.

Re-exportation The process of re-exporting a consignment.

Ref Reference; or ship fitted with refrigerated cargo equipment.

Refer carrier A vessel designed to carry refrigerated cargoes such as meat.

Refer container A container which can convey refrigerated cargoes.

Reference The submission of a matter in dispute to an arbitrator for his award – insurance definition.

Reference period Time intervals specified in Article 104c (2a) of the Treaty and in Protocol No. 6 of the convergence criteria for examining progress towards convergence. ECB term.

Reference value for the fiscal position Protocol No. 5 of the Treaty on the excessive deficit procedure sets explicit reference values for the deficit ratio (3% of GDP) and the debt ratio (60% of GDP), whereas Protocol No. 6 on the convergence criteria specifies the methodology for the computation of the reference values relevant for the examination of price and long-term interest rate convergence. ECB term

Refer ship Vessels which have refrigerated accommodation for cargo.

Refinery A processing plant which produces high purity metal either by electrolytic or fire refining. In copper and lead production, refining is always preceded by smelting (qv), but for tin, zinc and nickel smelting and refining blend into one process of recovering marketable quality metal from concentrates. Refiners (and smelters) may also treat scrap materials.

Refit The process of a vessel undergoing survey and repairs.

Reform process/programme The Uruguay Round agriculture Agreement starts a reform process. It sets out a first step in the process, i.e. a programme for reducing subsidies and protection and other reforms. Current negotiations launched under Article 20 are for continuing the reform process. WTO term.

Re-forwarding The process of a consignment/goods/merchandise being transported/re-forwarded from the airport/seaport of destination by a surface or air transportation agency, but not a carrier under the Air Waybill/Bill of Lading or other document.

Re-forwarding charge Charges paid or to be paid for subsequent surface or air transport from the airport of destination by a forwarder, but not by a carrier under the Air Waybill (air cargo).

Refrigerated box van A road vehicle which conveys refrigerated merchandise. It is a two axle road vehicle unit with a purpose build refrigerated unit.

Refrigerated container An ISO covered container of 20 feet (6.10m) or 40 feet (12.20m) length designed to transport cargoes requiring temperature control from frozen meat and fish to citrus fruits, bananas and vegetables, dairy produce and chocolate and photographic film. The container has an integral refrigeration unit and many are fitted with remote monitoring points. Many operate on dual voltage. Models with air and water cooling enable under deck operation, and all units are fitted with fresh air ventilation systems. Heaters provide defrost and container heating as required and meat rails for hanging carcass can be fitted without container modification. Many units are specially strengthened internally for the carriage of dry freight cargo.

Refrigerated swap body A swap body designed to convey refrigerated cargo at controlled temperatures. See also swap body entry.

Refrigerated vessel Ship built to convey chilled or frozen cargo, i.e. meat, butter, eggs, etc.

Refuge country A place/country where a person may chose to reside in order primarily to lessen the risk of any judicial proceedings which may arise consequent on any irregularity the person may have committed earlier in another country.

Refund The refund to the purchaser of all or a portion of a charge for unused carriage which may involve a passenger journey ticket or a consignment which was conveyed only a portion of the scheduled transit.

Refund policy Policy regarding reimbursement, especially following cancellation.

Refurbishment The process of modernising and revitalising a product with a view to extending its useful life. This maybe an hotel, lounge suite, interior of a passenger rolling stock, and so on. Its major advantage is that the cost tends to be economical when related to the benefits which are derived therefrom and the alternative of a new purchase. Each situation requires to be treated on its merits.

Regd Registered.

Regional agreements Agreements between countries in geographic proximity aiming to facilitate trade between them.

Regional airline See commuter airline.

Regional Airline Association A trade group representing commuter airlines.

Regional distribution centre (RDC) Term much in vogue in supermarket and retail distribution circles, sometimes crossing international boundaries. Usually implies a large distribution hub which receives products from a wide range of manufacturers or growers, breaks them down into loads for individual supermarkets and then reconsolidates loads for delivery in an efficient manner. Typically, serves several supermarkets across several counties or a complete region.

Regional organisation A Company whose organisation structure is based on a specific region/group of countries etc, with a Regional/Area manager in charge.

Regional quotas A situation/circumstance whereby a particular Free Trade region/area conducts trade on the basis of limiting the quantity of a particular commodity(ies) exported within the Trading Group from each country.

Regional trade Trade amongst neighbouring countries such as Far East region.

Regioncentric Tending to be oriented toward regions larger than individual countries as markets.

Register Book Lloyd's Register of Shipping book issued annually detailing vessels classified with the Society.

Registered Excise Dealers and Shippers A customs authorised trader who may be a principal importing excise goods purely for his own business or an agent/freight forwarder/Customs broker acting for one or more importers. Customs duties become payable on receipt of the goods.

Registered luggage See baggage checked entry.

Registered Manager Person responsible for statutory compliance with maritime regulations for specified ship(s).

Registered ton A 100 cubic feet or 2.8317 cubic metres of a vessel enclosed area viz, NRT or GRT.

Registrar of Companies Senior civil servant whose duty it is to monitor all companies in UK.

Register general of Shipping and Seamen The UK organisation responsible for the registration of British Ships.

Register of Shipping (Russia) The former Russian Register of Shipping Classification Society and based in St Petersburg, Russian Federation.

Registro Italiano Navale The Italian Ship Classification Society based in Genova, Italy.

Registry Attestation of a ship's registration in a country not necessarily the country of ownership; does not indicate the quality of the ship or the nationality of officers, crew or service personnel.

Registry, Certificate of A document giving all particulars of the vessel, including the names of the owner and the master.

Regression analysis The process of returning to the place or point of departure/origin in an analytical manner.

Re-groupage The process of splitting up shipments into various consignments (de-groupage) and combining these small consignments into other shipments (groupage).

Regular service A transport service which operates on a consistent/regular basis. This maybe a bus service which operates throughout the year each Tuesday and Thursday between A and B at hourly intervals during time band of 0930 hours and 1930 hours.

Regulated economy An economy where a large number of government regulators exist relative to the conduct of business and economy. This embraces labour law, fiscal measures, company law, social benefits and technology. Examples include France, Germany, Belgium, Italy and UK. It is found especially in fully developed economies.

Regulatory agency A local, state, federal or international agency with authority to approve or disapprove the actions of principals in an industry.

Regulatory Body Organisation set up by governments or other accredited authorities such as Trade Associations to oversee/regulate its members or specified companies within a code of practice. It could be the licensing authority of a government department issuing import licences, or a Trade Association ensuring goods sold conform to certain standards/quality.

Reheat plant (reheat cycle) Turbine plant where the steam exhausting from the high-pressure turbine is reheated in the boiler (or in a nuclear power station in a special heat exchanger) before entering the intermediate-pressure turbine. Reheating raises the cycle efficiency, i.e. more useful energy is obtained from the turbine per unit of heat supplied to the steam.

Re-hiring The process of hiring a piece of equipment such as a container following an interval when the hiring had ceased.

Reimburse To pay back.

Re-imbursement Payment of the expenses actually incurred as a loss covered by the policy. An insurance definition.

Reimbursing Bank The Bank nominated by the issuing bank to pay claims under the credit to the negotiating or paying bank.

Re-imported goods The process of re-importing goods previously exported.

Re-instatement In the event of loss, the repair or replacement of damaged property to a condition equal to, but no better than its original condition; or following a loss the restoration of the sum insured to the original figure.

Reinsurance A contract for reinsurance is to indemnity against losses which the original underwriter had suffered, but not against gifts. The re-insurers liability is limited to the original insurer's liability under the original policy. Overall, insurance protection taken out by an insurer to limit its exposure to losses on original business accepted. The reinsured maybe referred to as the original or primary insurer or the ceding party.

Reinsurance to close In syndicates, an agreement under which underwriting members (the reinsured members) who are members of a syndicate for a year of account (the closed year) agree with underwriting members who comprise that or another syndicate for a later year of account (the reinsuring members) that the reinsuring members will indemnify the reinsured members against all known and unknown liabilities of the reinsured members arising out of insurance business underwritten through that syndicate and allocated to the closed year, in consideration of: a premium; or the assignment to the reinsuring members of all the rights of the reinsured members arising out of or in connection with the insurance business (including without limitation the right to receive all future premium, recoveries and other monies receivable in connection with that insurance business).

Re-invoicing The procedure whereby goods shipped directly from a supplier to the customer are invoiced in two stages: at first by the supplier to an intermediary, and subsequently by the intermediary to the customer.

Rejection The non-acceptance of cargo for example.

Relational database A database in which data is held in a number of related files, the data structures of which obey normalisation rules. A computer term.

Relationship (database) The way in which entities in a database system are related to form a complete relational database. The relationships may be one to one, one to many or many to many. A computer term.

Relative Liquid Density The mass of a liquid at a given temperature compared with the mass of an equal volume of fresh water at the same temperature or at a different given temperature.

Relative Vapour Density The mass of a vapour compared with the mass of an equal volume of air, both at standard conditions of temperature and pressure.

Re-launch The process of updating a product/service to increase sales and market appeal/acceptance and subsequently re-launching it in terms of marketing strategies.

Relay port A port which acts as a feeder to a hub port. See hub port entry.

Release devices – explosive Articles consisting of a small charge of explosive with means of initiation. They sever rods or links to release equipment quickly.

Release note The receipt signed by the customer acknowledging delivery of the goods.

Reliability of delivery The reliability of a supplier concerning the agreed terms of delivery with regard to the quality, quantity and delivery time, conditions and price.

Relief goods Third country goods imported or delivered from a customs warehouse under inward processing relief arrangements.

Relief sailing An additional sailing usually provided to cope with additional traffic volume.

REM Remittance; or replacement equipment manufacturing (after sales).

Remaining on board Quantity of cargo or fuel on board a vessel. A chartering term.

Reminder advertising An advertising campaign devised to overcome the memory lapse and in so doing focus on product/service/entity awareness in the market place.

Remission The waiving of import or export duties which have not yet been paid (see also Repayment). A customs term.

Remitting Bank The Bank which sends the documents to the issuing bank under a documentary credit situation.

Remote access Pertaining to a system in which distant terminals are connected to a computer to enable them to use the facilities it provides.

Remote access (to an IFTS) The facility for a credit institution established in one country ('home country') to become a direct participant in an interbank funds transfer system (IFTS) established in another country ('host country') and, for that purpose, to have a settlement account in its own name with the central bank in the host country, if necessary, without having established a branch in the host country. An ECB term.

Removal deck Deck of a ship which is capable of being removed and stowed out of the way. This type of deck is found in some car carriers and is removed when the ship is carrying a bulk cargo.

Removal note A form issued by Customs following customs clearance of goods to permit authorised removal from the port (formerly known as the 'out-of-charge note'). Indicates goods clear of Customs.

REMS Remittance.

REMSA Joint Requirements for Emergency and Safety Airborne Equipment. (ECAC term).

Remanufactured packagings These include metal drums that are produced as a UN type from a non-UN type; are converted from one UN type to another UN type; and undergo the replacement of integral structural components (such as non-removable heads). Remanufactured drums are subject to the same requirements of these Regulations as apply to a new drum of the same type. Term associated with dangerous cargo shipments.

Rendu Extended form of cost, insurance, freight contract.

Renewal Continuation of a policy, contract, tenancy, etc, for a further term.

Renewal notice Notice sent to insured, tenant, etc, advising the approaching renewal date, and the premium rent for the next year.

Rent A periodic payment for the use of land, buildings, or equipment such as containers.

Rental finance Income from rent.

Rentcon Rent-a-Container.

Renunciation New issues of shares are usually made by means of allotment letters or renounceable certificates. During a period of renunciation, holders may dispose of their allotment to a new owner.

Repacking The process of packing cargo/merchandise having previously been packed but unpacked for a variety of reasons. It maybe customs examination, processing the commodity, larger/smaller packages and so on.

Repaircon The standard ship repair contract devised by BIMCO and introduced in 2003.

Repatriation Process of a person returning to one's own country following a period overseas.

Repatriation of funds The process of transferring funds/monies from one national currency in which an expatriate has been working to his national currency. For example, a UK resident may have worked in Thailand for eight months and plans to transfer all his/her Baht currency to sterling.

Repatriation of overseas funds The process of recovering/transferring overseas funds/capital/monies of the client/owners to his/her national country such as a UK citizen with investments in Switzerland transferring the funds to a UK based sterling investment.

Repayment The total or partial refunds of import or export duties which have been paid. Customs term.

Repeat purchasing Process of consumers buying on a regular basis. Such products are usually found in the mass market and the buying cycle may range from daily, weekly, yearly or three to five years. It features strongly Brand loyalty and embraces foodstuffs, newspapers, cars, clothes and service items such as insurance.

Replacement The substitution of insurance coverage from one policy contract to another – insurance term.

Replacement clause A clause limiting underwriter's liability for damage to machinery, cargo.

Replacement planning The process of formulating a programme/schedule to replace assets which may be life expired, obsolete, uneconomic and so on.

Replacement price The price at which material identical to that which is to be replaced could be purchased at the date of valuation (as distinct from actual cost price at actual date of purchase).

Reply card A postcard, usually prepaid, which responds to a direct marketing initiative. It maybe through a magazine/journal or mail shot.

Repo Sale and Repurchase Agreements. These are agreements to sell bonds with a simultaneous agreement to repurchase these bonds back at a later date, typically less than one year. A repo effectively enables the holder of a bond to borrow money whilst using the bond as collateral with the lender. An international banking definition.

Reporting day Date on which master indicated to charterer ship ready to commence loading or discharge – chartering definition.

Report, irregularity The document referred to in Resolution 603 (IATA) and is equivalent to the term 'Notice of Non-Delivery' which is forwarded by a participating airline upon discovery of any irregularity in a consignment or its documentation to the issuing airline or his handling agent at the airport of departure. An IATA definition. Similar report irregularities arise in other transport modes.

Report (ship's inward) Customs documents which must be completed and lodged with Customs for each vessel by the shipowner or agent before discharge can commence. This usually includes cargo manifest; details of vessel; voyage; crew; passengers and so on.

Reposition containers To move empty ISO shipping containers away from a location or area where there are no further loads, to location where demand exists for such containers.

Repositioning Changing the marketing mix of a product to reach a different segment of the market – marketing definition.

Representation A statement of fact usually made in negotiations, litigation etc, such as when made by the assured or his broker when negotiating insurance cover with the underwriter.

Repurchase agreement An arrangement whereby an asset is sold while the seller simultaneously obtains the right and obligation to repurchase it at a specific price on a future date or on demand. Such an agreement is similar to collateralised borrowing, with the exception that ownership of the securities is not retained by the seller. The Eurosystem uses repurchase agreements with a fixed maturity in its reverse transaction. An ECB term.

Repurchase operation (repo) A liquidity-providing reverse transaction based on a repurchase agreement. An ECB term.

REQ Requested.

Request-and-offer system Request lists for tariff concessions are first exchanged, followed by the tabling of offer lists. This bilateral process can take place several times before a final package of concessions is agreed. This approach has been used by some participants in the negotiations on tariffs, non-tariff measures and tropical products. A WTO term.

Re-routing The route to be followed as amended or altered from that originally specified on the consignment note or elsewhere.

RES Received for Shipment.

Res agent A person who takes reservations and/or sells tickets.

Resale value The value of a product on the second-hand market. See secondhand tonnage.

Rescheduling In financial terms the process of permitting a borrower to repay their borrowing over a longer period than had been originally agreed. This practice is adopted widely with countries with large debts; or in transportation terms the process of revising/amending a transit schedule due to bad weather, improved technology/facilities, rationalisation of ports of call and so on.

Rescission In legal terms the process of cancelling or unscrambling the contract. It is peculiar to situations where there has been a misrepresentation, and is one of the options open to the injured party in such a situation. Overall, the termination of an insurance contract by the insurer on the grounds of material misstatement on the application for insurance. The action of rescission must take place within the contestable period or time limit on certain defences, but takes effect as of the date of issue of the policy, thus voiding the contract from its inception.

Research and development aids These involve a government providing funds to aid research and development in the area of ship construction, particularly for new technology in the development of high-performance ships.

Research briefing See Market Research briefing.

Research cost (applied) The cost of original investigation undertaken in order to gain new scientific or technical knowledge and directed towards a specific practical aim or objective.

Research cost (basic) The cost of original investigation undertaken in order to gain new scientific or technical knowledge and understanding, not primarily directed towards any specific practical aim or application.

Research debriefing See market Research debriefing entry.

Research Director The Director responsible at Board level for Company Research policy and ultimate results. This usually involves technical research and development of the Company's products, and ancillary services including new technology the Company may be developing.

Research objectives The aims and objectives of a particular research programme/project within a particular company/entity.

Reservation The process of booking a seat on a train, theatre, airline, coach or linear/cubic capacity on a transport unit prior to the transit taking place. It maybe by the individual passenger/shipper, or the Agent/Freight Forwarder – the latter often having a block booking arrangement for a contracted period with the carrier on specified services and giving some form of traffic volume guarantee. Additionally it may be hotel reservation, passenger cruise, holiday package and so on.

Reservation form A form confirming the reservation has been effected which maybe a hotel booking, coach booking and so on.

Reservation of cargo The process of allocating cargo space for a particular consignment on a specified sailing/flight on a given date from named seaport/airport; or the process of a particular country reserving a particular percentage of commercial cargo volume/tonnage between two trading nations to the national fleet and not foreign tonnage. See also UNCTAD 40/40/20 entry.

Reservation of title A legal term which permits a reservation of the right of disposal relative to an international sales transaction. It enables the seller to reclaim the goods in the event of the buyer becoming bankrupt or going into receivership. See Romalpa clause.

Reservation system A system where accommodation/facilities have to be pre-booked to guarantee applicant their accommodation reservation. It maybe in a hotel, airline seats, cruise, cargo space in ship etc. Usually it is computer operated and passengers/cargo arriving without reservations would only be allocated unbooked accommodation or cancelled bookings.

Reserve base The sum of the balance sheet items (in particular liabilities) which constitute the basis for calculating the reserve requirement of a credit institution. ECB term.

Reserve currency A national currency which because of its widespread acceptability is used for making international payments and as part of a countries' foreign exchange reserve. The US dollar is the most important reserve currency.

Reserve price The minimum selling price of a product/service. It may arise when old stock is being cleared to make way for a new range of models.

Reserves That part of a company's capital which represents past profits retained in the business and not paid as dividends, or profit which has been obtained by selling shares at a price above their nominal value (the share premium).

Reserve requirements The practice is adopted in some countries, including the United States, whereby banks are required to invest in a specified proportion of funds taken on deposit in certain risk-free assets. These reserve assets, because they are risk free, tend to yield less than commercial lending or other investments.

Reserve tranche policies An IMF term. It permits a member to have a reserve tranche position in the IMF to the extent that its quota exceeds the IMF's holdings of its currency, excluding credits extended to it by the IMF.

Hence a member may draw up to the full amount of its reserve trance position at any time, subject only to balance of payments need. See credit tranche policies.

Re-shipped The process of a consignment such as a container being discharged from the first carrier (vessel) and subsequently loaded/reshipped onto the second carrier (vessel).

Residual fuel oil This term is sometimes used to describe heavy fuel oil and the labels can be used interchangeably. An IPE term.

Residual market A system through which insurance is made available to buyers that represent unusually high risks.

Residual value The value of an asset at the end of its useful life. For a ship, its scrap value. Depending upon the scrap price per ton assumed, this may be a substantial figure in the case of a large ship. The calculation of residual values has importance in relation to the size of balloon which a bank may consider appropriate in a loan repayment schedule. Attempts have been made to ease loan repayment terms by insuring residual values or arranging put options at prices above a straight forward scrap valuation. This obviously has validity for ships which, at the end of the loan repayment period, will not be at the end of their economic life. Progress so far as been limited, however.

Resigning Addendum An endorsement attaching to a Lloyd's policy changing a syndicate's participation.

Res ipsa liquitur (Latin) The thing speaks for itself.

Resort Generally, an area offering recreation and leisure possibilities along with accommodations.

Resort condo or condominium Usually, individually owned, but joint managed apartments, that are rented out to vacationers for short stays.

Resource allocation The process of allocating resources.

Resource availability The degree of resources available to undertake a particular assignment(s). This will involve labour, equipment, capital, time scale and so on.

Resourced based industry An industry which is located in the country in which the indigenous resources/raw materials are located. Examples include the French wine industry.

Resource planning The process of planning the effective use of a company/entity/Government resources on an optimum basis within a stated objective and environment.

Resources The actual available means by which to meet the demand. It maybe manpower, finance, equipment, materials or some other combination of various resources.

Res perii domino (Latin) The loss of a thing falls on its owner.

Respondeat superior (Latin) Let the principal answer (for his servants acts).

Respondentia Process of Master pledging his ship's cargo at a port to secure money to enable voyage to continue: A loan taken out on the cargo.

Respondent superior (Latin) Let him answer who exercises control or authority.

Response rate The number of personnel/companies etc, in percentage terms responding to a market survey and can be measured for example by the number of completed questionnaires returned.

Responsibility clause The section of a brochure that explains all the conditions of tour offering; should include the name of the financially responsible principals.

Responsible carrier The carrier responsible for the transport of goods as indicated in the transport document (air cargo).

Rest Restoration of (ship classification) class.

Restitutio in integrum (Latin) A reinstatement into the former position.

Restowage The process of re-locating and stowing a transport unit such as a container following it being emptied – partially or complete. This maybe due to a customs 'turn out'.

Restricted Articles See Dangerous Goods.

Restricted fare Airline fare that restricts the time period the holder can travel and imposes other restrictions, such as advance-purchase and minimum/maximum-stay requirements.

Restricted gauging A system employing a device which penetrates the tank and which, when in use, permits a small quantity of cargo vapour or liquid to be expelled to the atmosphere. When not in use, the device is kept completely closed.

Restricted goods Goods needing a licence or certificate before they can be imported into, or exported from the UK. A customs term.

Restricted Letter of Credit A Letter of Credit, the negotiation of which is restricted to a bank specifically mentioned.

Restrictive customs procedures Customs procedures resulting in onerous and lengthy import procedures. This includes product valuation, inspection, documentation, health and safety regulations, permits and licences. This is practised widely in less developed countries.

Restrictive practice Any practice which restricts a person's freedom to trade as he sees fit. This restriction may arise through laying down the price the seller must charge when selling the goods; preventing the seller from selling the goods to a certain type of customer; or laying down the terms on which the seller must resell the goods.

Retail analysis Process of analysing retail sales of a particular product or group of products.

Retail audit The process of evaluation the competence/creditability of a retail outlet maybe as a selling outlet for a product/service under consideration. See Marketing Audit entry.

Retailer The person/company which sells a product(s)/service(s) direct to the market place. In the travel business, a retail travel agent or agency – see retail outlet entry.

Retailg. Retailing.

Retail network A network of retail outlets in a country/region/trading bloc/area which is usually by commodity or generally. It maybe the retail outlets in the pharmacy or household furniture sectors. A feature of the network is the retail outlet profile.

Retail outlet The place where goods are sold to the consumer. It may be garage, supermarket, hypermarket, general store and so on.

Retail price index (RPI) The official statistic (in UK) used to monitor changes in the price of consumer goods and services. It is produced monthly and is a measure of inflation.

Retail Sales Analysis The process of analysing retail sales in a retail outlet (shop/supermarket) or group of retail outlets. This maybe for an individual commodity, or group of commodities; total sales in the retail outlet and so on. It would be for a particular period of one week, a day, or other specified period and one can compare one year's results with another. Also it can compare the results of one retail outlet with another.

Retention An amount retained by a reinsured when effecting a reinsurance. The proportion of a hull insurance premium that is retained by the underwriter, when returning part of the premium for a period during which the vessel is laid up; or the amount of premium retained by an insurer after paying claims and expenses, i.e. profit (US).

Retention bond A bond which guarantees refund of the released retention monies to the buyer, in the event of non-performance of the exporter's obligations after the contract has been completed. Such a bond is helpful to assist the cash flow of suppliers. An international banking term.

Retonnaging The process of reinvestment in the maritime Fleet either new or secondhand tonnage.

Retractable bogies A facility on a heavy vehicle equipped with rear bogie lifts which enables the second axle in a pair of tandem axles to be raised when running empty to reduce friction and tyre wear.

RETRO Retrospective rating or retrocession – insurance term.

Retrocedent The reinsurer placing a retrocession.

Retrocession Reinsurance of a line accepted as a reinsurance on a long term contract. The original insurer cedes the line and the reinsurer, having arranged reinsurance cover to protect his own interest, retrocedes the line.

Retrocessionaire A reinsurer who buys reinsurance on some of the risks he has agreed to bear.

Retrofitting of double hull The process of fitting a double hull to a vessel. Such a conversion represents a major conversion and additional safety requirements and usually is economically unfeasible.

Retrospective rating A plan or method which permits adjustment of the final reinsurance ceding commission or premium on the basis of the actual loss experience under the subject reinsurance treaty – subject to minimum and maximum limits.

Retrospective rebate A rebate granted on traffic which has already been conveyed by the shipowner.

Returned goods relief A customs facility whereby goods exported from and then returned to the UK in virtually the same condition within three years are admitted against declarations made by the original exporter and in certain cases his agent and detailing the particulars of the original exportation. Such goods are immune from Customs Import duty and include particularly a cancelled exporter order, damage in transit and refusal by the importer to accept the goods.

Return load The process of traffic being conveyed on the return journey of a transport unit.

Reused packagings These are packagings to be filled which have been examined and found free of defects affecting the ability to withstand the performance tests; the term includes those which are transported within distribution chains controlled by the shipper of the product. Term associated with dangerous cargo shipments.

REV Reversing.

Revalidation sticker Attachment to a flight coupon validating a change made in the original reservation.

Revaluation of exchange rate The process of an exchange rate which is fixed officially by a government and is subject to a change imposed by a government causing the price/value to rise in terms of other currencies. It may be a ten per cent revaluation of an exchange rate.

Revamping The process of up-dating/revising a particular situation/project which may, for example, have been dormant for a couple of years due to lack of investment resources and a depressed market. A further example is to update/revitalise a magazine to increase sales and market acceptance.

Revenue allocation The manner in which revenue obtained from a transport service is allocated. For example, a revenue of €100,000 may include five per cent (€5,000) to terminal charges to handle/process the traffic.

Revenue budget The process of formulating forecast and objectives relative to revenue including traffic volume during a specified future period for a particular service/trade or Company.

Revenue centre A centre devoted to raising revenue with no responsibility for production, e.g. a sales centre, often used in a 'not-for-profit organisation'.

Revenue dilution The process of introducing lower rates/tariffs thereby lowering the average yield per unit conveyed and in so doing dilutes/reduces revenue.

Revenue expenditure Expenditure in the supply and manufacture of goods and provision of services charged in the accounting period in which they are consumed. This includes repairs and depreciation of fixed assets as distinct from the provision of those assets.

Revenue passenger mile (RPM) One paying passenger carried one mile, a basic statistical unit in the airline industry. Also applied to rail, and bus/coach operation.

Revenue production The process of a Company generating income from a particular activity of its business such as a particular promotion for a new product, retail sales outlets, a bus service and so on.

Revenue projects Projects devised to increase income in a particular Company/Entity. This maybe an investment project involving the modernisation of a coach fleet; the provision of a number of inclusive tours extending the range of this product by a travel agent; or selective price increases of a number of products in a supermarket.

Reversible An option given to the charterer to add together the time allowed for loading and discharging relative to the terms of a particular charter party. Where the option is exercised the effect is the same as a total time being specified to cover both operations. A charter party term.

Reversible kingpin A trailer kingpin which is reversible or interchangeable thus allowing the trailer to be coupled to a variety of tractive units.

Reversible lay days An option given to the charterer to add together the time allowed for loading and discharging relative to the terms of a particular charter party. Where the option is exercised the effect is the same as a total time being specified to cover both operations.

Reversible laytime It is an option given to charterer to add together the time allowed for loading and discharging. Where the option is exercised the effect is the same as a total time being specified to cover both operations. A BIMCO term. It is the opposite of normal (non reversible) laytime in that the calculation for the loading port(s) is carried out and the result – e.g. 'days lost/saved' – held in abeyance to be added to/subtracted from the laytime allowed at the discharge port(s), before proceeding with the discharge port(s) calculations. An example arises with '16 days all purposes' (i.e. 16 total days for loading and discharging), although Reversible laytime conditions can be applied to the fixture – if the parties so negotiate. Once the vessel has finished at the load port, the days saved/lost are transferred to the discharge port allowance. See also normal laytime and non reversible laytime entries.

Reversible pallet – doubled decked pallet A flat pallet with similar top and bottom decks, either of which will take the full rated load.

Reversible stacking cone A single stacker that can be used either way up.

Revetment A stone wall or stack intended for diverting or breaking the force of the current in a river.

Revitalisation The process of rejuvenating/resuscitating/bringing up-to-date a particular situation. This could involve the injection of new ideas/methods/concepts to effectively revive the situation. This may involve in transport terms the provision of additional off-peak fares to encourage more travel in such periods.
Revocable credit A revocable credit can be cancelled or amended by the buyer without your supplier's prior knowledge, any time before seller presents the documents for payment. Once documents complying with the terms of the credit have been presented to the bank nominated in the credit, payment is guaranteed, but because of the possibility of cancellation prior to this, revocable credits are seldom used.
Revolving credits The revolving documentary credit may revolve in relation either to time or to value. Moreover it may either revolve automatically or the way in which it is allowed to revolve may be qualified. It means that when payment has been made for documents, the value of the credit is reinstated without the need for issuing an amendment. Documentary credits that revolve automatically without further qualification are extremely rare. Hence, revolving credits arise where there are shipments in a series at intervals and the parties wish the programme to proceed without interruption: accordingly a credit is established for a certain sum and as shipments are made the credit automatically becomes available as the documents are presented for subsequent payment to the exporter for the goods.
rf Rise of floor (or deadrise of a deck).
RFA Royal Fleet Auxiliary.
RFD Raised Fore deck.
RFR Required freight rate.
RFS Received for Shipment.
RFT Registered Floor Trader – IPE term.
RG VARIG SA (Viacao Aérea Rio Grandense) – an IATA Member.
RGE Range.
RGR Returned Goods Relief.
RGSS Register General of Shipping and Seaman.
RH Air Zimbabwe. The National airline of Zimbabwe.
RHA Road Haulage Association.
Rhine – Main – Danube This canal links the North Sea through Rotterdam with the Black Sea a distance of 3500 km.
Rht Engines fitted with reheaters.
RHT Radio hand held terminal – associated with container berths.
RHV Road haulage vehicle.
RI Registro Italiano Navale – Italian ship classification society.
R/I Reinsurance.
Rial The currency of Iran. See Appendix 'C'.
Rial invoice An invoice payable in Rial currency relative to an export sales invoice.
Rial Omani The currency of Oman. See Appendix 'C'.
Rial Omani invoice An invoice payable in Rial Omani currency relative to an export sales invoice.
RIBA Royal Institute of British Architects.
Rich market A market place where excessive wealth exists and therefore strong buying power to purchase products/services. Countries of immense wealth include USA, Japan and Brunei.
RICS Royal Institute of Chartered Surveyors.
RID Regulations concerning the International Carriage of Dangerous Goods by Rail; or The Royal Irrigation Department (Thailand).

Ridge A line or wall of broken ice forced up by pressure. May be fresh or weathered. The submerged volume of broken ice under a ridge forced downwards by pressure is termed an ice keel.
Riel The currency of Kampuchea. See appendix 'C'.
Riel invoice An invoice payable in riel currency relative to an export sales invoice.
Rigger A dockside labourer responsible for fitting lashing gear. Also termed Lasher.
Right-lock twist-lock A twist-lock which is locked when the handle is pointing to the right.
Right of establishment The right to establish an economic activity in any country normally implying free of controls not imposed also on domestic industry.
Rights issue The issue of new shares to existing shareholders in a fixed ratio to those already held at a price which is generally below the market price of the old shares.
Rigid vehicle Vehicle with a rigid chassis construction.
RIMS North American Risk & Insurance Management Society.
RIN Royal Institute of Navigation.
RINA Registro Italiano Navale – the Italian Ship Classification Society.
Rina certificate A certificate issued by the Italian government for carrying dangerous goods in Italian waters.
RING The official trading location of an exchange. A BIFFEX term; or a five minute trading period in a single metal on the LME.
Ring Dealing Broker An organisation which is a full member of a futures exchange, entitled to a seat on The Ring and able to deal directly across The Ring.
Ring fenced A business/commercial term to describe a situation which totally restricts access to market penetration, release of data and so on.
Ringi The decision making process in Japan embracing all involved in a project giving its approval. This involves obtaining agreement at all levels of management. The document is called Ringisho which carries all the data and recommendations.
Ringisho See Ringi entry.
R/IRS Reinsurers.
RISDA Rubber Industry Smallholder Development Authority (Malaysia).
Risk A fortuity, it does not embrace inevitable loss. The term is used to define causes of loss covered by a policy. Overall something which may happen but not something that is inevitable. Used in insurance in many senses, notably: the subject matter of insurance; uncertainty as to outcome of event; probability of loss; the hazard or peril insured against; and danger.
Risk analysis The process of analysing a particular investment situation or general business situation with a view to determining the risk areas, and how they can best be minimised. The risk may be technical, financial, commercial, political, and so on. It maybe, for example, the replacement of a fleet of vessels engaged in a particular trade and the risk arises whether the same company will carry the cargo or it maybe transferred to the National Flag. In so doing risk analysis may suggest the vessels be built to a multi purpose design to permit them to be used in another trade should such circumstances so arise.
Risk Assessment A process of reviewing the risks attached to operations at offshore installations, most often carried out along quantitative lines. In some countries this procedure is known as a safety case but it is mainly applicable to fixed installations.
Risk assessment of trading and investing This involves the process of evaluating the risk of trading and/or investing into a country or company. This is often

Risk aversion

overseas trading and/or investing and includes economic, political, technical, legal, and social factors.

Risk aversion To develop a strategy to minimise risk.

Risk basis Excess of loss reinsurance by which the reinsurer protects the reinsured against losses affecting individual exposure units.

Risk excess An excess of loss reinsurance where the contract is expressed "excess of each and every loss, each and every risk".

Risk factors Features that will influence the frequency or severity of losses.

Risk in the goods Risk in the goods means, the risk that the goods might be destroyed, following passing of title to the goods. If they are destroyed the question arises as to whether the buyer or the seller must stand the loss. A legal term.

Risk management The application of financial analysis and diverse financial instruments to the control and the reduction of selected types of risk.

Risk takers A person/company/entity prone to taking risks to realise an objective. It maybe financial, product development/launch, merger/acquisition, currency speculation, rights share issue and so on.

RIT Rangoon Institute of Technology.

River bus A bus service operating on a canal or river network, the most famous of which is found in Venice, Italy.

River port A port situated on a river.

River tonnage dues Charge raised on ship entering river or port usually based on her g.r.t.

Riyal The currency of Saudi Arabia, Qatar, and Yemen Arab Republic. See Appendix 'C'.

Riyal invoice An invoice payable in Riyal currency relative to an export sales invoice.

RJ Alia – The Royal Jordanian Airline – the national airline of Jordon and member of IATA.

RK Rubbing keel (of a ship); or Air Afrique – an IATA airline.

RKT Reichskraftverkehrstrif (Official German Road Transport Tariff).

RLR Report Loss Ratio.

RM Air Moldova International – IATA Member Airline.

RMC Refrigerated cargo installation classified with British Corporation.

RMD Rhine – Main – Danube canal.

RMI Resource Management International (Consultants in USA).

RMPM Rotterdam Municipal Port Management.

RMT National Union of Rail, Maritime and Transport workers.

RN Release note – receipt signed by customer acknowledging delivery of goods; Rubber non-continuous Liner; Euralair International – an IATA airline member; or Royal Navy.

RNAV Area Navigation – IATA term.

RNCs Raster navigational charts.

RND Round (voyage).

RNG Routes and NavAids Group – IATA term.

RNGE Range.

RNLI Royal National Lifeboat Institution.

RNMP Replacement of the Nautical Mile Panel – ICAO term.

RNPC Required Navigation Performance Capabilities – IATA term.

RNR Registru Naval Roman – Italian Ship Classification Society; or rate note reported.

RO Receiving order; Romania – Customs abbreviation; or Romanian Air Transport SA (TAROM) – an IATA Member Airline

R/O Routing order.

ROA Reinsurance offices association.

Road carrier The party undertaking transport by road from one point to another such as indicated in the contract.

ROADS The ports of Hampton Roads, i.e. Norfolk, Newport, News and Sewells Point or can be 'roads' leading into various ports.

Road Show A marketing term to describe the presentation to invited audiences of a Company product or service. Usually the road show involves a number of presentations at different key locations and may involve seven presentations in a week's programme.

Road Toll A charge/tariff raised on road users embracing motorists, pedestrians, motor cyclists, cyclists, coaches, lorries and so on.

Road Train The combination of two road haulage vehicles involving a rigid two or three axle unit hauling a two or three axle trailer. The road train overall length is 18 metres including the tow bar and has a gross weight of up to 45 tonnes (40 tonnes in UK).

Roadworthiness The process of providing a vehicle capable of undertaking the journey planned having regard to the commodity to be conveyed, the statutory obligations imposed on the owner and operator. Overall, the prime consideration is safety, vehicle stability and axle loading.

ROB Remaining on board i.e. cargo or fuel; or round of beam.

ROCE Return on capital employed.

Rocket motors Articles consisting of a solid, liquid or hypergolic fuel contained in a cylinder fitted with one or more nozzles. They are designed to propel a rocket or a guided missile. The term includes: rocket motors; rocket motors with hypergolic liquids, with or without expelling charge; and rocket motors, liquid fuelled. A term associated with dangerous cargo shipments.

Rockets Articles consisting of a rocket motor and a payload which may be an explosive warhead or other device. The term includes "Guided Missiles" and: rockets, line-throwing; rockets, liquid-fuelled, with bursting charge; rockets, with expelling charge; and rockets, with inert head. A term associated with dangerous cargo shipments.

Rod A lashing constructed out of a solid metal alloy.

Rodent Control Certificate Document renewable half yearly – issued to home trade ships indicating vessel free of rate infestation.

ROG Receipt of goods.

ROI Return on investment.

Rollback Commitment to progressively dismantle all trade restrictions which are inconsistent with the WTO within an agreed timeframe not later than by the end of the Uruguay Round. A WTO term.

Roller bed truck A truck equipped with a roller bed to facilitate the handling of unit load devices containing air freight.

Rolling contract An employment contract which offers an employee a contract of limited duration usually varying between three to five years. On expiry of the contract the employee is usually offered a renewal. Such renewal is subject to the employers strategy and overall circumstances.

Rolling market entry The process of entering one country and then another to sell/promote a product.

Rolling programme A term usually associated with investment indicating the provision of a continuous investment programme of a particular product. For example it maybe the fleet of 3 ton vehicles will be replaced on a seven year cycle basis, or a number of railway ticket offices will be subject to a face lift covering a ten year plan/programme.

Roll over The view of interest rates at agreed intervals, based on a formula agreed at the time the loan was made. A banking term; or the transfer of a futures or options position from one delivery/expiry month to another – always involving the purchase (sale) of the nearby month and simultaneous corresponding sale (purchase) of a further out delivery month. An International Petroleum Exchange term; or the phenomenon where the stability of two stratified layers of liquid of differing relative density is disturbed resulting in a spontaneous rapid mixing of the layers accompanied in the case of liquefied gases, by violent vapour evolution. A term associated with dangerous cargo shipments.

Roll-over loan A loan with a fixed repayment period but with provision for review of interest at regular intervals. The interest rate will usually be expressed as a fixed percentage or spread above a particular inter-bank offer rate.

Roll trailer Special trailer for terminal haulage and stowage on board Roll on – Roll off vessels. Also termed Mafi trailer.

Ro/Lo Roll-on, roll-off/lift-on, lift-off (type of vessel).

ROM Random Only Memory. Electronic, read only, memory which is not volatile. It does not lose its contents when power is removed. See 'Storage devices". Computer term.

Romalpa clause This clause enables the seller to reserve ownership rights over the goods until they have been paid for. It is used in international sales transaction.

Romalpa sale A clause which the seller puts into his sale contracts reserving title to his goods until they are paid for. The result of such a clause is to place the seller at the top of the list of creditors in the event of the insolvency of the buyer.

Rome Convention Treaty, administered by WIPO, UNESCO and ILO, for the protection of the works of performers, broadcasting organisations and producers of phonograms. A WTO term.

Roof load External static and dynamic loads imposed on the roof of a container.

Roof rail Longitudinal structural member situated at the top edge on either side of the container.

Room night One hotel room occupied by one or more guests for one night; see bed night.

Room type General room description by type, such as single, twin, suite.

ROP Return of Premium.

RoPax Roll On/Roll Off and Passengers.

RORCE Rate of return on capital employed.

Ro/Ro Roll on/roll off ship designed for conveyance of cars, road haulage units, and unitised cargo. A number of vessels so equipped have passenger accommodation for coaches, motorists and foot passengers; a vehicular ferry service onto which goods and containers can be driven, usually via a ramp.

Ro/Ro berth A purpose built berth situated at a seaport handling Ro/Ro tonnage. The berth is equipped with a ramp to enable the vehicles to be driven on or off the vessel.

Ro/Ro Cargo/Ro/Ro Passenger Vessel arranged for Roll-on Roll-off loading/discharging of vehicles (road and/or rail) as cargo and/or passenger conveyances. Cargo carried in wheeled containers or wheeled trailers aboard and moving on to the ship and off it on wheels, usually over ramps.

Ro/Ro Drivers cabins Cabins allocated to Ro/Ro drivers which accompany their vehicles on a particular vessel. Usually such cabins are available free on the ferry and are included in the overall freight tariff for the vehicle.

Ro/Ro Drivers Club A club available to Ro/Ro drivers operating through international road haulage services on a particular shipping route. Membership benefits usually include subsidised shipboard meals, free cabins, social events etc. It is under the aegis of the Shipping Company.

Ro/Ro Drivers restaurant A restaurant provided on a Ferry or at a seaport handling international road haulage traffic provided for exclusive use by Ro/Ro drivers.

Ro/Ro King 7 A multi purpose modern Roll on/Roll off cargo ship with an overall length of 150 m; breadth of 23 m; draught 6.90 m; deadweight tonnage of 7450 tonnes; and speed of 14.8 knots. She is a two deck vessel, equipped with two long lanes, stern-quarter ramp, four hoistable car decks and a scissors lift. The ship has a variable capacity of 470 TEU's, or 360 TEU's on MAFI trailers, or 111 TIR 12 m length road trailers, or 1485 cars of 4.2 m length or homogeneous cargo of 7450 tonnes. It is built by British Shipbuilders.

Ro/Ro rates A tariff devised for the shipment of freight vehicles conveyed on a Ro/Ro type of ship whereby the vehicle is driven on/off the vessel via a ramp. The rate is usually based on the length of the vehicle, whether accompanied or unaccompanied, and empty or loaded.

Ro/Ro vessel Roll on/roll off type of vessel – a vehicular ferry onto which wheeled cargo is conveyed and transhipped by being driven on/off the ship via a ramp.

ROS Route Orientation Scheme – IATA term.

Rossby wave In the atmosphere a wave in the general circulation in one of the principal zones of westerly winds, characterised by large wavelength (c.6000 km), significant amplitude (c.3000 km) and slow movement, which can be both eastward and westward relative to the Earth. In the ocean, similar waves have a wavelength of an order of a few hundred kilometres and nearly always move westward relative to the Earth.

Roster A work schedule detailing the date, place/time of duty and nature of the work. This may involve a coach driver, a factory employee, a clerk, and so on.

ROT Reference our telex; or Rate of turn.

Rotation The sequence in which a vessel calls at the ports on her itinerary.

Rotation Clause A provision in a large open cover allowing small declarations to be allocated to a small group of the subscribing underwriters instead of over the whole subscription. The subscribing underwriters are split into small groups, each group taking the declaration in rotation.

ROTC Return on Total Capital.

Rotten ice Sea ice which has become honeycombed and which is in an advanced state of disintegration.

Rouble The currency of USSR. See Appendix 'C'.

Rouble invoice An invoice payable in Russian rouble's currency relative to an export sales contract.

Round amounts Amounts of stock or numbers of shares in which jobbers are ready to deal on a normal basis as opposed to 'odd lot' amounts or numbers.

Round C/P Round charter party.

Round the world service A service which operates on a global basis as found in the Evergreen round the world deep sea container service.

Round trip A transit/journey which combines the outward and homeward transit such as Dover to Calais and Calais to Dover.

Round trip cost The total cost incurred by a transport operator to undertake a round trip.

Round trip fare The rate charged for a trip to a destination and return by the same route.

Round trip rate The rate/tariff embracing both the outward and homeward transit to produce an overall round trip rate. For example, the round trip rate from Harwich to Hook of Holland would be its Harwich-Hook of Holland portion plus the Hook of Holland-Harwich leg.

Round trip time The total transit time to complete a specified round trip by a transport unit such as a ship, airline, road vehicle and so on. An example arises on the basis of a departure time of 0700 hours from terminal A; an arrival time at terminal B of 0900 hours; a departure time of 1000 hours from Terminal B and arrival time at terminal A of 1200 hours. This produces a round trip time of 5 hours.

Round voyage See round trip.

Route Path taken by a vessel.

Route analysis The process of analysing various route options for a particular transport mode(s) and transit(s).

Route capacity The total volume of traffic which is capable of passing through a specified route within a particular period for a specified transport mode(s).

Route diversion The process of diverting a transport unit such as coach, train, from its customary route to an alternative route. This maybe due to road/rail accident causing a blockage; weather conditions involving a landslide; or engineering work such as bridge repairs.

Route option(s) A situation where a transport operator has more than one route from which to choose for a given journey(ies).

Route order The route a particular consignment must follow as prescribed/issued by the shipper.

Route planning The formulation of the most optimum route for a particular consignment or transport service which maybe road, rail, air and sea. It is particularly relevant in road and rail operation – especially the former – where numerous routes options can exist between two points. In such a road transport operation factors determining route selection include: traffic congestion; nature of commodity; type of vehicle; restrictions on height and weight; cost of transit; time of travel day or over night; location of two depots; route capacity; transport unit speed; and so on.

Route selection The process of evaluating the route options for a particular consignment/journey and deciding on the one most ideal and yielding the best value added benefit having regard to all the circumstances including convenience, cost, journey time and quality of service both to the exporter and more especially to the importer.

Routes section A department found usually in International Road Haulier/Freight Forwarders company dealing with the routing of consignment/vehicles engaged on international overland transits which may involve the route option of a sea crossing/ferry route

Route, through The total route from point of departure to point of destination.

Routine control Repetitive procedures of comparing feedback information with defined standards and acting to correct deviations.

Routing In Maritime terms a series of measures aimed to reduce the risk of casualties through the provision of traffic schemes such as two way routes, traffic separation schemes, inshore traffic zones, deep water routes, etc; or it is designation of a route of any transport mode for a specified consignment and recorded on the Bill of Lading, Air Waybill or consignment note.

Routing certificate A document indicating the routing of a specified consignment.

Rov Registry of Vehicles (Singapore); or Remotely Operated vehicle – employed as a diving/research vessel under water.

Row A transverse line of containers or stacks across the ship.

Royal P & O Ned Lloyd A mega container shipowner.

Royal Swazi National Airways The National Airline of Swaziland.

RP Return premium; radio phone – ships navigation aid; Portuguese Register of Shipping – the Portuguese Ship Classification Society; or Rinave Portuguesa – Customs abbreviation.

R/p Return to port for orders.

RPA The Rural Payments Agency. The UK body responsible for administering export and refunds.

RPG Regional Planning Group (ICAO term).

RPI Retail Price Index.

RPL Repetitive Flight Plan – IATA term.

RPM Revolutions per minute; or revenue passenger mile.

RPR Reglement de police pour la Navigation due Rhin; or Return of Premium Rider – insurance term.

RQD Raised quarter deck (on a ship).

RQRD Required.

RR Reverse reduction; Rumanian Register of Shipping – the Rumanian ship classification society; or retrospective rating – insurance term.

RRIM Rubber Research Institute of Malaysia.

RS Register of Shipping (Russia). The former Russian Ship Classification Society.

RSC Rescue Sub Centre.

RSCWG Runway Surface Conditions Working Group (ICAO term).

RSD Raised shelter deck (on a ship).

RSTRAM Regulations for the safe transport of radio active materials.

RSU Remote Switching Units.

RT Right time of ship departure/arrival; radio telephone; radio telegraph; rye terms; road transport, or rail transport.

RTA Regional Trade Agreements; or Road Traffic Act.

RTBA Rate to be agreed.

RTC Regional Technical Conference (IATA term).

RTCA Radio Technical Commission for Aeronautics (USA term).

RTD Regional Technical Director – IATA term.

RTG Rubber tyred gantry (crane).

RTGS Real-Time Gross Settlement: the gross settlement in real time of payments across settlement accounts maintained at the central bank.

RTh Radio-telephone (high frequency).

RTm Radio-telephone (medium frequency)

RTO Request to off-load; or Regional Technical Office – IATA term.

RTOL Reduced Take-off and Landing – IATA term.

RTP Returnable Transit Packaging.

RTR Regional Technical Representative – IATA term.

RTv Radio-telephone (very high frequency).

Ru Russia – Customs abbreviation.

Rubber tyred traffic A shipping term associated with conveyance of road vehicles which are classified as rubber tyred traffic. This will include cars, coaches, road haulage vehicles, trade cars and so on.

RUCATSE Runway Capacity to Serve the South East – IATA term.

Rudder The directional control of the vessel.

Rules of origin Laws, regulations and administrative procedures which determine a product's country of origin. Thus a decision by a customs authority on origin can determine whether a shipment falls within a quota limitation, qualified for a tariff preference or is affected by an anti-dumping duty. These rules can vary widely from country to country. A WTO term.

Rules of practice A set of rules drawn up by the Association of Average Adjusters to give their members guidance in assessing the relationship of circumstances to general average and particular average losses.

Run in The part of the ships hold or tween deck which is not directly under the hatch square opening.

Running cost A chartering term which specifies the items which make up the 'running cost' of the vessel. These include insurance, crew, fuel, radar, repairs and servicing, provision of spare parts for the ship, hotel cost for repatriating crew members and so on.

Running days Days which follow on immediately after the other relative to charter party terms. Alternatively, it maybe termed 'consecutive days'. Each 'running day' is of 24 hours each and includes Sundays or Fridays and holidays and all days are treated the same. Other interruptions, such as bad weather, can be assumed not to effect laytime unless expressly stated to do so in the charter party. The same criteria applies to 'running hours'.

Running Down Clause A term sometimes used in market practice to define the collision liability clause in a hull policy. It extends the hull policy to incorporate amounts paid by a shipowner as legal liability consequent upon collision. The Institute Clauses limit underwriters' liability to three quarters of the amount paid with a further limit of three quarters of the insured value, whichever is less. Cover extends only in respect of damage to the other ship, to property on the other ship and loss of use of the other ship.

Running hours The period specified in the charter party during which cargo operations (loading and/or discharging cargo) are to be performed. See 'running days' entry.

Running working days The period specified in the charter party during which cargo operations (loading and/or discharging cargo) are to be performed. See 'running days' entry.

Run Off Keeping open a cancelled or expired insurance or reinsurance to accommodate outstanding claims and adjustments.

Run-off account A year of account of a Lloyd's syndicate which has been left open after the date at which that account would normally have been closed by reinsurance.

Run-of-the-house rate A flat price at which a hotel agrees to offer any of its rooms to a group.

Run of the ship The narrowing of the ship at the bow and stern.

Rupee The currency of Seychelles, Nepal, India and Pakistan. See Appendix 'C'.

Rupee invoice An invoice payable in rupees relative to an export sales contract.

Rupiah The currency of Republic of Indonesia. See Appendix 'C'.

Rupiah invoice An invoice payable in rupiah currency relative to an export sales invoice.

Rush hour The peak period(s) of travel on a transport network. This is usually in the morning 07.30/08.30 hours and evening 1700/1800 hours periods when the volume of passengers travelling to and from work reach their peak.

Russcon Code name of Chamber of Shipping approved charter party used in grain trade.

Rust bucket A term sometimes applied to a vessel in very poor condition and more than fifteen years old.

Ruswood The Baltic and International Maritime Council Russian Wood Charter Party 1961 (Revised 1995) embracing Russian Baltic, White Sea, Barents Sea and Kara Sea Ports to the United Kingdom, the Republic of Ireland and other Countries.

Ruswoodbill Code name for Bill of Lading used for shipments chartered on Ruswood Charter Party issued 1995.

RV See recreational vehicle.

RVR Reglement de visit des bateaux du Rhin; or Runway Visual Range – IATA term.

RVSM Reduced Vertical Separation Minima – IATA term.

RVW Road vehicle weight.

RW Riveted and welded; or reduce weight.

Rwanda Franc The currency of the Republic of Rwanda. See Appendix 'C'.

Rwanda Franc invoice An invoice payable in Rwanda franc currency relative to an export sales contract.

Rws Round the world service – usually referred to within the container shipowner context of Evergreen Line.

Rx Receiver.

Ryokan A traditional inn in Japan.

S7 Siberia Airlines (joint Stock Company Siberia Airlines) – IATA member Airline.

s Starboard side; or second – unit of time.

S Steamship; summer; single expansion engines; stations; special; strength; or south – a point on the compass.

s/a Safe arrival; or subject to approval.

SA South African Airways. The National airline of South Africa; single acting (machinery) (on a ship); or selective availability – this is the option of the US Department of Defence to scramble GPS signals and alter positional accuracy of the GPS operation.

SAARC South Asia Association of Regional Co-operation.

Sabena Airways The Belgian National airline and member of IATA.

Sabre American Airlines' computerised reservations system.

s & c Shipper and carrier.

SACAA South American Civil Aviation Association – IATA term.

SACC Southern African Development Co-ordination Conference.

SACD South African Container depot.

Sacrifice The deliberate casting away or destruction of property to prevent greater loss. General average sacrifice is for the common good and saved interests make good the sacrifice in proportion to the saving enjoyed.

SACU Southern African Customs Union comprising Botswana, Lesotho, Namibia, South Africa and Swaziland.

SAD Single Administrative Document (UK); Customs Declaration Form for both import and export cargo – Form C88 (see separate entry).

SADC Southern African Development Community.

Saddle tanks Tanks usually for water ballast found in a ship which saddle or are fitted over the upper sides of the main cargo tanks.

SADIS Satellite Distribution System – IATA term.

SAE Stamped addressed envelope; or Society of Automotive Engineers (USA). National Standards Body in USA. See ISO.

SAECS Southern African Europe Container Service.

S & FA Shipping and Forwarding Agent.

SAFAC Safety Advisory Committee – IATA term.

Safari An expedition involving hunting animals and game especially prevalent in Central and East African. Tour operators offer and plan such holidays which involve professional qualified experienced personnel to lead them. Also available in the Sub Continents, Far East and South America.

SAFCON Certificate Document confirming a cargo ship meets minimum standards of construction, fire protection equipment, machinery etc. In a classed vessel, the certificate involving survey work is administered by an approved classification society such as Lloyd's Register of Shipping.

Safe aground Term in a charter party which allows the charterer to order the ship to a port or place where she may safely touch the bottom. The term is often part of the expression 'not always afloat but safe aground'.

Safe berth A berth which, during the relevant period of time, the ship can reach remain at and depart from without, in the absence of some abnormal occurrence, being exposed to danger which cannot be avoided by good navigation and seamanship.

Safeguard measures Action taken to protect a specific industry from an unexpected build-up of imports. Article XIX (of the GATT 1994) of the WTO requires that the imports should be causing actual damage to the domestic industry or, at least, threatening it. In these circumstances, the country affected may either increase tariffs or introduce quantitative restrictions. A WTO term.

Safeguards Sometimes referred to as the 'escape clause' or emergency measures clause, refers to Article XIX of the WTO which permits (under certain conditions) the imposition of temporary (or emergency) trade restrictions on imports which have caused or threatened to cause injury to domestic industry. A WTO definition.

Safe port A port which, during the relevant period of time, the ship can reach, enter, remain at and depart from without, in the absence of some abnormal occurrence, being exposed to danger which cannot be avoided by good navigation and seamanship. Overall, this embraces meteorologically, physically, and politically safe and free from temporary dangers such as ice blockage.

Safety equipment certificate Document issued in UK by the Department of Transport confirming the ship has observed the appropriate Merchant Shipping Acts.

Safety Management Certificate It is issued by an accredited Ship Classification Society, e.g. Lloyd's Register of Shipping London, to certify that the ship specified has been audited by an accredited classification society and that it complies with the requirements of the International Management Code for the Safe Operation of Ships and for Pollution Prevention (ISM Code) adopted by the organisation by resolution A741(18) as amended, following verification that the Document of Compliance (10.12) for the Company is applicable to this type of ship.

Safety Management System Under the ISM Code, shipowners are required to develop, implement, and maintain a safety management system with conformance of shore based management operations to standards validated by a Document of Compliance. It also requires audited compliance of the vessel(s) to retain mandatory Safety Management Certificates.

Safety stock A quantity of stock planned to be in inventory to protect against fluctuations in demand and/or supply; or in the context of production scheduling, safety stock can refer to additional inventory and/or capacity planned as protection against forecast errors and/or short term changes in the backlog. Also termed over planning and market hedge.

Safety surplus of bunkers The quantity of bunkers on board a ship which will not only enable the vessel to complete the particular leg of the voyage, but also have an adequate quantity on board so that in the event of unforeseen circumstances such as bad weather and/or engine difficulties, the ship can still safely make port and does not suffer the risk of running out of fuel at sea.

Safe working load The maximum load permitted in working specified equipment, such as lifting apparatus/ derricks etc., with a 20 tonne maximum. The equipment is usually tested regularly in accordance with Statutory regulations. Marking engraved on derricks, cranes and other lifting gear.

Safmarine South African Marine Corp. Ltd. A major container operator. A South African Shipping Line.

SAFTA South Asian Free Trade Agreement.

Sagging Excessive weight of cargo situated amidships thereby causing aft and forward to be higher than amidships portions. Overall, it is stress in which the ends of a ship are comparatively unsupported.

SAGT South Asia Gateway Terminal – the Port of Colombo container berth.

Said to contain A term in a Bill of Lading signifying that the master and the carrier are unaware of the nature or quantity of the contents in a carton, crate, container or bundle, and are relying on the description furnished by the shipper.

Sailing The operation of a ship from port of departure or other specified place, to port of arrival or other specified location.

Sailing cards Cards issued by shipbrokers to their customers, giving particulars of the ship, or ships they are about to load, the loading berth, date of departure, etc. Details also available on 'online' computers.

Sailing schedule Details of a ship's timetable for a particular service or trade giving times of departure and arrival at the various ports of call. Details also available on 'online' computers.

SAJ Shipbuilders Association of Japan.

Salami system A customs facility available relative to the single administration document whereby the document is split into its constituent parts so that formalities for export, transit documents, and import entry may be dealt with as separate elements. Usually termed split use option.

Salaried Agent An Agent resident in the country of operation and is paid a salary with or without expenses and fringe benefits, by his principal.

Sale and Purchase Broker Person who acts on behalf of the buyer or seller of ships thereby bringing the two parties together.

Sale Charges Costs incurred in selling goods. Underwriters are interested only when the sale is the result of insured damage or when they authorise the sale for their own account.

Sale form 1993 Code name of Norwegian Shipbrokers Association's Memorandum of Agreement for sale and purchase of ships. Adopted by BIMCO in 1956 and revised 1966, 1983, 1986/87 and 1993.

Sale of Vessel Clause A hull clause providing that the policy is cancelled automatically on the sale of the insured vessel or transfer of management.

Sales Agent An Agent engaged in selling his principal product(s)/service(s) usually on a Commission earning basis.

Sales aid Resources available to aid the process of selling/promoting a particular product(s) or service(s). This may include catalogue, brochures, visual aids, leaflets, posters, general handouts, gifts, testimonials, website, articles and so on. The important aspect relative to an effective sales aid is that it must be ideally 'tailor made' for the job after careful analysis, research and evaluation.

Sales and Marketing Director A member of a Board of the Directors reporting to the Managing Director. The duties would include, subject to the size of the Company, the ability to identify opportunities in the market place, and devise a strategy plan, initiate business development, leadership and motivation. The Director would be responsible for the Sales and Marketing Budget of the Company and its overall strategy.

Sales Budget A budget devised by the Sales/Marketing Director featuring income and expenditure within his department. The income/revenue would identify the sources and commodities/product(s)/service(s) and the expenditure detail e.g. sales department staff costs, agents commission, marketing and promotion cost embracing brochures, sponsorships, advertising and so on. In the larger Company there may exist a separate marketing/promotion budget. The budget will be for one year and be reviewed regularly to institute remedial measures to counter inadequacies such as a fall in sales results or need for greater selling effort. Agents and Sales personnel would work towards individual targets as made up on the budget.

Sales bulletin An information document/circular/paper prepared by a Company/Entity by its sales department circulated/distributed on a regular basis such as each month or three months. Many companies distribute it over the web. It will contain information of interest to the Company/Entity clients including potential clients and contain primary sales information including new promotions, price changes, personnel changes, matters relevant to after sales service, trade exhibitions and so on. It is basically a sales aid.

Sales by products The specific sales of individual products for a particular period/area/region/country.

Sales by territories The specific sales of one or more products in a particular area which maybe a region/country. It is likely to indicate the period such as monthly/quarterly/annually and identify individual products or services.

Sales calls Process of a sales person visiting existing and potential customers with a view to formulating a sale.

Sales conference A conference conducted to promote, for example, a product launch; 'teach in' to salesmen on future policy and so on.

Salescrap 87 Code name of BIMCO approved standard contract for the sale of vessels for demolition. See Demolishcon 2001 entry.

Sales Director A Company Director responsible for sales. This could embrace, depending on the size and nature of the Company business, the responsibility for the marketing and sales plan, advertising, agency policy, sales training, market research, customer relations, and all matters relative to the promotion and selling of the company products.

Sales Distributor agreement An agreement which reflects an importer who buys direct from a manufacturer or supplier and sells them in the market place. See distributor entry.

Sales drive Process of initiating a sales campaign to generate product/service awareness and thereby stimulate sales volume.

Sales Engineer A qualified engineer with responsibility for selling.

Sales Executive A managerial position in a Company/Entity involved in prime task of selling/promoting the Company product(s)/service(s).

Sales force Personnel forming the sales resources of a Company/Entity for a particular product(s)/service(s).

Sales forecasting The predicted sales income/volume over a specified period relative to a particular product(s) or service sector.

Sales lead Process of following up an item of market research data which could result in an ultimate sale being secured.

Sales Manager The executive in a particular company responsible for the sales personnel under his/her control. Responsibilities include sales management, budget training, execution of company sales policy, and internal sales service which includes sale analysis and customer relations department, etc.

Sales Negotiation plan A plan devised with a mission statement to secure a sale of a product/service on a domestic/local or international basis. It will embrace flexibility, decision making process, provisional time scale, costing/profit margins, terms of sale, product specification, culture, international regulations, currency and empathy. Overall, the strategy element of the plan will reflect the culture management environment and the nature of the product/service together with competition.

Sales Office An office/organisation within a Company/Entity dealing with sales matters.

Sales of goods law The legislation dealing with the rights of consumers who buy defective goods and other questions connected with the sale of goods.

Sales order processing The task of processing a sales order.

Sales penetration The process of conducting a sales strategy whereby the total market potential has been made aware of the entity product/service and in so doing the full market potential has been realised.

Sales pitch The place/point of sale where goods are sold.

Sales plan A plan devised by the Sales Manager which features sales objectives and the tactics/policy to realise them. It would include sales targets identified with members of the sales force.

Sales planning The process of devising the sales plan in accordance with company strategy/objectives and the sale budget. The plan would be reviewed regularly in the light of budget results, market conditions and company circumstances.

Sales programme The programme detailing the plan to realise the sales objectives. This will embrace field sales visits, trade fairs, sponsorship, advertising and so on.

Sales promotion The range of activities designed to promote sales but excluding face to face selling and advertising. In particular it embraces merchandising/brochures/in store displays and below the line advertising.

Sales publicity The process of publicising through the press and other media sales information. This could take the form of a press release to mark the sale of the millionth car of a particular model to a client.

Sales records Assembly of data relating to sales achieved in a particular company or regarding particular product/service or range thereof. In a modern sales office its sales records would be computerised.

Sales report A report prepared by a salesman on a particular visit to a company, client and so on. It will detail the nature of the visit, personnel seen, date and time, business secured and on what terms, action items 'in house' of the sales company, name of sales person, details of any order/time scale and price and so on. Alternatively, the sales executive can visit a new potential market maybe overseas and produce a comprehensive sales report thereon. An important document within a sales department.

Sales research The systematic gathering, recording and analysing of data of the selling activities of a particular entity/company. This includes the appropriateness of various selling methods, effectiveness of 'customer relations' department, adequacy of retail outlets and so on.

Sales strategy A sales plan which would identify the sales objectives and the tactics/policy to realise them. These include sales targets, selling technique, sales budget and client contact.

Sales target The target set covering a particular period and commodity/service to be attained by a person/entity/sales force and so on. It maybe 10,000 cars during the months of August/September/October.

Sales tax A tax raised by government based on the retail price of product(s) and/or service(s). It is usually raised on percentage basis such as four per cent on $100 would yield four dollars sales tax.

SALM Single Anchor Leg Mooring.

Saltcon Code name of charter party used in salt trade.

Salvage In maritime terms this relates to action carried out by a third party in the absence of contract whereby property in peril at sea is saved. The term is often used to define the money paid to a salvor as reward for such service, but this should be termed a 'salve award'. In marine insurance practice a salvage award, or a part thereof, which is recoverable under a marine insurance policy is termed a 'salve charge'. The term 'salve' may be used, in non-marine practice, to define property salved; as in the case of goods saved from a fire on land.

Salvage agreement Document completed by Master agreeing to the appointment of an arbitrator to decide salvage services remuneration.

Salvage Association An association, closely connected with the London insurance market, whose function is to take instructions from interested parties (e.g. underwriters) to investigate casualties and to make recommendations for the preservation and protection of property; also to determine the extent and proximate cause of loss when required.

Salvage award An amount awarded to a salvor for services rendered in the salvage of property in peril at sea. The award may be made by a court or by arbitration, depending on the terms of the salvage contract. Underwriters contribute towards a salvage award insofar as the award is in respect of insured property in peril from an insured risk, subject to any restrictions imposed by the policy (e.g. a policy deductible) and, except where the policy provides otherwise, subject to any restrictions to reflect under-insurance, if any.

Salvage charges The award due to a salvor for services rendered in saving the insured property.

Salvage loss A compromised settlement on a cargo policy, usually when the adventure has been terminated short of destination and damaged goods are sold at the intermediate port. The underwriter pays the difference between the sum insured by the policy and the proceeds of the sale.

Salving Ship Vessel engaged in maritime salvage operations. See Salvage.

Salvor The person claiming and receiving salvage for having saved a vessel and cargo, or any part thereof, from impending peril, or recovered after actual loss.

SAM South America(n) – IATA term.

SAMICK Sam ICK Lines – a shipowner.

SAML Security Assertion Mark up Language – logistics computer.

Sample A representative sample selection of a group which will be evaluated. A statistical sampling technique.

Sample licence Authorised temporary export of such as for example articles mounted or set with diamonds.

Sample survey The process of conducting a survey on a very limited scale as a forerunner to a major survey to be conducted.

Sampling The process of selecting a sub-group of a population for the collection of data which may be generalised to the whole population.

Sampling error A statistical term indicating the degree of error/bias found in the sample.

Sampling techniques The process of adopting one of the sampling techniques to identify the profile of the product(s) or service(s) sampled. It maybe the random or stratified sampling.

Samson port See Kingpost entry.

Samudera Shipping Line A South-East Asia container shipping line providing container feeder services in the region.

Samurai Bond A bond issued in Japan by a foreign company denominated in Japanese yen.

Sanitary and phytosanitary regulations Government standards to protect human, animal plant life and health, to help ensure that food is safe for consumption.

Sans recours Without recourse to me 'I accept no liability'.

SAOCLONL Ship and/or cargo lost or not lost.

SAPTA A trading block embracing Bangladesh, Bhutan, India, Maldives, Nepal, Pakistan and Sri Lanka; or South Asian Preferential Trade Agreement.

S & P Sale and Purchase Broker.

Saphir The computerised reservation and national distribution system of Belgium; partner of Galileo.

SAR Saudi Arabia Riyal – a major currency (International); Search & Rescue; or Special Administrative Region of the Peoples Republic of China, formerly Hong Kong and operative from 1st July 1997.

SARBE Search and rescue beacon.

SARPs Standards and Recommended Practices (ICAO term).

SART Search and Rescue Radar Transponder. A radar activated transponder which provides positional indication on a radar display.

SAS Scandinavian Airline Services. An international airline and member of IATA; or special annual survey – of a ship.

sa/sb Safe anchorage/safe berth.

SAT South Atlantic – IATA term.

Satang The currency of Thailand. See Appendix 'C'.

Satang invoice An invoice payable in the Satang currency of Thailand relative to an export sales invoice.

Sat Com Satellite Communication.

Satellite production The process of conducting production/assembly through a satellite. It is usually a logistic network and computer driven.

Satellite Sales Office The process of conducting sales through a satellite using videos in the sales office.

Satellite ticket printer (STP) A ticket printer placed on the premises of an agent's client to deliver airline tickets electronically to a corporate account.

Satellite warehouse Warehouse which is used as an intermediate distribution centre, typically used to service one or more towns in a specific region. May also be used as a 'buffer' warehouse to allow goods to be stored closer to point of final distribution.

Sat Nav Satellite Navigation.

Sato Travel A corporation owned by the US airlines that serves government travel accounts.

SATS South African Transport Services; or Singapore Airport Terminal Services Pte Ltd.

SATSAR Satellite Search and Rescue – IATA term.

SATSHEX Saturdays, Sundays, holidays excluded.

Saturated Vapour Pressure The pressure at which a vapour is in equilibrium with its liquid at a specified temperature.

Saturation coverage The process of advertising a product/service on a mass market basis thereby saturating the market. Usually it is undertaken on a limited period basis.

Saturation point The point at which sales have reached their ultimate level and the market is satisfied to the extent that further sales are unrealistic.

Saudi Arabian Airlines An international airline and member of IATA.

Savia The computerised reservation and national distribution system of Spain; partner of Amadeus.

SAWE Society of Aeronautical Weights Engineers (USA term).

SB Short Bill; single ended boiler(s) (on a ship); or safe berth.

SBA Small Business Administration.

SBAC Society of British Aerospace Companies.

SBC Singapore Broadcasting Corporation; or Strategic Business Centre.

SBE Safe berth (at both ends).

SBM Single Buoy Mooring (with facilities for crude oil loading/discharging).

SBP Solid non-ferrous propeller.

SBS Singapore Bus Service.

sbsp Safe berth safe port.

SBT Segregated Ballast tanks.

SBT-PL Segregated Ballast Tanks – Protected Location.

SBU Strategic Business Units, or small business unit.

SC Steel covers; slip coupling; special commodity rates (air freight); small craft; or Shandong Airlines Co Ltd – An IATA associate member Airline; or Suez Canal.

sc Salvage charges; or screw (propulsion) – on a ship.

SCA Suez Canal Authority.

Scab A person who disobeys union orders of which he or she is a member.

SCAC Standard Carrier Alpha Code.

Scale ten Freighting measurement used in certain trades for various commodities.

Scancon Code name of charter party used in Scandinavian trades.

Scanconbill Code name of BIMCO approved Bill of Lading for shipments on the Scancon Charter Party.

SCANDUTCH A Consortium of shipowners including A/S Det Ostasiatiske Kompagi Copenhagen; Brostrom Rederi AB Gothenburg; Wilh Wihelmsens, Oslo; Compagnie Generale maritime, Paris; Koninjlijke Nedlloyd NV, Rotterdam; and MISC, Kuala Lumpur.

Scanorecon Code name of charter party used in the ore trade.

Scantling draft Maximum permitted draft of a vessel as permitted by the shipowner.

Scantlings Thickness of plating, dimensions of ship frames, beams, girders etc.

SCAR Straight consignment – automatic release.

SCARAB Submersible Craft Assisting Repair and Burial.

Scatterometer An instrument which uses a beam of microwave radiation to measure the scattering properties of the ocean surface which can then be used to estimate the direction and size of the waves, and the associated wind field.

Scattershot The procedure of handling reservations/transactions by dealing with one or more offices independently of other offices and which results in one or more offices not having knowledge of the passengers complete remaining itinerary including the current status of each remaining segment. An ITA definition.

SCBA Self contained breathing apparatus.

SCC Spanish Chamber of Commerce.

SCCUK Swedish Chamber of Commerce of the UK.

Schedule Details of sailings in a particular trade or service; a list attached to a slip, open cover policy or other document usually detailing the rates of premium for various voyages, interest and risk; or 'Schedule of Specific Commitments'- a WTO member's list of commitments regarding market access and bindings regarding national treatment.

Schedule airline An airline operating passenger or cargo services on published schedules.

Scheduled services Scheduled flight, sailing, rail or road services or a combined transport operation performed according to a published timetable.

Schedule integrity The process of the aircraft departing and arriving on time. It measures the variation from the published schedule.

Schedule of Concessions List of bound tariff rates. A WTO term.

Schengen agreement The Schengen agreement was formed in 1985 and by 1997 had 13 member EU States. Its prime objective is to remove internal border checks and develop the concept of 'free movement of persons'. The measures adopted embrace removal of checks at common borders replacing them with external border checks, harmonisation of the rules regarding conditions of entry and visas for short stays, definition of role of carriers in the fight against illegal immigration; requirement of all non EU Nationals moving from one country to another to lodge a declaration; and the drawing up of rules for asylum seekers (Dublin Convention). The UK and Republic of Ireland are not members of the agreement. Since 11 September 2001 cross border immigration checks have been reinstated by some EU members to counter terrorism and illegal immigrants.

Schottel rudder A retractable rudder propeller powered by a diesel engine.

Schuldscheine An instrument which represents a participation in a German loan, on which interest is paid in full to foreign investors without deduction of withholding tax.

SCI Shipping Corporation of India Ltd – India's largest Shipping Line.

SCIA Sea Carrier Initiative Agreement.

SCINDIA Scindia Lines – a shipowner.

SCIP Solid cast iron propeller.

Scissor lift A platform device usually power operated, which may lift a load from ground level to vehicle or container floor height.

SCM Supply Chain Management.

SCNT Suez Canal net ton.

SCNZ Shipping Corporation of New Zealand Ltd.

SCOP Steering Committee on Pilotage.

Scotch derrick A stationary crane driven by any mechanical means. It is usually built up to a greater height than the normal quay crane, and so has greater radius for lifts, and the ability to handle heavier lifts. It pivots on a mast known as a king-post, and its slewing capacity is limited by the supports or stays necessary for this.

SCP & E Software Supply chain planning and execution software.

SCR Specific Commodity Rates.

Scrap and build A scheme or policy whereby a Government would offer substantial financial assistance in the form of a building subsidy, or very low interest rates usually to National shipowners who are willing to scrap some of their older vessels and have fewer new ones built in replacement. Usually it involves less vessels built in relation to those displaced such as two built for three scrapped or similarly three to four. The policy arises due to varying circumstances such as depressed shipyard employment; need for more modern and competitive National mercantile fleet; sustain and develop National mercantile fleet for varying reasons; low order book; and so on. The scheme is not in extensive use.

Scraping the bottom Removing weed and incrustation from the underside of a ship. Not covered by a hull policy; not even when necessary to repair damage recoverable under the policy.

Scratching Initialling an agreement. Also, a peril applying to tainted or similarly vulnerable cargo.

Screening Process of auditing/evaluating in a disciplined manner a new or modified product relative to its market acceptance and potential. In particular the screening will take account of competition and the environment in which the product/service will be sold. See vessel screening.

Screw A tensioner that is tightened by turning a threaded screw with left and right hand thread. Also termed bottlescrew or turnbuckle.

Scrip Specifically, a provisional document such as an allotment letter but more widely, any form of security.

SCS Solent Container Service; or Singapore Cosmos Shipping Co (Pte) Ltd – a shipowner.

SCSP Solid cast steel propeller.

Scuttle To let water into a vessel for the purpose of sinking her.

Scuttling The wilful casting away of a vessel with the privity or connivance of her owners.

SD Sea damaged (grain trade); standard design (of a vessel i.e. SD 14); or Sudan Airways Company Ltd. An International Airline and member of IATA; or standard displacement.

S/D Sailing date; or sight draft.

S/d Small damage.

SD 9 A general purpose tramp vessel of 9,000 dwt's with three or four holds each with tween decks, and speed of 14 knots suitable both for bulk dry or liner cargo shipments, of 183 TEU container units.

SD 14 A general purpose cargo tramp vessel of 14,000 dwt's with five holds – four of which have tween decks – and speed of 15 knots suitable for both bulk dry or liner cargo shipments.

SD 18 A general purpose cargo tramp vessel of 18,000 dwt's with four holds each with tween decks and speed of 15 knots suitable both for bulk dry or liner cargo shipments of 426 TEU container units.

S & D 'Special and differential treatment' provisions for developing countries. Contained in several WTO agreements.

SDAC Shipping Defence Advisory committee (of the Chamber of Shipping).

SDNR Screw down non return valve.

S d/k Shelter deck.

SD King 15 A versatile tween deck general cargo modern ship with an overall length of 144m, breadth of 20.42m, draught 8.87m, deadweight tonnage of 15,000 tonnes and speed of 14.50 knots. The ship has a cargo hold grain capacity of 21,380m³, or a hold bale capacity of 19,540m³, or container capacity of 196 TEU's. Cargo handling equipment includes one 30 tonne, one 90 tonne, six 10 tonne, and four 5 tonne derricks. Some eight 5 tonne and two 3 tonne winches are provided. The vessel has five holds which are suitable for general cargoes, dry bulk, ore, grain, timber and container shipments. It is built by British Shipbuilders.

SDP Simplified declaration procedure. See separate entry.

SDR The mean of a basket of currencies designed to 'iron out' currency exchange fluctuations in international valuations. (Now used to express limitation in the Hague Visby Rules and MSA Limitation Convention etc).

SDS Shipping Documentation system.

SD's Supplementary declarations – a customs term.

SDT Side tank(s) on a ship; or Shippers' declaration for the Transport of Dangerous Goods (IATA approved).

SDV Shallow draft vessel.

SDW Side wheel, or summer deadweight.

SE South East – a point on the compass; or Sweden – customs abbreviation.

SEA Ship event analysis; or South-East Asia – IATA term.

Sea air bridge A dedicated shipping/air service often involving through ISO container services. Examples include Singapore/Dubai/Frankfurt.

Sea anchor A device for keeping a small boat's head into the wind and sea.

SEABEE vessel An ocean going vessel designed to convey up to 38 fully loaded barges. Loading and discharge is achieved by a sophisticated elevator system which physically positions and lifts the barges from the water, elevates them to the proper deck and automatically transfers them to any predetermined location along the representative deck. Off loading the barges follows the same sequence in reverse. Some vessels also convey containers on the upper deck.

Sea carrier initiative agreement An agreement with the US Coast Guard and Customs authorities which was established in 2002 in response to the severe penalties imposed by the US authorities when drugs where inadvertently carried on board merchant ships.

SEADATA Lloyds Register of Shipping data base.

SEADD South-East Asia Development Division (ODA).

Seafarers Shipboard crew personnel involving Ship's Officers and seamen/ratings.

SEAFDEC Southeast Asian Fisheries Development Centre.

Sea going barges Vessels which possess special features such as low airdraught to maximise their capabilities for navigating inland waterways. Also termed as low profile coasters.

Sea going personnel Shipboard personnel forming part of the ship's crew.

Sea ice Any form of ice found at sea which has originated from the freezing of sea water.

Sea Island A pier structure with no direct connection to the shore, at which tankers can berth. Berthing can take place either on one or both sides of the structure.

SEAL A numbered and coded device used to seal doors of a trailer, railway wagon, or container whilst in transit through certain countries.

SEALAND Sea-land Services Inc – a shipowner.

Sea-land bridge A dedicated service involving a shipping/rail or road service often involving an ISO container.

Sea lane A navigational maritime route.

Sea leg A portion of an overall transit which may be multi modal such as road/sea/air.

SEALINE South East Asia Line – a shipowner.

Seal log A document used to record seal numbers.

Seaman Grade I A post in the Deck department of a vessel and formerly designated Able Seaman (AB).

Seaman Grade II A post in the Deck department of a vessel and formerly designated ordinary seaman.

Seamless transaction The use of a standard set of data to satisfy border release and clearance declaration requirements in both the exporting and importing countries.

Seamless transport The uninterrupted flow of cargo with a dedicated through service.

Seanotice The code name of a BIMCO approved charter party clause relative 'to notice of readiness' and entitled 'Baltic Conference Notice from Sea Clause 1970'.

Seaport A terminal and an area within which ships are loaded with and/or discharged of cargo and includes the usual places where ships wait for their turn or are ordered or obliged to wait for their turn no matter the distance from that area.

Seaport infrastructure All the facilities provided at a seaport embracing berth, warehouse, and dock handling equipment, security, rail/road/canal services/networks, customs, forwarding agents, shipbrokers, immigration, navigation systems and so on.

Sea protest See ship protest entry.

SEA(R) Ship event analysis continuous data recording capability.

Searching Searching through data to locate a given value or string of characters. Examples could be 'own' to find the word own or 'own' to find the names Brown and Crown.

SEARFS A group of shipowners including South East Asia Regional Feeder Service; the Neptune Orient Lines; the Pacific International Lines Pte Ltd; and the Regional Container Line.

Seasonal inventory An inventory built up in anticipation of a seasonal peak of demand in order to smooth production.

Seasonal variations A situation where the Spring, Summer, Autumn or Winter produce a seasonal variation demand such as for example a product(s) or service(s).

Season ticket A form of ticket permitting usually unlimited travel between the points specified on the ticket throughout its validity. This may be one week, one month, three months, one year and so on. Alternatively it maybe a season ticket for a theatre, concert hall, sporting events etc, permitting attendance at specified performances throughout the validity of the ticket. Usually season tickets are not transferable.

Season ticket valuation statements A customs facility available to UK importers for goods liable to customs duty where the value of the goods exceed £6,000 (2003). This facility is available to regular importers who comply with certain conditions and are required to complete a registration general valuation form every three years.

SEATEC South East Asia Technology Co Ltd (Thailand).

Seating etiquette The protocol/custom of a particular culture/nationality as found in Japan where guests sit furthest from the door at a meeting table.

Seat Kilometres used The product of number of revenue passengers carried on each stage flight multiplied by the stage distance equals the seat kilometres used. Overall, a seat kilometre is used when a revenue passenger is carried one kilometre.

Seat miles The number of seats provided on a transport unit multiplied by the distance travelled in miles.

Seat miles per engine hour The number of seats provided on a transport unit multiplied by the distance travelled in one hour.

SEATO South-East Asia Treaty Organisation.

Seat pitch The distance between the front edge of one seat in an aircraft, and the front edge of the seat immediately in front when both are in an upright position. An IATA definition.

SEATRAIN Seatrain Lines Inc – a shipowner.

SEATRANS Sea Transportation Pte Ltd – a shipowner.

Seat reservation system A facility where a client may reserve a seat such as on a specific service of a transport mode on payment of the appropriate fee. It maybe a train, flight, ferry, coach, theatre, etc.

Sea trials The process of a vessel – usually following her launch – undergoing sea trials to establish/monitor whether the vessel is fully operational as defined in the building contract specification. This includes speed, bunker consumption, navigational aids, manoeuvrability, stability, and so on. The sea trial may last for several days. Overall, it is the tests of ship and all equipment prior to handing over from builders to owners.

Seat rotation Schedule whereby passengers on a tour motor-coach change seats at each stage of the tour to afford everyone an equal opportunity for the best views.

Sea Waybill Non-negotiable document which evidences a contract for the carriage of goods by sea and the taking over or loading of the goods by the carrier, and by which the carrier undertakes to deliver the goods to the consignee named in the document.

SEAWIND Seawind Shipping Co Pte Ltd – a shipowner.

Seaworthiness The state of a vessel for a voyage (see Hague Rules etc). The fitness of a ship to encounter the hazards of the sea with reasonable safety. In addition to having a sound hull, the ship must be fully and competently crewed and sufficiently fuelled and provisioned for the next stage of voyage. All her equipment must be in proper working order and, if she carries cargo, she must be cargoworthy and properly trimmed and stowed. The MIA, 1906, incorporates an implied warranty in a hull voyage policy which requires the ship to be seaworthy at the commencement of the voyage and, if the voyage is in stages, at the commencement of each stage of the voyage. Breach of this warranty is not held covered, so in such case, the underwriters are discharged from all liability under the policy as from the date the ship sails in an unseaworthy condition. A similar implied warranty applies to seaworthiness of the carrying vessel in regard to a cargo policy; but underwriters waive breach of the warranty in this case, provided the assured and their servants are not privy thereto. The warranty in the MIA does not apply to a hull time policy, but if the ship sails in an unseaworthy condition, with the privity of the assured, the underwriter is not liable for any loss that is attributable to the unseaworthiness. This means the ship is tight, staunch and strong and in every way fitted for the voyage.

Seaworthy ship A ship is deemed to be seaworthy when she is reasonably fit in all respects to encounter the ordinary perils of the seas of the adventure insured. A marine insurance definition.

SEB Single Ended Boiler (on a ship).

SEBS Single ended Boiler survey (on a ship).

Sec Secant.

Secondary barrier The liquid-resisting outer element of a cargo containment system designed to provide temporary containment of a leakage of liquid cargo through the primary barrier and to prevent the lowering of the temperature of the ship's structure to an unsafe level.

Secondary data Data which already exists in the market place such as Company international agencies records, government statistics, trade statistics and other data to which the entrepreneur has access – a marketing term, see primary data entry; or the market in which securities are traded after issue when the initial distribution has finished.

Secondary industry A manufacturing industry involving the conversion of raw materials into semi finished or finished products. These maybe consumer or non-consumer goods.

Secondary key The attribute or attributes which are used as the secondary index key for an entity.

Secondary market The market in which securities are traded after issue when the initial distribution has finished. The buying and selling of a bond after its primary issue.

Secondary Metal Dealer A firm which specialises in buying and selling scrap metal. The metal is often in alloy form (such as brass) and often in a fabricated form, but is always at risk from fluctuations in the value of the basic metal.

Secondary readership The process of a newspaper/journal/magazine being read by persons other than the original subscriber. For example, the head of the household buys the publication which is read by other household members. It is especially common with technical journals which are Company purchased and circulated among key members in the Company or department.

Secondary research The process of obtaining market research data which is already available in the market place. This includes government statistics, UN trade statistics, Trade Associations data, Export Councils data and so on. It is usually referred to as Desk or Armchair research.

Secondary storage Using read/write data storage devices which are not part of the micro-processor. These usually use non-volatile magnetic media and are disc- or tape-based, e.g. hard and floppy discs and cassette tapes.

Second carrier The carrier who actually performs the second leg of the contracted transit. It maybe by Air (IATA schedule) or by sea.

Second class A class of accommodation found on a passenger transport unit especially rail, air, and sea transport. It is inferior to first class accommodation.

Second class hotel See hotel classifications.

Second Freedom The privilege to land in another state for non-traffic purposes (that is for refuelling, mechanical reasons, etc, but not for putting down or taking on a load). One of the six Freedoms of the air.

Second generation product A product/service which evolves from one already on the market and ultimately supersedes it.

Second hand market The volume of tonnage available for sale. This represents the first or subsequent owner of a vessel endeavouring to find a buyer.

Second hand tonnage Vessels purchased which have been previously in operational use in a particular trade.

Second register The process of a shipowner transferring the registration of vessels from one registry to another thereby adopting a flagging out strategy. See flagging out entry.

Second-year ice Old ice which has survived only one Summer's melt. Because it is thicker and less dense than first-year ice, it stands higher out of the water. In contrast to multi-year ice, Summer melting produces a regular pattern of numerous small puddles. Bare patches and puddles are usually greenish-blue.

Sectional rates The rate established by schedule air carrier(s) or International Train Ferry service for a section of a through rate involving an international transit. For example, the sectional air carrier rate represents each country's proportion of the overall rate through which the consignment is conveyed.

Sectoral annex/annotation An agreement, supplementing the GATS framework, which extends or modifies the application of principles set out in the framework with a view to addressing specificates of individual sectors. A WTO term.

SECU Storaenso cargo units.

Securing pad The fitting for connecting a lash to either corner casting or tensioner, in container securing practice. Also termed lash end fitting.

Securities Bank Notes, equity, loan stock, bonds or other debt instruments.

Securities settlement system A system which permits the transfer of securities either free of payment or against payment.

Securitisation A key product area and an important financing technique used to convert cash generating assets in to marketable securities for sale to investors. The objective of securitisation is to separate business risk from the risk of sold assets. This enables investors to assess the credit quality and the cash flow of the pool of assets rather than looking to the business risk. Securitisation of assets consists of the division of a single asset or a portfolio of assets into securities which can be sold to investors. A liquid market can then be made in these securities. The market in securitised assets or collateralised mortgage obligations has grown rapidly in recent years, and has been extended from residual mortgage to commercial mortgages to commercial mortgages, credit card receivables and other consumer loan receivables. The critical element in all such transactions is the quality and regularity of the cash flows produced by the loans in question. There has been some debate as to whether a CMO market can be developed for portfolios of ship mortgages. Major obstacles include the poor credit rating of many ship mortgage obligations and the frequency of early redemption of ship mortgages upon the sale of the related vessels. Little practical progress has been made to date in this area.

Security The underwriters subscribing a risk. A term used in marine insurance to define the insurers with whom a policy has been effected.

Security maintenance clause See asset protection clause.

Security of concessions Ensuring that non-tariff measures, which are not as quantifiable as tariffs, not be diluted or negated by other subsequent measures. A WTO term.

Secus (Latin) Otherwise on the contrary.

SED Shippers Export declaration.

SEDB Singapore Economic Development Board.

SEDCS State Economic Development Corporations (Malaysia).

SEED Southeast Europe Enterprise Development.

Segment A leg of an air itinerary from boarding to deplaning point.

Segmentation The process of dividing the market for a product into sub-groups which can then be targeted through marketing activity.

Segregated ballast tanks Tanks used for sea-water ballast only with own pumps and piping system.

Segregation The separation, for example, in the stowage of dangerous goods from other incompatible goods or from accommodation and machinery space or other working spaces occupied for the operation of the ship, aircraft or other transport unit.

Seiche Very long waves of small height generated by resonant oscillation within a partly closed harbour or other body of water. Strong horizontal currents can also be set up which may cause ship surging in adverse circumstances.

SEKNEG Indonesian Government State Secretariat.

SELA Caracas-based Latin American Economic System (SELA) which co-ordinates economic policy in the region and co-operates on technology, food etc.

Selecting Involves searching for and extracting data which matches the search. See 'Searching'.

Selection (programs) Program statements which cause the selection of one component or route from two or more possibilities.

Selectivity Trade actions taken against a selected country or group of countries, as opposed to action against all countries on a non-discriminatory basis (mfn treatment). Overall, the imposition of safeguard restrictions on a single supplier nation rather than non-discriminately against all suppliers. A WTO definition.

Self catering The provision of holiday accommodation which involves the client undertaking their own catering arrangements such as in a flat, chalet, caravan, etc.

Self demounting pallets In this system the container is carried on a steel pallet fitted with foldaway legs which can be extended hydraulically. There is sufficient clearance between the legs in their working position to accept a road trailer or a rail wagon.

Self drive hire The provision of a vehicle/car involving the person/entity hiring the vehicle and providing the driver, as distinct from a vehicle/car being chauffeur driven.

Self drive tour A tour that has a pre-planned itinerary with vouchers for meals, hotels and optional air travel arrangement; clients can rent a car or drive their own.

Self governing port A port which is controlled and operated under the direction of the port and other interested organisations including local public authorities, and state departments, all of which are represented on a governing body usually called a Board.

Self help passenger luggage trolley A piece of passenger baggage handling equipment available to passengers at passenger terminals to facilitate passenger luggage handling arrangements. It maybe at a railway terminal, airport or coach station. The equipment is manually operated by the passenger and enables baggage transfer to be effected between transport modes.

Self liquidating offer A marketing term indicating a special promotion offer is likely to yield/generate enough revenue to cover the cost of the offer/promotion cost.

Self liquidating trust A term used to describe the type of shipping fund developed originally by Bergvall + Hudner and floated on the NASDAQ and Amex exchanges in New York. The basic feature – a stated intention to liquidate the company at a specified or undefined point in the future – had previously been used in shipping investment funds such as Bulk Transport Ltd and Anangel-American Ship Holdings Ltd, but the B+H formula also incorporated certain elements which had become standard in real estate limited partnerships in the US. The most important were: the distribution to investors of virtually all the companies' operating cashflow after debt service, thereby offering the prospect of a high current yield; and an arrangement whereby management would participate substantially in such distributions and liquidation proceedings once the outside investors had received their original investment plus a 'preferred return'.

Self loading trailers A specially adapted road trailer gathers the container from ground level or from some form of docking stand or stillage. The container is tilted during acquisition from ground level, but the method would not be suitable for all cargoes. Trailer mechanisms typically include a tilting platform, and rails or guides along which one end of the container is drawn as the road vehicle reverses and possibly a pair of short-stroke rams at the rear end of the trailer.

Self reference criteria A form of check list adopted by a company.

Self sale Sale of airline tickets by an agency to a company with which it is financially affiliated; limited by some carriers.

Self trimming hatches Hatches so designed to allow the cargo to flow freely into the holds without the necessity for trimming.

Self trimming ship Ship with large hatches and with holds which are shaped in such a way that the cargo levels itself on loading.

Self-propelled Barge A barge possessing its own mechanical means of propulsion.

Sell and Report agreement A bilateral or multi-lateral agreement between airline carriers which requires signatories to exchange and maintain information relative to the availability of space on selected flights and between specified points. An IATA definition.

Sell by date The date by which a perishable consumable product must be sold in a retail outlet otherwise beyond that date quality quickly deteriorates. The date is recorded on the food product at the point of sale.

Seller-centric portal A place where a seller finds several buyers at the same time. The buyers are aware of the presence of other buyers. An E-commerce term.

Seller credit The process of the overseas supplier/manufacturer/exporter/seller allowing credit terms to the importer/buyer and obtains finance from a bank or export credit agency. See also supplier credit.

Seller credits A term associated with ship sale and purchase. It reflects the value of ship sale – in particular it embraces the liquid assets thereby permitting the ability to sell the ship quickly.

Sellers Lien In legal terms the right to hold onto goods until some charge or debt due on those goods has been paid. A sellers' lien on goods arises where the seller is still in possession of the goods, or the money outstanding is immediately due.

Sellers market A situation in a competitive market place where an excess demand over supply of a commodity creates an imbalance thereby generating the climate/environment for generally high prices with little need to undertake aggressive selling.

Selling The process of persuading a client to buy a product or service. It may be face to face selling, video sales, telephone sales and so on.

Selling Agent A person appointment by a Company/entity to represent them in the market place with a view to generating sales. Usually the Agent operates on a commission basis.

Selling group The banks and other financial institutions that have been brought together by a lead-manager to take up the securities being offered for sale to their customers.

Selling price The listed price of the product in the market place.

Selling rate of exchange The price at which a Bank or other specific organisation/company is prepared to sell any particular currency.

Selling service The process of selling a service in the market place such as insurance, banking, and tourism, rather than a manufactured product such as a car.

Selling targets A target set to be attained/reached by a sales unit. It maybe a sales representative, total sales force, a retail outlet such as a ship/ticket office, and so on. Usually the sales target(s) would be discussed/agreed with the sales unit(s) concerned. Budgetary controls feature prominently in this area.

Selling technique The technique to be used in selling the product(s)/service(s). It may be a mixture of direct selling, strong advertising, agents, sponsorship and trade exhibitions.

Sell-through Booking accepted by a hotel, although unavailable days maybe included, because of length or value.

Semi developed economy An economy – usually formerly an LDC – in the early stages of its economic development. See DMEC's.

Semi-elliptic spring Partly curved leaf spring as opposed to a coil-type spring.

Semi-portal crane See crane – semi portal entry.

Semi refrigerated gas carrier A purpose built tanker capable of conveying gas in a semi refrigerated condition at a controlled temperature.

Semi rigid receptacle One which will retain its shape when empty and uncompressed.

Semi-submersible A mobile drilling platform that floats and is held on location by anchors and/or thrusters.

Semi-trailer Trailer drawn by a tractive unit in a manner where a proportion of the weight of the trailer is superimposed on the tractive unit.

SEMKO Svenska Elektriska Material – Kontrallan – a Stalten. National Standards Body. See ISO.

SEN Ship entry notice.

Sen The currency of Kampuchea, Republic of Indonesia and Malaysia. See Appendix 'C'.

Sender See shipper.

Senes The currency of Western Samoa. See Appendix 'C'.

Senes invoice An invoice payable in Western Samoa senes relative to an export sales invoice.

SENGBEE Seng Bee Shipping Pte Ltd – a shipowner.

Sengi The currency of the Republic of Zaire. See Appendix 'C'.

Sengi invoice An invoice payable in the Republic of Zaire sengi relative to an export sales invoice.

Senior Export Clerk A senior clerical position in a Manufacturing Company/Freight Forwarder office involved in the many aspects of processing the export consignment. It will include documentation, customs, rates, booking of consignment, export finance, insurance, computerisation and so on. The role/responsibility of the job will vary by Company.

Senior fares Lower airline fares available from some US carriers for senior citizens; age qualifications and other elements vary from time to time.

Senior Import Clerk A senior clerical position in a manufacturing Company/Freight Forwarder/Receiver/Agents office involved in the many aspects of processing the import consignment. It will include documentation, customs, distribution, financial settlement and so on. It is likely to involve computerisation. The role/responsibility of the job will vary by Company.

Seniti The currency of the Kingdom of Tonga. See Appendix 'C'.

Seniti invoice An invoice payable in the Kingdom of Tonga seniti relative to an export sales invoice.

Sensible heat Heat energy given to or taken from a substance which raises or lowers its temperature.

Sentimental loss A market loss to goods brought about by fear that goods may have suffered from a known casualty whereas no such loss exists in fact.

Separate ballast tank Tank in a tanker tonnage which is used only for water ballast. There is no risk of cargo being mixed with ballast as the latter is pumped out by its own separate piping and pumping system with no risk of pollution.

Sequence list A list or plan which provides the numerical rotation in which containers or barges are to be loaded to or discharged from a vessel.

Sequence (programs) Program statements which are executed one after the other in list order.

Sequestrate To take possession for creditors.

Serang Senior Indian deck rating.

Seriatim (Latin) In strict order of previous mention.

Series All options of the same class which have the same exercise price and expiry date – an International Petroleum Exchange term.

SERS Ship Emergency Response Service – A Lloyd's Register of Shipping service.

Service Any activity or benefit one party can offer to another that is basically intangible. Examples include Banking, Health, Tourism, Transport and so on. Ownership of the service rest with the supplier and not with the beneficiary/user who are regarded as the consumer.

Service boat A small boat which provides victualling, stores materials, or a garbage collection service in a particular port/estuarial area to vessels at their moorings. Usually such ships are at buoys, or other anchorage and away from a berth/quay.

Service charge Additional charge levied for care of guests.

Service compris French term meaning a tip has already been included in the bill.

Service industries A range of services which serve industry and the market generally, but are not directly involved in the manufacturing sector. These include banking, insurance, transport, medical, finance, tourism, consulting, training and so on. Overall, a fast growing sector of business worldwide.

Service non compris French term for tip or gratuity not included.

Service pick up The carriage of outbound consignments from the point of 'pick up' to the airport of departure. An IATA definition.

Service providers The services such as transport, tourism, banking, freight forwarding, customs broker, agent, insurance, and so on. See service sector. A non tangible service. See seven P's.

Service quality The standard/quality of service provided by the seller to the client. See quality of service entry.

Service sector A part of the economy of a country/region/trading bloc which features the services such as transport, tourism, banking, insurance and so on. See service entry.

Services trade Items which as invisibles in the balance of payments of a country and embracing tourism, banking, insurance, consultancy and so on. See also invisible exports.

SES Stock Exchange of Singapore; Standard Exchange System, Surface effect ship, or Ship Earth Station – term employed with GMDSS communications.

Session of the Contracting Parties The most authoritative body of WTO, consisting of all Contracting Parties, which usually meets once a year.

SET Securities Exchange of Thailand.

Settlement The giving of consideration, i.e. payment of money or acceptance of a bill of exchange, in return for documents presented under a credit. An international trade banking term.

Settlement Agent An institution that manages the settlement (e.g. the determination of settlement positions, monitoring the exchange of payments, etc) for transfer systems or other arrangements that require settlement.

Settlement currency The currency in which a premium or claim is settled. Depending on the circumstances, this could be any of the so, called, 'hard' currencies (i.e. GBP, US$ or Can $).

Settlement price The last unfulfilled offer to sell cash at the close of the second morning ring in each metal. This price, which is officially announced on the LME floor, prevails as the accepted cash price for the metal for the succeeding twenty four hours; or IPE term. The official price established by the Exchange at the close of each trading day. This figure is used to calculate variation margin.

Settlement risk A general term used to designate the risk that settlement in a transfer system will not take place as expected. This risk may comprise both credit and liquidity risk.

Settling Agent An underwriter's representative who is authorised to settle claims.

Set up time The time required to prepare a work station from a standard condition to readiness to commence a specified operation.

Seven P's The marketing mix for the service sector embracing product, price, promotion, people, process (logistics) and physical aspects. See marketing mix.

SEVENSEAS Seven Seas Maritime Co Pte Ltd – a shipowner.

Seven 'S' framework A framework developed by McKinsey which helps to analyse the different aspects of the competitor organisation. These are strategy, skills, shared values, staff, systems, style and structure. They cover both 'hard' and 'soft' aspects of the organisation and should be consistent one with another.

Severance cost An example is found in the BIMCO Crewman B Lump Sum standard crew management agreement, whereby the crew managers are legally obliged to pay to the crew as a result of the early termination of fixed term employment contract for service on the vessel.

SEZ Special Economic Zones, e.g. Industrial Parks/Estates, FPZ, EPZ.

SF Sparing fitting (cargo battens or equivalent fitted); stowage factor; or Tassili Airlines – an IATA associate member Airline.

SFA Securities and futures authority.

SFD Simplified Frontier Declaration – a customs term.

SFP Structural fire protection.

SFR Swiss Franc – a major currency.

SFTC Standard Freight Trade Classification of the Council for Mutual Economic Assistance.

sg Steam heated steam generator(s) – on a ship.

SG Specific gravity; or Sempati Air – an IATA Airline Member.

Sgd Signed.

SGIA Seamen Grade 1A.

SGS Société Générale de Surveillance SA.

SG Policy The policy form adopted by Lloyd's in 1779, as a standard form for all marine insurance business. The form covered both ship and goods on a specified voyage and required considerable adaptation to be used for modern hull and cargo business. The London company market adopted separate policy forms for hull and cargo business, but these were very similar to Lloyd's SG form and still required adaptation. In January, 1982, the London cargo insurance market replaced the SG form with the MAR form of policy. The London hull insurance market took the same measure in October 1983. The Institute clauses published for use with the SG form and its company counterpart have been withdrawn.

SGTOAA Study Group on Take-Off Obstacle Accountability Areas (ICAO term).

SH Super heater.

S & H/exct Sundays and holidays excepted in laydays – a chartering term.

Shackle A U shaped piece of iron with a pin fitted through a hole on either side of the top, which is screwed in or secured with a forelock used to connect various sections of cargo gear.

Shadow Areas Minimal spaces within cargo tanks difficult to clean completely.

Shaft tunnel Accommodation housing the propeller shaft linking the propeller with the machinery (engine) in a vessel.

Shank The threaded bar of a turn buckle.

Share A share is a security representing a portion of the ownership of a Company and carrying corresponding property rights and rights of membership.

Share capital The name given to the total amount of money put into a company by the owners. It usually equals the nominal amount of each share, multiplied by the number of shares issued.

Shared airline designator operations When a carrier operates flights or legs for another carrier using the latter's airline designator; most frequently when commuter lines provide feeder service to major carrier's hub.

Shared statistics A circumstance which arises in some countries whereby one country would collect statistical data relative to an international consignment relative to other countries participating in the specified transit. In so doing the statistical data would be distributed to the interested countries. Overall, it centralises the statistical data collection.

Shareholder The owner of a share who is, incidentally, part owner of a company.

Share security It consists of an owner depositing stock of the shipping company for a specific period, or up to the final amortisation of the loan. This enables the bank or shipowner default to assume ownership of the vessel owning company, as opposed to the vessel itself, resulting in the bank acting as the new owning company as opposed to a mortgage.

Sharing boxes Usually termed Box sharing – automated system for swapping empty containers between carriers/shipowners.

Sharing risk The process of two or more persons/entity/company participating in a particular venture which usually is financial.

Shearing forces Stresses created on ship's frames leading to hull distortion at possible breaking points.

Shekel The currency of Israel. See Appendix 'C'.

Shekel invoice An invoice payable in Israeli shekels relative to an export sales invoice.

Shelf life The date by which a perishable consumable product on display in a retail outlet must be sold. The specified length of time prior to use for which items can be held in stock before deterioration is considered to take place. See also sell by date.

Shell A small company with stock-market quote but negligible assets or activities which is partially taken over by entrepreneur wanting a stock-market vehicle.

Shell and tube condenser A heat exchanger where one fluid circulates through tubes enclosed between two end-plates in a cylindrical shell and where the other fluid circulates inside the shell.

Shell plating The plates forming the watertight skin of the ship.

Shells See tour shells.

Shellvoy Code name for oil charter party.

Sheltered tonnage Vessels operating in coastal waters as found in West Africa Seaboard.

SHEX Sundays and holidays excepted – this excludes Sundays and holidays counting as laytime in the charter party.

SHIEHFU Shiehfu Navigation SA – a shipowner.

Shifting berth The process of moving/shifting a vessel from one berth to another usually within the same port/environs.

Shifting boards Boards and frames used to prevent excessive movement of bulk cargoes such as grain.

Shifting cargo The process of cargo shifting on a vessel during the course of the voyage thereby undermining the stability of the vessel.

SHINC Sundays and holidays included etc – a chartering term.

SHINHEUNG Shin Heung Shipping Korea – a shipowner.

Ship A seagoing vessel which is self-propelled and capable of carrying cargo or passengers.

Ship arresters Steel ropes to protect ships from hitting the lock gates.

Shipboard assessment A review of the systems and management control on a vessel. The review is sufficiently comprehensive to establish that all the ISM requirements of the Code have been met.

Shipboard conveyer belt and elevator system Cargo handling equipment operative at a seaport. It operates on the basis the commodity can drop out of the bottom of the ships hold onto a conveyor belt and transverse the length of the ship to be subsequently elevated to the shore link by either bucket wheel elevator or compressed belt system.

Shipboard entertainment The provision of entertainment resources during the course of the sailing to passengers. The scale/range of entertainment will depend on the size of the ship and nature of the voyage but may include films, live entertainment, gaming machines, clay pigeon shooting, dancing, pre-recorded music and so on.

Shipboard facilities Those facilities found on a vessel including passenger lounge, crew mess-room etc, and provided for the welfare and comfort of passengers and crew.

Shipboard Management The technique of Managing a fleet of ships whereby each vessel's Chief Officers have personal responsibility of not only the traditional navigational and operational aspects of the control of seaworthy tonnage, including the welfare of passengers and conveyance of cargo, but also the Management of the ship in terms of responding to an agreed budgeted expenditure and revenue level. The Master – termed the Managing Director – responds to a shore based Senior Manager relative to the budgetary control and other aspects including overall Management of the ship within the Company policy guidelines viz surveying, manning, victualling, bunkering etc. See International Ship Management Code.

Shipboard marine pollution emergency plans Regulation 26 of Annex I of MARPOL 73/78 and article 3(1)(a) of the OPRC Convention require that every oil tanker of 150 tons gross tonnage and above and every ship other than an oil tanker of 400 tons gross tonnage and above carry on board a shipboard oil pollution emergency plan.

Shipborne barge An independent non self propelled vessel specially designed and equipped to be lifted in a loaded condition and stowed aboard a barge carrying ship, i.e. BACAT or barge feeder vessel.

Shipbroker A person having one of several occupations: chartering agent or owner's broker, negotiating the terms of the charter of a ship on behalf of a charterer or shipowner respectively; sale and purchase broker, negotiating on behalf of buyer or seller of a ship; ship's agent, attending to the requirement of a ship, her master and crew while in port on behalf of the shipowner; loading broker, whose business is to attract cargoes to the ships of his principal.

Ship building Market Cycles The process of the ship building industry production being evaluated and thereby enable predictions to be made on future levels of ship building activity. These can be traced over a seven year or may be a longer period recording the cyclical nature of the business recording the peaks and troughs of the market and cyclical growth, decline or stabilisation of the market.

Ship chandler Merchant who supplies ships with stores and provisions.

Ship classification society See classification society.

Ship conversion The process of a vessel being converted from one specification to another such as a bulk carrier being converted to a container ship.

Ship productivity

Ship design strategy The process of devising/formulating a policy relative to a particular company/entity ship design. This may include consideration of ship viability; cabin design; navigational equipment; hull formation; trades in which the ships will ply involving possible standardisation of equipment/stores/victualling where practicable; and so on.

Ship double steel hull Vessel design with a double steel hull structure. See double hull.

Ship entry notice The process, for example, in Nigeria whereby all foreign registered vessels excluding those carrying petroleum or in ballast must be registered before entering Nigerian waters.

Ship Finance The process of funding a vessel which maybe through equity, debt, mezzanine or leasing.

Ship hardware The hull, machinery and equipment on a vessel.

Ship Inspection Reports Program A data base of ship inspection reports.

Shipman Code name of BIMCO approved standard ship management agreement.

Ship Management The technique of managing vessels from a shore based orientated organisation embracing all the revenue, expenditure, investment and overall policy including both day to day and long-term aspects of the business.

Shipment The process of loading cargo on a vessel for a particular maritime transit.

Ship mortgage bank Specialist ship financing banks, dating their origin from the 19th century in the Netherlands. Subsequently, similar institutions were set up in Germany, Norway, France and elsewhere. They are restricted from engaging in most other types of financing than ship mortgage lending. The leading institutions of this type today are the German ship mortgage banks. They finance their lending through the issue of long-term mortgage bonds in the domestic market, and are therefore able to advance medium- to long-term loans to their customers at fixed rates of interest. Their operations are similar to the practice of mortgage securitisation in the commercial and residential property market.

Shipowner A person(s)/company/entity which owns a particular ship and thereby accepts all the legal responsibilities/obligations in accordance with the country's maritime law to which the ship is registered.

Shipowner's Agent A person/company who acts on behalf of a shipowner. See Port Agent entry.

Shipped Bill A bill of lading which evidences that the goods stated thereon have been shipped.

Shipped Bill of Lading Acknowledgement that cargo accepted on board a vessel for shipment, as recorded on Bill of Lading.

Shipped on board Endorsement on a Bill of Lading confirming loading of cargo on a specified vessel.

Shipper The person/Company/Agent who provides the goods for despatch and whose name appears on the consignment note as the party contracting with the carrier for the carriage of the goods.

Shipper Councils A number of maritime related companies who have decided to become members of a Shipper Council either by application or nomination. These embrace Forwarding Agents, Logistic Operators, Customs, Port Police, Security Personnel, Rail Operators, Inland Waterway Operators, Customs Clearance Brokers and Road Haulage operators. Usually the Shipper Council is National such as Nigerian, Ghanian, European, Sri Lanka, Singapore, Thailand etc. Overall, a representative organisation/body.

Shipper's Export declaration A United States customs form to be completed for all exports to assist the government in compiling export statistics.

Shipper's letter of instruction A document containing instructions given by the shipper, or the shipper's Agent, for preparation of documents and forwarding of the specified consignment.

Shipper's Letter of Instruction for issuing Air Waybills Data prepared by the shipper outlining the information to be contained on the Air Waybill. Usually a 'standard transport instruction' form is used which includes details of consignor, and consignee; reference number; notifying party; carrier/carriers agent; terms of delivery/delivery time limit; points of origin and destination including routing; full description of cargo; remarks/special instructions; freight charges – prepaid/collect; other charges; insurance coverage; declared value; documents enclosed; number of original Air Waybills; place of issue; date and authentication.

Shipping cycles The periodic oscillation usually about seven years in the level of business embracing peak, decline, trough and growth. Each cycle phase has a profound impact on freight rates, profitability, scrapping, new building/investment, and merchant fleet productivity/growth/decline.

Shipping Documentation system All the shipping documents involved in the international distribution system/network embracing Bills of Lading, invoices, Delivery Orders, Manifests and so on.

Shipping documents Documents other than transportation receipts or transportation contracts required to enable shipments to be forwarded or received.

Shipping instruction A document advising details of cargo and exporter's requirements of its physical movement.

Shipping Invoice Document giving details of merchandise shipped.

Shipping label A label attached to a shipping unit containing certain data.

Shipping Lane A chartered shipping navigation route.

Shipping Magnate A most senior person in the global maritime industry.

Shipping market cycle The process by which the market co-ordinates supply with changes in demand by means of booms and slumps embracing peak, decline, trough and growth. See shipping cycles entry.

Shipping-Note A document provided by the Shipper or his Agent to the berth operator and/or shipping company giving information about export consignment(s) offered for shipment. It is a delivery or receipt note.

Shipping Officer Customs Officer exercising control over imported/exported cargo.

Shipping overcapacity A situation in the market place whereby too much shipping capacity exists in relation to demand thereby producing a surplus of shipping capacity.

Shipping supply It is the quantity of shipping service (total aggregated capacity) offered on the market at a particular freight over a given period of time.

Ship plan A general arrangement plan of a vessel.

Ship planning Process of planning utilisation of ship capacity. It is undertaken by the shipowner on maybe a day-to-day basis involving usually cargo manifest compilation, and/or longer term involving deployment of vessels and their adequacy to meet market capacity demand.

Ship productivity Key factors are the comparison of cargo generation, and fleet ownership, tons of cargo carried, and ton miles performed per dwt; and analysis of tonnage oversupply in the main shipping market sectors.

Ship recycling The process/procedures regarding the decommissioning of vessels with particular focus on protecting the environment and adopting a safe code of practice. This embraces a standard inventory of potentially harmful material on vessels sold for recycling, guidance on minimising residues and other substances prior to arrival at the yards, the pre-arrival condition of the vessel – including the condition of tanks, all designed to enhance safe working practices on the site.

Ship recycling yards A shipyard which undertakes the demolition and breaking up of vessels.

Ship repair facilities Purpose built accommodation/site where vessels may be repaired.

Ship's Agent The ship's agent represents the Shipowner/Master at a particular seaport either on a permanent or temporary basis. This includes notification of arrival and departure of the vessel; acceptance of vessel for loading, discharge, repairs, storing and victualling; arranging berths, tugs, harbour pilots, launches; ordering stevedores, cranes, equipment etc; and so on. 'In port' requirements include the needs of the Master embracing bunkers stores, provisions, crew mail and wages, cash, laundry, engine and deck repairs, and crew repatriation; completion of customs, immigration and port health formalities; hatch and cargo surveys; collection of freight; collection and issuing of Bills of Lading; completion of manifests; Notorial and Consular protests and so on.

Ship's articles The conditions and terms which seamen undertake to conform to, and abide by, on taking on board a merchant ship, and which are binding on shipowner and seamen.

Ship's bag Receptacle containing shipowners mail on specified vessel.

Ship's chandler A person/company/entity engaged in the sale of ships equipment particularly ships tackle such as ropes, paints, etc.

Ship's clerk A clerical position at a port under a Cargo Superintendent. The ship's clerk duties embrace the employment and supervision of checkers (or tally clerks); the liaison between the ships Master/Chief Officer and stevedore foreman, the consignees representatives, and cargo surveyors, ensuring adequate supply of lighterage for overside discharge, and so on. The ship's clerk is primarily found at a port/berth dealing with break bulk cargoes.

Ship's cubic capacity Total cubic capacity available to passenger/cargo/stores.

Ship's deadweight Total weight of cargo/passengers/stores etc permitted to be carried to the limit of ship loadline.

Ship's Delivery order See Delivery order entry.

Ship security plan Operational and physical security measures on a vessel. See International Ship and Port Security Code.

Ship's husband The person legally responsible on behalf of the shipowners, for the management of the ship; also the shipowner's agent who superintends the vessel when in port.

Shipside interface The facilities provided alongside the vessel whilst in the berth.

Ship's lightweight See lightweight and displacement light.

Ship's manifests Inventory of cargo shipped.

Ship's option The process of cargo being accepted for shipment on a specified sailing/vessel, subject to capacity available. Usually such cargo is accepted at a lower rate as no guarantee of shipment can be given, as if the vessel is full the cargo must wait its turn on the next available sailing where capacity exists. Ships option traffic is the lowest priority of shipment and is only accepted when all pre-booked traffic is stowed on board.

Ship's papers Documents/certificates required to be carried by a ship to conform to Statutory (International) Regulations.

Ship's protest Declaration made before a notary public or consular representative by the master of a ship on arrival in port, that the ship has encountered circumstances beyond his control, such as heavy weather, which may have caused damage to the ship or her cargo, and that the crew exercised all diligence to avoid such damage. This document maybe used in legal proceedings such as claims for loss or damage in transit, or to support the carriers defences against such claims.

Ship's rail A term found in INCO 2000 terms indicating the stage when the buyer/seller responsibility of risk and expenses emerges relative to cargo loading and discharge. See 'Free on Board', and 'Carriage Insurance Freight'.

Ship Register The document which provides fullest details of a particular ship to which the vessel is registered. See carving note and classification society entries.

Ship's Routeing The International Convention for the Safety of Life at Sea (SOLAS), IMO is recognised as the only international body for establishing and adopting measures of ship's routeing on an international level. Hence, Ship's Routeing embraces detailed information on all routeing measures adopted by IMO: traffic separation schemes, deep-water routes, mandatory ship reporting schemes and areas to be avoided by ships.

Ship specification The fullest details of the composition of a particular vessel including data on machinery, hull dimensions, crew/passenger accommodation, tonnage and so on.

Ship subsidies An operating or building subsidy granted to a particular shipping Company by their National Government usually under certain conditions.

Ship survey Examination of ship's hull and machinery by surveyor in shipwright department of Shipowners Classification Society organisation e.g. Lloyd's Register of Shipping, responsible for maintenance of hull, derricks and ship accommodation to a requisite standard.

Ship-to-ship transfer operations Transfer of crude oil or petroleum product between two ocean-going ships made fast alongside at anchor or underway. The transfer of petroleum to barges and estuarial craft is specifically excluded.

Ship turn round time The duration of a particular vessel stay at a seaport/berth based on the time the vessel is received at the berth to commence passenger disembarkation and/or cargo discharge, until the ship duly loaded leaves the berth to proceed on her voyage or other specified assignment. The vessel may arrive in a loaded or empty condition and likewise depart on a similar basis.

Ship vetting A ship evaluation, best undertaken by experienced professionals rather than using a computer model. This involves focusing on three or more areas, procedure systems, quality manning and superindency. It is often conducted prior to a second-hand ship purchase or chartering a vessel. Often referred to as dry vetting – dry bulk cargo tonnage, or wet vetting – wet bulk cargo tonnage.

Ship watch Crew personnel responsible for watch keeping duties on a particular vessel.

Shipwood load Timber demurrage scheme: Scandinavian/Continent.

Shipyard Purpose built site where ships are surveyed, repaired and built.

Shipyard credit Credit facilities offered by a shipyard in regard to work undertaken such as ship survey for a shipping company. It maybe the third survey on a vessel whereby the contractual payment cycle maybe spread over twelve months and not a one off payment following survey completion.

Shipyard subsidy A government subsidy granted to a shipyard for the construction of new tonnage usually under specific conditions.

SHOFUKU Shofuku Line – a shipowner.

Shoot yourself in the foot The process of a person/representative of a company/entity making a remark/comment/statement which is derogatory to the person/representative/company/entity and thereby attracts very adverse criticism/remarks.

Shop Place where goods/services are sold/retailed.

Shore-based assessment A review of the company's systems, which is sufficiently comprehensive to establish that all the requirements of this scheme have been met, for the purpose of issuing an initial certificate or verifying its continued validity. See ISM code.

Shore based gantry crane A gantry crane usually operative at a berth on the quayside.

Shore excursion A tour or trip off the ship at a port of call, usually for a charge.

Shores Stout timber seated to prevent cargo movement.

Short Starting a transaction by the sale of a contract – a BIFFEX term; a sale of a contract to open a position – an International Petroleum Exchange term; anyone who has sold securities he does not own in anticipation of buying them later; or an open sold position on a futures market. Thus 'to go short' : to start a transaction by the sale of a futures contract.

Short Bills Bills drawn for any period within ten days of their maturity.

Short Closing Closing a risk for less than the amount insured.

Short delivery The quantity of cargo delivered is less than the bill of lading quantity.

Shortfall The failure to attain a particular target/objective. It may be the sales target of €10,000 but the actual results were €8,500 producing a short fall of €1,500.

Shortfall Freight This arises when freight is calculated on a pro rata basis and fails to achieve the requisite minimum (as per tariff) and the carrier will increase the charge by imposing a 'shortfall freight' to cover the balance.

Short Form Bill of Lading A Bill of Lading where an abbreviated version of contract clauses – or reference to such clauses – has been printed on the back. Often referred to as Common Short Form Bill of Lading.

Short form ocean waybills A sea waybill where an abbreviated version of the clauses – or reference to such clauses – has been printed on the back. It is used in deep sea trades. See sea waybill.

Short haul The shorter distance transit which crosses international boundaries and in a time band between 24-48 hours for surface transport services which may be regarded as less than 300km.

Short haul airline An airline which offers services less than 2/2½ hours flight time.

Short in a currency A term found in the foreign exchange market whereby if a dealer is short in a currency, it means that liabilities exceed holdings for the currency.

Short interest The outstanding balance of the sum insured by a floating policy after all declarations have been made.

Short international voyage A short international voyage is one in which a ship is never more than 200 miles from a port or place in which the passengers and crew could be placed in safety and which does not exceed 600 miles in length.

Short landed The process of cargo being unloaded/discharged from a vessel and in so doing a quantity is missing such as eight tonnes discharged instead of ten tonnes. The cargo manifest is thereby 'over stated'. See short shipped.

Short notice of cancellation Right by the insurer to cancel cover for voyages that have not yet commenced (usually due to an unexpected change in circumstances).

Short run A time period within which a firm is not able to vary all its factors of production.

Short run industry supply curve A curve that shows how the relative quantity supplied by the industry varies as the market price varies assuming the plant size of each firm and the number of firms in the industry remain the same. In perfect competition it is the horizontal sum of marginal cost curves (MC) above the level of average variable costs (AVC) of all firms in an industry.

Shorts Gilt edged and similar stocks due for repayment within five years.

Short Sea Shipping Short Sea Shipping is that part of sea borne trade between countries which does not cross oceans.

Short Sea Trade International Trade between United Kingdom and Continent (of Europe).

Short shipment Goods 'shut-out' of a vessel usually through lack of space or cargo late arrival.

Short Shipped Cargo which is included in a manifest but which is erroneously not loaded or forwarded.

Short Tail A term used to describe a risk in respect of which all claims are likely to be advised within the period of cover or shortly after the cover ceases to attach.

Short term A time policy effected for a period of less than 12 months.

Short term commitment An undertaking extending to up to seven years.

Short term forecast A market forecast which ranges from one to three years.

Shoulder period The period when the market demand for a product(s) or service(s) falls/occurs immediately before and immediately after the peak period(s).

SHOWA Showa Line – a shipowner.

shp Shaft horse-power (steam turbine engine) – of a ship.

Shrink wrapping A form of packaging which is provided by placing the goods on a covered base – such as a pallet – and covering it with a film of plastic which is shrunk to enclose the items by the use of hot air blowers (thermo-guns). It is relatively cheap and its configuration follows the outline of the goods.

SHSV Super heater safety valve.

SHUNCHEONG Shun Gheong S N Co Ltd – a shipowner.

Shut out Cargo/merchandise not carried on transport unit which maybe a ship, aircraft and so on. In such circumstances the transport unit (ship) is fully loaded and no capacity exists for the 'shut out' cargo to be carried and the consignment has to wait for the first available schedules (sailing) on which capacity exists. A similar criteria applies to other transport modes.

Shut out cargo Merchandise/cargo not carried on a transport unit – which maybe a ship, aircraft and so on – due to the non availability of space/capacity. See also 'shut out'.

Shuttle tanker See off take vessel.
SI Shipping Inspectorate (Scheepwaart Inspectie); or Statutory instrument – legal term.
SI (Systeme International) Units An internationally accepted system of units modelled on the metric system consisting of units of length (metre), mass (kilogram), time (second), electric current (ampere), temperature (degrees Kelvin), and amount of substance (mole).
SIA Singapore International Airlines; or Simplified authorisation to use IPR – customs term.
SIB Securities and Investment Board.
SIC Standard Industrial Classification; Said to contain; or Sea Farers identification cards.
Sic (Latin) Thus, so given.
SICASP SSR Improvements and Collision Avoidance Systems Panel (ICAO term).
SICC Singapore International Chamber of Commerce.
SID Single decker.
SIDAP New Zealand Committee/Organisation on Trade Procedures and Facilitation based in Wellington.
Siding Place where railway wagons are stabled/accommodated.
Side door Water tight barrier which seals on opening in the side of a roll on/roll off ship through which rolling cargo is wheeled or drive along a ramp into or out of the ship.
Side frame handling A fort lift truck attachment capable of handling empty and fully loaded 20 feet (6.10m) and 40 feet (12.20m) ISO containers by lifting from only one side of the unit.
Sideline point The sideline point arises in the air cargo tariff and represents a 'city' published on a page below a headline point opposite which the applicable rates are shown. See also headline entry. An IATA term.
Side shift mechanism type of fork lift truck A fork lift truck equipped with sideshaft mechanism to handle a wide variety of cargoes.
Side thrusters A thruster situated at the stern of the vessel on both the port and starboard sides to aid ship manoeuvrability in confined water and fulfils a similar role to the bow thruster.
Sidetrip See excursion.
SIECA General Treaty of Economic Integration, Central American States.
Sievert (Radioactive Material Only) The sievert is the standard unit of measure for radiation dose-equivalent used in these Regulations; it is represented by the symbol 'Sv'. Other multiples of the sievert are frequently used in these Regulations (see B.2.2.3). The sievert replaces the older unit for dose-equivalent the 'rem'. One Sv is equal to 100 Rem. An IATA term air freight shipments.
SIF Standard Interchange Facilities – IATA term.
Sift-proof Packagings are packagings impermeable to dry contents including find solid material produced during transport. An IATA term.
Sift proof receptacle One which is impermeable to dry contents.
Sight The time when a draft is presented to the drawee for payment or acceptance. Under a sight credit the bank should settle immediately if it is satisfied with the documents (unless the sight terms of the credit in some way is modified by its contents). Banks are allowing a reasonable time to check documents.
Sight Bill A Bill of Exchange calling for payment as soon as it is presented to the drawee.
Sight draft A bill of exchange payable at sight, i.e. on demand or presentation, but with respect to which three days of grace are allowed for payment.

Sighting Bills of Lading Process of Master or Agent scrutinising Bill of Lading at port of discharge.
Sighting the bottom Examining the underside of a ship for damage following an accident. The Institute hull clauses provide that the underwriters will pay the cost of sighting the bottom without applying the policy deductible and even if no damage is found, but only when the cost is incurred specially to examine the bottom following stranding.
Sight payment Method of trade finance payment whereby payment is made to the seller locally upon presentation of conforming documents.
Sigma The computerised reservation and national distribution system of Italy; partner of Galileo.
SIGMET Significant Meteorological (Report: Forecast) – IATA term.
Signals Articles containing pyrotechnic substances designed to produce signals by means of sound, flame or smoke or any combinations thereof. The term includes signal devices, hand; signals, distress, ship; signals railway track, explosive; and signals, smoke. An IATA term air freight shipments.
Significant wave height The average height of the highest third of the waves in any sea state, which equates to visual estimates by observers of the average wave height: the average height is actually just under two thirds of the estimate made by observers.
Signing and Accounting A procedure at LPSO whereby accounting procedures are conducted concurrently with the policy signing procedure.
Signing slip The broker's slip that is used for submitting details to LPSO for signing and accounting. This can be the original slip, certified photocopy of the original slip, or an initialled off slip.
SIGTTO Society of International Gas Tanker and Terminal Operators Ltd.
SIGWX Significant Weather – IATA term.
SIL Singapore Islands Line – a shipowner; or Specific individual licence.
Silent language A gesture, a smile, eye contact etc. Overall, it is an expression of a culture in communication. See non verbal message and culture.
Silica gel A chemical used in driers to absorb moisture.
Sill of a dock The timber at the base against which the dock gates shut.
SIMA Scientific Instrument Manufacturers Association of Great Britain.
SIMEX Singapore International Monetary Exchange – a futures exchange.
Simple Deferment An imports accounting system relative to VAT.
Simpler Trade Procedures Board A UK government funded Agency responsible for the development of trade facilitation embracing documentation and procedures including electronic data interchange. Its prime aim is to assist in the more efficient distribution/flow of cargo.
Simplex Commendatio non obligat (Latin) A simple recommendation is not binding.
Simplified declaration procedure A customs term. It is available to UK exporters and is a simplified export procedure involving a two part declaration. The first part requires the exporter to submit brief details of the export consignment. Once the goods have been presented to customs, and CHIEF has accepted the declaration and released the goods, they may be loaded for export shipment. The second part of the declaration must take place within 14 days of the acceptance by CHIEF of the 'goods departed message' or date of export, whichever is the earlier. The exporter must

submit both parts of the declaration to customs electronically using one of the available transmission routes. It is applicable to the UK seaports and airports. It is only available to approved traders by HM Customs. The system can also be used for local clearance procedure where goods for export maybe declared at the traders own or other nominated inland premises. See NES and CHIEF.

Simplified procedure for import clearance An HM Customs facility as a simplified procedure for declaring and clearing certain low value goods instead of completing SAD. Customs charge can be deferred against the deferment number of either the importer or the person making the declaration. See SAD and Period entry traders

SIMPROFRANCE The trade facilitation organisation of France.

SIMS Sims Consolidated Ltd – a shipowner.

SIMTRA South Korea Committee/Commission on Trade Procedures and Facilitation based in Seoul.

Simultaneous Payments Clause A reinsurance clause which provides that the reinsurance shall pay a claim at the same time as the original claim is paid by the reinsured.

SIN Singapore Dollar – a major currency.

Sin Sine.

SINCHIAO Sin Chiao Shipping Pte Ltd – a shipowner.

Sine die (Latin) Indefinitely – without a day being appointed.

Sine qua non (Latin) An indispensable condition.

Singapore Airline The National Airline of Singapore.

Singapore dollar The currency of Singapore. See Appendix 'C'.

Singapore dollar invoice An invoice payable in Singapore dollars relative to an export sales invoice.

Single Any reservation, facility or service to be used by one person.

Single Administrative Document A document used for trade throughout the Community, EFTA and Visegrad countries. The single administrative document has seven copies and involves a data input for seventy boxes. This includes consignor, consignee, cargo description, gross weight, net weight, calculation of taxes and much statistical data. The object of the single document is that it will be raised by the exporter and used as an export declaration; carried by the transport operator and used as a transit document, and completed by the importer and used as an import declaration. It has increasing use outside the EU. Also called Form C88.

Single back The movement of an empty road trailer on the return leg following the completion of the loaded journey.

Single berth cabin Passenger sleeping cabin accommodation on a ship or train which accommodates one person.

Single buoy mooring A fixed structure or floating mooring positioned outside a shallow draught harbour, but in deep waters to enable large tankers to moor by the bows and enable them to load or unload their cargoes of oil.

Single commodity ship Vessel specifically designed to convey a single commodity such as iron ore, oil, LNG, etc.

Single contract finance This is finance for contracts between one overseas buyer and one seller.

Single country The focus on one country. See single country data capture.

Single country data capture A circumstance whereby one country would collect statistics at importation on behalf of the exporting Member State relative to the despatch of an international consignment. This usually arises within a Free Trade area.

Single country international market research An international market research programme embracing a single country.

Single deck A vessel having no deck below the weather deck.

Single decker A one deck or single deck bus which has a seating capacity of about 50 passengers.

Single deck link span A bridge ramp on which the vessel's ramp can rest and thus connects a vessel with the quay or shore and allows for the working of ro/ro trailers and cargo. In so doing the wheeled cargo which may include coaches and motorists are transhipped at one level over the stern or bow of the vessel via the link span.

Single entity charter An air charter sponsored and paid for by a single company, organisation or person.

Single hull A vessel constructed with a single hull. See double hull.

Single manning The process of one man or woman operation of a bus thus combining the driver and conductor duties, or a train with one train driver and no assistant.

Single market The twenty five countries forming the European Union and permitting the free circulation of goods. See European Union.

Single nationality crew The crew are all of one nationality such as a French registered ship have a crew of French officers and seamen.

Single packagings These are packagings which do not require any inner packaging in order to perform their containment function during transport.

Single-plane An arrangement in which the turbine(s), condenser and gearbox are all in one plane. This is particularly advantageous for shipboard use.

Single point mooring A facility whereby the tanker is secured by the bow to a single buoy or structure and is free to swing with the prevailing wind and current.

Single Port Operation The situation where a major shipping company will opt to serve one port in an area/region rather than a series of ports and rely on feeder services to develop trade through the port. See also one stop ports and hub and spoke system.

Single product sales strategy The process of focusing exclusive/dedicated resources on one single product in the whole ambit of sales resources by a company/client.

Single sourcing Process of an entity obtaining supplies/raw materials/component parts from only one country/company/supplier. A Procurement term.

Single supplement Additional charge to one person occupying a hotel room accommodating two or more; usually charged on a tour.

Single transport document A commercial document indicating a contract for the carriage of goods between two member states within EEC. See Single Administrative document.

Single user port A port used by only one operator/shipowner/consortia.

Single voyage contract A contract usually a charter party for a single voyage Felixstowe-Singapore or one trip.

Single wall A packaging term relative to the fibreboard carton with a single wall structure comprising of one layer of fibreboard.

Single width (passenger) gangway A piece of port equipment linking the ship with the shore (quay) of single passenger width – about one metre – thereby enabling passengers/crew personnel to embark or disembark in single file.

Sinking fund Moneys set aside by a borrower, normally at regular intervals from revenue for the purpose of establishing a reserve to repay a loan at maturity.

SIO Senior Information officer.

SIPP ATC Standard Interline Passenger Procedures Manual – IATA term.
SIPROCOM Committee for the Simplification of International Trade Procedures based in Belgium.
SIRA Sira Institute Ltd (was Scientific Instrument Research Association).
SIRE Ship inspection reports programme. See Separate entry.
SIRIM Standards and International Research Institute of Malaysia.
SIS Special Intermediate Survey.
SISC South India Shipping Corporation Ltd Madras – a shipowner.
SISIR Singapore Institute of Standards & Industrial Research.
Sistership Clause A clause in a hull policy whereby the underwriters agree to treat sisterships as if they were separately owned (and capable of legal liability, one to the other) in regard to collision liability claims and claims for salvage charges.
Sister ships Vessels built to an identical or near identical specification for the same owner consortia and usually from the same shipyard.
SITA Société Internationale de Téléommunications Aéronautiques.
SITPROSA South African organisation for the simplification of International Trade Procedures based in Johannesburg.
SITC/SITC(R)/SITC (RZ) The standard international trade classification.
SITPRO Simpler Trade Procedures Board.
SITPRONETH The trade facilitation organisation of the Netherlands.
Sitting Meal serving times on a ship for breakfast, lunch and dinner, usually two sittings each and often an hour and a half apart.
Situation analysis The process of analysing a particular situation at a specified time. The analysis may involve a Company, product, patient, student etc and would identify the salient features of relevance at the time of the analysis.
SIVs Standard Import values. See separate entry.
Sixth Freedom The privilege where an airline registered in a state may take on revenue passengers, mail and freight in a second state, transport them via the state of registration and put them down in a third state. One of the six Freedoms of the air.
SK Scandinavian Airlines Systems (SAS). An international airline and member of IATA.
Sk Smack.
SKA Station Keeping Assistance.
SKD Semi knocked down.
Skeletal trailer A road trailer without decks designed solely for the purpose of carrying containers with twistlock securing.
Skeleton case Method of cargo packing for shipment such as a wooden case affording limited protection.
Skeleton crates See wooden crates.
Skeleton service A limited service to the public/market usually embracing just the essentials.
Skeleton trailer A road trailer consisting of a frame and wheels specially designed to carry containers.
Skids Battens fitted beneath stillages, boxes, or packages to raise them clear of the floor and allow easy access of fork lift trucks, slings, or other handling equipment.
Skimming price The process of adopting a pricing strategy whereby a high price is fixed of a product/service to target the high income groups socioeconomic categories 'A' and 'B' (those with high buying power). This is usually a short term strategy to recover develop-ment cost and generate high profit. See market skimming.
Skim pricing The introduction of a new product at a high initial price to 'skim' the market.
Skin The inside of the outer shell of a ship's hull.
SKR Swedish Krona – a major currency.
SKU Semi knocked down unit.
Sky bridge The provision of a through international dedicated service operating on a regular basis. It maybe Far East/European service in sea/air/road with the airline linking the sea and road service. See land bridge.
Sky market 'In flight' products for sale such as duty free products, drinks etc.
Sky marshal An in-flight security officer operative on passenger flights on some services/airlines to combat potential terrorism.
SL Surveillance licence; or Salvage loss.
sl & c Shippers' load and count.
SL/NL Ship lost or not lost.
S/LC Sue and Labour clause.
Slack water The period between the flood or ebb tide, or between the cessation of the tidal stream in one direction, and its commencement in the opposite direction.
Slamming Impact of water on the bows of the ship. Force created when fore-part of ballasted ship surfaces and pounds into rough sea.
Slave pallet A pallet which can be lifted by the hook and is not used for loose cargo, but for made up pallet loads. One means of conveying a pallet with cargo between ship and shore. Also termed master pallet.
SLC Synchronous Link/Network Control Procedures – IATA term.
SLEPT Social Legal Economic Political Technical – a marketing term.
Slewing crane A crane with slewing platform capable of rotating with the load in a plane in relation to its undercarriage or base.
Slewing ramp Ramp of a roll-on roll-off ship which is capable of being swung to either side to enable cargo to be rolled on and off from a variety of berthing positions at different ports.
SLI Shippers Letter of Instruction.
Sliding fifth wheel Fifth-wheel plate mounted on a tractive unit, which can be moved forwards and rearwards to accommodate different lengths of semitrailer.
Sliding tandem An undercarriage with a subframe having provision for convenient fore and aft adjustment of its position on the chassis/semi trailer. The purpose being to be able to shift part of the load to either the king pin or the suspension to maximise legally permitted axle loads (road cargo).
Sling Cargo handling equipment suitable for bag or case transhipments.
Sling loss Cargo lost by falling from ship's lifting tackle during loading or unloading.
Slings Rope, artificial fibre strap, or wire for lifting cargo particularly for bagged merchandise and case transhipments.
Slip agreement An agreement expressed by an underwriter on the broker's slip.
Slip policy A broker's slip which is used as the policy document. Subject to certain requirements slip policies can be used, where practicable and where no formal policy is required, for cargo insurance, for some types of hull insurance and for reinsurance.
Slips Notification of marine insurance risk acceptance.
Slip sheet A plastic sheet used as an alternative to a pallet which allows entrance of fork truck spades.

Slip tube A device used to determine the liquid-vapour interface during the ullaging of semi and fully pressurised tanks. See also Restricted Gauging.

Slipway A recess provided with a sloping bed, leading off from the main dock or other water area. Slipways are provided at many ports for vessels which are small enough to be hauled clear of the water and for which in consequence the provision of expensive graving or floating dock facilities is not necessary.

Slop reception A facility usually available in a seaport for the reception of slop, sludge and dirty ballast emerging from ship calling at the seaport.

Slop tank An oil tanker into which slops are pumped. These represent a residue of the ship's cargo of oil together with the water used to clear the cargo tanks. They are left to separate out in the slop tank.

Slot The section in a cell (of a containership) that will take a container; a berth availability at a particular seaport; or a landing/take off flight path at an airport.

Slot capacity The number of slots allocated to shippers/container operators on a particular sailing. See slot.

Slot charters The process of allocating charters on a slot basis; or an agreed number of cells on a container ship.

Slothire Code name of BIMCO approved standard slot charter party for use in the container trade.

Slot swaps The process of two container shipping companies swapping a number of container slots on particular sailings. Overall, it is a vessel sharing agreement. See slot.

Slow steaming The process of a vessel moving slowly during the voyage and well below her usual speed.

SLPA Sri Lanka ports authority.

SLS Strategic Logistics System.

SLSA St Lawrence Seaway Authority.

SLSC Sri Lanka Shippers Council.

Sludge Semi-solidified residue of Crude Oil cargo deposited on frames etc within tanks – maritime definition.

Sludge acid The acid waste resulting from oil refining, or from nitrating processes. It generally has somewhat the same hazards as the original acid. IATA definition.

SMA Society of Maritime Arbitrators (New York).

Smart and Secure Trade Lanes A partnership of three of the largest terminal operating companies Hutchison Port Holdings, PSA Corporation Ltd and P & O ports. Its objective is to provide an automatic tracking system for containers entering US ports that involves tagging containers electronically. The electronic seal contains full details of the box content. It operates between Singapore, Hong Kong, and Seattle/Tacoma. It establishes partnerships with importers, carriers, brokers, warehouse operators, and manufacturers to improve container security along the entire supply chain. Overall, it produces 'fast lane' cargo clearance. It was set up in 2002. See multilateral security measures.

Smart card A card operative in the international transport of goods by road which covers all payments, documentary and reporting requirements and Customs formalities for the international transit operative in Europe. It incorporates a miniaturised processing system (micro-chip). Overall, it is a small computerised portable file and payment device as well as a means of accessing a network.

SMC Safety Management Certificate.

SMCG Slow moving consumer goods. See FMCG.

Smelter A processing plant which produces crude metal by treating mine feed (concentrate).

SME's Small and medium sized companies.

SMG Speed made good.

SMGCS Surface Movement Guidance and Control Systems (ICAO term).

SMGS Agreement on International Goods Traffic by rail.

SMIT Bracket A fitting for securing the end link of a chafing chain, consisting of two parallel vertical plates mounted on a base with a sliding bolt passing through the plates.

SMM Shipping Market Modes.

SMMT Society of Motor Manufacturers and Traders.

SMS Safety Management System. See ISM Code; SAR Mission Co-ordination; or Seaworthy Management System.

SN SNB Brussels Airlines (Delta Air Transport dba SN Brussels Airlines). IATA member airline.

S/N Shipping Note.

SNCM Société Nationale Maritime Corse-Mediterranee.

Snood Polythene sleeve or cover on pallets or blocks of cartons.

Snotter Cargo handling equipment suitable for bag or case transhipment.

SNSC Singapore national Shippers Council.

SO Seller's option; senior officer; or indicates that the ship is provided with special quality steel cable – Lloyd's Register of Shipping abbreviation.

SOAP Simple Object Access Protocol – logistics computer term.

SOB Shipped on board (a specified vessel).

Social Customs The customs of a particular culture which can be classified as social embracing protocol, dress, geniality and so on.

Social sanctions Measures taken by the International Community or individual governments on a bilateral or multilateral basis which penalises/legalises restrictions/constraints on unimpeded international travel. It isolates such a country and hampers tourism. It may apply to certain sectors of a particular country community.

Social stratification The division of a particular population into classes.

Société Générale de Surveillance SA Organisation and its affiliates which are appointed by shippers/various authorities in countries throughout the world to carry out pre-shipment inspection on the quality, quantity of goods being exported and on the prices proposed and price comparison. See CRF and NRF entries.

Society for worldwide Interbank Financial Telecommunications A computer system for interbank transfers, operated by the four major UK banks, to achieve fast money transfers between banks worldwide which are linked to the system. It can assist the importer in reducing to a minimum any delays in money transfers. Through the system money is transferred by coded interbank telex and as long as the supplier makes it clear to the buyer exactly to which bank and account the remittance should be made, the supplier should receive very speedy payment.

Socio cultural environment The attitudes, sociology, behaviour, psychology and culture development of and within a country as a whole, and the various sub populations which go to make up that country.

Socio cultural factors A wide variety of patterns of living including behaviour norms such as those regarding diet and styles of dress.

Socio-economic groups The segmentation of a population into sections to represent the main sub-sections of a community in accordance with the selected economic and social criteria. The groups are identified by letter ranging from A & B representing the higher income groups to C1, C2, D & E. The C1 and C2 reflect the middle stream income categories whilst the D & E category identify the lower end of the economic and social category. The analysis of the population into socio-economic groups is widely used in market research and the evolvement of marketing strategies. The spread of the socio-economic groups will vary by market/country/region.

Soc-oil Soc-Oil Corporation Berhad – a shipowner.

Soda lime A mixture of calcium oxide or calcium hydroxide with sodium hydroxide.

Sodium metal liquid alloy A metal mixture or alloy of sodium and another metal, existing as a liquid at ordinary temperatures and which flows more or less readily depending upon composition. Avoid contact with moisture since such contact may cause the mixture to become ignited and burn.

SOEC Statistical Office of the European Communities.

SOF Statement of Facts – a chartering term.

SOFIA Standard Off-line Internet Applications.

Soft currency A currency which is consistently depreciating in terms of other currencies over the long term.

Soft data Data obtained from a secondary source and not regarded as absolutely factual or even reliable.

Soft economy Non-essential areas of an economy which arise in a wealth focused society particularly in the areas of leisure.

Soft loan A loan granted below the market level of interest rates and subject to the most favourable repayment arrangement in comparison with the current market practice.

Soft selling A low key and unobtrusive method of selling with emphasis on subtlety and subconscious in terms of presentation. Overall, selling conducted in a very relaxed atmosphere/environment with no pressure being placed significantly on the buyer to complete a sale. Such a policy tends to court the potential buyer in a very subtle approach.

Soft top container Container with a removable waterproof 'tilt'. Also termed a top loader or open top.

Software The programs which enable computers to operate; instructions to a computer. Software can generally be classified according to the following different types: operating systems, applications, utilities (editors, diagnostics, file management) user interface, program generators, system services (database management, translators) videotext systems, coding and programming languages, system analysis (CASE), local and wide area networking, training and learning (CBL), testing and assessment, games and leisure applications.

Software facilities Facilities within software packages which enable user to perform tasks more effectively and efficiently. Examples include macros, mail merge, pagination and page headers or footers.

SoG Speed over Ground.

SOIR Simultaneous Operations on Parallel or Near-Parallel Runways (Study Group – ICAO term).

SOL Shipowners liability.

Sol The currency of Peru. See Appendix 'C'.

SOLAIR Solomon Islands Airways. A national airline and member of IATA.

Solar radiation The amount of radiation or energy received from the sun at any given point.

SOLAS International Convention for Safety of Life at Sea, 1974; as amended.

Sold message The report of a sale under a Sell and Report, or Free Sale agreement. See Sell and Report and Free Sale Agreement entries. An IATA definition and relative to airline traffic.

Solent Container Service A Southampton Terminal Operator.

Sole rights agency An Agency with exclusive rights.

Solid frame securing system A rigid structure for restraining containers such as cell guide or buttress.

Sol invoice An invoice payable in the Peruvian sol currency relative to an export sales invoice.

Solomon Islands Airways A national airline and member of IATA.

Solomon Islands Dollar The currency of Solomon Islands. See Appendix 'C'.

Solomon Islands Dollar invoice An invoice payable in the Solomon Islands Dollar currency relative to an export sales invoice.

Solution sales Encompasses corporate sales activities with the addition of tailored investment opportunities. An international banking term.

Solvent Person/Company able to pay all debts contracted.

SOMS Suez Odense Marine Service – a joint venture between the Suez Canal Authority and Odense Steel Shipyard of Denmark.

SONIA Sterling OverNight Index Average: the weighted average rate to four decimal places of all unsecured sterling overnight cash transactions brokered in London; calculated and published by the WMBA.

SOONKIAT Soon Kiat Shipping Pte Ltd – a shipowner.

SOONSIANG Soon Siang Shipping Pte Ltd – a shipowner.

SOPEP Ships oil pollution emergency plan.

sor(s) Starboard.

Sorting Ordering data in numerical or alphabetical order. Sometimes sorting is undertaken on two fields so that where one field recurs that data is in order of a second field. Examples of this are the telephone directory where Smith is the primary field but a second field of 'initial' or 'second name' is used.

SOTA Shannon Oceanic Transition Area – IATA term.

Sounding Measurement of water (usually) found in any tank.

Sour crude A crude oil of which contains a relatively high concentration of sulphur compounds.

Sour palan (French) Under the hook (ship's tackle) and arises when loading or discharging cargo into lighters.

Sour palan au sous vergues (French) A French expression indicating that discharge will be made 'under ship's derrick', 'under ship's tackle', 'under ship's slings', 'over ship's side', or 'at ship's rail'.

South African Airways An international airline and member of IATA.

Southern Africa Europe Container Service The consortium in which the Overseas Container Ltd trades in the Europe-Southern Africa.

South Africa Transport Services The organisation which operates container freight stations and terminals in South Africa.

South Asia Gateway Terminal See SAGT.

South Atlantic rates Such Air Freight rates apply between the South Atlantic Area and points in IATA area 2.

Southbound The process of proceeding in a southbound direction. It maybe a consignment, transport unit, passenger, transport route and so on.

South East Asia Trade Advisory Group Organisation responsible for developing/facilitating trade between UK and South East Asia.

South grain The code name of a BIMCO approved charter party for the South American grain trade.

South Pacific rates Such Air Freight rates apply between points in IATA Area 1 and points in the South West Pacific Area.

SOUTHSEA South Sea Freighters Ltd – a shipowner.

South Yemen Dinar The currency of the Peoples Democratic Republic of Yemen. See Appendix 'C'.

South Yemen dinar invoice An invoice payable in South Yemen dinars currency relative to an export sales invoice.
Sovcoal Code name of charter party used in coal trade.
Sovcoalbill Code name of BIMCO approved Bill of Lading for shipments on the Sovcoal Charter Party.
Sovcoaltic Code name of charter party used in coal trade.
Sovconround Code name of BIMCO approved Soviet Roundwood charter party for pulpwood, pitwood, roundwood and logs from Baltic and White Sea Ports of the CIS.
Sovconroundbill Code name of BIMCO approved Bill of Lading for shipments on the Sovconround charter party.
Sovereign risk This is a term applied to debt whereby a country's government/central bank is the borrower/guarantor. A WTO term.
Sovietwood Soviet/UK timber trade Charter Party 1961 – code name of charter party used in timber trade from CIS.
Sovorecon Code name of charter party for shipment of ores and ore concentrates from CIS ports.
Sovoreconbill Code name of BIMCO approved Bill of Lading for shipments on the Sovorecon charter party.
S/P Submersible pontoon.
SP Sales promotion; special price; SATA Air Acores – an IATA associate member airline; sailing plan; safe port; special performance or Spain.
Sp Sloop.
SPA Subject to particular average.
SPACDAR Specialist Panel on Conflict Detection and Resolution (EURO-CONTROL) – IATA term.
Space Charter A voyage charter for a certain quantity of cargo, whereby the shipowner has the liberty to compete with other (space charter) cargoes.
Spacer A framework or other device to maintain a space between parts of a composite package.
Spanfrucon Code name of charter party used in Spanish fruit trade.
Span Gas A vapour sample of known composition and concentration used to calibrate gas detection equipment.
Spanish Chamber of Commerce The organisation based in London representing Spain whose basic role is to develop, promote and facilitate international trade between UK and Spain and vice versa
Spanish Shippers Council The organisation in Spain representing the interest of Spanish Shippers in the development of international trade.
Spare capacity Unused/under utilised capacity. It maybe an airliner with 100 seats and 20 unoccupied – the latter representing spare capacity.
Sparred Describes a system of packing hollow-ware cargo whereby wooden supports hold the goods rigid during handling.
SPARTECA South Pacific Regional Trade and Economic Co-operation. It involves Australia, New Zealand and other Forum Island countries.
Spatial resolution (In the case of satellite observations) the lower limit to which the distance between features in an image can be resolved.
SPC Self polishing copolymer – a patent anti-fouling paint; or Scheduling Procedures Committee (IATA term).
spd Steamer pays dues.
SPDC Southern Philippine Development Corporation.
SPEC Special; specification; or Society of Philippine Electrical Contractors.
Special Administrative Region On 1st July 1997 Hong Kong became a special Administrative Region of the Peoples Republic of China.
Special Agent In legal terms, an Agent who has authority to do only one act, although he can do that act many times, but that act is the limit of his authority. If the Agent exceeds this authority in anyway, the principal is not bound.
Special and different treatment The principle included in part IV of the GATT and under the Enabling clause recognises tariff and non tariff preferential treatment in favour of and among developing countries.
Special Drawing Rights (SDR) The SDR was created by the International Monetary Fund in 1969 as an official reserve asset to alleviate the shortages in relation to world trade of gold and other reserve currencies. It does not exist physically, but it is recognised as the IMF's official unit of account. SDR's are allocated to members of the IMF according to set quotas which are based on their economic size in relation to other members of the fund. Members of the IMF can use SDR's to finance balance of payments deficits and to settle international financial transactions. In addition the SDR has been promoted as a standard unit of account for international transactions in both official and private sectors. To allow its use in international payments and as a unit of account, the SDR is valued in terms of a basket of currencies. Initially there were sixteen currencies in the basket but in 1981 this was reduced to five – the US Dollar, French Franc, Deutsche Mark, Japanese Yen and Sterling. Today (2004), the IMF uses SDR's for book-keeping purposes and issues them to member countries. Their value is based on a basket of the US dollar (with a weight of 45%), the Euro (29%), the Japanese Yen (15%) and the Pound Sterling (11%). The composition of the basket is reviewed every 5 years. It is now used to express limitation in the Hague-Visby rules and MSA convention following the Euro launch.
Special Economic Zones Areas in particular host countries which offer special incentives and privileges to foreign investors including tax concessions, easier repatriation of earnings, and more favourable government regulations regarding employment. Such zones exist in China.
Special feature A publication which embraces/devotes part of its content to a particular event or item of special interest to its readers. It maybe a separate publication; a 'pull out' or supplement to an existing form of media such as a journal magazine/newspaper.
Special fields The form fields designed to contain special data such as date, time, page number or calculation like column totals – computer term.
Specialised Carrier (Special Ship) A cargo vessel specially designed for the carriage of particular cargoes.
Special programming language Facilities now available in many applications, software, e.g. applications generator, macros, templates – computer term.
Special rate A rate which is usually a discounted contract rate negotiated between carrier and shipper, or one devised by the carrier for special goods on particular services/routes.
Special settlement Normally, settlement of premium and claims is arranged on balance on a specified date each month. Where early settlement is required it can be arranged so that settlement takes place outside the usual monthly settlement, usually within seven days.
Special survey Classification Society survey imposed by Lloyd's Register of Shipping on vessels every 4 years.
Special Types Vehicles which are permitted to operate outside the requirements of the Motor Vehicles (Construction & Use) Regulations for the purpose of carrying abnormal loads. The vehicles are controlled by the Motor Vehicles (Authorisation of Special Types) General Order (ASTGO).

Specie Valuable cargo such as money, precious metal, jewellery, etc (see also Loss of Specie).

Specification The fullest details of a product(s)/service(s). It is often associated with a price quotation.

Specific Bank Guarantee A guarantee given to a bank as security for finance made available to an exporter of capital goods covered by specific insurance or supplemental extended terms of cover.

Specific Commitments Negotiated commitments on market access and national treatment listed by countries in their national schedules.

Specific commodity rates A rate applicable to the carriage of specifically designated commodities to/from specifically named points/airports under IATA regulations.

Specific duty Imported cargo subject to a prescribed duty.

Specific gravity The ratio of the density of a liquid at a given temperature to the density of fresh water at a standard temperature. Temperature will affect volume and the comparison temperature must therefore be stated; e.g. specific gravity 60/60°F – substance and water at 60°F; specific gravity 15/4°C – substance at 15°C, water at 4°C. The use of the term is being superseded – See Relative Liquid Density.

Specific heat This is the quantity of energy in kilojoules required to change the temperature of 1 kg mass of the substance by 1°C. For a gas the specific heat at constant pressure is greater than that at constant volume.

Specific individual licence An import licence permitting the import of a stated quantity, or value, of a specific commodity. It is usually valid for a set period and issued by the Government.

Specific performance An order by the court to a party in breach of a contract to carry it out. It will only be awarded in certain very limited circumstances when damages would not be enough to compensate the injured party. Hence, the remedy sought by a plaintiff who, instead of damages for a breach of contract, seeks the enforcement of the terms of the contract.

Speculation Transactions based on the expectation that the exchange rate of a particular currency will change in response to some political or national event. Whether the operator buys or sells a currency depends on whether he anticipates a rise or fall in value. An international financial term.

Speculation in the Foreign Exchange Markets The process of trading in foreign currencies undertaken with the expectation of securing a profit yield from short run change/fluctuations in exchange rates.

SPEX SITPRO Export Consignment Processing and Invoicing System.

SPG Systems Planning Group (NAT – ICAO term).

SPHL Self propelled hyperbaric Lifeboat.

SPIC Simplified procedure for import clearance.

SPILPR Sri Lanka Committee/organisations on the Trade Procedures and Facilitation based in Colombo.

Spiral bevel axle Design of different gear in which the gear teeth are curved and therefore remain in mesh longer. Thus a number of teeth on the crown wheel and pinion are always engaged, relieving the strain which would be imposed if only one tooth were engaged.

Split charter This arises primarily in the air freight market whereby an aircraft will be split chartered by a number of separate shippers who each take a part of the available space of the aircraft through a charter broker for a particular flight.

Splitter gearbox A means of increasing the number of gear ratios available from a standard gearbox by splitting each full gear in half to give progressive but narrowly spaced steps across the ratio range.

Split use option A customs facility available relative to the Single Administrative Document whereby the document is split into its constituent parts so that formalities for export, transit documents and import entry may be dealt with as separate elements. Sometimes termed Salami system.

SPM Single Point Mooring.

SPNFZ South Pacific Nuclear Free Zone.

Spoke port See hub and spoke entry.

Sponson Large buoyancy tanks attached to the outer hull of a vessel usually Ro/Ro vessels.

Sponsorship The process of a Company/Entity selecting a particular sporting/entertainment event(s) and in return for maximum publicity of the sponsoring company product(s)/service(s) through advertising, contribute financially to the running of the sporting/entertainment fixture(s). It maybe exclusive sponsorship for one particular event over say five years, or a series of events such as Cricket/Football matches in one year or more with other participating sponsors. The sponsorship arrangements vary. The prime objective is to increase sales of the sponsors product(s)/Services(s).

Spontaneous combustion The ignition of material brought about by a heat-producing chemical reaction within the material itself without exposure to an external source of ignition.

Spontaneous ignition temperature It is defined as the lowest temperature at which a substance will ignite spontaneously without an external source of ignition.

Sporadic maintenance The process of maintaining a facility on an irregular basis and not to any scheduled pattern.

Sporadic service A transport service which operates irregularly and not to a scheduled pattern.

Spore Singapore.

Spot A term which described a one time open market transaction where a commodity is traded 'on the spot' at current market rates – an International Petroleum Exchange term. See cash market.

Spot against forward Currency transactions involving transactions between multinationals and their banks, between banks on an international scale and between national state governments when it is advantageous to move out of one currency into another for a limited time without exposure to a foreign exchange risk. See currency swap.

Spot business A type of business opportunity which is available/concluded outside the general market situation/strategy, usually at a much discounted price below the market norm. This maybe the purchase of oil on the spot market in Rotterdam; the conclusion of a business deal involving a property purchase whereby the seller decided to accept a low offer to conclude a quick sale to raise cash; or a spot charter.

Spot charter The charter of a vessel usually at a discounted price below the market norm of the current fixture rate.

Spot Cotton Cotton which is in a warehouse ready for immediate delivery.

Spot dealings An open market transaction. See spot transaction.

Spot earnings Revenue from the spot market. See spot market and spot charter.

Spot market An open market transaction. In the currency market, the market rates are quoted in the press and is applied to those transactions for completion immediately or within two days from the date of the deal.

Spot month The first trading month for which a quotation is available in futures relative to the freight market. A BIFFEX term.

Spot rates A currency rate which is for immediate completion or at least within two working days from the date of the deal. An exchange rate quoted on the spot date. It reflects the relative strengths of the economies and the respective buying power of the currencies at that date. Eg £1 = €1.492. A bank will sell currency at a lower rate and buy at a higher rate.

Spot transaction A foreign exchange deal where the currency involved must be delivered immediately i.e. within two days of agreement.

Spout A fitting usually supplied in conjunction with a pouring orifice of a drum to direct the flow of the contents of the receptacle when it is being emptied.

SPR Sun protection required – flagging system to ensure that containers with heat sensitive cargoes are not unduly exposed to direct solar radiation.

Spread Buying in one quoted month and simultaneously selling the same quantity in another delivery month – for futures, the price difference between contracts or contract months (e.g. April/July BIFFEX) – a BIFFEX term; or in Banking terms the difference between a dealers buying and selling rate for a currency; the difference between a jobbers buying price and selling price, or an options or futures position in which two or more open positions are established to offset risk. Future spreads involve buying and selling in differing delivery months or products. Option spreads may use different exercise prices and/or expiry dates – an International Petroleum Exchange term; or a trade in which two or more open positions are established in order to trade the differentials and offset risk. Option spreads may use different exercise prices and/or expiry dates.

Spreader A device used for lifting containers and unitised cargo from the top; the same size plan area as the container or unit, and as the guidance system, it has to fit. Also a beam or frame that 'spreads' the wire legs of the multi-leg sling.

Spreader in foreign exchange transactions The difference between the rate of exchange applied when a customer is selling currency to the bank and that applied when she/he is buying the same currency for them. Quotations found in the daily financial press will indicate on the previous day USD were quoted on the London Foreign Exchange Market at 1.5710 – 1.5718 spot against sterling.

Spreading the Risk The practice of obtaining subscriptions over a wide market so that the liability of the individual underwriter is reduced for each risk.

SPREP South Pacific Regional Environment Programme. It is based in Apia Western Samoa.

Springing Vertical movement of main-deck of VLCC when loaded and in a seaway.

Spring lines Mooring lines leading in a nearly fore and aft direction, the purpose of which are to prevent longitudinal movement (surge) of the ship while in berth. Headsprings prevent forward motion and backsprings aft motion.

Spring tide Highest tide.

SPS Software Products Scheme or Seakeeping Prediction System.

SPS regulations Sanitary and Phytosanitary regulations – government standards to protect human, animal and plant life and health, to help ensure that food is safe for consumption. WTO term.

Spt Spritsail or; superheater(s) – ship's machinery; or spot.

Spudding Settling of the legs of a drilling rig in the seabed.

Spurling pipe Leading of anchor cable from chain locker over windlass to hawse pipe.

SQ SIA (Singapore Airlines Ltd). The National airline of Singapore and member of IATA; or signifies the specified vessels is equipped with special quality steel cable.

SQL Structured query language. A special language designed to enable trained users to enter English-type queries onto an electronic database to extract the information they require.

Square of the hatch The floor of the hold immediately below the hatch.

Squat Tendency of a VLCC to sit lower in the water when under way in shallows.

Squeegeeing A chartering term. See sweeping entry.

Squeeze Pressure on a particular delivery period pushing the price up against the rest of the market – a BIFFEX term.

Squeeze clamp An attachment on fork lift trucks which enables a load to be picked up by the pressure of two plates on each side of the load.

sq ft Square foot.

sq in Square inch.

SR Swiss Air Transport Co Ltd (SWISSAIR). The National Airline of Switzerland and member of IATA; or Soviet Register – ship classification society.

S/R Signing/releasing (Bills of Lading).

Sr Schooner; or single reduction gearing (machinery).

SR & CC Strikes, riots and civil commotions.

SRC Slops receiving station or; self reference criteria – see separate entry.

SRII Stanford Research Institute International.

SRNA Shipbuilders and Repairers National Association.

SRR's Search and rescue regions.

SRT The State Railway of Thailand.

ss Steamship; or screw steamer.

SS Steamship; stainless steel; suspended solids; special survey (on a ship); or Corsair (Corse Air International) (Limited) – IATA member Airline.

SSA Special survey automated controls; Singapore Ship-owners Association, Shipbuilders and Ship repairers Association, or Stevedoring Services of America.

SS & C Same sea and country or coast.

SSB Single Side Band – IATA term.

SSC Swedish Shippers Council; Swiss Shippers Council; Spanish Shippers Council; or Singapore Sports Council.

SSCC Serial Shipping Container Code.

SSCL Swire Straits Container Line – a shipowner.

SSCV Semi-submersible crane vessel.

SSD Supplementary Statistical Declaration – a custom's term operative in the European Community.

SSDs Supplementary statistical declarations.

SSH Special survey of the hull.

SSHEX Saturdays, Sundays and Holidays excepted.

SSHINC Saturdays, Sundays and Holidays included.

SSIA Ship repairers and Shipbuilders Independent Association (UK).

SSIGS Special survey of inert gas system.

SSIs Standard settlement instructions: used by firms in the wholesale market to facilitate the payment and settlement process, by making available to counter-parties standardised instructions regarding relevant bank or settlement account details.

SSM Special survey of the machinery; or PT Susinma Line – a shipowner.

SSN Standard shipping note.

SSOs Ship Security Officers.

SSP Stainless steel propeller.

SSR Special survey of refrigerated machinery; or Secondary Surveillance Radar – IATA term.

SSR MODE S SSR with Discrete Addressing – IATA term.

SSS Single skin ship. Basically single hull ship.

SST Supplementary service tariff; SST Lines (India) – a shipowner; or Smart and Secure Trade lanes – see separate entry.

SSV Support supply vessel.

SSW Sawn soft wood.

SS with date Ship's special survey with date.

ST Stern thruster; side tank; self trimmer; short term; or super transport.

Stabiliser In appearance like a fish fin, situated amidships below the loadline – to counter ship rolling in heavy seas.

Stability A vessel's ability to right herself when heeled over from any outside cause.

Stability and Growth Pact Consists of two EU Council Regulations on the strengthening of the surveillance of budgetary positions and the surveillance and co-ordination of economic policies and on speeding up and clarifying the implementation of the excessive deficit procedure and of an EU Council Resolution on the Stability and Growth Pact adopted at the Amsterdam summit on 17 June 1997. It is intended to serve as a means of safeguarding sound government finances in Stage Three of EMU in order to strengthen the conditions for price stability and for strong sustainable growth conducive to employment creation. More specifically, budgetary positions close to balance or in surplus are mentioned as the medium-term objective for Member States, which would allow them to deal with normal cyclical fluctuations while keeping the government deficit below the referenced value of 3% of GDP. In accordance with the Stability and Growth Pact, countries participating in EMU will report stability programmes, while non-participating countries will continue to provide convergence programmes. An ECB term.

Stable currency An exchange rate of specific currency which is subject to the minimum degree of variation over a particular period due to market forces and other circumstances. A variation of more than one per cent over a three month period can be regarded as fluctuating currency.

Stable market A market place where the volume of business/sales displays little change despite any price variation.

Stable rate An exchange rate of a specific currency or a freight rate of a particular commodity in a shipping/air service which is subject to the minimum degree of variation over a particular period. See also stable currency.

STAC Standard Multi Modal Carrier and Tariff Agents Codes.

Stacker A one piece stacking fitting with non-locking parts.

Stacking To stockpile cargo contained in boxes/sacks/containers or on pallets.

Stacking fitting A fitting that connects into corner castings of top and bottom containers. It maybe locking or non-locking.

Stacking of cargo The process of placing cargo units on top of one another. It is particularly relevant to containers which can be stacked up to three high in a container standage area.

Stack measurements The process adopted by some shipowners to measure for freight calculation purposes involving shipments in bags or sacks a block of say twelve bags stacked on the ground two by two and three tiers high. In such circumstances the measurements are treated as one complete package.

Stackweight The total weight of the containers and cargo in a certain row.

Staff Appraisal A management development technique whereby personnel are appraised annually relative to their job performances, potential leadership, initiative, and so on. It is a method of facilitating the Management development of a Company's resources especially relative to career development of its executive personnel.

Staff Captain Officer responsible to the Master for passenger administration on a large passenger liner.

Stag A speculator who applies for new issues not as a permanent investor but in the hope that an allotment can be sold at a profit when dealings commence.

Stagflation A combination of high inflation and industrial recession.

Stakrak cargo units A type of container.

Stale Bill of Lading In banking practice a Bill of Lading presented so late that consignee could be involved in difficulties.

Stanchions A support for a ship's deck.

Stand A facility at an exhibition, trade fair etc, exhibiting the merchandise a particular company/entity/group of manufacturers has to sell to the market. The stand can also feature service sector companies.

Standage area A place usually at a port or inland clearance depot where vehicles, trailers, and containers are parked for varying reasons. This maybe due to awaiting customs clearance due to inadequate documentation currently available; awaiting shipment; awaiting forward despatch inland, and so on. The Port Authority would raise demurrage charges on standage area traffic which had been used beyond the prescribed free standage.

Stand-alone computer A computer system which is complete in itself and requires no other devices to operate satisfactorily. Used to describe micro-computers which are not connected to networks or other communicating devices.

Standard Measurement used in timber shipments equivalent to 165 cubic feet.

Standard Absorption clause This BIMCO clause now adopted in shipping documents – particularly charter parties and marine insurance – was introduced in 2002 and designed to eliminate the practice of declaring general average for small and uneconomic claims in all sectors of the industry. It is targeted for use in hull and machinery policies covering all types of vessels from container ships, bulk carriers, and tankers to cruise ships.

Standard Bunker Contract The bunker purchasing general terms and conditions contract devised by BIMCO in 2001. It displaced the Fuelcon Standard marine Fuels Purchasing contract of 1995.

Standard displacement The weight (tonnage) of the vessel when fully manned and equipped with stores, cargo, passengers, but without fuel or reserve feed water. See displacement loaded.

Standard Exchange system A customs facility whereby duty relief is provided on goods imported as replacement of goods exported for repair. See prior importation.

Standard export levy A charge payable when certain agriculture goods in free circulation are exported to non Community (EEC) countries.

Standard Import values (SIVs) Standard Import values are a special system for valuing certain fresh fruit and vegetables, and grape juice, at certain times of the year, when the entry price system is in force. SIVs do not operate concurrently with SPVs.

Standard Industrial Classification A method used in the United States to categorise companies into different industrial groupings.

Standard International Trade Classification The United Nations statistical classification of the commodities entering external trade designed to provide the commodity aggregates needed for purposes of economic analysis and to facilitate the international comparison of trade by commodity data.,

Standardisation or differentiation pricing strategy The process of adopting a pricing strategy throughout a country/region/economic bloc at either a standard price thereby offering parity, or a different price in each market place. See differential pricing and pricing strategy.

Standardised codes The codes required to represent each data element.

Standard of living The level of material affluence of a group or nation. It maybe measured as a composite of quantities of goods and their quality.

Standard order The order form used as an official receipt of all incoming orders onto the market floor. These forms must be immediately completed and time-stamped before communication to the trading pit for execution. IPE term.

Standard Shipping Note A document used when delivering cargo to a port, container base, or other freight terminal. It is used primarily for container groupage reception points and only goods for shipment to one port of discharge on one sailing and sometimes relating to only one Bill of Lading may be grouped on one shipping note. Shipping note covering delivery of non-hazardous cargo to any British port. A Dangerous Goods Note is used for hazardous cargo.

Standard tender A tender procedure used by the Eurosystem in its regular open market operations. Standard tenders are carried out within a time frame of 24 hours. All counterparties fulfilling the general eligibility criteria are entitled to submit bids in standard tenders.

Standby cargo Cargo which is not pre-booked on a particular scheduled sailing and will only be shipped if capacity becomes available. See filler traffic.

Standby credit An arrangement to lend money in case of need, usually at market rates and sometimes with a commitment fee; overdraft facilities are sometimes used as standbys by corporate borrowers; or issued in favour of the beneficiary, it stipulates that a penal sum will be paid to the beneficiary on demand in the event the beneficiary submits a signed statement setting forth that there has been default or non performance. This type of credit lacks the stability provided by the goods underlying the commercial credit transaction.

Standby power plant Power plant available for connection to a power system to meet unexpected breakdowns.

Standby ticket A ticket issued to passengers on the basis they only travel in the event of pre-booked passengers not presenting themselves at the time of departure. It is particularly common with airlines and the ticket is usually discounted as it offers no guarantee of travel on a specific flight.

Standby vessel A vessel which assists with the mooring of offtake vessels and can also have other functions to support oil field operations such as emergency towage. These vessels may also be known as anchor handling tug supply vessels or, if used to combine this duty with standby tasks, they may be called Multi-Role Vessels (MRVs).

Standing deposit Money deposited by Agent with Customs to meet Customs charges on continuous imported merchandise flow in UK.

Standing facility A central bank facility available to counterparties on their own initiative. The Eurosystem offers two overnight standing facilities: the marginal lending facility and the deposit facility. An ECB term.

Standstill Commitment not to introduce any new trade restrictions, during the Uruguay Round, which are inconsistent with WTO obligations. A WTO term.

Star A marketing term to describe a very successful and profitable product/service which has established a very competitive track record.

Starboard The right hand side of a vessel/aircraft facing forward.

Star fitting A 'D' ring designed to accept two lashings. Usually termed a cloverleaf deck fitting.

Star ratings A measure of the effectiveness of the advertising campaign particularly size of audience which saw the advertisement or media programme.

Star Ship A major container operator.

Statement of affairs A statement in a prescribed form showing the estimated financial position of a debtor, or of a company which maybe unable to meet its debts. It contains a summary of the debtors total assets and liabilities. The assets are shown at their estimated realisable values. The various classes of creditors, such as preferential, secured, partly secured and unsecured are shown separately. Usually prepared by a Receiver.

Statement of Claim The first step in the pleadings to an action, in which the plaintiff particularises his claim and the legal grounds on which it is based.

Statement of compliance A document issued by a Classification Society in accordance with the shipowner request following survey of specified ship types. See International Liquefied Gas Carrier Code and International Bulk Chemical Code entries.

Statement of recognised gains and losses See STRGL.

State of Origin The state in the territory in which the cargo was first loaded on board of an aircraft or ship.

State of registry The state/country in whose register the ship or aircraft is entered.

State of the operator The state in which the operator has his principal place of business, or if he has no such place of business, his permanent residence (air cargo).

State rooms The highest category of passenger accommodation on a passenger liner embracing a suite.

State Trading The process of conducting International Trade through Government Agencies or quasi Government Agencies.

State Trading Organisation Foreign state-owned buying organisations set up to buy goods and services for their governments, etc.

Static electricity Static electricity is the electrical charge produced on dissimilar materials caused by relative motion between each when in contact.

Static market A market demand for a particular product(s)/service(s) whose constituents remain unchanged over a period of time from any fluctuation especially in volume terms.

Static Stresses Forces acting upon hull of ship when stationary – loaded, unloaded or in ballast.

Status control The continuous monitoring of the credibility of buyers/creditors within a company sphere of business to ensure especially the creditor does not exceed their credit limits/terms of sale.

Status enquiry The process of a seller taking an audit on a potential client/company to determine the credit worthiness. It can also be termed 'credit rating' or status report.

Status quo (ante) (Latin) The former position.

Status report See Status enquiry.

Statute-Barred A debt the claim to which is barred by lapse of time under Statutes of Limitation.

Statute Law Law enacted by Parliament.

Statutory exclusion Policy exclusions specified in the Marine Insurance Act (1906), Section 55.

Statutory instrument Secondary legislation.

Statutory obligation Any obligation, liability or direction imposed by any legislative enactment, decree order or regulation having the force of law in any country.

STBC Self trimming bulk carrier.

STBD Starboard.

STBY Standby/rescue.

STC Straddle carrier operations, or said to contain – a container term.

STCA Short Term Conflict Alert – IATA term.

STCCD Singapore Trade Classification and Customs Duties.

STCW Standard of Training, Certification and Watchkeeping (IMO Convention); (1995 code for seafarers).

Std Standard (timber trade measurement).

STD Scheduled time of departure.

Stds Standards (timber trade measurement).

Steam data Steam data describes the condition of the steam at a particular moment, e.g. pressure, temperature, dryness.

Steel terminal A terminal linking two modes of transport where steel is loaded and/or discharged by mechanical means. It maybe a berth at a seaport, a railway terminal, or road depot.

Steering of containers The function with the aid of specific software for tracking and forecasting (IRMA, MINKA), to direct empty containers to demanding areas at minimum costs.

STEM Subject to existence of merchandise.

Stem Foremost part of a vessel which forms bow unit.

Stemming Arranging bunkers.

Step-frame trailer Trailer with a stepped frame to reduce the height of the load platform.

Sterling account An account which is conducted/managed in sterling.

Sterling area A group of countries which use sterling as a reserve currency and in the conduct of their external trade.

Sterling invoice An invoice payable in sterling relative to an export sales invoice.

Stern The after end of a vessel.

Stern discharge An oil tanker capable of discharging cargo over the stern.

Stern door The watertight barrier which seals an opening in the after end of a roll-on roll-off ship through which rolling cargo is wheeled or driven along a ramp into or off the ship. This door is often made up of the ramp itself which is operated hydraulically.

Stern Frame The substantial member often combining rudder post, propeller post and their extensions. The after ends of plating are secured to it. It frequently carries the rudder and propeller.

Stern lines Mooring lines leading ashore from the after end of poop of a ship, often an angle of about 45 degrees to the fore and aft line.

Stern ramp An inclined plane which connects the after end of a roll-on roll-off ship with the shore or quay on which rolling cargo is wheeled or driven onto or off the ship. The ramp is designed to a watertight door to cover the opening in the ship.

Stern thruster A propeller or water jet system set in the stern of the ship and positioned to give a sideways thrust to assist manoeuvring in a confined space.

STET (Latin) Let it stand.

Stevedores Dockers engaged on cargo/baggage transhipment.

Stevedoring department Shipowners organisation responsible for planning of stowage and cargo transhipment arrangements.

Stevedoring gang Manual workers (labourers) at a port engaged in the handling/transhipment of cargo to/from a ship. An average number would be six persons, but much will depend on whether it is a manual gang physically handling the cargo, or mechanical/computerised gang using mechanised cargo handling equipment. Other factors include the nature of the cargo operation being performed; the custom of the port; the country in which the port is situated, and union agreements.

Stevenson shelter A standard housing for ground-level meteorological instruments designed to ensure that reliable shade temperatures are measured.

STF Security Task Force – IATA term.

STG United Kingdom sterling pound – major currency; or salvage tug.

Stiff vessel Ship with low centre of gravity and consequently will not tend to roll excessively in heavy seas. It is usually caused by excessive weight in the bottom of the ship.

Stillage An open top cage of about one metre or more cubed made of metal with bearers or legs of 10 cms or more used for the conveyance in unitised form of a wide variety of commodities especially ideal for items of a heavy weight to cube ratio such as components parts. They are stackable and are moved about usually by fork lift trucks. It has an uninterrupted space between the bearers or legs for the entry of a stillage track.

STIR Short-Term Interest Rate. Normally used in the context of a three-month interest rate futures contract.

Stl Steel – (material for decks).

STL Submerged Turret loading. See separate entry.

STM Steam.

STN Special Traffic Notice.

Stockbroker A person who deals in stocks and shares.

Stock control The process of keeping a vigilant eye on the quantity of stock held by a Company/Entity to ensure the optimum quantity is available at all times compatible with the demand. Budgetary control and computers feature predominantly in modern techniques of stock control.

Stock Exchange A market place within the International Trade context for the purchase and sale of securities on foreign exchanges (bourses) by private or corporate investors in order to yield a return or hold for capital appreciation.

Stock Exchange operations The purchase of securities on the foreign stock exchanges (bourses) by private or corporate investors in order to yield a return, or in the expectation of a capital appreciation. These are portfolio investments as opposed to industrial investments which represent capital placed by manufacturers in subsidiary or associated enterprises abroad.

Stock flow adjustments See deficit-debt adjustment.

Stockist A stockist holds supplies of semi-fabricated products ready for sale to users. Stockists usually specialise either in steel/stainless steel/aluminium or in copper/brass/high-value metals. A stockist function is also ful-filled by, e.g. a builders' merchant holding large stocks of copper tubing or plumbing fitments. See Fabricator).

Stockist Agent An Agent which sells on commission but undertakes to buy for his own account, or to share in the cost of stocks of some of the items within the range.

Stock Keeping Unit The description of the unit of measurement by which the stock items are recorded on the stock record.

Stock Locator system A system in which all places within a warehouse are named or numbered.

Stock optimisation The process to ensure that the stock holding policy provides effective customer service at a minimum cost.

Stocktaking A process whereby stocks (which may comprise direct and indirect materials, work in progress and finished goods) are physically counted and are then valued item by item. This maybe as at a point in time (periodic) or by counting and valuing a number of items at different times (continuous).

STOL Short takeoff and landing; an aircraft with those capabilities.

Stop-loss order An order to buy or sell a futures contract at market price when protection level is breached within the freight market. A BIFFEX term. Also an order which becomes a market order to buy only if the market advances to a specified level or to sell only if the market declines to a specified level. As soon as this specified level is traded the order is executed for the client at the next obtainable price. There is no guarantee the order will be executed at the price specified. A stop-loss order, as its name implies, is instituted to prevent or minimise losses from either a short or long position.

Stop Loss Reinsurance A reinsurance to protect the reinsured from losses in excess of this net premium income.

Stop-over A break of journey. In so doing, it results in a deliberate interruption of a journey by the passenger – agreed to in advance by the carrier – at a point between the place of departure and the place of destination. An IATA definition.

Stoppage in Transitu Right of a seller to give instruction to a carrier or other bailee to withhold delivery to the buyer, usually due to non-payment for the goods.

Stopper A device for securing a mooring line temporarily at the ship whilst the free end is made fast to a ship's bitt. A Carpenter's stopper is a device with opening jaws to receive wire and shaped wedges to hold line when tension is applied.

Stopper (glass, earthenware, porcelain receptacles) An unthreaded piece of material inserted in the orifice of the receptacle and held by friction.

Storage The process of keeping goods in a warehouse.

Storage charge The fee for keeping goods in a warehouse.

Storage devices Include: RAM, ROM (CD, electronic), magnetic disc, CD disc, magnetic tape, magnetic card/strip. Computer term.

Stores Provisions and supplies on board a vessel required for running a ship.

Stores promotion The process of a particular large shop/retailer promoting specific product(s) relative to a special event/promotion. An example is a large retail store in Switzerland promoting all the Danish products it imports/retails for one week.

Stotinki The currency of Bulgaria. See Appendix 'C'.

Stotinki invoice An invoice payable in Stotinki relative to an export sales invoice.

Stove An arched tapering component forming the walls of a wooden barrel.

Stow The process of accommodating cargo in a specific area usually on a transport unit, quay, warehouse and so on. A good stow is one where there is little room for movement of cargo and is economical in the use of space.

Stowage The process of accommodating cargo in a specific area. See stowage area.

Stowage area An area designated for the storage of merchandise. It maybe open or covered accommodation and typically found at a seaport, airport, road/rail depot and so on. The storage area maybe for general cargo or purpose built for a particular merchandise such as cold store for foodstuff, or tank form for oil/liquid products.

Stowage factor Cubic space (measurement ton) occupied by one ton (2240 lbs), or one tonne (1000 kgs) of cargo. Cotton has high stowage factor whilst steel has a low one.

Stowage instructions Details provided by the shipper/agent about the way certain cargo is to be stowed.

Stowage order A document issued by the shipowner indicating acceptance of a specified cargo on a particular ship sailing from a port and featuring the shipowner's reference. It is usually associated with the movement of dangerous classified cargo by sea. The stowage order will indicate where the cargo is to be stowed on the ship. The Master has the right to refuse to ship the dangerous cargo if the circumstances warrant such a decision. See dangerous goods authority form.

Stowage plan Plan depicting location of cargo stowed in ship.

Stowaway A person who hides on board a vessel/container/truck/railway wagon/aircraft to get a free passage to evade immigration at the destination seaport/airport/frontier point.

STP Satellite ticket printer.

STPB Singapore Tourist Promotion Board.

STPL Steeple.

Str Steamer; or strengthened.

Straddle Buying one quoted delivery month and simultaneously selling the same quantity in another delivery month – a BIFFEX term.

Straddle carrier A machine capable of lifting a container within its own framework and travelling with the load in a dock or container terminal area.

Straight Bill of Lading An American term for a Way Bill to which reference is made in US Pomerene Act. It is a Bill of Lading where the consignee is clearly nominated by name without any qualification ('to the order of' or similar). It is the opposite of an 'Order Bill' which is one where the Consignee is shown as being 'to order', or with a name qualified by the word 'to the order of' or similar. Worldwide a 'Straight Bill' is not negotiable and cannot be used to transfer title by endorsement (or passage of a blank endorsed Bill) in the way an 'Order Bill' can. The delivery of the goods is not contingent on surrendering the Bill of Lading but only on its being produced by the consignee as means of identification. Hence, while a straight Bill of Lading is non negotiable it differs from the ordinary Way Bill in being transferable subject to existing equities.

Straight Bonds Bond issues with a fixed coupon.

Straight consigned Consigned directly to the order of the Named Party.

Straight debt The fixed interest rate Eurobond without the option to convert to equity.

Straight lift A method of slinging in which the hook(s) of a sling engage directly onto lifting points on the load, and leg(s) of chain lie in a straight line.

Straight tankers Vessels only capable of conveying only oil as compared with OBOs. See OBOs.

STRAITS Straits Shipping – a shipowner.

Stranding A ship being caught on rocks or a beach.

Strapping Any material (string, plastic tape, sticky tape, etc) used to secure pallet loads.

Strategic alliance A joint venture or marketing co-operation are two examples of a strategic alliance whereby a company joins forces with another operator with identical objectives. In so doing they improve their business objective which maybe motivated by market entry mode, market share improvement, technology advancement, profitability, investment and so on.

Strategic Business Unit A unit/department in a company with complete focus to devise business strategy and to monitor its execution.

Strategic located industry An industry which is strategically located to take full advantage of its optimum development. This would embrace not only geographical locations and country credit rating, but access to markets, competent labour, production resources, raw materials, tax benefits, good infrastructure, legal environment, political stability etc. The importance/significance of each would vary by the nature of the industry which maybe in the service sector, airport/seaport or manufacturing car production.

Strategic Management See Strategic Management accounting.

Strategic Management accounting The provision and analysis of management accounting data relating to a business strategy, particularly the relative levels and trends in real costs and prices, volumes, market share, cash flow and the demands on a firm's total resources.

Strategic Planning The process of planning within a company the available resources on an optimum basis including investment project(s) with a view to taking maximum advantage of business opportunities based on profit motivation.

Strategic pricing The process of establishing and offering the optimum price of a product/service in the market place.

Strategic Trade Policy A policy laid down by a company/business to realise an objective. It may be control of exports to favour trade development within an economic region/bloc, or discriminating import custom tariffs.

Strategy A course of action including the specification of the resources required, to achieve a specific objective.

Strategy of empathy A strategy of a Company/entity where the wishes/needs of the buyer/end user are respected in regard to the provision and specification of the product/service provided.

STRATH P & O Strath Services – a shipowner.

Stratified sampling A statistical term indicating that the sampling is based in sub-dividing the situation into groups and selecting from each group in the relevant ratio.

Stratosphere A region of the upper atmosphere, which extends from the tropopause to about 50 kilometers above the Earth's surface, and where the temperature rises slowly with altitude. The properties of the stratosphere include very little vertical mixing, strong horizontal motions and low water vapour content compared to the troposphere. The ozone layer is located in the stratosphere.

Strawboard A low quality fibreboard made entirely or almost entirely of straw, having a tendency to brittleness and recognisable by its characteristic straw colour.

Strengthened WTO rules and disciplines The negotiating group's mandate states that the integration of the textiles sector into the WTO should be on the basis of "strengthened WTO rules and disciplines". To several major textile importing countries this refers to new rules on anti-dumping, countervailing duties, intellectual property and safeguards. A WTO term.

Strength of the market An evaluation of a market with particular emphasis on the salient features of the market of interest to the entrepreneur. This will extend to the general stability and viability of the market including its political aspects/stability, the market profile, and opportunities to the entrepreneur.

Stretch wrapping A form of packaging which is provided by enveloping the goods in cellophane.

STRGL Statement of recognised gains and losses. Accountancy term embracing the financial gains and losses on assets and liabilities found in the Balance Sheet. See assets and liabilities entries.

Strike A concerted industrial action by workmen causing a complete stoppage of their work which directly interferes with the working of the vessel. Refusal to work overtime, go-slow or working to rule and comparable actions not causing a complete stoppage shall not be considered a strike. A strike shall be understood to exclude its consequences when it has ended, such as congestion in the port or effects upon the means of transportation bringing or taking the cargo to or from the port. A BIMCO term.

Strike Expenses Expenses incurred as a result of a strike, such as forwarding costs for goods that cannot be discharged at the scheduled destination port or extra freight charged for overcarriage to another port when the scheduled discharge port is strikebound. These expenses are not covered by the marine policy with the standard strikes clauses attached thereto.

Strike price The market price agreed between the buyer and seller at the time of the deal being struck. A BIFFEX term.

Strikes cover Limited to damage caused to insured property by strikers, locked out workmen and persons involved in a labour dispute. Does not include loss or expenses incurred as a result of strikes etc.

Striking price Price at which shares are allotted in a tender offer for sale – a form of new issue. Everyone tendering at or above the striking price receives some shares at the striking price.

Stripped package A package that includes the minimum ingredients to qualify for an IT number; a package or tour offering lesser accommodations and features.

Stripping The process of unloading a container.

Stripping Pump Used in final stages of tank cleaning to clear residue of cargo from tanks.

Strong manufacturing based economy An economy in a particular country/region/area which has a very competitive industry and in so doing is able to compete effectively/decisively for overseas markets/tenders thereby producing a favourable external manufacturing Balance of Trade surplus.

Strong market A market which is very stable and has a strong economy.

Strops Ropes or wires with an eye at each end used for slinging cargo.

Structural reform The process of conducting/undertaking a structural reform which may focus on one or more of the following: economic, political, social, technical and legal.

Structured data Data which is placed within a structure such as a record, usually within a database – computer term.

Structured design A combination of Top Down Design and Structured Programming – computer term.

Structure diagram Also called a structure chart. A diagram showing the structure of data (as in JSP) or the structure of a program and the relationship between modules of the program. The structure of program modules often reflects the structure of data – computer term.

Structured English A shorthand form of the English Language which is used to define elements of a computer program. Sometimes called 'Pseudo code'.

Structured interview In market research terms an interview conducted between a skilled professional marketeer researcher and a respondent using a specific range of questions in a prescribed sequence/manner.

Structured programming A method of programming which uses the principles of a hierarchy of components related to the three basic constructs, a modular approach and modules which have restricted entry and exit points.

Structure of the market The composition/make up of a market and how it functions/operates.

Strugglers A social/economic category term with the lowest incomes and too few resources to be included in any consumer class.

STS Ship-to-ship transfer operations.

STs Side Tanks (on a ship).

Stuffing The process of loading a container.

Stuffing/Stripping The process of packing/unpacking a container.

S Turb Steam turbine (of a ship).

stvdrs Stevedores.

STVP Short term vehicle park.

STW Stern wheel.

SU Aeroflot – Russian International Airlines – an IATA Airline Member.

SUA Suppression of unlawful acts against the safety of Maritime Navigation (SUA treaties). USA legislation.

Sub contract The process of a contractor delegating an element of the contractual work to a third party – the sub contractor who usually has some specialisms in that particular area which the contractor cannot provide. Alternatively the work may be sub-contracted due to efficiency or other reasons.

Sub contractor A local operator that provides service for a wholesaler.

Sub stem See subject stem entry.

Subject approved No Risk The abbreviations SANR appears on a slip where the assured is required to approve the condition etc for the risk to attach.

Subject details A fixture maybe agreed in respect of the main term but the parties agree that certain points need to be clarified, for example, the length of vessel, whether the necessary documents/certificates can be obtained or the fact there may have to be alterations to the terms of the standard charter form in a manner which is clear and acceptable to both parties, and this, particularly in the case of a period charter, might take some time. A chartering term.

Subject matter insured The subject matter exposed to risk to which the assureds' insurable interest attaches. In a cargo policy it would be the 'goods', a hull policy the 'ship and machinery', and a freight policy the 'freight'. In a policy the subject matter must be designated with reasonable certainty so that it is readily identifiable.

Subject receivers The Charterer is required to check with the receiver or cargo buyer that the vessel is acceptable at the Receiver discharge terminal, the sale of the cargo is finalised etc. The opposite to subject stem. A chartering term.

Subject stem A chartering term. It applies to shippers and/or suppliers agreement to make a cargo available for specified dates. In the case of stem not granted as required, no other ship can be fixed by the charterers before the one initially fixed subject stem has received the first refusal to accept the amended dates and/or quantity provided they are reasonably near.

Subject to contract A legal term indicating the contract does not become binding until certain formalities have been concluded.

Subject to management approval There are many valid reasons to check the charterers company. For example, in the area of vessel environmental performance, and a more senior person can improve the charter party terms as it is ultra vires to the executive originally handling it. A chartering term. See vessel screening.

Sub judice (Latin) Under (judicial) consideration.

Sub letting Right of permitting in a charter party the charterer to transfer the document.

Sub mariner A person engaged in under water activities such as a diver salvage operator, submarine etc.

Submerged pump A type of centrifugal cargo pump commonly installed on gas carriers and in terminals in the bottom of a cargo tank. It comprises a drive motor, impeller and bearings totally submerged by the cargo when the tank contains bulk liquid.

Submerged turret loading A sub-sea device, hanging above the sea-bed at neutral buoyancy over oil wells. It is connected to the wells by flexible hoses or flowlines. Connection to the offtake vessels is accomplished through a vertical trunking designed into the ship. The offtake vessel manoeuvres into position over the STL and hauls it into the ship so forming a temporary cargo connection and mooring.

Suboptimising The process of striving for optimum performance in one element of an organisation disregarding the effects this may cause to the performance of the other elements. In other words, a solution for a problem that is best from a narrow point of view but not from a higher or overall company point of view.

Subordinated loan A loan which, in the event of a liquidation, ranks after the claims of ordinary creditors and prior only to the interests of the ordinary shareholders.

Subordination A clause sometimes inserted in the terms of a capital loan stock issue whereby the rights of the stockholders rank after the rights of some or all unsecured creditors of the borrower in the event of his liquidation.

Subrogation The right of the underwriter to inherit the rights and liabilities of the assured following payment of a claim to recover the payment from a third party responsible for the loss. Limited to the amount paid on the policy.

Sub rosa (Latin) Secretly, in confidence.

Subscription The extent of the underwriter's liability in an insurance. The line signed by the insurer.

Subsidiaries guarantee The process of providing insurance cover to a sale by an overseas subsidiary of goods sold to it by its parent or associated company. An example would be the parent or associated company being situated in the UK.

Subsidiary Company A company shall be deemed to be a subsidiary of another if, but only if (a) that other either: (i) is a member of it and controls the composition of its Board of Directors, or (ii) holds more than half in nominal value of its equity share capital, or (b) the first mentioned company is a subsidiary of any company which is that other's subsidiary and it otherwise comes within the terms of Section 154 of the Companies Act, 1948.

Subsidies Government financial grant to shipowners for shipbuilding, operating, or making good losses incurred in providing a service.

Subsidised export credits The process of a government or Economic bloc subsidising export credits. See export credits.

Subsidised travel The process of a passenger being financially assisted in his/her travel arrangements. It maybe an employer's contribution; a local authority grant such as providing free travel to school/college; or the provision of free or discounted travel for senior citizens offered by a bus company, rail operator etc.

Subsidy There are two general types of subsidies: export and domestic. An export subsidy is a benefit conferred on a firm by the government that is contingent on exports. A domestic subsidy is a benefit not directly linked to exports. A WTO term.

Substituted Expenses

Substituted Expenses Expenses incurred in place of loss or expense which would be allowed as general average (e.g. cost of removal of a ship, with general average damage, to a place where repairs would cost less).

Successful export strategy This embraces following key areas: identify company objectives in business plan; undertake feasibility study; examine legal and political constraints, review political and economic stability in markets to enter; understand and empathise with the culture and continuously evaluate buyers needs; research market to decide which market to enter; visit market to produce an evaluation; prior to visit personnel to undergo culture briefing; adapt/modify product or service to buyers' needs; choose carefully overseas representative; recruit experienced professional and multilingual personnel; undertake a cost analysis and focus on risk areas; examine competition; undertake credit insurance; evaluate/determine pricing structure and terms; evolve a trade payment cycle analysis; attend trade fairs/exhibitions and feature in outward trade missions; display strong commitment to the buyer; undertake buyer screening analysis; ensure exporter has ISO 9000 quality assurance accreditation; make full use of computer technology and satellite communication; provide adequate resources within the company; formulate export/international marketing plan; exercise control systems through budgetary management; develop international logistic driven distribution strategy; develop multilingual brochure/promotional material; prepare negotiating plan; establish country/buyer credit rating; and develop good customer relations/after sales/spares/servicing provisions. See export market malaise strategy.

Successful procurement strategy This embraces the following areas. Identify company objectives in business plan; determine product specification; research the market regarding potential suppliers; examine legal and political constraints; review political and economic stability in procurement markets; recruit experience professional procurement personnel with multilingual skills; evaluate import regulations; visit the market to produce an evaluation and discuss with potential suppliers; undertake screening of potential suppliers; choose carefully overseas representative; develop a logistic culture with a supply chain; ensure suppliers have ISO 9000 quality assurance accreditation; visit trade fairs/exhibitions and feature in inward trade missions to identify potential suppliers; provide adequate in-house resources; evolve a buying price structure and terms; ideally obtain product samples from potential suppliers and testing evaluation; study the competition and their procurement strategy; undertake networking; develop payment cycle strategy; evaluate product cost analysis of procurement strategy; undertake culture briefing before visiting overseas market; develop procurement distribution strategy; exercise control system through budgetary management; continuously review procurement strategy in all areas; especially financial and value added in product specification; formulate a procurement plan; evaluate risk areas; prepare negotiating plan; establish country/supplier credit rating; and make full use of computer technology and satellite communication. See procurement malaise strategy.

Sucre The currency of Ecuador. See Appendix 'C'.

Sucre invoice An invoice payable in Ecuadorian sucres relative to an export sales invoice.

Suction hopper dredger A vessel designed to recover silt etc from the bed of a river/port/estuary etc and in so doing undertakes the work by means of suction.

Sudan Airway The National Sudanese Airline and member of IATA.

Sudanese pound The currency of the Republic of Sudan. See Appendix 'C'.

Sudanese pound invoice An invoice payable in Sudanese pounds relative to an export sales invoice.

Sue and Labour Sue and Labour is action taken by the assured, his representative or assignee to prevent or reduce loss for which the underwriter would be liable. The action relates to the Sue and Labour Clause, excluding General Average or Salvage.

Sue and Labour Charges Expenses incurred by a marine assured to prevent or minimise loss for which underwriters would have been liable. Expenditure incurred as part of a GA act is not allowed as sue and labour, nor are expenses incurred in a salvage operation involving a salvage award. Sue and Labour charges are recoverable under a marine insurance policy whether or not the purpose for which they were incurred was achieved. They are recoverable in addition to any other loss under the policy, even a total loss. Under a hull policy, except where they are associated with a total loss, S & L charges are subject to the policy deductible. Claims for S & L charges under a hull policy are subject to reduction to reflect underinsurance, if any.

Sue and Labour Clause This clause formed part of the SG policy form and was the basis for applying to the policy expenses incurred by the assured to prevent loss for which underwriters would have been liable. The clause was not retained when the MAR form of policy was introduced to replace the SG policy in 1982 (cargo) and 1983 (hull), and was, therefore, omitted from the new cargo and hull clauses drafted for attachment to the MAR form of policy. Nevertheless, the effect of the sue and labour clause is incorporated in the 'duty of the assured' clause in the new clauses.

Suezmax A vessel which is capable of passing through the Suez canal having regard to its draught and overall dimensions especially length and beam. A Suezmax crude oil tanker has a capacity range between 120,000 and 200,000 dwt.

Suez stop The code name of BIMCO approved charter party clause relative to the Suez canal and entitled 'Baltic Conference Stoppage of Suez Canal Traffic clause, 1956'.

Sugar ship Purpose built vessel to convey sugar in bulk shipments.

Suite Hotel unit of at least two rooms, one a parlour; may include kitchen facilities.

SUK State Secretariat (Brunei).

SUL Singapore Union Line – a shipowner.

Sulcon Code name of Chamber of Shipping approved charter party for grain trade.

Sulphuryl Fluoride Fumigant used to control pests to timber. Often to fumigate containers.

Summer deadweight The deadweight of a ship when loaded to summer marks.

Summer marks The summer loadline mark and centre of Plimsoll disc on the ship's side.

Summer tanks In early oil tankers, the expansion tanks served as spaces that could be filled with oil to bring a vessel to the loading mark in the summer.

Sundays and Holidays excepted This excludes Sundays and holidays counting as laytime in the charter party.

Sun protection required A 'flagging' system to ensure that containers with heat sensitive goods are not unduly exposed to direct solar radiation.

Sunset Clause Specific time limits within which specific national trade measures, for safeguards, anti-dumping actions, countervailing measure etc must expire.

SUNUNION Sun Union Line – a shipowner.

Super Superstructure.

Supercargo An experienced officer assigned by the charterer of a vessel to advise the management of the vessel and protect the interests of the charterer.

Super cube container An ISO container of the latest generation in terms of specification with a very high cube to weight ratio and as found in the container of 2.59m high, 2.44m wide and 12.19m long. Also termed High Cube Container or High Capacity Container.

Super Handymax A vessel in excess of the Handymax dry bulk carrier capacity. See dry bulk carrier fleet.

Superheated vapour Vapour removed from contact with its liquid and heated beyond its boiling temperature.

Superstructure The decked structure above the upper or main deck and outboard sides of which are formed by the shell plating on a vessel. Examples include bridge, poop, forecastle etc.

Super tanker An oil tanker in excess of 200,000 tons.

Super Trailer A road trailer provided with small rear road wheels enabling its rear half platform height to be much lower giving a greater total cube and enlarged internal floor to ceiling height without exceeding the overall four metre maximum permitted height. The weight carrying capacity of the trailer is not increased but merely the volume load available.

Supplement The price for an extra or better grade of service; see single supplement.

Supplemental liability insurance Liability coverage for accident claims against a car rented by a third party.

Supplementary declaration An electronic message sent to CHIEF used to declare the fiscal, statistical and control information for all consignments imported under CFSP. A customs term.

Supplementary service tariff Tariff for services other than freight.

Supplementary Stocks Guarantee The process of providing to an exporter insurance cover of goods while held in stock overseas against certain events mainly of a political nature.

Supplementary statistical declarations A monthly return which must be completed by traders above a certain annual threshold and submitted to Customs. It records all intra-community despatches and arrivals. Such data is required under the EC system for collecting intra-community trade statistics termed INTRA-STAT.

Supplier An entity offering travel products or services for sale, generally through retail travel agents and often directly to the public.

Supplier client relationship A marketing term involving the relationship between the client buying the goods/service and the supplier/manufacturer.

Supplier credit The process of the supplier/manufacturer/exporter/seller allowing credit terms to the importer/buyer, and obtaining finance from a bank or export credit agency. Payments are made to the supplier/exporter/seller by the importer/buyer over the period of the contract and in the currency and under the terms agreed. The importer/buyer liability is to the supplier/exporter/seller until payment is completed. It maybe termed Exporter credit or Sellers credit.

Supplier integration A logistics term. This is the process of the supplier at an early stage of product design/development and forecasting to liaise with the customer(s) needs. This enables logistic cost savings to be realised in supply chain management performance.

Supply chain The sequence of events in a goods flow which adds to the value of a specific good. These events may include conversion assembling and/or disassembling and movements and placements.

Supply chain management The Management mechanism/structure and strategy/objective to supply goods on an international trade basis involving not only the exporter/seller/supplier, and buyer/consumer/end user, but also all the intermediaries and their management in the supply chain.

Supplytime 89 Code name of BIMCO approved Uniform Time Charter Party for Offshore service vessels.

Supply vessel A vessel which carries stock and stores to drilling rigs.

Supporting document Additional evidence needed to verify or identify for a transaction; driver's licence, birth certificate, health card, visa and passport are examples for the former, cancelled receipts for the latter.

Surcharge An additional charge on the basic tariff such as the currency, bunkering fuel surcharges.

Surcharge value A surcharge for the carriage of cargo having a value in excess of a specified amount per kilogram.

Surface consolidators An Agent usually who offers regular services on schedule sailings and in so doing despatches as one overall consignment in the truck, ISO container, or swap body a number of individual compatible consignments from various consignees to various consignors. It could involve a dedicated rail service and usually is multi-modal.

Surface freight rates Freight rates associated with road and rail conveyance of goods.

Surface transport Provision of transport services overland such as rail and road.

Surf days Charter party term indicating when a heavy swell of surf prevents loading/discharging at the port; it will not count as laydays.

Surge pressure A phenomenon generated in a pipeline system when there is a change in the rate of flow of liquid in the line. Surge pressures can be dangerously high if the change of flow rate is too rapid and the resultant shock waves can damage pumping equipment and cause rupture of pipelines and associated equipment.

Surinam Guilder (Florin) The currency of Surinam. See Appendix 'C'.

Surinam Guilder (Florin) invoice An invoice payable in Surinam Guilders (Florins) relative to an export sales invoice.

Surplus capacity The surplus/spare manufacturing/production capacity of a particular company/entity for a specific product/service based on the company specified market demand. Hence, production levels maybe 70% leaving 30% for surplus capacity.

Surplus production
The process of manufacturing/producing a product(s)/service(s) in excess of the company/entity defined market demand. Hence, planned/anticipated market demand in production terms maybe 70%, leaving 10% for surplus production based on a production level of 80%.

SURT Simple Universally Recognised and Truthful. An advertising term.

Surveillance A system of monitoring safeguard and other trade policy measures to ensure transparency and consistency with WTO rules. A WTO definition. The concept of surveillance applies in a wide variety of other situations worldwide where consensus exists.

Surveillance Body Established in 1986 to oversee the two commitments on standstill and rollback. A WTO term.

Surveillance licence An import licence issued for goods on which import levels are being monitored, such as textile articles. It is usually valid for three months and issued without restriction. The licence is issued by the Government.

Surv Survey.

Survey In marketing terms the process of conducting field research usually in the transportation areas/disciplines. See ship survey, traffic survey and field research entries.

Surveyor Marine Engineer responsible for examining/surveying/classifying ships usually employed by Ship Classification Society or Government Maritime department; or in insurance terms, person who inspects property etc to advise the underwriter about the physical aspects of the risk.

Survey report A detailed document, compiled by a surveyor, of the cause of loss, valuation, etc, quantifying the amount of the loss. Overall, it is a report of an expert, issued by an independent party such as an assessment and cause of damaged cargo on which claim has been lodged.

Susie Air supply coupling pipe between tractive unit and semi-trailer.

Suspense account An account in which debits or credits are held temporarily until sufficient information is available for them to be posted to the correct accounts.

Suspension A customs term whereby customs duties/charges are suspended when the goods are imported such as exist in the UK under the inward processing relief system for traders authorised by customs.

Suspension seat Driving or passenger seat incorporating a spring suspension system to give the occupant a bump-free ride, often adjustable to take account of the occupant's weight.

SV Saudi Arabian Airlines Corp (SAUDIA) – an international airline and member of IATA; sales voucher; or sailing vessel.

Sv Sievert.

SVQ Scottish Vocational Qualification.

SVS Snap shot valuation service.

SW Shipper weights; salt water; South West – a compass point; or Air Namibia – an IATA airline member.

Swan neck or goose neck General description for the actual stepped portion of step-frame trailer, so called because of its resemblance to a swan or goose neck.

Swap An International Banking term which refers to the purchase/sale of a currency in the spot market combined with the simultaneous sale/purchase in the forward market. A hedging mechanism whereby a spot transaction is wholly or partially matched by an opposite forward transaction. Shipowners and charterers have made increasing use of swaps to limit their risk exposure in such areas as interest rates, currencies and bunker prices.

Swapbody An intermodal transport unit which takes the form of a container. It is without wheels but relies on the skeletal trailer for road movement and the flat bed wagon for rail operation. The swapbody differs from the ISO freight container in that it is bottom lift only. It is not stackable and many are equipped with skeleton legs to facilitate intermodal transfer. A typical swapbody would be 2.777m overall height 2.420m internal width, 2.55m internal height, 13.6m overall length and pay load 29 tonnes. Overall, it offers a door to door service and involves road/rail/sea.

Swap rate A type of exchange rate whereby a standard swap is an agreement between two counter parties in which the cash flows from two assets are exchange as they are received for a fixed time period (tenor), with the terms initially set so that its present value is zero. It is a negotiable instrument between the Bank and Customer. See also two types of exchange rates, Spot and Forward.

Swaps Exchange of interest obligations between two parties, either fixed to floating rate, or vice versa, or a given notional period for an agreed time period. Currency swaps exchange interest flows in two currencies, and also include an exchange of principals in the two currencies.

Swap shot valuation service The process of offering a service which merely provides an overview of the market situation at a particular time.

SWATH Small waterplane area twin hull.

SWBM Still Water Bending Movement.

SWD Salt water draft.

Sweat damage Damage caused to cargo by condensation, usually due to lack of ventilation in a ship's hold.

Swedish Chamber of Commerce for the UK The organisation based in London representing Sweden whose role is to develop, promote and facilitate international trade between UK and Sweden and vice versa.

Swedish Corrosion Institute International organisation specifying standards of steel preparation prior to protective painting.

Swedish Shippers Council The organisation in Sweden representing the interest of Swedish shippers in the development of international trade.

Sweeping A chartering term associated with discharging tanks on an oil tanker. It features in the charter party contract and may be termed puddling or Squeegeeing

Sweet crude A crude oil which contains a low sulphur content.

Swell Waves generated by distant or earlier weather systems which travel large distances with little attenuation.

SWEPRO Swedish organisation for the simplification of International Trade Procedures based in Gothenburg.

Swettenham Fender One of a series of fenders fitted to a smaller tanker used when lightening a VLCC.

SWIFT The Society for worldwide Interbank Financial Telecommunications.

Swing When an exporting country transfers part of a quota from one product to another restrained product. A WTO term.

Swing clearance The clear arc, for example, swept by the front corners of a trailer when turning in relation to the towing vehicle.

Swire Straits Container Line An Overseas Containers Ltd associate company trading between Japan and the Malacca Straits area.

Swissair The National Swiss Airline and member of IATA.

Swiss Franc The currency of Switzerland. See Appendix 'C'.

Swiss Franc invoice An invoice payable in Swiss Francs relative to an export sales invoice.

SWISSPRO Swiss organisation of the simplification of International Trade Procedures based in Lausanne.

SWISS Shippers Council The organisation in Switzerland representing the interest of Swiss shippers in the development of international trade.

Switch Bills A Bill of Lading which seeks to conceal source of the cargo, shipment date, destination and avoid customs duties. See switch bill of lading.

Switch Bill of Lading A switch Bill of Lading is issued by, or on behalf of, the carrier in substitution for the Bill of Lading at the time of shipment. The prime reason why the switch bill of lading is issued is that from the point of view of the holder of the bills, the first set of bills is unsuitable under one of the sale and purchase contracts of the goods in question. Carriers often feel under commercial pressure to issue switch bills in order to satisfy the requirements of their customers.

Switching Transferring an open position into a different month for the same commodity within the freight market. A BIFFEX term; or (a) on the LME, the exchange of metal in one warehouse for that in another, e.g. Rotterdam for London or (b) the movement out of one futures contract into another, usually further forward.

Switch Trading A counter trade technique. It arises in imbalances in long-term bilateral trading agreements usually between East European countries and developing nations. It can sometimes lead to the accumulation of un-cleared surpluses in one or other country; for example Brazil at one stage had a large credit surplus with Poland. These surpluses can sometimes be tapped by third parties so that, for example, UK exports to Brazil might be financed from the sale of Polish goods to the UK or elsewhere. A sophisticated development of switch trading involves actual triangular trading with hard currency payments through switch dealers.

SWL Safe working load.

SWOPS Single Well Oil Production System.

Swot analysis A marketing term involving the process of analysing a product to determine its strengths, weaknesses, market opportunities, and competition threats.

SWR Steel wire ropes.

Sy Synchronous system (on a ship).

Syli The currency of the Republic of Guinea. See Appendix 'C'.

Syli invoice An invoice payable in the Republic of Guinea syli relative to an export sales invoice.

Sympathetic Damage Damage to property which arises because other property in close proximity has been damaged. An example would be where fumes given off by water damaged cargo cause taint to another cargo.

Synaxomex 90 Code name of BIMCO approved Continent Grain voyage charter party.

Syndicated Bank Credits The process of providing medium-term loans extended by a syndicate of banks, normally at a fixed margin over the (variable) interbank cost of funds. A distinction is often made between international loans, financed, in whole or in part, from funds raised on the eurocurrency markets, and foreign loans. The latter are loans by a group of banks in one country, denominated in the currency of that country, to a non-resident borrower.

Syndicated Euro credit A means of borrowing in a Euro currency from a syndicate of banks at floating interest rates.

Syndicate (Lloyd's) A group of underwriting members at Lloyd's whose acceptances and liabilities are handled jointly by an underwriting agency acting on their behalf, while each member remains legally liable solely for his/her own share of the syndicate's liability.

Syndication of Shipping loans Process of a consortium of banks under a lead Bank providing a large loan for a new build.

Synergy The simultaneous joint action of separate parties which together have greater total effect than the sum of their individual effects.

Syntax The rules governing the structure of an EDI message.

Synthetic based muds Drilling muds that have synthetic oils in place of mineral oil. Synthetics mainly include chemicals such as esters, ethers, and polyalpha-olephins, but other types can include other various chemical forms of olephins, as well as acetals and paraffin. A term associated with exploring for offshore oil and gas.

Syrian Pound The currency of Syria. See Appendix 'C'.

Syrian Pound invoice An invoice payable in the Syrian Pound relative to an export sales invoice.

System configuration (LANs) The setting of passwords, rights, access to servers and printer queues.

System flow chart A diagrammatic model of the movement and interaction of documents, data and storage in a computer system.

System One Automated reservations system owned by System One Corp, which is affiliated with Continental Airlines.

Systems analysis Investigating and analysing user needs for an information system and proposals for a computer system. It often includes the structured design and implementation of the system.

System specification Comprises the data and information required to enable analysts and programmers to design and produce a system which meets user requirements. The specification may comprise some or all of the following: data definition-model, process specifications, input specifications, output specifications, resource requirements.

SZ China Southwest Airlines – IATA Member Airline.

T1 Status Goods not in free circulation – a customs term.
T5 Turkmenistan Airlines.
T Twin screw; ton(nes); thrusters; tropical; triple expansion engine; tapered hatchway – with narrow breadth; or tug; or metric ton(s) or tonne(s); unit of mass, but often also used for forces (sometimes expressed as 'tf'); ltf = 9.81kN.
ta Tank tainers or Tanks.
TA Trade Association; Technical Assistant – a person engaged on technical work within a company; or TACA International Airlines SA – an IATA Airline Member.
Ta Tanks.
TAA Transferable account area; or Trans Atlantic Agreement.
TAAG Tropical African Advisory Group. An HM Government Trade Advisory group.
TAB Technical Assistance Bureau (ICAO term).
Table Entity types in a relational model are represented by a table of values where the columns represent attributes of the entity and each row of the table corresponds to an entity occupance. The header and body parts of the table form a relation containing attribute values (cells), attribute domains (columns-fields), records (rows) and keys. Computer term.
Table d'hote A full-course meal served at a fixed price.
Tabloid Newspaper with a small page area.
Taboo An unacceptable strategy, for example, when exporting to Japan to discuss with buyers the second world war.
Tabulate To arrange or display in pre-determined position; to set out in a systematic grouping. Data processing or statistical term.
TAC The Australian Conference. The Conference for the Europe-Australia trade; Thai Airways Co Ltd; or Transport Agencies Corporation (Burma).
TACA Trans Atlantic conference Agreement; or Total Access Communications Co Ltd.
TACAN Tactical Air Navigation – IATA term.
Tachograph A speedometer and mileage counter fitted with a clock and recording mechanism. It is obligatory for virtually all road haulage vehicles licensed within EU countries and its prime purpose is to measure time at the wheel so as to obtain compliance with the restrictions on drivers' hours. Distance and speed are therefore recorded and gives the driver a warning light to remind the driver that a pre-set speed has been exceeded.
Tackle Device or arrangement used between the hook effecting the lift and the package or packages lifted.
Tactics The most efficient deployment of resources in an agreed strategy.
TAD Transit accompanying document.
TAF Aerodrome (terminal or alternate) forecast – in abbreviated form – IATA term.
TAFA Thai Air Freight Forwarders Association.
TAFOR Aerodrome (terminal or alternate) Forecast – in full form – IATA term.
Tag A cardboard, plastic, or metal marker used for identification or classification.
T & G Tongued and grooved (timber trade).
TAIGA Taiga Navigation Co Ltd – a shipowner.
Tail A short length of synthetic rope attached to the end of a wire mooring line to provide increased elasticity and also ease of handling.

Tail-lift Hydraulically or electrically operated loading platform fitted to the rear of a road vehicle body to assist in loading and unloading heavy objects or unit loads.
Tail shaft The part of the propeller shaft that passes through the stern tube and takes the propeller in a vessel.
Tainted vessels Ships carrying goods loaded in/to be discharged in a third country port calling at one or more community ports; vessels which call at third country ports between the Community port of departure and destination; vessels calling at freeports/zones; goods transiting through non EU countries; goods transported by air via a non EU country.
Tainting Process of cargo being soiled by atmospheric conditions arising for example from merchandise in close proximity giving off odours such as oranges tainting tea. This problem arises with consolidated consignments.
Taka The currency used in Bangladesh. See Appendix 'C'.
Taka invoice An invoice payable in Bangladesh taka's relative to an export sales contract.
Take off The process of an aircraft leaving the ground or aircraft carrier etc and in so doing commences the flight.
Takeover bid An offer by an outside company to shareholders to buy their shares in order to take control of the existing company.
Taker The buyer or purchaser of an option. A BIFFEX definition.
Taking inward pilot Frequently used in a time charter to determine the time and place of delivery of a ship by the owner to the charterer. The hire charge commences at the moment the pilot embarks.
Talisman Transfer Accounting Lodgement for investors.
Tally clerks Personnel engaged in the process of checking or taking account of the goods loaded placed into a ship or unloaded from a vessel.
Tallying The process of checking, or taking an account of the goods loaded placed into a ship or unloaded from a ship.
Tallyman A person who records the number of cargo items together with the condition thereof at the time it is loaded into or discharged from a vessel.
Tally sheet Written record of the count of the number of pieces of cargo, their description, marks and numbers, carried out at the time they are loaded into, or discharged from a ship.
Tambala The currency of Malawi. See Appendix 'C'
Tambala Invoice The payment in Malawi Tambala currency relative to an export sales contract.
TAMS Thruster Assisted Mooring System.
TAN Transit Advice Note; Total Acid Number; Lloyd's Register of Shipping term; or Tangent.
Tandem The process of working together of two cranes heavy gear to increase maximum lifting capacity.
Tandem axle Two axles of a road vehicle, located close together and usually having a linked suspension system.
Tandem configuration This is the most common mooring arrangement found offshore. It comprises a mooring hawser between the bow or the offtake vessel and the FPSO. The two units lie in 'line ahead'.
Tangible variables A marketing term used in market forecasting whereby tangible products can be identified in levels of production/consumption and on that basis a judgement made with further research of the likely level of production/consumption. Examples include car production, oil, grain, iron ore, etc.

Tank barge River barge designed for the carriage of liquid bulk cargoes.

Tank car Container used for carrying liquids in bulk by rail.

Tank cleaning The process of cleaning out and clearing of gases from a container tanker unit, road tanker, railway tanker or the cargo tanks on a tanker. It may be oil, chemical or other liquid cargo.

Tank container A container especially built for transporting liquids and gases in bulk.

Tanker Single-deck vessel constructed and arranged for the carriage of liquid cargoes in tanks mainly crude oil, chemicals, or non-hazardous refined products. Tankers are sometimes used to carry grain

Tanker broker A person who deals with tankers on the tramp market.

Tanker cargo chartering It embraces the chartering of tanker tonnage which has been designed particularly to convey one of the following commodity products and related consumables: crude oil; petroleum products; chemicals; gas; vegetable oil and wine.

Tanker chartering The process of chartering a tanker.

Tanker freight index Published monthly by Lloyd's Ship Manager it is compiled in four elements: VLCC/ULCC; medium size crude carriers; handy size clean (refined); and handy size dirty (not refined). It measures/records the tanker freight movement in each of the four categories monthly. See liner freight index.

Tanker/MX A computer software package enabling cargo posting to be placed electronically. The package can also alert brokers to possible deals.

Tanker owner's voluntary agreement An agreement concerning liability for oil pollution. An agreement subscribed by most of the world's tanker operators whereby the Owners agree to reimburse Governments for oil pollution clean up costs. Each member insures his potential liability under the agreement. It is located in London.

Tankers Accommodation for liquid cargo in a road vehicle, railway wagon, container, or ship.

Tankervoy 87 Code name of BIMCO approved International Association of Independent Tanker Owners tanker voyage Charter Party document.

Tank farm A large tank storage installation for liquid cargoes such as oil, chemical, serviced by pipeline, or road, rail, and/or sea services.

Tankradar Highly accurate and non-dangerous. Tank Gauging system.

Tanks deep Tanks on a vessel used for cargo or ballast purposes.

Tanks wing Tanks on a vessel used for cargo or ballast purposes and situated on the starboard and port sides.

Tanktainer Container for bulk liquids.

Tank top Upper plating of the double bottom in a ship. Floor of the hold so called it is the top of the double bottom tank.

Tankwaybill '81' A non negotiable tanker waybill formulated by the International Association of Independent Tankers Owners in 1981.

Tanolith C Trade name given for Copper Chrome Arsenate solution used in immunising timber against wood-boring insects.

Tanolith U A fluoride Chrome Arsenate solution used in immunising timber against wood-boring insects.

Tanzanian Shilling The currency of Tanzania. See Appendix 'C'.

Tanzanian Shilling invoice An invoice payable in Tanzanian shillings relative to an export sales contract.

Tap The security such as a certificate of deposit issued on an 'as required' basis by the borrower; it is not a 'managed issue' (see 'tranche'). An International Banking term.

TAPS Turret Anchored Production System.

Tap stock A Government bond or gilt issue, not offered all at once but as and when there is demand, i.e. on tap.

Tare Weight of packing in a consignment or unladen weight in a vehicle container or other carrier of goods.

Tare mass of container See tare weight of container entry.

Tare weight The unladen weight of a transport unit. In a road vehicle, tare weight excludes weight of driver and any passenger prior to loading.

Tare weight of container The mass of an empty container including all fittings and appliances associated with that particular type of container on its normal operating condition.

Target The market segment at which activity is aimed. Targeting is a strategy for positioning goods to reach selected segments; or the Trans-European Automated Real-time Gross settlement Express Transfer system: a payment system consisting of one RTGS system in each of the Member States participating in the euro area. The national TRGS systems are interconnected through the Interlinking mechanism so as to allow same-day cross-border transfers throughout the euro area. TRGS systems of non-euro area Member States may also be connected to TARGET, provided that they are able to process euro.

Target audience The audience/market to which the media plan will direct/target their advertising strategy/efforts. This will embrace advertising/selling/merchandise. Overall, a group of people or segment of the market to whom a sales person/promoter will direct their advertising/selling efforts.

Target market The market to which the promoter/salesman will direct their selling efforts.

Target marketing The process of directing/aiming for a specific segment of the market in the promotion of a particular product(s)/service(s).

Target risk A very large risk shared by most of the insurance market.

TARIC The Community's Integrated Tariff. (An acronym for Tarif Integré des Communautés Européenes). EC term.

Tariff List of duties payable on imported or exported cargo imposed by the government; published passenger fares/freight rates/charges and/or related conditions of carriage of a carrier; or standard insurance premium rates applied by all insurers.

Tariff barrier An impediment to trade between two nations and expressed in a high level of import duty to protect the home industry and/or discourage imports to preserve hard currency. It may be the import tariff, preferential tariff, ad valorem duty, specific duty, compound duty, anti dumping duty, and countervailing duty, temporary import levies and compensating import taxes.

Tariff binding Commitment not to increase a rate of duty beyond an agreed level. Once a rate of duty is bound, it may not be raised without compensating the affected parties. Article II of WTO relates to bound tariff schedules, while Article XXVIII covers modification of schedules. A WTO definition.

Tariff escalation Higher import duties on semi-processed products than on raw materials, and higher still on finished products. This widespread practice protects domestic processing industries and discourages the development of processing activity in the countries where the raw materials originate. Overall, the application of higher duties as the degree of processing of a product increases. A WTO definition.

Tariff free zones A zone immune/free of any import duty tariffs.
Tariffication Procedures relating to the agricultural market-access provision in which all non-tariff measures are converted into tariffs. WTO definition.
Tariff peaks Relatively high tariffs, usually on "sensitive" products, amidst generally low tariff levels. For industrialised countries, tariffs of 15 per cent and above are generally recognised as "tariff peaks". A WTO definition.
Tariff quotas Goods eligible for importation only on a quota basis.
Tariff rates Tariff concerned with rates and related charges.
Tariffs Customs duties on merchandise imports. Levies either on an ad valorem basis (percentage of value) or on a specific basis (e.g. 7 USD per 100 kgs). Tariffs give a price advantage to similar locally-produced goods and raise revenues for the State. See Tariff barrier.
TAS Telecommunication Authority of Singapore.
TASA Telecommunicacoes Aeronauticas SA (Brazil) – IATA term.
TASC Technical Advisory Sub-Committee.
TASG Ticketing Automation Strategy Group – IATA term.
TAT Transitional Automated Ticket – IATA term; or The Tourism Authority of Thailand.
TAWB Through air waybill.
Tax Arbitrage Trading that takes advantage of a difference in tax rates or tax systems as the basis for profit.
Tax avoidance The process of avoiding to pay tax indicated by a deliberate policy to evolve/develop strategies which do not qualify tax payment. See Freight Tax and Harbour Maintenance Tax.
Tax break A situation in a country/region whereby foreign investors are given preferential tax concessions on their profits. An example arises in Hamriyah Free Zone – Shariah – UAE which provides no corporate taxes or personal income taxes, and complete exemption from import and export taxes, all commercial levies, together with a range of other investment incentives. See free Trade Zones and Special Economic Zones.
Tax evasion The process of evading/escaping/contriving to pay tax levied by Government/State/entity. See freight tax and Harbour Maintenance Tax.
Taxtainer Group A major container leasing company.
Tax treaty A contract between states relative to taxation policies.
TBN Total Base Number – Lloyd's Register of Shipping Term; or to be Named or Nominated.
TBNC Thames Barrier Navigation Centre.
TBS Tubeshaft.
TBT Tribulyfin.
T/C Till countermanded; or Time Charter.
TC Air Tanzania Corporation – an IATA airline; or time charter.
TCAS Traffic Alert and Collision Avoidance System – IATA term.
TCC Technical Co-operation Committee; or Thai Chamber of Commerce.
TCCS Trans Caucasian Container Service. It operates on the spur of the Trans-Siberian railroad from Leningrad to the Iranian port of Djulfa.
TCDC Technical Co-operation amongst Developing Countries (UN/ODA).
TCIM Tank Control Inventory Management system.
TCPA Time of closest point of approach. A term used exclusively to radar plotting.
TCS Tilbury Container Service.
TCT Time charter trip.

TDA Transport Distribution Analysis.
TDB Trade Development Board (Singapore).
TDC Top dead centre; or Trade development cycle.
TDCC Transportation Data Co-ordinating Committee.
TDED Trade Data Elements Directory.
TDG Training development guidelines.
TDI Trade data interchange.
TDID Trade Data Interchange Directory.
TDMA Time Division Multiple Access.
TDRI Thailand Development Research Institute Foundation.
TDW Tons deadweight.
TE Turbo electric; triple expansion; or Lithuanian Airlines, the National Airline – a member of IATA.
Teach in The process of acquainting a person(s) with a particular circumstance, procedure, scheme and so on. For example, the Manager may call a meeting to inform all the Sales Executives of a product and in so doing obtain general comments thereon.
teak s Teak sheathed deck.
Tear proof Material which is proofed against tearing during normal handling.
TEC Department of Technical and Economic Co-operation (Thailand), or Trans European Combined Rail Service.
TECE Trans Europe Express Rail Service.
TECH Cargo Toxic, Explosive, Corrosive, Hazardous cargo – a dangerous classified cargo.
Technical and Clauses Committee A group of Lloyd's and company underwriters who meet to discuss, formulate and amend marine insurance clauses which are then recommended to the London market for general use.
Technical barriers to trade The existence of non-harmonised standards for product and service sector areas in different countries or regions. Examples include packaging, shipbuilding, textiles, banking, distribution and information processing and communications. Overall, it jeopardises the free flow of trade as it is contra to international standardisation
Technical co-operation The process whereby two parties agree to co-operate on technical issues. It may be between Companies/Economic Trading blocs/governments. Overall, it could result in the exchange of technical data and/or the construction of products jointly involving high technology.
Technical culture An attitude/focus adopted by Government/Company which focus primarily on the technical evaluation in the strategy/management/development of the business. See accountancy and marketing culture.
Technical drawing A graphic representation produced using vector-based software which enables entities to be edited individually and the whole drawing to be scaled and manipulated. Such drawings could be layouts such as a building floor or car park layout or detailed drawings of artefacts used for manufacture. A computer term.
Technical drivers for marine propulsion improvement/development Technical advancement in marine propulsion focuses on need to improve power output, power to weight ratio, control systems, design of hull and propeller and more efficient bunker consumption. These contribute to added value to the maritime service and better ship productivity overall.
Technical Help to Exporters A branch of the British Standards Institute which helps exporters identify any rules, regulations and standards which might apply to their product in a specific country.
Technical Policy The technical policy of a particular company or in a particular set of circumstances.

Technical press Periodicals which deal specifically with technical subjects and usually not read by the public consumer market. In the main it serves the trade and technical readership markets.

Technical reserves A collective term for the amounts set aside in respect of underwriting liabilities: unearned premium reserves, unexpired risk reserves, claims outstanding and the funds held for three-year and treaty business.

Technical stop A stop en route, planned or unplanned, for refuelling crew change or other operational need, but not for discharging or taking passengers on board an aircraft.

Technological variables A marketing term used in market forecasting whereby technological products can be identified in levels of production/consumption and on that basis a judgement can be made with further research of the likely level of production/consumption. An example is energy consumption which maybe oil, gas, nuclear and related to the degree of efficiency. Such forecasting will be influenced by further progress in technology and innovation coupled with market conditions.

Technology Technology is a perishable resource comprising knowledge, skills and the means for using and controlling factors of production for the purpose of producing, delivering to users and maintaining goods and services for which there is an economic and/or social demand. See also transfer of technology.

Technology transfer The process of transferring the technical specification/manufacturing skills/management on a bilateral or multilateral company basis.

TECO Technical Co-operation Committee (OECD).

TEDIS Trade Electronic Data Interchange Systems.

TEEM Trans Europe Express Merchandises Rail Service.

Tel Telephone.

Tel add Telegraphic address.

Telegraphic transfer A debtors instruction to his bank to request its correspondent bank in the exporter country to pay the specific amount to the exporter. Such instruction would be cabled using telex, telegram or facsimile method of communication.

Telematics Telecommunication and Data Information.

Telemetry A system to two-way radio communications which interlinks data from the offtake vessel and installation. This data is often related to ESD and Emergency Release activities, but can also include information on valve positions.

Telementry system An export software facility which allows exporters to measure different levels of inventory in warehouses and thereby reduce supply chain cost in a logistics environment.

Telemotor Machinery which operates the steering system of a ship.

Telemotor system An arrangement of hydraulic pipes for controlling a ship's rudder.

Telephone interviews In market research terms, the process of the skilled professional marketeer researcher conducting an interview with a respondent by telephone using a questionnaire. The questionnaire will be designed to obtain from the respondent specific data within the market research survey remit/objective and could include up to 30 questions covering a five to fifteen minute interview.

Telephone network See survey.

Telephone sales The process selling/promoting a product by initiating telephone calls on a selected basis and thereby endeavouring to conclude a product(s)/service(s) sale.

Telephone survey To conduct market research using the telephone network. See survey.

Tele sales The process of conducting sales using the telephone network as the media and thereby conducting/negotiating a sale over the telephone. It is also more commonly employed as a means of identifying a potential customer, identifying customer needs, and subsequently agree a date on which to conduct a face to face sales visit. See also cold call and personal selling.

Tele sales video A training video for tele sales trainees.

Teletex The method of carrying pages of information 'piggyback' on a broadcast (over air or wide-band cable) television signal, for display on any TV receiver which incorporates the necessary decoder. The standard international term for teletext is broadcast videotext. Teletex (without the 't') refers to communication word processors.

Teleticketing Automated procedure that permits a machine to print airline tickets at an agency when the booking is made by telephone.

Television rating The size of the audience viewing a particular programme or advertisement. It maybe expressed in percentage terms or in volume.

TEM Thailand Exhibitions and Management Co Ltd.

Temperature clause This features in a Bill of Lading whereby shippers insist the good product be conveyed, for example, at -18°C. It arises through the increasingly strict food regulations relative international trade shipments.

Temperature – controlled transport Transport units conveying goods under controlled temperature conditions which primarily involve refrigerated units. It may involve a road vehicle, railway wagon, container or ship.

Template Mould or pattern. An electronic file which holds a standardised document layout or screen format, e.g. letter type, position of references and location of addressee details. The template can also hold style data, variable data or macros. Templates also refer to overlays for keyboard keys to indicate their action when used with a particular application. Computer term.

Temporary Importation Goods imported temporarily such as for a trade fair, etc.

Temporary imports Goods imported on a temporary basis. These include advertising films, aircraft, commercial samples, goods imported for the promotion of the export trade, for technical examination or testing as an export service, goods destined for approved exhibitions; certain machinery imported on hire or loan, recorded material for use, amendment or further recording; used personal and household effects; professional effects, returnable containers and road vehicles, tourist publicity material, goods in transit or for transhipment.

Temporary storage Place approved for the deposit of imported goods which have not been cleared out of customs control. A customs term; or storage which allows deletion of data or files. Magnetic tape and disc are the most common but laser discs and writable CD discs are also available. Computer term.

Tender In an offer by tender, buyers of shares specify the price at which they are willing to buy. The shares are allocated to these buyers offering at least the "striking price", which is the highest price at which all the shares are brought. They all pay this striking price, unlike an auction where each successful buyer pays the price he or she offered; or an offer by a supplier to supply goods for a defined period of time on specific price and delivery terms; excessive weight of cargo on the upper deck of a ship; a document for delivery against an expiring futures contract; or issues offered for subscription at minimum prices which are allotted to the highest bidders or at the minimum prices if under subscribed, a BIFFEX term.

Tender bond A bond provided to support the exporter's offer to supply the goods or services. It is an indication

Tender clause

to the buyer of the serious interest of the exporter in bidding for the contract. The bond is usually required for amounts of two to five per cent of the tender. Also termed bid bond. An international banking term.

Tender clause A clause in a policy covering hull and machinery of a ship. It specifies the measures the assured must take in the event of an accident, whereby a claim may arise under the policy, regarding advice to underwriters and Lloyd's agents and survey arrangements. The clause also gives underwriters rights regarding the taking of tenders for repairs, deciding the place and firm of repair and acceptance of the successful tender. It details the assured's rights in regard to delay experienced in awaiting further tenders requested by underwriters. A penalty is applied to the assured's claim for non compliance with the terms of the clause.

Tender documents Documents with respect to a bid for a contract.

Tender to contract cover An insurance cover facility available to exporters quoting prices in foreign currency. The cover protects against loss due to the adverse movement of exchange rates between the date the tender is offered and the date of acceptance by the overseas customer.

Tender vessel A ship with a high centre of gravity and consequently will tend to roll excessively in heavy seas. It is usually caused by excessive weight of cargo in the upper decks thereby resulting in an unstable vessel.

Tenor The period of time to maturity of a bill of exchange. Also termed usance.

Tenor or Usance Bill A bill of exchange calling for payments at a fixed or determinable future time.

TERFN Trans European Rail Freight Network.

Term A definition or word; or the period of validity of a contract such as the length of time for which a mortgage arrangement will run.

Term Bill The legal definition as per the Bills of Exchange Act 1882, Section 3 of a Bill of Exchange is: 'An unconditional order in writing addressed by one person to another, signed by the person giving it, requiring the person to whom it is addressed to pay on demand, or at a fixed or determinable future time, a sum certain in money to, or to the order of, a specified person or to bearer'. Bills of exchange are widely used in international trade as a means of claiming payment by the supplier (drawer) from the buyer (drawee) and can be used as a vehicle for obtaining trade finance. A 'term' bill calls for payment either upon a fixed future date or upon some future determinable date, as opposed to a 'sight' bill which calls for payment immediately upon presentation to the drawee. By including a term bill of exchange as one of the documents required within a letter of credit, the importer is obtaining credit. Normal periods range from 30 to 180 days. Where a Term Bill is drawn on and accepted by a UK bank under a Letter of Credit, the bank undertakes to pay at maturity. The 'Bank Bill' is therefore capable of being discounted at the prime market rate for the relevant currency to provide immediate finance and there is no recourse to the exporter.

Term draft A Bill of Exchange which is made out payable 'at a fixed or determinable future time'. It is called term draft because the buyer/importer is receiving a period of credit which is identified by the tenor of the Bill of Exchange. The buyer/importer signifies an agreement to pay on the due date by writing an acceptance across the Bill of Exchange.

Terminal A link in a transport chain at which interchange facilities exist between common and different modes of transport. It maybe a seaport, railway depot, station, airport, heliport, road haulage depot, coach station etc.

Terminal emulation The type of terminal selected will set the character and control codes used in transmission and reception. Three of four standards are commonly used. Computer term.

Terminal Manager A Manager in a railway or road haulage company – engaged in the management control, and operation of a particular rail or road transport depot. This maybe road haulage/parcels depot which could also be rail served. It maybe also a bus station terminal manager.

Terminal throughput Total volume of traffic passing through a terminal.

Terminal traffic Traffic passing through a railway, road, or other transport terminal including a sea or air port; or telecommunications traffic which terminates at a particular point. For example, telephone calls from London to receivers in Dublin are Dublin's terminal traffic.

Termination The Customs procedure formally ending the carnet's validity. A custom term.

Term Policy Policy written for a period in excess of 12 months.

Terms It usually describes the conditions of the credit and the dispatch of goods and the basis on which the payment amount is fixed. An international trade Banking term.

Terms and conditions A term generally refers to a period covered by an insurance, but can also mean a condition or stipulation in a contract. A 'condition' is imposed in the contract by the insurer and must be literally complied with unless it is waived by the insurer. A condition goes to the root of a contract and noncompliance by the assured enables the insurer to avoid the policy from inception.

Terms of access All the conditions which apply to the importation of goods manufactured in a foreign country embracing import duties, foreign exchange regulations, import restrictions or quotas, and preference arrangements.

Terms of credit Terms on which business is conducted. For example, at Lloyd's the broker has a maximum period of credit during which he must pay the premium. It is the responsibility of the broker to make arrangements for payment by his client in sufficient time for him to meet the terms of credit requirements.

Terms of delivery The terms under which the goods are delivered to the buyer/importer and this is found in the export sales contract which embraces INCO-TERMS 2000.

Terms of (export) sale The terms of the export sale contract agreed between buyer (importer) and seller (exporter).

Terms of sale The terms of the sale's contract agreed between buyer (importer) and seller (exporter).

Terms of shipment The negotiations with the supplier (exporter) which will lead to an agreement on who is to meet the various charges involved in shipping the goods, e.g. insurance, freight, duty, stowage. The invoice will show the price of the goods and indicate by the inclusion of a standard abbreviation, known as a shipping term, i.e., CIF, FOB, which additional charges the price covers. There are thirteen different 'INCO terms 2000' in use. See separate entries i.e. CIF, CFR EXW, CIP, FCA.

Terms of Trade An economic term determined by dividing the index of import prices into the index of export prices multiplied by 100 as given below:–

Terms of Trade = $\dfrac{\text{Index of Export Prices}}{\text{Index of Import Prices}} \times 100$

Overall, the rate at which goods are exchanged. It is the ratio of the average price of a country's exports to the average price of its imports.

Terotechnology A combination of management, financial, engineering and other practices, applied to physical assets in pursuit of economic life cycle cost, i.e. its aim is to obtain the best use of physical assets at the lowest total cost to the entity.

Territorial franchise A franchise which operates in a specified area/country/region. See Franchise.

Territorial Waters Those waters which form a twelve mile band within a coast line and exclude river estuaries. The country who controls the coast line likewise extends its control to the territorial waters.

Territory Manager A Manager responsible for a particular area/territory within a Company/Entity organisation structure involving a particular segment(s) of the Company business. It maybe for example the Northern or Southern Territory.

Tertiary industry A service sector industry. See primary and secondary industries.

Tertiary readership The casual readership of a particular journal/magazine/newspaper.

Test condition Means the physical conditions imposed on the sample during the process of testing. A term associated with dangerous cargo shipments.

Test marketing The process of promoting/retailing a product(s) in a particular area, region, group of persons or other situation with a view to establishing the adequacy of the product(s) and its marketing. Such an analysis will help determine whether any alterations are required to the product(s) and/or its marketing before the overall promotion is undertaken.

Test transit A consignment which is monitored at specified points of its transit and in so doing enables the operator/shipper to gauge the adequacy of the overall transit time for the market having regard to the prescribed schedule.

TETOC Technical Education and Training Organisation for Overseas Countries (ODA).

TEU Twenty foot/feet equivalent units Technique of quantifying ISO containers, ie 1 x 40 ft = 2TEU; 1 x 20 ft = 1 TUE.

Textile bag A bag made of jute.

Textile bag – paper lined A crepe or plain paper bag lined with bitumen or other water resistant adhesive, suitable for textile products.

Textiles Surveillance Body An organisation set up by WTO to oversee the operation of the multi fibre arrangement. See Multi fibre arrangement.

Textile waste-wet It maybe subject to spontaneous heating and ignition caused by physical or chemical reactions producing heat energy on the surface of the material and has therefore to be classed as flammable solid. A term associated with the movement of dangerous classified cargo.

TF Tropical Fresh Water; or Malmˆ Aviation AB – IATA Airline member.

TFG Traffic Forecasting Group (of EANPG) – IATA term.

TFPA Thai Food Processors' Association.

TFR Transfer.

'T' forms Customs transit Forms T2, T2ES, or T2PT required for non-Community goods under the Community transit system involving minimum of customs formalities. See NCTS.

TG Thai Airways International Ltd. The National airline of Thailand and member of IATA; or TUG.

TGB Tongued, grooved and beaded.

TGV Train a Grande Vitesse, France's ultra-fast passenger train.

TH LAR Transegional (Linhas Regionais SA) – an IATA Member Airline.

THAIAGRO Thai Agro Navigation Co – a shipowner.

Thai Airways The National Airline of Thailand and member of IATA.

Thames Oil Spill Clearance Association. See TOSCA.

THC Terminal Handling Charge. For example, a charge for handling FCLs at ocean terminals. Also known as a Container/Port Service Charge.

THD Truck height dimensions.

THE Technical help to exporters. A facility available to UK exporters on a subscription basis which gives UK manufacturers details of foreign technical requirements of specified goods.

Theatre warehouse Sometimes referred to as postponement warehousing. This type of facility is used for 'kitting', whereby the owners of the warehouse provide services for the assembly, bagging, labelling or other processing of goods, in order to prepare them for the market.

Thebe The currency of Botswana. See also Appendix 'C'.

Thebe invoice An invoice payable in Botswana Thebe currency relative to an exporter's invoice.

The Ghan A passenger train launched in January 2004 between Melbourne – Darwin involving a journey of 47 hours of 1851 miles. The Transcontinental rail route was opened in January 2004 and conveys passenger and freight trains embracing bulk cargo and containers. It serves the deep sea port of Darwin which will develop quickly as container and bulk cargo tonnage shipowners reschedule there away from the Queensland and New South Wales seaports to the rail served deep seaport of Darwin. It will provide a stimulus to the Australian economy.

The Hague Rules Code of minimum conditions under which cargo shipped under Bill of Lading. See also Hague – Visby Rules and Hamburg Rules entries.

Theme park A large amusement facility with rides, shows, restaurants, shops and other attractions; architecture, decoration, uniforms, music and other features suggest a theme or image for the entire park or for designated sections of it.

Thermal container A container built with insulating walls, doors, floor, and roof which retard the rate of heat transmission between the inside and outside of the container.

Thermocline In the ocean a region of rapidly changing temperature between the warm upper layer (the epilimnion) and the colder deeper water (the hypolimnion).

Thermohaline circulation The deep-water circulation of the oceans driven by density contrasts due to variations in salinity and temperature.

Thermostatic fan Electrically operated engine cooling fan which is activated by a thermostat pre-set to open and close at specific temperatures to keep engine temperature within the most efficient range.

The Royal Dutch Airlines The Dutch National Airline and member of IATA.

Think tank A group of people/a department with a common focus who meet to encourage both lateral and horizontal thinking with emphasis on creative/innovative thinking with a view to generating new ideas/strategies/technologies/ideologies.

Thin trading A market place where limited trading is being undertaken due to sluggish demand or other circumstance.

Third carrier The carrier who actually performs the third leg of the contracted transit. Usually associated with IATA Air Freight consignment.

Third class A class of accommodation found on a passenger transport unit especially rail or estuarial transport. It is usually found on a service offering first and third class, or first, second and third class accommodation.

Third countries Countries which are outside the customs territory of the Community, also referred to as non-Community countries. An EU definition.

Third country A State or country which is not the bilateral trading nation in the conduct of overseas trade. Hence the two trading nations of UK and USA would identify Brazil as the third country if UK was the exporter, USA the importer and Brazil the manufacturer/supplier.

Third country co-operation The process of two trading nations engaging a third nation/country in a trading deal or other circumstance/objective.

Third country currency See third currency entry.

Third country free goods Merchandise which is non-Community goods which have not been put into circulation. An EU definition.

Third country goods Merchandise which originates/processed/manufactured/assembled in neither of the bilateral trading nations, but in a third country. An example arises where the export sales, contract originates in UK with the importer in USA, but the goods are supplied from Switzerland.

Third country port Cargo which is shipped between two trading nations but is routed through a third country as a transhipment/hub seaport. Hence goods from France destined for Malaysia being routed through Singapore.

Third country production The process of bilateral trading situation and a third country being the manufacturing base supplying the goods. Hence Japan may be the exporter, India the importer, and Taiwan the place of manufacture which supplies the goods to India.

Third country sourcing The process of bilateral trading situation and a third country being the supply source of raw materials/componentised products. See also third country products.

Third country transfer risk An increasing volume of business is now conducted on a third country basis. For example, an order placed with a Japanese company can be supplied to the EU from a manufacturing plant in Taiwan. Hence, the risk is transferred in order execution from Japan to Taiwan.

Third currency An export/import contract quotation being transacted in a currency that is foreign to both the exporter or the importer. This is termed a third currency. See also invoices under a third currency.

Third dimension The height dimension which can be utilised to effect additional storage capacity by installing high racking of mezzanine floors.

Third flag carrier A foreign shipowner operating a service between two trading nations whereby such a vessel being of another country's registration.

Third freedom The privilege to put down in another state revenue passengers, mail and freight taken on in the state of airline registration. One of the six freedoms of the air.

Third freedom flight Where cargo is carried by an airline, from the country in which it is based, to a foreign country.

Third party A person claiming against an insured. In insurance terminology the first party is the insurer and the second party is the insured. Insurance definition.

Third party agent An Agent which originates traffic and in so doing retails it through another Agent to obtain a more favourable tariff. For example, the transport operator obtains traffic from an Agent with the latter obtaining traffic from third party Agents. Also termed Umbrella Agent.

Third party currency This arises where the specific currency used in an export sales contract is a currency of neither the importer or suppliers country. For example, goods exported from UK to India with the currency being in Swiss Francs.

Third party liability Liability incurred by the assured to another party but excluding contractual liability. Legal liability to anyone other than another party to a contract (e.g. liability of one ship to another consequent upon a collision).

Third party logistics The process among manufacturers to outsource logistics thereby taking cost advantage of consolidation and other benefits.

Third party logistics management The process of a company contracting another company to undertake the logistics management. See outsourcing.

Third party service provider The process of a company contracting another company – a third party to provide a service. An example is a shipowner engaging a logistic management firm. See outsourcing.

Third party ship management The process of a shipowner contracting another company specialising in ship management to undertake the ship management elements of the business which maybe crewing survey programme, marine insurance and so on. See outsourcing and operating alliance.

Third person Person other than those detailed in a two party contract.

Third world country A country with a very low GDP per Capita. Such countries are termed less developed countries and represent the poorest countries in the world such as Tanzania.

Third world debt The international debt outstanding with the less developed countries. See international debt.

THONGSOON Thong Soon Shipping Pte Ltd – a shipowner.

Thorne Ship Management Pte Ltd Singapore A ship management company based in Singapore.

Three axle flat trailer A tiltable three axle trailer with a length of 12.2m and having a pay-load of 24,000 kilogrammes.

Three axle insulated trailer A three axle thin wall insulated body trailer with a length of 12.2m and having a capacity of 72 cubic metres and 22,500 kilogrammes payload.

Three axle refrigerated trailer A three axle trailer with a cooling unit and having a length of 12.34m. It has a capacity of 68 cubic metres and payload of 20,000 kilogrammes.

Three axle tilt trailer A tiltable three axle trailer with a length of 12.12m and having a capacity of 75 cubic metres and 23,500 kilogrammes payload.

Three axle vehicle A road vehicle having three axles.

Three berth cabin A cabin/passenger berths on a vessel accommodating three persons.

Three island ship Single deck vessel on which are superimposed forecastle, bridge, and poop.

Three Leader Agreement A condition in a slip or policy whereby the subscribing underwriters allow the leading underwriters to agree amendments etc to the insurance on their behalf. Where both Lloyd's and ILU companies share the slip it is understood that the first two underwriters in each market subscribing the risk shall indicate their agreement to amendments, etc. It is also understood that the leaders' agreement relates only to changes which do not materially affect the cover (e.g. they cannot increase written lines).

Three Leader Agreement (Cargo) A standard Institute clause, for cargo slip use only, which binds following underwriters to acknowledge amendments agreed by the three leading underwriters on the slip, or, if both Lloyd's and Institute Companies are on the slip, by the first two Lloyd's underwriters and the first two companies.

Three market expansion This involves product development, market penetration, and market development.

Threshold A business term to indicate/describe the starting point or benchmark in a business situation.

Threshold (no fault) The point, measured in money, time or other ways, beyond which tort liability can be established. Until that point is reached, reparations must be paid within the provisions of the no-fault plan, with no recourse to the courts.

Through air waybill An air waybill covering the entire transportation from departure to destination airports of shipment.

Through Bill A contract of affreightment that covers goods throughout the period of transit, including both overland and sea transit.

Through Bill of Lading Acknowledgment of Bill of Lading that cargo shipped between two specified ports (loading and discharge) plus inland portion of the transit.

Through charge The total inclusive charge of a specified consignment between two points involving two or more co-ordinated transport services often of different modes such as rail/ship/road thereby involving a through service.

Through fare The total fare from points of departure to point of destination. It may involve one form of transport and/or operator, or it could involve two or more transport operators embracing rail and sea.

Throughput The total volume of business in a specified period and circumstances, e.g. total tonnage passing through a warehouse or terminal; total passengers passing through a terminal; or total number of cars produced at a factory in 24 hours.

Through rate The total inclusive charge of a specified consignment between two points involving two or more co-ordinated transport services often of different modes such as rail/ship/road thereby involving a through service.

Through route The total route from point of departure to point of destination.

Through Transport Operator A carrier who contracts to carry goods only part of which carriage he undertakes himself on the basis that he is a 'principal' whilst the goods are in his personal care and an 'agent' only whilst they are not.

Through transport system A transport schedule offering a door to door or warehouse to warehouse service such as a container involving rail/ship/road transport modes.

Thrust block Unit that takes the propeller thrust on a vessel.

Thruster-Assist A system of thrusters fitted to an offshore installation which help to limit fishtailing or, say, in the event of a break-out, can control its drift.

Thw Thwartship.

Thwartship Crosswise of (across) the vessel.

THY THY (Turkish Airlines Inc) – IATA Member Airline.

TI Temporary Importation, the bringing into customs territory of goods (including vehicles and other means of transport) which will be re-exported, subject to certain conditions.

TIB Trimmed in bunkers.

TIBA Traffic Information Broadcast by Aircraft (ICAO term).

TIC Textile Industries Corporation (Burma).

Tick The smallest price movement measurable in the freight market. For BIFFEX this equals one full index point (value $10); or the standardised minimum price movement of a futures or options contract – an International Petroleum Exchange term.

Ticket A facility giving authority for the passenger to travel on a passenger transport mode(s) journey and issued subject to the terms and conditions of the carrier. The ticket is usually not transferable and is only valid during the period recognised within that particular ticket type. It maybe a season ticket, day excursion, ordinary single, weekend ticket and so on.

Ticket Agency A retailer selling primarily theatre tickets and other areas of live entertainment such as sporting events.

Ticket agent Office or counter authorised to make reservations and prepare tickets for a travel supplier, the employee authorised so to do.

Ticket analysis The process of analysing tickets which maybe by type, journey length, origin and destination, classification, date, service, child or adult, and aggregate revenue generated from each flight or sailing.

Ticket Office A point of sale dispensing travel and/or theatre etc tickets.

Ticket stock Blank ticket forms held by travel agents and travel industry services to be filled out and validated, at which point they become tickets that can be exchanged for travel services. This is now computerised in many countries.

Tick size This is the minimum price movement for the futures contract (e.g. 0.01% of the nominal value of the trading unit). A LIFFE term.

Tidal dock A dock with no lock and therefore subject to tidal variation with the constant rise and fall of the water level.

Tidal fall The difference between the height of high water and low water which maybe for example 5.5m.

Tidal range The difference between the height of high water and low water which maybe for example 4.2m.

Tidal River Navigations/Estuaries Tidal stretches of rivers and estuaries, without locks.

Tidal seaport A seaport subject to tidal conditions.

Tied aid The process of a donor country giving financial aid to an overseas territory or globally, and in so doing insisting the beneficiary obtains/purchases services or goods from the donor country and not from another State.

Tie downs Device used for securing a container or trailer to a ship's deck.

Tier A level in a container stack – bottom container equals first tier; next one up equals second tier and so on.

TIF International Transit by Rail.

TIL Taiwan International Line Ltd – a shipowner.

Tilt Canvas or other waterproof material used to cover or protect the interior of an open-top or open-sided container.

Tilt trailers A road trailer with a tilt facility to enable unloading of its contents.

Tilt transport See tilt entry.

Timas The computerised reservation and national distribution system of Ireland; partner of Galileo.

Timber ship Vessel designed to convey timber probably with a three island formation.

Time agreement Agreement between shipowner and crew for a specified period.

Time and sales The bids, offers, traded prices and estimated volume as observed by Exchange Pit Observer and then recorded in the sequence in which they occurred. IPE definition.

Time band A specified period. For example the time band during which you may use a discounted travel ticket maybe during time band period of 1000/1600 hours and 1800/2400 hours which is outside the peak travel period each weekday.

Time bar clause A clause found in a contract such as a charter party whereby all claims must be submitted within a specified period of 12 months from final discharge date. Arbitration tribunals tend to construe such clauses quite strictly.

Time Charter Vessel hired for a specified period of time. The shipowner is remunerated by payment of hire at an agreed daily rate.

Time Charter Party Charter Party hiring a vessel for specified period of time.

Time Charter trip A vessel contracted on a time charter basis undertaking a specific voyage and for the carriage of a specific cargo.

Time Clauses Insurances that remain in force for a specified period of time (one year).

Time C/P Charter Party hiring a vessel for specified period of time.

Time delivery A service offered by some carriers especially airlines which will specify the time band the goods will be delivered to major industrial/commercial/city/countries.

Time frame The period allocated for a particular task or series of manufacturing processes. Likewise it can apply to the service industry such as a product launch.

Time lost waiting for berth to count as loading or discharging time or as laytime If no loading or discharging berth is available and the vessel is unable to tender notice of readiness at the waiting-place, then any time lost to the vessels shall count as if laytime were running, or as time on demurrage if laytime has expired such time shall cease to count once the berth becomes available. When the vessel reaches a place where she is able to tender notice of readiness laytime or time on demurrage shall resume after such tender and, in respect of laytime, on expiry of any notice time provided in the charter party. A BIMCO term.

Time on Risk A period, during which insurance has applied, used for the calculation of premium when for some reason the insurance has been discontinued. Where a non-marine risk has expired before the policy is signed and no claims attach, it may be agreed to dispense with the issue of a policy noting the slip with 'TOR' (time on risk). The term TOR is not normally applicable to marine insurance policies.

Time Phasing Adding the dimensions of TIME to inventory status data by recording and storing the information on either specific dates or planning periods with which the respective quantities are associated.

Time Policy Underwriting subject matter insured for a specified period.

(Times) uncovered The amount of money a company has available for distribution as dividend is termed cover and is divided by the amount actually paid. If the result is less than one the dividend is uncovered.

Time service analysis A statistical term whereby using historical/existing data and statistical techniques one can project future trends. It is used in market forecast, budgeting projects and so on.

Timetable Details of a transport mode timetable which may be rail, coach, air and so on.

Timetable analysis The process of analysing a timetable to establish average transport unit speed, utilisation of staff and transport units and so on.

Timetable planning The process of planning a timetable of a transport mode which may be rail, coach, bus, ship, air and so on. It will take account of market demand, revenue; manpower resources, cost and utilisation; transport unit availability utilisation, capacity, and speed; terminal capacity and availability; transit times; route capacity and so on.

Time value The amount which an equity or index option premium exceeds its intrinsic value. The time value reflects the remaining life of an option – a financial definition; or the part of an option's premium over and above its intrinsic value (this may be zero) – insurance definition.

Time zones The surface of the earth is divided into 24 time zones. Each zone represent $15°$ of longitude or one hour of time. Hence when it is noon 12.00 in London it is 17.00 in Bombay, 13.00 in Venice and 20.00 in Beijing.

Timon Code name for time charter party.

Tin plate Light gauge mild steel with an alloy coating of tin.

TIP Taking inward Pilot (on a ship).

TIPEE Technical Investigation Propulsion and Environmental Engineering – a Lloyd's register of shipping service.

Tipping The overturning of a container about one bottom edge.

TIR Transport International Routier. Bond conditions under which containerised/vehicular merchandise conveyed internationally under the convention. The vehicles are not subject to any intermediate frontier Customs inspection; or (US) Trailer Interchange Receipt. Receipt to shipper delivering a loaded export container to a terminal.

TIR CARNET An international Customs transit document, issued by a guaranteeing association approved by the Customs authorities, under the cover of which goods are carried, in most cases under seal, in road vehicles and containers in compliance with the requirements of the Customs Convention on the International Transport of Goods under cover of TIR Carnets. A system which allows loaded vehicles to cross national frontiers with minimal customs formalities.

TIR Container An (ISO) container which complies with the TIR provisions and thereby enables the container to operate under an international customs transit document permitting the merchandise to be carried under customs seal/bond. This obviates the need for customs examination in transit countries coming under the aegis of the TIR convention.

TI relief Temporary importation. Goods imported for example to the EU subject to certain conditions, e.g. exhibitions etc which maybe re-exported. See IPR.

TIR trailer A road trailer which complies with the TIR provisions and thereby enables the trailer to operate under an international customs transit document permitting the merchandise to be carried under customs seal/bond. This obviates the need for customs examination in transit countries coming under the aegis of the TIR convention.

TI'S Technical Instructions for the safe transport of dangerous goods by air.

TIS International declaration of goods carried by rail (Customs).

TISI Thai Industrial Standards Institute.

TISTR Thailand Institute of Scientific and Technological Research.
Title XI A section of the United State Merchant Marine Act covering the provision of government guarantees for the issue by US-flag shipowners of long-term ship financing bonds.
Title documents Documents within international trade which give virtual control of the goods to the document holder, by possession or assignment of the document. These include Ocean Bill of Lading, Forwarders Bill of Lading and Forwarders delivery order.
Title to the goods The title to the goods is ownership. A legal term.
TK Tank.
TL Truck load.
T/L Total loss.
TL Trans-Mediterranean Airways SAL (TMA) – an IATA Member Airline; or top layer.
TLD Truck length dimensions.
TLO Total loss only.
TLP Tension Leg Platform.
TLR Terminal Loss Ratio – insurance term.
TLS Target Level of Safety – IATA term; or Transport Logistics Service.
TLV This is the abbreviation for threshold limit value. It is the concentration of gases in air to which personnel may be exposed 8 hours per day or 40 hours per week throughout their working life without adverse effects. The basic TLV is a time-weighted average (TWA). This may be supplemented by a TLV-STEL (Short-Term Exposure Limit) or TLV-C (Ceiling exposure limit) which should not be exceeded even instantaneously.
TM Traffic Manager; Turkmenia – customs abbreviation; LAM – Linhas Aéreas de Moçambique – IATA Airline Member.
TMA Terminal control area.
TMAP Textile Mills Association of the Philippines.
TMB The Textiles Monitoring Body, consisting of a chairman plus ten members acting in a personal capacity, oversees the implementation of ATC commitments. WTO definition.
Textiles Monitoring Body It supervises the WTO agreement on Textiles and Clothing and any unresolved disputes are referred to the Dispute Settlement body. See WTO Dispute Settlement Procedure.
TMC Transmitting magnetic compass.
TMESS Tele message.
TMK Tonnage mark – found on each side of the ship aft.
TML Transportable Moisture Limit.
TMM Transportation Maritime Mexicana SA – a shipowner.
TMNC Thai Maritime Navigation Co Ltd.
TMO Telegraphic Money order.
TMS Telex Message Switch.
TMS Software Transportation Management systems software.
TMV Tanker motor vessel.
TN Air Tahiti Nui – IATA Member Airline.
TNC A trade negotiations committee set up to oversee and co-ordinate negotiations on both goods and services involving the eighth round of the Uruguay WTO negotiations.
TNCs Transnational Corporations.
TNGE Tonnage.
TNK Tank type stabilizer.
TNS Thames Navigation Service; or Tons.
TO Tonnage opening (relative to a ship's deck access).
To average laytime Separate calculations are to be made for loading and discharging and that any time saved in one operation is to be set off against any excess time used in the other. A BIMCO term.
TOCA Transfer of Class Agreement. A ship classification term whereby the vessel is transferred from one classification to another.
Toea The currency of Papua New Guinea. See Appendix 'C'.
TOEPFER Alfred C. Toepfer Line – a shipowner.
Toe Puffs They are box toe boards used in the manufacture of boots and shoes and may consist of several layers of fabric impregnated with celluloid solvent, resin and dye. They are liable to spontaneous combustion. A term associated with the movement of dangerous classified cargo.
TOFC Trailer on flat car system/network.
TOH Thermal oil heater.
TOKO Toko Line – a shipowner.
Tokyo round The seventh round of the WTO multi trade negotiations 1973 – 1979.
Tolls Charges raised by Port/Dock Authority on cargo or passengers passing over the quay following or prior to shipment.
Tombstone The advertisement which lists the managers and underwriters and sometimes the providers of a recently floated issue.
Tomming Off Using wedges between cargo and the ship's side or a bulkhead to prevent movement of the goods during transit.
Toms Devices such as wire strainers or other wire or chain tighteners for securing cargo.
Ton Imperial English ton 20 cwt or 2240 lbs; American ton 2000 lbs; or Metric tonne 1000 kilogrammes. See Appendix 'F'.
Tonga Dollar (Palanga) The currency of the Kingdom of Tonga. See Appendix 'C'.
Tongs Scissor like devices for lifting cargo.
Ton miles A measure of a unit of transport service, calculated by dividing the cargo tonnage over distance covered. For example three tons of cargo conveyed 75 nautical miles will produce 225 ton miles.
Tonnage Measurement of ship's (Gross Registered Tonnage, Net Registered Tonnage and Dead Weight tonnage) size of a vessel; or cargo tonnage.
Tonnage calculations The most common type of calculable laytime arises where a charter party indicates that a vessel will load and/or discharge at the rate of certain number of tons per day or hour. A chartering term.
Tonnage deck The upper continuous deck in merchant ships less than three decks and second continuous deck from below in all other vessels.
Tonnage dues ship Document issued by harbour authority indicating that the required harbour dues, dock rent etc, has or will be paid on ship's NRT.
Tonnage Mark Mark on ship's hull situated aft.
Tonnage measurement The type of tonnage measurement used in shipping/international trade embracing cargo, deadweight, displacement, gross and net tonnages. See separate entries.
Tonnage on order The total dwt on order in shipyards embracing all types of merchant and passenger ships at various stages of their construction.
Tonnage opening Continuously open cargo access situated aft on shelter deck ship.
Tonnage oversupply A surplus of tonnage (ships) for which no immediate employment is available.
Tonnage tax Introduced by UK Government in 2000, the tonnage based corporation tax encourages expansion of the UK registered fleet. A key condition is minimum training obligation requiring each shipping company to

train sufficient UK seafarers to match their future manpower needs. The UK register is open to any vessel whose owner is a British Citizen or company incorporated in an EEA member state.

Tonne (metric) 1000 kilogrammes or 2204.6223 lbs. See Appendix 'F'.

Tonne miles The total number of tonnes conveyed on a transport unit multiplied by the distance conveyed.

Tonne miles per vehicle hour The total number of tonnes conveyed on a transport unit multiplied by the distance conveyed in one hour.

TOP Taking outward Pilot (on a ship).

To pay as cargo Used in ancillary insurances relating to the cargo (e.g. increased value) when the assured is not required to show evidence of loss or interest and can claim on the policy if he can show that a corresponding loss has been settled on the main cargo policy.

Top layer A layer of excess of loss reinsurance arranged to protect the reinsured against the occasional exceptionally large loss.

Top loader container An ISO container designed with a pull back plastic waterproof cover/roof thereby permitting cargo to be loaded from the top. Ideal for large awkwardly shaped freight such as heavy pieces of machinery. Also termed soft top container.

Top of line range The product/service which is the ultimate/best in a specific range. It maybe a washing machine or car and the top of the range would be the most expensive and most sophisticated.

Topology The topology of a network refers to the way it is organised. Examples are bus, ring, star, mesh etc – computer definition.

Topping lift gear Arrangement of blocks and wires for raising derricks.

Topping-off Final stages of tanker loading operation.

Topping-up clause A requirement upon the borrower to ensure that in the event of the margin of security (eg the excess of the secondhand value of a ship over the outstanding balance of the mortgage loan) falling below a stipulated percentage, he will deposit further security to make good the shortfall, should the bank so request. Failure to comply with such a request triggers a demand for full repayment of the loan.

Tops-down The opposite approach to "bottoms-up" in which participants are assumed to be committed fully to the application of the framework except where exclusions are explicitly listed in national schedules. These would be reduced over time. A WTO definition.

Top twenty container seaports The top twenty container seaports as at 2002 include: Hong Kong, China, Singapore, Busan, Kaohsiung, Shanghai, Rotterdam, Los Angeles, Shezhen, Hamburg, Long Beach, Antwerp, Port Klang, Dubai, New York, Bremerhaven, Felixtowe, Manila, Tokyo, Quingdao and Gioia Tauro.

TOR Telex over radio; Time on risk – insurance term; or transfer of residence, the permanent transfer into a customs territory of a person's personal and household effects including vehicles, subject to certain conditions – customs abbreviation.

TORC Thai Oil Refinery Co Ltd.

Torpedoes Articles containing an explosive or non-explosive propulsion system and designed to be propelled through water. They contain an inert head or a warhead. The term includes: Torpedoes, liquid-fuelled, with inert head; Torpedoes, liquid-fuelled, with or without bursting charge; and Torpedoes, with bursting charge. Term associated with dangerous cargo shipment.

Torpedoes with bursting charge These are devices containing a means of propulsion and a change of secondary detonating explosive. A term associated with the movement of dangerous classified cargo.

Torquay round The third round of the WTO multilateral trade negotiations held during period 1950/51.

Tort An injury or wrong, independent of contract as by assault, liable, malicious prosecution, negligence, slander, or trespass.

Tort feasor (Latin) A person guilty of tortious conduct.

TOS Traffic Orientation Scheme – IATA term.

TOSCA Thames Oil Spill Clearance Association. A service provided by the Port of London Authority to deal with any incident likely to lead to pollution from oil or other hazardous and noxious substances within the port area.

TOT The Telephone Organisation of Thailand; Tourist Organisation of Thailand; or type of transport.

Total cost T/C The market value of all resources used to produce a good or service. Usually divided into fixed and variable costs.

Total distribution cost The aggregated cost of distributing a product/commodity. In export terms it embraces freight charges, packaging, import duty, handling charges at the terminal – airport/seaport, insurance, interest on capital tied up in transit, collection/delivery cost, customs clearance, agents commission etc.

Total fixed costs (TFC) The total of a firm's costs that do not vary with output in the short-run.

Total freight costs as a percentage of import value This represents the total freight costs as a percentage of import value (cif) relative to an imported product. In 2001 the percentage for the world was 6.11%, for developed market economy countries 5.12% and for developing countries 8.70% – the latter reflecting the less efficient infrastructure to gain access to such countries.

Total freight costs for imports This represents the total freight paid for transport services to enter an overseas market crossing international boundaries relative to an imported product.

Total logistics A strategy adopted by an increasing number of international entrepreneurs which examine their distribution network continuously on a total distribution concept. Overall, the art of logistics is the ability to get the right product at the right place at the right time. See also total distribution concept.

Total loss Total loss – this can be actual total loss or constructive total loss. In hull insurance it may include arranged or compromised total loss.

Total product All the ingredients when the buyer accepts the product/service from the manufacturer/supplier, and as inferred in a relative media programme. This embraces product specification, packaging, company brand image, after sales, warranty, spares, technical manual, any training as found in industrial products, service and guarantee.

Total Quality Management TQM The total involvement of an organisations workforce into quality achievement and excellence both relating to the final product or service and the departmental operations within the firm.

Total Revenue The price of each unit multiplied by the quantity sold as detailed below:-

TR = Price x quantity sold.

Overall, the value of a firm's sales of its product. It is calculated as the price of the good multiplied by the quantity sold in a given time period.

Total solvency ratio % The total solvency ratio measures the total capital against the size of its risk weighted balance sheet expressed as a percentage. Total capital consists of the general reserve, subscribed capital, the revaluation reserve, general loss provisions and subordinated liabilities and net of intangible fixed assets.

Total utility The total satisfaction or benefit that a person derived from consuming some amount of a good or service.

Total variable costs (TVC) The total of those of the firm's costs that do vary in the short-run the cost of variable resources.

Tour A prearranged and usually prepaid journey to and from one or more destinations and usually including transportation meals accommodations and other elements.

Tour-basing fare A reduced rate excursions fare available to purchasers of prepaid tours or packages, including group inclusive incentive, and inclusive tours; any fare offered by a carrier on which a travel agent may claim a higher commission by selling specified arrangements at the same time.

Tour brochure See brochure.

Tour departure Date, time and place for the beginning of a tour and by extension the entire operation of a single tour.

Tourism All those activities aligned to persons from overseas visiting/touring a country. It includes income from travel, hotels, and so on.

Tourism deficit A nation where money spent by the indigenous population travelling/touring overseas exceed the income generated by tourist visiting the country. See Balance of Payments.

Tourism surplus A nation where the income from tourists visiting the country exceeds the money spent by the indigenous population travelling/touring overseas. See Balance of Payments.

Tourist card Document issued to prospective tourists as a prerequisite for entry and exit; the card maybe the only travel document required by the issuing country.

Tourist class A class of passenger accommodation found on a transport unit especially in sea transport and airlines.

Tourist fare A type of fare found on ships and airlines. It is sometimes called economy fare.

Tourist hotel See hotel classification.

Tourist Information office An office dispensing tourist information. This includes maps, restaurant details, forthcoming events in the town/area/region, general transport information and items of interest to the tourist in the town/area/region including historical places/areas of interest.

Tour manager A person who leads a trip from beginning to end or at least the land portion; also known as escort.

Tour operator A company that develops tour packages with all the transportation accommodation and other elements and usually markets the tours; tour operators sell and pay commission to travel agents and some sell directly to the public.

Tour Operator Study Group (TOSG) A trade association of major UK tour operators.

Tour organiser An individual, sometimes a travel agent, who forms a group of passengers to participate in a special, prepaid tour; an organiser does not necessarily have conference appointments, nor does he usually pay commissions.

Tour package See package.

Tour shells Brochures containing artwork and graphics but no printed copy which subsequently is printed separately by wholesalers or tour operators.

Tour wholesaler See wholesaler.

TOVALOP Tanker owners' voluntary agreement concerning liability for oil pollution. An agreement subscribed by most of the world's tanker operators whereby the Owners agree to reimburse Governments for oil pollution clean up costs. Each member insures his potential liability under the agreement. It is located in London.

Tow A ship being towed plus the tug.

Towage Bill An account rendered for towage expenses.

Towage dues A charge raised on a vessel requiring the services of a tug(s).

Towcon Code name of BIMCO approved international ocean towage agreement. (lump sum).

Tower's Liability Liability incurred by any ship or vessel when she is towing another ship, vessel or other object.

Towhire Code name of BIMCO approved international ocean towage agreement (daily hire).

Towing dolly A road transport facility with two or more wheels and specifically designed to support a superimposed semi trailer.

Towing tractor A tractor capable of towing wheeled transport units. It is found at a seaport, container base, transport terminal.

Toxicity detector An instrument used for the detection of gases or vapours. It works on the principle of a reaction occurring between the gas being sampled and a chemical agent in the apparatus.

TP Tank pressure; TAP – AIR Portugal – an IATA Member Airline; Third Party or Teleprocessing.

TPA Trade Practices Act.

TPC Tonnes per centimetre.

TPFT Third party fire and theft – insurance abbreviation.

TPI Tropical Products Institute (ODA).

TPND Theft pilferage and non delivery – insurance term.

TPRB The Trade Policy Review Body is General Council operating under special procedures for meetings to review trade policies and practices of individual WTO members under the Trade Policy Review Mechanism. WTO definition.

TPRM (Trade Policy Review Mechanism) Regular reviews conducted by the WTO Council of trade policies of individual WTO members permitting a collective evaluation of their effects and impact on the multilateral trading system. Introduced at the Mid-Term Review, the TPRM has so far covered 37 countries. A WTO definition.

TQ Tale quale (as found in grain trade).

TQC Total quality control.

TQM Total quality management.

TR Twin screw; triple reduction; triple screw; turkey – a customs abbreviation; or Transbrasil SA Linhas AÉreas (TRANS BRASIL) – an IATA Member Airline.

Tr Triple screw; or trawler.

Trace To locate a consignment lost or believed lost in transit.

Trace metals Metals occurring in minute amounts in water or sediment. Elevated levels of some trace metals can be harmful to marine life. A term associated with exploring for offshore oil and gas.

Tracers for ammunition Sealed articles containing pyrotechnic substances, designed to reveal the trajectory of a projectile. A term associated with dangerous cargo shipments.

Tracing The action of retrieving information concerning the whereabouts of cargo, cargo items, consignments or equipment.

Track Charter route particularly the route followed by a series of back-to-back charters; or railway line/route.

Track gauge The distance between the rails which, in Britain and most European countries, is a standard width of 4'8½" (1.43m). Main exceptions are Spain, Portugal the CIS and Ireland.

Tracking A facility offered by most major airlines and shipping companies which record the progress of a consignment throughout its transit whilst in the responsibility of the carrier. Electronic Data Interchange Computer technology features strongly in the system and the carrier can at any time tell the shipper the location of the consignment and the ETA – expected time of arrival at the destination point. The function of maintaining status information including current location of cargo, cargo items, consignments or containers either full or empty.

Track record The case history of a particular circumstance. For example, a person seeking a job may stress he/she has a good track record as a Sales representative indicating that he/she has produced good sales results.

TRACON Terminal Radar Approach Control.

Traction winch A special winch fitted at the mooring station on board the offtake vessel and used for heaving up the mooring hawser by means of the messenger and pick-up rope.

Tractive unit Vehicle not designed to carry a load but intended to form part of an articulated road vehicle. Overall, it is a motorised unit.

Trade The exchange of goods, funds, services or information with value to the parties involved. This value is either previously agreed or established during business; use of a vessel to carry cargo (or passengers) or a commercial connection between two or more individual markets.

Trade agreement An agreement between two or more trading nations which specifies the conditions under which a specific area of trade will be conducted or it maybe of a more general nature.

Trade and distribution centre A facility usually found in the environs of a major seaport where cargo is received, processed and distributed. Usually the facility is located in the business park area and much of the cargo would be despatched to neighbouring or more distant markets. A typical complex found in Rotterdam would cover a total area of 4000m^2 with a commercial building containing offices, a showroom and 2500m^2 of storage space. Merchandise is sold to traders clients from such trading centres.

Trade and Port Promotion Council A group of representatives who are users and operators at a particular seaport. The Council meets regularly to help develop and expand the seaport and the industries which serve it with a view to promoting the port and expanding its resources to meet the market needs both short and long-term

Trade association An Association representing the interest of a particular manufacturing industry at both national and sometimes international level.

Trade balance The trade performance of a country/state/region. See trade surplus and trade deficit.

Trade-balancing measure Requirement that the investor use earnings from exports to pay for imports. A WTO definition.

Trade ban A ban on trade between two or more countries. It maybe defence equipment or a range of goods for political reasons.

Trade barrier A technique adopted by a country or state to protect their trade such as through the issue of import licences; provision of high customs tariffs; complex import documentation procedures; import quotas, and so on. It is the converse of open market trading principles.

Trade Bill A bill of exchange issued in connection with a trade transaction, contrary to a draft issued merely for financing purposes.

Trade brief A bulletin produced at regular intervals usually involving maybe an International Bank outlining trading developments and marketing opportunities for companies trading around the world.

Trade Caravans The term applied to describe caravans when they are in a bulk transit – which may be national or international – between the factory and distribution depot. On arrival at the distribution depot the trade caravans are despatched to the retailer for sale for example through the retailers new caravan showroom.

Trade Cars The term applied to describe cars and small commercial vehicles such as vans when they are in bulk transit which maybe national or international between the factory and distribution depot. On arrival at the distribution depot the trade cars are despatched to the retailer for sale, for example, through the new car showroom.

Trade Commission An organisation representing a country or a particular product or activity which exists for the prime purpose to promote its interest. It may be the French Trade Commission situated at the Port of Dubai promoting/facilitation French exports.

Trade Custom How things are done in a particular trade. For example, the trade custom maybe to raise 50% surcharge on all dangerous classified merchandise on a particular service.

Trade cycle A periodic oscillation usually of about seven years in the level of business activity spanning the booms/peaks to the troughs/depressions. It is particularly evident on a global basis.

Trade deficit The amount by which the value of a country of import exceeds the value of its exports, in a given time period.

Trade development cycle The periodic oscillation in the level of global business/trade embracing peak, decline trough and growth. It embraces both the dry and wet bulk cargo markets together with industrial/consumer goods. It is influenced by both cultural and political characteristics. Overall, the trade cycle has a major impact on the international transport industry. See shipping cycle.

Trade Directory A directory of a particular Trade/Profession detailing the Company details, personnel, turnover, products and other data. It maybe on automobiles, shipbuilders, aircraft manufacture and so on. Overall, the directory may be national or global and be fully computerised with on line access to members.

Trade discount The discounted rate the manufacturer sells the product below the recommended selling price to the wholesaler/retailer. Hence the recommended selling price may be €100 and the distributor/retailer/wholesaler obtains it from manufacturer at 30% discount which is €70.

Trade dispute Where two traders or more disagree about a particular trade – IPE definition; or dispute over trade between two nations.

Trade distribution centre A centre where trade is conducted and goods distributed to neighbouring/distant market(s). An increasing number of major world ports are now developing in this manner especially Singapore, Rotterdam and Dubai.

Trade distribution concept This involves all the elements which go to make up the international distribution of a specific product as a complete entity including packaging, modes of transport, overall transit time, route, freight, documentation, customs clearance, handling, import tariff, duration of capital tied up in transit, warehousing, etc and the overall perceived benefits to the shipper.

Trade diversion The cost to a particular country when a group of countries trade a product freely among themselves while maintaining common barriers to trade with non-members.

Traded option An option is a right (but not an obligation) to buy or sell a security or commodity at a given price within a certain period. Investors can trade (i.e. sell) the option before expiry to make a profit or reduce a loss, depending on which way prices move.

Traded options The buyer (taker) acquires the right to buy or sell a futures contract at one of several fixed base (or strike) prices for a fixed delivery month. The premium he pays will vary accordingly to a (i) the length of time he is buying, and (ii) the distance from the current price to the base (or strike) price of the option. A BIFFEX term.

Trade elasticity This is measured by the percentage growth of international trade divided by the percentage growth in industrial production. When trade grows faster than production elasticity is positive and negative when the reverse applies.

Trade embargo An embargo/stoppage of trade between two or more former trading nations due to political, economic, strategic or fiscal reasons.

Trade Exhibitions The promotion of a particular product/commodity usually involving International Exhibitions at an established commercial centre such as the International Book Fair in Frankfurt.

Trade Facilitation The process of adopting/progressing measures whose objective is to facilitate/make more easy the conduct of trade with other countries. This maybe on a bilateral or multilateral country basis and embraces reduction/elimination of tariff barriers, export/import licences, documentation rationalisation and/or simplification, and so on; or removing obstacles to the movement of goods across borders (e.g. simplification of customs procedures). WTO definition.

Trade facilitator An entrepreneur/Company/Government Agency which endeavour to facilitate trade and thereby encourage international trade growth. See trade facilitation entry.

Trade Fair The promotion of a particular product/community usually involving International Exhibition at an established commercial centre such as the International Book Fair in Frankfurt.

Trade finance The process of funding international trade transactions. This involves the financial resources available to the seller/buyer to conduct international trade and their availability. See funding options.

Trade/GDP Ratio Exports expressed as a share of GDP of a country economy. It may be for example, 20.6% in 1992 and 29.6% in 2002. Overall, it represents a measure of trade growth in ten years.

Trade liberalisation The liberalisation of trade thereby exploiting the concept of free trade permitting the free flow of goods and services without inhibition by government interference across national frontiers. See free trade.

Trade margins The mark up/profit margin on a product sale.

Trade mark The distinctive identification of a product or service. It may take the form of a name, logo, letter, picture, or other device to distinguish it from similar offerings. It protects the exclusive rights to use the brand name and mark.

Trade mark licensing A form of licensing which permits the names or logos of recognisable individuals or groups to be used on products.

Trademarks and trade names Terms of designations identifying an enterprise and distinguishing its activities from those of other enterprises.

Trade mission A marketing technique to facilitate an exporter develop/promote his product/service. It is usually organised by a trade association/government/chamber of commerce and involves groups of businessmen/women representing companies. Overall, a co-ordinated overseas visit designed to help a number of businesses to meet potential buyers and/or agents. Often arranged by Chambers of Commerce.

Trade names The name given to a product/service. An example is Aerosol.

Trade Negotiations Committee (TNC) The most senior body under the auspices of the Uruguay Round which oversees the work of the two major groups and the working of the Round as a whole. A WTO term.

TRADENET The code name given to a computer facility available in the Port of Singapore. Overall, it is an electronic data interchange system linking the Port of Singapore with government agencies, traders, transport companies, shipping lines, freight forwarders and airlines. It deals with container ship planning, berth allocation and yard planning.

Trade off The process of evaluating one option against another and deciding on the most favourable option within a prescribed criteria. For example, in supply chain management one can calculate the total cost of where the cost of long supply pipelines may outweigh the production cost saving.

Trade press Periodical/journals which serve the trade market. See also technical press.

Trade price The price agreed between buyer and seller in the trading pit at which the transaction is executed. IPE definition.

Trade protectionism A technique adopted by a country or state to protect their trade which is the converse of the open market trading principles.

Trader A business entrepreneur.

Trade receivables See managing trade receivables.

Trade receivable portfolios evaluation Process of providing a credit rating for diversified receivable portfolios. See managing trade receivables.

Trade related investment measures A specific topic in the eighth round of the Uruguay Gatt multilateral trade negotiations which examines where investment measures distort trade flows.

Trade round A multilateral negotiation held to a prescribed timetable and plan, aimed at improving trade rules and exchanging concessions in a variety of areas.

Trader Unique Reference Number A number allocated by Customs to importers/forwarders which is normally based upon the existing VAT Registration number with additional digits.

Trade sanctions The process of a Government placing an embargo on trade with a specific country or range of countries. It maybe on all trade activities or specific items.

Trade surplus The amount by which the value of a country's exports exceeds the value of imports, in a given time period.

Trade ullage An allowance for natural loss to cargo (e.g. evaporation).

Trade vehicle rate The freight rate applicable to an export vehicle such as trade car. See also trade car.

Trade wars A dispute between two or more countries relative to the conduct of international trade. This may embrace dumping policy, tariff barriers, subsidies, repatriation of funds, inwards investment and so on.

Trade week The promotion of a particular country's products at a trade centre involving exhibitors displaying their products such as cars, machinery, textiles, etc for one week such as the British Week in Tokyo, Japan.

Tradient A business to business market exchange for freight transportation that brings together carriers, shippers, and freight forwarders. It enables carriers and shippers to participate in a global online market for containerised cargo.

Trading bloc A group/cluster of countries which have economic and/or political ties/allegiance. Accordingly, members of the trading bloc are encouraged to trade with each other usually on a free trade basis and can assert themselves as one trading bloc in global politics and trade negotiations rather than on an individual country basis.

Trading down The process of lowering the category/class/level of a particular product/service available in the market place. It maybe changing business class accommodation to economy class on an aircraft or a retail food outlet targeting the social economic category 'C' rather than 'A' & 'B' in their marketing strategy with a new range of commodities at lower prices.

Trading House An organisation involved in the conduct of international trade embracing exporters (sellers) and importers (buyers).

Trading limits Geographical maritime area usually specified by range of ports in which a vessel may operate.

Trading Nation A nation which engages in overseas trade such as Belgium trading with France.

Trading partner A country/buyer/importer with whom you do trade.

Trading restrictions The maritime area usually specified by a range of ports or region/area in which a vessel may officially operate.

Trading up The process of raising the category/class/level of a particular product/service available in the market place. It maybe converting the economy class on an aircraft to business class, or a hotel moving from two to three star status through the provisions of improved facilities and higher standards coupled with higher prices; or the process of selling at high prices to secure special custom and high profits.

Trade law The international law governing the conduct of trade embracing goods/merchandise/services which cross international boundaries. See Vienna convention.

Traffic In transport terms the number/quantity of passengers, freight etc conveyed on a service, trade route and so on.

Traffic analysis An analysis of traffic statistics and other data.

Traffic by flag An analysis of traffic/ship by flag of registration passing through a seaport/maritime canal/inland waterway.

Traffic by ship type An analysis of traffic/type of ship (tankers/bulk carriers/Ro-Ro ships/Passenger Ships/Car carriers/Container ships etc) passing through a seaport/maritime canal/inland waterway.

Traffic conferences Within IATA, the three areas into which the organisation has divided the world.

Traffic control The process of controlling/regulating/operating a transport system over a specific area which maybe, rail road, and air. This concerns particularly regulating traffic flow; maintaining schedules and inter-transport modal connections; instituting alternative services in an emergency including an accident situation or major breakdown of equipment; marshalling transport units and crews to best advantage practicable conducive with current traffic demand and operating technical circumstances; and finally to maintain safety at all times coupled with service quality.

Traffic control tower The accommodation housing and controlling/regulating all the operations of aircraft or ship movements at an airport or seaport. Its operations are usually computer and/or radar controlled.

Traffic flow The volume of traffic on a specified route/trade/service or through a terminal for a particular period which maybe per hour, day, month or year. This maybe passengers, containers, road trailers, cargo tonnage, and so on.

Traffic forecasts The process of estimating the volume of traffic for a particular trade, service, route, on a particular transport mode. It is usually undertaken on a budgetary control criteria and prepared on an annual basis spread over each month by the type of traffic involved.

Traffic lane An area with defined limits inside which one way traffic is established.

Traffic Manager Managerial position in a transport department or transport company responsible for the commercial and operational aspects of the transport mode which may be road, rail, air or shipping.

Traffic mixture The composition of the traffic units in a particular situation such as cars, lorries/trucks, containers, cargo, passengers, etc. See cargo mixture.

Traffic potential The anticipated carryings of a particular flow of traffic over a specific period. For example, one may take the reasonable view that following the introduction of a new vessel it may produce an increase of up to 15% more traffic. This is the traffic potential from such an investment.

Traffic rights Government controls on the type and frequency of commercial flights into or out of its territory. See Freedom of the Air.

Traffic Separation Scheme A transport mode scheme which separates traffic proceeding in opposite directions by the use of separation zone or line, traffic lane or by other means. It applies to road, rail, air and sea transport. See ship routeing.

Traffic Superintendent Managerial position in the transport department or transport company responsible for the day to day commercial and operational aspects of the transport mode which maybe road, rail air or shipping.

Traffic Supervisor A person who is involved in the day to day manning of a particular segment of a transport operation. It maybe road vehicles, container control, rail operation in a marshalling yard, coach operation, air traffic control and so on. The nature of the job specification will vary, but basically it is one of general supervision under the guidance of a Manager.

Traffic survey The process of conducting field research into transportation relative its user; the profile of the passenger/shipper; the adequacy/convenience of the schedules/information; the quality of the service/management and so on. See market research.

Traffic trends An analysis of traffic carryings and related derivatives including cost and revenue over a particular period which maybe each month or annually. For example, the average road vehicle load maybe as follows:- 1999 10 tons; 2000 11 tons; 2001 12 tons, 2002 14 tons and 2003 15 tons. This suggests improved loading results of the vehicle fleet coupled with the fact the smaller capacity vehicles are being scrapped and larger vehicles taking their place on a more intensive schedule basis.

Traffic – 'what will the traffic bear' A term associated with the pricing of a transport service(s) or product(s) – primarily the latter. It emerges through the considerations involved in the determination of the freight rate or passenger fare especially where competition is intense. Overall, the fare or freight rate charged should encourage maximum traffic flow and revenue generation, but if the tariff is too high the traffic may not pass. A good example is found in commuter rail travel whereby to journey daily 100 miles (50 miles in each direction) may cost €5000 per year per annual season ticket, but by motorway coach €1400. The stage may have been reached by the rail passenger that unless he is on a high income, he cannot afford to travel such long distance daily by rail, but can do so by coach albeit slower journey times and lower quality of service.

Traffic yield The income a particular form of traffic would generate during a particular period. This may be €4.0 million per month from 400,000 passengers' journeys producing a yield of €10 per passenger carried.

Trailer A road transport unit which only becomes operational when linked to motorised tractor unit thereby forming an articulated vehicle combination.

Trailer on flat car Carriage of piggy back highway trailers on specially equipped railway wagons.

Trailer operations The operation of trailers in the road transport fleet. The trailer only becomes operational when linked to the motorised transport unit.

Trailer park A place where road trailers maybe parked. Usually associated with a seaport; depot; container base; inland clearance depot where the TIR trailers can be parked.

Trailer trains An intermodal trailer which can be attached to a road tractor unit or become in effect a rail wagon. On the road the trailer is hauled by a road tractor whilst on rail the road wheels are retracted and the units ride on

Trailer unladen weight The weight of the trailer inclusive of the body and all parts (the heavier weight being taken where alternative bodies or parts are used) which are necessary to or ordinarily used with the trailer when working on a road, but exclusive of any vehicle by which the trailer is drawn and of loose tools and loose equipment.

Trailer utilisation The overall utilisation of road trailers usually measured in the total number of loaded ton miles generated by the trailer fleet in a particular period and divided by the number of trailers to obtain the individual trailer performance.

Train ferry Vessel capable of conveying loaded rail freight and/or passenger rolling stock.

Train ferry deck The deck(s) provided on a Train Ferry vessel accommodating/stowing Train Ferry wagons.

Training officer A post responsible for all aspects of training within a company. This may involve formulation of training programme, provision of an adequate number of suitably trained personnel to meet technological and business opportunities within the company/market place, and so on.

Training packs Portable structured training sessions that trainers or topic specialists can present themselves with minimum personal preparation.

Train loading plan The allocation/location of freight on each of the wagons forming the train. Also applies to passenger/rolling stock.

Train Number The numerical designation of a train schedule as found in the public timetable.

Tramp Vessel engaged in bulk cargo shipments or time chartering business. A ship which will call at any port to carry whatever cargoes available, normally on the basis of a charter or part charter. The opposite of a liner vessel. See tramp ship.

Tramp ship A dry cargo ship which is not being used for the provision of a regular service and usually finds employment in the carriage of homogeneous dry cargoes in bulk or tankers. The two main sub-divisions are general purpose tramps (usually multi-deck) and bulk carriers. They are non-schedules, non-conference vessels. Overall, dry cargo ships and tankers trading irregularly between any of the world's ports on a time or time charter basis.

Tranche The 'managed' issue of securities, often in relatively small denominations. Typically refers to different classes of securities e.g. 'A' Tranche and 'B' Tranche in a securitisation. Can also more generally mean an issue of a certain amount of securities e.g. a €100m tranche.

TRANS Transverse.

Transaction fee Charges for certain types of services, such as the making or cancelling of bookings, delivery of tickets and providing insurance or visas; it has been proposed as an alternative to commissions.

Transaction loss The loss resulting from a foreign exchange transaction covering the movement of goods or services. An international Trade Banking term.

Transaction processing The type of 'real time' data processing system which handles one transaction at a time. The system ensures that other users are locked out of the records being used, so that when a transaction is completed it is secure. A computer term.

Transaction profit The profit resulting from a foreign exchange transaction covering the movement of goods or services. An international Trade Banking term.

Transaction risk The flow of cash in and out of the Company to suppliers and customers crossing international frontiers and inherent risk in currency fluctuation from date deal agreed until final payment made.

Transaction value The basis of the transaction value emerging under the WTO valuation code is determined by the price actually paid or payable for the goods. It is to determine the level of import duty payable. See WTO valuation.

Transamerica Leasing A major container leasing company.

Transatlantic Freight Conference Liner conference embracing trade between Great Britain and Canada including the Northern ports of the United States.

Transatlantic trade Trade between the Atlantic and European seaboard country markets involving the Atlantic Maritime and Air Trade networks.

Trans border data The process of transmitting/obtaining data which crosses international boundaries.

Trans-European Networks A European Community all transport modes network featuring especially combined transport operation.

TRANSCLINE Trans-C Line – a shipowner.

Transfer Local transportation and porterage from one airline terminal to another, from a terminal or station to a hotel or from a hotel to a theatre or restaurant, all as part of a tour contract which also states the mode of transfer; or the form signed by the seller of a stock or share, authorising the company to remove his name from the register and to substitute that of the buyer.

Transferable credit The transferable credit is designed to meet the requirements of international trade. It enables a middleman who is receiving payment from a buyer under a documentary credit to transfer his claim under that credit to his own supplier. In this way he can carry out transactions with only a limited outlay of his own funds. The middle man gets his buyer to apply for an irrevocable credit in the middleman's favour. The issuing bank must expressly designate it as transferable.

Transferable credit transhipment entry One which may be transferred by the first beneficiary. Customs entry for cargo imported for immediate re-exportation.

Transfer buyers This arises in transfer pricing (see separate entry) whereby the company purchasing/receiving the goods/services is termed the transfer buyer.

Transfer cargo Cargo arriving at a point by one flight and continuing therefrom by another flight. (Air cargo).

Transfer cost The cost incurred in transferring technology to a licensee and all ongoing expenses of maintaining the licensing agreement.

Transfer manifest A list of cargo executed by the transferring carrier which is handled and endorsed by another airline, the endorsed list serving as a receipt of transfer of a consignment.

Transfer of class See change of class entry.

Transfer of marketing mix The process of transferring the constituents of the marketing mix from a domestic/home market environment to one overseas which crosses international boundaries. It requires substantial modification including price, specification, design, culture, literacy, distribution, language and so on. See Marketing Mix.

Transfer of portfolio The substitution of a new insurer for the original one in respect of all insurance business of a particular class.

Transfer of product mix The process of transferring the product mix from the domestic/home market to an overseas market thereby crossing international boundaries. See product mix.

Transfer of technology The process of transferring technology from one country to another for the purpose of its development/production and so on. It maybe undertaken on the basis of foreign direct investment; turnkey projects; trade in goods and services involving sale of equipment tools etc; contracts and agreements involving licensing of patents, trademarks etc; research and development; personnel involving employment of nationals by foreign firms or employment of foreign technicians; acquisition of foreign companies, and transfer through international tender invitations. See also technology.

Transfer of technology methods Some eight basic methods exist regarding transfer of technology including product purchases; local assembly; marketing arrangements; local production by overseas firms; licensed production; licensed technology; two way agreements, and copy.

Transfer price A price related to goods or other services transferred from one process or department to another, or from one member of a group to another. The extent to which cost and profit are covered by the price is a matter of policy. A transfer price may, for example, be based upon marginal cost, full cost, market price or negotiated price.

Transfer pricing The process of a multinational company/entity undertaking inter-company transactions of goods and services, and in the process thereof determining the 'transfer price' of such goods and services. In such situations involving transactions across 'international frontiers' within the corporate family of the multi-national company/entity, the goods/services are subject to customs duties, currency risks, foreign exchange controls and so on. Transfer pricing presents the multinational with the opportunity to manipulate/adjust the transfer price to further the goods/aims of the enterprise.

Transferred credit A letter of credit which has been transferred to another party. These credits can only be transferred once unless the credit specifically permits it.

Transferring carrier A participating carrier who delivers the consignment to another carrier at a transfer point (air cargo).

Transhipment Process of transferring (loading or discharging) cargo from one transport mode to another involving a vessel, aircraft, railway wagon, road vehicles, or barge. See hub and spoke system.

Transhipment Bill of Lading Bill of Lading permitting cargo to be transhipped from one vessel to another en route to reach final destination.

Transhipment entry Customs entry for cargo imported for immediate re-exportation.

Transhipment Freight The freight rate raised to tranship merchandise from one vessel to another to enable the commodity to complete the transit such as Felixstowe/Rotterdam – Rotterdam/Aqaba.

Transhipment trade The international trade sector which passes through a country/trading bloc during the course of a commodity transit and usually would involve the goods being transferred from one ship to another or one aircraft to another.

Transire A Customs document used when a vessel is coasting, giving full cargo details. It serves as clearance from the port of issue.

Transit Bond note National Customs document providing authority for goods to be conveyed in Customs transit without pre-payment of import duties and taxes.

Transit Boxes Containers in transit.

Transit breaks The time spent by a commodity/goods in transit waiting for an onward schedule.

Transit card A form of ticket enabling the passenger to leave the aircraft and later rejoin it at a 'stopover' schedule airport thereby enabling the passenger to visit the passenger transit lounge. An example arises with Thai Airline London/Frankfurt and Frankfurt/Bangkok involving London originating passengers to have a transit card for the 45 minutes stop at Frankfurt.

Transit cargo The process of cargo passing through a country.

Transit clause A clause in the Institute Cargo Clauses, specifying the attachment and termination of cover.

Transit credit A credit issued by a bank in one country, and advised or confirmed by a bank in a second country to a beneficiary in a third country, sometimes through a third bank in the beneficiary's country.

Transit Department A department dealing with transits as found, for example, in the Suez Canal Authority organisation handling multi-type tugs used for towing, salvage, fire fighting and berthing of ships. See transit cargo.

Transit document A customs document often referred to as a movement certificate for the throughout transit. An example arises in movement certificate form EUR 1.

Transit passenger A passenger en route to his/her ultimate destination and in so doing would leave the first carrier aircraft at a particular airport and then proceed to join the second carrier aircraft to complete the journey. It maybe Bangkok to Copenhagen, and Copenhagen to London with a three hour stopover/transit time at Copenhagen. Transit passengers usually do not go outside the transit lounge and its environs area including duty free shop and therefore are not subject to customs/immigration procedures/control but the customary security check/control.

Transit population A population not in permanent residence but temporary such as a tourist.

Transit Shed Port (quay) accommodation for cargo pending shipment or inland transport distribution.

Transitime Code name of time charter party.

Transitional safeguard mechanism Allows members to impose restrictions against individual exporting countries if the importing country can show that both overall imports of a product and imports from the individual countries are entering the country in such increased quantities as to cause – or threaten – serious damage to the relevant domestic industry. WTO term.

Transit time The time taken to complete a specified journey involving merchandise or passenger traffic between two specified points.

Translation loss Estimated loss resulting from the revaluation of foreign assets and liabilities in company balance sheets. An international Trade Banking term.

Translation profit Estimated profit resulting from the revaluation of foreign assets and liabilities on company balance sheets. An international banking term.

Translation risk Risk incurred by Companies making investments resulting in them earning either equity in overseas companies or fixed assets in the form of buildings or plant. An international trade banking term.

Translator A program language processor which converts program statements into another form, e.g. machine code. Three types of translator are common: Interpreters, Compilers and Assemblers. Computer definition.

Transmission modes There are several configuration settings which affect the mode of transmission. Flow control and cable provisions affect the type of exchange between stations. Simplex is a one-way mode of transmission; duplex is two-way; half duplex enables two-way transmission, but not for each station simultaneously; asymmetric duplex is two-way but with different speeds each way. Asynchronous transmission sends a character at a time with markers either end; synchronous systems are used to speed up this process and send whole blocks of data at a time by timing the start and end of the block. Cable types and facilities affect the type of transmission. Serial transmission sends each data bit down the same cable in sequence; parallel transmission enables the eight data bits to travel along separate cables simultaneously. Computer term.

Transmission rate Expressed in bits per second (bps), this rate depends on the transmission media, e.g. cable. Baud is the signalling rate and is sometimes the same as the bps, but the bps can be three or four times the baud rate. Computer term.

Trans National Company A company whose activities cross international boundaries.

Transpacific Stabilisation Agreement A maritime agreement featuring freight rates between the Pacific Rim of countries and the Americas. It focuses on the stabilisation of container rates.

Transpacific trade Trade between the Pacific Rim of countries and the Americas involving the Pacific Maritime and Air trade networks.

Transparency The principle that national policy measures which affect international trade, should be clearly visible and open to scrutiny by trading partners. The same criteria can apply to a wide range of other activities in the international business world where a consensus exists.

Transparent decision making The process of undertaking/conducting decision making with a rationale which is clearly visible and open with no confidentially or constraints on access to data relative to the factors/data determining the decision arrived at. See logistics and supply chain management.

Transpacific Stabilisation Agreement A maritime agreement featuring freight rates between the Pacific Rim of countries and the Americas. It focuses on the stabilisation of container rates.

Transportation documents Transportation receipts or transportation contracts issued by carriers.

Transport chain The ensemble of operations undertaken in the transport of goods from the point of origin to the point of destination. It embraces not only the carriers sea – road air – rail, but also the freight forwarders, handlers, packers, insurance companies, banks, ports, airports, exporter, importer and so on. Overall, it embraces all the infrastructure of the overall transit chain. See supply chain management.

Transport controller A managerial post responsible for the optimum development control of the transport fleet and this extends to manning levels, operating schedules, maintaining provision, loading-unloading arrangements, vehicle utilisation and so on. It is usually found in road transport.

Transport documents Formerly and still usually referred to as shipping documents, the documents which demonstrate that goods have been dispatched from one place to another. It maybe Bill of Lading, Air Waybill, CMR/CIM consignment note and so on.

Transport Distribution Analysis A technique by which alternative methods of cargo distribution are analysed and the optimum pattern of transportation selected.

Transport Emergency Card Often referred to as 'Trem card'. It is a road transport scheme intended to standardise and improve the means by which written instructions are given to drivers of vehicles carrying hazardous chemicals. Additionally, it is of assistance to manufacturers in complying with regulations which require written instructions to be provided.

Transport facilitation The process of adopting procedures/policies which enables a transport network and their traffic to operate unimpeded across/between border frontiers. An example arises with merchandise operating under a customs bond in transit countries.

Transport hub The centre of a transport system/network which would involve more than one form of transport. Modern airports and seaports are usually regarded as transport hub which provides an interchange/interface with other transport services providing dedicated and integrated services. Examples include, Frankfurt, Paris, Rotterdam and Singapore. See hub and spoke.

Transport International Routier Goods conveyed on a road haulage vehicle displaying a TIR plate indicate the vehicle complies, for customs purposes, with the internationally agreed vehicle construction requirements. Once TIR approved, the vehicle can operate under the authority of a TIR carnet. Overall, it is the Bond conditions under which containerised/vehicular merchandise is conveyed internationally under the convention. See also TIR carnet.

Transport International Routier Carnet An international Customs transit document, issued by a guaranteeing association approved by the Customs authorities, under the cover of which goods are carried, in most cases under seal, in road vehicles and containers in compliance with the requirements of the Customs Convention on the International Transport of Goods under cover of TIR carnets.

Transport Internationaux Routiers Bond conditions under which containerised/vehicular merchandise is conveyed internationally under the convention. The vehicles are not subject to any intermediate frontier customs inspection.

Transport instructions (form) A document approved by IATA containing data prepared by the shipper containing all particulars required by any carrier, on whatever mode, in order that the appropriate transport document for the particular mode or combination of modes could be produced. This includes details of consignor/

consignee; reference number; notify party; carrier/carriers agent; terms of delivery/delivery time limit; point of origin and destination including routing, full description of cargo, remarks/special instructions; freight charges – prepaid/collect, other charges; insurance coverage, declared values; documents enclosed, and so on.

Transport Manager A post found in a Transport Company or other entity where transport services are provided. The content of the post responsibilities varies by individual company and the nature of the business. Ideally the person should be professionally qualified and the responsibilities could include maintenance and provision of vehicles, appointment of personnel, operating schedules, route planning, tariff matters, industrial relations, marketing and so on. The smaller the Company, the wider the job role. In many companies it is confined to the provision and operation of the schedules and equipment for their maintenance in a road worthy condition.

Transport index (TI) (Radioactive Material Only) A single number assigned to a package, overpack or freight container to provide control over both nuclear criticality safety and radiation exposure. Term associated with dangerous cargo shipment

Transport mode Any type of transport which maybe rail, road, ship, air, hovercraft, and so on.

Transport security To focus on the security of a transport network.

Transport units Transport units comprise of road tank or freight vehicles, rail tank or freight wagons, and tank or freight containers which are loaded, stowed and discharged as one piece provided they give additional protection to the dangerous goods contained therein. Flat bed units are not to be considered as falling within this definition. A maritime transport definition used in association with dangerous cargo.

Trans Siberian Landbridge Overland route from Europe to the Far East via the Trans Siberian Railway.

Transtainer A vehicle used for carrying cargo containers during loading or discharge operations or within port of terminal areas. See Straddle Carrier

Transversal Frame extending width (or beam) of ship.

Transverse propulsion unit Propeller located usually in ships bows to aid manoeuvrability in confined waters.

Transversal lashing system A method of below-deck securing where containers are connected laterally by bridge fittings to each other or to the ship's structure.

Transverse Across a ship at right angles to a line drawn from bow to stern.

Transverse members Transverse structural members at the top and bottom of an ISO container joining the corner fitting of the end in question.

TRANSWORLD Transworld Line – a shipowner.

Trans World A major airline of the United States.

Travaux Prearatoires (Latin) Preparatory work, in the case of legislation, drafts, and discussion in the legislative chamber, recommendations of Commissions, learned writers and experts.

Travel Agent A retailer dealing in travel and inclusive tours.

Travel and Tourism Research Association (TTRA) A professional society of US travel industry market research specialists.

Travel documents Documents relevant to a passenger's journey. These may include reservation boarding pass, port tax ticket, passport, ticket, visa, international health certificate and so on.

Travel Incentives Promotional transport market products encouraging passengers to travel such as redeemable vouchers, free cabins, free meals and so on.

Travel Industry Association of America (TIA) A non-profit organisation of government and private organisations dedicated to promoting travel to and around the US.

Travel Industry Association of Canada (TIAC) A trade association including the entire range of Canadian travel, transportation and related industries and governmental organisations.

Travellers Cheques A cheque permitting the bearer to cash it for a foreign currency in accordance with the cheque valuation at a specified bank(s) or other official outlets. A facility used by tourist and overseas businessmen.

Travelling exhibition A mobile exhibition usually by road or rail.

Travel restrictions Restrictions imposed by the carrier on passengers or freight which may for example limit excursion travel to certain periods. It can apply to any transport mode.

Travel samples The process of the passenger (overseas sales person/business person) taking samples of the product intended to promote in the overseas territory. This attracts customs documentation and the need for an ATA carnet to obtain customs duty relief in some markets.

Travel Shopper A version of the PARS computerised reservations system that is available to the public through commercial networks.

Travel voucher A credit voucher which for example, can be exchanged in part or full payment for a ticket. It is usually associated with a special promotion of the transport mode product and can include excursion trips, meals, and so on.

Travolator A moving walkway usually found in passenger terminals.

TRAWSEC Transport Security Section (of the UK Department for Transport).

Trb Trabaccolo – small coasting vessel of the Adriatic.

TRC Type rating certificate.

Trdg Trading.

Treaty A long term reinsurance contract usually effected to cover the whole or a certain section of the reinsured's business. Accounting is usually effected on a monthly or quarterly basis; or refers to the Treaty establishing the European Community. The Treaty was signed in Rome on 25 March 1957 and entered into force on 1 January 1958. It established the European Economic Community (EEC) and was often referred to as the "Treaty of Rome". The Treaty on European Union was signed in Maastricht (therefore often referred to as the "Maastricht Treaty") on 7 February 1992 and entered into force on 1 November 1993. It amended the EEC Treaty, which is now referred to as the Treaty establishing the European Community. The Treaty on European Union has been amended by the "Amsterdam Treaty", which was signed in Amsterdam on 2 October 1997 and later ratified. An ECB definition.

Treaty of Rome The formulation of the European Economic Community when six European Nations signed the Treaty of Rome in 1957. It came into existence in 1958 and its objective was to remove customs barriers between member states and introduce a Common External tariff to third countries.

Trem card A card prepared in conformity with the European agreement concerning the international carriage of dangerous goods by road (ADR) in Europe. Its aim is to standardise a code of practice across Western Europe for all bulk transport of chemicals by road.

Tret Allowance for wear and tear and deterioration during transit.

TRIAD A group of three countries: the United States, Japan and European Union.

Triangle service A transport service between three seaports.

Triangular compensation (with financial switch) A counter trade technique. It arises when a bilateral clearing agreement exists between countries B and C. Country A requires payment in hard currency. Country C sells the clearing dollars to a switch dealer in exchange for hard currency, but pays a discount or disagio. Country C pays country A in the hard currency. The switch dealer then sells the clearing currency to country D, again at a disagio. The switch dealer's profit is the difference between the disagio granted and the disagio earned.

Triangular compensation (with hard currency goods) A counter trade technique. Triangular compensation involves the movement of goods between three countries. It is appropriate when the hard currency goods offered in compensation by country B are readily saleable, but not necessarily in country A. The goods are sold in country C who then settle in foreign exchange with country A.

Triangular compensation (with weak currency goods) A counter trade technique. Triangular compensation involves the movement of goods between three countries. It is appropriate when the weak currency goods offered in compensation by country B are saleable, but not necessarily in country A. However, as the goods are less easily saleable as they are not a hard currency goods country C will accept them only where a subsidy is paid by country A. Allowance for this subsidy will have been made by country A in pricing their hard goods exports to country B.

Triangulation A customs term used to describe a chain of suppliers of goods involving three or more parties instead of a bilateral trading situation involving two parties. An example arises where the supply of the goods/services originate from UK to Germany, but the invoice being charged to a company in Switzerland. Hence, a system whereby goods are despatched from the Community by being exported from one member state and returned to the Community by being imported into another member state. Customs term.

Tri-axle The axles of a road vehicle or trailer in line at right angles to the chassis frame.

Tribacks The process of an articulated vehicle – platform type – conveying two further trailers. The trailers are placed one above another thereby producing a tier of three trailers.

TRIESTINO Lloyd Triestino – a shipowner.

TRIKORA P T 'Trikora Lloyd' – a shipowner.

Trilateral Trade The conduct of trade involving three trading nations.

Tri lingual The ability to speak or communicate in three languages such as French, German and English.

Trim Difference between forward and aft draft of a ship.

Trim a cargo To level a bulk cargo in the hold of a ship in order to contribute to her stability at sea.

Trimmed by the bow The practice of having a ship with a deeper draft bow compared with the stern.

Trimmed by the stern Practice of having ship with deeper draft aft compared with bow draft.

Trimming The operation of shovelling coal, grain, or similar dry bulk cargoes to the wings or ends of the hold when loading a vessel.

Trimming tank A watertight compartment that can be filled with water ballast to change the trim of the ship.

Tri-modal infrastructure An infrastructure featuring three transport modes such as road, rail and waterways.

TRIMS Trade related investment measures.

Trinidad dollar The currency of Trinidad and Tobago. See Appendix 'C'.

Trinity House Association concerned with licensing pilots and having charge of aids to navigation, buoys, lighthouses etc in the United Kingdom.

TRIOCEAN Tri Ocean Shipping Co – a shipowner.

Trio Tonnage centre The co-ordinating body for the trio consortium.

Trip analysis The process of evaluating/analysing a number of railways/flights etc to determine payload, average/flight cost, average income per voyage/flight, composition of the traffic and so on.

Trip charter The process of chartering a vessel for a single voyage.

Trip charters A chartering term where the vessel delivers on a time charter at her previous voyage discharge port, and where the trip ends back in a relative position to other voyage employment alternatives. Overall, a combination of time chartering and voyage chartering. See trip time chartering.

Trip duration The duration of a voyage.

Trip hire The process of hiring a particular from of transport unit for single journey such as a railway wagon or road vehicle.

Triple A hotel room suitable for occupancy by three persons.

Triple back The process of conveying three road trailers as one unit involving one trailer conveying two empty trailers.

Triple lane link bridge A bridge with three lanes of traffic capacity and found at Ro/Ro berths thereby permitting loading/discharging over three lanes.

Triple wall A packaging term relative to the fibre board carton with a triple wall structure comprising of three layers of fibre board to afford improved protection to the goods.

TRIPS Trade related aspects of intellectual property rights. An area embraced in the 8th multilateral trade negotiations of the GATT Uruguay round where the distortions caused to international trade by an inadequate protection of intellectual property rights was under examination.

Trip time chartering The process of a vessel being chartered for a trip which may be a single voyage or a round trip. Hence the trip may be Hull/Rotterdam or the round trip Hull/Rotterdam/Hull.

Triton Container Intl A major container leasing company.

TRK Trunk.

TRO Trailer operations.

Tropical products Products under the following seven groups in their raw and processed forms have constituted a base for negotiations on tropical products: tropical beverages (i.e. coffee, tea, cocoa); spices, flowers and plants; oilseeds, vegetable oil and oilcakes (e.g. palm and coconut oils); tobacco, rice and tropical roots (e.g. manioc); tropical fruit and nuts (e.g. bananas, oranges and pineapples); tropical wood and rubber; and jute and hard fibres. However, this list was interpreted flexibly. For example, rice has not been treated as a tropical product in certain countries producing rice. A WTO term.

Tropopause The boundary between the troposphere and the stratosphere.

Troposphere The lower atmosphere, from the ground to an altitude of about 8 kms at the poles, about 12 kms in mid-latitudes, and about 16 kms in the tropics. Clouds and weather systems as experienced by people, take place in the troposphere.

TRS Tropical revolving storm.
TRSA Terminal Radar Service Area – IATA term.
Trs Shed Transit Shed.
TRT Turn round time.
Truck A road vehicle or railway wagon designed to convey goods/merchandise.
Truck carrier A road transport operator.
Trucker A road transport operator (a haulier in the UK).
Truck Waybill An inventory of cargo conveyed on a road vehicle.
Trunking The movement of containers or road haulage vehicles between terminals involving either container freight depots or container bases.
Trustee Institution in which the rights of the bond or note holders may be vested.
TS Traffic Superintendent; Tailshaft; or Transhipment; or Tons.
Ts Tippers.
T/S Time Sheet – chartering term.
TSA Tonnage Sharing Agreement; or Thai Shipowners' Association.
T's & P's Temporary and preliminary notices to mariners.
TSB Textiles Surveillance Body.
TSC Technical Services Corporation (Burma).
TSCS Trans-Siberian Container Service.
TS D/W Tons deadweight.
TSH Tanker ship.
TSIC Thai Standard Industrial Classification.
TSK TSK Line – a shipowner.
TSL Trans Siberian Landbridge.
TSPC Tropical Stored Products Centre (TPI).
TSR Trans-Siberian Railway.
TSS Turbine steamship twinscrew steamer; Terminal stacking space; or Traffic separation scheme.
TSWG Telecommunications Strategy Working Group – IATA term.
Tt Tilt (road) trailers.
TT Timetable; Tank top; telegraphic transfer – a remittance purchase by the debtor from his banker in international trade; or Technology Transfer.
TTC Trio Tonnage centre; or the tender to contract.
TTC/FES cover The Tender to contract and Forward Exchange Supplement. A facility available from the ECGD.
TTM Thailand Tobacco Monopoly.
TTO Through Transport Operator: A Carrier who contracts to carry goods (only part of which carriage he undertakes himself) on the basis that he is a principal whilst the goods are in his personal care and an agent only whilst they are not.
TT&T Thai Telegraph and Telephone.
TTT Thai Telephone & Telecommunication Co Ltd.
TU Tunis Airline. An International Airline and member of IATA; or Tadzhikistan – customs abbreviations.
Tug A vessel which tows another vessel, the other vessel being termed the 'tow'.
Tugrik The currency of the Peoples Republic of Mongolia.
Tugrik invoice An invoice payable in Tugrik currency relative to an export sales contract.
Tug boat A towing vessel. An American term. A tug boat tows dumb barges using a line connection.
Tugger Winch A small usually air-operated auxiliary winch.
Tugmaster Proprietary name for tractor unit used in ports and elsewhere, equipped with fifth wheel type of coupling. It enables/permits trailers to be transhipped from and to the Ro/Ro vessel via the Ro/Ro ramp.
Tugs A vessel available at a port to give assistance to ships entering or departing from their berth or mooring often in bad weather conditions.
Tunis Airline The international Airline of Tunisia and member of IATA.

Tunisian Dinar The currency of the Republic of Tunisia. See Appendix 'C'.
Tunnel Toll A charge/tariff raised on tunnel users embracing motorist, motor cyclist, lorries, coaches and so on.
Turbofan Jet aircraft powered by turbojet engines, the thrust of which has been increased by the addition of a low-pressure fan.
Turbojet Aircraft powered by engines incorporating a turbine-driven air compressor to take in and compress air for fuel combustion, the combustion gases and/or heated air being used both to rotate the turbine and create a thrust-producing jet.
Turboprop Aircraft powered by engines in which the propulsive force is supplied by a gas turbine that drives a propeller.
TURB(S) Turbines.
Turkish pound (Lira) The currency of Turkey. See Appendix 'C'.
TURN Trader Unique Reference Number – customs term.
Turnbuckle A tensioner that is tightened by turning a threaded screw with left and right hand thread. Also termed a screw or bottlescrew.
Turning back Emptying or filling a lock for a waiting vessel going in the same direction.
Turning basin An area of water for turning vessels usually found on a seaport/harbour.
Turning circle The arc swept by the wheels of a vehicle when moving in a circle; or measurement of manoeuvrability when navigating any class of ship.
Turnkey An export sales corporate contract provision where the seller (exporter) undertakes to supply not only the goods, but also commission it on the site specified by the importer. It thus involves the exporter providing the facility and also setting it up on the site. It is usually on a consortia basis.
Turnkey ship operation The most complete ship management service as it combines the operational and commercial management of the vessel. The ship management company takes responsibility for the ship as a business venture and provides the shipowner with complete accounting package.
Turnaround of a ship The period of time that a ship spends in port. An important measure of the efficiency of the port management and ship operators.
Turn out The process of examination(s) in a container, road vehicle, railway wagon, pallet etc being examined by a Customs Officer. This may arise at a seaport, airport, inland clearance depot or other prescribed area.
Turn round time The time taken to discharge and/or load a vehicle, wagon, ship, aircraft at a terminal or other transhipment point.
Turn time It arises at certain ports where it is customary for ships to queue to wait their turn either to load or discharge. When a ship is waiting her turn, laytime will not normally count. A port and chartering term.
Turn turtle Denotes a vessel capsizing.
Turpentine Substitutes A petroleum distillate which might contain some aromatic components and which usually has a flash point of approximately 40°C. White spirit is a synonym for turpentine substitute. A term associated with dangerous cargo shipments.
Turret Mooring A method of permanently mooring an offshore installation to the sea bed in which the upper most ends of the anchor chains terminate in a swivel arrangement attached to the vessel. The swivel may be internal, in a trunking built through the hull; or external built, for example, as bow extension. Turret moorings allow the installation of weathervane.
TVR Television rating.

TW Twin screw; Trans World Airlines Inc (TWA) – an IATA Airline Member; or Tween decker.

TWD Tween deck tank(s) between deck space (on a ship); or truck width dimensions.

TWEEN Tweendecker – type of dry cargo ship.

Tween deck Deck inserted between two continuous decks (on a ship). The first continuous deck below the main deck of the ship. A ship with more than three decks has an upper and lower 'tween deck. The gap between the 'tween deck and the main deck above is known as the 'tween deck.

Tween deck space A closed space between two consecutive decks and bounded by permanent bulkheads.

Twenty fourths MO A basis for calculation of unearned premiums involving the assumption that the average date of issue for all policies written during a month is the middle of that month.

TWI The Welding Institute.

Twin A hotel room with two single beds.

Twin deck link span A bridge ramp on which the vessel's ramp can rest and thus connects a vessel with the quay or shore and allows for the working of Ro/Ro trailers and cargo. In so doing the wheeled cargo which may include coaches and motorist are transhipped (driven to and from the two decks simultaneously) over the stern or bow of the vessel via the link span. The link span is at two level upper and lower deck.

Twin deficits The deficit on the Balance of Payments and the Government current account occurring simultaneously in a particular country.

Twin double A hotel room with two double beds.

Twin lock Two locks operative alongside each other and thereby permitting two-way operation as found in the St Lawrence Seaway.

Twistlock Container lifting securing device which locks into the corner castings.

Two axle vehicle A vehicle having two axles.

Two berth cabin Passengers sleeping cabin accommodation on a ship or train which accommodates two persons.

Two class vessel Passenger ship offering two classes of accommodation e.g. first and tourist.

Two lane deck A deck on a passenger or cargo vessel which has two lanes for vehicular traffic. Each lane would be of lorry width.

Two way pallet A pallet of which the frame permits the entry of forks at opposite sides.

Two Way Scheme An area within definite limits inside which 'two way' maritime traffic is operated.

TWRA Transpacific Westbound Rate Agreement.

Tx/Rx Transmit/Receive.

Types of intellectual property In WTO terms under the TRIPS agreement, embrace copyright and related rights; trademarks, including service marks; geographical indications; industrial designs; patents; layout-designs (topographics) of integrated circuits and undisclosed information, including trade secrets. See TRIPS.

Types of letters of credit There are eight types of letters of credit embracing documentary, red clause, advance payment, green clause, performance, standby, revolving and combination. See separate entries.

Type of tickets The range of tickets available in the market place and these include season ticket, ordinary single, ordinary return, weekend, awayday, party ticket, and so on.

Typhoon A name of Chinese origin (meaning 'great wind') applied to tropical cyclones which occur in the western Pacific Ocean. They are essentially the same as hurricanes in the Atlantic and cyclones in the Bay of Bengal.

U1, U2, U3 Grades of chain cable – ship classification term.
6U Air Ukraine – IATA member airline.
8U Afriqiyah Airways – IATA member airline.
U Unit Loads. A transport term involving the conveyance of goods in a unitised way on a door to door, or warehouse to warehouse basis involving a pallet, container and so on; Unaflow or uniflow – Ship Classification Society term; or TAT European Airlines – an IATA Airline Member.
u Universal Containers.
UA United Airlines – a member of IATA; unit of account (EEC); or Ukraine – customs abbreviations.
U/A Underwriting account or Underwriting agent.
UAC Dutch EAN Numbering Association.
UAE United Arab Emirates.
UASC United Arab Shipping Co (SAG) – a shipowner; or Union of African Shippers' Councils.
UAWB Universal Air Waybill.
UBB Union of Burma Bank.
Uberrimae Fidei (Latin) 'Utmost good faith'. Insurance contracts are one of a limited class that requires the parties (insurer and insured) to exercise the utmost good faith in their dealings with each other. Specifically the proposer of an insurance must disclose all material facts. The contract is voidable if this is not complied with.
Ubi remedium ibi jus (Latin) Where there is a remedy there is a right.
Ubi supra (Latin) Where above mentioned.
UC LADECO SA – an IATA associated Airline Member.
UCAP United Coconut Associated of the Philippines.
UCC Ultimate Container Carrier; User Charges Committee (IATA term); or (US) Uniform Commercial Code.
UCE Unforeseen circumstances.
UCOM United Communications Industry Co Ltd (of Thailand).
UCP 500 Uniform Customs and Practice for Documentary Credits and UCP 600. A code of practice developed by the International Chamber of Commerce and introduced in 1993. A supplement termed as UCP – electronic presentation was introduced in 2003 to accommodate trade which is either completely electronic or where a mixture of electronic and paper documentation is used. Overall, UCP 500 provides a uniform interpretation of Documentary Credits terms and conditions. See documentary credits.
UCP 600 Uniform Customs and Practise for Documentary Credits. Currently (2005) in process of formulation by ICC it will displace UCP 500 (1994).
UCPC ULD Control and Procedures Committee.
UCR Unique Consignment Reference Number – customs term.
UDA Urban Development Authority (Malaysia); or Upper Advisory Service
UDDI Universal Description Discovery and Integration – a computer logistics term.
UDEAC Central African Customs and Economic Union.
Udk Upper deck.
UEO L'Union de l'Europe Occidentale (Western European Union).
UEPR Unearned premium reserve.
UER Union Europeenne de Radiodiffusion (Eurovision).
UFL Upper flammable limit. See separate entry.
UFS Universal Freight Services – a shipowner.
Uganda Shilling The currency of the Republic of Uganda. See Appendix 'C'.

Uganda Shilling invoice An invoice payable in the Republic of Uganda shillings relative to an export sales invoice.
UGS Union of Greek Shipowners.
UHF Ultra High Frequency.
UHMWPE Ultra high molecular weight polyethylene. See High-modules polyethylene.
UIA Union of International Associations.
UIC Union Internationale des Chemins de Fer (International Union of Railways). Based in Paris, its role is to co-ordinate and improve conditions regarding construction and operation of railways international traffic.
UINF International Union for Inland Navigation.
UIR Upper Flight Information Region – IATA term.
UIRR International Union of combined Road-Rail transport companies.
UITP Union Internationale des Transports Publics.
UIs Unified interpretations – an IACS term.
UK United Kingdom.
UKAEA United Kingdom Atomic Energy Authority.
UKASTA United Kingdom Agricultural Supply Trade Association.
UKC Under keel clearance – that measurement obtained from the echo sounding machine.
UK CNTRL CHIEF (EDI) error response message. Customs term – see CHIEF entry.
UDDI Universal Description Discovery and Integration – a computer logistics term.
UK/Cont United Kingdom or Continent with limits – port destination location.
UK/Cont (BH) United Kingdom or Continent (Bordeaux – Hamburg range of ports).
UK/Cont (GH) United Kingdom or Continent (Gibraltar – Hamburg range of ports).
UK/Cont (HH) United Kingdom or Continent (Le Havre – Hamburg range of ports).
UKCOSA United Kingdom Council for Overseas Student Affairs.
UK fo United Kingdom for orders.
UKHAD United Kingdom and Le Havre, Antwerp, and Dunkirk range of ports.
UKHH United Kingdom and Le Havre – Hamburg range of ports.
UKHO United Kingdom Hydrographic Office.
UKOOA United Kingdom Offshore Operators' Association.
UKPA United Kingdom Pilots Association.
UKPx UK Power Exchange.
UKREP United Kingdom Permanent representative to the European Economic Community.
UKSATA United Kingdom South Africa Trade Association.
UK South Africa Trade Association Organisation responsible for developing/facilitating international trade between UK and South Africa.
UKTOTC United Kingdom tariff and overseas trade classification.
UKWA United Kingdom Warehousing Association.
ULCC Ultra large crude carrier – Tanker in excess of 320,000 sdwt.
ULCS Ultra large container ship.
ULD Unit Load Device.
Ullage The difference between the total capacity and the actual content of a cask or package. In a tank container it is the free space allowed for expansion of a liquid; usually about 5% of the total capacity. It is the space between surface of oil cargo and under-deck level at top

of tank's ullage point. Ullage is required to allow for the expansion of the contents.

Ullage report A tanker document – see ullage.

ULR 'Ultimate Loss Ratio' the ratio on net losses to net premiums assessed quarterly in respect of the open years for all lines of Business representing the ratio anticipated when all premiums are earned and all expired.

Ultra large container ship A container ship with a capacity between 12,000 to 14,000 TEUs.

Ultra large crude carrier A crude oil tanker of 400,000 dwt plus.

Ultra vires (Latin) Beyond their legal powers.

Ultimate customers The end user such as a person or organisation which will use a company's particular product(s) or service(s).

Ultimo (Latin) In or of the last month.

UM Air Zimbabwe Corporation – an IATA airline.

Umbrella Agent An Agent who generates traffic for an operator, viz shipowner, airline etc, and in so doing has other Agents working with him in the Umbrella Agent's name, thereby enabling the Umbrella Agent to claim commission on all the traffic despatched in the Umbrella Agent's name.

Umbrella Branding One brand which is known for all the Company products such as Sony covering cameras, television, hi fi systems and so on.

Umbrella Cover Cover providing excess limits over the normal limits of liability policies and giving additional excess cover for perils not insured by the primary liability policies and in reinsurance, cover against an accumulation of losses under one or more classes of insurance arising out of a single event.

Umbrella Liability Insures losses in excess of amounts covered by other liability insurance policies; also protects the insured in many situations not covered by the usual liability policies.

Umbrella Organisation An organisation which embraces all the activities of the company to one control unit.

UMG Underwriting (UK Property) Management Group.

UMS Unattended (unmanned) machinery spaces or unmanned engine room.

UN United Nations; or Transaero (Transaero Airlines) (limited) – IATA Member Airline.

UNA Unaflow. See 'U' entry; or United Nations Association.

Unaccompanied baggage Baggage which is not accompanied by the passenger and in some transport modes conveyed as cargo.

Unaccompanied load The process of a trailer being conveyed on a Ro/Ro vessel without the driver accompanying the shipment. Usually the trailers are loaded on and discharged from the vessel via the ramp using the tugmaster operated by the Port Authority or Shipowner.

Uncalled capital The part of a company's capital which has not been 'called up'.

UNCED United Nations Conference on Environment and Development.

Uncertainty A situation in which more than one individual event might occur but it is not known which of them will occur – insurance definition.

UNCID The Uniform Rules of Conduct for Interchange of Trade Data by Teletransmission 1987. It governs the conduct between parties and embraces the CMI Rules for Electronic Bills of Lading.

UNCITRAL United Nations Commission for International Trade Law, drafts model laws such as the one on government procurement.

Unclean Bill A bill of lading that has been claused by the carrier to show that the goods were not in sound condition when received. See Clean Bill.

UNCLOS UN Convention on the Law of the Sea.

Uncompressed gas Gas at a pressure not exceeding ambient atmospheric pressure at the time the containment system is closed. Term associated with dangerous cargo shipments.

UNCON Uncontainerable goods. Goods which, because of their dimensions, cannot be containerised and which are therefore carried other than in a container.

Unconditional In legal terms a contract which has no strings attached or that the contract must have been finalised without any possibility that it might not go through.

Unconfirmed Letter of Credit/Confirmed Letter of Credit An Unconfirmed Credit is where the advising bank has not been asked by the issuing bank to add any financial commitment by way of confirmation. Notwithstanding that the beneficiary may have presented documents in full compliance of the credit, the advising bank has no obligation to pay the beneficiary until reimbursement has been obtained from the issuing bank. When trading with a country experiencing political or economic difficulties a beneficiary may desire the additional security of the legal and binding undertaking of the advising bank in his own country. If this additional undertaking is given by the advising bank, the Documentary Credit becomes a Confirmed Documentary Letter of Credit, and the advising bank does have a financial interest. Under this arrangement, the confirming bank undertakes that payment will be made irrespective of what may happen to the applicant or issuing bank regardless of any changed political or economic situation in the country of the applicant, providing that the documents submitted by the beneficiary comply fully with the terms of the credit. Confirmations involve payment of an additional fee and the advising bank may not always be willing to enter into such arrangement.

Uncontainerable goods Goods/merchandise which because of their dimensions, cannot be containerised and which are therefore carried other than in a container.

UN Convention on Conditions for Registration of Ships 1986 Not yet in force (2004) – requires 40 contracting parties with a least 25 per cent of the world's tonnage as per ANNEX III to the Convention.

UN Convention on International Multi-modal Transport of Goods 1980 Not yet in force (2004) – requires 30 contracting parties.

Uncovered/Naked position An option position where the writer has no cover against a market move in the holders favour. The writer of a naked call does not have the underlying commodity to deliver and the writer of a naked put has not sold the commodity short – an International Petroleum Exchange term.

UNCTAD United Nations Conference on Trade and Development. Based in Geneva, its aim is to promote international trade especially with a view to accelerating economic development.

UNCTAD 40/40/20 The process of two trading nations reserving/sharing the cargo volume to the two national fleets on a basis of 40/40, and the residue of the trade of 20 being shipped on foreign flag tonnage.

UNCTAD V Conference Manila, 1979 Conference which agreed to the Liner conference code of 40/40/20 and proposed cargo sharing between trading Nations of bulk shipments of both dry and liquid cargoes. Adopted from 1983.

UNCTAD NMO UNCTAD multi modal Transport Convention.

UNCTC United Nations Centre for International Corporation.

Under carrier A carrier who carries less cargo than the allotment distributed to him.

Under cost freight rate A situation whereby the freight rate quoted/adopted for a particular transit/schedule is below the operator's cost of conveying the merchandise.

Under Deck Tonnage The capacity of the ship below the tonnage deck calculated in cubic tons.

Under developed market A market situation whereby only a limited area of the total market is using the facility/service due to price, consumer resistance and so on. An example is that in some countries only social category B's use a washing machine, whilst a mass market would exist throughout most households if the price of the washing machine was lower; or overall an under developed market is one below its economic potential level.

Under funding The process of guaranteeing an investment.

Under insurance Insurance that is not adequate in terms of the sum insured to provide for full payment of a loss.

Under invoicing An underpriced item on an invoice.

Under load A transport unit or cargo handling equipment conveying merchandise.

Underlying The basic insurance on which a supplementary insurance has been effected.

Underlying asset The asset from which the future or option's price is derived. In the case of BIFFEX, this is the Baltic Freight Index.

Undermanning Below the full complement (of a crew or stevedores gang).

Undermining market confidence A situation which undermines market confidence. This may arise through a rapid increase in the price of oil thereby causing the market to look at alternative energy sources.

Under priced A product/service which is under priced in the market place. It maybe a packaged holiday, whereby due to heavy demand, the volume of business exceeds considerably the available holiday accommodation, but overall, market volume can be reduced through a price increase to equate demand and supply.

Under provision The under resourcing of a situation which may be production capacity to meet the needs of the market.

Under reserve Under certain circumstances involving documentary credit, the bank will be prepared to credit a sellers account with the amount immediately 'under reserve'. In so doing the bank puts the funds at the sellers disposal on condition that the other party will accept the documents despite the discrepancy. If the correspondent bank or the buyer refuses to accept the documents, the seller will be debited with the amount paid out plus all costs, interest and exchange rate differences, if any. When there are discrepancies in the documents the decision whether to release the funds under the reserve is made solely by the paying bank.

Under resourced The provision of an inadequate degree of resources available to undertake a particular assignment. See resource availability.

Underspend The process of not fulfilling expectations/budget in a spending/buying situation. It maybe a budget figure of $2,000 and the amount of $1,800 actually spent. An underspend of $200.

Underspent The process of not spending the allocated budget such as €100 instead of €150 – an underspending of €50.

Under subscription An issue is under subscribed when applications total less than the amount offered for sale for subscription. If the issue is underwritten, the underwriters have to take up the amount not subscribed pro rata to their commitments. In such cases it is not unusual for dealings in the security to begin at a discount on the issue price.

Under tonnaging The under provision/resourcing of shipping capacity/fleet complement to meet the needs of the market or other criteria.

Undervaluation An exchange rate is under-valued when it is below its purchasing power parity. The consequence of undervaluation is that the goods produced in the undervalued country will be too cheap in the export markets.

Undervalued currency Currency whose rate of exchange is persistently above the parity rate.

Undervalued freight rate A freight rate which is under priced in the market place. See under priced.

Underwriter The person who is authorised to accept insurance business on behalf of an insurance company or Lloyd's syndicate. An insurer who in return for an underwriting commission, undertakes to apply for, or to find, other applicants for all or part of an issue, or offer which is not taken up by the public. An underwriter usually passes part or all of his commitment to sub-underwriters such as stock brokers, jobbers, pension funds, insurance offices, merchant banks, investment trusts, finance houses and individuals who share commission.

Underwriters There are three categories as follows: (a) Insurers (e.g. companies of Lloyd's syndicates or Names); or (b) Those individuals performing the underwriting for a company, or underwriting agency (the active underwriters); or (c) A person soliciting business on behalf of an insurer (US).

Underwriters (Lloyd's) Active underwriters are the people or organisations who provide insurance cover. The name dates back to the early days of insurance, when the person seeking the insurance would write down the details of the risk and then take it to merchants who in turn wrote their names and the amount they would pay in the event of a claim underneath. Today, insurance is underwritten in two ways – insurance companies use their corporate wealth to meet claims, while in the Lloyd's market it is the private means of syndicate members which provide the capital base to meet claims. The underwriters seen in the Room at Lloyd's are the active underwriters representing and doing business on behalf of the underwriting members of their syndicates.

Underwriters (Stock Exchange) Market-makers or broker/dealers who agree to buy a new issue of a stock or share if other investors do not subscribe for the whole issue. For this service, the underwriters charge a fee, payable whether or not the whole issue is bought by outside investors.

Underwriting An agreement by which a company is guaranteed that an issue of shares will raise a given amount of cash because the underwriters for a small commission undertake to subscribe for any of the issue not taken up by the public; or subscribing ones name to a policy – to become answerable for loss or damage – for a premium. The acceptance of insurance business by an underwriter. One who agrees to compensate another person for loss from an insured peril in consideration of payment of a premium.

Underwriting Agent A person authorised to accept insurances on behalf of an insurance company as its agent; or an agent appointed by a Name under the terms of an underwriting agency agreement to manage his underwriting affairs at Lloyd's on his behalf. There is a distinction between a managing agent (who manages one or more syndicates) and a members' agent (who acts for his Names in other capacities and may place them on syndicates run by a managing agent), although many agents undertake both functions. An underwriting agent must have complied with the requirements of the Council of Lloyd's for conducting an underwriting agency at Lloyd's (see Agent – Lloyd's Underwriting).

Undeveloped market A market which has no evidence of it having reached any stage of development. See also developed and under developed market.

Undiscovered Loss Clause A cargo insurance condition which provides that the assured shall not be prejudiced because a loss is not discovered until cases are opened some time after termination of the risk.

Un Dk Under deck tank(s) – on a ship.

UNDP United Nations Development Programme.

Undue influence In legal terms the existence of a special kind of relationship as a result of which one party can exert subtle pressure on the other such as doctor-patient, solicitor-client, or teacher-pupil.

Unearned Premium That part of the premium (if any) which relates to a period during which the underwriter is not on risk. Overall, it relates to the portion of a risk which has not yet expired.

Unearned premium reserve The fund set aside as at the end of the financial year by a non-life insurer out of premiums in respect of risks to be borne by the company after the end of that year under contracts of insurance entered into before the end of that year.

UNECE United Nations Economic Commission for Europe.

UNECLAC United Nations Economic Commission for Latin America and the Caribbean.

UN/EDIFACT United Nations Electronic Data Interchange for Administration Commerce and Transport; the International multi-sector standard for EDI.

UN Embargo All goods prohibited under UN embargo.

UNEP United Nations Environment Programme.

UNEPTA United Nations Expended Programme of Technical Assistance.

Unequal trade Trade that is not in balance in quantity and/or fiscal terms between two ports/countries or specified shipping/air service or trade.

UNESCO United Nations Education, Scientific and Cultural Organisation.

Unfriendly end user A client – the buyer/consumer which is obstructive to dialogue and co-operation in relation to the vendor and the product/service bought. See end user.

UNGA United Nations General Assembly.

UNHCR United Nations High Commissioner for refugees.

Unibulk A system which makes it possible to carry both solid and liquid bulk cargoes in containers.

UNIC Unique Identification Code Methodology.

UNICE The Union des Industries de la Communaute Europeenne (French). An umbrella group of national employers' associations, including the Confederation of British Industry.

UNICEF United Nations International Children Emergency Fund.

UNIDF United Nations Industrial Development Fund.

UNIDO United Nations Industrial Development Organisation.

UNIDROIT International Institute of Unification of Private Law.

UNIFOR United Fortune Shipping (Pte) Ltd – a shipowner.

Uniform Customs and Practice for Documentary Credits (1994) A standardise code of practice formulated by the International Chamber of Commerce regarding the processing of Documentary Credits. See UCP 600. It is an ICC publication No UCP 500. Changes in the new code affect new developments in transport industry; and in data transferred communications technology.

Uniform rules for collections A code of practice detailing the main rules to be applied by the banks in connection with the collection operations for drafts, their payment or non-payment, protest and for documentary collections. The code of practice was formulated by the International Chamber of Commerce.

Unilateral aid The process of giving aid to one recipient country/organisation/agency.

Unilateral approval (Radioactive Material Only) An approval of a design which is required to be given by the competent authority of the State of origin of the design only. A term associated with dangerous cargo.

Unimodal transport The transport of goods by one mode of transport by one or more carriers.

UN Import Sanctions All goods subject to UN import sanctions unless specifically licensed by the UN.

UNIONOCEAN Union Ocean Shipping (Pte) Ltd – a shipowner.

Union of Greek Shipowners Formed in 1916 and based in Piraeus, it represents 95 per cent of the Greek flag. It represents the Greek Shipping Community internationally.

Unionised Company A company which recognises unions as the vehicle to represent the work force's views in the areas especially of consultation and negotiation of work practices and conditions of employment especially pay negotiation.

Unionised work force A work force relying on unions to represent their views and negotiate with the management on terms and conditions of employment especially pay negotiations.

Union purchase A rig in which a pair of derricks is used in combination, the derricks being fixed, and the cargo runners coupled in such a way that the load is swung from a position vertically under one derrick to a position vertically under the other (sometimes known as burtoning).

UNIPACIFIC Uni-Pacific Container Lines Ltd – a shipowner.

UNIQUE Unique Shipping & Trading Co (Pte) Ltd – a shipowner.

Unique Consignment Reference (UCR) A reference, allocated by the CFSP authorised trader to each import (or Customs Warehouse removal) consignment, which can be used to trace the consignment through all that traders' records. A customs term.

Unique Identification Code Methodology A UN Committee engaged in trade facilitation whose aim is to reduce the current number and variety of references used in both national and international trade.

Unique selling point USP is the main selling point based on a distinctive feature of the product or on the way it is promoted.

Unique selling proposition A marketing term indicating the perceived benefits of the product/service can be regarded as unique as a primary selling argument.

Unirradiated uranium Uranium containing not more than 10?6 gram of plutonium per gram of uranium-235 and not more than 9 MBq (0.2 mCi) of fission products per gram of uranium-235. Term associated with dangerous cargo shipment.

UNISCAN United Nations Anglo-Scandinavian Economic Committee.

UNISON Unison Shipping (Pte) Ltd – a shipowner.

UNITAR United Nations Institute for Training and Research.

Unit cargo Merchandise shipped in bulk but handled in units. These include steel and forest products. Additionally it embraces palletised cargo which is applicable to all forms of transport and includes general cargo. Unitised cargo is handled primarily by fork lift trucks.

United Arab Emirates The following states combined to form a single independent state UAE: Abu Dhabi, Ajman, Dubai, Fujairah, Sharjah, Umm al-Qaiwain, and Ras al-Khaimah.

United Arab Shipping Co (UASC) A major liner shipowner and container operator.

United Container Systems A major container leasing company.

United Nations Conference on Trade and Development Based in Geneva, its aim is to promote international trade especially with a view to accelerating economic development.

United Nations Convention for the International Sale of Goods The international conditions relative to the sale of goods as prescribed by the United Nations involving the Vienna convention.

United Nations Convention on a Code of Conduct for Liner Conferences 1974 Entered into force 6 October 1983 and has 78 contracting states. See liner conference code.

United Nations Convention on Conditions for Registration of Ships 1986 Not yet in force – 40 contracting parties with at least 25 per cent of the world's tonnage required for adoption as per annex III to the Convention – presently eleven contracting parties (2002).

United Nations Convention on International Multimodal Transport of Goods 1980 Not yet in force. Requires 30 contracting states for adoption, presently nine (2002).

United Nations Convention on the Carriage of Goods by Sea 1978 (Hamburg Rules) Entered into force 1st November 1992 and has 26 contracting states. See Hamburg Rules.

United Nations Industrial Development Organisation Based in Vienna, its aim is to promote and accelerate the industrialisation of the developing countries through the mobilisation of national and international resources.

United Nations Number Most dangerous goods have been allocated a UN number based on the generic name reflecting the correct technical name of the chemical or hazardous substance. The UN number will enable the shipper/sender to identify the correct classification of the merchandise.

United States Tour Operators Association (USTOA) A US organisation of tour firms devoted to promoting professionalism in the industry.

United States Travel & Tourism Administration The US Commerce Department agency (USTTA) for the promotion of tourism to the US.

UNITHAI United Thai Shipping Corporation Ltd – a shipowner.

Unitised freight Goods normally conveyed in containers, closed wagons and road trailers or lorries, e.g. manufactured products and perishable goods.

Unitised Load Device Rates Air freight rates devised by IATA relative to the conveyance of a container or pallet within the equipment profile use of the Unitised Load Device.

Unitised fleet This embraces fully cellular container ships, partly cellular container ships, ro/ro ships, and barge carriers.

Unit Load A number of individual packages bonded, palletised or strapped together to form a single unit for more efficient handling by mechanical equipment. It usually involves palletised or containerised cargo. Overall, it is a transport term involving the conveyance of goods in a unitised way on a door to door, or warehouse to warehouse basis concerning a pallet, container and so on.

Unit loading The process of stowing standard sized units thereby maximising the cubic capacity in a vessel/aircraft/trailer/railway wagon. It may be pallets, containers or trailers.

Unit Load Device (ULD) Any type of container with integral pallet, aircraft container or aircraft pallet, whether or not owned by an airline. The specification of IATA registered unit load devices are published in the IATA Register of Containers and Pallets and airline's tariffs.

Unit of freight The factor in which the freight rate is to be calculated which maybe gross weight, measurement, volume, value per head (livestock) etc.

Units of account The composite units designed to reduce exchange exposures of both borrower and investor.

Unit trusts Collective investment schemes whereby banks, insurance companies and other investment firms sell "units" to investor and invest the proceeds in shares. Selling and repurchase prices of trusts are published in many newspapers.

Universal Agent A person who has power to do anything the Principal could have done. Such an Agency is usually created by deed and is called a 'General Power of Attorney'. A legal term.

Universal fairlead A fairlead with three or more cylindrical rollers.

Universal Federation of Travel Agents Associations (UFTAA) A world organisation of national travel agents trade associations.

Universal Postal Union Based in Berne, it is a liaison and advisory body to national postal administrations.

UNL Ultimate net loss.

Unladen weight The weight of the road vehicle inclusive of body and all parts normally used in operation (the heavier of any alternative body being counted) but exclusive of the weight of water, fuel, loose tools and equipment and batteries where these are used for propelling the vehicle. A similar criteria maybe applied to other transport modes.

Unlawful trading The process of conducting international trade on an unlawful/illegal basis thereby contravening legislation/regulations operative in the importing country. See dumping and maritime fraud.

Unless Caused by Unless caused by the vessel being stranded, sunk, on fire, in collision or contact with any external substance (including ice but excluding water).

Unless General Other than general average.

Unless sooner commenced If laytime has not commenced by loading or discharging is carried out, time used shall count against laytime. BIMCO definition.

Unless used (UU) If laytime has commenced but loading or discharging is carried out during periods excepted from it, such time shall count. BIMCO term.

Unlimited liability A situation where the responsibility for paying all the debts of the business lies with an individual – say, a sole trader or partner.

Unlimited profit potential The profit potential in holding options is the same as having long or short positions in the underlying shares (except for the cost of the options). A LIFFE term.

Unlisted Securities Market A junior market for shares in companies which cannot fulfil all the requirements of Stock Exchange for a full quotation.

UNLK UN. Layout Key.

Unloader cranes Cranes specifically provided/used for the purpose of discharging merchandise from a vessel onto the quay or ligherage.

Unmooring The process of releasing a vessel's cables or ropes from quayside bollards and/or buoys the latter which would involve mooring boats; or to release the ship's anchors from the seabed to enable the vessel to proceed.

UN No United Nations Number

Un number The four or six digit number assigned by the UN to identify a commodity on a Bill of Lading, Air Waybill, or other form of consignment note. For example with regard to the movement of dangerous cargo by air, it would identify a substance or a particular group of substances. The prefix "UN" must always be used in conjunction with these numbers.

Unorganised market A market which is fragmented in terms of its promotion, channels of distribution, retail outlets and so on. In so doing, it does not produce any coherent pattern of the market profile.

Unpaid trade receivables Losses incurred due to buyers insolvency. See managing trade receivables.

UNPFA The United Nations organisation which is the leading international provider of development funding for population and reproductive health programmes. It was formed in 1969, and provides funds in excess of 6 billion dollars in assistance in developing countries.

Unquoted securities Stocks and shares not officially listed on any recognised stock exchange.

Unrebated oil Oil upon which the full rate of excise duty has been paid and upon which no rebate has been allowed.

Unregulated labour market A labour market which is not influenced by trade unionism, and mobility of labour and flexibility in job application is paramount.

Unrepaired Damage This term relates to damage to an insured ship that has not been repaired at the time the policy expires. It applies to both time and voyage hull policies, and provides that underwriters are not liable for unrepaired damage if the ship becomes a total loss from any cause (insured or not) before the expiry of the policy period. If the ship was not a total loss, the assured could normally claim a depreciation allowance for unrepaired damage, caused by an insured peril, on expiry of the policy.

Unrestricted letter or credit A letter of credit which maybe negotiated through any bank at beneficiaries choice

UNRISD United Nations Research Institute for Social Development.

UNSD United Nations Statistics Division.

Unseaworthy The condition of a vessel where from any cause it is unsafe to send her to sea.

Unsecured loan stock A stock which is not secured in any or all of the assets of a company and which therefore ranks as an ordinary creditor in a winding up.

UNSM United Nations Standard Message.

Unstructured interview In market research terms an interview conducted between a skilled professional marketeer researcher and respondent using no specific sequence of questions but simply adopting a rambling attitude jumping from one question to another with no specific relevance/continuity.

UNTAC United Nations Transitional Authority in Cambodia.

UNTAO United Nations Technical Assistance Organisation.

UNTDED United Nations Trade Data Elements Directory.

UNTDID United Nations Trade Data Interchange Directory.

Unvalued policy A policy that does not express the insured value of the property insured, as distinct from the sum insured by the policy. A claim under an unvalued policy is limited to the insurable value of the subject matter insured at the time of loss, even though the sum insured may be more than the insurable value. Unvalued policies are normally used for policies covering such hull interest as 'excess liabilities' and 'anticipated freight'. Overall, it is subject to the limit of the sum insured, leaves the insurable value to be subsequently ascertained.

UOB United Overseas Bank.

UOFP Under water optical fibre project.

up Under proof.

UPC Uniform Product Code.

Up front The process of an entrepreneur having money immediately available to support his/her offer/bid. Usually associated with the term 'money up front'.

Upgrade To move to a better accommodation or class of services; to make improvements to a facility.

Up market The top end of the market relative to consumer demand involving the A and B social/economic classifications.

Up market product/service The top end or near top end of a product/service range which is regarded as above the norm for quality of service, design, durability, specification and general market appeal.

UPOV International Union for the Protection of New Varieties of Plants (Union internationale pour la protection des obtentions végétales).

Upper deck Uppermost continuous deck (of a ship).

Upper flammable limit (UFL) The concentration of a hydrocarbon gas in air above which there is insufficient air to support combustion.

Upper lakes The four great lakes above Lake Ontario embracing Lakes Superior Huron, Erie and Michigan.

Upper tween-deck Space for carrying cargo below the main deck of a ship above the deck which divides the upper hold.

Upr Upper (deck) (on a ship).

UPR Underwriting Performance Report.

UPS Uninterrupted power supply – a means of continuous power guaranteed, over a limited period of time.

Upturn The process of an improvement relative to a particular situation. This would arise if the volume of business in a company rose resulting in an upturn in sales volume.

UPU Universal Postal Union.

URA Urban Redevelopment Authority (Singapore).

Uranium – natural, depleted, enriched Natural Uranium – chemically separated uranium with the naturally occurring distribution of uranium isotopes (approximately 99.28% uranium-238, 0.72% uranium-235 by mass). Depleted Uranium – uranium containing less than 0.72% uranium-235, with the remainder being uranium-238. Term associated with dangerous cargo.

URC Uniform rules for collection No 522. See separate entry.

URDG Uniform Rules for Demand Guarantees – ICC term.

URL The uniform resource locator used to address World Wide Web resources – a cyberspace term.

URN Unique Reference Number. The manifest or rotation number and the item number on the manifest uniquely identifies, for example, a vehicle.

Uruguayan Pesos The currency of Uruguay. See Appendix 'C'.

Uruguayan Pesos invoice An invoice payable in Uruguayan Pesos relative to an export sales invoice.

Uruguay Round Multilateral trade negotiations launched at Punta del Este, Uruguay in September 1986 and concluded in Geneva in December 1993. Signed by Ministers in Marrakesh, Morocco, in April 1994. WTO definition.

u/s Unsorted (timber trade).

US US Air Inc – an IATA Member Airline.

USA United States of America.
USA/GUL North America/Middle East/South Asia container service.
USAID United States Agency for International Development.
Usance The period time to maturity of a bill of exchange. The credit period granted by the exporter to the importer and reflected in the documentary credit terms, e.g. 30 days after Bill of Lading date, 60 days sight, 90 days after shipment date. Also termed tenor bill or term bill.
Usance Bill A bill of exchange called for payment at a fixed or determinable future time.
Usance credit Under this document credit, it authorises the beneficiary to draw a draft on a specialised bank together with the documents to be remitted. In return for the documents surrendered, the beneficiary receives the draft after it has been accepted by the drawee bank. If the beneficiary wishes to have the amount of the draft at his disposal prior to maturity, he can usually discount his bank at favourable terms. This puts the sales price at the seller's disposal immediately. On the other hand, the buyer will be charged with the draft amount at maturity only.
USB Unless sooner berthed.
USC Unless sooner commenced.
USCG United States Coast Guard.
USCS United States Customs Service.
US$ United States dollar – a major currency.
USD United States Dollar.
US Dollar The currency of the United States of America. See Appendix 'C'.
US Dollar invoice An invoice payable in US dollars relative to an export sales invoice.
USDV Ultra shallow draft vessel.
USEA United States East Coast.
User An information technology professional, technician or operator, including members of the general public when they are accessing information through information technology system. Computer term.
User friendly A facility which is easy to understand/operate/control/maintain/compatible and encourages a relaxed atmosphere/environment to the user thereby placing the person in favourable at ease situation.
User friendly product/service A product/service with a specification and design features enabling the person/company to be at ease when using the product/service.
User interface system A system used to improve the simplicity with which users can interact with another system, e.g. a graphic user interface (WIMP), or a menu selection system. Computer term.
User needs The needs/requirements of a particular user/clientele for a particular product(s)/service(s) in a particular area/region/country.
User profile The profile of the end user of a product or service.

User orientated organisation A form of company/entity organisation usually found in the marketing department whereby each phase of its operation is geared to meet the needs of the company product(s) or service(s) users.
USM Unlisted Securities Market. The Stock Exchange's second market for shares of medium-sized companies which either do not want or do not qualify for a full listing.
USMC United States Maritime Commission.
USNH United States North of Hatteras – ports of north Cape Hatteras sometimes called 'Northern Range'.
USP Unique selling proposition.
USPH United States Public Health (Standards).
USPSPI Uniform system of Port Statistics and Performance Indicators.
USS Uniform Symbology Specification.
USSR Union of Soviet Socialist Republics – now termed CIS. See CIS.
USTDP United States Trade Development Programme.
USWE United States West Coast.
UT Utility twin.
UTC Universal Co-ordinated Time; or Uniform Transport Code Foundation (the Netherlands).
Utilities (software) Software which performs common tasks such as file management, editor facilities and diagnostic routines.
Utility The ability of a good or service to satisfy a want. The satisfaction that an individual or household receives from consumption.
Utility models Patents granted for inventions that embody a lower level of inventive step and for a shorter period of time, mainly in the mechanical field.
Utmost good faith A fundamental principle of insurance with which strict compliance is required during negotiations to conclude an insurance contract. Its intention is to ensure that, by proper disclosure and representation, both parties are aware of all circumstances material to the risk at the time the contract is concluded between them. Breach of good faith on the part of the assured or his broker entitles the insurer to avoid the contract from inception.
Ut res magis valeat quam pereat (Latin) It is better for a thing to have effect than to be made void.
UU Unless used – time spent carrying out cargo work during excepted periods is to count as laytime. A chartering term; or Air Austral – an IATA Member Airline.
UV Ultraviolet.
U/W Underwriter or Underwriting.
UX Air Europea (Air Europa Líneas Aéreas, SA) – IATA Member Airline.
UY Cameroon Airlines – an IATA Member Airline.
UZ Uzbekistan – customs abbreviation.

3V TNT Airways SA – an IATA Member Airline.
V7 Air Sénégal International – an IATA Member Airline.
V Volt, or van containers.
V or vs (versus) (Latin) Against.
VA Value analysis; or Volare Airlines (Volare Airlines SpA) – an IATA Member Airline.
VAC Voyage Cash Flow analysis.
VAHD Veterinary & Animal Husbandry Department (Burma).
Valenti non fit injuria (Latin) A person cannot complain of an injury to which he gave consent.
Validate To check that data falls within certain prescribed limits.
Validation Imprinting a piece of generic airline ticket stock with the stamp of a particular airline, which makes it a ticket; the Customs procedure formally approving the carnet at the office of departure. A customs term; or checking a data entry to confirm that it is within the acceptable range and that it is not incomplete or unreasonable. Computer term.
VALMET A barge handling system involving ocean going tonnage.
Valuable cargo A consignment which contains one or more valuable articles (air cargo).
Valuation charge A charge for the carriage of goods based on the value declared for the carriage of such merchandise.
Valuation clause A clause in a hull policy, whereby a constructive total loss based on cost of repairs cannot be claimed unless such costs would exceed the insured value.
Valuation declaration A form on which an importer (or other eligible person) declares the method of valuation and the elements involved in calculating the value for import duty purposes – customs term.
Valuation report An assessment of the value of property or goods for mortgage, insurance or other prescribed purposes.
Value An LME term referring to a price which is traded in a volume sufficient to satisfy all the current buyers and sellers at that price.
Value added The increase in realisable value resulting from an alteration in form, location or availability of a product or service, excluding the cost of purchased materials or services. Unlike conversion costs, value added includes profit. Overall, the process of adding value to the service or product on offer to the buyer. See value added selling and value added services.
Value added benefit The accrued benefit to the user/buyer when purchasing a product/service or processing of a product/service. The buyer of a colour television obtains pleasure, education and above all communication from such a facility. Likewise the distributor importing bagged rice will package for the retail market and thereby provide a value added benefit to the product in the distribution chain process.
Value added chain The process of adding value-improve the benefits to the shipper from the supply chain. See value chain.
Value added selling The enhanced benefits to the supplier/sales company effective selling yields based on the external focus when developing good buyer/end user relationship. This embraces empathy, market positioning, relationships, forecasting/planning, pre sell/buy agreements, product application/development/cost analysis and sales training/management.

Value added services The process of increasing the perceived benefits of a product/service to the buyer. This involves developing customer focused benefits and added value rather than through price competition. It includes quality control, customs packaging, product development, warranty, after sales, technical manuals, training, maintenance, spares replacement and so on. Moreover, the goods may arrive in bulk and are packaged; oriental furniture may have cushions fitted to reflect local culture needs; cars may be fitted with air conditioners and so on.
Value added tax (VAT) Part of a country's tax system that imposes a levy at each stage of production of a product or service.
Value analysis A systematic procedure aimed at ensuring that necessary functions are achieved at minimum cost without detriment to quality, reliability, performance and delivery of a particular product(s).
Value chain A logistic term defined by the Harvard Business school professor as follows: 'Competitive advantage cannot be understood by looking at a firm as a whole. It stems from the many discrete activities a firm performs in designing, producing, marketing, delivering and supporting its product. Each of these activities can contribute to a firm's relative cost position and create a basis for differentiation. Hence a supply chain that adds value to its product or service between the supplier and buyer.
Value chain activities These can be categorised into two types: primarily activities (inbound logistics, operations, outbound logistics, marketing and sales service), and support activities/infrastructure, human resource management, technology development and procurement.
Value date This arises under a 'spot' transaction involving a foreign exchange where the currency involved must be declared immediately, i.e. within two days of agreement. The actual date of settlement is termed the value date.
Valued Policy A policy which specifies the insured value of the ship or goods insured, as the case may be. The value so expressed is conclusive of the insurable value of the ship or goods, as applicable, in the absence of fraud; but is not to be used to establish a constructive total loss, except where the policy so provides (as in hull policies). A policy that does not specify the insured value as such is an 'unvalued' policy.
Value Engineering A formal and systematic procedure for analysing product designs in order to cut costs and improve value. The function can be undertaken during initial product design or performed on existing products.
Value of materials The customs value at the time of importation of the non-originating materials used, or if this is not known, the first known price paid for those materials. This definition of value is also to be used where the value of any originating materials needs to be established. A customs term.
Value received for Often included on bills of exchange as a confirmation that a transaction exists and that the bill is not being used for financing purposes without there being a transaction.
Value surcharge A surcharge for the carriage of cargo having a value in excess of a specified amount per kilogramme.
Value to weight ratio The value of the goods in relation to its weight. High value goods like jewellery have low weight ratio, whilst coal is of low value but high weight ratio.

Valve bag A bag provided with an opening in a corner through which the bag maybe filled. It maybe a textile, multi-wall paper or plastic film bag.
Valve pressure A one way outlet on a receptacle to reduce internal pressure.
Vanning An American term sometimes used for the packing of a container.
VAN Value added network.
Vanilla The most basic type of any product e.g. vanilla swap, vanilla option. Distinguishes it from more complex types ('exotics') of swap and option. International Banking term.
VANS Value added network services.
VAP Visual Aids Panel (ICAO term).
Vaporising oil A blended kerosine type of petroleum distillate used in certain types of spark-ignition engines such as those used for agricultural purposes, stationary engines, and boats.
Vapour Density The density of a gas or vapour under specified conditions of temperature and pressure.
Variable cost Those expenses which will vary according to the frequency of the user of the transport unit such as fuel, day to day maintenance; or cost of production that will vary with the output level. It is the cost of a variable input.
Variation Margin In the event of a bought contract declining in value, or a sold contract increasing in value, the broker will request a payment from the client to meet the changing value of the contract, also payment by floor member to the clearing house. A BIFFEX term; or profits and losses on open positions which are calculated daily by the market-to-market process, which are then paid or collected daily – an International Petroleum Exchange term.
Variable levy Customs duty rate which varies in response to domestic price criterion. A WTO definition.
Variable overhead cost Overhead cost which tends to vary with changes in the level of activity.
Variable rate A rate which will vary according to the market forces or other circumstances. It may be an exchange rate or specific currency or a freight rate of a particular commodity in a shipping/air service.
Variometer Device for varying the inductance in an electric circuit composed of two or more coils.
VASI Visual Approach Slope Indicator – IATA term.
VAT Value Added Tax.
VAT Based system A situation to describe/recognise a VAT system exists in a particular country and derivative use is being made of such data.
VATOS Valid at time of shipment.
VAT Zero Rating A VAT commodity or service classification attracting a nil VAT level.
Vatu The currency of the Republic of Vanuatu. See Appendix 'C'.
Vatu invoice An invoice payable in the Republic of Vanuatu vatu relative to an export sales invoice.
VB Vertical main boiler(s) on a ship; or Maersk Air Ltd – an IATA Airline Member.
V BLT Vee built. American ship classification term.
VBPERKINS V B Perkins & Co Pty Ltd – a shipowner.
VC Valuation clause; variable charge – a special charge, comprising a fixed and a variable element, applicable to goods processed from certain basic agricultural products; ventilated containers symbol used in the import of goods subject to a Common Agricultural Policy under EEC; or voyage cost.
VCF Voyage Cash Flow analysis.
VCG Vertical centre of gravity.
VCM Vinyl Chloride Monomer.

VD Air Liberté SA – an IATA Airline Member.
VDI Verein Deutscher Ingenieure (Association of German Engineers) – its role is similar to the BSI codes of practices.
VDR Verbawd Deutscher Reeder (German Shipowners Association); or Voyage Data Recorder.
VDU Visual Display Unit or video display unit. A terminal linked to a computer consisting of a keyboard and a television-type screen. Information to and from the computer is displayed on the screen. Used extensively with electronic data storage equipment.
VE Aerovias Venezolanas SA (AVENSA) – an IATA Airline Member.
Vector graphic A graphic image where the graphic elements are defined using co-ordinate geometry, enabling them to be scaled each time they are used without loss of resolution. Each of the entities (e.g. lines, circle etc) in a vector graphic can be manipulated individually.
VEE BTM Vee Bottom.
veg Vegetable.
Vegetable oil tank farm A farm consisting of a number of tanks accommodating vegetable oil and usually adjacent to a Maritime terminal embracing a jetty berthing tanker vessels. Alternatively it may be situated inland, and road and/or rail served.
Vehicle leasing An agreement whereby a vehicle or vehicles are made available by one party – often a vehicle manufacturer, commercial vehicle dealer or vehicle rental specialist – to another who operated the vehicle for an agreed fixed period. Under the terms of the lease, the lessee will be responsible for maintaining and insuring the goods at his expense.
Vehicles carrier Ship built to convey vehicular cargo.
Vehicle processing centre The hub for processing import/export trade cars/vehicles by adding value to them.
Vehicle sling Cargo handling equipment for vehicle/car transhipment.
Vehicle transhipment In customs terms, the process of transferring of customs cleared goods between two vehicles in the dock area.
Vehicle turn round time The duration of a particular vehicle stay at a depot/terminal etc based on the time the vehicle is received at the loading bay to commence unloading of merchandise, until the vehicle duly loaded leaves the loading bay to proceed on her journey. The vehicle may arrive in a loaded or empty condition and likewise depart on a similar basis.
Vehicle unladen weight The weight of the vehicle inclusive of the body and all parts (the heavier being taken where alternative bodies or parts are used) which are necessary to or ordinarily used with the vehicle when working on the road but exclusive of the weight of water, fuel, or batteries used for the purpose of the supply of power for the vehicle, and of loose tools and loose equipment.
Vehicular cargo Vehicular traffic driven on/off the vessel. See wheeled cargo.
Vehicular Ferry Vessel capable of conveying vehicular traffic which is usually transhipped by being driven on or off over a ramp.
Vendor One that sells – a person or a company.
Vendor management The management responsible for selling activities. See vendor activities.
Vendor placing A placing of shares held by persons who have acquired them as payment for the purchase of their company. A form of financing which substitutes for a rights issue, but gives no pre-emptive rights of subscription to existing shareholders.

Vent A small outlet to permit the escape of air or vapour from a receptacle to reduce internal pressure.

Ventilated container A container with openings in the side and/or end walls to allow the ingress of outside air when the doors are shut. Ventilated containers are used for cargoes which have a high humidity or where there is a risk of condensation.

Venture Term embracing the three elements of any commercial voyage – hull, cargo and freight/passengers.

Venture capital Capital raised for the start-up of a new business venture. The high risks associated with such investments are reflected in the high rates of return which are targeted. It has been thought that shipping projects might attract venture capital, but the required rates of return are generally beyond those which shipping is capable of generating.

Verbatim ac litteratim (Latin) Word for word, letter for letter.

Verification Checking the accuracy of a data entry. Usually carried out by forcing an operator to read back and check the entry or by double entry of the data to check accuracy. A computer term.

Vers Voluntary export restraints.

Verso (Latin) The left hand page.

vert Vertical.

Vertical access Access to the vessel cargo accommodation is primarily vertical such as opening up the hatch covers and using cranes.

Vertical integration Company A Company which both manufactures the product and also sells it thereby becoming completely involved in the production and channels of distribution chain. An example is a dairy with various farms producing all the dairies products, and having a number of factories processing the dairy products, and ultimately retailing the dairy products through a chain of retail shops.

Verticalised cargo spaces Spaces in which the natural shape of the ship has been squared off to prevent stows collapsing.

Vertical market place Its aim is to encourage, facilitate and realise transactions between parties that belong to the same value chain.

Vertical mergers The process of companies merging to embrace manufacture and selling the product. See vertical integration company and horizontal mergers.

Vertical organisation A vertical organisation structure found in an export office dealing with the processing of an export consignment. In so doing each clerk deals independently with an individual aspect of processing the export consignment including order processing, credit control, production/assembly, packing, transportation and insurance, invoicing, documentation and so on. The structure encourages specialisation by each clerk involved in order processing technique. See also horizontal organisation entry.

Vertical spar A ceiling erected in certain spaces to square off the ship's space.

VERTREP Vertical replenishment.

Very large crude carrier A crude oil tanker of 150,000 – 299,999 dwt.

Vessel A ship.

Vessel being in free pratique and/or Having been entered at the Custom House The completion of these formalities shall not be a condition precedent to tendering notice of readiness, but any time lost by reason of delay in the vessel's completion of either of these formalities shall not count as laytime or time on demurrage. BIMCO term.

Vessel crossing A vessel proceeding across a fairway/traffic lane/route.

Vessel exclusion zones A maritime area where boat traffic and fishing are normally excluded. For example, in Atlantic waters boat traffic and fishing are normally excluded 500m from an oil drilling rig or production platform or 50m beyond its anchors, whichever is greater.

Vessel planning work station See VPWS.

Vessel screening The process of vetting a vessel on line. See vetting ship inspections.

Veterinary certificate A document required when livestock/domestic animals are exported and signed by the appropriate health authority in the exporter's country.

VFR traffic Visiting friends and relatives, a statistical category of traveller.

VGI Vertical gyroscopic indicator.

VH Aeropostal Alas de Venezuela CA – IATA Member Airline.

VHF Very high frequency.

VI Volga-Dnepr Airlies (JSC Volga-Dnepr Airlines) – IATA Member Airline.

Via En route. For example container shipment maybe Hamburg to Amman via Aqaba indicating the consignment will reach its destination Amman through the Port of Aqaba.

Via (Latin) By way of, Lit road.

VIC Very important cargo.

Vice Propre Inherent vice.

Vice versa (Latin) The position being reversed.

Victualling Ships catering.

Victualling bill Document detailing ship's stores at time of inward clearance. A document showing bonded stores for the vessel's use.

Victualling department Shipowners department responsible for ships catering and ancillary activities.

Victualling officer Catering officer.

Video conference The process of conducting a conference/meeting via satellite using videos.

Video display terminal (VDT) See cathode ray tube.

Video display unit (VDU) See cathode ray tube.

Video presentation A marketing aid involving a pre-recorded video tape which is shown usually to an invited audience or customer.

Videotex The internationally agreed generic term for viewdata and teletext services. Viewdata is interactive or wired videotex and teletext is broadcast Videotex.

Viditel The trade mark of the viewdata service of PTT Netherlands.

Vienna Convention A 1980 United Nations Convention on contracts for the International Sale of goods which came into force on 1st January 1988.

VIETRANS Vietnamese Forwarding and Warehousing Corporation.

Viewdata The original term used for the method, invented by the British Telecom, of carrying pages of information, stored in computers, over ordinary telephone lines for display on modified TV receivers and computer terminals. The standard international term for viewdata is interactive (or wired) videotex.

Vikane Trade name for sulphuryl fluoride. It is used as a fumigant to control insect pest or timber – often to fumigate containers.

VIO Very important object.

VIP Very important person.

Virtute officii (Latin) By virtue of his office.

Visa A document issued by a Consulate/Embassy etc to a person authorising/approving/permitting his/her entry/stay in their National country for a specified purpose and duration. Overall, an official authorisation appended to a passport that permits entry and travel within a country for a certain time.

Visconbill Code name of a BIMCO approved liner Bill of Lading for use in trades where Hague-Visby Rules are compulsory. English, French, German and Spanish editions are available.

Viscon booking Code name of BIMCO approved liner booking note used with Viscon-bill Liner Bill of Lading shipments.

Viscosity Measure of an oil's or liquid's resistance to flow.

Visegrad countries General name for the countries which have acceded to the Common Transit convention. Czech Republic, Hungary, Poland and Slovak Republic.

Visible exports Income received by a country from its overseas trade.

Visiting cards Usually termed business cards – see business card entry.

Visiting overseas market The process of visiting an overseas market which maybe in the role of either selling (exporting) or buying (importing) goods or some related purpose such as appointing an agent or simply to gather market intelligence data. See overseas trade fairs and seminars.

VITB Vocational & Industrial Training Board (Singapore).

Viva voce (Latin) Oral academic exam, Lit with the living voice.

Viz (videlicet) (Latin) By substitution, namely.

VK Air Tungaru Corporation – an IATA Member Airline.

V/L Vessel.

VLCC Very Large Crude Oil Carrier-Tanker in excess of 160,000 sdwt.

VLCS Very large container ships.

VLF Very Low Frequency.

VLLNGCs/VLNGCs Very large liquefied natural gas carriers.

VLOC Virtual logistics operation centre.

VMP Value as a marine policy; or Vertical Measurements Panel (ICAO term).

VMS Voyage management system.

Voidable Policy Where the underwriter has the right to avoid a policy (e.g. in event of a breach of good faith) the policy is termed 'voidable'. See avoidance.

Void Policy A policy which is not acceptable to a court of law as a valid legal document. Examples expressed in the MIA, 1906, are a gaming and wagering contract, a contract entered into where the assured had no insurable interest and no expectation of acquiring such interest, a policy which states 'interest or not interest', a PPI policy and a policy which contains terms whereby the underwriter waives any benefit of salvage to the insurer, or any similar terms.

Void Space An enclosed space in the cargo area external to a cargo containment system, other than a hold space, ballast space, fuel oil tank, cargo pump or compressor room or any space in normal use by personnel.

Volatile Subject to change. Used to refer to electronic memory which loses its contents when its power source is removed. Computer term.

Volcoa The code name of the BIMCO approved standard volume contract of affreightment for the transportation of bulk dry cargoes.

Volte-face (Latin) Reversal of position in opinion.

Volume The total weight of a commodity traded in one day, expressed either in tonnes or lots. LME volumes for silver are quoted in lots of 10,000 or 2,000 troy oz. Also referred to as turnover.

Volume charge A charge for carriage of goods based on their volume.

Volume market A market where a large amount of business opportunity obtains and a quasi mass market. See mass market.

Volume market sales A market which has a high volume of sale for a particular product(s)/service(s).

Volume rebates A rebate geared to traffic volume.

Volume trade route An international trade (shipping) route which has a high volume of trade such as the three major liner trade routes. See liner trade routes.

Voluntary excess An excess which the insured agrees to bear in consideration of a reduction in premium.

Voluntary export restraints A WTO initiative which introduced bilaterally voluntary export restraints to avoid disruption to a particular market.

Voluntary restraint agreements A WTO initiative which introduced bilaterally voluntary restraint agreements to avoid disruption to a particular market.

Voluntary restraint arrangement (VRA) Bilateral arrangement whereby an exporting country (government or industry) agrees to reduce or restrict exports without the importing country having to make use of quotas, tariffs or other import controls.

VO-MTO Vessel – operating multi modal transport operator.

VOP Value as in original policy.

VOR VHF Omnidirectional Radio Range – IATA term.

VOSA Vietnam Ocean Shipping Company.

VOSTRO An international banking term meaning 'your'. An example is found within an American Bank situation in New York having an (vostro) account with UK Bank Barclays. It would be a sterling based account.

Voucher Any documentary evidence in support of an accounting entry.

Vouchers, tour Document issued by tour operators to be exchanged for accommodations, meals, sightseeing and other services.

voy Voyage.

Voyage A maritime transit from one seaport (seaboard) to another seaport (seaboard).

Voyage account report A report produced by the shipowner detailing the expenses incurred and payable by various parties primarily the charterer. This includes daily fixture rate, crew overtime, port charges, disbursements and so on.

Voyage charter Voyage charter for a specified voyage. Usually the voyage charter is a non demise with the shipowner remunerated by freight – normally calculated on an agreed amount per ton of cargo carried. However, it may be under demise conditions. It maybe a port charter, berth charter, wharf charter, or dock charter. See relevant entries.

Voyage clauses Insurances that remain in force for the duration of a specified voyage or voyages.

Voyage cost The cost of performing a particular voyage which embraces fuel consumption, port charges, canal dues, tugs, crew cost, maintenance insurance, stores, lubricants and so on.

Voyage C/P Charter Party for specified voyage.

Voyage data recorder Sequential records of pre-selected data items relating to the status and output of ship's equipment and command and control of the ship. Also termed Black box. A mandatory requirement under regulation 20 of the SOLAS chapter V introduced 2002/4.

Voyage estimate The object of voyage estimating, whether provided by a shipowner, charterer, or their brokers, is to calculate the financial return a ship will realise on a particular voyage after deducting from the income the expenses of a particular voyage. Expenses include crew, victualling, bunkering, port dues, cargo loading/discharging, insurance, disbursements and so on. Income arises from the conveyance of cargo or fixture of the vessel.

Voyage expenses Those expenses directly attributable to performing a voyage which embrace crew cost, bunkering, pilotage, towage, light dues and so on. They do not embrace expenses at those ports which are naturally incurred in the normal operations of the ship such as radar repairs, hotel cost for repatriating crew members, cash advance to the charterer to meet ship board expenses and so on.

Voyage fixture The process of chartering a vessel for a particular voyage.

Voyage management The process of managing a voyage. See voyage plan.

Voyage number The allocated voyage number for a particular sailing.

Voyage plan The plan of a particular voyage embracing all its ingredients/elements such as bunkering, cargo, crewing schedules, ports of call, voyage schedule, financial input, and so on.

Voyage policy Contract of marine insurance covering a voyage from one seaport to another, or others.

VOYAGER Voyager Marine Corporation, SA – a shipowner.

Voyage report A report produced by the Master of a vessel, usually to the shipowner giving details of the voyage just completed. It will include cargo loading/arrangements and period in departure port; duration of voyage, average ship speed and fuel consumption; weather conditions throughout the voyage; any special incidents during voyage such as going to the aid of a vessel in distress, or exceptionally bad weather resulting in cargo damage; cargo discharge/arrangements and period of arrival in port; bunkering/victualling arrangements at both departure/arrival ports, and so on. The report is signed by the Master.

Voyage time The transit time for a particular voyage between specified seaports/seaboards.

Voylayrules 93 Code name of BIMCO approved document providing the voyage charter party laytime interpretation of the rules 1993.

VP Viaçao Aérea Sao Paulo SA (VASP) – an IATA Member Airline.

VPWS Vessel planning work station – computer system for ship planning usually based at a seaport.

VRAs Voluntary restraint agreements.

VRA, VER, OMA Voluntary restraint arrangement, voluntary export restraint, orderly marketing arrangement. Bilateral arrangements whereby an exporting country (government or industry) agrees to reduce or restrict exports without the importing country having to make use of quotas, tariffs or other import controls. WTO term.

VRM Variable range marker. One of the range controls incorporated on a marine radar.

VS Virgin Atlantic Airways – an IATA Member Airline.

VSA Vietnam Shipowners Association.

VSAT Very small aperture terminal.

VSG Vertical Studies Group (ICAO term).

VSOE Venice Simplon Orient Express. Pullman railway service.

VSSG Vertical Separation Study Group (IATA and EUROCONTROL).

VT Air Tahiti – an IATA associate Member Airline.

VTM Vessel traffic management.

VTMS Vessel traffic management system.

VTOL Vertical Take-off and Landing.

VTS Vessel traffic services/system.

VV Aerosvit Airlines.

VX ACES Aerolíneas Centrales de Colombia SA (Aerolíneas Centrales de Colombia SA) – IATA Member Airline.

W5 Mahan Air (Mahan Airlines Services Company) – IATA Member Airline.

9W Jet Airways (Jet Airways (India) Private Ltd) – IATA associate Member Airline.

w Wood; winches; West-compass point; winter; or worldscale.

WA With average.

WAC Whole account cover.

WACCC Worldwide Air Cargo Commodity Classification.

WA Cover WA Cover – the form of cover provided by the SG cargo policy with the Institute Cargo Clauses attached thereto (now obsolete).

WAEC West African Economic Community.

WAEMU Western African Economic and Monetary Union.

WAFC World Area Forecast Centre – IATA term.

WAFS World Area Forecast System – IATA term.

Wagering policy A policy in respect of which the assured has no insurable interest.

Wages The price paid to workers for the use of their labour, usually expressed as so much money per hour, day or week.

Wait berth The code name of a BIMCO approved charter party clause relative to a 'ready berth' and entitled 'Baltic Conference waiting for berth clause, 1962'.

Waiting a berth A ship awaiting a berth and in so doing for example, is anchored in the outer harbour or other convenient safe mooring place.

Waiting time The period of time for which an operator is available for production but is prevented from working by shortage of material or tooling, machine breakdown. An accounting definition.

Wait list The process of consignments or passengers being listed on a wait list in anticipation that the goods/passengers will be conveyed on the next available transport service. It arises because the transport service is fully booked and hopefully the clients will be conveyed on the service of their choice following a cancellation of a booking. The same criteria can apply to a theatre, hotel, inclusive tour, cruise and so on.

Waiver An agreement attached to a policy which exempts from coverage certain disabilities or injuries that otherwise would be covered by the policy; or permission granted by WTO members allowing a WTO member not to comply with normal commitments. Waivers have time limits and extensions have to be justified. WTO term.

Waiver Clause Waiver Clause – a clause which entitles both underwriter and assured to take measures to prevent or reduce loss, without prejudice to the rights of either party.

Waiver of premium A provision in some policies to relieve the insured of premium payments falling due during a period of continuous total disability that has lasted for a specified length of time, such as three or six months.

Wale A thick plate, on a ship, designed to withstand heavy impact.

Walk-ins Hotel guests who arrive without reserving.

Wallem Ship Management Ltd A ship management company based in Hong Kong.

Wall Street Trader An executive who works on the New York Stock Exchange and deals with the process of buying and selling stocks and shares on behalf of clients.

WAN Wide area network.

Wan Hail A major container operator.

Wan Hai Lines Ltd A major cargo liner operator.

WAP Wireless application protocol.

WARC World Administrative Radio Conference (ITU term).

War Clause A clause found in a charter party or insurance provision dealing with the risk of hostilities emerging from a specified shipment/trade/service etc.

Warehouse Any building or structure used for the storage of goods.

Warehouse entry Customs entry detailing cargo housed in a Bonded warehouse pending withdrawal.

Warehousekeeper The occupier of a warehouse approved for the deposit of goods chargeable with customs and/or excise duty that has not been paid or has been drawn back, who has entered into a bond with the Commissioners for the payment of duties chargeable on goods deposited in his warehouse, and is responsible for the observance of the law and any conditions applicable to the warehouse and to the goods warehoused.

Warehouseman The person responsible for the care of goods while they are in store.

Warehouse Officer Customs officer responsible for exercising control over goods kept in a Bonded warehouse.

Warehouse Receipt Document issued by a warehouse/shed operator (port administration etc) acknowledging receipt in a warehouse/shed of goods specified therein on conditions stated or referred to in the document.

Warehouse to Warehouse clause An insurance policy clause indicating the goods are insured from the warehouse of the seller to the warehouse of the buyer.

Warehouse warrant A document of title to the goods and issued by the warehouse in which they are stored.

Warheads Articles consisting of detonating explosives. They are designed to be fitted to a rocket, guided missile or torpedo. They may contain a burster or expelling charge or bursting charge. The term includes: warheads, rocket, with burster or expelling charge; warheads, rocket, with bursting charge and warheads, torpedo, with bursting charge. Term associated with dangerous cargo shipments.

Warlike confiscation Seizure of a vessel by the legal authorities during hostilities.

Warm front The boundary line between advancing warm air and a mass of colder air over which it rises.

Warm sector In the early stages of the life of many depressions in temperature latitudes, there is a sector of warm air, which disappears as the system deepens and the cold front catches up the warm front.

Warping Using ropes or cables to manoeuvre a ship.

Warrant The attachment to a fixed interest security that gives the holder certain rights, such as the right to purchase a given security price: the warrant can normally be detached from the bond and traded separately. Overall, it is a special kind of option given by a company to holders of a particular security giving them the right to subscribe for future issues either of the same or some other security. The document of title to metal stored in an LME registered warehouse. The warrant is a bearer instrument and states the brand of metal, its weight, the number of pieces and the rent payable.

Warranty A warranty in a marine insurance policy is an undertaking by the assured, whereby he promises that a certain thing shall be done or shall not be done, or that some condition will be fulfilled, or whereby he affirms or negatives the existence of a particular state of facts. A warranty may be express or implied, and must be complied with literally. Failure to comply with the warranty is termed 'breach of warranty', which, if not excused, discharges the underwriter from all liability under the policy as from the date of the breach.

Warranty bond A bond which provides a financial guarantee to cover the satisfactory performance for the equipment supplied during a maintenance or warranty period specified in the contract between the buyer and exporter. An international banking term.

War Risk The risk a shipper may encounter as a consequence of war in despatching the specified consignment.

Warsaw convention An international agreement for the unification of carriers responsibilities for the international carriage of goods by air which was signed by various countries in Warsaw in 1929. It was subsequently amended by the Hague Protocol of 1955 and the Guadalajara Convention of 1961. These are incorporated into the Carriage of Air Act, 1961, and Carriage by Air (Supplementary Provisions) Act 1962. Additional Protocol No 1 of Montreal 1975; as amended at The Hague 1955 and by Additional Protocol No 2 of Montreal 1975; as amended at The Hague 1955 and by Protocol No 4 of Montreal 1975. This legislation prescribes maximum limits of liability for carrier negligence and a code of uniformity for the carriage of goods by air. Currently the per-passenger liability limit is $75,000. This was suspended by Montreal Convention 1999. See separate entry.

WASA West African Shipper's Association.

Washing Overboard Loss of cargo, equipment (or crew) as a result of waves that pass over the deck of a vessel.

Washington treaty Treaty for the protection of intellectual property in respect of lay-out designs of integrated circuits. WTO term.

WASP Wind Assisted Ship Propulsion.

Waste cube Wasted capacity when cargo does not completely fill the interior of the container due to bad packing, awkward shapes, weight limitations, or lack of cargo.

Waste disposal The process of disposing of waste from a ship. Stringent restrictions and regulations exist as found in MARPOL 73/78.

Waste regulations Legislation governing waste as found in the European Packaging Waste 2001 Directive which requires member states to increase the proportion of packaging to be recovered and recycled after use. This has led to a range of innovative cost effective solutions to importers/exporters distribution problems.

Wasting asset Any asset of a fixed nature which is gradually consumed or exhausted in the process of earning income e.g. mines or quarries.

WAT Wing assisted tri-maran.

Watch man Artisans responsible for security activities at a seaport/container base/container freight station or elsewhere to prevent/lessen risk of any theft or related activity.

Water ballast tank Heavy weight of water – usually seawater – carried by ship to increase her stability and safety at sea. Usually carried in double bottom, deep and peak tanks.

Waterborne Agreement An understanding in the British marine insurance market whereby underwriters will cover goods against war risks only while they are on board an overseas vessel. Limited cover is allowed while goods are in craft en route between ship and shore and, also, during transhipment.

Waterborne property Items being carried on water (including the vessel itself).

Waterfront Container Leasing A major container leasing company.

Waterguard Service HM Customs organisation responsible for examination of passenger/crew effects, duty free stores and prevention of smuggling.

Waterproof Offering extreme resistance to the passage or absorption of liquid water.

Water resistant As applied to materials: having a degree of resistance to damage and permeation by water.

Water taxi A waterborne taxi service. Overall, it provides a taxi service in a waterborne environment as found in Venice and Baltimore.

Water vapour proof Offering extreme resistance to the passage of water vapour.

Waterway A canal such as inland waterways (canals) in Europe or Suez, Panama or Kiel waterways (canals).

Waterways Bill of Lading Negotiable transport document made out to a named person, to order or to bearer, signed by the courier and handed to the sender after acceptance of the goods.

WATOG ATA/IATA/ICCAIA Airline Industry Standard World Airlines Technical Operations Glossary.

WATRS Western Atlantic Route System – IATA term.

Wave oscillation The rise and fall in the sea level due to the action of waves.

Waybill List of passengers conveyed or inventory of cargo shipped. See Waybill (sea).

Waybill (sea) A non negotiable document used as an alternative to a negotiable bill of lading when all the facilities of a bill of lading are not required and when the owner of the goods does not wish to retain title to them. Waybills are made out to a named consignee and cannot be drawn 'to order'. A waybill may also be a list of passengers conveyed.

WB Waybill; water ballast; or warehouse book.

WB/EI West Britain/East Ireland.

wbs Without benefit of salvage (usually contained in ppi policies).

WC Wood covers.

wccon Whether cleared customs or not.

WCO World Customs Organisation, which studies questions of co-operation in customs matters to obtain harmony and uniformity. Governments of most countries are members.

WCSA West coast of South America.

W/d Warranted; or working day(s).

WD Working days – a chartering term.

W dk Weather deck.

WEAA Western European Airports Association – IATA term.

Weather deck Uppermost continuous deck.

Weather hindrances A chartering term usually quoted to indicate when adverse weather conditions prevent loading/discharge of cargo into/from the vessel and in so doing laytime does not count.

"Weather permitting" (WP) Any time when weather prevents the loading or discharging of the vessel shall not count as laytime. A BIMCO term.

Weather routeing Process of routeing a vessel having regard to vessel speed, fuel consumption and weather encountered during voyage. Also optimise routes based on a fixed ETA; track the weather impact on a vessel throughout her voyage; route vessels around hazardous weather conditions; reconstruct voyages for use if conflicts arose due to vessel poor performance, age of vessel; type of vessel; cargo loadability; type of cargo; convincing documentation of weather conditions encountered during the voyage for use in a claim which goes to arbitration, and offer potential of lower insurance premiums. The service is likely to be offered by a government department or private company whereby a shipowner or ship operator is provided with a route for his ship, devised by means of up-to-date weather predictions, which avoids severe weather conditions such as storms, fog and ice.

Weather routeing clause Clause found in a time charter indicating weather routeing data will be supplied by charterer to Master. Hence Master must comply with the reporting procedure of the charterers weather routeing service. See weather routeing.

"Weather working day (WWD) or "Weather working day of 24 hours" or "Weather working day of 24 consecutive hours" A working day of 24 consecutive hours except for any time when weather prevents the loading or discharging of the vessel or would have prevented it, had work been in progress. A BIMCO term.

Website A site on the internet system which features information Company/Government/Entity/individual wishes to display. Also a place where a buyer finds a single seller or vice versa. An E commerce term.

Website listing The process of keeping the website on a specialist company computer.

WEC World Energy Council.

WECM Warranted existing class maintained.

Wedge A structure used to provide a container stack with a horizontal base on a sloping deck or hatch cover.

Weekend freight tariff A freight tariff which is only valid at weekends. It is primarily found in the Shipping Ro/Ro market where in some trades Ro/Ro traffic is at a low ebb and the cheaper weekend freight tariff is designed to increase trade.

Weekend rates Special rates often used to attract leisure business to business hotels.

Weekend return A reduced return fare by rail, coach or air ticket permitting travel only at the weekend.

wef With effect from.

Weigh anchor To raise a ship's anchor.

Weighbridge A facility situated at a terminal which weighs road vehicles or railway wagons in a loaded or empty condition. The weighbridge tends to be situated in a depot, sea port, airport and so on.

Weight charge The charge for carriage of goods based on their weight.

Weight chargeable The weight upon which charges are assessed.

Weight dues Dues raised by a port authority on cargo handled/passing through a berth.

Weight gross The total weight of a shipment including the weight of the packaging and container.

Weight/less economy An economic value depends upon the trading of information rather than the trading of goods.

Weight/measurement The option of the carrier (shipowner/agent) to calculate the sea freight on a weight/measurement basis for an international transit whichever produces the greater revenue. The rate may be quoted on weight basis applied either per metric ton of 1,000 kg (2,205 lb) or measurement volumetric basis per tonne of 1.133m^3. Hence steel rails would be charged on a weight basis, and bales of cotton on a measurement basis. In most trades cargo measuring under 1.133m^3 per tonne weight would be charged on a weight basis, whilst cargo measuring 1.133m^3 or more per ton is charged on a measurement basis.

Weight/measurement ships option A freight rate term permitting the shipowner to charge the highest rate yield when evaluating the cargo which maybe based on a weight or measurement (cubic volume). For example, steel rails would be based on weight whilst bales of cotton on measurement.

Weight net Weight of the shipment except the weight of the packaging and container.

Weight note The weight/quantity of landed cargo at the port of discharge.

Weight of cargo The quantity/weight of cargo which is usually detailed on the Bill of Lading, Air Waybill, or other consignment note.

Weight over pivot The weight in excess of the pivot weight.

Weight pivot Minimum chargeable ULD weight.

Weight rated cargo Cargo whose freight is payable on the basis of weight by means of a rate per tonne.

Weight tare The weight of the packaging and/or container.

Weight volume Refers to the low density cargo for which the transportation charges are based on cubic dimensions of the shipment rather than upon actual gross weight. In such cases the charges are established on the basis that each 6,000 cubic centimetres (366 cubic inches) equal one kilogram or 166 cubic inches (equal one pound).

Welcon Code name of Chamber of Shipping approved charter party for coal trade.

Well The open deck space between erections on a vessel; or oil well service.

Well deck Area between continuous deck – fore and/or aft – which has not been developed.

WESA West Europe Airports Association.

Westbound The process of proceeding in a westbound direction. It maybe a consignment, transport unit, passenger, transport route and so on.

Westernised In broad terms, a culture/fashion/protocol/attitude reflecting the populace taste etc in the Western Hemisphere, especially among economically developed countries.

Wet bulk cargo Merchandise conveyed in bulk and of a liquid nature/specification such as oil, wine, gas, and so on.

Wet barrel A barrel constructed to hold liquids without leaking.

Wet barrels Trading in crude oil or products with the intention of making physical delivery – an IPE term.

Wet barrel trading Trading in crude oil or products with the intention of making physical delivery – an IPE term.

Wet dock A dock which has a constant level of water access which is gained via a lock.

Wet lease The rental of a vehicle, particularly an aircraft, including crew, and may include full operational and cabin crew supplies, fuel and maintenance services.

Wet vetting The process of vetting wet bulk cargo tonnage. See ship vetting.

WEU Western European Union.

WF Widerøe Flyveselskap A/S – an IATA member airline.

WFP World Food Programme.

WFUNA World Federation of United Nations Association (based in Geneva).

wg Weight guaranteed.

WGS World geodetic system, a datum reference.

Wgt Weight.

WH Workable hatch – a charter party term.

Wharf A place where vessels are moored and usually load and/or discharge their cargo. It is usually situated in a port or on a river.

Wharfage Port tariff – the charge for using a berth alongside a wharf. Also known as 'quayage'. See wharfage charges.

Wharfage charges Such charges are basically handling tariffs at a port relative to the movement of import or export cargo over the quay from or to a vessel. It covers a wide variety of cargo/commodity types including road haulage vehicles, containers, coaches, cars; loose cargo on a per tonne basis; livestock usually on a per head basis and so on.

Wharf Apron That portion of a wharf or pier lying between the waterfront edge and the (transit) shed, or that portion of the wharf carried on piles beyond the solid fill.

Wharf charter A voyage charter party which specifies the vessel to discharge her cargo at a specified wharf. In so doing, the vessel will not have officially arrived until she actually reaches the named wharf where cargo operations are to be performed.

Wharf dues Charges raised by the wharf operators on a vessel calling at the wharf which maybe for the purpose of cargo loading and/or discharging, or merely to bunker, victual the vessel and so on. It is based on the net registered tonnage of the vessel, with additional charges for example for mooring and unmooring.

Wharfinger The person in charge of a wharf.

What the market will bear The degree to which a particular market will bear/accept/absorb any price increase on a specified product(s)/service(s) without customer/client resistance and loss of market share/business.

WHB Waste Heat Boiler (of a ship).

WHBS Waste heat boiler survey (of a ship).

WHD Wagon height dimensions; or Workable hatch per day – a chartering term.

Wheeled cargo Cars, lorries, tractors, caravans, coaches and other wheeled cargo embracing both agricultural and industrial equipment. Usually associated with maritime transport.

Wheel house Accommodation located on the navigating bridge deck which forms the centre of navigational controls and wherein is located the steering wheel controlling rudder movement of the vessel.

"Whether in berth or not" (WIBON) or "berth or no berth" If no loading or discharging berth is available on her arrival the vessel, on reaching any usual waiting-place at or off the port, shall be entitled to tender notice of readiness from it and laytime shall commence in accordance with the charter party. Laytime or time on demurrage shall cease to count once the berth becomes available and shall resume when the vessel is ready to load or discharge at the berth. A BIMCO term.

WHFTB Waste heat fire tube boiler (of a ship).

WHFTBS Waste heat fire tube boiler survey (of a ship).

Whipping Longitudinal Force created on loaded VLCC by heavy seas causing deck movement.

Whistle blower A company employee who breaks silence by communication with the media regarding a company/government misdemeanour.

White durables Furniture usually found in the kitchen and classified as consumer durables such as dish washer, washing machine, refrigerator, cooker, deep freeze unit and so on.

White furniture See white durables.

WHITELINE Whiteline Navigation Co Ltd – a shipowner.

White list A list of maritime nations provided by the International Maritime Organisations which provide approved training establishments for safety training of crew personnel including catering personnel.

White lists A medium risks. See Black and Grey list entries.

White products This term applies to gasolines, naphthas, kerosenes and gas oils. See also Black oil. An IPE term.

White spirit A highly refined distillate with a boiling range of about 150° to 200° used as a paint solvent and for dry cleaning purposes etc.

WHO World Health Organisation (based in Geneva).

Wholesale market prices The price of oil products sold ex-refinery or from independent storage – an IPE term.

Wholesaler An intermediary between the producer and the final seller, usually buying large quantities for resale in smaller quantities to retailers; or a company that develops tour packages with all the components for sale through travel agents.

Wholesalers Middlemen who sell to other middlemen or industrial users but not to consumers and who take title to the goods they sell.

Whole trader control Customs control for all of a trader's customs-related activities by a designated "Account Manager" based at the local control office. Control will be carried out by working with the trader to understand and audit its systems and procedures. Customs term.

Wholly-owned foreign enterprises An enterprise/company which is owned by a foreign investor under a joint venture as found in China.

Whse Warehouse.

WHWTB Waste heat water tube boiler (of a ship).

WHWTBS Waste heat water tube boiler survey (of a ship).

WI Wrought iron cable – ship's equipment.

WIAS West Indies Associated States.

WIBON Whether in berth or not – laytime to commence and count whether the vessel is berthed or not depending upon other formalities in the charter party being complied with.

Wide area network The interconnection of items of computing and telecommunications equipment over a very large geographical area, e.g. countrywide or internationally. Such systems enable the sharing and transmission of data and information between LANs and individual users of the system. Computer term.

Wide bodied aircraft A high capacity passenger/freight aircraft as found in the McDonnell Douglas DC10-30, BOEING 747 100/200 and BOEING 747-400. Such aircraft with a high capacity passenger load can convey up to 20 tonnes of cargo usually in the larger type of Containers/Unit Load Devices. Such aircraft are on long distance routes and capable of conveying the IATA containers.

Wide-body Any of a number of large-capacity aircraft that can seat up to 10 passengers abreast, such as 707s, DC10s and L-1011s.

Wide body container A container wider than the standard width of 8ft (2.43m) such as 8ft 2.5ins (2.5m) or 8ft 6 ins (2.59m).

Wide-cut gasoline Light hydro carbons intended for use in aviation gas turbine power units.

Wide load A consignment which is in excess of the regulatory/standard width. It usually requires special transit arrangements and attracts higher freight charges. See out of gauge entry.

WIG Wing in Ground, system.

Wilhelmensen SA A major cargo liner operator.

WILLINE Willine Express – a shipowner.

Winding up An alternative for liquidation.

Window of change A gap in a situation/market place to develop an opportunity/idea/proposal with a view to change.

Window of opportunity A gap in a situation/market place to develop/explore/discuss an opportunity/idea/proposal.

Wine Import Agent A procurement/import agent which deals with the importation of wine.

Wine warehouse A warehouse accommodating wine for distribution to the market embracing wholesalers and/or retailers.

Winged pallet A pallet in which top and bottom planks extend beyond bearers to facilitate lifting by use of bar slings.
Wings The sides of the holds in a vessel.
Wing span The overall length of the wings of an aircraft.
Wing tank Tank situated at the side of a ship such as a tanker, bulk carrier, or combination carrier.
WIPO World Intellectual Property Organisation.
WIPON Whether in port or not.
Wire lashing Lashing constructed out of wire rope.
Wire lay A rope or textile structure formed by a stranding process similar to the fabrication of steel wire rope.
Wireless certificate Document certifying serviceable ships wireless.
Wireless log book Ships wireless diary. See Radio log book entry.
With Average A WA policy covers partial loss if the franchise is exceeded but is otherwise similar to an FPA policy.
Withholding tax Deducted in certain countries from interest payable to non-residents; depending on whether a double taxation agreement exists between the investor's country and that of the investment, the investment may or may not prove profitable.
Without Benefit of Salvage A term in a marine insurance policy whereby the underwriter forgoes his subrogation rights. A marine insurance policy that incorporates such term is deemed by the Marine Insurance Act to be a gambling policy and, as such, is invalid in a court of law.
Without Prejudice The claim is paid on this occasion, although the underwriter feels it does not attach to the policy, but this action must not be treated as a precedent for future similar claims.
Without recourse The process of placing a liability on the supplier for any funds not repaid by the buyer. It arises under export credit finance.
With particular average Accidental partial loss proximately caused by an insured peril. See Particular Average entry.
With recourse The process of placing no liability on the supplier for any funds not repaid by the buyer. It arises under export credit finance.
WKBT Utility/workboat.
WL Ward Leonard system.
WLD Wagon length dimensions.
WLTHC Water line/top of hatch coaming.
W/M Weight and/or measurements at ships option.
WM Water monitors.
W & M War and Marine.
WMARC World Maritime Administrative Radio Conference.
WMO World Meteorological Organisation (based in Geneva).
WMS Warehouse management systems.
WMS Software Warehouse management systems software.
WNA Winter North Atlantic.
WNG Wing.
Wob Washed overboard.
WOFEs Wholly-owned foreign enterprises
WOG Without guarantee – a chartering term.
WOL Wharfowners liability.
Won The currency of the Democratic Peoples Republic of Korea. See Appendix 'C'.
Won invoice An invoice payable in Won currency relative to an export sales contract.
Wooden cases A long established form of outer packing constructed of wood and affording complete protection to the cargo. Aids handling, stowage, prevents pilferage and usually permits stacking. Usually purpose built.
Wooden crates A long established form of outer packing constructed of wood and affording only limited protection to the cargo as unlike the wooden case, the crate is not close boarded. Many crates are mounted on skids. Often termed skeleton cases/crates.
Wooden pallet Mounted wooden platform of about one metre square designed to accommodate and facilitate cargo shipment and through cargo movement.
Wool Exchange A market place in London where imported wool particularly from Australasia is auctioned and sold.
Workable hatch A chartering term involving the availabililty of a ship's hatch to load/discharge cargo and sometimes the actual volume of cargo capable of passing through the hatch daily is given.
Work experience The process of a person or item of equipment having the experience of work in a particular circumstances. It maybe any export clerk in a manufacturing company processing export documentation or a fork lift truck operator handling palletised merchandise in an import shed.
Working capital The capital available for conducting the day to day operations of an organisation; the excess of current assets over current liabilities.
"Working days" (WD) Days not expressly excluded from laytime. A BIMCO term.
Working layer A layer of excess of loss reinsurance in which frequent claims are likely to arise.
Working luncheon A lunch at which the participants meet to discuss business matters of mutual interest.
Working Party A group of people representing different interests/departments formed for the purpose of examining/resolving/recommending a particular course of action in response to a remit submitted to them.
Working practice The customary practice at which a work assignment is performed.
Working time saved A chartering term indicating despatch money to be paid by the shipowner to the charterer for working time saved.
Working Time Table An operational schedule for rail, bus, airline or road haulage services.
Work in progress Partly-completed manufactured goods or contracts – e.g. a half-finished ship or bridge.
Work-out The process of negotiation between borrower and creditors in the event of serious financial difficulties or outright insolvency. The aim is to arrive at agreement on capital reconstruction and debt rescheduling.
Work Station Total Computerised Cargo Handling system; or generally a stand-alone computer but could be a dumb terminal – computer definition.
Work study A management study based on measuring employees' work against their output.
World Bank Agency associated with the United Nations that provides financing for development projects through its affiliate, The International Bank for Reconstruction and Development.
World bulk fleet The total dwt of world bulk cargo tonnage embracing both dry and wet bulk cargoes.
World Business Environment The culture, legal, infrastructure, technical, economic, financial, regulatory, political and social environment in which business is conducted worldwide.
World exports The components of the world trade exports which in 2003 embraced machinery and transport equipment 2584; manufactured products 1634; fuels 596; chemicals 567; food, drink and tobacco 388; raw materials 213 and producing a total including others of 6,180. All figures in $bn fob

World food 99 A charter party developed between the United Nations World Food Programme and BIMCO in 1999.

World food receipt Code name of BIMCO approved non-negotiable cargo receipt for use with world food charter party.

World food waybill Code name of BIMCO approved the world food programme non-negotiable liner waybill.

World Intellectual Property Organisation The international organisation which administers the protection of industrial property rights such as trade marks, patents, designs and copyright.

WORLDLINE World Line – a shipowner.

Worldscale The Worldwide Tanker 'nominal' Freight scale. It is a schedule of freight rates applying to tankers carrying oil in bulk. The freight rates which Worldscale provides are only intended solely as a standard of reference.

Worldscale (NYC) Inc of New York See London Tanker Brokers Panel Ltd and Worldscale entries.

World seaborne trade The total tonnage of cargo conveyed by maritime services globally usually compiled on an annual basis. It includes crude oil and products, the five major bulk cargoes and other dry cargoes.

World Shipping Council Based in Washington and formed in 2000. It has over 25 founding liner operator members.

Worldspan See DATAS II or PARS entries.

World's principal trading currencies US dollar, Sterling, Japanese yen, Euro, Norway Krone, Denmark Krone, India Rupee, Hong Kong dollar, Singapore dollar, Swiss franc, Canadian dollar, Australian dollar.

World Tourism Organisation (WTO) A worldwide group of government travel organisations involved in promoting tourism; formerly the International Union of Official Travel Organisations.

WORLDTRADE World Trade Shipping, New York – a shipowner.

World Trade Organisation The successor to GATT. It has 148 members (2002).

World trade routes Shipping routes which convey a very substantial volume of international trade.

World Travel and Tourism Council Organisation of travel industry executives established to ensure that governments understand the economic role of tourism and to promote the expansion of the tourist industry.

World Travel Market Annual global travel and tourism event in London featuring exhibition stands by hotel groups, airlines, car rental companies, cruise lines, ferry companies, tourist organisations, attractions, tour operators and business and incentive travel operators.

W/P Working Party.

WP Working Pressure; or weather permitting – a Chartering term.

wp Without prejudice.

WPA with particular average An insurance clause which means the items insured are only covered for partial loss.

WPC World Power Conference. Based in London, its role is to promote the regulation and co-ordination of the energy sources of individual countries; or Wave piercing catamaran.

WPI Written premium income.

WPT Way point. A term used when passage planning with an electronic chart. Usually defined as a point of course alteration, but not always so.

Wrap-around The use of one type of financing, e.g. fixed rate shipbuilding credit, as the basis for a larger composite credit, usually involving floating interest rates and a longer term.

Wreck This includes jetsam, flotsam, lagan and derelicts found in or on the shores of the sea or any tidal water.

Wreckcon Code name of BIMCO approved international wreck removal and marine services agreement (lump sum).

Wreck fixed 99 Code name of BIMCO approved international wreck removal and marine services agreement. It is based on no cure no pay agreement.

Wreck hire Code name of BIMCO approved international wreck removal and marine services agreement (daily hire).

Wreck line 99 Code name of BIMCO approved international wreck removal and marine service agreement. It is based on no cure no pay agreement.

Wreck stage 99 Code name of BIMCO approved international wreck removal and marine services agreement. It is based on no cure no pay agreement.

Write down A reduction in the recorded value of an item of stock to comply with the principle of valuing stock at the lower of cost or net realisable value; may also be applied to investments and fixed assets.

Writer The person who sells an option contract to open a position (also called 'Grantor') – an International Petroleum Exchange term.

Written line The amount written on the broker's slip by an underwriter accepting the risk.

Written premium The cumulative amount of premium income due to be received by an insurer.

wro War risk only.

WS Wood sheathed (decks); or Worldscale – see separate entry.

WSh Warren Spring Laboratory.

WSHTC Worldscale Hours terms and conditions – chartering term.

WSP Water spray protection.

WST Wholesale town.

WT Nigeria Airways Ltd – an International Airline and Member of IATA; or watertight.

W/T Wireless telegraphy.

WT Aux B Water tube Auxiliary boiler(s) (on a ship).

WTB Water tube boiler(s) (on a ship).

WTBS Water tube boiler survey (on a ship).

WTC World Trade Centre (Singapore).

wtdb Water tube domestic boilers (on a ship).

WTDR World Traders Data Report – a USA resource which provides an overview and its status of overseas companies.

WTO World Trade Organisation; or World Tourism Organisation.

WTO Dispute Settlement Procedure A code of practice prescribed by the World Trade Organisation to settle disputes relative to the global multilateral trading system. It involves ten stages: Preamble first hearing; first hearing; rebuttals; experts; first draft; interim report; review; final report; ruling and appeals. The time scale is one year or a further three months with appeal.

WTRPP Water pump propeller.

WTS Working Time saved – despatch to be paid for working time saved – Chartering term.

WTY Warranty.

WUS World University Service.

WVNS Within Vessel's Natural Segregation.

WW Warehouse warrant.

wwd Weather working days. A day of 24 hours on which work is not prevented by bad weather, and not any days excepted under the terms of the agreement (ex Saturdays, legal and local holidays).

wwdFHEx Weather working days, Friday and holidays excluded – a chartering term.

wwdSHEx Weather working days, Sundays and holidays excluded.
WWR When and where ready.
WWW World Weather Watch (WMO).
www World wide web.
WX CityJet – an IATA Airline Member.
WY Oman Air (Oman Aviation Services Co [SAOG]).
WYNSTANLEY4 Wynstanley Line – a shipowner.

X400 E-mail protocol standard.
5X UPS Airlines (United Parcel Service Company) – IATA Airline Member.
XD Ex Directory.
xd (ex-dividend) Shares so marked indicate that a buyer is not entitled to the latest declared dividend. Shares marked "xr" and "xc" give similar indications regarding rights and capitalisation issues recently announced. Most stocks and shares are bought and sold "cum-dividend".
XGS/GDP(%) Exports of goods and services: Debt servicing as a percentage of exports of goods and services. A debt indicator found in LDCs representing in financial terms the cost of servicing the external debt in repayment/interest charges as a percentage of the exports of goods and services.
XK Compagnie Aérienne Corse Méditerannée – an IATA Airline Member.
X/L Excess of Loss.
xl & ul Exclusive of loading and unloading.

XML Extensible Mark-up Language – a computer logistics term.
XO China Xinjiang Airlines – IATA Airline Member.
Xpired Risks Reserve The amount set aside as at the end of a financial period in addition to unearned premium in respect of risks to be borne by the company after the end of that year under contracts of insurance entered into before the end of that year, in order to provide for additional costs of claims. This may be necessary because certain business is expected to be loss making, or for example, to cover claims whenever they may arise in the future under contingency insurances paid for by single premiums.
x pri Without privileges.
XREF Cross reference.
X rf Rise of floor (or deadrise); or ribbed furnaces (on a ship).
XS Excess.

Y2 Alliance Air (African Joint Air Services)
Y4 Eagle Aviation (Eagle Aviation Ltd) – IATA associate Member Airline.
5Y Atlas Air, (Atlas Air, Inc) – IATA Member Airline.
9Y Air Kazakstan – IATA Member Airline.
Yacht broker A person who buys and sells yachts in the market place.
Yang Ming Marine Transport Corp Ltd A major liner shipowner and container operator.
Yankee bond US domestic issue for a foreign borrower.
Yaounde Treaty A treaty which seeks to improve co-operation and integration among Africa airlines.
YAR York Antwerp Rules.
Yard dolly A wheeled device used to move containers around a yard or area pulled by a tow tractor.
Yard operation computer system See YOCS.
Yard planning A term associated with a container berth and its infrastructure whereby the Port Management plan/allocate the container import, export, and transhipment to a specified location in the nearby quayside/container stacking area/container yard.
YARN A generic term for a continous strand of textile fibres, filaments or material in a form suitable for intertwining to form a textile structure via any one of a number of textile processes.
Yatres The computerised reservation and national distribution system of Yugoslavia; partner of Amadeus.
Yaw The process of a vessel starting to roll. In such a situation the keel depth may increase up to fifty centimetres with every degree of yaw.
y/c Your cable.
YC Flight West Airlines Pty Ltd – an IATA Member Airline.
Year closed The calendar year in which the premium is first advised to the Company by a broker irrespective of the year or period time when the Policy incepted. (Also known as closing year).
Year entered Year during which premiums are entered on the Company's books. Losses are recorded within the same year. (Booked, Closed and Signed are all synonymous.
Year free of premium Some insurers offer insurance, especially household, under which a free year is granted from time to time, usually as a no claims bonus.
Year inception The Calendar Year in which a risk commences to run.
Year of construction Year in which a vessel was built.
Year of origin The Financial Year in which a claim made against a Company is known to have originated.
Year of refitting Year in which a vessel is given a major overhaul.
Year open An Insurance Year of Account which is not closed or finalised, common with Lloyd's who keep a year of account open until the 36th month.
Year policy The year commencing with the effective date of the Policy or with an anniversary of that date.
YEF Year End Forecast.
Yen The currency of Japan. See Appendix 'C'.
Yen account An account which is managed in Yen.
Yen invoice An invoice payable in Japanese Yen relative to an export sales contract.
YEWLIAN Yew Lian Shipping Co (Pte) Ltd – a shipowner.
YGC Yard gantry crane (operations).
YGS Year of grace survey.
Yield The annual return on an investment based on the current price of the security on the assumption that the next dividend paid will be the same as the last one. Flat yield is the income on a fixed interest stock, ignoring any capital gain that may be made if the stock is due to be redeemed at par at some future date. Redemption yield is the same, but allowing for the expected capital gain if the price is below par.
Yield, current The ratio of the coupon (interest) to the market price expressed as a percentage; the maturity and frequency of interest payments are not considered. A banking term.
Yield curve The relationship between the maturity dates of an instrument and the yield; a positive yield curve is one where the shortest maturity bears the least interest and the longest the most. A banking term.
Yield, direct The discount to yield based on the total life of the asset at simple interest. A banking term.
Yield, effective Calculated on the annual, semi-annual or quarterly compounded interest basis over the period of the actual investment. A banking term.
Yield load A load at which irreversible deformation begins.
Yield management Tourist industry practice of controlling the mix of bookings by computer to achieve highest revenue and profit.
Yield to maturity The return on a security held to maturity at a given price; a variety of this is the 'yield to average life' when the average life (qv) is substituted for final maturity in the formula. A banking term.
YM Montenegro Airlines – IATA Member Airline.
YOA Year of Account.
YOCS Yard (container) Operation Computer System.
York/Antwerp Rules The rules relating/governing general average. The present draft is found in the 'York Antwerp rules 1974 as amended 1990'. A set of internationally accepted rules for application to general average circumstances. Most contracts of affreightment provide for general average to be adjusted in accordance with these rules. In the absence of such agreement adjustment is made in accordance with the law of the place where the adventure is terminated.
Young society A situation where the demographics are such that a high proportion of the population are relatively young.
Youth culture The culture associated with a young person expressed in dress code/fashion, attitude, taste and life style.
YTD Year to Date.
Yu Yugoslavia – customs abbreviation.
Yuan (Renminbi) The currency of China. See also Appendix 'C'.
Yuan (Renminbi) invoice An invoice payable in the Chinese currency of Yuan (Renminbi) relative to an export sales contract.
YUG Yugoslavia.
YW Air Nostrum (Air Nostrum, Líneas Aéreas del Mediterr·neo SA) – IATA Member Airline.

4Z SA Airlink (SA Airlink (Pty)) Ltd – IATA Member Airline.

7Z Laker Airways (Bahamas) Ltd – IATA Member Airline.

Z9 Aero Zambia (Aero Zambia Ltd) – IATA Member Airline.

ZA ZAS Airline of Egypt – an IATA Airline Member.

Zaire The currency of the Republic of Zaire. See Appendix 'C'.

Zaire invoice An invoice payable in the Republic of Zaire – currency of Zaire – relative to an export sales contract.

Zambia Airways The Zambian National Airline and member of IATA.

ZC Chinese Register of Shipping – vessels classified by the Peoples Republic of China Ship Classification Society; Royal Swazi National Airways Corp Ltd, the National Airline of Swaziland; or Zianian Chuen – Chinese Ship Classification Society.

ZE Líneas Aéreas Azteca (SA de CV) – IATA associate Member Airline.

Zener Barrier Safety interface connector between an electronic cargo handling system and hazardous deck area on tanker.

Zero base budgeting A method of budgeting whereby all activities are re-evaluated each time a budget is formulated. Each functional budget starts with the assumption that the function does not exist and is at zero cost. Increments of cost are compared with increments of benefit, culminating in the planned maximum benefit for a given budgeted cost.

Zetu-Coupon bonds Bonds issued at a heavy discount and paying no interest but redeemable at par at a future date. They are designed to exploit the different rates of tax on interest income and capital gains.

Zero defect A situation whereby no deficiencies arise and a perfect situation obtains. For example, a total of 100 items – all coloured black – arrived in Japan on 1st January which conforms precisely with the contract of sale and therefore has a zero defect.

Zero failure rate The aim of every exporter/distributor/forwarder to have a nil failure rate in the execution of the contract which could involve obviating, for example, late delivery, damaged merchandise, incorrect completion of documents and wrong specification.

Zero flight time A situation emerging in flight simulation embracing the use of simulation for training airline pilots.

Zero-for-zero The complete elimination of tariffs and non-tariff measures on certain products. WTO term.

Zero option No option/alternative available.

Zero or no cost claims Personal Lines – claims which are settled or closed without indemnity payment to the Insured or Claimant except for possible service fees. All other areas fall in line with the DOT definition that a no cost claim or zero claim is where no payment has been made.

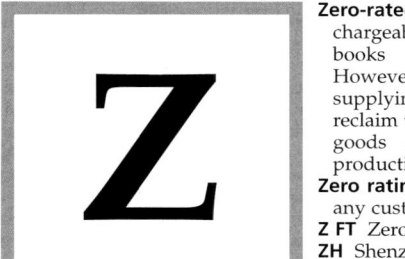

Zero-rated Goods on which no VAT is chargeable to the general public (e.g. books and children's clothing). However, VAT – registered businesses supplying zero-rated goods may still reclaim the VAT they have paid for any goods and services used in the production process.

Zero rating Goods which do not attract any customs duty.

Z FT Zero Flight Time.

ZH Shenzhen Airlines Co Ltd – IATA associate Member Airline.

Zimbabwe dollar The currency of Zimbabwe. See Appendix 'C'.

Zimbabwe dollar invoice An invoice payable in the Zimbabwe currency of Zimbabwe dollar relative to an export sales invoice.

Zim Israel Navigation Co Ltd (ZIM) A major liner shipowner and container operator and registered in Israel.

ZIPP Zonal improvement project plan.

Zirconium suspended in a flammable liquid It consists of very finely divided metallic zirconium which is usually suspended in some highly volatile and flammable liquid. If spilled, the material is liable to self-ignition and therefore can be shipped only in very limited quantities when specially packed. A term associated with dangerous classified cargo shipments.

ZL Hazelton Airlines – an IATA Airline Member; or Affretair (PVT) – an IATA Airline Member.

Zloty The currency of Poland. See Appendix 'C'.

Zloty invoice An invoice payable in the Polish currency of Zloty relative to an export sales contract.

Zone charge A pre-shipment or post shipment charge in the ocean carriers tariff for collection or delivery haulage of LCL or FCL traffic. Sometimes termed 'carriers haulage'; or the basis of an express parcel/small package tariff system whereby the territories served are split up into zones and the rates per package applied on that basis. The zonal charging system with minor variations is widely used in transport.

Zone rate See zone charge.

ZQ Ansett New Zealand – an IATA Member Airline.

ZS AZZURRAair SpA – IATA Member Airline.

ZT Zone Time.

ZU Helios Airways Ltd – IATA Airline Member.

ZY ADA Air – an IATA Member Airline.

Appendix A

EXPORTMASTER OPERATIONS FLOWCHART

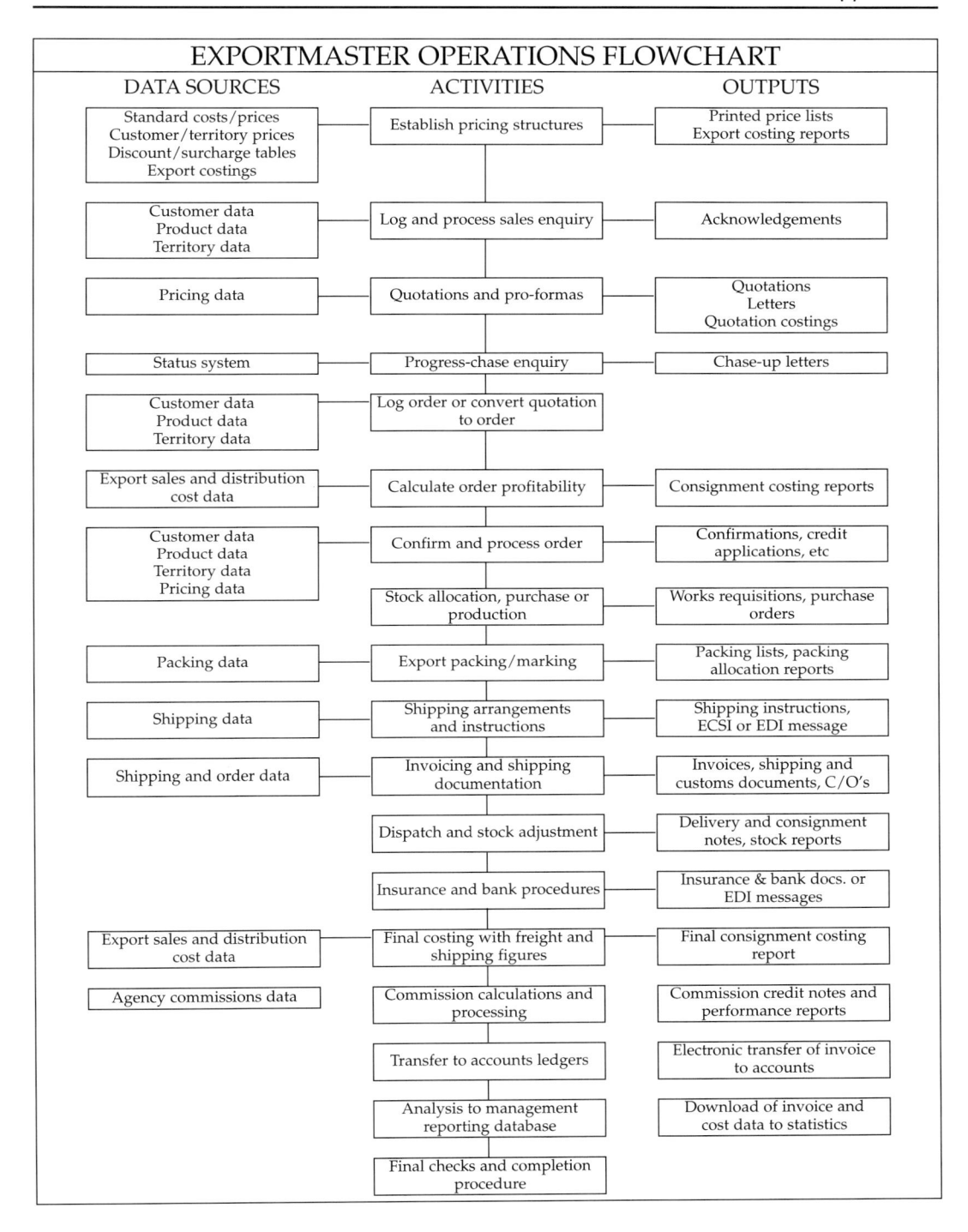

DATA SOURCES	ACTIVITIES	OUTPUTS
Standard costs/prices Customer/territory prices Discount/surcharge tables Export costings	Establish pricing structures	Printed price lists Export costing reports
Customer data Product data Territory data	Log and process sales enquiry	Acknowledgements
Pricing data	Quotations and pro-formas	Quotations Letters Quotation costings
Status system	Progress-chase enquiry	Chase-up letters
Customer data Product data Territory data	Log order or convert quotation to order	
Export sales and distribution cost data	Calculate order profitability	Consignment costing reports
Customer data Product data Territory data Pricing data	Confirm and process order	Confirmations, credit applications, etc
	Stock allocation, purchase or production	Works requisitions, purchase orders
Packing data	Export packing/marking	Packing lists, packing allocation reports
Shipping data	Shipping arrangements and instructions	Shipping instructions, ECSI or EDI message
Shipping and order data	Invoicing and shipping documentation	Invoices, shipping and customs documents, C/O's
	Dispatch and stock adjustment	Delivery and consignment notes, stock reports
	Insurance and bank procedures	Insurance & bank docs. or EDI messages
Export sales and distribution cost data	Final costing with freight and shipping figures	Final consignment costing report
Agency commissions data	Commission calculations and processing	Commission credit notes and performance reports
	Transfer to accounts ledgers	Electronic transfer of invoice to accounts
	Analysis to management reporting database	Download of invoice and cost data to statistics
	Final checks and completion procedure	

1. Progressing Export Consignment – Diagram A

Reproduced by kind permission of Export Master Ltd
33 St Peter's Street
South Croydon CR2 7DG
Tel +44(0)20 8681 2321
Fax +44(0)20 8687 1816

E-mail info@exportmaster.co.uk
Web-site www.exportmaster.co.uk

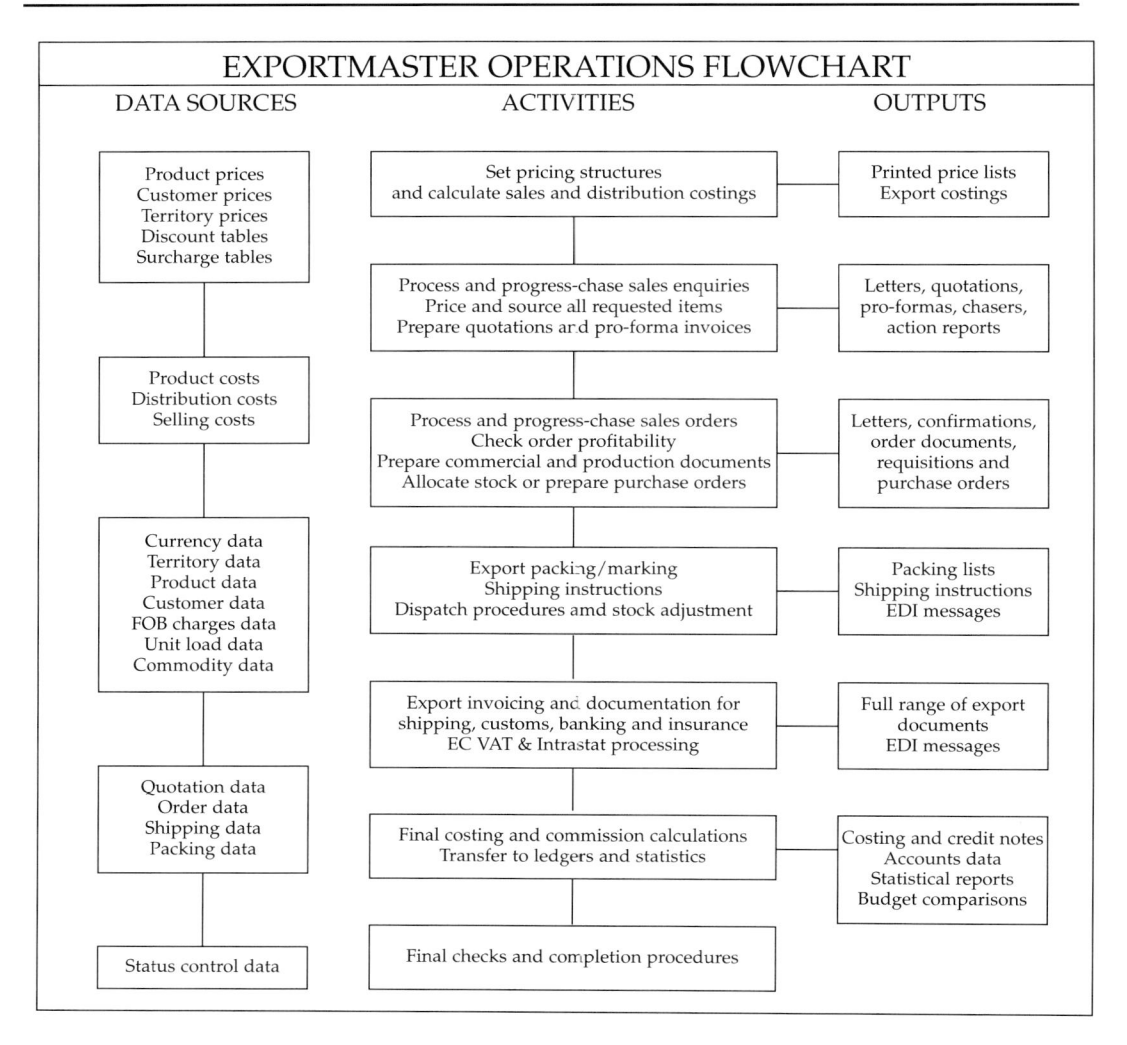

Appendix A

EXPORTMASTER OPERATIONS FLOWCHART

DATA SOURCES	ACTIVITIES	OUTPUTS
Product prices Customer prices Territory prices Discount tables Surcharge tables	Set pricing structures and calculate sales and distribution costings	Printed price lists Export costings
	Process and progress-chase sales enquiries Price and source all requested items Prepare quotations and pro-forma invoices	Letters, quotations, pro-formas, chasers, action reports
Product costs Distribution costs Selling costs	Process and progress-chase sales orders Check order profitability Prepare commercial and production documents Allocate stock or prepare purchase orders	Letters, confirmations, order documents, requisitions and purchase orders
Currency data Territory data Product data Customer data FOB charges data Unit load data Commodity data	Export packing/marking Shipping instructions Dispatch procedures amd stock adjustment	Packing lists Shipping instructions EDI messages
	Export invoicing and documentation for shipping, customs, banking and insurance EC VAT & Intrastat processing	Full range of export documents EDI messages
Quotation data Order data Shipping data Packing data	Final costing and commission calculations Transfer to ledgers and statistics	Costing and credit notes Accounts data Statistical reports Budget comparisons
Status control data	Final checks and completion procedures	

2. Progressing Export Consignment – Diagram B

Reproduced by kind permission of Export Master Ltd
33 St Peter's Street
South Croydon CR2 7DG
Tel +44(0)20 8681 2321
Fax +44(0)20 8687 1816

E-mail info@exportmaster.co.uk
Web-site www.exportmaster.co.uk

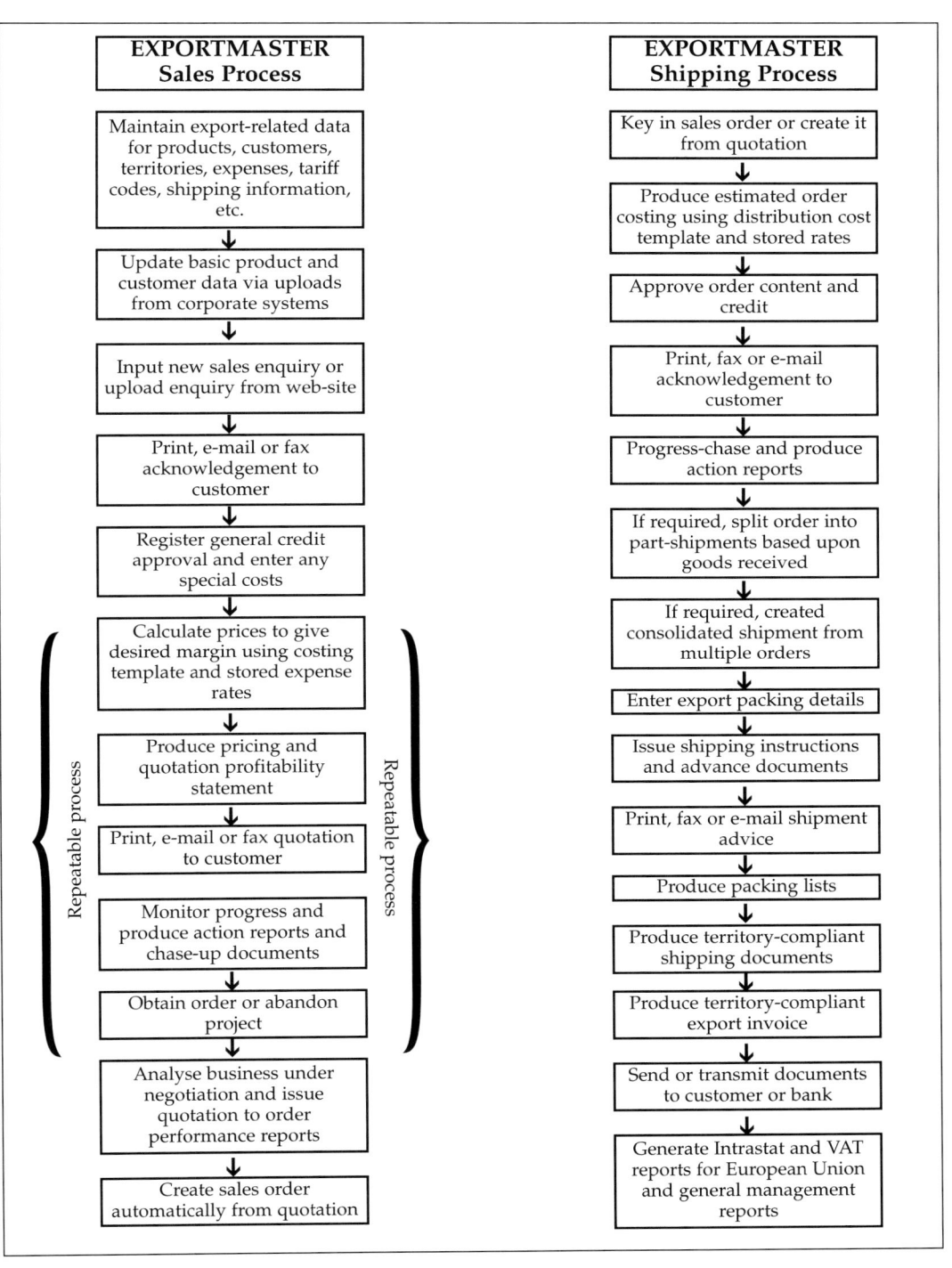

Appendix A

EXPORTMASTER Sales Process

- Maintain export-related data for products, customers, territories, expenses, tariff codes, shipping information, etc.
- ↓
- Update basic product and customer data via uploads from corporate systems
- ↓
- Input new sales enquiry or upload enquiry from web-site
- ↓
- Print, e-mail or fax acknowledgement to customer
- ↓
- Register general credit approval and enter any special costs
- ↓
- *(Repeatable process begins)*
- Calculate prices to give desired margin using costing template and stored expense rates
- ↓
- Produce pricing and quotation profitability statement
- ↓
- Print, e-mail or fax quotation to customer
- ↓
- Monitor progress and produce action reports and chase-up documents
- ↓
- Obtain order or abandon project
- *(Repeatable process ends)*
- ↓
- Analyse business under negotiation and issue quotation to order performance reports
- ↓
- Create sales order automatically from quotation

EXPORTMASTER Shipping Process

- Key in sales order or create it from quotation
- ↓
- Produce estimated order costing using distribution cost template and stored rates
- ↓
- Approve order content and credit
- ↓
- Print, fax or e-mail acknowledgement to customer
- ↓
- Progress-chase and produce action reports
- ↓
- If required, split order into part-shipments based upon goods received
- ↓
- If required, created consolidated shipment from multiple orders
- ↓
- Enter export packing details
- ↓
- Issue shipping instructions and advance documents
- ↓
- Print, fax or e-mail shipment advice
- ↓
- Produce packing lists
- ↓
- Produce territory-compliant shipping documents
- ↓
- Produce territory-compliant export invoice
- ↓
- Send or transmit documents to customer or bank
- ↓
- Generate Intrastat and VAT reports for European Union and general management reports

3. Sales Process and Shipping Process – Diagram C

Reproduced by kind permission of Export Master Ltd
33 St Peter's Street
South Croydon CR2 7DG
Tel +44(0)20 8681 2321
Fax +44(0)20 8687 1816
E-mail info@exportmaster.co.uk
Web-site www.exportmaster.co.uk

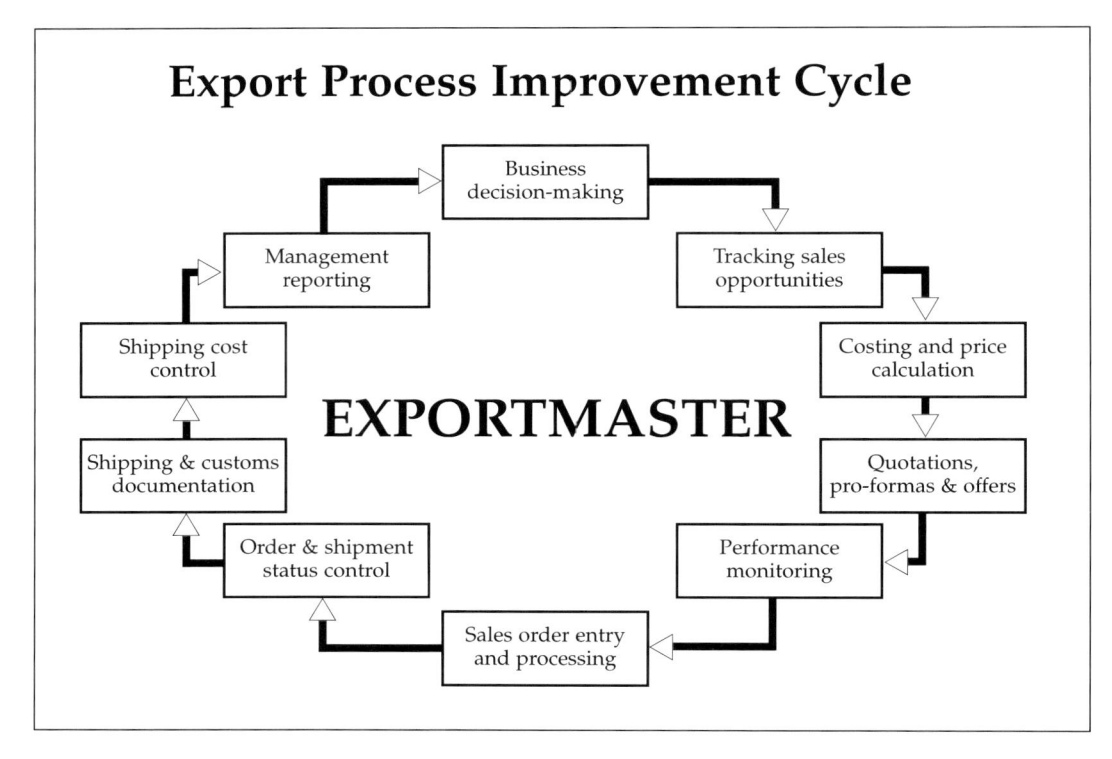

Appendix A

Export Process Improvement Cycle

EXPORTMASTER

Cycle (clockwise from top):
- Business decision-making
- Tracking sales opportunities
- Costing and price calculation
- Quotations, pro-formas & offers
- Performance monitoring
- Sales order entry and processing
- Order & shipment status control
- Shipping & customs documentation
- Shipping cost control
- Management reporting

4. Export Process Improvement Cycle – Diagram D

Reproduced by kind permission of Export Master Ltd
33 St Peter's Street
South Croydon CR2 7DG
Tel +44(0)20 8681 2321
Fax +44(0)20 8687 1816

E-mail info@exportmaster.co.uk
Web-site www.exportmaster.co.uk

Appendix B

Major ports of the World and their location

A

Aabenraa – Denmark
Aalborg – Denmark
Aalesund – Norway
Aarhus – Denmark
Abidjan – Ivory Coast
Abu al bu Khoosh – United Arab Emirates
Acajutla – El Salvador
Adelaide – Australia
Aden – Yemen, Republic of
Affarsverken – Sweden
Agioi Theodoroi – Greece
Ain Sukhna – Egypt
Airlie Island – Australia
Akita – Japan
Akrotiri – Cyprus
Al Muajjiz Terminal – Saudi Arabia
Al Shaheen Terminal – Qatar
Albany (NY) – USA
Albany (WA) – Australia
Alexandria – Egypt
Algeciras – Spain
Algiers – Algeria
Aliaga – Turkey
Alicante – Spain
Alicante Bay
Aliveri – Greece
Alliance – USA
Alpha Zone
Altona – Australia
Ambarli – Turkey
Ambes – France
Amesville – USA
Amorco – USA
Amoy – China, People's Republic of
Amsterdam – Netherlands
Amuay Bay – Paraguana Refinery Centre – Venezuela
Anacortes – USA
Anchorage – USA
Ancona – Italy
Anegasaki – Japan
Anewa Bay – Papua New Guinea
Angle Bay – United Kingdom
Ango Ango – Zaire
Angra Dos Reis – Brazil
Annaba – Algeria
Anoa Natuna – Indonesia
Antan Terminal – Nigeria
Antifer – France
Antigua – Caribbean Islands
Antilla – Cuba
Antofagasta – Chile
Antsiranana – Madagascar
Antwerp – Belgium
Aokata – Japan
Apapa – Nigeria
Apia – Pacific Ocean Islands
Aqaba – Jordan
Aransas Pass – USA
Ardjuna – Indonesia
Arendal – Norway
Argyll Field – North Sea Installations
Arica – Chile
Arroyo (PR) – Puerto Rico
Aruba – Netherlands Antilles
Arun – Indonesia
Arzew – Algeria
Ash Shihr – Yemen, Republic of
Ashdod – Israel
Ashkelon – Israel
Ashtart – Tunisia
Aspra Spitia – Greece
Aspropyrgos – Greece
Assab – Ethiopia
Atlas Cove – Nigeria
Atreco – USA
Atsumi – Japan
Auckland – New Zealand
Augusta – Italy
Auk Field – North Sea Installations
Aviles – Spain
Avon (Cal) – USA
Avondale – USA
Avonmouth – United Kingdom
Axelsvik – Sweden
Azzawia – Libya

B

Bach ho – Vietnam
Bahia Blanca – Argentina
Bahrain – Bahrain
Bahregansar – Iran
Baie Comeau – Canada
Bajo Grande – Venezuela
Bakar – Croatia
Balao – Ecuador
Balboa – Panama, Republic of
Balikpapan – Indonesia
Baltimore – USA
Banana – Zaire
Bandar Abbas – Iran
Bandar Bushire – Iran
Bandar Imam Khomeini – Iran
Bandar Mahshahr – Iran
Bangkok – Thailand
Banias – Syria
Banjul – Gambia
Bar – Yugoslavia
Barbados – Caribbean Islands
Barbers Point – Hawaiian Islands (USA)
Barcelona (Spain) – Spain
Barrow Island – Australia
Barry – United Kingdom
Bataan – Philippines
Batangas – Philippines
Batangas Bay – Philippines
Bathurst (Gambia) – Gambia
Baton Rouge – USA
Batum – USSR (Former)
Batumi – USSR (Former)
Bayonne (France) – France
Bayonne (NJ) – USA
Bayport – USA
Baytown – USA
Bayway (NJ) – USA
Beaufort – USA
Beaumont – USA
Bec D'Ambes – France
Beira – Mocambique
Beirut – Lebanon
Bejaia – Algeria
Belawan – Indonesia
Belem (Brazil) – Brazil
Belfast (UK) – United Kingdom
Bell Bay – Australia
Bellingham – USA
Benghazi – Libya
Benicia – USA
Berbera – Somali Republic
Bergen – Norway
Bergs Oljehamn – Sweden
Berville – France
Beryl Field – North Sea Installations
Bethioua – Algeria
Beyrouth – Lebanon
Big Stone Beach
Bilbao – Spain
Bima Terminal – Indonesia
Bintulu Terminal – Malaysia
Birch Field – North Sea Installations
Birkenhead (S Aust) – Australia
Bizerta – Tunisia
Blang Lancang – Indonesia
Blexen – Germany
Bluff – New Zealand
Bodo – Norway
Bohai BZ Terminals – China, People's Republic of
Bonaire – Netherlands Antilles
Bone – Algeria
Bonny – Nigeria
Bonny Anchorage – Nigeria
Bordeaux – France
Borga – Finland
Borsele – Netherlands
Boston (Mass) – USA
Botany Bay – Australia
Botas (Ceyhan) Terminal – Turkey
Botlek – Netherlands
Botwood – Canada
Bougainville Island – Papua New Guinea
Bougie – Algeria
Bourgas – Bulgaria
Bouri Terminal – Libya
Brake – Germany
Brass Terminal – Nigeria
Bravo Zone
Brayton Point – USA
Bremen – Germany
Bremerhaven – Germany
Bremerton – USA
Brent Field – North Sea Installations
Brest – France
Bridgeport – USA
Bridgetown – Caribbean Islands
Brindisi – Italy
Brisbane – Australia

Appendix B

Bristol – United Kingdom
Brofjorden – Sweden
Broome – Australia
Brownsville – USA
Brunsbuttel – Germany
Buchanan – Liberia
Buchan Field – North Sea Installations
Bucksport – USA
Budge Budge – India
Buenaventura – Colombia
Buenos Aires – Argentina
Bufadero – Cuba
Bula – Indonesia
Bullen Bay – Netherlands Antilles
Bundaberg – Australia
Burnaby – Canada
Burnie – Australia
BZ 28 Terminal – China, People's Republic of
BZ 34 Terminal – China, People's Republic of

Cabinda – Angola
Cabo San Antonio
Cadiz – Spain
Caen – France
Cagliari – Italy
Cairns – Australia
Calais – France
Calcutta – India
Caldera (Chile) – Chile
Caleta Cordova – Argentina
Caleta Olivares – Argentina
Caleta Olivia – Argentina
Callao – Peru
Camar Marine Terminal – Indonesia
Canaport – Canada
Canvey Island – United Kingdom
Cap Lomboh – Cameroun
Cap Lopez – Gabon
Cape Canaveral – USA
Cape Lambert – Australia
Cape Lopez – Gabon
Cape Town – South Africa
Cardiff – United Kingdom
Caripito – Venezuela
Cartagena (Col) – Colombia
Cartagena (Spain) – Spain
Carteret (NJ) – USA
Carthagena (Spain) – Spain
Casablanca – Morocco
Castellon – Spain
Catania – Italy
Cayo Arcas – Mexico
Cebu – Philippines
Cekmece – Turkey
Ceuta – Spanish Morocco
Challis Venture – Australia
Chalmette – USA
Champ Moudi – Cameroun
Chanaral – Chile
Chandler – Canada

Charleston (SC) – USA
Charlie Zone (See Cabo San Antonia)
Charlottetown (PEI) – Canada
Chelsea – USA
Cherbourg – France
Cherry Point – USA
Chiba – Japan
Chimu Wan – Japan
Chinwangtao – China, People's Republic of
Chita – Japan
Chittagong – Bangladesh
Cienfuegos – Cuba
Cilacap – Indonesia
Cinta – Indonesia
Ciudad Madero – Mexico
Civitavecchia – Italy
Cliffe – United Kingdom
Clifton Pier – Bahamas
Cloghan Point – United Kingdom
Coatzacoalcos – Mexico
Cochin – India
Colombo – Sri Lanka
Come By Chance – Canada
Comeau Bay – Canada
Common Zone
Comodoro Rivadavia – Argentina
Conakry – Guinea Republic
Conchan – Peru
Constantza – Romania
Convent – USA
Coos Bay – USA
Copenhagen – Denmark
Cordemais – France
Corinto – Nicaragua
Corner Brook – Canada
Corpus Christi – USA
Corunna – Spain
Coryton – United Kingdom
Cotonou – Benin, People's Republic of
Cousins Island – USA
Covenass – Colombia
Craney Island – USA
Cristobal – Panama, Republic of
Crofton – Canada
Curacao – Netherlands Antilles
Curacao (Emmastad Terminal, Willemstad) – Netherlands Antilles

Dai Hung (STS Load)
Dai Hung (Tandem Load)
Dae San – Korea, South
Dagenham – United Kingdom
Dairen – China, People's Republic of
Dakar – Senegal
Dalhousie (NB) – Canada
Dalian – China, People's Republic of
Dampier – Australia
Danang – Vietnam
Dar-es-Salaam – Tanzania
Darwin – Australia
Das Island – United Arab Emirates

Deer Park – USA
Delair – USA
Delaware City – USA
Delta Yesil Terminal – Turkey
Delta Zone
Derince – Turkey
Destrehan – USA
Devonport (Tasmania) – Australia
Dhekelia – Cyprus
Diego Garcia – Indian Ocean Islands
Diego Suarez – Madagascar
Djakarta – Indonesia
Djeno – Congo Republic
Djibouti – Djibouti
Domsjo – Sweden
Donges – France
Doornzelle – Belgium
Dora – Lebanon
Dordrecht – Netherlands
Dortyol – Turkey
Dos Bocas – Mexico
Douala – Cameroun
Drapetzona Bay – Greece
Drift River – USA
Duala – Cameroun
Dublin – Irish Republic
Dulang Marine Terminal – Malaysia
Dumai – Indonesia
Dundee – United Kingdom
Dunedin – New Zealand
Dunkirk – France
Durban – South Africa
Durres – Albania
Dutch Harbour – USA

Eagle Point – USA
East Braintree – USA
East London – South Africa
Eastham – United Kingdom
Ebeye – Pacific Ocean Islands
Eden – Australia
Edmonds – USA
Egersund – Norway
Eilat – Israel
Einswarden – Germany
Ekeberg – Norway
El Chaure – Venezuela
El Palito – Venezuela
El Sequndo – USA
El Tablazo – Venezuela
Eleusis – Greece
Ellesmere Port – United Kingdom
Emden – Germany
Emmastad Terminal – Netherlands Antilles
Empire – USA
Ensted – Denmark
Erawan – Thailand
Ertvelde – Belgium
Es Sider – Libya
Esbjerg – Denmark
Escombreras – Spain
Escravos – Nigeria

Appendix B

Esmeraldas – Ecuador
Esperance – Australia
Espiritu Santo Island – Pacific Ocean Islands
Essungo Terminal – Angola
Essvik – Sweden
Etajima – Japan
Europoort Rotterdam – Netherlands
Everett (Mass) – USA
Everett (Wash) – USA

Fagerstrand – Norway
Fagervik – Sweden
Falconara – Italy
Fall River – USA
Falmouth (UK) – United Kingdom
Farge – Germany
Fateh Terminal – United Arab Emirates
Fawley – United Kingdom
Felixstowe – United Kingdom
Ferndale – USA
Ferrol – Spain
Finnart – United Kingdom
Fisher Island – USA
Fiumicino – Itlay
Flaxenvik – Sweden
Floating Storage Unit "Tuma" – Nigeria
Flotta Terminal – United Kingdom
Flushing – Netherlands
Forcados – Nigeria
Fort de France – Martinique
Fort Miflin – USA
Fortaleza – Brazil
Fos – France
Foynes Harbour – Irish Republic
Fredericia – Denmark
Frederikshavn – Denmark
Fredrikstad – Norway
Freeport (Bahamas) – Bahamas
Freeport (Texas) – USA
Freetown – Sierra Leone
Fremantle – Australia
Fulmar Field – North Sea Installations
Funakawa – Japan
Funchal – Madeira
Fuzhou – China, People's Republic of

Gabes – Tunisia
Gaeta – Italy
Galena Park – USA
Galeota Point – Trinidad & Tobago
Galveston – USA
Gamba (Gabon) – Gabon
Gan Island – Maldive Islands

Garyville – USA
Gashaga – Sweden
Gastgivarehagen – Sweden
Gavle – Sweden
Gdansk – Poland
Gdynia – Poland
Geelong – Australia
Gefle – Sweden
Gela – Italy
Gemlik – Turkey
Genes – Italy
Genoa – Italy
Georgetown (Gd Cayman) – Caribbean Islands
Geraldton – Australia
Ghent – Belgium
Gibraltar – Gibraltar
Gijon – Spain
Gisborne – New Zealand
Gladstone (Queensland) – Australia
Goa – India
Goi – Japan
Golfo San Matias
Gonfreville – France
Good Hope (La) – USA
Goteburg – Sweden
Gothenburg – Sweden
Gove – Australia
Grangemouth – United kingdom
Gravelines – France
Gravenchon – France
Grays – United Kingdom
Gretna – USA
Griffin Venture – Australia
Grimsby – United Kingdom
Groote Eylandt – Australia
Groton – USA
Guam – Pacific Ocean Islands – USA
Guanta – Venezuela
Guantanamo Bay (US Base) – Cuba
Guaraguao Terminal – Venezuela
Guayama – Puerto Rico
Guayanilla – Puerto Rico
Guayaquil – Ecuador
Guaymas – Mexico
Guiria – Venezuela
Gulf of Izmit – Turkey
Gulfhavn – Denmark
Gulfmex No 1
Gullfaks – North Sea Installations

Hafnarfjordur – Iceland
Haifa – Israel
Haiphong – Vietnam
Hakodate – Japan
Halat al Mubarras – United Arab Emirates
Haldia – India
Halifax – Canada
Hallstavik – Sweden
Halmstad – Sweden
Halul Island – Qatar
Halvorshavn – Norway

Hamble – United Kingdom
Hamburg – Germany
Hamilton (Ont) – Canada
Hamina – Finland
Hammerfest – Norway
Hamriyah – United Arab Emirates
Haramidere – Turkey
Harbor Island – USA
Harburg – Germany
Harmac – Canada
Harstad – Norway
Hastings – Australia
Hastings-on-Hudson – USA
Haugesund – Norway
Havana – Cuba
Havre – France
Haydarpasa – Turkey
Hazira – India
Helsingborg – Sweden
Helsinki – Finland
Hemixem – Belgium
Hernosand – Sweden
Heroya – Norway
Hilo – Hawaiian Islands (USA)
Himeji – Japan
Hjartholmen – Sweden
Ho Chi Minh City – Vietnam
Hobart – Australia
Hodeidah – Yemen, Republic of
Hoheschaar – Germany
Holehaven – United Kingdom
Holmsund – Sweden
Holyrood – Canada
Honfleur – France
Hong Kong – Hong Kong
Honiara – Pacific Ocean Islands
Honningsvaag – Norway
Honolulu – Hawaiian Islands (USA)
Hopa – Turkey
Horta – Azores (Portugal)
Hound Point – United Kingdom
Houston – USA
Huangpu – China, People's Republic of
Huasco – Chile
Hudiksvall – Sweden
Huelva – Spain
Hulaylah Terminal – United Arab Emirates
Hull – United Kingdom
Humber River – United Kingdom
Huntingdon Beach – USA
Husum – Sweden
Hvalfjordur – Iceland

Ilha Grande Bay – Brazil
Ilichevsk – USSR (Former)
Iligan – Philippines
Ilo – Peru
Iloilo – Philippines
Immingham (East and West Jetties) – United Kingdom
Ince Terminal – United Kingdom

Appendix B

Inchon – Korea, South
Ingleside – USA
Inkoo – Finland
Inshore Bonny – Nigeria
Ioco – Canada
Iquique – Chile
Isdemir Payas – Turkey
Iskenderun – Turkey
Isle of Grain – United Kingdom
Istanbul – Turkey
Iwakuni – Japan
Izmir – Turkey
Izmit – Turkey

Jabiru – Australia
Jacksonville – USA
Jacobstad – Finland
Jakarta – Indonesia
Jamestown Anchorage
Jarrow – United Kingdom
Jebel Ali – United Arab Emirates
Jebel Dhanna – United Arab Emirates
Jeddah – Saudi Arabia
Jesselton – Malaysia
Jobos – Puerto Rico
Johannedal – Sweden
Jose Ignacio Terminal – Uruguay
Jose Monobuoy – Venezuela
Jose Platform – TAECJ – Venezuela
Jose Terminal – Venezuela
Juaymah – Saudi Arabia
Jubail – Saudi Arabia
Jurong – Singapore

Kainan – Japan
Kakap Marine Terminal – Indonesia
Kalmar – Sweden
Kalundborg – Denmark
Kamigoto – Japan
Kamsar – Guinea Republic
Kandla – India
Kanmon – Japan
Kaohsiung – Taiwan
Karachi – Pakistan
Kardeljevo (Ploce) – Croatia
Karlshamn – Sweden
Karlskrona – Sweden
Karsto – Norway
Kartal – Turkey
Kashima – Japan
Kasim – Indonesia
Kavalla – Greece
Kawasaki – Japan
Kearney – USA
Keelung – Taiwan
Kemi – Finland
Kenai – USA
Keoje – Korea, South

Kerteh – Malaysia
Kharg Island – Iran
Kiire – Japan
Kikuma – Japan
Killingholme – United Kingdom
Kimuwan – Japan
Kin (See Chimu Wan) – Japan
King Fahd Industrial Port (Yanbu) – Saudi Arabia
Kingsnorth – United Kingdom
Kingston (Jamaica) – Jamaica
Kingston (NY) – USA
Kingston (Ont) – Canada
Kinwan – Japan
Kittiwake Field – North Sea Installations
Klaipeda – USSR (Former)
Kobe – Japan
Kokura – Japan
Kole Terminal – Cameroun
Koper – Slovenia
Kopmanholmen – Sweden
Koshiba – Japan
Kota Kinabalu – Malaysia
Kotka – Finland
Kramfors – Sweden
Kristiansand S – Norway
Kristinestad – Finland
Kuala Beukah – Indonesia
Kuching – Malaysia
Kudamatsu – Japan
Kumul Marine Terminal – Papua New Guinea
Kurnell – Australia
Kvarnholmen – Sweden
Kwajalein – Pacific Ocean Islands
Kwinana – Australia

La Corunna – Spain
La Goulette – Tunisia
La Libertad (Ecuador) – Ecuador
La Nouvelle – France
La Pallice – France
La Pampilla – Peru
La Plata – Argentina
L Plata Roads
La Romana – Dominican Republic
La Salina – Venezuela
La Skhirra – Tunisia
La Spezia – Italy
Labuan – Malaysia
Lae (New Guinea) – Papua New Guinea
Lagos (Nigeria) – Nigeria
Lake Charles – USA
Lalang – Indonesia
Lan Shui Terminal – China, People's Republic of
Larkollen – Norway
Larnaca – Cyprus
Larne – United Kingdom
Las Minas – Panama, Republic of
Las Palmas – Canary Islands

Las Piedras Paraguana Refinery Centre – Venezuela
Lattakia – Syria
Laugarnes – Iceland
Lavan Island – Iran
Lavera – France
Lavrion – Greece
Lawi Lawi – Indonesia
Leghorn – Italy
Le Havre – France
Leixoes – Portugal
Leningrad – USSR (Former)
Lerwick – United Kingdom
Levuka – Fiji
Lewisporte – Canada
Limerick – Irish Republic
Limhamn – Sweden
Linden – USA
Linnton – USA
Lisbon – Portugal
Littlebrook – United Kingdom
Liuhua Terminal – China, People's Republic of
Liverpool Bay – United Kingdom
Livorno – Italy
Lobit – Angola
Loire River – France
Lokele – Cameroun
Lome – Togo
London – United Kingdom
Londonderry – United Kingdom
Long Beach – USA
Longview – USA
LOOP Terminal – USA
Lorient – France
Los Angeles – USA
Loudden – Sweden
Lourenco Marques – Mozambique
Luanda – Angola
Lucina Gabon
Lufeng – China, People's Republic of
Lulea – Sweden
Lutong – Malaysia
Lyme Bay – United Kingdom
Lysaker – Norway
Lyttelton – New Zealand

Macassar – Indonesia
Maceio – Brazil
Mackay – Australia
Madang – Papua New Guinea
Madras – India
Madre de Deus – Brazil
Madura Terminal – Indonesia
Magnisi – Italy
Magpetco – USA
Mahajanga – Madagascar
Mai Liao – Taiwan
Majunga – Madagascar
Malacca -Malaysia
Malaga – Spain
Malmo – Sweden
Malongo Terminal – Angola

Appendix B

Malta – Malta
Mamonal – Colombia
Manaus – Brazil
Manchester (Wash) – USA
Manchester Ship Canal – United Kingdom
Mandal – Norway
Mandalay Beach (Cal) – USA
Manfredonia – Italy
Manila – Philippines
Mantua – USA
Manus Island – Papua New Guinea
Manzanillo (Mexico) – Mexico
Map Ta Phut – Thailand
Maputo – Mocambique
Marcus Hook – USA
Mariel – Cuba
Marifu (See Iwakuni) – Japan
Marrero – USA
Marsa el Brega – Libya
Marsa el Hariga – Libya
Marsaxlokk Bay – Malta
Marsden Point – New Zealand
Martinez – USA
Massawa – Ethiopia
Matanzas (Cuba) – Cuba
Matias de Galvez – Guatemala
Matsuyama – Japan
Mayaguez – Puerto Rico
M'Bya – Gabon
Medway River – United Kingdom
Megara – Greece
Melbourne – Australia
Melilli – Italy
Melville – USA
Meraux – USA
Mersin – Turkey
Mesaieed – Qatar
Miami – USA
Middlesbrough – United Kingdom
Milazzo – Italy
Milford Haven – United Kingdom
Mina Abdulla – Kuwait
Mina al Ahmadi – Kuwait
Mina al Fahal – Oman
Mina Khalid – United Arab Emirates
Mina Saqr – United Arab Emirates
Mina Saud – Kuwait
Mina Zayed – United Arab Emirates
Miri – Malaysia
Mizushima – Japan
Mobile – USA
Mocambique – Mocambique
Mogadiscio – Somali Republic
Mohammedia – Morocco
Moji – Japan
Mollendo – Peru
Mombassa – Kenya
Monfalcone – Italy
Mongstad – Norway
Moni – Cyprus
Monrovia – Lliberia
Montego Bay – Jamaica
Montevideo – Uruguay
Montreal – Canada
Montrose Field – North Sea Installations
Morehead City – USA
Mormugao – India

Moron – Venezuela
Mossel Bay – South Africa
Moudi – Cameroun
Mount Airy – USA
Mount Maunganui – New Zealand
Mozambique – Mocambique
Mtwara – Tanzania
Muanda Terminal – Zaire
Mubarek – United Arab Emirates
Mukalla – Yemen, Republic of
Mukilteo – USA
Mumbai – India
Munkedal – Sweden
Murmansk – USSR (Former)
Muroran – Japan
Mutsure – Japan

Naantali – Finland
Nacal – Mocambique
Nagasaki – Japan
Nagoya – Japan
Nakagusuku – Japan
Nakhodka – USSR (Former)
Namikata Terminal – Japan
Namsos – Norway
Nan Hai Fa Xian – China, People's Republic of
Nanaimo – Canada
Nanchital – Mexico
Nanjing – China, People's Republic of
Nanking – China, People's Republic of
Nantes – France
Nantong – China, People's Republic of
Napier – New Zealand
Naples – Italy
Nassau – Bahamas
Natal (Brazil) – Brazil
Nederland (Texas) – USA
Negishi – Japan
Nelson (NZ) – New Zealand
Nemrut Bay – Turkey
New Bedford (Mass) – USA
New Haven (Conn) – USA
New Lond (Conn) – USA
New Mangalore – India
New Orleans – USA
New Plymouth – New Zealand
New Sendai – Japan
New York – USA
Newark – USA
Newburgh – USA
Newcastle (NSW) – Australia
Newcastle (UK) – United Kingdom
Newington (NH) – USA
Newport News – USA
Nicaro – Cuba
Nigg Terminal – United Kingdom
Niigata – Japan
Nikishka – USA
Nikiski – USA

Ningpo – China, People's Republic of
Nishihara – Japan
Nonoc Island – Philippines
Norco – USA
Nordenham – Germany
Norfolk (Va) – USA
Norresundby – Denmark
Norrkoping – Sweden
North Tees Jetties – United Kingdom
North West Cape – Australia
North West Malacca Strait No 1
North West Malacca Strait No 2
North Weymouth (Mass) – USA
Northfleet – United Kingdom
Northport – USA
Northville – USA
Nossi Be – Madagascar
Nosy Be – Madagascar
Nouadhibou – Mauritania
Nouakchott – Mauritania
Noumea – New Caledonia
Novorossisk – USSR (Former)
Nuevitas – Cuba
Nyborg – Denmark
Nynashamn – Sweden

Oak Harbour – USA
Oak Point – USA
Oakland – USA
Obbola – Sweden
Ocho-Rios – Jamaica
Odessa – USSR (Former)
Odudu Terminal – Nigeria
Offshore Ain Sukhna
Offshore Ambrose Light
Offshore Bombay
Offshore Cochin
Offshore Corpus Christi No 1
Offshore Corpus Christi No 2
Offshore Curacao
Offshore Dakar
Offshore Delaware No 1
Offshore Freeport (Bahamas)
Offshore Freeport (Texas)
Offshore Fujairah
Offshore Galveston No 1
Offshore Galveston No 2
Offshore Great Isaac Island
Offshore Jeddah
Offshore Kaohsiung
Offshore Khor Fakkan
Offshore Malta
Offshore Mesaieed
Offshore Pascagoula No 1
Offshore Sikka
Offshore Singapore No 1
Offshore Singapore No 2
Offshore Terminal, Bonny – Nigeria
Offshore Umm Said
Ofunato – Japan
Oguendjo Terminal – Gabon
Oita – Japan

Appendix B

Okrika – Nigeria
Oleum – USA
Olympia – USA
Omisalj – Croatia
Onahama – Japan
Onsan – Korea, South
Oran – Algeria
Orange – USA
Orangemund – Namibia
Orfirisey – Iceland
Ornskoldsvik – Sweden
Ortviken – Sweden
Osaka – Japan
Oskarshamn – Sweden
Oslo – Norway
Ostermoor – Germany
Ostrand – Sweden
Ostrica – USA
Oulu – Finland
Owase – Japan
Oxelosund – Sweden

Pachi – Greece
Padang – Indonesia
Pago Pago – Pacific Ocean Islands – USA
Pajaritos – Mexico
Palanca Terminal – Angola
Palembang – Indonesia
Palermo – Italy
Palm Beach – USA
Palma (Majorca) – Spain
Pamatacual – Venezuela
Pangkalan Susu – Indonesia
Papeete – Pacific Ocean Islands
Paradip – India
Paraguana Refinery Centre – CRP – Venezuela
Paranagua – Brazil
Pasadena – USA
Pascagoula – USA
Pasir Panjang – Singapore
Pastelillo – Cuba
Patras – Greece
Pauillac – France
Paulsboro – USA
Pearl Harbour – Hawaiian Islands (USA)
Peekskill – USA
Pembroke Dock – United Kingdom
Penang – Malaysia
Pennington – Nigeria
Pennsauken – USA
Pensacola – USA
Perama – Greece
Pernis – Netherlands
Perth Amboy – USA
Pesaro – Italy
Peterhead – United Kingdom
Petit Couronne – France
Petrozuata SPM – Venezuela
Philadelphia – USA
Phillippeville – Algeria
Pilot Town – USA

Piney Point – USA
Piombino – Italy
Piraeus – Greece
Pisco – Peru
Pitea Sweden
Pladju – Indonesia
Ploce – Croatia
Plymouth (UK) – United Kingdom
Point Breeze – USA
Point Comfort – USA
Point Lisas – Trinidad and Tobago
Point Orient – USA
Point Tupper – Canada
Point Wells – USA
Pointe a Pierre – Trinidad and Tobago
Pointe a Pitre – Guadaloupe
Pointe Limboh (Cameroun) – Cameroun
Pointe Noire – Canada
Pointe Noire (Congo Rep) – Congo Republic
Ponce – Puerto Rico
Ponta Delgada – Azores (Portugal)
Pori – Finland
Poro (Luzon Island, Philippines) – Philippines
Porsgrunn – Norway
Port Alfred (Quebec) – Canada
Port Allen (Baton Rouge rates apply) – USA
Port Alma – Australia
Port Arthur (Tex) USA
Port au Prince – Haiti
Port Balongan – Indonesia
Port Blair – Indian Ocean Islands
Port Bonython – Australia
Port Botany – Australia
Port Canaveral – USA
Port Cartier – Canada
Port Costa – USA
Port Dickson – Malaysia
Port Elizabeth – South Africa
Port Esquivel – Jamaica
Port Etienne – Mauritania
Port Everglades – USA
Port Gentil – Gabon
Port Harcourt – Nigeria
Port Hawkesbury – Canada
Port Hedland – Australia
Port Hueneme – USA
Port Isabel – USA
Port Jefferson – USA
Port Jerome – France
Port Kaiser – Jamaica
Port Kelang – Malaysia
Port Kembla – Australia
Port Latta – Australia
Port Limon – Coast Rica
Port Lincoln – Australia
Port Louis (Mauritius) – Mauritius
Port Manatee – USA
Port Moody – Canada
Port Moresby – Papua New Guinea
Port Neches – USA
Port Okha – India
Port Raschid – United Arab Emirates
Port Rhoades (Jamaica) – Jamaica
Port Said – Egypt
Port Stanvac – Australia

Port Sudan – Sudan
Port Swettenahm – Malaysia
Port Tampa – USA
Port Victoria (Mahe Island) – Seychelles
Port Walcott – Australia
Portland (Me) – USA
Portland (Ore) – USA
Portland (Victoria) – Australia
Porto Corsini – Italy
Porto Foxi – Italy
Porto Marghera – Italy
Porto San Stefano – Italy
Porto Torres – Italy
Porto Vesme – Italy
Portsmouth (NH) – USA
Portsmouth (Va) – USA
Porvoo – Finland
Poti – USSR (Former)
Poughkeepsie – USA
Powell River – Canada
Pozos Colorados – Colombia
Prai – Malaysia
Preston – Cuba
Prinos Terminal – Greece
Priolo (Isab Terminal) – Italy
Priolo (Agip Terminal) – Italy
Priolo G – Italy
Providence (RI) – USA
Puerto Barrios – Guatemala
Puerto Bayovar – Peru
Puerto Cabello – Venezuela
Puerto de las Mareas – Puerto Rico
Puerto Galvan – Argentina
Puerto la Cruz – Venezuela
Puerto las Minas – Panama, Republic of
Puerto Libertad (Mexico) – Mexico
Puerto Mexico – Mexico
Puerto Miranda – Venezuela
Puerto Ordaz – Venezuela
Puerto Plata – Dominican Republic
Puerto Rosales – Argentina
Puerto Sandino – Nicaragua
Pulau Ayer Chawan – Singapore
Pulau Merimau – Singapore
Pulo Bukom – Singapore
Pulo Bunyu – Indonesia
Pulo Busing – Singapore
Pulo Sambu – Indonesia
Pulo Sebarok – Singapore
Punta Cardon – Paraguana Refinery Centre – Venezuela
Punta Cuchillo – Venezuela
Punta Fijo – Paraguana Center – Venezuela
Punta Palmas – Venezuela
Purfleet – United Kingdom
Pusan – Korea, South

Qingdao – China, People's Republic of
Qinhuangdao – China, People's Republic of

Appendix B

Qua Iboe – Nigeria
Quebec – Canada
Quincy – USA
Quinfuquena – Angola
Quintero Bay – Chile

Rabaul – Papua New Guinea
Rabigh – Saudi Arabia
Rabon Grande – Mexico
Ramshall – Sweden
Rangoon – Burma
Ras al Khafji – Saudi Arabia
Ras Bahregan – Iran
Ras Budran – Egypt
Ras es Sider – Libya
Ras Gharib – Egypt
Ras Isa Terminal – Yemen, Republic of
Ras Laffan – Qatar
Ras Lanuf – Libya
Ras Shukheir – Egypt
Ras Tanura – Saudi Arabia
Rasta – Sweden
Rauma – Finland
Ravena (NY) – USA
Ravenna – Italy
Recife (Brazil) – Brazil
Rensselaer – USA
Resaro – Sweden
Reunion – reunion
Revere – USA
Reykjavik – Iceland
Rhyahamen – Sweden
Richards Bay – South Africa
Richmond (Cal) – USA
Rijeka – Croatia
Rio de Janeiro – Brazil
Rio Grande (Brazil) – Brazil
Rio Haina – Dominican Republic
Riverhead – USA
Riverton – USA
Rocky Point – Jamaica
Rong Field – Vietnam
Ronneby – Sweden
Rosario (Philippines) – Philippines
Rosarito – Mexico
Roseton – USA
Rospo Mare Terminal – Italy
Rostock – Germany
Rotterdam Europoort – Netherlands
Rouen – France
Ruwais – United Arab Emirates

Sabine – USA
Saigon – Vietnam
Sakai – Japan
Sakaide – Japan
Sal – Cape Verde Islands

Saladin Marine Terminal – Australia
Salawati – Indonesia
Saldanha Bay – South Africa
Salem (Mass) – USA
Salina Cruz – Mexico
Salonica – Greece
Saltend – United Kingdom
Salvador (Brazil) – Brazil
Samson Point – Australia
Samsun – Turkey
San Ciprian – Spain
San Deigo – USA
San Francisco – USA
San Jose (Guatemala) – Guatemala
San Juan (PR) – Puerto Rico
San Nicholas (Peru) – Peru
San Pedro – USA
San Pedro de Macoris – Dominican Republic
San Vicente (Chile) – Chile
Sandakan – Malaysia
Sandarne – Sweden
Sandefjord – Norway
Sandviken – Sweden
Sandwich (Mass) – USA
Santa Catalina Gulf
Santa Maria Island (Azores) – Portugal
Santa Marta – Colombia
Santa Panagia Bay – Italy
Santan – Indonesia
Santander – Spain
Santiago de Cuba – Cuba
Santo – Pacific Ocean Islands
Santo Domingo – Dominican Republic
Santo Tomas de Castilla – Guatemala
Santos – Brazil
Sao Francisco do Sul – Brazil
Sao Luiz de Maranhao – Brazil
Sao Sebastiao – Brazil
Sariseki – Turkey
Sarnia – Canada
Sarroch – Italy
Savannah – USA
Savona – Italy
Scapa Flow
Seal Sands Terminals – United Kingdom
Searsport – USA
Seattle – USA
Seine Bay
Seme Terminal – Benin, People's Republic of
Sendai – Japan
Senipah – Indonesia
Seria – Brunei
Serviburnu – Turkey
Sete – France
Setubal – Portugal
Seven Islands – Canada
Sewaren – USA
Sewell's Point – USA
Sfax – Tunisia
Sha Lung – Taiwan
Shanghai – China, People's Republic of
Shekou – China, People's Republic of
Shellburn – Canada

Shellhaven – United Kingdom
Shenzhen Floating Terminal – China, People's Republic of
Shimizu – Japan
Shimonoseki – Japan
Shimotsu – Japan
Shuaiba – Kuwait
Shuidong – China, People's Republic of
Sibuko Bay – Indonesia
Sidi Kerir – Egypt
Sidon – Lebanon
Sines – Portugal
Singapore – Singapore
Siracha – Thailand
Sirri Island – Iran
Sitra (Bahrain)
Sjursoya – Norway
Skaerbaek – Denmark
Skarvikshamnen – Sweden
Skelleftehamn – Sweden
Skerjafjordur – Iceland
Skikda – Algeria
Skredsvik – Sweden
Slagen – Norway
Slemmestad – Norway
Slite – Sweden
Smiths Bluff – USA
Soderhamn – Sweden
Sodertalje – Sweden
Sofala – Mocambique
Sola – Norway
Solin – Croatia
Solvesborg – Sweden
Somerset – USA
Sorel – Canada
Sorong – Indonesia
Sourabaya – Indonesia
Sousse – Tunisia
South Brewer – USA
South Riding Point – Bahamas
South Sabine Point
South West Point
Southampton – United Kingdom
Spezia – Italy
Split – Croatia
Sriracha – Thailand
St Croiz (Virgin Islands) – USA
St Eustatius Island – Netherlands Antilles
St George's (Bermuda) – Bermuda
St George's (Grenada) – Caribbean Islands
St George Harbour – Greece
St Helens – USA
St Hoggarn – Sweden
St James (La) – USA
St John (NB) – Canada
St Johns (Antigua) – Caribbean Islands
St John's (NF) – Canada
St Louis du Rhone – France
St Lucia – Caribbean Islands
St Nazaire – France
St Petersburg – USA
St Petersburg – USSR (Former)
St Romauld – Canada
St Rose – USA
St Vincent (CV) – Cape Verde Islands

Appendix B

Stade – Germany
Stadersand – Germany
Stanlow – United Kingdom
Statjford – North Sea Installations
Stavanger – Norway
Steilene – Norway
Stenungsund – Sweden
Stephenville – Canada
Stettin – Poland
Stigsnaes – Denmark
Stockholm – Sweden
Stoney Point – Australia
Studstrup – Denmark
Stugsund – Sweden
Sture – Norway
Suape – Brazil
Subic Bay – Philippines
Suez – Egypt
Sullom Voe – United Kingdom
Sunderland – United Kingdom
Sundsvall – Sweden
Sungai Udang – Malaysia
Sungei Gerong – Indonesia
Sungei Pakning – Indonesia
Sunshine – USA
Supsa – USSR (Former)
Surabaya – Indonesia
Susak – Croatia
Susu – Indonesia
Suva – Fiji
Svano – Sweden
Svartvik – Sweden
Svolvaer – Norway
Swansea – United Kingdom
Swinemunde – Poland
Swinoujscie – Poland
Sydney (NS) – Canada
Sydney (NSW) – Australia
Szczecin – Poland

Tabangao – Philippines
Tacoma – USA
Taft – USA
Takao – Taiwan
Takoradi – Ghana
Takula – Angola
Talara – Peru
Talcahuano Bay
Talien – China, People's Republic of
Tallaboa Bay – Puerto Rico
Tallinn – USSR (Former)
Tamatave – Madagascar
Tampa – USA
Tampico – Mexico
Tandjung Priok – Indonesia
Tandjung Uban – Indonesia
Tangier – Morocco
Tanjong Pagar – Singapore
Tanjong Penjuru – Singapore
Tanjung Gerem – Indonesia
Tarafa – Cuba
Tarakan – Indonesia
Taranto – Italy

Tarbert (Irish Rep) – Irish Republic
Tarragona – Spain
Tartous – Syria
Tauranga – New Zealand
Tawau – Malaysia
Tazerka – Tunisia
Tees River – United Kingdom
Teesport – United Kingdom
Tees-Side – United Kingdom
Telegrafberget – Sweden
Teluk Semangka – Indonesia
Tema – Ghana
Teneriffe – Canary Islands
Terceira Island – Azores (Portugal)
Termini Imerese – Italy
Terneuzen – Netherlands
Tetney Oil Terminal – United Kingdom
Texas City – USA
Thameshaven – United Kingdom
Theodosia – USSR (Former)
Thessaloniki – Greece
Thistle Field – North Sea Installations
Thorshavn – Faroe Islands
Thrace – Turkey
Three Rivers – Canada
Timaru – New Zealand
Tiverton – USA
Tjilatjap – Indonesia
Toamasina – Madagascar
Tobata – Japan
Tocopilla – Chile
Tokuyama – Japan
Tokyo – Japan
Tomakomai – Japan
Toronto – Canada
Toros (Ceyhan) Terminal – Turkey
Toulon – France
Townsville – Australia
Toyama – Japan
Trabzon – Turkey
Tracey – Canada
Tramandai – Brazil
Tranmere – United Kingdom
Trelleborg – Sweden
Trieste – Italy
Trincomalee – Sri Lanka
Trinidad – Trinidad and Tobago
Tripoli (Lebanon) – Lebanon
Tromso – Norway
Trondheim – Norway
Tsamkong – China, People's Republic of
Tsingtao – China, People's Republic of
Tuapse – USSR (Former)
Tucupita – Venezuela
Tuma – Nigeria
Tumaco – Colomgia
Turku – Finland
Tuticorin – India
Tutunciftlik – Turkey
Tuxpan – Mexico
Tyne River – United Kingdom

Ube – Japan
Udang Terminal – Indonesia
Uddevalla – Sweden
Ulsan – Korea, South
Umea – Sweden
Umm al Nar – Untied Arab Emirates
Umm Said – Qatar
Urinj – Croatia
Utansjo – Sweden

Vaasa – Finland
Vadinar Terminal – India
Vado – Italy
Vaja – Sweden
Valdez – USA
Valencia – Spain
Valletta – Malta
Vallvik – Sweden
Valparaiso – Chile
Vancouver (BC) – Canada
Vancouver (Wash) – USA
Varanus Island – Australia
Varberg – Sweden
Varna – Bulgaria
Vartan – Sweden
Vassiliko – Cyprus
Vatia – Fiji
Venice – Italy
Ventspils – USSR (Former)
Vera Cruz – Mexico
Vesteras – Sweden
Vigo – Spain
Visakhapatnam – India
Vitorio – Brazil
Vlaardingen – Netherlands
Vlissingen – Netherlands
Vlore – Albania
Vuda – Fiji
Vung Tau – Vietnam

Wadi Feiran – Egypt
Wakamatsu – Japan
Wakayama – Japan
Walvis Bay – Namibia
Warri – Nigeria
Waterford – Irish Republic
Weipa – Australia
Weizhou – China, People's Republic of
Wellington – New Zealand
Westernport – Australia
Westervik – Sweden
Westville – USA
Westwego – USA

Whampoa – China, People's Republic of
Whangarei – New Zealand
Whiffen Head (NF) – Canada
Whitegate – Irish Republic
Widuri Marine Terminal – Indonesia
Wifstavarf – Sweden
Wilhelmsburg – Germany
Wilhelmshaven – Germany
Willbridge – USA
Williamstown – Australia
Wilmington (Cal) – USA
Wilmington (Del) – USA
Wilmington (NC) – USA
Wismar – Germany
Withnell Bay – Australia
Woodlands – Singapore
Wulsan – Korea, South
Wyndham – Australia

Xeros – Cyprus
Xiamen – China, People's Republic of

Yabucoa – Puerto Rico
Yanbu – Saudi Arabia
Yantai – China, People's Republic of
Yantian – China, People's Republic of
Yarmouth (Me) – USA
Yawata – Japan
Yesil Terminal – Turkey
Yokkaichi – Japan
Yokohama – Japan
Yokosuko – Japan
Yombo Oil Terminal – Congo Republic
Yonkers – USA
Yorktown – USA
Yosu – Korea, South
Yxpila – Finland

Zaafarana Terminal – Egypt
Zadar – Croatia
Zafiro Producer – Guinea Equatorial
Zahrani – Lebanon
Zaire Estuary
Zarzis – Tunisia
Zawia – Libya
Zeebrugge – Belgium
Zeit Bay – Egypt
Zhanjiang – China, People's Republic of
Zirku Island – United Arab Emirates
Zona Comun
Zueitina – Libya

World Currencies

Country **Currency**

A

Afgahanistan	One Afghani = 100 Puls
Albania	One Lek = 100 Qindarka
Algeria (Republic of)	One Algerian Dinar – 100 Centimes
Andorra	One Euro – 100 Cents
Angola	One Kwanza = 100 Lwei
Antigua	One E Carb Dollar =100 Cents
Argentina	One Argentina Peso = 100 Centavos
Australia	One Australian Dollar = 100 Cents
Austria	One Euro = 100 Cents
Azores	One Euro = 100 Cents

B

Bahamas	One Bahamian Dollar = 100 Cents
Bahrain	One Bahrain Dinar = 1000 Fils
Balearic Isles	One Euro = 100 Cents
Bangladesh	One Taka = 100 Poisha
Barbados	One Barbados Dollar = 100 Cents
Belgium	One Euro = 100 Cents
Belize	One Belize Dollar = 100 Cents
Benin (Republic of)	Communauté Financiére Africaine (CFA) Franc
Bermuda	One Bermuda Dollar = 100 Cents
Bolivia	One Bolivian Peso = 100 Centavos
Botswana	One Pula = 100 Thebe
Brazil	One Real = 100 Centavos
Brunei	One Brunei Dollar = 100 Cents
Bulgaria	One Lev = 100 Stotinki
Burundi (Republic of)	One Burundi Franc = 100 Centimes

C

Cameroon (United Republic of)	Communauté Financiére Africaine (CFA) Franc
Canada	One Canadian Dollar = 100 Cents
Canary Islands	One Euro = 100 Cents
Cape Verde Islands	One Cape Verde Escudo = 100 Centavos
Cayman Islands	One Cayman Dollar = 100 Cents
Central African Republic	Communauté Financiére Africaine (CFA) Franc
Chad (Republic of)	Communauté Financiére Africaine (CFA) Franc
Chile	One Peso = 100 Centavos
China (People's Republic of)	One Yuan (Renminbi) = 10 Jiao and 10 Jiao = 100 Fen
Colombia	One Peso = 100 Centavos
Costa Rica	One Costa Rica Colon = 100 Centimos
Côte d'Ivorie	Communauté Financiére Africaine (CFA) Franc
Croatia	One Kuna = 100 Cents
Cuba	One Cuban Peso = 100 Centavos
Cyprus (Republic of)	One Cyprus Pound – 100 Cents
Czech Republic	One Koruna = 100 Haléru

Appendix C (i)

Denmark (including the Faroe Islands and Greenland)	One Danish Crown (Krone) = 100 Oere
Djibouti (Republic of)	Djibouti Francs (related to the USA Dollar)
Dominican Republic	One Domincan Peso = 100 Centavos
Ecuador	One US Dollar = 100 Cents
Egypt (Arab Republic of)	One Egyptian Pound = 100 Piastres = 1000 Milliëmes
Equatorial Guinea (Rio Muni and Fernando Poo)	One CFA Fr = 100 Centimes
Estonia	One Kroon = 100 Cents
Ethiopia	One Birr = 100 Cents
Falkland Islands and Dependencies	One Falkland Islands Pound = 100 pence
Fiji Islands	One Fiji Dollar = 100 Cents
Finland	One Euro = 100 Cents
Formentura	One Euro = 100 Cents
France	One Euro = 100 Cents
French Antilles	One Euro = 100 Cents
French Guyana	One Euro = 100 Cents
Gabon (Republic of)	Communauté Financiére Africaine (CFA) Franc
Gambia	One Dalasi = 100 Bututs
Germany	One Euro = 100 Cents
Ghana (Republic of)	One Cedi = 100 Psewas
Gibraltar	One Gibraltar Pound = 100 Pence
Greece	One Euro = 100 Cents
Greenland	One Danish Crown (Krone) = 100 Oere
Grenada	One East Caribbean Dollar = 100 Cents
Guadaloupe	One Euro = 100 Cents
Guam	One US Dollar = 100 Cents
Guatemala	One Quetzal = 100 Centavos
Guinea (Republic of)	One Guinea Franc = 100 Centimes
Guinea Bissau	Communauté Financiére Africaine (CFA) Franc
Guyana	One Guyanan Dollar = 100 Cents
Haiti	One Gourde = 100 Centimes
Honduras (Republic of)	One Lempira = 100 Centavos
Hong Kong	One Hong Kong Dollar = 100 Cents
Hungary	One Forint = 100 Fillér

Appendix C (i)

Ibiza	One Euro = 100 Cents
Iceland	One Icelandic Crown (Krona) = 100 Aurar (Eyrir)
India	One Indian Rupee = 100 Paise (singular – Paisa)
Indonesia (Republic of)	One Rupiah = 100 Sen
Iran	10 Rials = 1 Toman
Iraq	One Dinar = 1000 Fils
Ireland (Republic of)	One Euro = 100 Cents
Israel	One New Shekel = 100 New Agorot (singular – Agora)
Italy	One Euro = 100 Cents
Ivory Coast (Republic of) (Cote d'Ivoire)	Communauté Financiére Africaine (CFA) Franc
Jamaica	One Jamaican Dollar = 100 Cents
Japan	Yen
Jordan	One Jordan Dinar = 1000 Fils
Kenya	One Kenya Shilling = 100 Cents
Kiribati (Republic of) (formerly Gilbert Islands)	Australian Dollar
Korea North (Democratic People's Republic of)	One Won = 100 Chon
Republic of Korea (South)	One Won = 100 Chon
Kuwait	One Kuwati Dinar = 1000 Fils
Laos (People's Democratic Republic of)	One New Kip = 100 Cents (at)
Lebanon	One Lebanese Pound = 100 Piastres
Lesotho	One Loti = 100 Lisente
Liberia (Republic of)	One Liberian Dollar = 100 Cents
Libya (Libyan Arab Republic)	One Libyan Dinar = 1000 Dirhams
Liechtenstein (Principality of)	One Swiss Franc = 100 Centimes (Rappen)
Luxembourg (Grand-Duchy of)	One Euro = 100 Cents
Macao	One Pataca = 100 Avos
Madeira	One Euro = 100 Cents
Majorca	One Euro = 100 Cents
Malagasy Republic (Madagascar)	Madagascar Francs (MG Frs)
Malawi	One Kwacha = 100 Tambala
Malaysia	One Malaysian Ringgitt = 100 Sen
Maldives (Republic of)	One Maldivian Rufiyaa = 100 Iari
Mali (Republic of)	One CFA Franc = 100 Centimes
Malta	One Lira = 100 Cents
Martinque	One Euro = 100 Cents
Mauritania (Islamic Republic of Mauritania)	One Ouguiya = 5 Khoums
Mauritius	One Mauritius Rupee = 100 Cents
Mayotte	One Euro = 100 Cents
Melilla	One Euro = 100 Cents

Appendix C (i)

Mexico	One Mexican Peso = 100 Centavos
Minorca	One Euro = 100 Cents
Miquelon	One Euro = 100 Cents
Monaco	One Euro = 100 Cents
Mongolia (People's Republic of)	One Tug = 100 Mongos
Montserrat	One E Carb. Dollar = 100 Cents
Morocco (Kingdom of)	One Dirham = 100 Centimes
Mozambique	One Metical = 100 Centavos
Myanmar (adopted by Burma in 1989)	One Kyat = 100 Pyas

Nambia	One Dollar = 100 Cents
Nauru Islands (Republic of)	One Australian Dollar = 100 Cents
Nepal (Kingdom of)	One Nepalese Rupee = 100 Paise
Netherlands (Holland)	One Euro = 100 Cents
Netherlands Antilles (Netherlands West Indies)	Antilles Guilder = 100 Cents
New Zealand (including Ross Dependency)	One New Zealand Dollar = 100 Cents
Nicaragua	One Cordoba = 10 Centavos
Niger (Republic of)	Communauté Financiére Africaine (CFA) Franc
Nigeria (Federation of)	One Naira = 100 Kobo
Norway	One Krone = 100 Ore

Oman	One Rial Omani = 1000 Baizas

Pakistan	One Pakistan Rupee = 100 Paisa
Panama	One Balboa = 100 Centesimos
Papua New Guinea	One Kina = 100 Toeas
Paraguay	One Guarani = 100 Centimos
Peru	One New Sol = 100 Centimos
Philippines (Republic of)	One Philippine Peso = 100 Centavos
Picairn Island	Currency rates issued by Picairn Island Currency Board (with Sterling parity)
Poland	One Zloty = 100 Groszy
Portugal (including Azores and Madeira)	One Euro = 100 Cents
Puerto Rico (Commonwealth of)	One US Dollar = 100 Cents

Qatar	One Qatar Riyal = 100 Dirhams

Reunion Island	One Euro = 100 Cents
Romania	One Leu = 100 Bani
Rwanda (Republic of)	One Rwanda Franc = 100 Centimes

Appendix C (i)

St Christopher	One Carb. Dollar = 100 Cents
St Helena Island (including Ascension Island)	Currency rates issued by St Helena Currency Board (with Sterling parity)
St Lucia	One E Carb. Dollar = 100 Cents
St Pierre	One Euro = 100 Cents
St Vincent	One E Carb. Dollar = 100 Cents
San Marino	One Euro = 100 Cents
Sao Tome and Principe	One Dobra = 100 Centimos
Saudi Arabia	One Riyal = 100 Halalat
Senegal (Republic of)	Communauté Financiére Africaine (CFA) Franc
Seychelles	One Seychelles Rupee = 100 Cents
Sierra Leone	One Leone = 100 Cents
Singapore (Republic of)	One Singapore Dollar = 100 Cents
Solomon Islands	One Solomon Islands Dollar = 100 Cents
South Africa (Republic of)	One Rand = 100 Cents
Spain (including the Canary Islands, Ceuta and Melilla)	One Euro = 100 Cents
Spanish Ports North Africa	One Euro = 100 Cents
Sri Lanka	One Rupee = 100 Cents
Sudan (Republic of the)	One Dinar = 100 Piastres
Suriname	One Suriname Guilder (Florin) = 100 Cents
Swaziland	One Lilangeni = 100 Cents
Sweden	One Krona = 100 öre
Switzerland	One Swiss Franc = 100 Centimes (Rappen)
Syria (Syrian Arab Republic)	One Syrian Pound = 100 Piastres

Taiwan (Republic of China)	One New Taiwan Dollar = 100 Cents
Tanzania (United Republic of Tanganyika, Zanzibar and Pemba)	One Tanzanian Shilling = 100 Cents
Thailand	One Baht = 100 Stang
Togo (Republic of)	Communauté Financiére Africaine (CFA) Franc
Tonga (Kingdom of) (Friendly Islands)	One Tonga Dollar (Pa'anga) = 100 Seniti
Trinidad and Tobago	One Trinidad Dollar = 100 Cents
Tunisia (Republic of)	One Tunisian Dinar = 1000 Millimes
Turkey	One Turkish Pound (Lira) = 100 Kurus
Turkmenistan	One Manat = 100 Tenga
Tuvalu	Australian Dollar

Uganda (Republic of)	One Uganda Shilling = 100 Cents
Ukraine	One Hryvnia = 100 Kopiykas
United Arab Emirates	One Dirham = 100 Fils
United Kingdom	One Pound (Sterling) = 100 Pence
United States of America (including Virgin Islands of the USA, Guam, American Samoa and Panama Canal Zone)	One US Dollar = 100 Cents
Uruguay	1000 Uruguayan Pesos = One New Uruguayan Peso =100 Centesimos

Appendix C (i)

Vanuatu (Republic of) — Vatu
Vatican City State — One Euro = 100 Cents
Venezuela — One Bolivar = 100 Centimos
Vietnam (Socialist Republic of) — One Dong = 100 Xu
Virgin Islands (British) and (US) — One US Dollar = 100 Cents

West Indies (East Caribbean Area) — One East Caribbean Dollar = 100 Cents
Western Samoa — One Tala = 100 Senes

Yemen Arab Republic — One Riyal = 100 fils
Yemen (People's Democratic Republic of) — One South Yemen Dinar = 1000 Fils
Yugoslavia — One New Dinar = 100 Paras

Zaire (Republic of) — One Zaire = 100 Makuta = 10000 Sengi
Zambia — One Kwacha = 100 Ngwee

Euro based Currency Countries:-

Andorra
Austria
Azores
Balearic Isles
Belgium
Canary Islands
Ceuta
Corsica
Crete
Finland
Formentera
France
French Antilles
French Guyana
Germany
Greece
Guadaloupe
Ibiza
Ireland
Italy
Luxembourge
Madeira
Majorca
Martinique
Mayotte
Melilla
Minorca
Miquelon
Monaco
Netherlands
Portugal
Reunion Isles
San Marino
Spain
St Pierre
Vatican City

Conversion of Weights and Measures
Conversion Factors

To Convert	Into	Multiply By

C

Centigrade or Celsius	Fahrenheit	9/5ths and add 32F
Centimetres	Inches	0.3937
Centimetres	Metres	0.0100
Centimetres	Millimetres	10.0000
Cubic feet	Cubic centimetres	28320.0000
Cubic feet	Cubic inches	1728.0000
Cubic feet	Litres	28.3200
Cubic inches	Cubic centimetres	16.3900
Cubic metres	Cubic feet	35.3100
Cubic metres	Cubic inches	61023.3779
Cubic inches	Fluid ounces (Imperial)	0.4614
Cubic inches	Fluid ounces (US)	0.5541

F

Fahrenheit	Centigrade or Celsius	Subtract 32F and multiply by 5/9ths
Feet	Centimetres	30.4800
Feet	Metres	0.3048
Feet	Inches	12.0000
Fluid ounces (Imperial)	Millilitres	28.4123
Fluid ounces (Imperial)	Pints (Imperial)	0.0500
Fluid ounces (US)	Millilitres	29.5729
Fluid ounces (US)	Pints (US)	0.0625

G

Grams	Ounces (avoirdupois)	0.0353

I

Imperial gallons	Litres	4.5460
Imperial gallons	US gallons	1.2009
Imperial pints	Fluid ounces (Imperial)	20.0000
Imperial pints	Litres	0.5682
Imperial quarts	Litres	1.1365
Inches	Centimetres	2.5400

K

Kilograms	Pounds (16 ounces)	2.2046
Kilograms per square centimetre	Pounds per square inch	14.2234

Appendix D

L

Litres	Fluid ounces (Imperial)	35.1960
Litres	Fluid ounces (US)	33.8147
Litres	Imperial gallons	0.2200
Litres	Imperial pints	1.7598
Litres	Imperial quarts	0.8799
Litres	US gallons	0.2642
Litres	US Pints	2.1136
Litres	US quarts	1.0567

M

Metres	Centimetres	100.0000
Metres	Feet	3.2808
Metres	Inches	39.3700
Millimetres	Centimetres	0.1000
Millimetres	Inches	0.0394

O

Ounces (avoirdupois)	Pounds	0.0625
Ounces (avoirdupois)	Grams	28.3495

P

Pounds	Ounces	16.000
Pounds	Grams	453.5924
Pounds (16 ounces)	Kilograms	0.4536
Pounds per square inch	Kilograms per sq centimetre	0.0703

T

Tons (long)	Kilograms	1016.000
Tons (long)	Pounds	2240.0000
Tons (long)	Tons (short)	1.1200
Tons (metric)	Kilograms	1000.0000
Tons (metric)	Pounds	2205.0000
Tons (short)	Pounds	2000.0000
Tons (short)	Kilograms	907.1849
Tons (short)	Tons (long)	0.8929
Tons (short)	Tons (metric)	0.9072

U

US gallons	Imperial gallons	0.8327
US gallons	Litres	3.7853
US pints	Fluid ounces (US)	16.0000
US pints	Litres	0.4732
US quarts	Litres	0.9463

Appendix E

LABELS, PLACARDS, MARKS AND SIGNS

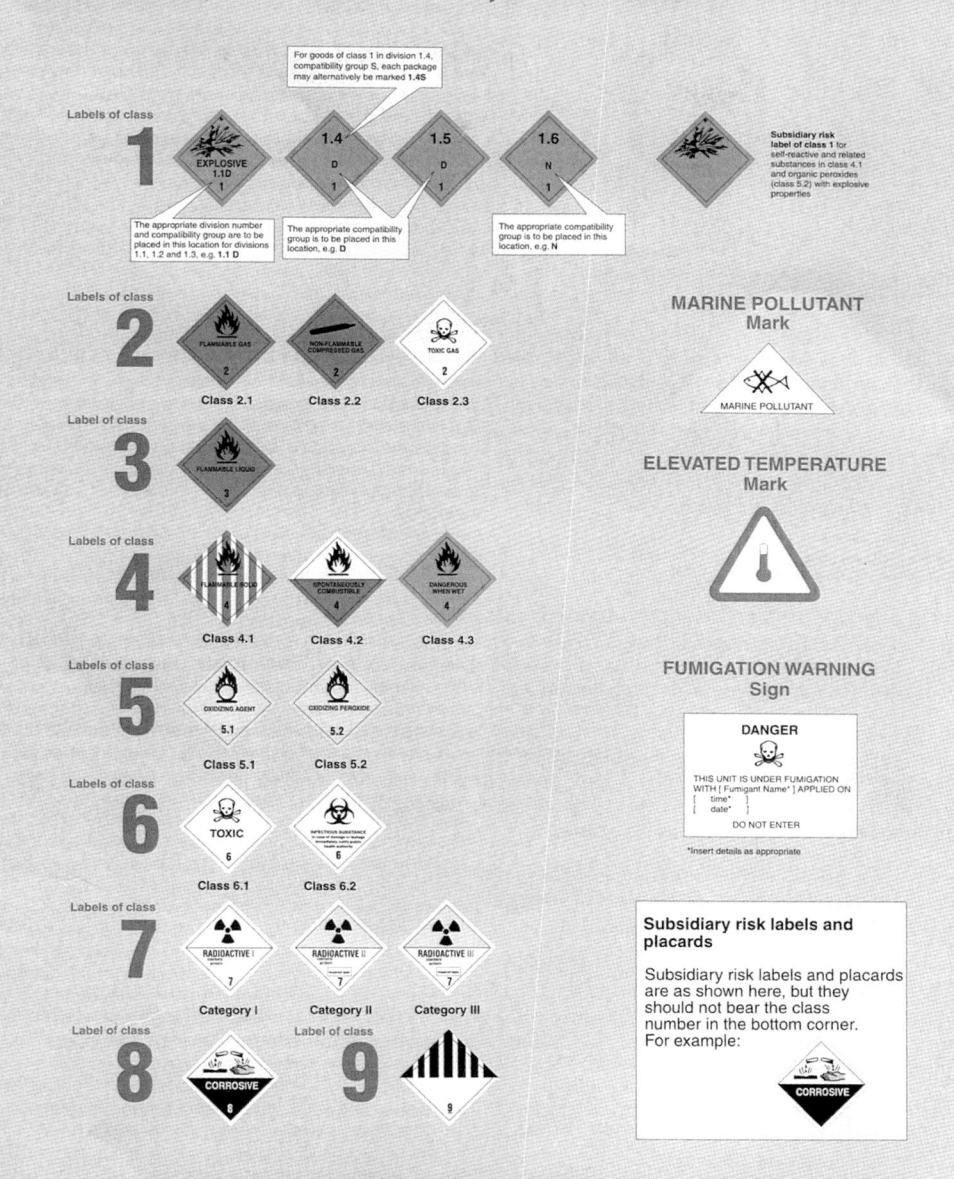

Labels of class 1

For goods of class 1 in division 1.4, compatibility group S, each package may alternatively be marked **1.4S**

- **EXPLOSIVE 1.1D** — The appropriate division number and compatibility group are to be placed in this location for divisions 1.1, 1.2 and 1.3, e.g. **1.1 D**
- **1.4 D** — The appropriate compatibility group is to be placed in this location, e.g. **D**
- **1.5 D**
- **1.6 N** — The appropriate compatibility group is to be placed in this location, e.g. **N**

Subsidiary risk label of class 1 for self-reactive and related substances in class 4.1 and organic peroxides (class 5.2) with explosive properties

Labels of class 2
- Class 2.1 — FLAMMABLE GAS
- Class 2.2 — NON-FLAMMABLE COMPRESSED GAS
- Class 2.3 — TOXIC GAS

MARINE POLLUTANT Mark

Label of class 3
- FLAMMABLE LIQUID

ELEVATED TEMPERATURE Mark

Labels of class 4
- Class 4.1 — FLAMMABLE SOLID
- Class 4.2 — SPONTANEOUSLY COMBUSTIBLE
- Class 4.3 — DANGEROUS WHEN WET

Labels of class 5
- Class 5.1 — OXIDIZING AGENT
- Class 5.2 — OXIDIZING PEROXIDE

FUMIGATION WARNING Sign

DANGER

THIS UNIT IS UNDER FUMIGATION WITH [Fumigant Name*] APPLIED ON
[time*]
[date*]
DO NOT ENTER

*Insert details as appropriate

Labels of class 6
- Class 6.1 — TOXIC
- Class 6.2 — INFECTIOUS SUBSTANCE

Labels of class 7
- Category I — RADIOACTIVE I
- Category II — RADIOACTIVE II
- Category III — RADIOACTIVE III

Subsidiary risk labels and placards

Subsidiary risk labels and placards are as shown here, but they should not bear the class number in the bottom corner. For example: CORROSIVE

Label of class 8 — CORROSIVE

Label of class 9

For the requirements on labels, placards, marks, signs and orange panels, refer to sections 7 and 8 of the General Introduction to the IMDG Code.

To purchase the IMDG Code, please contact:
Publications Section
International Maritime Organization
4 Albert Embankment
London SE1 7SR
Tel: +44(0)171 735 7611
Fax:+44(0)171 587 3241
E-mail: publications.sales@imo.org

IMO dangerous goods labels
Reproduced by kind permission of International Maritime Organisation

Appendix F

Further Recommended Reading

Journals

A
ABC – Air Cargo Guide

B
Baltic Exchange
Barclays Bank – Country Reports
BIMCO annual review

C
Chartered Institute of Logistics and Transport
Containerisation International

E
Export Focus

F
Fair Play
Freight Management International
Freight News

H
Handy Shipping Guide
HSBC – Country Report

I
Institute of Export Members Annual Handbook
International Freighting Weekly
International Trade Today

K
Kellys Export Services

L
Lloyd's Freight Transport Buyer
Lloyd's List and Shipping Gazette
Lloyd's Shipping Economist

M
Marketing
Merchant's Guide (P & O Containers)
Motorship

O
Overseas Trade

P
Pipeline

R
Review of Maritime Transport
UNCTAD Annual Review

S
Sea Trade Review
Shipping Intelligence Weekly
Shipping Review and Outlook
SITPRO – Annual Report

T
The Economist
The Ship Broker